SPRINGER PROCEEDINGS IN PHYSICS

J. M. L. M. Palma
A. Silva Lopes (Eds.)

Advances in Turbulence XI

Proceedings of the 11th EUROMECH European
Turbulence Conference, June 25-28, 2007
Porto, Portugal

With 555 Figures and 11 Tables

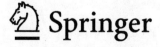 Springer

J. M. L. M. Palma
Faculdade de Engenharia
da Universidade do Porto
Rua Dr. Roberto Frias, s/n
4200-465 Porto
Portugal
jpalma@fe.up.pt

A. Silva Lopes
Faculdade de Engenharia
da Universidade do Porto
Rua Dr. Roberto Frias, s/n
4200-465 Porto
Portugal

Library of Congress Control Number: 2007928308

ISSN 0930-8989
ISBN 978-3-540-72603-6 Springer Berlin Heidelberg New York

Springer is a part of Springer Science+Business Media
springer.com

Typesetting: Integra Software Services Pvt. Ltd., India
Cover design: WMX Design, Heidelberg, Germany

Printed on acid-free paper   SPIN: 11675709   45/3100/Integra   5 4 3 2 1 0

Preface

This book comprises the written versions of all communications presented at ETC11, the 11th EUROMECH European Turbulence Conference, held at FEUP — Faculdade de Engenharia da Universidade do Porto, in Porto, Portugal, during June 25–28, 2007.

The history of the ETC conference series goes back to 1986, in Lyon, France. Latest editions were held in Barcelona (2000), Southampton (2002) and Trondheim (2004). The present edition is taking place after a three year gap, in order that, in the future, the two largest EUROMECH conferences in fluid mechanics and turbulence, EFMC and ETC, can occur in different years. During the last three years, there was time for announcing or loosing the élan of previous editions of ETC11. The conference committee had many reasons to be apprehensive, but their concerns were put to rest by the large response of the community to the call of papers for ETC11. This also shows the need for a conference on turbulence along the lines of the previous edition of ETC and the establishment and appreciation of the conference by the research community in fluid mechanics and turbulence.

Conference and Book Organisation

The book is organised as the conference programme. Chapters are called sessions and papers follow an order identical to their order of presentation within the session. A list of authors is provided at the end. In total, the book contains the written versions of 8 invited talks, 197 oral and 65 poster presentations.

Paper Selection

A total of 443 abstracts, originating from 39 countries, were submitted for possible presentation at ETC11. Each abstract was scrutinised by three reviewers and marked on a scale of A to D. Considering the number of submissions and available space, only abstracts marked with a majority of A and Bs could be accepted. A total of 225 oral and 90 poster presentations were selected. Both the abstract submission and reviewing were carried out using web technologies, which facilitated much of the work and enabled us to inform the authors of the acceptance of their abstracts before the deadline. Final decisions and assessment of the reviewers criteria were made at a meeting of the Conference Committee. Great care was put into the abstract selection, and we hope that this will be apparent during the four days of the conference from the quality of the presentations and from many fruitful meetings and discussions among the participants.

Turbulence is a topic of interest to a large community of researchers of different backgrounds. It is our hope that, apart from perpetuating the spirit of the ETC11, this book can also provide a source of knowledge and inspiration for some years to come.

Acknowledgements

Several teams were put together in charge of aspects such as the conference web page, the electronic paper submission and evaluation, the conference registration, etc. Because it would be almost impossible to name them all, we express here our gratitude to those at different units within FEUP, namely the multimedia unit for producing all documents related to the conference announcements and graphic image, the computer centre for their support and assistance in the conference rooms and the economical and financial unit for taking care of all procedures required by the sponsors that also made possible the organisation of ETC11.

One person should be mentioned here to whom we are particularly grateful: our colleague João Correia Lopes, for his enthusiasm, availability at all times, support and experience in the use of Cyberchair, the web-based system used in electronic submission.

We are most grateful to the European Commission through the Marie Curie programme, by making possible the award of 162 grants that lowered the financial burden of many participants. The University of Porto (the Rector) and the Portuguese-American Foundation (FLAD) partly supported the expenses of some of the invited speakers.

Last but not least, we must thank all the authors for their contribution, which is the main purpose of the conference.

Porto, March 2007

José M. Laginha M. Palma
Alexandre Silva Lopes

Additional Reviewers

Jean-Pierre Bertoglio
Fabien S. Godeferd
Mikhael Gorokhovski
Lionel Le Penven
Benoit Pier

Frédéric Plaza
Florence Raynal
Liang Shao
Serge Simoens

Contents

Session 4.1: Wall bounded flows (I)

Invited Speaker II

Session 1.2: Lagrangian dynamics in turbulence/inertial particles (II)

Session 2.2: MHD turbulence (II)

Session 3.2: Instability and transition (II)

Session 4.2: Wall bounded flows (II)

Session 1.3: Lagrangian dynamics in turbulence/inertial particles (III)

Session 2.3: MHD turbulence (III)

Session 3.3: Instability and transition (III)

Session 4.3: Wall bounded flows (III)

Invited Speaker III

Session 1.4: Transport in quasi-2D-turbulence (I)

Session 2.4: Instability and transition (IV)

Session 3.4: Wall bounded flows (IV)

Session 4.4: Control of turbulent flows (I)

Session 1.5: Intermittency and scaling (I)

Session 2.5: Instability and transition (V)

Session 3.5: Wall bounded flows (V)

Session 4.5: Vortex dynamics and structure formation (I)

Invited Speaker IV

Session 1.6: Transport in quasi-2D-turbulence (II)

Session 2.6: Turbulence in multiphase and non-Newtonian flows (I)

Session 3.6: Wall bounded flows (VI)

Session 4.6: Vortex dynamics and structure formation (II)

Session 1.7: Large eddy simulation and related techniques (I)

Session 2.7: Turbulence in multiphase and non-Newtonian flows (II)

Session 3.7: Wall bounded flows (VII)

Session 4.7: Vortex dynamics and structure formation (III)

Invited Speaker V

Session 1.9: Lagrangian dynamics in turbulence/inertial particles (IV)

Session 2.9: Geophysical and astrophysical turbulence (I)

Session 3.9: Transport and mixing (II)

Session 4.9: Atmospheric turbulence (I)

Invited Speaker VI

Session 1.10: Lagrangian dynamics in turbulence/inertial particles (V)

Session 2.10: Intermittency and scaling (II)

Session 3.10: Transport and mixing (III)

Session 4.10: Atmospheric turbulence (II)

Session 1.11: Lagrangian dynamics in turbulence/inertial particles (VI)

Session 2.11: Intermittency and scaling (III)

Session 3.11: Transport and mixing (IV)

Session 4.11: Stability and flow control

Invited Speaker VII

Session 1.12: Turbulence in multiphase and non-Newtonian flows (III)

Session 2.12: Atmospheric turbulence (III)

Session 3.12: Transport and mixing (V)

Session 4.12: Wall bounded flows (VIII)

Session 1.13: Geophysical and astrophysical turbulence (II)

Session 2.13: Large eddy simulation and related techniques (III)

Session 3.13: Transport and mixing (VI)

Session 4.13: Control of turbulent flows (II)

Invited Speaker VIII

Session 1.14: Geophysical and astrophysical turbulence (III)

Session 2.14: Atmospheric turbulence (IV)

Session 3.14: Transport and mixing (VII)

Session 4.14: Control of turbulent flows (III)

Posters I: Acoustics of turbulent flows

Posters II: Atmospheric turbulence

Posters III: Control of turbulent flows

Posters IV: Instability and transition

Posters V: Intermittency and scaling

Posters VI: Large eddy simulation and related techniques

Posters VII: MHD turbulence

Posters VIII: Reacting and compressible turbulence

Posters IX: Transport and mixing

Posters X: Turbulence in multiphase and non-Newtonian flows

Posters XI: Vortex dynamics and structure formation

Posters XII: Wall bounded flows

Experimental Measurements of Lagrangian Statistics in Intense Turbulence

H. Xu[1,3], N. T. Ouellette[1,3],H. Nobach[1,3], and E. Bodenschatz[1,2,3]

[1] MPI for Dynamics and Self-Organization, 37077 Göttingen, Germany
eberhard.bodenschatz@ds.mpg.de
[2] LASSP and M&AE, Cornell University, Ithaca, NY 14853, USA
[3] International Collaboration for Turbulence Research (ICTR)

Summary. We report Lagrangian laboratory measurements of fluid turbulence at high Reynolds numbers. First, the experimental techniques that have made Lagrangian measurements possible, including both optical and acoustic particle tracking, are reviewed. Then some of the laboratory flows used in Lagrangian measurements are described and a selection of new experimental results are presented.

1 Introduction

Even after more than a century of inquiry, turbulence in an incompressible, Newtonian fluid still lacks a satisfactory understanding. This may be the more surprising, as the physical description based on the Navier-Stokes equations is well understood. The complexity of turbulence lies in its highly disordered, spatio-temporally chaotic nature, involving phenomena over many decades of spatial and temporal scales. For sufficiently strong turbulence, the small–scale statistical behavior of a turbulent flow is expected to be independent of its initial and boundary conditions. Theoretical descriptions of turbulence typically rely on phenomenological models employing the 1941 hypotheses of Kolmogorov (K41) [1], as well as more sophisticated extensions [2]. Kolmogorov's hypotheses (and most subsequent theories) were originally formulated in the Eulerian framework, where flow statistics are considered at fixed spatial locations. Like everything in fluid mechanics, however, turbulence and the Kolmogorov hypotheses can also be cast in the Lagrangian framework [3], where statistics are measured along the trajectories of individual fluid elements. Experimentally, much more is known about the Eulerian characteristics of turbulence than its Lagrangian counterpart. Lagrangian experiments have historically been extremely difficult, if not impossible for highly turbulent flows. Since a general mapping between Eulerian and Lagrangian statistics remains elusive, however, we cannot fully understand turbulence without a full characterization of its Lagrangian nature as well as its Eulerian counterpart. In addition, many problems, including mixing and transport, are inherently Lagrangian.

Lagrangian experiments have been carried out for many decades in field measurements in the atmosphere and ocean using balloons and floaters, but have typically been limited to probing large scales. Such measurements are also often significantly influenced by the non-stationarity, anisotropy, and inhomogeneity of natural flows. Over the past twenty years or so, however, Lagrangian turbulence has become the subject of laboratory experiments with the development of powerful experimental techniques based on digital imaging and signal processing, and data comparable in quality with the best Eulerian results have been obtained.

Here, we report laboratory measurements of fluid turbulence at high Reynolds numbers, and as such we will largely omit some interesting research topics, *e.g.*, numerical simulations, inertial particle dynamics, and turbulence in non-Newtonian fluids. The manuscript is organized as follows: First, in section 2, the experimental techniques that have made Lagrangian measurements possible, including both optical and acoustic particle tracking, are presented. Section 3 describes some of the laboratory flows used in Lagrangian measurements. In section 4, we present a selection of new experimental results, and compare them with previous measurements when available. We show measurements of the statistics of single Lagrangian particles, including their velocity structure functions and acceleration, and of multi-particle statistics. We conclude with a brief discussion of the future of Lagrangian measurements.

2 Lagrangian Particle Tracking

The Lagrangian framework describes fluid flow in terms of the motion of individual fluid elements. Although some progress has been made in tracking marked fluid molecules [4], Lagrangian measurement techniques usually approximate true fluid elements with small particles. To follow a turbulent flow faithfully, a tracer particle must have a size on the order of the Kolmogorov length scale $\eta \equiv \epsilon^{3/4}\nu^{-1/4}$ and a response time smaller than the Kolmogorov time scale $\tau_\eta \equiv (\nu/\epsilon)^{1/2}$ [5, 6, 7]. Here, ν is the kinematic viscosity and ϵ is the mean energy dissipation rate per unit mass. Equivalently, for particles with sizes of order η, we require that the Stokes number $\mathrm{St} \equiv \frac{1}{18}\frac{\rho_p - \rho_f}{\rho_f}\left(\frac{d}{\eta}\right)^2$, be much smaller than unity, where ρ_p and ρ_f are the densities of the particle and the fluid, respectively, and d is the particle diameter. Particles with higher Stokes numbers show important inertial effects and deviate from the motion of fluid elements[8].

2.1 Optical Techniques

Early particle tracking measurements used strobe lights and photographic film to record the motion of particles in the flow [9]. The first application of optical Lagrangian Particle Tracking (LPT) to turbulence with Taylor-scale Reynolds $R_\lambda > 300$ used a position-sensitive photodiode, but the results were limited by temporal resolution [10]. This difficulty was later overcome by the use of silicon strip detectors, originally developed for particle tracking in high-energy physics [11, 5]. The design of the silicon strip detectors allowed the simultaneous tracking of only one tracer particle, but the high spatial resolution (1/5000) and fast response time (up to 70 kHz) enabled particle accelerations to be measured accurately [6]. In these experiments, particles motions were recorded in three dimensions (3D-LPT) by using multiple detectors [5, 6, 7]. In general, 3D-LPT requires simultaneous imaging of the tracer particles viewed from multiple directions. Under special conditions images from a single camera may be used to extract the three-dimensional positions of the tracer particles, by using digital inline holography (see, *e.g.* [12]) or by measuring the defocusing ring of individual particles [13]. A 3D-LPT technique usually consists of three components, as laid out in [14, 15]. First, the images from each camera are processed to obtain the coordinates of particle centers on the two-dimensional camera image planes. Next, the particle coordinates measured from each camera must be matched together to produce three-dimensional coordinates in the laboratory frame. Finally, the particles must be tracked in time. In most of the 3D-LPT algorithms, the three steps are carried out in the order given above [16, 17, 6, 18]. While the stereo-matching and tracking steps may be logically interchanged [19], it is typically easier to track in three dimensions, since the particles will then have an entire extra dimension in which to distribute themselves, lowering the particle number density for the same number of seeding particles [18]. Recently, the tomographic technique developed in the particle imaging velocimetry community [20] proposed an interchange of the first two steps: the images from each camera are first used to reconstruct the three-dimensional optical field, from which the three-dimensional particle centers are then obtained. This technique can handle occlusion better, but the increase in computational load may limit its use. A promising new idea is hybrid tomographic 3D-LPT. A traditional 3D-LPT algorithm is used to process most of the images. Only the regions with possible occlusion, which can be detected by the size of the particle images, for example, are processed using tomographic technique.

2.2 Acoustic Techniques

Typical digital two-dimensional imaging sensors are divided into a limited number of individual photosensors: strips, in the case of silicon strip detectors, or pixels, for cameras. Therefore the range of resolvable spatial scales is necessarily limited. A system that resolves the smallest length scales of intense turbulence well will not simultaneously be able to resolve the largest scales, and vice versa. Acoustic techniques circumvent this issue [21, 22, 23]. Instead of using scattered light to image tracer particles, scattered ultrasound is recorded by an array of transducers, and the phase shift between the different transducers allows the determination of the particle position [23]. By using several arrays of transducers, three-dimensional particle positions of a single particle can be measured. Acoustic particle tracking also provides a direct measurement of the particle velocity due to the Doppler shift of the scattered signal [24, 25]. However, only one particle can be in the measurement volume at any time, similar to the limitation of the silicon strip detectors discussed above. Additionally, the particle size must be large. Typical tracer particles have been larger than the Kolmogorov length (as in [21, 22]). This may influence the results, as was found, for example, for measurements of acceleration [11, 6]. The large measurement volume possible in the acoustic technique does allow the particles to be tracked for long times. Additionally, acoustic techniques can be used to track particles in fluids that are not optically transparent. It is worth noting that the traditional Laser Doppler Velocimetry (LDV) operates on the same principle as the acoustic technique, but because of the smaller tracer particles used in the laser Doppler technique, the small scales of turbulence can be resolved. Recently, this technique has been extended to measure particle accelerations in intense turbulence [26, 27].

Apparatus	P (bar)	ν $(\mathrm{m}^2/\mathrm{s})$	u' (m/s)	ϵ $(\mathrm{m}^2/\mathrm{s}^3)$	L (m)	λ $(\mu\mathrm{m})$	η $(\mu\mathrm{m})$	τ_η (ms)	R_λ
SF$_6$ tunnel	15	1.5×10^{-7}	1.0	1.2	0.45	1400	7.3	0.36	9600
air tunnel	1	1.5×10^{-5}	1.2	2.5	0.76	11790	190	2.5	974
SF$_6$ tank	15	1.5×10^{-7}	1.0	5.5	0.094	648	5.0	0.17	4360
water tank I	1	1×10^{-6}	0.87	9.2	0.071	1110	18	0.33	970
water tank II	1	8×10^{-7}	1.3	30	0.08	820	11.4	0.16	1300

Table 1. Upper limit for flow parameters in experimental facilities, where P is the pressure, ν is the kinematic viscosity, u' is the rms velocity fluctuation, ϵ is the dissipation rate, L is the integral length scale, $\eta \equiv \epsilon^{3/4}\nu^{-1/4}$ is the Kolmogorov length scale, $\tau_\eta \equiv (\nu/\epsilon)^{1/2}$ is the Kolmogorov time scale and $R_\lambda \equiv u'\lambda/\nu$ is the Reynolds number. "Tunnel" refers to active-grid wind tunnels and "tank" refers to the von Karman swirling flows. Note that the parameters for the facilities are either estimates based on the input power, facility dimensions, and fluid properties or measured and previously published.

3 Laboratory Flows for LPT Measurements

Wind tunnels have provided the canonical flows for experimental turbulence research. By using local probes with multiple hotwires and assuming Taylor's frozen flow hypothesis the Eulerian statistics of the turbulence has been well-characterized. Wind tunnels have strong mean flows. Mechanical systems can be constructed to move the cameras along with the flow [28, 29, 8], but such sled systems are complex.Therefore, small enclosed chambers with negligible mean flows have been preferred for Lagrangian measurements. Various driving mechanisms have been developed.Von Kármán swirling flows between two counter-rotating disks produce very large Reynolds numbers at the cost of introducing anisotropic and inhomogeneous large-scale flows. The highest Reynolds number Lagrangian data has all come from these flows [10, 11, 5, 22, 21, 6, 7, 25, 30, 23, 31]. Flows forced with oscillating grids have less complex large-scale flows than swirling flows, but typically produce lower Reynolds numbers [17]. Flows forced electromagnetically alleviate some of these difficulties [32]. They can be configured to have very week mean flows, unlike a swirling flow. The Reynolds number, however, is limited. The same is true of other recently-developed low-mean-flow devices, driven by audio speakers [33] or random jet arrays [34]. An overview of parameters attainable in different flows are shown in table 1. The two von Kármán swirling water flow devices are being used at the MPIDS and at Cornell U., the atmospheric active grid wind tunnel with a sled system at Cornell U. [8], and the high-pressure von Kármán swirling flow device and active grid wind tunnel with sled are under construction at the MPIDS. In the latter devices, the working fluid will be SF_6 compressed to 15 bar. The table shows that for Lagrangian measurements in these devices very high-speed cameras with good spatial resolution are needed.

4 Experimental Results

4.1 Single-Particle Single-Time Statistics

Many early Eulerian measurements confirmed that the probability density function
(PDF) of turbulent velocity fluctuations is close to Gaussian, a result also found in
3D-LPT measurements using three cameras [18, 35]. The first high-resolution measurements of the PDF of turbulent acceleration using silicon-strip detectors have
shown the highly intermittent nature of acceleration [11, 5, 6, 7]. In Figure 1, acceleration PDFs from three different techniques are compared. The silicon strip detector
measurements were carried out in water tank I, listed in Table 1, at $R_\lambda = 690$, while
the camera and the laser Doppler measurements were carried out in water tank II
and at $R_\lambda = 600$ and 500, respectively. The four silicon strip detectors were operated
at 70kHz and gave a temporal resolution of 65 frames per τ_η and a spatial resolution
of $8\mu m$ per pixel. The frame rates of the cameras were 37kHz giving a temporal resolution of 67 frames per τ_η with a spatial resolution of $40\mu m$ per pixel. Even though
the spatial resolution of the camera system was lower, the fourth moment is still
resolved up to $|a^+| \leq 30$, where $a^+ \equiv a/\langle a^2 \rangle^{1/2}$ is the normalized acceleration. The
agreement of the laser Doppler measurement with both the silicon strip detector
and the camera data is very encouraging.

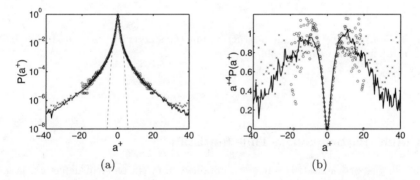

(a)   (b)

Fig. 1. Probability density functions of the normalized acceleration from different
techniques. (a) $P(a^+)$; (b) $a^{+4}P(a^+)$. The solid lines are data from silicon strip
detectors, the crosses are from cameras, the circles are from laser Doppler technique,
and the dashed line is the standard Gaussian. For experimental details see text.

4.2 Single-Particle Multi-Time Statistics

The Lagrangian velocity structure functions $D_n^L(\tau) \equiv < (u(t + \tau) - u(t))^n >$ are
some of the most important quantities for characterizing the Lagrangian nature of
turbulence. The measured inertial-range scaling exponents of $D_n^L(\tau)$ deviate from

the simple predictions from K41 theory much faster than their Eulerian counterparts [36]. The second-order structure function $D_2^L(\tau)$ is expected to scale as $C_0\epsilon\tau$ in the inertial range. C_0 is an important parameter in stochastic models and is related to other fundamental constants [37, 38]. In Figure 2, we show the compensated structure function $D_2^L(\tau)/(\epsilon\tau)$ measured in water tank I with $200 \leq R_\lambda \leq 815$ (see [39]). All curves have a peak that may be identified as the value of C_0. With increasing R_λ the peak height appears to approach an asymptotic value and the peak-width, $i.e.$, the extent of the Lagrangian inertial range, increases slowly. These features have also been observed in recent direct numerical simulations [40]. A detailed investigation of the dependence of C_0 on R_λ and the effects of large-scale anisotropy can be found in [39].

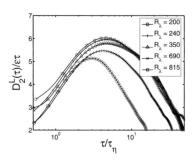

Fig. 2. Compensated Lagrangian structure functions $D_2^L(\tau)/\epsilon\tau$ measured from 3D-LPT using multiple cameras [39].

4.3 Multi-Particle Single-Time Statistics

While single-particle statistics may be obtained from silicon-strip detectors, laser Doppler and acoustic techniques, multi-particle statistics are exclusively measured using multiple cameras. As an example of single-time multipoint statistics, Eulerian structure functions are presented in Sec. 4.3, and respectively, as an example of multi-time multipoint statistics turbulent relative dispersion measurements are shown in Sec. 4.4. The measurement of Eulerian structure functions also serve as a way to determine the energy dissipation rate ϵ of the turbulent flow. As we discussed in Sec. 3, most laboratory flows for LPT are not homogeneous and have non-trivial mean flows. Even though there is lack of rigorous theoretical justification, it has been found that the inertial-range scaling laws derived for homogeneous, isotropic turbulence work well for swirling flows, especially if the local mean velocities are removed. Figure 3 shows the measured longitudinal and transverse second-order Eulerian structure functions $D_{LL}(r)$ and $D_{NN}(r)$, and the third-order longitudinal structure function $D_{LLL}(r)$. These were measured at $R_\lambda = 690$ in water tank I with a camera frame rate of 27 kHz, which is equal to 24 frames / τ_η. Within the measurement volume of approx. $(2cm)^3$ centered in the middle of the apparatus, the mean velocity was at most 20% of the fluctuation velocity. As shown in Fig. 3, the two second-order structure functions have well-developed inertial-range plateaus and are insensitive to the presence of the mean flow. The mean flow, on the other hand, has a stronger effect on the third order structure function.

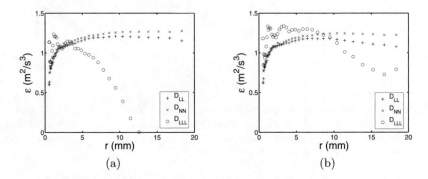

Fig. 3. Eulerian structure functions measured from 3D-LPT using multiple cameras. (a) with local mean velocity; (b) after subtraction off local mean velocity.

4.4 Multi-Particle Multi-Time Statistics

Turbulent relative dispersion – the separation of two particles in a turbulent flow – has wide applications in mixing and transport. Despite intense effort both in experiments and in numerical simulations, a conclusive observation of the long-standing Richardson t^3 law [37] has yet to be conducted. Recent measurements suggest that very intense turbulent flows with R_λ as high as 10^4 might be required in order to observe the t^3 law convincingly [31]. Berg et al. [41] studied the backward dispersion, i.e., how fast two fluid particles come together. They found from their experimental data at relatively low Reynolds number that backward dispersion was faster than the usually considered forward dispersion. In Figure 4, the forward and the backward dispersion in water tank I with $R_\lambda = 815$ are compared (for details see [31]). In the range of measurements, backward dispersion is found to be faster; the ratio between the two was, however, found to be smaller than that reported by Berg et al. [41]. In addition, a significant effect of the initial separation was observed, similarly to the case of forward dispersion [31].

5 Outlook

Even though it has been used to great effect, Lagrangian particle tracking is still a relatively new tool in experimental turbulence research. As discussed in Sect. 2, when considered in isolation each step in the algorithm has significant challenges. The entire technique may potentially be improved and made more robust by combining the

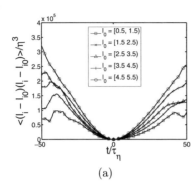 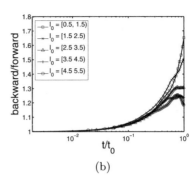

(a)   (b)

Fig. 4. (a) forward and backward relative dispersion measured from 3D-LPT using multiple cameras ; (b) ratio of the backward/forward dispersion. l_i is the separation vector between particle pairs, l_{i0} is the initial separation vector, l_0 is the magnitude of l_{i0}, and $t_0 \equiv (l_0^2/\epsilon)^{1/3}$ is the Batchelor timescale.

three main steps iteratively, at the cost of speed. A careful analysis of the limitations of LPT is needed not only for improvement of the method but also for the proper interpretation of existing LPT results. As LPT continues to improve, we expect it to become a tool as prevalent in turbulence research as the hot-wire anemometer. In the short term, we expect that LPT will be fruitfully applied to problems ranging from the temporal evolution of multi-particle clusters (see e.g. [42]), including the dynamics of material volumes and of the turbulent velocity gradients at many scales, to the dynamics of particles in turbulence, including both the motion of particles with finite inertia and the modification of turbulence in densely particle-laden flows, to turbulence in complex, non-Newtonian fluids. In recent years, there has been growing interest in these and other topics in Lagrangian turbulence in the scientific community. Turbulence, however, is a challenging problem, and progress requires ever more complex experimental, numerical, and analytical tools. In that light, we are pleased to announce the recent formation of the International Collaboration for Turbulence Research (ICTR), a partnership of experimentalists, simulators, and theorists from around the world dedicated to the understanding of the Lagrangian nature of turbulence. As this collaboration matures, we are optimistic that combining the skills of these individual scientists will lead to rapid progress and exciting new discoveries. We thank Z. Warhaft and L. Collins for many fruitful discussions and the MPG and the NSF for financial support.

References

1. A. N. Kolmogorov. The local structure of turbulence in incompressible viscous fluid for very large Reynolds numbers. *Dokl. Akad. Nauk SSSR*, 30:301–305, 1941a.
2. U. Frisch. *Turbulence: The Legacy of A. N. Kolmogorov*. Cambridge University Press, Cambridge, England, 1995.

3. A. S. Monin and A. M. Yaglom. *Statistical Fluid Mechanics*, volume 1. MIT Press, Cambridge, MA, 1971.
4. M. Pashtrapanshka and W. van de Water. The dispersion of microscopic molecular clouds in turbulence. *Bull. Am. Phys. Soc.*, 50:153–154, 2005.
5. A. La Porta, G. A. Voth, A. M. Crawford, J. Alexander, and E. Bodenschatz. Fluid particle accelerations in fully developed turbulence. *Nature*, 409:1017–1019, 2001.
6. G. A. Voth, A. La Porta, A. M. Crawford, J. Alexander, and E. Bodenschatz. Measurement of particle accelerations in fully developed turbulence. *J. Fluid Mech.*, 469:121–160, 2002.
7. N. Mordant, A. M. Crawford, and E. Bodenschatz. Experimental Lagrangian acceleration probability density function measurement. *Physica D*, 193:245–251, 2004.
8. S. Ayyalasomayajula, A. Gylfason, L. R. Collins, E. Bodenschatz, and Z. Warhaft. Lagrangian measurements of inertial particle accelerations in grid generated wind tunnel turbulence. *Phys. Rev. Lett.*, 97:144507, 2006.
9. W.-C. Chiu and L. N. Rib. The rate of dissipation of energy and the energy spectrum in a low-speed turbulent jet. *T. Am. Geophys. Union*, 37:13–26, 1956.
10. G. A. Voth, K. Satyanarayan, and E. Bodenschatz. Lagrangian acceleration measurements at large Reynolds numbers. *Phys. Fluids*, 10:2268–2280, 1998.
11. G. A. Voth. *Lagrangian acceleration measurements in turbulence at large Reynolds numbers*. PhD thesis, Cornell University, 2000.
12. G. Pan and H. Meng. Digital holography of particle fields: Reconstruction by use of complex amplitude. *Appl. Opt.*, 42:827–833, 2003.
13. M. Wu, J. W. Roberts, and M. Buckley. Three-dimensional fluorescent particle tracking at micron-scale using a single camera. *Exp. Fluids*, 38:461–465, 2005.
14. H.-G. Maas, A. Gruen, and D. Papantoniou. Particle tracking velocimetry in three-dimensional flows—Part 1. Photogrammetric determination of particle coordinates. *Exp. Fluids*, 15:133–146, 1993.
15. N. A. Malik, Th. Dracos, and D. A. Papantoniou. Particle tracking velocimetry in three-dimensional flows—Part 2. Particle tracking. *Exp. Fluids*, 15:279–294, 1993.
16. Th. Dracos. Particle tracking in three-dimensional space. In Th. Dracos, editor, *Three-Dimensional Velocity and Vorticity Measuring and Image Analysis Techniques*, pages 129–152. Kluwer Academic Publishers, Dordrecht, the Netherlands, 1996.
17. J. Mann, S. Ott, and J. S. Andersen. Experimental study of relative, turbulent diffusion. Technical Report Risø-R-1036(EN), Risø National Laboratory, 1999.
18. N. T. Ouellette, H. Xu, and E. Bodenschatz. A quantitative study of three-dimensional Lagrangian particle tracking algorithms. *Exp. Fluids*, 40:301–313, 2006.
19. Y. G. Guezennec, R. S. Brodkey, N. Trigui, and J. C. Kent. Algorithms for fully automated three-dimensional particle tracking velocimetry. *Exp. Fluids*, 17:209–219, 1994.
20. G. E. Elsinga, F. Scarano, B. Wieneke, and B. W. van Oudheusden. Tomographic particle image velocimetry. *Exp. Fluids*, 41:933–947, 2006.
21. N. Mordant, P. Metz, O. Michel, and J.-F. Pinton. Measurement of Lagrangian velocity in fully developed turbulence. *Phys. Rev. Lett.*, 87:214501, 2001.
22. N. Mordant. *Mesure lagrangienne en turbulence: mise en œuvre et analyse*. PhD thesis, École Normale Supérieure de Lyon, 2001.

23. N. Mordant, P. Metz, J.-F. Pinton, and O. Michel. Acoustical technique for Lagrangian velocity measurement. *Rev. Sci. Instr.*, 76:025105, 2005.

24. N. Mordant, J.-F. Pinton, and O. Michel. Time-resolved tracking of a sound scatterer in a complex flow: Nonstationary signal analysis and applications. *J. Acoustical Soc. Am.*, 112:108–118, 2002.

25. N. Mordant, E. Lévêque, and J.-F. Pinton. Experimental and numerical study of the Lagrangian dynamics of high Reynolds turbulence. *New J. Phys.*, 6:116, 2004.

26. M. Kinzel, H. Nobach, C. Tropea, and E. Bodenschatz. Measurement of Lagrangian acceleration using the laser Doppler technique. In *Proc. of the 13th International Symposium on Applications of Laser Techniques to Fluid Mechanics, June 26-29, 2006, Lisbon, Portugal.* 2006.

27. N. Mordant, R. Volk, A. Petrosyan, and J.-F. Pinton. Lagrangian measurements using doppler techniques: Laser and ultrasound. *Stirring and mixing: The Lagrangian approach workshop, Leiden, The Netherlands*, 2006.

28. Y. Sato and K. Yamamoto. Lagrangian measurement of fluid-particle motion in an isotropic turbulent field. *J. Fluid Mech.*, 175:183–199, February 1987.

29. M. Virant and Th. Dracos. 3D PTV and its application on Lagrangian motion. *Meas. Sci. Technol.*, 8:1539–1552, 1997.

30. A. M. Crawford, N. Mordant, and E. Bodenschatz. Joint statistics of the Lagrangian acceleration and velocity in fully developed turbulence. *Phys. Rev. Lett.*, 94:024501, 2005.

31. M. Bourgoin, N. T. Ouellette, H. Xu, J. Berg, and E. Bodenschatz. The role of pair dispersion in turbulent flow. *Science*, 311:835–838, 2006.

32. B. Lüthi, A. Tsinober, and W. Kinzelbach. Lagrangian measurements of vorticity dynamics in turbulent flow. *J. Fluid Mech.*, 528:87–118, 2005.

33. W. Hwang and J. K. Eaton. Creating homogeneous and isotropic turbulence without a mean flow. *Exp. Fluids*, 36:444–454, 2004.

34. E. A. Variano, E. Bodenschatz, and E. A. Cowen. A random synthetic jet array driven turbulence tank. *Exp. Fluids*, 37:613–615, 2004.

35. N. T. Ouellette. *Probing the statistical structure of turbulence with measurements of tracer particle tracks*. PhD thesis, Cornell University, 2006.

36. H. Xu, M. Bourgoin, N. T. Ouellette, and E. Bodenschatz. High order Lagrangian velocity statistics in turbulence. *Phys. Rev. Lett.*, 96:024503, 2006.

37. B. L. Sawford. Turbulent relative dispersion. *Annu. Rev. Fluid Mech.*, 33:289–317, 2001.

38. P. K. Yeung. Lagrangian investigations of turbulence. *Annu. Rev. Fluid Mech.*, 34:115–142, 2002.

39. N. T. Ouellette, H. Xu, M. Bourgoin, and E. Bodenschatz. Small-scale anisotropy in Lagrangian turbulence. *New J. Phys.*, 8:102, 2006.

40. P. K. Yeung, S. B. Pope, and B. L. Sawford. Reynolds number dependence of Lagrangian statistics in large numerical simulations of isotropic turbulence. *J. Turbul.*, 7:58, 2006.

41. J. Berg, B. Lüthi, J. Mann, and S. Ott. Backwards and forwards relative dispersion in turbulent flow: An experimental investigation. *Phys. Rev. E*, 74:016304, 2006.

42. L. Biferale, G. Boffetta, A. Celani, B. J. Devenish, A. Lanotte, and F. Toschi. Multiparticle dispersion in fully developed turbulence. *Phys. Fluids*, 17:111701, 2005.

Tracer particles in turbulent superfluid helium

Y. A. Sergeev[1], C. F. Barenghi[2] and D. Kivotides[2]

[1] School of Mechanical and Systems Engineering, Newcastle University, Newcastle upon Tyne, NE1 7RU, UK (contact: `yuri.sergeev@ncl.ac.uk`)
[2] School of Mathematics, Newcastle University, Newcastle upon Tyne, NE1 7RU, UK

1 Motivations

Much of the interest in the turbulence of superfluid helium is motivated by the simplicity of the vortex structures: the core of a vortex line has atomic thickness, the circulation, $\kappa \approx 10^{-3}\,\mathrm{cm}^2/\mathrm{s}$ is quantized, and the superfluid has zero viscosity. The recent success [1, 2] of implementing the PIV technique in liquid helium has opened up the possibility of visualizing flow patterns. This technique consists of tracking the motion of micron-size inertial particles using lasers. However, the interpretation of PIV data is complicated by the presence in He II of two fluid components: the viscous normal fluid and the actual inviscid superfluid. Our aim is to provide the theoretical understanding of the dynamics of small particles which is necessary to interpret the data.

2 Equations of motion and particle dynamics

We consider first the motion of spherical solid particles under the assumptions that they do not modify the turbulence, trapping of particles by superfluid vortices does not occur, and the particle radius, a_p is smaller than the Kolmogorov length in the normal fluid and the intervortex distance in the superfluid. The Lagrangian equations of particle motion are [3]:

$$\frac{d\mathbf{r}_\mathrm{p}}{dt} = \mathbf{v}_\mathrm{p}, \quad \frac{d\mathbf{v}_\mathrm{p}}{dt} = \frac{1}{\tau}(\mathbf{v}_\mathrm{n} - \mathbf{v}_\mathrm{p}) + \frac{3}{2\rho_0}\left(\rho_\mathrm{n}\frac{D\mathbf{v}_\mathrm{n}}{Dt} + \rho_\mathrm{s}\frac{D\mathbf{v}_\mathrm{s}}{Dt}\right) + \frac{\rho_\mathrm{p} - \rho}{\rho_0}\mathbf{g}, \quad (1)$$

where ρ is the density (subscripts n, s and p refer to the normal fluid, superfluid, and solid particle, respectively), $\rho = \rho_\mathrm{n} + \rho_\mathrm{s}$, $\rho_0 = \rho_\mathrm{p} + \frac{1}{2}\rho$, and $\tau = 2a_\mathrm{p}^2\rho_0/(9\mu_\mathrm{n})$ is the particle relaxation time. Analysis of Eqs. (1) shows that for neutrally buoyant particles a number of different regimes [3] can be identified: in some regimes the particles trace the normal fluid, in others the superfluid, in others the total mass current. However, an instability of particle trajectories may require modification of these conclusions. We find that in pure superfluid (at $T < 1$ Kelvin), due to the instability and the mismatch of initial velocities of particles and the fluid, small particles cannot be used to visualize the full superfluid velocity field [4]. However, it is possible to obtain information from the observation of particles that are trapped on vortex lines.

In the case of finite temperature superfluid, we analyze collisions of particles with superfluid vortices taking into account normal fluid disturbances induced by the mutual friction between the normal fluid and superfluid. We find that these disturbances can deflect the particle which otherwise could have been trapped by the vortex. We also perform calculations of particle interaction with the superfluid vortex ring propagating against a particulate sheet and show that particle trajectories collapse to the normal-fluid path lines. We propose an experiment in which, by measuring velocities of particles, direct information could be obtained about the normal-fluid velocity [5].

3 Interaction of particles and superfluid vortex lines

Particle trapping on vortex lines can lead to a modification of our results. To address this issue, we analyze a close interaction between a quantized vortex and a particle. Our calculation is self-consistent and takes into account an influence of the particle on the superfluid vortex (including reconnection of the vortex line with the particle surface.) We find that trapping occurs only in the presence of a dissipative force acting on the particle. At finite temperatures such a force is provided by the viscous drag exerted by the normal fluid. However, even at temperatures below 1 K, when the normal fluid is absent, there exists a dissipative force caused by phonon scattering.

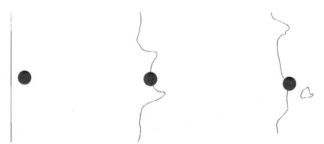

Fig. 1. Trapping of the particle by the initially straight vortex line.

A sequence shown on Fig. 1 illustrates trapping of the neutrally buoyant $1\,\mu$m particle by the initially straight vortex line. Fig. 1 (middle) shows the reconnection of the superfluid vortex with the particle surface, accompanied by excitation of Kelvin waves propagating along the vortex line. Fig. 1 (right) shows the moment of trapping accompanied by the emission of small vortex ring reducing the total energy of the particle-vortex system.

Based on the mechanism of particle-vortex interaction, we develop a theory of particle motion in turbulent thermal counterflow in helium II. In counterflow, the vortex tangle can be dense so that several vortices can be attached to the particle surface. We find that the nonuniform pressure distribution, caused by N vortices attached to the sphere, yields the body force

$$\mathbf{F} = \frac{\rho_s \kappa^2}{4\pi} \ln \frac{a_p}{\xi} \sum_{n=1}^{N} \mathbf{n}_i \, , \qquad (2)$$

where $\xi \approx 10^{-8}$ cm is the core radius, and $\mathbf{n}_i$ the unit vector along the ith attached strand. In symmetric configuration (Fig. 2 (a) and (b)) $\mathbf{F} = \mathbf{0}$. As the particle moves through the tangle, the vortex loops will be attached asymmetrically, as in Fig. 2 (c) and (d), causing a net body force. This force enables us to explain the surprising result of a recent experiment [2] that in turbulent counterflow the tracer particles move with about half the speed of the normal fluid. Fig. 2 (right) shows the calculated particle velocity vs the normal fluid velocity. The results compare well with experimental data [2].

This research is funded by EPSRC grant GR/T08876/01.

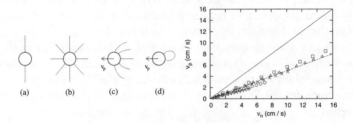

Fig. 2. Left ((a)-(d)): Sphere-vortex configurations. Right: Particle velocity vs the normal fluid velocity; dashed line – calculation based on Eq. (2).

References

1. R.J. Donnelly, A.N. Karpetis, J.J. Niemela, K.R. Sreenivasan, and W.F. Vinen, J. Low Temp. Physics **126**, 327 (2002); T. Zhang and S.W. Van Sciver, Nature Physics **1**, 36 (2005). G.P. Bewley, D.P. Lathrop, and K.R. Sreenivasan, Nature **441**, 588 (2006).
2. T. Zhang and S.W. Van Sciver, J. Low Temp. Phys. **138**, 865 (2005).
3. D.R. Poole, C.F. Barenghi, Y.A. Sergeev, and W.F. Vinen, Phys. Rev. B **71**, 064514 (2005).
4. Y.A. Sergeev. C.F. Barenghi, D. Kivotides and W.F. Vinen, Phys. Rev. B **73**, 052502 (2006).
5. D. Kivotides, C.F. Barenghi and Y.A. Sergeev, Phys. Rev. Lett. **95**, 215302 (2005)

Particle dispersion in stably stratified turbulence

M. van Aartrijk and H.J.H. Clercx

J.M. Burgerscentre, Fluid Dynamics Laboratory, Eindhoven University of Technology, The Netherlands m.v.aartrijk@tue.nl, h.j.h.clercx@tue.nl

The effect of stable stratification on turbulent dispersion plays an important role in nature, for example in plankton bloom formation in shallow coastal environments and in aerosol distribution in the atmospheric boundary layer. Including all biological and physical parameters in modelling this particle dispersion is complicated and as a starting point in this work the effect of stratification on dispersion of both fluid particles and more realistic particles (including mass and inertial effects) will be studied.

A natural way to describe turbulent dispersion is the Lagrangian frame of reference, in which the observer is moving with the particle. This is performed by first calculating the Eulerian velocity field using a 3D pseudo-spectral Direct Numerical Simulation (DNS) code. Next, particle trajectories are obtained by numerical integration of $\frac{d\mathbf{x}_p}{dt} = \mathbf{u}_p$. For fluid particles their velocities are derived from cubic spline interpolation of the velocity field at the particle position. To get the velocities of more realistic particles the Maxey-Riley equation [1] is solved in the limit of heavy particles ($\rho_p \gg \rho_f$) in which drag forces and gravity are taken into account: $\frac{d\mathbf{u}_p}{dt} = \frac{1}{\tau_p}(\mathbf{u} - \mathbf{u}_p) + \mathbf{g}$. A constant linear stable background stratification is applied and the turbulence is forced by injecting energy at the largest scales to reach a statistically stationary state. Particles are released when this stationary state is reached and velocity and position time series of $O(10^5)$ particles are collected in simulations with initial $Re_\lambda \approx 85$ and $Sc = \nu/\kappa = 1$.

Stratified flows display strong anisotropy on the large scales, and horizontally layered, pancake-like structures can be identified. The anisotropy of the flow is expressed in the dispersion characteristics. For fluid particle dispersion in stably stratified turbulence some results are shown in figure 1. On the left the trapping of particles within a horizontal layer can be seen and on the right the vertical dispersion is shown for five different values of the stratification. Dispersion in vertical direction is clearly suppressed by stratification. Three successive regimes can be identified for the vertical mean square displacement:

the classical t^2-regime, a plateau which scales as N^{-2} and the diffusion limit displaying the well-known linear scaling with time. The initial growth and the appearance of the plateau in these forced stratified turbulence simulations correspond with previous work by Kimura & Herring [2] (decaying stratified turbulence, DNS) and Nicolleau & Vassilicos [3] (non-decaying, Kinematic Simulations). By applying forcing we have been able to follow particles for sufficiently long times to find also a long-time diffusion regime. This diffusion of fluid particles away from the original equilibrium position is caused by molecular diffusion of the active scalar (density), what we checked by changing the Schmidt-number Sc. Horizontal dispersion (not shown) initially displays behaviour similar to the classical results for isotropic turbulence. However, the data on the long-time growth rates provide indications for possible super-diffusive behaviour. It shows scaling behaviour as t^α with α generally in the range $2 - 2.5$.

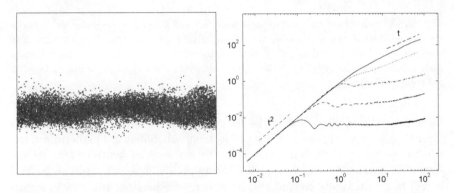

Fig. 1. a) Snapshot of the particle positions projected in the x-z plane for strong stratification ($N^2 \approx 0.1$). Fluid particles released at the same height are trapped within a layer. b) Vertical dispersion $\overline{(z - z(0))^2}$ scaled with the rms-velocity $\overline{w'^2}$ and the Lagrangian timescale T_L^2 as a function of time t/T_L for five different values of the stratification. A plateau can be seen around $t = 2\pi/N$ which scales as $1/N^2$.

The next - and new - step is the study of Lagrangian dispersion of heavy particles in stratified turbulence. When considering heavy particles, drag forces and gravitational forces influence their dispersion. In homogeneous isotropic turbulence preferential concentration occurs when $\tau_p = O(\tau_K)$ [4]. This is caused by a centrifugal effect that forces particles with $\rho_p > \rho_f$ away from vortex cores into strain dominated regions. As a result, particles will no longer be uniformly distributed over the domain. Some regions are depleted, whereas other regions are more closely packed. The tendency towards preferential concentration can be evaluated by looking at the average distance between a given particle and its closest neighbour [5]. As a preliminary result, for both isotropic turbulence and strongly stratified turbulence this distance is shown in figure 2 as function of time after particle release. From this graph it can be concluded that it

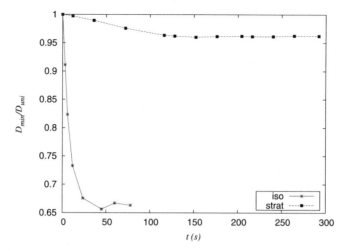

Fig. 2. Averaged distance between a particle and its closest neighbouring particle, scaled with this distance for a uniform distribution, as function of time. For strongly stratified turbulence ($N^2 \approx 1$) it takes longer to reach a a quasi-steady distribution and the effect of preferential concentration is less pronounced than in isotropic turbulence.

takes more time for heavy particles to reach a quasi-stationary distribution in strongly stratified turbulence than in isotropic turbulence. Furthermore, the effect of preferential concentration is less pronounced in stratified turbulence. In this work it will be addressed what causes these differences. More in general it will be studied how inertial effects influence dispersion statistics in stratified flows. Furthermore, particle-pair dispersion and the effect of gravitational settling on heavy particle dispersion will be discussed.

Acknowledgements

This programme is funded by the Netherlands Organisation for Scientific Research (NWO) and Technology Foundation (STW) under the Innovational Research Incentives Scheme grant ESF.6239. The use of supercomputer facilities for this work is sponsored by NCF (National Computing Facilities). We would like to thank Kraig Winters for kindly providing the (original version) of the numerical code.

References

1. M.R. Maxey, J.J. Riley: Phys. Fluids **26**, 4 (1983) pp 883–889
2. Y. Kimura, J.R. Herring: J. Fluid Mech. **328**, (1996) pp 253–269
3. F. Nicolleau, J.C. Vassilicos: J. Fluid Mech. **410**, (2000) pp 123–146
4. K.D. Squires, J.K. Eaton: Phys. Fluids **3**, 5 (1991) pp 1169–1178
5. S.W. Coppen, V.P. Manno, C.B. Rogers: Comp. Fluids **30**, (2001) pp 257–270

Is it possible to study Euler (or inviscid/purely inertial) evolution in low Reynolds number flows?

B. Galanti, D. Gendler-Fishman and A. Tsinober

Department of Fluid Mechanics and Heat Transfer, School of Mechanical Engineering, Tel-Aviv University, P.O. Box 39040, Tel-Aviv 69978, Israel tsinober@eng.tau.ac.il

The main issue we are concerned in this communication relates to the difference in the evolution of vorticity and material lines due to nonlinearity of the equation governing the evolution of vorticity. In other words, the fact that vorticity is frozen in the inviscid flow field does not mean that vorticity behaves in the same way as material lines, but the other way around: those material lines which coincide with vorticity behave like vorticity, because they are not passive anymore as are all the other material lines (Tsinober, 2001). In order to observe such a difference ideally one needs to compare a purely inviscid evolution of vorticity (and without external forcing) with that of material lines. While the latter is intrinsically nondiffusive and can be obtained from Lagrangian particle tracking, the latter is impossible especially at low Reynolds numbers. However, this impossibility refers to the whole flow field and can be circumvented locally (both in space and time) in the following sense. One can look for regions in which the forcing and the viscous terms are (approximately) balancing each other and use the Lagrangian tracking of fluid particles originating from these regions for time periods during which the above balance between the forcing and the viscous terms remain valid. Along with standard quantities as vorticity, enstrophy and similar we looked at the so called modified helicity $\tilde{h}_\omega$ obeying the equation:

$$
\frac{D\tilde{h}_\omega}{Dt} = v\left(\nabla^2 \tilde{h}_\omega - 2\frac{\partial \omega_i}{\partial x_k}\frac{\partial v_i}{\partial x_k} \right) + v_i\{\operatorname{curl}\mathbf{f}\}_i + \omega_i f_i. \tag{1}
$$

The modified helicity is defined as $\tilde{h}_\omega \equiv \boldsymbol{\omega}\cdot\mathbf{v}$ with $\mathbf{v} = \mathbf{u} + \nabla\varphi$ and $\mathbf{u}$ being the fluid particle velocity. It is known (Kuz'min, 1983; Oseledets, 1989) that there exists a particular choice of φ such that $\tilde{h}_\omega$ is a pointwise

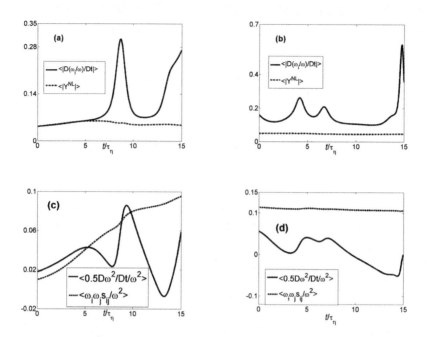

Fig. 1. Comparison of the behavior of vorticity related quantities for particle trajectories originating from regions in which the forcing and the viscous terms are balancing each other (a, c) and with those for randomly chosen trajectories (b, d). $Y^{NL} = \omega^{-1}\omega_k S_{ik} - \omega^{-3}\omega_i\omega_k\omega_j S_{jk}$.

inviscid Lagrangian invariant (i.e. it is unchanged along a particle trajectory in an inviscid and unforced flow) as seen from the equation (1). Thus using modified helicity is a more elaborate way of assessing the roles of viscosity and forcing due to absence of production (nonlinear) terms in the equation (1). The scalar function φ satisfies the following parabolic equation which is solved in parallel with the NSE:

$$\frac{D\varphi}{Dt} = p - \frac{\mathbf{u}^2}{2} + \nu\nabla^2\varphi \tag{2}$$

At some initial moment we have located flow regions in which the viscous and forcing terms are approximately balancing each other in the NSE, the equation for vorticity, enstrophy and additionally requiring the RHS of (1) is small. Figures 1 exemplifies comparison of the behavior of vorticity related quantities for particle trajectories originating from regions in which the forcing and the viscous terms are balancing each other with those for randomly chosen trajectories. The main result is that the vorticity evolution remains approximately inertial during at least 5 Kolmogorov time scales.

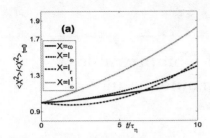 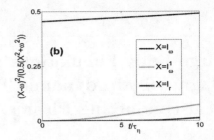

Fig. 2. An example of comparison of evolution of vorticity and material lines. Dashed lines present evolution of material lines parallel to vorticity, and dashed–dotted lines present randomly chosen lines. On the figure: $l_\omega \| \omega$ for the regions in which the forcing and the viscous terms are balancing each other, $l^l_\omega \| \omega$ for the randomly chosen positions.

The next figure represents comparison of the evolution of vorticity and material elements for the two sets as in figure 1. The main result shown in fig. 2 is consistent with the previous one. It shows that evolution of vorticity and material lines is very similar for particle trajectories originating from locations in which the forcing and the viscous terms are balancing each other. On the conceptual level the main result of this communication is the feasibility of comparative study of evolution of material line elements and that of vorticity which is almost pure inertial (locally in time and space) in the regions where the forcing and the viscous terms are in approximate balance for at least 5 Kolmogorov time microscales. The success of such an approach most probably is due to the persistency of Lagrangian evolution more evidenced from much larger characteristic times of Lagrangian correlations than corresponding Eulerian ones. We have shown that generally this evolution is qualitatively different even for the material lines which initially coincide with vorticity which becomes drastic for other material line elements. An important technical point is that when addressing issues involving vorticity (and its Laplacian) adequate resolution with a grid 128^3 is achieved $Re_\lambda \approx 50$ only. Looking at modified helicity at this Reynolds number for adequate resolution requires 256^3.

References

1. Galanti, B, Gendler-Fishman, D. and Tsinober, A. (2006) In: *Turbulence, Heat and Mass Transfer 5*. K. Hanjalic, Y. Nagano and S. Jakirlic (Editors), pp.105—108.
2. Kuz'min G.A. (1983) *Phys.Lett.*, **A96**, 88—90.
3. Oseledets, V.I. (1989) *Russ. Math. Surveys*, **44**, 210—211.
4. Tsinober A. *An informal introduction to turbulence*. Kluwer (2001).

Large Eddy Simulations of Compressible Magnetohydrodynamic Turbulence in Heat-conducting Plasma

A.A. Chernyshov[1], K.V. Karelsky[1] and A.S. Petrosyan[1]

Theoretical section, Space Research Institute of Russian Academy of Sciences, Profsoyuznaya 84/32, 117997, Moscow, Russia apetrosy@iki.rssi.ru

In this paper we present the large eddy simulation (LES) technique for study of compressible magnetohydrodynamic (MHD) turbulence in heat-conducting plasma. We consider the complete system of compressible MHD equations for heat-conducting plasma. We develop parameterizations of extra terms appearing in energy equation due to presence of the magnetic field.

In order to simplify the equations describing turbulent MHD flow with variable density, we use the Favre filtering to avoid additional SGS terms. A Favre-filtered variable is defined as: $\tilde{F} = \overline{\rho F}/\bar{\rho}$. To denote the filtration in this relation we use two symbols, namely the overbar indicates the ordinary filtering and the tilde specifies the mass-weighted filtering. Applying the filtering operation, we obtain the resolved system of equations of compressible magnetohydrodynamics.

In the following we shall use notations: ρ is the density; T is the temperature; u_j is the velocity in direction x_j; B_j is the magnetic field in direction x_j; $P = p + \frac{1}{2} B_i B_i / M_a^2$ is the total pressure; diffusive fluxes are given by $\tilde{\sigma}_{ij} = 2\mu \tilde{S}_{ij} - \frac{2}{3}\mu \tilde{S}_{kk}\delta_{ij}$ and $\tilde{q}_j = -k\partial \tilde{T}/\partial x_j$, where $S_{ij} = 1/2 \left(\partial u_i/\partial x_j + \partial u_j/\partial x_i\right)$ is the strain rate tensor; μ is the coefficient of molecular viscosity; η is the coefficient of magnetic diffusivity.

The effect of the SGSs appears on the right-hand side of governing equations through the SGS stresses τ_{ij}^u; magnetic SGS stress tensor τ_{ij}^b; SGS heat flux Q_j; SGS turbulent diffusion J_j; SGS magnetic energy flux V_j; SGS energy of the interaction between the magnetic tension and velocity G_j. These quantities are defined as:

$$\tau_{ij}^u = \bar{\rho}\left(\widetilde{u_i u_j} - \tilde{u}_i \tilde{u}_j\right) - \frac{1}{M_a^2}\left(\overline{B_i B_j} - \bar{B}_i \bar{B}_j\right);\tag{1}$$

$$\tau_{ij}^b = \left(\overline{u_i B_j} - \tilde{u}_i \bar{B}_j\right) - \left(\overline{B_i u_j} - \bar{B}_i \tilde{u}_j\right);\tag{2}$$

$$Q_j = \bar{\rho}\left(\widetilde{u_j T} - \tilde{u}_j \tilde{T}\right);\tag{3}$$

$$J_j = \bar{\rho}\left(u_j\widetilde{u_k u_k} - \tilde{u}_j\widetilde{u_k u_k}\right); \tag{4}$$

$$V_j = \left(\overline{B_k B_k u_j} - \overline{B_k B_k}\tilde{u}_j\right); \tag{5}$$

$$G_j = \left(\overline{u_k B_k B_j} - \tilde{u}_k \bar{B}_k \bar{B}_j\right). \tag{6}$$

Here the diffusion terms connected with magnetic energy are also disregarded. SGS viscous diffusion $D_j = \overline{\sigma_{ij}u_i} - \tilde{\sigma}_{ij}\tilde{u}_i$ is a small term in the total energy equation and is about 5% of the divergence of Q_j. Therefore, we have neglected term D_j.

The SGS stresses (1) and (2) are the only unclosed terms that appear in the momentum and the induction equations. For these SGS terms we use Smagorinsky model for MHD-case:

$$\tau_{ij}^u = -2C_s\bar{\rho}\tilde{\triangle}^2|\tilde{S}^u|\left(\tilde{S}_{ij} - \frac{1}{3}\tilde{S}_{kk}\delta_{ij}\right) + \frac{2}{3}Y_s\bar{\rho}\tilde{\triangle}^2|\tilde{S}^u|^2\delta_{ij}; \tag{7}$$

$$\tau_{ij}^b = -2D_s\tilde{\triangle}^2|\bar{j}|\bar{J}_{ij}. \tag{8}$$

Here, $|\tilde{S}^u| = \left(2\tilde{S}_{ij}\tilde{S}_{ij}\right)^{1/2}$ is the filtered strain-rate magnitude; $\bar{j}$ is the resolved electric current density; and $\bar{J}_{ij} = \frac{1}{2}\left(\partial\bar{B}_i/\partial x_j - \partial\bar{B}_j/\partial x_i\right)$ is the large-scale magnetic rotation tensor. Parameters C_s, Y_s and D_s are model coefficients, their values being self-consistently computed during run time with the use of the dynamic procedure has been applied for MHD.

In filtered energy equation several SGS terms appear that need to be parameterized. Consider first the subgrid-scale heat flux Q_j. The eddy diffusivity model is used for the closure of this term:

$$Q_j = -C_s\frac{\tilde{\triangle}^2\bar{\rho}|\tilde{S}^u|}{Pr_T}\frac{\partial\tilde{T}}{\partial x_j}, \tag{9}$$

where C_s is the coefficient used earlier in the Smagorinsky model (7). The Pr_T is the turbulent Prandtl number which also is calculated dynamically.

The model for J_j was is based on analogy to Reynolds-averaged Navier-Stokes equations and on the assumption that $\tilde{u}_i \simeq \tilde{\tilde{u}}_i$. Here we use the model in the form: $J_j \simeq \tilde{u}_k\tau_{jk}^u$ where tensor τ_{jk}^u was found above.

For the final closure of the full system of compressible magnetohydrodynamics equations, it is necessary to parameterize the new SGS terms in energy equation arising from the presence of the magnetic field. In order to derive these SGS terms we use an approach based on generalized central moments, extended here for MHD case. We obtain for G_j:

$$G_j = \varphi\left(B_j, B_k, u_k\right) + \bar{B}_k\varphi\left(u_k, B_j\right) + \tag{10}$$
$$\tilde{u}_k\varphi\left(B_j, B_k\right) + \bar{B}_j\varphi\left(u_k, B_k\right).$$

$$\frac{1}{2}V_j - G_j \simeq \bar{B}_k\tau_{jk}^b. \tag{11}$$

In order to assess the LES technique we have performed simulations of decaying isotropic turbulence using the second order of accuracy for MHD equations written in the conservative form as the reference DNS, the same integration domain (a cube of linear extension 4π) and the same set of parameters ($Re = 75$, $Re_m = 10$, $M_s = M_a = 0.6$, $Pr = 0.01$, $\gamma = 5/3$). In order to separate turbulent flow into large and small eddies we have applied Gaussian filter.

In the following (Fig. 1,2) asterisk line represents the filtered DNS results, solid line is LES results without any SGS closure, that is, DNS on the coarse LES-grid starting from filtered initial conditions. Dashed line presents LES results. In order to measure the effect of modelling SGS terms in energy equation, additional simulation is performed which omit the energy SGS terms. The dot line represents LES of decaying turbulence where only SGS tensors τ_{ij}^u and τ_{ij}^b are calculated.

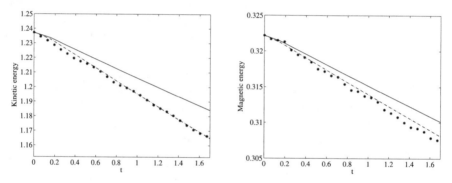

Fig. 1. Time evolution of the kinetik energy (left) and magnetic energy (right).

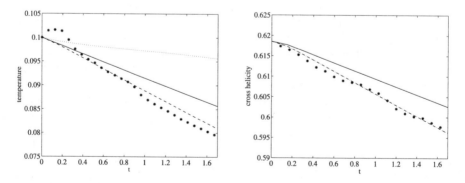

Fig. 2. Time evolution of the temperature (left) and the cross-helicity (right).

The effect of a finite cascade time on the normalized energy dissipation

Wouter Bos, Liang Shao, Jean-Pierre Bertoglio

LMFA, Ecole Centrale de Lyon - CNRS - UCBL -INSA, Ecully, France
wouter.bos@ec-lyon.fr

The normalized dissipation rate C_ϵ is a quantity that has been extensively studied in the literature. C_ϵ is defined as:

$$C_\epsilon = \frac{\epsilon \mathcal{L}}{\mathcal{U}^3},\tag{1}$$

in which ϵ is the viscous dissipation of turbulent kinetic energy, $\mathcal{L}$ is the integral lengthscale and $\mathcal{U}$ is the root-mean-square velocity.

An important and thoroughly discussed issue is the possibility of a universal value of C_ϵ for isotropic turbulence at high Reynolds number. Recently Burattini, Lavoie and Antonia [1] revisited the behavior of C_ϵ and in particular discussed its dependency with the Reynolds number. Considerable scatter was observed in the data of various experiments. The same scatter was already observed by Sreenivasan [2]. The Direct Numerical Simulations (DNS) of isotropic turbulence Kaneda *et al.* [3], exhibited significantly less scatter at large Reynolds number, which seemed to corroborate the general belief that, at high Reynolds numbers, C_ϵ tends to a constant value, at least for isotropic turbulence.

An important problem when evaluating the value of C_ϵ is that experiments of isotropic turbulence are generally done in decaying grid turbulence, whereas the value of C_ϵ from DNS are generally corresponding to forced turbulence, since high Reynolds number fully developed decaying turbulence is still extremely expensive to simulate.

In the present work we address the issue of the high Reynolds number universality of the value of C_ϵ by performing LES, DNS and closure calculations of forced and decaying isotropic turbulence. The simultaneous use of these different methods allows to compare decaying and forced turbulence and to carefully evaluate the effect of initial conditions, forcing scheme and Reynolds number on the value of C_ϵ.

In figure 1 the effect of the initial conditions for decaying turbulence is illustrated. After a transient period of a few turnover times, the LES computations yield values of C_ϵ close to one for decaying turbulence for all the

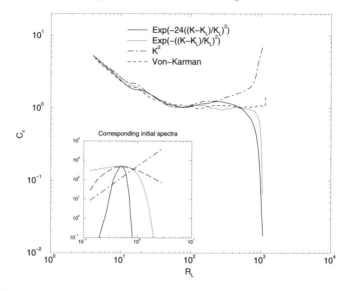

Fig. 1. Starting from entirely different initial energy distributions, all computations yield values of C_ϵ close to one for decaying turbulence.

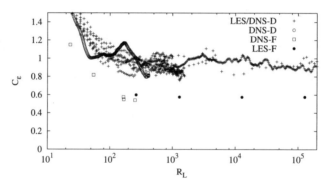

Fig. 2. DNS and LES results for the normalized dissipation rate as a function of the Reynolds number

different initial distributions at intermediate Reynolds number. In figure 2 DNS and LES results for the normalized dissipation rate as a function of the Reynolds number are shown. For comparable Reynolds number, the values for decaying turbulence (indicated by D in the figure) seem to be larger than the values of C_ϵ for the forced turbulence (indicated by F).

A scenario, based on simple phenomenological arguments, can be proposed which explains the observations in Figure 2. These arguments are based on two concepts. First, we define $\epsilon_f(t)$, the flux of energy that enters the *energy cascade* at time t. It can be assumed that this flux is determined by the large scale quantities $\mathcal{U}$ and $\mathcal{L}$ so that dimensional analysis gives,

$$\epsilon_f = C_{\epsilon_f} \frac{\mathcal{U}^3}{\mathcal{L}}. \tag{2}$$

It is stressed that this expression is different from (1), as it involves the energy flux entering the cascade and not the viscous dissipation. In stationary turbulence these quantities can supposed to be equal ($\epsilon_f(t) = \epsilon(t)$) so that for forced turbulence $C_\epsilon(Forced) = C_{\epsilon_f}$. In general this must not be true. We therefore introduce the second concept, the cascade time $\mathcal{T}_c$, which is the time it takes for the energy entering the cascade at time t to reach the viscous dissipation range:

$$\epsilon_f(t) = \epsilon(t + \mathcal{T}_c). \tag{3}$$

Combining equation (1), (2) and (3) and using that $C_\epsilon(Forced) = C_{\epsilon_f}$ the following expression is obtained:

$$\frac{C_\epsilon}{C_\epsilon(Forced)} = \frac{\epsilon(t)}{\epsilon(t + \mathcal{T}_c)}. \tag{4}$$

If we consider self-similar decaying turbulence, $\epsilon(t)$ will be a decreasing function of time so that the ratio in (4) is larger than unity, as observed in figure 2. A more detailed study of this issue, using dimensional arguments for the cascade time (Lumley [4]) and a model spectrum proposed by Comte-Bellot and Corrsin [5], is performed in [6]. It is shown comparing LES, DNS and EDQNM results combined with these phenomenological arguments, that for isotropic turbulence the ratio $C_\epsilon(Decaying)/C_\epsilon(Forced)$ tends to a constant value of order 2 at high Reynolds numbers.

References

1. P. Burattini, P. Lavoie, and R. A. Antonia. On the normalized turbulent energy dissipation rate. *Phys. Fluids*, 17:098103, 2005.
2. K.R. Sreenivasan. On the scaling of the turbulence energy dissipation rate. *Phys. Fluids*, 27:1048, 1984.
3. Y. Kaneda, T. Ishihara, M. Yokokawa, K. Itakura, and A. Uno. Energy dissipation rate and energy spectrum in high resolution direct numerical simulations of turbulence in a periodic box. *Phys. Fluids*, 15(L21), 2003.
4. J.L. Lumley. Some comments on turbulence. *Phys. Fluids*, 4:206, 1992.
5. G. Comte-Bellot and S. Corrsin. The use of contraction to improve the isotropy of grid-generated turbulence. *J. Fluid Mech.*, 25:657–682, 1966.
6. W.J.T. Bos, L. Shao, and J.-P. Bertoglio. Spectral imbalance and the normalized dissipation rate of turbulence. *Phys. Fluids*, 19, 2007.

Anisotropy in three-dimensional MHD turbulence

Bigot B.[1,2], Galtier S.[1], and Politano H.[2]

[1] Institut d'Astrophysique Spatiale, UMR 8617, Université de Paris-Sud - Bât. 121, 91405 Orsay Cedex
[2] Laboratoire Cassiopée, UMR 6202, Observatoire de la Côte d'Azur, 06304 Nice Cedex

We perform numerical simulations to characterize the transition from strong to weak magnetohydrodynamic (MHD) turbulence for freely decaying incompressible flows in presence of a uniform magnetic field $\mathbf{B}_0$. Due to reduction of energetic transfers along $\mathbf{B}_0$, the flow anisotropy, as measured for example by Shebalin angles, increases together with $\mathbf{B}_0$ intensity. At high $\mathbf{B}_0$, ratios of Alfvén to nonlinear turnover timescales stay below unity allowing for an Iroshnikov-Kraichnan description of anisotropic flows.

1 Introduction

In order to better understand the solar wind turbulence and its anisotropic properties due to the presence of a strong component of solar magnetic field [1], we perform tri-dimensional simulations of incompressible MHD flows as first approximation to astrophysical plasmas. The conductive flow is thus described by the coupled Navier-Stokes and induction equations. The kinetic and magnetic energy transfers create structures of smaller and smaller size up to dissipative and diffusion scales. The external magnetic field induces the development of Alfvén waves and flow anisotropy along the $\mathbf{B}_0$ direction. At $\mathbf{B}_0$ intensity strong enough, this can even lead to bi-dimensionalization of the flow. For B_0 intensities well above the *rms* level of kinetic and magnetic fluctuations, the flow dynamics becomes dominated by the Alfvén waves dynamics [2], leading to weak turbulence as opposed to turbulence also known as strong. For different intensities of the background magnetic field, $B_0 = 0, 1, 5$ and 15, we integrate numerically the MHD equations, in a 2π-periodic box using a pseudo-spectral method with 256^3 collocation points, and a second-order finite-difference scheme in time. The initial velocity, $\mathbf{v}$, and magnetic, $\mathbf{b}$, fields correspond to spectra proportional to $k^2 exp(-k/2)^2$ for $k = [1, 8]$, *i.e.* a flat modal spectrum up to $k = 2$, with equal kinetic and magnetic energies $E_v(t = 0) = E_b(t = 0) = 1/2$. At scale injection, the initial kinetic and magnetic Reynolds numbers are about 800 for flows at $\nu = \eta \sim 4 \times 10^{-3}$. The dynamics of the flow then freely decays.

2 Temporal analysis

To quantify the degree of anisotropy associated with flows, we use the generalized Shebalin angles [3] defined as $\tan^2 \theta_\mathbf{q} = \sum k_\perp^2 |\mathbf{q}(\mathbf{k}, t)|^2 / \sum k_z^2 |\mathbf{q}(\mathbf{k}, t)|^2$, with $k_\perp^2 = k_x^2 + k_y^2$, and $\mathbf{q}$ stands for $\mathbf{v}$ or $\mathbf{b}$ fluctuations (see Fig. 1). Note that here $\mathbf{B_0}$ is in the $\mathbf{z}$-direction. Initially, $\theta_{\mathbf{v},\mathbf{b}} \sim 54, 74°$ corresponding to an isotropic $3D$ flow. The angle depends only on the intensity of $\mathbf{B_0}$. For $B_0 = 0$ the energy transfer is similar in all directions and the temporal evolution of Shebalin angles remains almost constant, close to its initial value. For $B_0 = 5$ and $B_0 = 15$, the Shebalin angles quickly increase and stabilize around $78°$. Thus, as expected, anisotropy develops along $\mathbf{B_0}$-direction. However, the flow is not totaly confined in planes perpendicular to $\mathbf{B_0}$, which should be the case of a purely bi-dimensional flow with current and vortex lines perpendicular to flow plane. This $2D$ configuration compares to asymptotic value of $90°$ of Shebalin angles: spectra have all their energy in modes perpendicular to $\mathbf{B_0}$. Flows with highest values of B_0 behave quite similarly while for $B_0 = 1$, a transitional regime from case with no background magnetic field is observed.

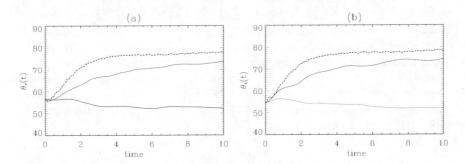

Fig. 1. Temporal evolutions of Shebalin angles of velocity (a) and magnetic (b) field for $B_0 = 0$ (black), $B_0 = 1$ (blue), $B_0 = 5$ (dash black) and $B_0 = 15$ (red).

To characterize flow timescales, we investigate ratios, $\chi^\pm(t) = \tau_A^\pm(t)/\tau_{NL}^\pm(t)$, between Alfvén times $\tau_A^\pm$ of shear Alfvén waves and eddy turnover times $\tau_{NL}^\pm$, based on toroïdal components of the Elsässer variables $\mathbf{z}^\pm = \mathbf{v} \pm \mathbf{b}$. At characteristic flow scales, $\chi^\pm(t) = \tau_A^\pm(t)/\tau_{NL}^\pm(t) = L_z^\pm(t)/B_0 \cdot (L_\perp^\pm(t)/z_{rms}^\mp(t))^{-1}$ is defined on parallel $L_z^\pm(t) = \int \int E^\pm(k_\perp, k_z)k_z^{-1}dk_\perp dk_z/E^\pm$ and perpendicular $L_\perp^\pm(t) = \int \int E^\pm(k_\perp, k_z)k_\perp^{-1}dk_z dk_\perp/E^\pm$ integral scales. At these flow scales, the ratio $\chi^+(t)$ stays well below unity (see Fig. 2a). Similarly, at smaller scales, $\chi_l^+(t)$, estimated as $k_\perp z_l^-/k_z B_0$ (with $z_l^- \sim \sqrt{E^-(k_\perp, k_z)k_\perp k_z}$), are smaller than unity at all times, as shown in Fig. 2b for given values of $(k_\perp, k_z)$. These values of time ratios are typical of a wave dominated regime and allow for an Iroshnikov-Kraichnan description of anisotropic flows rather than for a Kolmogorov one [4]. For some specific $(k_\perp, k_z)$, $\chi_l(t)$ is almost constant between

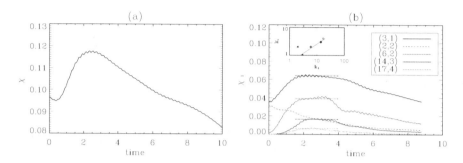

Fig. 2. Temporal evolutions of $\chi^+(t)$ based on integral scales L_z^+ and $L_\perp^+$ (a) and $\chi_l^+(t)$ based on $(k_\perp, k_z)$ for different values (b) (see text for definitions). Insert frame: plot of of k_z against $k_\perp$ fitted by a 2/3 scaling law.

$t = 2.$ and $t = 3.5$, a time interval after generation of all small scales and within intense nonlinear interaction. Note that this is not true for all $(k_\perp, k_z)$, see for example plot for $(2, 2)$ mode in Fig. 2b. The observed plateau behavior could reveal an underlying relationship $k_z \sim k_\perp^{2/3}$ as stated in [4], although plateau levels are not at the same value, probably due to absence of well established constant fluxes in energy cascades in our simulation. However, plot of k_z against $k_\perp$ reveals a 2/3 scaling law when the corresponding $\chi_l^+(t)$ displays a plateau (see Fig 2.b, insert frame). Same behaviors are observed for time ratios $\chi^-(t)$ and $\chi_l^-(t)$ (defined as $k_\perp \sqrt{E^+(k_\perp, k_z)k_\perp k_z}/k_z B_0$), when dealing with $\mathbf{z}^-$ field dynamics.

References

1. Dasso S., Milano L.J., Matthaeus W., Smith C.W.: Astro. J. **635**, 181 (2005)
2. Galtier S., Nazarenko S.V., Newell A.C. & Pouquet A. : J. Plasma Phys. **63**, 447 (2000)
3. Oughton S., Priest E.R. & Matthaeus W.H.: J. Fluid Mech. **280**, 95 (1994)
4. Galtier S., Pouquet A. & Mangeney A.: Phys. Plasmas **12**, 092310-1 (2005)

Receptivity to Roughness and Vortical Free-stream Modes

Lars-Uve Schrader, Luca Brandt, and Dan S. Henningson

Linné Flow Centre, KTH Mechanics, 100 44 Stockholm, Sweden
schrader@mech.kth.se

Transition in boundary layers is due to the growth of unstable disturbances finally breaking down to turbulence. Numerous experimental investigations have shown that the initial conditions of these unstable waves depend strongly on the external perturbation environment. Therefore, this study is devoted to the initial receptivity phase.

In particular, the Falkner-Skan-Cooke boundary layer is considered here. The misalignment between pressure gradient and free-stream velocity induces an inflectional wall-normal profile supporting inviscid instability. A number of papers in literature deal with the receptivity of swept boundary layers to surface roughness, e.g. [1, 2]. This study also includes the receptivity to vortical free-stream perturbations.

Simulation Approach. The presented results are obtained using a spectral method to solve the three-dimensional, time-dependent, incompressible Navier-Stokes equations. Streamwise periodic boundary conditions are combined with the spatially developing boundary layer by the implementation of a "fringe region", where the flow is forced to the prescribed inlet velocity field. The inflow consists of a Falkner-Skan-Cooke profile at $Re_{\delta^*} = 337.9$ with a favorable pressure gradient ($\beta_H = 0.51$) and a sweep angle of 55.26°, as in [3]. The flow field is resolved by $n_x \times n_y \times n_z = 576(512) \times 65(145) \times 16(8)$ grid points; the values in brackets apply to the simulations employing vortical free-stream modes.

SURFACE ROUGHNESS. Since the spectral framework of the DNS requires uniform boundaries, roughness elements on the plate surface need to be modelled by modified boundary conditions. A Taylor expansion of fourth order is used to project the no-slip conditions along the desired bump contour onto the wall plane. This model has turned out to provide an accurate approximation of the flow over a bump.

VORTICAL FREE-STREAM MODES. The vortical disturbances in the outer flow are modelled by modes of the continuous spectrum of the Orr-Sommerfeld and Squire operators and are forced in the fringe region. Since these modes behave sinusoidally in the free stream and decay towards the wall, they provide a suitable model for free-stream perturbations.

Results. Fig. 1 shows the base flow at the inflow plane in free-stream aligned coordinates, in particular the inflectional cross-flow profile W_{FS} (left plot), and the neutral curve for the inflow conditions (right plot). The contour lines represent the angular frequency of the eigenmodes, showing that there is an unstable stationary mode for a large range of wavenumber vectors $(\alpha, \beta)^T$. The crosses mark the stationary and the most unstable travelling wave at $\beta = -0.25$, giving $\alpha_0 = 0.315$ and $\alpha_1 = 0.235$, respectively.

SURFACE ROUGHNESS. The left plot of Fig. 2 displays the chordwise shapes of three different spanwise periodic bumps (inserted figure) and their spectral representation $H(\alpha)$. The bumps are centered around $x_R = 21$ and feature a chordwise and spanwise length-scale of 10.0 and 25.14, respectively, corresponding to $2\alpha_0$ and $-\beta$. Downstream of $x = 80$ the disturbance evolution is determined solely by the most unstable stationary wave with $\alpha_0 = 0.315$. The disturbance-growth curves collapse, if they are re-scaled based on $H(\alpha_0)$, as shown in the right plot. Hence, the receptivity of the boundary layer to roughness depends only on the the spectral content of the bumps at α_0, but not on their specific shapes. A receptivity coefficient defined as $C_R = u_m'^2(x_R)/H(\alpha_0)^2$ turns then out to be constant, see Tab. 1, where "m" denotes the wall-normal maximum and $u_m'^2(x_R)$ is obtained from a PSE simulation.

VORTICAL FREE-STREAM MODES. The left plot of Fig. 3 shows the shape of three stationary modes with different wall-normal length scales l_y and of a travelling mode. The chordwise evolution of $u_m'^2$ is depicted in the right figure. Downstream of $x = 230$ the growing cross-flow disturbance starts to dominate over the decaying free-stream perturbation. The receptivity coefficient is taken as the ratio between the amplitudes of the cross-flow mode and the free-stream mode at x_R. In comparison with surface roughness free-stream modes are less effective in exciting stationary cross-flow vortices. In contrast, the most unstable travelling mode is more excited by free-stream modes.

Table 1. Receptivity coefficients. Left part: Roughness, $C_R = A_0^{CF}/H(\alpha_0)^2$, where $A_0^{CF} = u_{m,CF}'^2(x_R)$. Right part: Free-stream mode, $C_{FS} = A_0^{CF}/A_0^{FS}$, where $A_0^{FS} = u_{m,FS}'^2(x_R)$. "CF"=cross flow; "FS"= free stream

Bump	$H(\alpha_0) \cdot 10^4$	$A_0^{CF} \cdot 10^8$	C_R	Mode	ω	l_y	$A_0^{FS} \cdot 10^5$	$A_0^{CF} \cdot 10^7$	$C_{FS} \cdot 10^3$
A	1.184	10.43	7.440	D	0	10.5	24.48	5.843	2.387
B	1.038	9.146	7.440	E	0	6.3	10.24	2.776	2.711
C	1.775	15.64	7.440	F	0	4.5	9.461	1.633	1.726

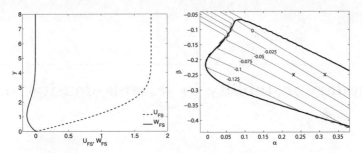

Fig. 1. Falkner-Skan-Cooke boundary layer flow for $Re_{\delta*} = 337.9$, $\beta_H = 0.51$ and $\phi_0 = 55.26°$. Left: Base flow in swept coordinates. Right: Neutral curve

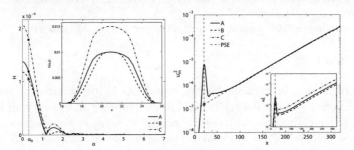

Fig. 2. Roughness-induced perturbation. Left: Bumps in physical and spectral space. Right: Evolution of u'^2_m and scaled curves. Comparison with PSE

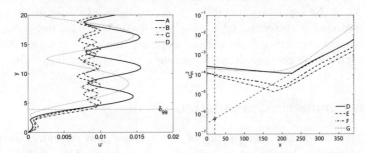

Fig. 3. Perturbation of the boundary layer by free-stream modes. Left: Mode shapes at $x_R = 21$. Right: Streamwise evolution of u'^2_m. Comparison with PSE

References

1. F. P. Bertolotti. Receptivity of three-dimensional boundary layers to localized wall roughness and suction. *Phys. Fluids*, 12:1799–1809, 2000.
2. M. Choudhari. Roughness-induced generation of crossflow boundary vortices in three-dimensional boundary layers. *Theor. Comput. Fluid Dyn.*, 6:1–30, 1994.
3. M. Högberg and D. Henningson. Secondary instability of cross-flow vortices in falkner-skan-cooke boundary layers. *J. Fluid Mech.*, 368:339–357, 1998.

Secondary instability in variable-density round jets

Joseph W. Nichols[1,2], Jean-Marc Chomaz[1], and Peter J. Schmid[1]

[1] Laboratoire d'Hydrodynamique (LadHyX), CNRS-École Polytechnique, 91128
Palaiseau cedex, France.
[2] Corresponding author: joseph.nichols@ladhyx.polytechnique.fr

1 Abstract

Side jet formation in variable-density round jets is investigated by means of direct numerical simulation (DNS) and linear stability analysis. From DNS, it is observed that a light jet with density ratio $S = \rho_0/\rho_j = 4$ supports sustained side jets which eject fluid from the center of the jet in a star-shaped pattern. These side jets persist over an axial distance of approximately 5 jet diameters before the jet transitions to turbulence, and do not precess around the jet. It is conjectured that this behavior can be explained by a change in the local properties of the secondary instability from convective to absolute in nature. This hypothesis is tested by examining the spatio-temporal development of the wavepacket resulting from a small impulse, taken about a non-diffusing periodic base state corresponding to the saturated primary instability.

2 Direct numerical simulation results

Figure 1(a) shows the relationship of the primary and secondary instabilities as a variable-density jet transitions to turbulence. The jet fluid enters as a laminar flow from the left, and then quickly undergoes an axisymmetric primary instability which is self-sustaining for sufficient values of S [2]. Further downstream, side jets form, which are visible as coherent axially aligned structures near the middle of the figure (see also [3]). These are seen to have a star-shaped cross section as shown in figure 1(c). Finally, near the outlet of the domain, the jet transitions to a fully turbulent flow. In figure 1(b), vortex tubes are visualized by plotting an isosurface of $Q = 1/2\left(\Omega_{ij}\Omega_{ij} - S_{ij}S_{ij}\right)$ for the same field. The primary instability is evident as a train of Kelvin-Helmholtz rings in the left half of this figure. As these vortex rings convect downstream, they develop a secondary instability which manifests itself as azimuthal oscillations on each ring. In the figure, these azimuthal oscillations have grown enough to be visible on the fifth vortex ring from the left.

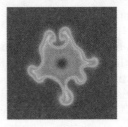

Fig. 1. (a) Three dimensional isosurface ($z = 0.18$) of a conserved scalar, where $z = 0, 1$ signify pure ambient/jet fluid, respectively. (b) Vortex visualization of the same field, using the Q-criterion. (c) Radial cross section of the scalar field taken at an axial location of 7.39 jet diameters downstream of the nozzle.

3 Linear impulse results

To test the hypothesis that the side jets are due to an absolute secondary instability, we consider the evolution of an infinitesimal localized perturbation superimposed on a base state consisting of the saturated primary instability.

3.1 Base state

The saturated base state was calculated using an axisymmetric domain, periodic in x, of length L equal to the axial wavelength of the primary instability observed from the DNS. The flow is allowed to develop about a fixed parallel mean flow. This method captures one saturated vortex ring within the period of the domain.

3.2 Wavepacket

The calculated base state forms the axisymmetric component of a three dimensional simulation to which a perturbation of azimuthal wavenumber n is applied. We have chosen $n = 6$ in the current study because six side jets were observed in the full DNS, but the reader should note that in general, n may be any integer. The axisymmetric mode is frozen by means of a body force which exactly compensates for the slow diffusion of the saturated primary instability. (Since this diffusion is small compared to the saturation process in the previous section, we can be reasonably sure that the obtained base state is representative of the primary instability observed in the full DNS. To accurately observe the development of the wavepacket, however, we must eliminate this time dependence due to diffusion which otherwise might obscure the small amplitudes in which we are interested.) To include the interaction of axial harmonics in this secondary stability analysis, the base state is repeated 8 times in a computational domain of length $8L$.

Figure 2(a) displays the wavepacket amplitude (see [1]) at various times as a function of group velocity $c_g = x/t$ resulting from a initial perturbation applied at $x = 0$ and $t = 0$. Since there exists positive growth for negative group velocities as shown in figure 2(a), we conclude that the flow is absolutely unstable with respect to secondary perturbations of azimuthal wave number $n = 6$.

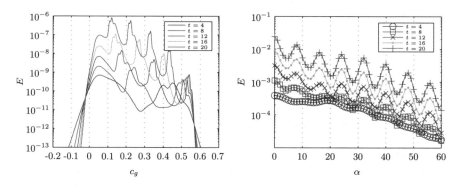

Fig. 2. (a) Time development of wavepacket, (b) axial spectrum of wavepacket.

Figure 2(b) shows the wavepacket amplitude at various times as a function of axial wavenumber α. The Floquet modes consist of a sum of fundamental and harmonic wavenumbers separated by intevals of 8. We find that the Floquet mode corresponding to wavenumbers $\alpha = 0, 8, 16, \ldots$, exhibits the strongest growth. This agrees with the stationary side jets observed in the full DNS which were characterized by long axial wavelengths.

References

1. F. Gallaire and J.-M. Chomaz. Mode selection in swirling jet experiments: a linear stability analysis. Journal of Fluid Mechanics, **494**:223-253, 2003.
2. P.A. Monkewitz and K.D. Sohn. Absolute instability in hot jets. AIAA Journal, **26**:911-916, 1988.
3. J.W. Nichols, P.J. Schmid, and J.J. Riley. Self-sustained oscillations in variable-density round jets. Journal of Fluid Mechanics, 2007 (*in press*).

Axisymmetric absolute instability of swirling jets

J. J. Healey

Department of Mathematics, Keele University, Keele, Staffs. ST5 5BG, UK
j.j.healey@maths.keele.ac.uk

Summary. The spatio-temporal stability of axisymmetric inviscid disturbances to swirling jets is investigated. There is a known difficulty associated with the pinch-point controlling absolute instability. Previous authors have noted that for axisymmetric waves the pinch-point moves into the left half of the complex wavenumber plane as the swirl increases. This behaviour appears unphysical because such waves diverge exponentially in the radial direction. However, our recent understanding of this phenomenon in the context of the rotating disk boundary layer tells us that it implies that confining the flow by placing it inside a cylinder concentric with the jet axis, will be destabilizing for absolute instabilities, even when this cylinder has very large radius. This prediction is confirmed, and it is shown for the first time that swirling jets can become absolutely unstable to axisymmetric waves. The possible connection with vortex breakdown is discussed.

1 Introduction

The flow separating from a wing-tip undergoes a vigorous swirling motion, and for large aeroplanes these trailing vortices are strong and persistent and represent a hazard to following aircraft. The resulting flow is characterized by the presence of both axial and azimuthal velocity components, i.e. swirling jets. In this paper the absolute/convective instability characteristics of axisymmetric inviscid disturbances to a family of model swirling jets are investigated. The new result is that these flows can be absolutely unstable to axisymmetric waves. If the onset of axisymmetric absolute instability turns out to be related to the appearance of axisymmetric vortex breakdown, then these results could provide guidance for how a trailing vortex might be modified to promote vortex breakdown and thereby greatly enhance its dissipation.

The classification of a flow as either absolutely unstable, in which case disturbances propagate both upstream and downstream, and grow in time, or convectively unstable, when disturbances only grow downstream, can be very helpful. Absolutely unstable flows can act as self-excited ocillators with their own intrinsic dynamics, while convectively unstable flows act as spatial amplifiers of externally introduced disturbances. It has been suggested by [1] and [2] and others that vortex breakdown of a swirling jet might be associated with transition from convective to absolute instability. This idea represents a generalization of Benjamin's theory of a transition from super-critical to sub-critical flow states (but which was based on neutral waves).

The absolute/convective character of a flow can be elegantly determined from a spatio-temporal analysis of the dispersion relation in which both wavenumbers and frequencies are complex, as shown by [3]. Central to the method is the identification of 'pinch-points' at which upstream and down-stream modes coalesce in the complex wavenumber plane.

Although non-axisymmetric waves are known to produce absolute insta-bility (AI) in swirling jets, and have been successfully related to helical modes of vortex breakdown, a difficulty in the application of Briggs' method to ax-isymmetric waves has been noted by [1] and [2]. The 'pinch-point' can cross the imaginary axis of the complex wavenumber plane, leading to solutions that grow, instead of decay, with distance outside the jet. This behaviour of the pinch-point was recently found in the rotating disk boundary layer [4], and the physical consequences were studied in detail in that context [5]. The effects of confinement for this type of problem is counter-intuitive: it turns out that placing a plate parallel to the boundary layer and well outside the boundary layer has a destabilizing effect on the absolute instability, and can convert a convectively unstable flow into an absolutely unstable flow [6].

2 Results

Our new understanding of this behaviour in the rotating disk problem has been used to help us to calculate for the first time AI of inviscid axisymmetric waves in swirling jets. It is found that these waves can grow with distance in the radial direction outside the jet, and that the addition of a bounding cylinder far from the jet converts this spatial growth into an AI. As the distance from the jet to the outer cylinder is reduced, the AI increases, and less swirl is needed to produce AI, see figure 1. It is interesting to note that many experiments on axisymmetric vortex breakdown involve swirling flows in pipes, and it may be that the presence of the pipe is instrumental in creating the AI, which perhaps leads in turn to vortex breakdown. Following previous studies, the model profiles have uniform axial flow, and constant angular velocity, within the jet and both velocity components drop discontinuously to zero outside the jet. A more realistic family of profiles with finite thickness smooth shear layers at the jet edge has also been studied. It is found that increasing the shear layer thickness produces AI, even in the unconfined jet, via an enhancement of the centrifugal instability, see figure 2. For a given swirl, the shear layer at the jet's edge thickens with downstream distance, causing the flow to change from CI to AI at a critical distance from the jet nozzle, possibly providing the location for axisymmetric vortex breakdown.

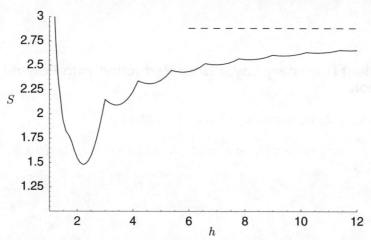

Fig. 1. Neutral curve for axisymmetric AI for confined jets: swirl vs. outer cylinder radius, h, (dashed line is limit as $h \to \infty$). Swirl, S, is maximum azimuthal velocity divided by centre-line axial velocity.

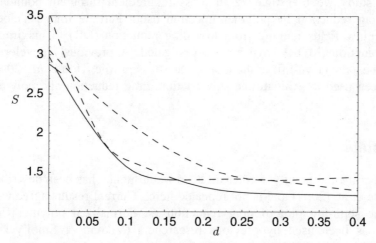

Fig. 2. Neutral curves for axisymmetric AI for smooth jets: swirl vs. shear layer thickness (dashed lines are non-pinching saddles).

References

1. T. Loiseleux, J.-M. Chomaz, P. Huerre: Phys. Fluids **10**, 1120 (1998)
2. D.W. Lim, L.G. Redekopp: Eur. J. Mech. B/Fluids **17**, 165 (1998)
3. R.J. Briggs: *Electron-Stream Interaction with Plasmas.*(MIT Press 1964)
4. J.J. Healey: Proc. R. Soc. Lond. A **462**, 1467 (2006)
5. J.J. Healey: J. Fluid Mech. **560**, 279 (2006)
6. J.J. Healey: J. Fluid Mech. Accepted, in press (2007)

Turbulent Boundary Layer Drag Reduction with Polymer Injection

Y. X. Hou, V. S. R. Somandepalli and M. G. Mungal

Mechanical Engineering Department, Stanford University, Stanford, CA 94305, USA

1 Introduction

In this study we investigate zero pressure gradient turbulent boundary layer drag reduction by polymer injection using Particle Image Velocimetry. Flow fields ranging from low drag reduction (DR) to maximum drag reduction (MDR) have been investigated. A previously developed technique - the $(1-y/\delta)$ fit to the total shear stress profile (Hou *et al.*, 2006) - has been used to evaluate the skin friction, drag reduction and polymer stress.

2 Results

The experimental facility, a constant head water tunnel, has been described in White et al. (2004) and is not repeated here. Current results agree well with the semi-log plot of drag reduction *vs.* normalized polymer flux which has been used by previous researchers (Vdovin & Smol'yakov 1981, Petrie *et al.* 2003) and can be used as a guide to optimize the usage of polymer from a single injector.

The mean velocity profile in polymer flow is reduced near the wall, increases slightly further away from the wall and then merges with that of the Newtonian flow. The mean velocity profiles in wall units, Fig. 1, vary from the classical Newtonian flow line, first due to a shift in the profile, then due to a change in the slope of the profile, to the final MDR limit (Virk, 1975) as drag reduction increases.

The u_{rms} and v_{rms} profiles are in good agreement with previous studies of drag reduced boundary layers (Fontaine *et al.*, 1992) as well as homogeneous channel flows (Warholic *et al.*, 1999).

The Reynolds shear stress in polymer flow is reduced and the reduction is greater when the DR is higher. Sometimes, the Reynolds shear stress for polymer flow is only reduced at smaller y/δ (where y is the wall normal coordinate and δ is the boundary layer thickness) and not affected further outward. The polymer effect on Reynolds shear stress spreads outward from the wall to the outer part of the boundary layer gradually with downstream distance. It is seen that the mean velocity responds quickly to the flow condition change, i.e., the suddenly reduced wall shear stress at injection, however, it takes a much longer time for the entire Reynolds shear stress profile to adjust to the same change. The Reynolds shear stress profiles in wall units, Fig. 2, can sometimes be higher than unity, for high polymer concentrations, and this unique feature can be used to judge whether the flow is in equilibrium. Here the stress profile measured just after the injection station (position $x02$) reflects this non-equilibrium behavior while other stations (positions $x2$ and $x4$) appear to be in equilibrium as the flow evolves downstream.

3 Conclusions

A comprehensive dataset of turbulence quantities for turbulent boundary layer drag reduction with single slot polymer injection has been generated and shows the evolution of mean and fluctuating quantities and Reynolds stresses. The flow evolves from non-equilibrium to equilibrium for high polymer concentration. The data is useful for understanding of DR flows and for comparison to numerical predictions.

References

Fontaine, A. A., Petrie, H. L. & Brungart, T. A. 1992 Velocity profile statistics in a turbulent boundary layer with slot-injected polymer. *J. Fluid Mech.* **238**:435 - 466.

Hou, Y. X., Somandepalli, V. S. R. & Mungal, M. G. 2006 A technique to determine total shear stress and polymer stress profiles in drag reduced boundary layer flows. *Experiments in Fluids*, **40**, 589-600.

Petrie, H. L., Deutsch, S., Brungart, T. A. & Fontaine, A. A. 2003 Polymer drag reduction with surface roughness in flat-plate turbulent boundary layer flow. *Exp. Fluids* **35**:8–23.

Vdovin, A. V. & Smol'yakov, A. V. 1981 Turbulent diffusion of polymers in a boundary layer. *Zh. Prikl. Mekh. Tekh. Fiz.* **4**:98-104 (transl. in UDC532.526 (1982) 526-531, Plenum).

Virk, P. S. 1975 Drag reduction fundamentals. *AIChE J.* **22**:625-656.

Warholic, M. D., Massah, H. & Hanratty, T. J. 1999 Influence of drag-reducing
polymers on turbulence: effects of Reynolds number, concentration and mixing. Exp. *Fluids* **27**:461–472.
White, C. M., Somandepalli, V. S. R. & Mungal M. G. 2004 The turbulence structure of drag-reduced boundary layer flow. *Exp. Fluids* **36**:62-69.

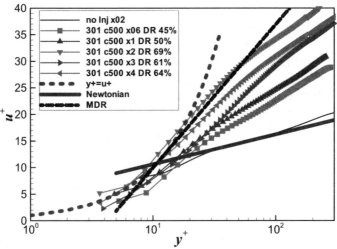

Figure 1. Examples of mean streamwise velocity in wall units at intermediate DR
(45 – 69%) for 500 ppm concentration injection of PEO WSR 301 polymer.

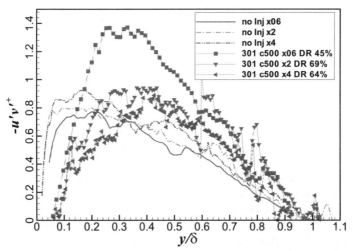

Figure 2. Reynolds shear stress in wall units, showing three examples of no
polymer injection (lines) and three examples of polymer injection (symbols) with
one case not in equilibrium (case x06, close to the injector) for 500 ppm concentration injection of PEO WSR 301 polymer.

Characterisation of Marginally Turbulent Square Duct Flow

M. Uhlmann[1], A. Pinelli[1], A. Sekimoto[2], and G. Kawahara[2]

[1] Modeling and Numerical Simulation Unit, CIEMAT, 28040 Madrid, Spain
{markus.uhlmann,alfredo.pinelli}@ciemat.es
[2] Department of Mechanical Science, Osaka University, 560-8531 Osaka, Japan
{sekimoto,kawahara}@me.es.osaka-u.ac.jp

1 Introduction

We have performed direct numerical simulation of fully developed turbulent flow in a straight duct with square cross-section. The main objective of the present study is to determine the minimal requirements for maintaining turbulence in duct flows [1]. A detailed analysis of this limit regime allows to elucidate the dominant mechanisms governing the marginally turbulent state, where the self-sustaining coherent structures, i.e. streamwise vortices and streaks, have a cross-streamwise length scale comparable with the duct width. It is therefore expected that the coherent structures are of direct relevance to the appearance of secondary flow of Prandtl's second kind.

2 Methodology and Results

For the numerical simulations we have used a pseudo-spectral method applied to the primitive variable formulation of the incompressible Navier-Stokes equations. The time advancement is based on a three-step Runge-Kutta scheme with implicit viscous terms, and continuity is imposed by means of a pressure-correction method. A dealiased Fourier expansion is employed in the streamwise (x) direction, while Chebyshev-polynomial expansions are used in the cross-streamwise (y, z) directions. Turbulence statistics resulting from the present simulations are in good agreement with those obtained in former finite-difference simulations [2] as well as experimental measurements [3].

In order to determine the critical Reynolds number and the minimum streamwise period allowing for self-sustained turbulence, the parameter values have been gradually reduced from their initial values $Re_b \equiv u_b h/\nu = 2205$ (based on bulk velocity u_b and duct half-width h) and $L_x/h = 4\pi$, respectively. It is found that turbulence can be maintained above $Re_b \simeq 1080$ (cf. Fig. 1). If we use the mean friction velocity u_τ as a velocity scale, the corresponding

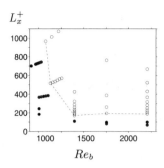

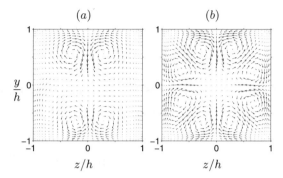

Fig. 1. Map of the critical values for the bulk Reynolds number Re_b and the streamwise box length in wall units L_x^+. (•) laminar flow; (◦) turbulent.

Fig. 2. Mean velocity in the cross-sectional plane of a marginally turbulent square duct at $Re_b = 1180$. (a) Data averaged along the streamwise direction and over a time period of $100\,h/u_b$; (b) average over a much longer temporal interval (approximately $3300\,h/U_b$).

lowest Reynolds number is $Re_\tau \equiv u_\tau h/\nu \simeq 75$. In viscous length units the minimal streamwise period has a value of $L_x^+ \equiv L_x u_\tau/\nu \simeq 200$, approximately independent of the Reynolds number. These critical values are comparable to their counterparts in plane channel flow [1, 4].

Focusing upon one of the marginal cases ($Re_b = 1180$, $L_x/h = 2\pi$), we observe persistent periods of time characterized by two almost quiescent walls (opposite to each other) while turbulence activity is concentrated on the other two walls. During those intervals, a single low-velocity streak is located around the bisector of each one of the active walls whereas no streak is on the other pair of parallel walls. Each streak is flanked by staggered streamwise vortices of alternating signs. Therefore, the mean flow exhibits a pair of counter-rotating streamwise vortices, when the flow field is averaged over intervals of $\mathcal{O}(100)$ bulk flow time units, as shown in Fig. 2a (active walls being located at $y/h = \pm 1$ in this example). It is observed that the pairs of active walls alternate in time, thus leading to a long-time average secondary flow consisting of a superposition of the two possible states, as shown in Fig. 2b. The latter image shows an 8-vortex pattern similar to the well-known secondary flow observed at higher Reynolds number. The tendency to exhibit the 4-vortex state can be quantified by evaluating the distribution of the streamwise vorticity in the cross-stream plane, i.e. by integrating (the square of) its streamwise average separately over the four sectors delimited by the diagonals and comparing the sums of the values pertaining to pairs of opposite walls. We have found that this measure of inequality is largest for the lowest Reynolds numbers and for the shortest domains. Similarly, the mean interval between successive reorientations of the instantaneous streamwise-averaged vortex pattern strongly decreases with the Reynolds number (figures omitted).

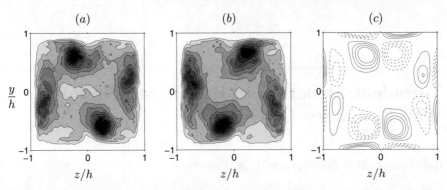

Fig. 3. Statistical data for a case with $Re_b = 1143$ and $L_x/h = 4\pi$, accumulated from 960 instantaneous flow fields over a time interval of $915h/u_b$. (a) Contours indicating .1(.1).9 times the maximum probability of occurrence of vortex centers with positive streamwise vorticity (increasing from white to black) ; (b) the probability for vortices with negative streamwise vorticity; (c) the average streamwise vorticity over the same interval (negative values dashed).

Our conjecture is that the secondary flow in this marginal state is a footprint of the quasi-streamwise vortices associated with the near-wall turbulence regeneration cycle. In order to confirm this scenario, we have identified the central axes of the streamwise vortices by means of the criterion of Kida and Miura [5]. The accumulated data has been used to determine the p.d.f. of the position of vortex centers in the cross-plane. Fig. 3 shows the result for one low-Reynolds-number case, averaged over an interval during which turbulence activity is primarily found near the walls at $y/h = \pm 1$. Comparing the most probable positions for the coherent vortex centers associated with positive/negative streamwise vorticity on the one hand (cf. Fig. 3a-b) to the mean secondary vorticity on the other hand (Fig. 3c) a striking correspondence can be noted. In this case, where the coherent structures are highly constrained by the geometry, the instantaneous streamwise vortices are practically locked into their positions. For higher Reynolds numbers, the p.d.f.s associated with positive/negative streamwise vorticity are expected to overlap near the wall-bisector, restricting the selectivity to the corner region.

References

1. J. Jiménez, P. Moin: J. Fluid Mech. **225**, 213 (1991)
2. S. Gavrilakis: J. Fluid Mech. **244**, 101 (1992)
3. G. Kawahara, K. Ayukawa, J. Ochi, et al: Trans. JSME B **66**, 95 (2000) *(in Japanese)*.
4. M. Nishioka, M. Asai: J. Fluid Mech. **150**, 441 (1985)
5. S. Kida, H. Miura: J. Phys. Soc. Japan **67**, 2166 (1998)

Turbulent Structure in Rough and Smooth Wall Boundary Layers

R.J. Volino[1] M.P. Schultz[2] and K.A. Flack[1]

[1] Mechanical Engineering Department, United States Naval Academy, Annapolis, MD, USA volino@usna.edu

[2] Naval Architecture and Ocean Engineering Department, United States Naval Academy, Annapolis, MD, USA mschultz@usna.edu

1 Introduction

Surface roughness plays an important role in many practical applications, and the study of roughness effects can provide insight into the nature of turbulent boundary layers. Raupach et al.[1] noted a consensus among several studies that the flow in the outer region of both rough and smooth wall boundary layers scale similarly with the boundary layer thickness, δ, and the wall friction velocity, u_τ. Hence, the wall sets the boundary condition via the wall shear, and the outer region responds similarly in all cases. This behavior is consistent with the Perry and Marusic[2] model of detached eddies which are only indirectly influenced by the near wall flow. Krogstad et al.[3], however, noted significant differences in the outer flow. More recently, Kunkel and Marusic[4] and Flack et al.[5] presented turbulence statistics showing outer region similarity for a variety of rough surfaces. The present work considers turbulence structure, as shown through spectra, swirl strength and two point correlations of turbulence quantities.

2 Experiments

Experiments were conducted in a water tunnel with a 2 m long, 0.2 m wide and 0.1 m tall test section. The upper wall was set to maintain a uniform freestream velocity of 1.255 m/s. The lower wall was an acrylic test plate for the smooth wall cases. A wire mesh screen was affixed to a similar plate for the rough wall cases. The mesh spacing, mesh wire diameter and roughness height were t=1.69 mm, d=0.26 mm and k=0.52 mm respectively. A trip near the leading edge ensured a turbulent boundary layer. A two component LDV system was used to acquire velocity profiles to determine δ and u_τ, and also to acquire long time records (500,000 readings at 2200 Hz) at y/δ=0.1 and 0.4 for spectral processing. Sets of 2000 image pairs were acquired for PIV processing at the spanwise centerline of the test section (xy, streamwise-wall normal, planes) and at y/δ=0.1 and 0.4 (xz, steamwise-spanwise, planes).

3 Results and Discussion

Comparisons are made between the smooth and rough wall cases at high and similar momentum thickness Reynolds numbers (Re_θ=6069 and 7663, respectively). For the rough wall, δ/k=71, and δ/k_s=22, where k_s is the equivalent sand grain roughness. The roughness Reynolds number was k_s^+=112, and δ/t=22. The conditions were similar to those of Krogstad and Antonia[6] who had δ/k=109, δ/k_s=15, k_s^+=331, and δ/t=24 at Re_θ=12800. The present mean velocity profiles and Reynolds stresses normalized in outer coordinates showed excellent agreement between the rough and smooth wall cases, in agreement with the study of Flack et al.[5].

Spectra of the streamwise (u^2), and wall normal (v^2) Reynolds stresses and the Reynolds shear stress ($-uv$) are shown in Fig. 1 in premultiplied coordinates for the y/δ=0.4 data. Agreement between the rough and smooth wall data is excellent. Agreement was similar at y/δ=0.1.

Similarity in velocity vector fields is quantified by two point correlations of the streamwise velocity, R_{uu} shown in Fig. 2. The angle of inclination of the correlation is related to the inclination of hairpin packets, and is equal to 9.3° and 9.6° for the rough and smooth walls respectively. The results differ from those of Krogstad and Antonia[6] who saw a large rise in the angle for the rough wall case to 38°. R_{uu} in the xz plane gives an average sense of the size of the hairpin packets, and agreement between the rough and smooth cases was good.

The swirl strength, λ, can be used to locate vortices. It is defined as the imaginary part of the complex eigenvalue of the local velocity gradient tensor. A sign can be assigned to λ based on the local vorticity. Figure 3 shows the probability density function of signed λ in the xy plane at y/δ=0.4. The high pdf point at zero has been removed for clarity. The positive peak is significantly larger than the negative peak since many of the vortices are the heads of the hairpins, which have a positive direction of rotation. At all wall-normal locations there was good agreement between the rough and smooth wall cases. Similar good agreement can be seen in the xz plane, where λ identifies the legs of hairpins.

In conclusion, good agreement between the rough and smooth wall cases was observed in the outer part of the boundary layer for all quantities when scaled using the boundary layer thickness and wall friction velocity. Correlations of other quantities and cross correlations between the various quantities were also computed, and the agreement between rough and smooth was good in all cases.

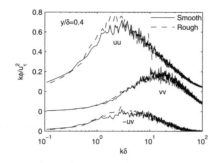

Fig. 1. Spectra from smooth and rough wall boundary layers

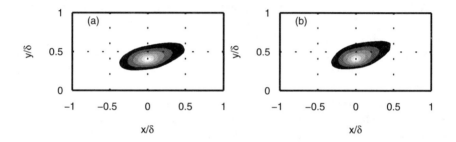

Fig. 2. R_{uu} correlation in xy plane, a) smooth wall, b) rough wall

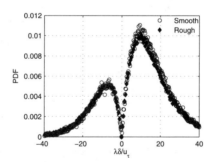

Fig. 3. Probability density function of signed swirl strength in xy plane at $y/\delta=0.4$

References

1. M.R. Raupach, R.A. Antonia, S. Rajagopalan: Appl. Mech. Rev. **44**, 1 (1991)
2. A.E. Perry, I. Marusic: J. Fluid Mech. **298**, 361 (1995)
3. P.-Å. Krogstad, R.A. Antonia, L.W.B. Browne: J. Fluid Mech. **245**, 599 (1992)
4. G.J. Kunkel, I. Marusic: J. Fluid Mech. **548**, 375 (2006)
5. K.A. Flack, M.P. Schultz, T.A. Shapiro: Phys. Fluids **17**, 035102 (2005)
6. P.-Å. Krogstad, R.A. Antonia: J. Fluid Mech. **2 77**, 1 (1994)

Wall-layer models for large-eddy simulations of high Reynolds number non-equilibrium flows

Ugo Piomelli,[1] Senthil Radhakrishnan,[1] Liejun Zhong[2] and Ming Li[2]

[1] Department of Mechanical Engineering, University of Maryland, College Park, MD (USA) ugo@umd.edu
[2] Horn Point Laboratory, University of Maryland Center for Environmental Science, Cambridge, MD (USA)

1 Introduction

One of the main obstacles to the use of large-eddy simulation (LES) to practical applications is the size of the turbulent eddies near the solid boundary, which decreases with Reynolds number Re, so that the resolution requirements (and the cost) of the simulation increase. While the cost of a calculation that only resolves the outer layer of a wall-bounded flow is proportional to $Re^{0.6}$ (where the Reynolds number is defined using the outer-flow velocity and some global length-scale of the problem), if the inner layer is resolved the cost of an LES scales like $Re^{2.4}$ [1, 2, 3].

Because of this constraint, wall-resolving LES can only be used at moderate Re, and are not suited for the high Reynolds-number flow that occur in engineering devices, in geophysical or environmental flow. Alternative treatments that model the wall layer have, therefore, been used since the earliest applications of LES. A review of wall-layer models can be found in [3]; here we only highlight a few points of interest to the present work.

The most common class of wall-layer models postulates the existence of a constant-stress region and a logarithmic velocity profile, which can be used to relate the velocity in the outer layer (where the LES resolves the flow) to the wall stress [4, 5]. Modifications can be made to the logarithmic law to account for roughness or stratification. This approach is simple and inexpensive, but has two main shortcomings: first, it cannot give accurate flow predictions if the logarithmic law does not apply (strong acceleration or deceleration, streamline curvature, mean-flow three-dimensionality, separation). The second one is due to the fact that, at the first grid point, the integral scale is of the order of the distance from the wall, which is comparable to the grid spacing itself. There, the critical assumption that the integral scale be well-resolved fails, often resulting in significant modeling errors. Despite these deficiencies, wall-layer models based on the logarithmic law have been in widespread use, especially in meteorology, physical oceanography and environmental sciences, in applications in which separation does not occur and the wall layer does not play a major role in the flow dynamics.

In engineering, on the other hand, geometry plays an important role, and separation, mean-flow three-dimensionality, streamline curvature and flow acceleration often occur. The limitations of models based on the logarithmic law has spurred the development of hybrid approaches that couple the solution of the Reynolds-averaged Navier-Stokes (RANS) equations in the near-wall region with LES away from the wall. These methods are based on the assumption that, at high Re, a near-wall grid cell includes a very large number of eddies, and the time step is much longer than a near-wall eddy turnover time. As a result, each near-wall cell contains a large ensemble of eddies, and one needs to account only for their average effect on the mean velocity profile and wall stress. Hybrid methods have some desirable properties: the use of a turbulence model to represent the inner layer makes it possible to account for more complex physical phenomena, compared with the constant-stress layer approach. Furthermore, the RANS region can extend far enough from the wall that the grid becomes smaller than the integral scale in the LES zone. One shortcoming of this approach, however, is the mismatch in length- and time-scales between the LES and RANS zones, which results in an adjustment region in which the flow is unphysical. Symptoms of this problem are under-prediction of the wall stress, and the presence of unphysically long, streamwise-coherent structures (super-streaks) in the near-wall region [3]. Space limitations prevent us from discussing all the hybrid methods that have been proposed, or the modifications made to accelerate the development of eddies capable of supporting the Reynolds shear stresses; we will only mention a representative hybrid method based on the Spalart-Allmaras model [6, 7, 8], and two modifications designed to amplify the instabilities in the RANS/LES interface region, the first based on stochastic forcing to perturb the flow, the second on decreasing the eddy viscosity in the RANS/LES interface region to enhance the amplification of existing flow instabilities.

Several researchers suggest adding perturbations in the RANS/LES interface region to facilitate the generation of eddies. Keating & Piomelli [9] performed calculations with the wall-layer model based on the Detached-Eddy Simulation (DES) technique [7, 8], with the addition of a stochastic forcing term in the interface region. They used a control system to determine the magnitude of the stochastic force, which required the resolved Reynolds shear stresses to be equal (on average) to the modeled ones in the interface region. This method gave very good results in channel flows [9].

Another wall-layer model based on the DES technique was recently proposed by Travin et al. [11], who modify the blending functions and the length scale in the Spalart-Allmaras (SA) model [6] to decrease the eddy viscosity in the RANS/LES interface region. This results in more accurate prediction of the skin friction in calculations of plane channel flow for a range of Reynolds numbers.

While the development of more advanced wall-layer models was motivated by engineering applications, their use in the geophysical sciences (and, in particular, in physical oceanography) can be beneficial. Examples of flows in which the solid boundary plays an important role are wave-generated boundary layers, estuarine dynamics, and many problems in coastal oceanography. Thus, it is our goal to compare wall-layer treatments in two configurations relevant to oceanographic studies. In the following, we will first briefly present the relevant methodologies, referring to the appropriate articles for an in-depth treatment. We will then show some representative results, and conclude with recommendations for future research directions.

2 Problem formulation

In LES the velocity field is separated into a resolved (large-scale) and a subgrid (small-scale) field, by a spatial filtering operation. In the unsteady RANS (URANS) approach, on the other hand, the Reynolds decomposition is used to separate the mean from the fluctuating part of the velocity. Despite this difference, the governing equations for LES and URANS have the same form:

$$\frac{\partial \overline{u}_i}{\partial x_i} = 0 , \tag{1}$$

$$\frac{\partial \overline{u}_i}{\partial t} + \frac{\partial (\overline{u}_j \overline{u}_i)}{\partial x_j} = \nu \frac{\partial^2 \overline{u}_i}{\partial x_j \partial x_j} - \frac{\partial \tau_{ij}}{\partial x_j} - \frac{1}{\rho} \frac{\partial \overline{p}}{\partial x_i} . \tag{2}$$

In these equations, an overline denotes a quantity averaged over one grid cell and time step, and $\tau_{ij} = \overline{u_i u_j} - \overline{u}_i \overline{u}_j$ are either the subgrid-scale (SGS) stresses or the Reynolds stresses that must be modeled.

Equations (1–2) are solved with the curvilinear finite-volume code described in [12, 13]. The code is second-order accurate in time and space, and is parallelized using the MPI message-passing library and the domain-decomposition technique.

In all calculations periodic boundary conditions were used in the spanwise (z) direction (note that in this work the coordinate normal to the solid surface is y). In the oscillating boundary layer case, periodic conditions in the streamwise direction (x) were used; to simulate the free surface in the estuaries and coastal oceans a free-slip condition was applied at the top boundary of the domain. In the calculations of the flow over a bump the convective condition $\partial \overline{u}_i / \partial t + U_b \, \partial \overline{u}_i / \partial x = 0$ [14] was used at the outlet, and a velocity plane computed from a separate calculation was assigned at the inflow at each time-step.

At solid surfaces the no-slip condition is imposed in hybrid calculations. When the inner layer is bypassed, in order to relate the wall stress to the outer-region velocity we postulate the existence of a constant-stress layer, which results in a logarithmic velocity profile that is assumed to hold at the first grid point in the outer region:

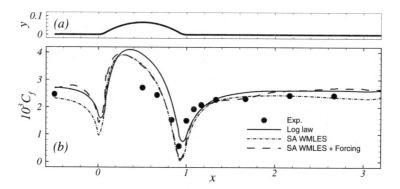

Fig. 1. *(a)* Sketch of the bump geometry and *(b)* profiles of the skin-friction coefficient.

$$\frac{\overline{u}}{u_\tau} = \frac{1}{\kappa} \log \frac{y u_\tau}{\nu} + B, \tag{3}$$

where $\kappa = 0.41$ and $B = 5.0$ are constants, y is the coordinate normal to the wall, and $u_\tau = (\tau_w/\rho)^{1/2}$. Given $\overline{u}$, one can solve (3) for u_τ, and thus recover the wall stress [4, 5].

Two models were used to parametrize the unresolved stresses τ_{ij}: the Smagorinsky model [15] was used in calculations that employed the logarithmic law-of-the-wall; in hybrid calculations we used the Spalart-Allmaras (SA) model [6], both in the detached-eddy simulation formulation [7, 8] and in the Improved Delayed DES (IDDES) form [11].

3 Results

The first case examined is the flow past the two-dimensional bump shown at the bottom of Figure 1, formed by tangentially connecting two short concave arcs to a longer convex arc. The flow is characterized by a sequence of favorable and adverse pressure gradients due to the curvature and the constriction. The momentum Reynolds number at the inlet is $Re_\theta = 12,170$. All lengths are non-dimensionalized by the chord length ($L_c = 305$ mm) of the bump. At the inlet, the boundary layer thickness is $0.09718L_c$ and height of the bump is $0.0659L_c$. A second wall is located $0.498L_c$ above the bottom wall. The simulation domain is $3\delta_{ref}$ wide in the spanwise direction and extends two chord lengths downstream of the trailing edge, which are followed by a buffer region. The grid uses $538 \times 150 \times 36$ nodes in the streamwise, wall-normal and spanwise directions, respectively for the hybrid RANS/LES cases, but only 72 point in y when the logarithmic-law boundary condition is used. Simulation results are compared to the experiments of DeGraaff [16].

Figure 1 shows the skin-friction coefficient, $C_f = 2\tau_w/\rho U_\infty^2$. The hybrid wall-modeled LES based on the SA model (SA WMLES) under-predicts the

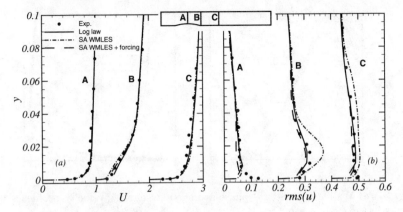

Fig. 2. *(a)* Mean horizontal velocity profiles and *(b)* streamwise velocity fluctuation rms at the locations shown in the inset.

skin-friction in the upstream flat-plate region (a well-known feature of this model), but shows reasonable agreement elsewhere. With the addition of the stochastic forcing this error is corrected. The model based on the logarithmic law also gives acceptable results. In the downstream region, which is dominated by non-equilibrium effects, we see a good agreement with the experimental values for all cases (see also [10] for further discussion of this case).

Figure 2*(a)* shows the mean horizontal velocity normalized by the local free stream velocity at select locations. The predicted mean velocity shows good agreement with the experimental values at all the locations. Figure 2*(b)* shows the rms of the fluctuations of x-component of the velocity. Notice that at the trailing edge of the bump the flow is mostly attached, so one should not expect the logarithmic law to result in the significant errors observed, for instance, by Silva Lopes et al [17]. The turbulent fluctuations decrease in the first half of the bump due to the favorable pressure gradient, and increase again in the adverse pressure-gradient region beyond the summit. The SA WMLES over-predicts the rms of the fluctuations by as much as 40% in the recovery region; the addition of the stochastic forcing results in better agreement with the experimental values.

The second case we examine is a boundary layer over a flat plate forced by an oscillating pressure gradient that causes the freestream velocity to vary sinusoidally, $U_\infty = U_o \sin \omega t$. This acceleration may result in flow transition, and causes flow reversal without separation. Experimental results [18] and direct simulation data for a low-Re case [19] are available. Two cases are examined: one is transitional, the other fully turbulent (respectively, cases 6 and 10 in [18]). In Case 6 the Reynolds number based on Stokes layer thickness $\delta_s = (\nu/\omega)^{1/2}$ and U_o is 700, while in Case 10 it is 2,315. The Reynolds number based on the thickness of the turbulent layer and the friction velocity goes up to 1,000 for the low-Re case, and 7,000 for the high-Re one. The grids used

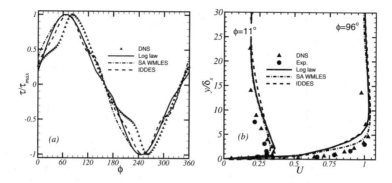

Fig. 3. *(a)* Wall stress and *(b)* phase-averaged streamwise velocity profiles at two phases in the cycle. Case 6.

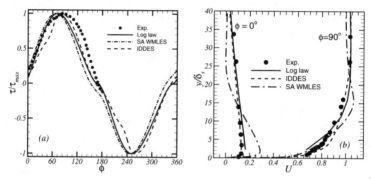

Fig. 4. *(a)* Wall stress and *(b)* phase-averaged streamwise velocity profiles at two phases in the cycle. Case 10.

for Case 6 have 144×96×96 points for the hybrid models, 120×32×60 for the logarithmic-law boundary condition. For Case 10, in the hybrid calculations the number of points in y was increased to 144. The parameters of Case 10 are close to those of a wave-generated bottom boundary layer. The low-Re case is particularly challenging for wall-layer models, since the flow undergoes transition, which none of the existing models is designed to capture.

All models reproduce the shape of the wall-stress curve (Figures 3 and 4). At low Re, however, no model predicts the transitional behavior correctly, and the maximum wall stress is under-predicted by 30% for the SA WMLES, 20% for the other cases. The phase lag between the maximum velocity and the maximum wall stress is also incorrect: the simulations consistently predict an earlier occurrence of the peak τ_w. Despite these shortcomings, the mean velocity profile is predicted reasonably well by all models at low Re; at high Re the SA WMLES predicts a slower response of the inner layer to the outer layer acceleration, resulting in errors in the near-wall region. The IDDES gives better predictions, indicating that the interaction between the

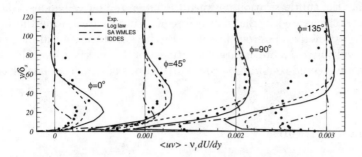

Fig. 5. Total shear stresses. Case 10. Each profile is shifted by 0.001 units.

turbulent eddies in the RANS/LES interface region play an important role in exchanging information between inner and outer layer. Paradoxically, the simpler logarithmic-law model gives better results than the SA WMLES. This may be attributed to the fact that in the hybrid calculation the long inner-layer scales adversely affect the prediction of the inner-outer layer interaction, while in the logarithmic law model the outer layer imposes its scales directly on the inner one, and no intermediate RANS region exists.

Both the logarithmic law and the IDDES give reasonable prediction of the Reynolds shear stresses (Figure 5), except during the deceleration ($\phi = 135^o$) phase. The models predict the general shape of the normal stresses reasonably well (not shown), but the levels are overestimated by the logarithmic law model, underestimated by the SA WMLES. In stratified flows this error is expected to be important, as it will significantly affect the mixing.

4 Conclusions

We have applied several wall-layer models to two flows: one, the flow over a bump, is characterized by a succession of favorable- and adverse-pressure-gradient regions, and unsteady but mild separation. In the second, the flow in an oscillating boundary layer, the unsteady (but uniform) pressure gradient results in acceleration and deceleration of the flow, with a phase lag between the inner- and outer-layer response to the perturbation.

In flows of this type, with mild separation and pressure gradients, the mean velocity profile can be predicted reasonably well by all models; however, a more accurate treatment of the RANS/LES interface appears very beneficial. In the oscillating boundary layer, the RANS treatment of the inner layer affects significantly the interaction between the inner and outer layer. The SA WMLES model results in significant errors in this flow. The simplest (and most economical) approach, the logarithmic law, gives reasonable results, perhaps because the outer-layer time-scales are imposed directly on the wall stress. In the SA WMLES the inner layer RANS does not mimic adequately the subtle relation between the inner and outer regions. Allowing eddies to be generated in the interface region (in this case using the IDDES approach) again gives improved results. Prediction of the second moments is more difficult; here a more accurate treatment of the interface region is clearly beneficial, and could result in improved predictions of flows that include scalar transport.

Acknowledgments

UP and SR were supported by the Office of Naval Research under Grant No. N00014-03-1-0491, monitored by Dr. R. D. Joslin, and by the National Science Foundation under Grant No. OCE0452380. LZ and ML were supported by the National Science Foundation under Grant No. OCE0451699.

References

1. Chapman, D. R.: *AIAA J.* **17**, 1293 (1979).
2. Reynolds, W. C.: In *Whither turbulence? Turbulence at the crossroads,* edited by J. L. Lumley (Springer-Verlag, Heidelberg), 313 (1990).
3. Piomelli U., and Balaras, E.: *Annu. Rev. Fluid Mech.* **34**, 349–374 (2002).
4. Deardorff, J. W.: *J. Fluid Mech.* **41**, 453 (1970).
5. Schumann, U.: *J. Comput. Phys.* **18**, 376 (1975).
6. Spalart, P.R. and Allmaras, S. R.: *La Recherche Aérospatiale,* **1**, 5 (1994).
7. Spalart, P.R., Jou, W.H., Strelets, M. and Allmaras, S.R. In: Liu, C., Liu, Z. (Eds.), *Advances in DNS/LES*, Greyden Press, Columbus, 137 (1997).
8. Nikitin, N.V., Nicoud, F., Wasistho, B., Squires, K.D., and Spalart, P.R.: *Phys. Fluids* **12**, 1629 (2000).
9. Keating, A. and Piomelli, U.: *J. of Turbulence,* **7**(12), 1 (2006).
10. Radhakrishnan, S., Keating, A., and Piomelli U.: *J. of Turbulence,* **7** 63, 1 (2006).
11. Travin, A.K., Shur, M.L., Spalart, P.R., and Strelets, M. Kh.: In *European Conference on Computational Fluid Dynamics ECCOMAS CFD 2006* P. Wesseling, E. Onate, J. Pèriaux.) (2006).
12. Silva Lopes., A., and Palma, J.M.L.M.: *J. Comput. Phys.,* **175(2)** 713 (2002).
13. Silva Lopes, A., Piomelli, U. and Palma, J.M.L.M.: *J. of Turbulence,* **7** 1 (2006).
14. Orlanski, I.: *Journal of Computational Physics,* **21**, 251 (1976).
15. Smagorinsky, J.: *Mon. Weather Rev.* **91**, 99 (1963).
16. DeGraaff, D.G.: PhD Dissertation, Dept of Mechanical Engg, Stanford University (1999).
17. Silva Lopes, A., Piomelli, U. and Palma, J.M.L.M.: In *Advances in turbulence X*, eds. H.I. Andersson and P.A. Krogstad (Kluwer, Dordrecht), 249, 2004.
18. Jensen, B.L., Sumer, B.M., and Fredsøe, J.: *J. Fluid Mech.,* **206**, 265 (1989).
19. Spalart, P.R., and Baldwin, B.S.: In *Turbulent Shear Flows 6*, edited by J.-C. Andre, J. Cousteix, F. Durst, B.E. Launder,and F.W. Schmidt, (Springer-Verlag, Heidelberg), 417 (1989).

Trajectories of solid particles in a tangle of vortex-filaments

A.J. Mee[1], D. Kivotides[1], C.F. Barenghi[1] and Y.A. Sergeev[2]

[1] School of Mathematics, University of Newcastle, Newcastle upon Tyne, NE1 7RU, UK (contact: `C.F.Barenghi@ncl.ac.uk.uk`)
[2] School of Mechanical and Systems Engineering, University of Newcastle, Newcastle upon Tyne, NE1 7RU, UK

1 Motivations

Understanding the dynamics of small particles advected by incompressible turbulent flows is an important problem with applications ranging from atmospheric physics[1] to industrial processes. The problem is often tackled by generating the turbulence using direct numerical simulations (DNS) of the Navier–Stokes equations[2] or Kinematic Simulation models[3]. It is known that homogeneous isotropic turbulence consists of coherent vortex structures or filaments[4] superimposed to a more incoherent background. The coherent structures carry most of the energy and the enstrophy and correspond to the observed Kolmogorov spectrum of the total (coherent plus incoherent) velocity field[5]. Our aim is to determine the role of these structures on the particles' trajectories. To achieve this aim we use a model of turbulence based entirely on vortex filaments.

2 Model

Our model[6] consists of three–dimensional, time–dependent vortex filaments with finite viscous cores which evolve under the action of inertial and viscous forces and reconnect with each other. The central line $\mathbf{r}(s,t)$ of each filament moves according to $d\mathbf{r}/dt = \mathbf{V}(\mathbf{r},t)$ where t is time, s is arclength measured along the filament, and the velocity $\mathbf{V}(\mathbf{r},t)$ is given by the Biot–Savart law[7]

$$\mathbf{V}(\mathbf{r},t) = -\frac{1}{4\pi} \int \frac{(\mathbf{r} - \mathbf{r}') \times \boldsymbol{\omega}(\mathbf{r}')d^3\mathbf{r}'}{|\mathbf{r} - \mathbf{r}'|^3}. \tag{1}$$

The vorticity distribution $\boldsymbol{\omega}$ is Gaussian; the radius of the vortex filaments (defined as the standard deviation of the Gaussian) decreases because of vortex stretching and increases because of viscous diffusion. A second dissipation mechanism arises from vortex reconnection events; these events introduce a cascade of energy to small scales which is damped by the model[8].

3 Results

We make the variables dimensionless using the size of the (periodic) computational box, b, as unit of length, and b^2/Γ as unit of time, where Γ is the circulation around each vortex and ν is the kinematic viscosity. We start from an arbitrary configuration of vortex loops. The loops interact with each other, distort and reconnect. The total vortex length increases and soon a tangle of vortex filaments is produced - see Fig. (1). At this point the effective Reynolds number is $Re = bV_{rms}/\nu \approx 10^4$ where V_{rms} the rms velocity (at later stages, since there is no forcing in our model, the vortex tangle decays). The velocity field $\mathbf{V}$ has properties which agree well with DNS[6], such as the Kolmogorov spectrum and the third order longitudinal structure function. Then we introduce a large number of small, neutrally buoyant solid particles of radius a_p, position $\mathbf{r}_p(t)$, velocity $\mathbf{v}_p(t)$, and solve the equations of motion of each particle[9], $d\mathbf{r}_p/dt = \mathbf{v}_p$ and

$$\frac{d\mathbf{v}_p}{dt} = \frac{(\mathbf{V} - \mathbf{v}_p)}{\tau} + \frac{\partial \mathbf{V}}{\partial t} + (\mathbf{V} \cdot \nabla)\mathbf{V}, \qquad (2)$$

where $\tau = a_p^2/3\nu$ is the Stokes relaxation time. The last two terms at the right hand side of Eq. (2) represent inertial effects which have been neglected in the literature[2] concerned with heavy particles (rather than our neutrally buoyant particles) and have been considered only for simple flows[10]. Particles' trajectories are computed in the two cases of (i) time–frozen vortex tangle (for simplicity), and (ii) time–evolving vortex tangle (more realistically). In both cases we notice that particles can become trapped in the vicinity of one or more vortex filaments, as in Fig. (1). Fig. (2) shows that the acceleration of fluid particles is larger than the acceleration of solid particles.

Work is in progress to compare the statistics of the three contributions to the acceleration of solid particles in Eq. (2) as a function of τ.

This research is funded by EPSRC grant GR/T08876/01.

References

1. G. Falkovich, A. Fouxon and M. Stepanov, Nature (London), **419**, 151 (2002).
2. G. Boffetta, F. De Lillo, and A. Gamba, Phys. Fluids **16**, L20 (2004); J. Bec, L. Biferale, M. Cencini, A. Lanotte and F. Toschi, Phys. Fluids **18**, 081702 (2006).
3. J. C. H. Fung and J. C. Vassilicos, Phys. Rev. E **68**, 046309 (2003).
4. Z.-S. She and S.A. Orszag, Nature, **344**, 226 (1990); A. Vincent and M. Meneguzzi, J. Fluid Mech. **258**, 245 (1994).
5. M. Farge, G. Pellegrino and K. Schneider, Phys. Rev. Lett. **87**, 054501 (2001).

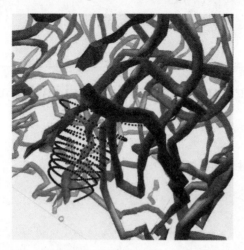

Fig. 1. Red: snapshot of vortex filaments. Green: position of a solid particle at different times as it becomes trapped and spirals around a vortex filament.

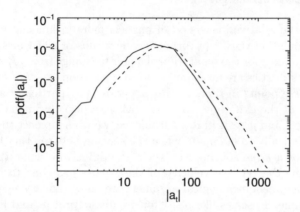

Fig. 2. Probability density function of the magnitude $|a_t|$ of the total acceleration of solid particles (solid line) for $\tau = 0.74 \times 10^{-3}$ and fluid particles (dashed line).

6. D. Kivotides and A. Leonard, Phys Rev. Lett. **90**, 234503 (2003); Europhys. Lett. **63**, 354 (2003); Europhys. Lett. **65**, 344 (2004) and Europhys. Lett. **66**, 69 (2004).
7. P.G. Saffman, *Vortex Dynamics*, Cambridge University Press (1992).
8. P. Chatelain, D. Kivotides and A. Leonard, Phys. Rev. Lett. **90**, 054501 (2003).
9. D.R. Poole, C.F. Barenghi, Y.A. Sergeev and W.F. Vinen, Phys. Rev.B **71**, 064514 (2005); Y.A. Sergeev, C.F. Barenghi, D. Kivotides and W.F. Vinen, Phys. Rev. B **73** 0525021 (2006).
10. A. Babiano, J.H.E. Cartwright, O. Piro and A. Provenzale, Phys. Rev. Lett. **84**, 5764 (2000).

Quantifying turbulent dispersion by means of exit times

B.J. Devenish and D.J. Thomson

Met Office
Fitzroy Road
Exeter
EX1 2SR
U.K.
ben.devenish@metoffice.gov.uk
david.thomson@metoffice.gov.uk

We consider the relative dispersion of particle pairs in kinematic and direct numerical simulation (DNS) of turbulence by means of exit times, defined to be the time, t, taken for the separation of a pair to change from r/ρ to r. In this approach the statistics are measured at fixed separations, or *thresholds*, rather than at discrete points in time. In this paper, we pay particular attention to the variation of the exit time statistics with ρ, especially in the asymptotic limits $\rho - 1 \ll 1$ and $\rho \gg 1$. In real turbulence, we would expect the pairs that separate ballistically to dominate when the spacing between the thresholds is small while for a large spacing we expect the diffusively separating pairs to dominate since the relative velocity of these pairs has time to decorrelate. Using a simple phenomenological argument we show that for $\rho - 1 \ll 1$ the typical exit time, $\hat{T}$, scales like $\rho - 1$ for a ballistic process and like $(\rho - 1)^2$ for a diffusive process (indicated by a subscript D). Analysis of the diffusion equation with a separation-dependent diffusivity proportional to r^m shows that for $\rho - 1 \ll 1$ the probability density function (pdf) of the exit times scales like $1/t^{3/2}$ for $\hat{T}_D \ll t \ll \hat{T}$, where $\tilde{T}$ is the typical time for the pair to separate from zero to separation r. In this limit, all the positive moments scale like $\rho - 1$ and the negative moments, $\langle t^{-n} \rangle$, like $(\rho - 1)^{-2n}$. For $\rho \gg 1$, the condition $\hat{T}_D \ll t \ll \tilde{T}$ can no longer be satisfied and there is no power law scaling range. In this limit, the pdf and its positive moments are independent of ρ. We use these results to analyse the exit time statistics in the two flow fields.

Fig. 1 shows the exit time pdf for kinematic simulation for $\rho = 1.075$ and $\rho = 2$ which are taken to be representative of the two asymptotic cases described above. For $\rho = 1.075$ it is clear that the pdf scales like $t^{3/2}$ with the scaling range centred about the mean. For $\rho = 2$ this scaling range has disappeared as the condition $\rho - 1 \ll 1$ can no longer be satisfied. As ρ

increases, the difference between $\hat{T}_D$ and $\tilde{T}$ decreases until they coincide. For large times and both values of ρ, the pdf decays at a faster rate, consistent with the exponential decay predicted analytically from the diffusion equation. The lack of sweeping of the large scales in kinematic simulation implies that particle pairs reach their diffusive limit much earlier than in real turbulence. When this lack of sweeping is exaggerated by adding a large mean velocity to the flow (as is the case here), the mean exit time grows like $r^{1/3}$ in the inertial subrange in contrast with $r^{2/3}$ in real turbulence. This result is consistent with Thomson and Devenish[1] who provide a phenomenological explanation for this scaling.

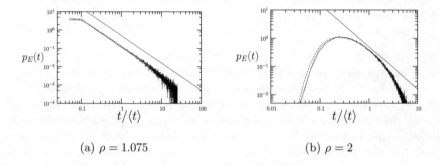

(a) $\rho = 1.075$ (b) $\rho = 2$

Fig. 1: The exit time pdf for kinematic simulation for three thresholds in the inertial subrange. The straight line is proportional to $1/t^{3/2}$.

Fig. 2 shows the exit time pdf for DNS for $\rho = 1.075$ and $\rho = 2$. In fig. 2(a) no $t^{3/2}$ scaling is observed in contrast with kinematic simulation. In real turbulence, when $\rho - 1 \ll 1$ only the slowly separating pairs are reasonably approximated by the diffusion equation and which are observed with a low probability. These pairs account for the long tail of the pdf which decays exponentially, as predicted analytically, for exit times larger than the relative velocity decorrelation time scale which is of order $\tilde{T}$. For large ρ, we would expect the majority of pairs to separate diffusively and no power law scaling range in the pdf. The presence of the power law in fig. 2(b) then suggests that $\rho = 2$ is both large enough for the majority of pairs to separate diffusively and small enough to satisfy the conditions required for the diffusion equation to yield a power law. Even though theoretically this is unexpected (formally requiring both $\rho - 1 \ll 1$ and $\rho \gg 1$) it indicates that here the pairs are decorrelating faster than $\tilde{T}$. For $\rho = 2$ we also find that the finite size of the inertial subrange affects the tail of the pdf.

Finally, we examine how Richardson's constant, g, calculated from the exit time moments via the diffusion equation varies with ρ and the order of the moment. We find that there is more variation of g calculated from the

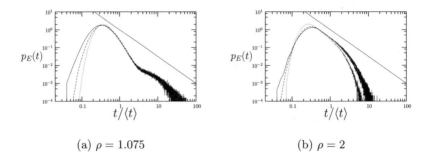

(a) $\rho = 1.075$ (b) $\rho = 2$

Fig. 2: The exit time pdf for DNS for three thresholds in the inertial subrange. The straight line is proportional to $1/t^{3/2}$.

negative moments than from the positive moments, reflecting the fact that these moments are dominated by the rapid (ballistic) separators. For $\rho - 1 \ll 1$ we use a simple phenomenological argument to show that g calculated from the negative moments scales like $(\rho - 1)^3$. This is illustrated in fig. 3. We find that g calculated from $\langle t \rangle$ does not vary with ρ even for $\rho - 1 \ll 1$ where the diffusive approximation does not hold. Using a simple Lagrangian stochastic model to illustrate our arguments, we show that this is because the ballistic and diffusive scalings coincide for $\langle t \rangle$ for $\rho - 1 \ll 1$. As a result, we argue that a large value of ρ is most appropriate for calculating g from the exit time statistics as it is in this case that the diffusion equation best approximates a turbulent flow. On the other hand, for large ρ, the finite size of the inertial subrange affects the statistics which makes it difficult to give a precise value of g using this method.

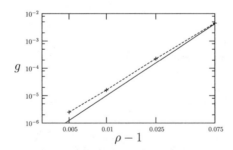

Fig. 3: Richardson's constant calculated from the mean inverse exit time as a function of ρ. The straight line is proportional to $(\rho - 1)^3$.

Reference

1. D.J. Thomson, B.J. Devenish: J. Fluid Mech. **506**, 277 (2005).

Settling velocity of inertial particles

Antonio Celani[1], Marco Martins Afonso[2], and Andrea Mazzino[3]

[1] CNRS, INLN, 1361 Route des Lucioles, 06560 Valbonne, France
[2] Department of Physics of Complex Systems, The Weizmann Institute of Science, Rehovot 76100, Israel   marcomar@fisica.unige.it
[3] Department of Physics - University of Genova & CNISM and INFN - Genova Section, via Dodecaneso 33, 16146 Genova, Italy

1 Introduction and general equations

Our work gathers some analytical results (also supported by numerical computations) on the settling velocity of a single small, rigid, spherical inertial particle in a given incompressible flow $\boldsymbol{u}(\boldsymbol{x}, t)$. Such flow may be known deterministically or only in its statistical properties (e.g. in the presence of turbulence), but for the sake of simplicity in this paper we will confine ourselves to either steady or periodic cellular flows, possessing odd parity with respect to reflections in the vertical direction.

We focus our attention on the Stokes regime, in which the surrounding flow is differentiable on scales of the order of the particle radius and the mean free path is negligible. We moreover neglect the feedback of the particle on the flow and any possible interactions with other particles or boundaries. However, besides inertia, we keep into account both gravity and molecular diffusivity.

The motion of the particle [1] is thus influenced by viscous drag, buoyancy and thermal fluctuations (through Brownian motion). These three effects reflect themselves in the presence of as many relevant adimensional parameters, namely the Stokes (St), Froude (Fr) and Péclet (Pe) numbers, which can be varied independently.

It is moreover customary to introduce the adimensional coefficient β, which depends on the ratio between the fluid and particle densities and keeps into account the added-mass term. It ranges from 0 (for heavy particles, like drops in gases or solid powders) to 3 (for light particles, like bubbles in liquids), and becomes 1 when the two densities are equal (and inertial effects thus absent). Neglecting the Basset, Faxen, Oseen and Saffman corrections, the adimensionalized evolution equation for the particle covelocity, $\boldsymbol{v}(t) \equiv \dot{\boldsymbol{x}}(t) - \beta\boldsymbol{u}(\boldsymbol{x}(t), t)$, reads [1]:

$$\dot{\boldsymbol{v}} = -St^{-1}[\boldsymbol{v} - (1 - \beta)\boldsymbol{u}(\boldsymbol{x}(t), t)] + (1 - \beta)Fr^{-2}\hat{\boldsymbol{e}}_g + \sqrt{2}St^{-1}Pe^{-1/2}\boldsymbol{\eta}(t) \, ,$$

where $\hat{e}_g$ denotes the direction of gravity and $\boldsymbol{\eta}$ represents the adimensionalized standard white noise.

The consequent Fokker–Planck equation for the phase-space density $\rho(\boldsymbol{x}, \boldsymbol{v}, t)$ is the starting point for our analysis. For the sake of simplicity we do not report it here, but we simply point out that the highest-order derivative is represented by the diffusive term, carrying a prefactor $St^{-2}Pe^{-1}$ in front.

2 Multiscale expansion and second-quantization formalism

We introduce a multiscale expansion [2, 3, 4] for the space and time variables, while considering the velocity exclusively as a fast variable. Plugging this development in the Fokker–Planck equation, the zeroth order gives the evolution equation for ρ as a function of the fast variables only. At the first order in the scale-separation parameter,[4] one gets the expression for the adimensionalized terminal settling velocity [5], which, in general, may differ from the "bare" value (corresponding to falling or rising in still fluid)

$$(1 - \beta)StFr^{-2} . \tag{1}$$

We study the deviation from this value, which amounts to compute the integral (over time, space and covelocity variables) of $\boldsymbol{u}(\boldsymbol{x}, t)\rho(\boldsymbol{x}, \boldsymbol{v}, t)$. Because of the assumptions imposed on the flow, every (eventual) nonzero result must originate from the component of the density antisymmetric in the vertical coordinate and is to be interpreted as due to preferential concentration in areas of rising or falling fluid.

The following step consists in the introduction of a second-quantization formalism for the covelocity, namely in the Hermitian reformulation of the problem and in the definition of creation and annihilation operators. At this point, we focus on particles deviating little from the fluid-particle behaviour, and we perform a power-series expansion of the phase-space density in $St^{1/2}$ (the presence of half-integer orders is necessary for a consistent development, but for the final result only integer orders are relevant). The vacuum state is represented by a Gaussian (in $\boldsymbol{v}$), which is the solution of the zeroth-order (Ornstein–Uhlenbeck) equation, times a constant given by the spatial normalization. At the following orders, we obtain a set of semi-coupled partial differential equations, which can be solved (at least numerically) once the basic flow is known. At every order, indeed, the unknown is acted upon by the classical advection-diffusion operator and is forced by the lower-order, already-solved densities.

---

[4] It is worth noticing that the second order in this expansion gives the auxiliary equation which describes the eddy-diffusivity tensor [4]: this will be the subject of our future investigation.

3 Analytical and numerical results

From the analytical point of view, the following conclusions can be drawn.

1. With respect to (1), a correction in the settling velocity at the order St^2 is, in general, possible. Its actual presence and its eventual sign depend on the flow.
2. The aforementioned correction scales with β as $(1-\beta)^2$, which means that it has the same sign for both heavy and light particles. In other words, if an increase in falling is found for the former, a decrease in rising takes place for the latter in the same flow, or viceversa.
3. At this order in St, the deviation scales as Fr^{-2} for any value of Fr, i.e. gravity turns out to be a simple prefactor in the settling velocity. In general, this is not the case for the corrections at higher orders in St.
4. The dependence on Pe is more complicated. However, at small Pe, it can be shown that the asymptotic scaling is at least Pe^2, and for some simple synthetic flows a behaviour $\propto Pe^3$ can be proven.
5. To capture the behaviour at large Pe, a matching-asymptotic expansion is required. Analogously, to consider large St, it is necessary to exploit techniques such as coloured-noise approximation. This is due to the fact that, as pointed out at the end of Sect. 1, both developments are singular, and will constitute the object of future work.

Let us turn to the numerical analysis of the problem. In particular, we made some preliminary tests on the 3D ABC flow and on its 2D restriction (BC flow), and then we focused on the well-known 2D Gollub flow, with one axis parallel to the gravity direction, by performing both Lagrangian simulations and DNS with very heavy particles. As a result, we found an increase in the falling velocity at the order St^2 in perfect agreement with our analytical predictions. More interestingly, at growing $St \sim O(1)$, we observed first a maximum increase, followed by a sign inversion and then by a maximum decrease. At large St, such a decrease in the falling velocity vanishes asymptotically. Moreover, the behaviour of the correction as a function of Pe shows two well-identified regimes, namely $\propto Pe^3$ at small Pe (which could also be shown analytically for this flow, see point 4. above) and $\propto Pe$ at large Pe.

References

1. M.R. Maxey, J.J. Riley: Phys. Fluids **26**, 883 (1983)
2. A. Bensoussan, J.L. Lions, G. Papanicolaou: *Asymptotic Analysis of Periodic Structures* (North-Holland, Amsterdam 1978)
3. M. Vergassola, M. Avellaneda: Physica D **106**, 148 (1997)
4. G.A. Pavliotis, A.M. Stuart: Physica D **204**, 161 (2005)
5. M.R. Maxey: J. Fluid Mech. **174**, 441 (1987)

Backwards/forwards dispersion and inertial range stretching rates

Jacob Berg

Wind Energy Dept., Risø National Laboratory/DTU, Roskilde, Denmark
jacob.berg.joergensen@risoe.dk

We want to shed light on the difference between forwards and backwards dispersion. As reported by [2] and [1] there is a distinct difference between the speed of separation in the separation of two nearby particles in a turbulent flow when time is running backwards compared to forwards. The time asymmetry is a trademark of turbulence and is justified by the lack of time irreversibility in the Navier-Stokes equation. In this contribution we investigate the difference by studying the eigenvalues of coarse-grained strain $\langle \widetilde{s}_{ij} \rangle$ and Lagrangian stretching rates in a turbulent flow at $Re_\lambda \sim 172$ by using Particle Tracking Velocimetry (PTV). The experimental technique as well as the properties of the mean flow is well described in [1].

We define the coarse-grained velocity derivative tensor $\widetilde{A}_{ij} = \widetilde{\partial_i u_j}$ as a least square fit

$$\min \left[\int_{B(r)} d^3\mathbf{x} \, (u_i - \widetilde{A}_{ij} x_j)^2 \right] \tag{1}$$

to the given number of particles with position $\mathbf{x}$ and velocity $\mathbf{u}$ present inside an artificially ball with radius r located in the flow. It is not the only reasonable definition of a coarse grained velocity derivative. Preliminary investigations show in fact that a definition based on projections may be statistically more well behaved when analyzing particle tracking data.

The antisymmetric and symmetric part of $\widetilde{A}_{ij}$ is denoted vorticity and strain, respectively. The latter has mean eigenvalues $\langle \widetilde{\Lambda}_i \rangle$ with $\sum_i \langle \widetilde{\Lambda}_i \rangle = 0$ due to incompressibility. In Figure 1 $\langle \widetilde{\Lambda}_i \rangle$ are shown as a function of the coarse-grained ball size r compensated with a characteristic time scale $t_\star(r) = \sqrt{\frac{2r^2}{15f(r)}}$, where $f(r)$ is the Eulerian longitudinal second order structure function $\langle \delta u_\parallel^2(r) \rangle$.

We observe constant compensated eigenvalues for all values of r. The observation suggests that $t_\star$ is a characteristic time scale for coarse-grained strain. $|\langle \widetilde{\Lambda}_3 \rangle|/\langle \widetilde{\Lambda}_1 \rangle \sim 1.13$. If separation of nearby particle trajectories would align perfectly with the eigenvector associated with the highest eigenvalue, we could

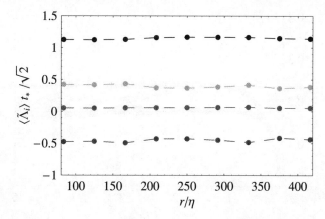

Fig. 1. Eigenvalues; green, $i = 2$: blue and $i = 3$: red. The black curve is the ratio $|\langle \widetilde{\Lambda}_3 \rangle| / \langle \widetilde{\Lambda}_1 \rangle$.

perhaps argue that backwards separation would be fastest by a factor of 1.13^3 since $r^2(t) \sim t^3$, where $r(t)$ here denotes the distance between pairs starting with a separation r_0 at $t = 0$. This is however not the case. We therefore define the stretching rate

$$L(t) = \left\langle \frac{1}{2r^2} \frac{dr^2}{dt} \right\rangle. \tag{2}$$

in order to measure the stretching directly. By sampling the particles in our Lagrangian database pairwise we can calculate $L(t)$ for the forward case (time running forwards) and the backward case (reversed time). The result is shown in Fig. 2. For small times the forwards and backwards curves follow each other indicating an initial ballistic regime. For longer times the curve representing backwards separation saturates at a higher level than the curve representing forward separation. Both values being smaller than the corresponding eigenvalues $\langle \widetilde{\Lambda}_3 \rangle$ and $\langle \widetilde{\Lambda}_3 \rangle$, respectively, meaning that the separation is not perfectly aligned with the eigenvector associated with the largest eigenvalue. The ratio of $L^b(t)/L^f(t)$ is 1.31. Using again the simple argument of stretched times this would give a ratio between backwards and forwards dispersion of $1.31^3 \sim 2.23 \pm 0.33$ within errors of the result obtained from a strict Lagrangian analysis by [1]. How exactly the time scale $t_\star$ is related to the speed of separation is still to be investigated. Including coarse-grained strain in the analysis of pair separation adds an extra dimension to the much more common Lagrangian description.

We are currently investigating relative dispersion in two-dimensional turbulence. Here forwards and backwards dispersion seem quite similar with forwards being slightly larger as opposed to the three-dimensional case. The fact that in 2D turbulence the compressing and stretching eigenvalues are of the

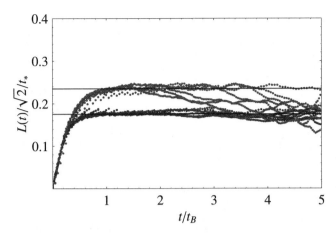

Fig. 2. Stretching rates $L(t)$ rescaled by $t_\star$. The blue curves correspond to forwards separation while the red curves corresponds to backwards separation for different initial separations r_0. Particle pairs with r_0 from $3 - 4$mm to $9 - 10$mm in steps of $1mm$ are shown. The x-axis is divided by the Batchelor time scale $t_B = (r_0^2/\varepsilon)^{1/3}$ indicating the length of the ballistic regime.

exact same magnitude (but with opposite sign) supports the stronger similarity between forwards and backwards relative dispersion.

References

[1] Jacob Berg, Beat Lüthi, Jakob Mann, and Søren Ott. Backwards and forwards relative dispersion in turbulent flow: An experimental investigation. *Phys. Rev. E*, 74(1):Art. No. 016304, 2006.

[2] B. L. Sawford, P. K. Yeung, and M. S. Borgas. Comparison of backwards and forwards relative dispersion in turbulence. *Phys. Fluids*, 17:095109, 2005.

Multiscale Analysis of Convective Magnetic Systems in a Horizontal Layer

M. Baptista[1], S. M. A. Gama[1], V. A. Zheligovsky[2,3]

[1] CMUP & DMA, University of Porto, R. C. Alegre 687, 4169-007 Porto, Portugal
(mbaptist@fc.up.pt)
[2] I.I.E.P.T.M.G., 79 bldg.2, Warshavskoe ave, 117556 Moscow, Russian Federation
[3] O.C.A., CNRS - U.M.R. 6529, BP 4229, 06304 Nice Cedex 4, France

1 Introduction

Boussinesq hydromagnetic convection obeys the Navier-Stokes equation with Lorentz and Archimedes forces for the flow, the magnetic induction equation for the magnetic field, and the heat transfer equation for temperature (ν, η, k, α are, respectively, the molecular viscosity, magnetic diffusivity, thermal diffusivity and thermal compressibility; $\tilde{\mathbf{F}}$, $\tilde{\mathbf{R}}$, $\tilde{S}$ are external source terms):

$$\partial_t \mathbf{V} = \mathbf{V} \times (\partial \times \mathbf{V}) - \partial p + \nu \partial^2 \mathbf{V} - \mathbf{H} \times (\partial \times \mathbf{H}) - \alpha (T - T_0)\mathbf{G} + \tilde{\mathbf{F}},$$
$$\partial_t \mathbf{H} = \partial \times (\mathbf{V} \times \mathbf{H}) + \eta \partial^2 \mathbf{H} + \tilde{\mathbf{R}}, \qquad \partial \cdot \mathbf{H} = 0, \qquad \partial \cdot \mathbf{V} = 0, \qquad (1)$$
$$\partial_t T = -(\mathbf{V} \cdot \partial)T + k \partial^2 T + \tilde{S}.$$

They may be used to simulate the evolution of astrophysical convective hydromagnetic (CHM) systems. Accurate simulations for geo- and astrophysical parameter values are close to impossible, due to the limited power of available computers. A semi-analytic approach, based on multiscale analysis, can be applied to this system, in particular to evaluate eddy diffusivity [1, 2, 4, 3]. Negative values of eddy diffusivity indicate that the system is unstable to perturbations involving large scales. This instability is a possible mechanism for kinematic magnetic field generation (non-convective dynamos exploiting this mechanism were studied in [5, 6]).

2 Multiscale analysis

The modes of perturbations of steady CHM states obey the equation $\mathbf{PAW} = \lambda \mathbf{W}$, where λ is the growth rate of perturbations, $\mathbf{W} = \begin{bmatrix} \mathbf{W^V} & \mathbf{W^H} & \mathbf{W}^T \end{bmatrix}^t$ is a block vector representing their spatial profile, $\mathbf{A}$ is the linearised CHM operator and $\mathbf{P}$ is the projection onto the subspace of solenoidal vector fields [1].

In the two-scale expansion, slow variables $\mathbf{X} = \varepsilon\mathbf{x}$ are introduced to describe large-scale dynamics. Expanding perturbations and growth rates in power series in the scale ratio ε, substituting them in the eigenvalue equation and equating the terms in ε^n at each order n, a hierarchy of equations of the form $\mathbf{PA}^{(0)}\mathbf{Pf} = \mathbf{Pg}$ is obtained. Such an equation has a solution, as long as its right hand side is orthogonal to the kernel of the adjoint operator, $\mathbf{PA}^{(0)*}\mathbf{P}$. For $n = 0$ and $n = 1$, this solvability condition is trivially satisfied, if the perturbed CHM state possesses certain symmetries. For $n = 2$, it yields an eigenvalue equation for the eddy diffusivity operator in slow variables.

3 Numerical results and discussion

To model magnetic instabilities in turbulent convective flows, periodic steady CHM states can be randomly generated in the Fourier space [1, 6]. These states satisfy (1) for the appropriate source terms. Symmetry and solenoidality conditions must be imposed on the Fourier coefficients, which are normalised afterwards to have a prescribed decaying energy spectrum and the r.m.s. average one. Auxiliary problems are solved numerically in the Fourier space by pseudo-spectral methods (sine or cosine transforms are applied in the vertical direction, in accordance with boundary conditions for respective components of vector fields). Algebraic ($E(k) \sim k^{-\xi}$) or exponential ($E(k) \sim \exp(-\xi k)$) spectra were considered in [6], where flows with exponentially decaying spectra were found to be statistically better dynamos.

Simulations have been carried out for $\nu = \eta = k = 0.5$, and $\alpha = 1$ for the periodicity box of size $2\pi \times 2\pi \times \pi$, with the resolution of $32 \times 32 \times 16$ Fourier harmonics. An ensemble of 1000 instances of CHM steady states, involving Fourier harmonics with wave numbers not exceeding 7, has been generated for both algebraic and exponentially decaying spectra, assuming $\xi = 4$ in both cases. For algebraic spectra, it turns out that 110 out of 1000 (11%) generated flows exhibit negative combined eddy diffusivity. The number rises to 131 (13%) for exponential spectra (see Fig. 1 (a)-(b)). Steady states leading to negative eddy diffusivity are unstable to large-scale perturbations. The growth rate of the perturbation is quadratic in the scale ratio ε. Therefore, this instability can be observed only if the considered CHM steady state is stable to short-scale perturbations, which would have larger growth rates otherwise.

For one of the generated CHM states, one of the molecular diffusivities has been varied, keeping all the other parameters equal to the previously used values (see Fig. 1 (c)-(d)). The combined eddy diffusivity depends explicitly on the molecular diffusivities ν and η, and on a correction involving the solutions of the auxiliary problems. Molecular diffusivities ν, η and k are also present in the linearised operator and affect the solutions of these auxiliary problems. If no correction is present, the maximum growth rate is negative and equal to $-\min(\nu, \eta)$. For large molecular diffusivities, the growth rate remains negative, since the correction is of the order of $1/\min(\nu, \eta, k)$ and cannot outweigh the additive contribution of ν and η. Beyond the interval of this asymptotic behaviour, the correction may grow in amplitude, leading eventually to positive growth rates λ_2. Whether this happens depends in particular on the spectral properties of the linearisation of (1.1). Thus the influence of molecular diffusivities on the growth rate of the dominant mode of large-scale perturbations is difficult to predict.

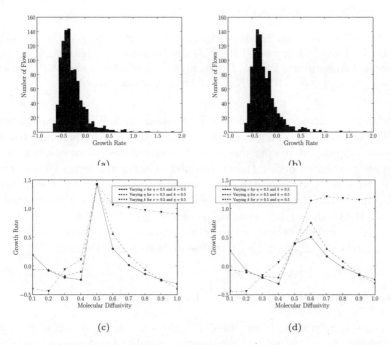

Fig. 1. Statistics of growth rates (opposite of eddy diffusivity), for algebraic (a) and exponential (b) spectra; growth rates as function of molecular diffusivities, for algebraic (c) and exponential (d) spectra.

References

[1] M. Baptista, S.M.A. Gama, and V.A. Zheligovsky. Submitted to the Euro. Phys. Jour. B., 2007.

[2] V.A. Zheligovsky. *Izvestiya, Phys. Solid Earth*, 42(3):244–253, 2006.

[3] B. Dubrulle and U. Frisch. *Phys. Rev. A*, 43:5355, 1991.

[4] S. Gama, M. Vergassola, and U. Frisch. *J. Fluid Mech.*, 260:95–126, 1994.

[5] A. Lanotte, A. Noullez, M. Vergassola, and A. Wirth. *Geophys. Astrophys. Fluid Dynam.*, 91:131, 1999.

[6] V. A. Zheligovsky, O. M. Podvigina, and U. Frisch. *Geophys. Astrophys. Fluid Dynam.*, 95:227, 2001.

Hall-MHD turbulence in the solar wind

Sébastien Galtier[1] and Éric Buchlin[2]

[1] Institut d'Astrophysique Spatiale (IAS), Bâtiment 121, F-91405 Orsay (France); Université Paris-Sud 11 and CNRS (UMR 8617) sebastien.galtier@ias.fr
[2] Space and Atmospheric Physics Department, The Blackett Laboratory, Imperial College, London SW7 2BW, UK e.buchlin@imperial.ac.uk

Waves and turbulence are ubiquitous in the solar wind. Whereas Alfvén waves and Kolmogorov-type energy spectra are found at low frequencies, whistler waves and significant steeper magnetic fluctuation power law spectra are detected at frequencies higher than a fraction of hertz at 1 astronomical units. This multi-scale turbulence behavior may be investigated in the framework of 3D Hall magnetohydrodynamics. We show that both wave turbulence analysis and high Reynolds number simulations of strong turbulent flows converge towards a steepening of magnetic spectra which may be attributed to dispersive nonlinear processes rather than pure dissipation as often stated.

1 Waves and turbulence in the solar wind

From the very beginning of *in situ* observations it was realized that the interplanetary medium was not quiet but rather highly turbulent and permeated by fluctuations of plasma flow velocity and magnetic field on a wide range of scales, from 10^{-6} Hz up to several hundred hertz. The detailed analyses revealed that these fluctuations are mainly characterized (at 1AU) by power law energy spectra around $f^{-1.7}$ at low frequency ($f < 1$ Hz), which are generally interpreted directly as wavenumber spectra by using the Taylor "frozen-in flow" hypothesis. This spectral index is somewhat closer to the Kolmogorov prediction for neutral fluids ($-5/3$) than the Iroshnikov–Kraichnan prediction for magnetohydrodynamic (MHD) ($-3/2$). Both these heuristic predictions are built, in particular, on the isotropic turbulence hypothesis which is questionable for the inner interplanetary medium since apparent signatures of anisotropy are found through, for example, the detection of Alfvén waves (Belcher & Davis 1971) or the variance analysis of the magnetic field components and magnitude.

For timescales shorter than few seconds ($f > 1$ Hz), the statistical properties of the solar wind change drastically with, in particular, a steepening of the magnetic fluctuation power law spectra over more than two decades (Leamon et al. 1998) with a spectral index on average around -3. This new inertial range – often called dissipation range – is characterized by a bias of the polarization suggesting that these fluctuations are likely to be right-hand polarized with a proton cyclotron damping of Alfvén left circularly polarized fluctuations. This proposed scenario seems to be supported by DNS of compressible $2\frac{1}{2}$D Hall-MHD turbulence (Ghosh et al. 1996) where a steepening of the spectra is found – although on a narrow range of wavenumbers – and associated with the appearance of right circularly polarized fluctuations. Therefore, it is likely that what has been conventionally thought of as a dissipation range is a second – dispersive – inertial range.

2 Hall Magnetohydrodynamics turbulence

2.1 Wave turbulence

Spacecraft measurements made in the interplanetary medium suggest a nonlinear dispersive mechanism that may be modeled by the 3D incompressible Hall-MHD equations. In that context, a rigorous analysis of nonlinear transfers in the wave turbulence regime has been proposed by Galtier (2006) where Alfvén, ion cyclotron and whistler waves are taken into account. The main rigorous result derived is a steepening of the anisotropic magnetic fluctuation spectrum at scales smaller than the ion inertial length d_i with anisotropies of different strength, large scale anisotropy being stronger than at small scales. The Hall MHD turbulence spectrum is characterized by two inertial ranges, which are exact solutions of the wave kinetic equations, separated by a knee as in the solar wind. The position of the knee corresponds to the scale where the Hall term becomes sub/dominant. The single anisotropic phenomenology proposed recovers the power law solutions found and makes the link continuously in wavenumbers between the two scaling laws such that

$$E(k_\perp, k_\parallel) \sim k_\perp^{-2} k_\parallel^{-1/2} (1 + k_\perp^2 d_i^2)^{-1/4} . \tag{1}$$

We see how the large scale prediction is affected by the Hall effect through the term proportional to d_i: only the scaling in the perpendicular (to the mean magnetic field) wavenumber $k_\perp$ is modified. It is precisely the direction along which the non linear transfer is dominant. This regime is particularly relevant close to the Sun (< 0.5 AU) where strong anisotropy is expected.

2.2 Cascade model for strong solar wind turbulence

DNS of turbulent flows at very large (magnetic) Reynolds numbers are still well beyond today's computing resources. Therefore, any reasonable simplification of corresponding equations is particularly attractive. In the case of the solar wind, for which the Reynolds number is as large as 10^9, simplified models are currently the only way to investigate the multi-scale behavior described above. Following this idea, a description of solar wind turbulence in terms of a cascade (shell) model based on the 3D incompressible Hall-MHD equations is presented in Galtier & Buchlin (2007). In spite of the simplifications made, shell models remain highly non trivial and are able to reproduce several aspects of turbulent flows like intermittency (Buchlin & Velli 2007). These simulations reveal that the large-scale magnetic fluctuations are characterized by a $k^{-5/3}$-type spectrum which steepens at scales smaller than the ion inertial length d_i, to $k^{-7/3}$ if the magnetic energy overtakes the kinetic energy, or to $k^{-11/3}$ in the opposite case. These results are in agreement both with a heuristic description *à la* Kolmogorov, and with the most recent range of power law indices found in the solar wind (Smith et al. 2006).

3 Conclusion

Hall-MHD may be seen as a natural nonlinear model for explaining the strong steepening of the magnetic fluctuation spectra observed in the solar wind at different distances from the Sun. Of course, the scalings predicted by the models may be altered by effects not included. For example density variations – although weak in the pure/polar wind – could modify slightly these results as well as intermittency whose effects is mainly measured in higher order moments. Nonlocal effects (Mininni et al. 2007) and anisotropy are also important ingredients. In the latter case a recent analysis made with Cluster, a multi-spacecraft mission dedicated to the Earth's magnetosphere, shows only a slight difference in the power law index between the frequency magnetic spectrum and the 3D spatial one although a strong anisotropy is detected in this medium (Sahraoui et al. 2006). The predominance of outward propagating Alfvén *and* whistler waves has also certainly an influence on the spectral laws but it has never been studied in a multi-scale model. It is likely that such asymmetric wave flux (imbalanced turbulence) leads for the scaling exponents to a range of values centered around the exponents discussed here as it is found in other fluids like MHD flows.

References

1. J.W. Belcher & L. Davis: J. Geophys. Res. **76**, 3534 (1971)
2. É. Buchlin & M. Velli: Astrophys. J. , in press
3. S. Galtier: J. Plasma Physics **72**, 721 (2006)
4. S. Galtier & É. Buchlin: Astrophys. J. **656**, 560 (2007)
5. S. Ghosh et al.: J. Geophys. Res. **101**, 2493 (1996)
6. R.J. Leamon et al.: J. Geophys. Res. **103**, 4775 (1998)
7. P. Mininni, A. Alexakis & A. Pouquet: J. Plasma Phys., in press
8. F. Sahraoui, G. Belmont, L. Rezeau et al.: Phys. Rev. Lett. **96**, 075002 (2006)
9. C.W. Smith, K. Hamilton, B.J. Vasquez et al.: Astrophys. J. **645**, L85 (2006)

Transition to turbulence in plane channel flow with spanwise magnetic field

Thomas Boeck[1], Dmitry Krasnov[1], Maurice Rossi[2], and Oleg Zikanov[3]

[1] Fakultät Maschinenbau, Technische Universität Ilmenau,
   Postfach 100565, 98684 Ilmenau, Germany Thomas.Boeck@tu-ilmenau.de
[2] Institut Jean Le Rond D'Alembert, Université Pierre et Marie Curie,
   4 place Jussieu, F-75252 Paris Cedex 05, France maur@ccr.jussieu.fr
[3] Department of Mechanical Engineering, University of Michigan - Dearborn,
   4901 Evergreen Road, Dearborn, MI 48128-1491, USA zikanov@umich.edu

1 Introduction

We consider pressure-driven flow of an incompressible, electrically conducting fluid in an infinite plane channel between insulating walls with a magnetic field B in the spanwise direction. Such flows can be found in numerous metallurgical and materials processing applications, e.g. in electromagnetic flow control in continuous steel casting and in growth of large silicon crystals. Another area of applications is the liquid metal (Li or Pb-17Li) cooling blankets of breeder type for fusion reactors. Our numerical study is performed within the quasi-static approximation, whereby the governing equations reduce to the Navier-Stokes system with the additional Lorentz force. The length and velocity scales are the laminar centerline velocity U and the channel half width L. The non-dimensional basic velocity profile is the parabolic Poiseuille profile. Nondimensional parameters are the Reynolds number $Re = UL/\nu$ and the Hartmann number $Ha = BL\sqrt{\sigma/\rho\nu}$, where σ denotes the electric conductivity and B the magnetic field strength.

We perform a systematic study of the optimal linear perturbations providing the strongest transient growth for different Ha at a fixed, subcritical Reynolds number $Re = 5000$. Among the possible transition scenarios, we then focus on the one based on the algebraic transient growth of optimal perturbations and their subsequent three-dimensional breakdown.

2 Transient growth of linear perturbations

In the linear analysis we study the evolution of decoupled monochromatic Fourier modes with the wavenumbers α and β in the streamwise and spanwise

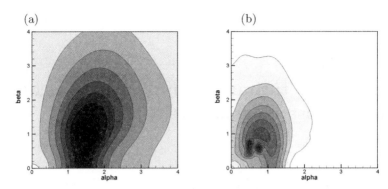

Fig. 1. Isolevels of energy amplification $\hat{G}(\alpha, \beta)$ for $Ha = 50$ at different moments in time T. (a) Global maximum at $T = 15$, (b) three local peaks at $T \approx 28$.

directions. The flow is linearly stable, i.e. all eigensolutions decay exponentially. However, linear combinations of eigenmodes can experience substantial transient algebraic growth before they eventually decay. To quantify the amplification at time T we use the kinetic energy E of the perturbations, whereby the individual contributions of each wavenumber pair (α, β) can be considered independently. The amplification gain of any given mode at time T is the ratio $E(T)/E(0)$. This ratio is maximized over all possible initial vertical shapes by an optimization procedure [1] to give the maximum amplification $\hat{G}(T, \alpha, \beta)$ of the disturbances with the wavenumbers (α, β) at the time T.

For $Ha = 0$, streamwise vortices with $\alpha = 0$ provide the largest transient amplification. These modes are strongly damped by the magnetic field, and are therefore supplanted by oblique modes with $\alpha > 0$ for $Ha \geq 5$. Fig. 1 shows isolines of $\hat{G}$ for $Ha = 50$ for different times T. The contours indicate variation of $\hat{G}$ from low (white regions) to high (black regions) values. The highest amplification occurs for $T \approx 15$ (Fig. 1a), but there may be several co-existing local maxima, e.g. at $T \approx 28$ (Fig. 1b). Maximization with respect to T, α and β provides the maximum amplification $\hat{M}$ and the corresponding wavevector (α, β) as functions of Ha, which are shown in Fig. 2. The transient growth of oblique perturbations is also reduced for $Ha > 0$, and the oblique angle of the optimal modes increases monotonically with Ha. For $Ha \geq 100$, the spanwise Orr-mechanism modes unaffected by the magnetic field become the modes with strongest transient amplification. In Fig. 2(b), these modes provide the constant amplification level for $Ha \geq 100$. We also show the maximum amplification of streamwise rolls with $\alpha = 0$ for comparison. For these modes, the amplification $\hat{M}_{stream}$ eventually reduces to unity, i.e. they experience no transient growth for sufficently large Ha. Remarkably, we find scaling relations $\beta \sim Ha^{-1}$ and $\hat{M}_{stream} \sim Ha^{-2}$ for these modes.

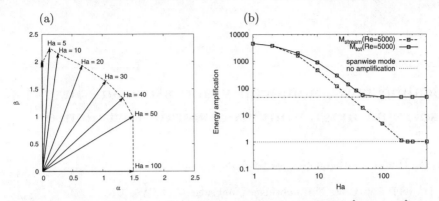

Fig. 2. Optimal wavevector vs. Ha (a) and amplification gains $\hat{M}_{tot}$ and $\hat{M}_{stream}$ for global and streamwise optimal modes (b).

3 Direct simulations of transition

In the transition simulations we study the evolution of the basic flow modulated by the optimal linear mode of a specified amplitude. The initial kinetic energy of the perturbations $E(0)$ varies between 10^{-5} and 10^{-2} relative to that of the basic flow. Weak three-dimensional noise with energy $E_{3D} = 0.01E(0)$ is added at the moment of maximum linear amplification. This approach corresponds to the so-called two-step scenario, which was shown to be pertinent for other parallel shear flows, such as the plane Poiseuille [2], pipe Poiseuille [3], and Hartmann [4] flows. For $Ha = 10$, the maximum energy amplification is provided by an oblique mode. By comparison, the amplification of streamwise vortices is about twice lower. Streamwise vortices do not induce transition to turbulence if $E(0) < 10^{-4}$. For the oblique optimal mode, even $E(0) = 10^{-5}$ is sufficient for the transition when 3D noise is added. Moreover, the transition occurs earlier than in the case of streamwise vortices.

TB, DK and MR acknowledge financial support from the Deutsche Forschungsgemeinschaft in the framework of the Emmy–Noether Program (grant Bo 1668/2-2). OZ's work is supported by the grant DE FG02 03 ER46062 from the U.S. Department of Energy.

References

1. P. J. Schmid, D. S. Henningson: *Stability and Transition in Shear Flows* (Springer, Berlin Heidelberg New York 2001)
2. S. C. Reddy et al: J. Fluid Mech. **365**, 269 (1993)
3. O. Zikanov: Phys. Fluids **8**, 2923 (1996).
4. D. S. Krasnov et al: J. Fluid Mech. **504**, 183 (2004)

Scaling laws, nonlocality and structure in isotropic magnetohydrodynamic turbulence

T. A. Yousef[1], F. Rincon[1] and A. A. Schekochihin[1,2,3]

[1] DAMTP, University of Cambridge, Cambridge CB3 0WA, UK.
[2] Department of Physics, Imperial College London, London SW7 2W7, UK.
[3] King's College, Cambridge CB2 1ST, UK.

Statistically stationary and isotropic magnetohydrodynamic (MHD) turbulence satisfies exact scaling laws analogous to Kolmogorov's 4/5 and Yaglom's 4/3 laws:

$$\left\langle \delta u_L^3 \right\rangle - 6 \left\langle B_L^2 \delta u_L \right\rangle - 6\nu\partial_r \left\langle \delta u_L^2 \right\rangle = -\frac{4}{5}\,\epsilon r, \quad (1)$$

$$\left\langle \delta z^{\mp} |\delta z^{\pm}|^2 \right\rangle - \partial_r \left[(\nu + \eta) \left\langle |\delta z^{\pm}|^2 \right\rangle + (\nu - \eta) \left\langle \delta z^+ \cdot \delta z^- \right\rangle \right] = -\frac{4}{3}\,\epsilon r, \quad (2)$$

where ν and η are the fluid's viscosity and resistivity, ϵ is the average injected power, $z^{\pm} = u \pm B$ are the Elsasser variables, the subscript L stands for "longitudinal" and $\delta f = f(x+r) - f(x)$ [1, 2]. Since Kolmogorov's 1941 picture of hydrodynamic turbulence is consistent with the 4/5 law, it may be tempting to conclude that the phenomenology of isotropic MHD turbulence is similar to that of Kolmogorov turbulence and notably that nonlinear interactions in MHD turbulence are predominantly local. However, the exact laws do not imply this: they are only constraints that can be satisfied by a local cascade or by the formation of structures. Actually, the nonlocal nature of MHD turbulence is evident in the *small-scale dynamo* problem, where turbulent motions of a conducting fluid amplify magnetic fluctuations to dynamically strong levels by means of random stretching [3]. The right panel of Fig. 1 depicts a cross section of $|B|$ from a numerical simulation of the saturated state of small-scale dynamo and shows that the fields are dominated by long folded structures, whose length is comparable to the outer scale of the turbulence $\ell_{\|} \sim \ell_0$, with direction reversals on scales comparable to the Ohmic dissipation scale $\ell_{\perp} \sim (\eta^3/\epsilon)^{1/4}$. The magnetic field saturates in a state where the magnetic tension in the folds is strong enough to oppose stretching by turbulent stresses that are characterized by the forcing scale of the turbulence. This balance is clearly nonlocal. In this study, we attempt to show *how* the saturated state of the small-scale dynamo, unlike Kolmogorov turbulence, manages to comply simultaneously with the exact laws and nonlocality.

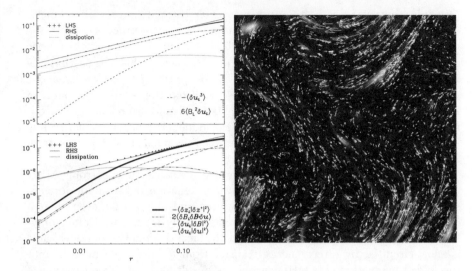

Fig. 1. *Right*: Cross section of $|B|$ from a simulation with Pm = 1250 and Re ~ 1. *Top left*: 4/5 law for simulation with Pm = 1 and Rm $\sim$ Re ~ 400. *Bottom left*: 4/3 law for the same simulation. Terms not shown are small. See Ref. [2] for details.

To this end, let us start by considering a simple theoretical model that captures the basic characteristics of the folded field. We assume that the magnetic field is straight and parallel, but oscillates with a spatial period of $2\ell_\eta \ll \ell_0$ across itself. For scales $r \gg \ell_\eta$, this *stripy field* model implies that the magnetic field increments δB will, with equal probability, be either 0 or $-2B(x)$. For clarity, we first concentrate on the asymptotic limit Pm $\gg 1$ and examine the subviscous scales, $r \ll \ell_\nu \sim (\nu^3/\epsilon)^{1/4}$. The velocity field is smooth on these scales, so velocity increments can be approximated by $\delta u \simeq r \cdot \nabla u(x)$. If we substitute these estimates for δu and δB into Eqs. (1) and (2) we see that both laws reduce to the power balance, $\epsilon = \nu \langle |\nabla u|^2 \rangle + \eta \langle |\nabla B|^2 \rangle$ [2], but it is important to stress that to achieve this seemingly trivial result, we had to make assumptions about the *spatial structure* of the magnetic field. Thus, we have shown that this *stripy field model is consistent with the exact laws*. The model also predicts that the mixed third-order structure functions $\langle \delta z_L^\mp |\delta z^\pm|^2 \rangle = \langle \delta u_L |\delta u|^2 \rangle \pm 2 \langle \delta u_L \delta u \cdot \delta B \rangle \mp \langle \delta B_L |\delta u^2| \rangle + \langle \delta u_L |\delta B|^2 \rangle - 2 \langle \delta B_L \delta u \cdot \delta B \rangle \mp \langle \delta B_L |\delta B|^2 \rangle$ in Eq.(2) are dominated by $2 \langle \delta B_L \delta u \cdot \delta B \rangle$, which scales linearly with r, and that the magnetic interaction term $6 \langle \delta B_L^2 \delta u_L \rangle$ dominates over $\langle \delta u^3 \rangle$ in Eq.(1) [2].

It turns out that the predictions of the stripy field model are consistent with numerical simulations of high Pm isotropic MHD [2] and are to a large extent also valid for the nonasymptotic case of Pm = 1. The top left panel in Fig. 1 shows the different terms in Eq. (1) for a numerical simulation with Re $\sim$ Rm ~ 400. We see that the viscous term is subdominant for $r \gtrsim \ell_\nu$. In principle, this leaves some room for a possible inertial range with a kinetic energy cascade. Nonetheless, the dominant balance in this range is between the magnetic interaction term $6 \langle B_L^2 \delta u_L \rangle$ and $(4/5)\epsilon r$ and not between the kinetic energy "cascade" term $\langle \delta u_L^3 \rangle$ and $(4/5)\epsilon r$. Physically this means that the *kinetic energy cascade is short circuited by energy transfer into the magnetic field*. The 4/3 law indicates the same. We see from the bottom left panel of Fig. 1 that the main contribution to $\langle \delta z_L^{\mp} | \delta \boldsymbol{z}^{\pm} |^2 \rangle$ comes from $2 \langle \delta B_L \delta \boldsymbol{u} \cdot \delta \boldsymbol{B} \rangle$, much like the prediction of the stripy field model. Furthermore, the scaling law approximately reduces to a balance between the magnetic-interaction term $2 \langle \delta B_L \delta \boldsymbol{u} \cdot \delta \boldsymbol{B} \rangle$ and $(4/3)\epsilon r$.

Our study therefore adds to the weight of evidence that isotropic MHD turbulence is controlled by direct nonlocal interaction between the forcing-scale motions and small magnetic structures and shows that even though certain third-order structure functions scale linearly with r, no Kolmogorov-like inertial range or energy cascade exist. In fact, the kinetic-energy cascade is short-circuited by direct energy transfer from forcing scale motions into the folded magnetic field. MHD turbulence resulting from the saturation of small-scale dynamo manages to satisfy the linear scaling prescribed by the exact laws by means of organizing the magnetic field into a folded structure with field reversals at small scales. This emphasizes that the existence of exact laws is not by itself a proof of a local cascade. Note that the short-circuiting of the kinetic-energy cascade is not peculiar to isotropic MHD turbulence. It also appears to be a feature of turbulence in polymer solutions where elastic polymer chains are stretched by turbulence much in the same way magnetic fields are in MHD turbulence [4]. It may be worth enquiring whether the break between the exacts laws and the assumption of a local cascade also occurs in other types of turbulence.

References

1. S. Chandrasekhar: Proc. R. Soc. London, Ser. A **204**, 435–449 (1951). H. Politano & A. Pouquet: Phys. Rev. E **57**, R21–R24 (1998)
2. T. A. Yousef, F. Rincon & A. A. Schekochihin: J. Fluid Mech. **575**, 111–120 (2007) (eprint astro-ph/0611692)
3. A. A. Schekochihin & S. C. Cowley: Turbulence and magnetic fields in astrophysical plasmas. In *Magnetohydrodynamics: Historical Evolution and Trends* ed by S. Molokov, R. Moreau & H. K. Moffatt (Springer, Berlin 2007) (eprint astro-ph/0507686) and references therein. A. A. Schekochihin, C. S. Cowley, S. F. Taylor et al: Astrophys. J. 612, 276–307 (2004)
4. E. De Angelis, C. M. Casciola, R. Benzi et al: J. Fluid Mech. **531**, 1–10 (2005)

Absolute Instabilities and Transition to Turbulence in a Rotating Cavity

B. Viaud[1], E. Serre[2], and J.M. Chomaz[3]

[1] CReA Ecole de l'Air 13661 Salon France bviaud@cr-ea.net
[2] CNRS MSNM-GP IMT Chateau-Gombert 13451 Marseille France
  Eric.Serre@L3m.univ-mrs.fr
[3] CNRS LadHyX Ecole Polytechnique 91128 Palaiseau France
  chomaz@ladhyx.polytechnique.fr

1 Focus of the Work

The flow over a rotating disk has long served as a prototype flow for 3D boundary layer. So, many works have been devoted to the study of the unstable waves found in this configuration, and the associated process of transition toward turbulence. Results of linear stability analysis have shown that the Ekman layer is subject to at least two types of instability. For a review of the subject see [4].

More recent insight on the transition process has been gained by considering the impulse response of the unstable boundary layer.

The interest here is particularly great since [1] discovered that one family of waves is subject to absolute instability at a critical Reynolds number close to the one associated to breakdown to turbulence.

Further studies have been conducted recently, see [2] and [3], focusing on the possible role played by this absolute mode in the last stage of transition, which remains an open question.

Here highly-accurate Direct Numerical Simulations, based on spectral methods, have been performed in order to provide well controled numerical experiments in the case of an open rotating cavity with forced throughflow. The results completely characterize the convective modes of instabilities and bring new elements in the analysis of the role played by the absolute one in the breakdown to turbulence.

2 Modelling and Method

The numerical method is based on pseudo-spectral collocation-Chebyshev approximation in radial and axial directions while a Galerkin-Fourier expansion is applied in the tangential direction. The pseudospectral discretization ensures exponential convergence of the solution, and the use of Gauss Lobatto points improves accuracy near the boundaries. The time scheme is second order accurate.

Appropriate boundary condition are used, especially at the outflow where convective conditions are implemented in order to avoid spurious reflections.

3 Results

First of all, the laminar base flow was favorably compared with Ekman analytical solution, then the flow was locally disturbed at the start of each computation, to record its impulse response.

Multi-dimensionnal FFT of the spatio-temporal recording of the perturbation velocity field enabled precise measure of the characteristic parameters of the wave-packets, namely the radial wave-length λ/δ scaled by the Ekman length-scale, the time frequency ω_r, the azimuthal wave-number β corresponding to the number of spiraling arms, and the wave angle ϵ between the vortices axis and the azimuthal direction. Critical Reynolds and parameter values have been found in good agreement with previous experimental studies.

Then the properties expected from a convective mode have been observed: advection of the wave packet and amplifier-like behaviour. Below a saturation threshold, the growth rate is independent of the perturbation amplitude, while increasing with the distance to the critical Reynolds. The processes of saturation and de-tunning have been clearly identified, thus illustrating the non-linear and non-parallel effects.

Simulations above the expected absolute critical Reynolds led to two kinds of results. If, by means of filtering in the azimuthal direction, the occurence of an absolute mode is prohibited, the flow exhibits only convective behaviour and eventually returns to stability. Non-linear effects limit the amplitude of the wave packet, while its advection does not give time for the higher harmonics to develop and very small scales to appear. On the other hand, if an absolute mode is stimulated, either directly or by energy transfert via saturation of a lower convective mode, the wave packet stays in the computational domain long enough for full saturation to occur (fig. 2). We then observe an harmonic cascade in the sole azimuthal direction, see fig. 1. It has been fully solved, down to its cut-off at approximately 20 Kolmogorov length-scale. It is not associated with any other non-linear effects in the radial or axial directions, so that the flow remains coherent. During these simulations we made sure that the absolute ($\beta = 68$) mode and its harmonics were alone allowed to develop. The fact that transition did not occur seems to indicate that, whereas the absolute mode is necessary for transition to turbulence, it is not sufficient, and another mechanism must be invoked to achieve spatio-temporal disorder.

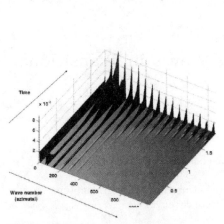

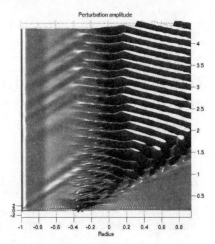

Fig. 1. Absolute type I mode: Time evolution of the azimuthal spectrum showing progressive saturation of higher harmonics.

Fig. 2. Absolute type I mode: spatiotemporal recording of the axial perturbation velocity, from the inflow $(r = -1)$ to the outflow $(r = +1)$

4 Conclusion

High resolution spectral Direct Numerical Simulation has been applied to the study of the transitionnal regime for the boundary layer flow in a rotating cavity. Comparison of the results with the literature in the convective range of control parameters validates the approach. Computations above the critical Reynolds for absolute instability have shown that such a mode is a pre-requisite for transition, but that it is not sufficient in itself to promote turbulence. In this prospect, the secondary instability described by [3] may be the additionnal mechanism required. Further computations are needed, first to investigate the sustained presence of a non-linear global mode, taken as base flow by [3], then with less restriction on the authorized azimuthal wave-numbers, to study the effects of a secondary perturbation.

References

1. R.J. Lingwood:J Fluid Mech **299**, 17 (1995)
2. C. Davies,P.W. Carpenter:J Fluid Mech **486**, 287 (2003)
3. B. Pier: J Fluid Mech **487**, 315 (2003)
4. E. Crespo del Arco, E. Serre, P. Bontoux, B.E. Launder: In *Advances in Fluid Mechanics*, vol 41,(WIT Press 2005) pp 141

Recurrence of Travelling Waves in Transitional Pipe Flow

R.R. Kerswell[1] and O.R. Tutty[2]

[1] Department of Mathematics, Bristol University, Bristol, BS8 1TW, UK
   R.R.Kerswell@bris.ac.uk
[2] Department of Aeronautics & Astronautics, University of Southampton,
   Southampton, SO17 1BJ, UK O.R.Tutty@soton.ac.uk

Wall-bounded shear flows are of tremendous practical importance yet their transition to turbulence is still poorly understood. A new direction in rationalising this phenomenon revolves around identifying alternative solutions (beyond the laminar state) to the governing Navier-Stokes equations. Such solutions which take the form of travelling waves (TWs) have only recently been found in pipe flow [1,2]. Despite being unstable, experimental observations [3] have nevertheless indicated that these solutions are transiently realised. To quantify this, we perform a series of numerical experiments in which the spatial signatures of these TWs are sought in transitional pipe flows [4].

1 Formulation

The governing Navier–Stokes equations for the flow of an incompressible, Newtonian fluid along a circular, straight pipe under the conditions of constant mass flux are solved in primitive variables and cylindrical coordinates (s, θ, z) using finite differences in s and Fourier modes in θ and z. Along with the standard non-slip conditions at the pipe radius, periodic boundary conditions are imposed over a pipe length $5D$. The mean axial speed U, the pipe diameter D and the kinematic viscosity ν define a Reynolds number $Re := UD/\nu$. A working value of $Re = 2400$ is chosen as this is high enough to give turbulent flows but low enough to allow travelling waves to be adequately resolved numerically. At this Re, the TWs identified so far are arranged into symmetry classes of 2, 3- and 4-fold rotational symmetry about the axis and a total of 37 can fit integral wavelengths into a $5D$ periodic pipe. The simulations were started by using a TW with a small perturbation along its most unstable direction as the initial condition (e.g. see figure 1). Two correlation functions,

$$I_{tot}(t) := \max_{\theta_0, z_0} \left[\frac{\langle \mathbf{v}_{DNS} \cdot \mathbf{v}_{TW} \rangle}{\sqrt{\langle \mathbf{v}_{DNS} \cdot \mathbf{v}_{DNS} \rangle} \sqrt{\langle \mathbf{v}_{TW} \cdot \mathbf{v}_{TW} \rangle}} \right], \tag{1}$$

(where $\langle \; \rangle$ represents a pipe volume average, $\mathbf{v}_{DNS}$ and $\mathbf{v}_{TW}$ are the directly numerically simulated (DNS) flow and TW velocity field *after* the mean profile of the TW has been removed and θ_0 and z_0 encompass all phase possibilities) and I_{uv} (similarly defined but including only the cross-pipe velocities), were used to measure how close the instantaneous flow field was to any one travelling wave.

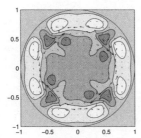 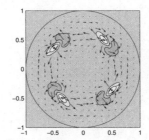

Fig. 1. A 4-fold rotationally symmetric TW (left) and its most unstable eigenfunction (right) at $Re = 2400$. Light(dark) shading indicates $+(-)$ streamwise velocity for the eigenfunction and differences relative to the underlying laminar state for the TW. Arrows indicate size of cross-stream speeds in both.

2 Results & Conclusions

For all the low shear stress (lower branch) TWs considered, starting the run in one sense along the TW's most unstable manifold invariably led to an uneventful gradual relaminarisation, whereas starting in the other sense always produced a turbulent evolution: see figure 2. Both signs of perturbation, in contrast, produced a turbulent trajectory for the high shear stress (upper branch) TWs. During the turbulent episodes, which at $Re = 2400$ and a periodic pipe length of $5D$ are always transient, there is clear evidence that TWs are recurrently visited (e.g. see figure 3). Using the visit criterion that $I_{tot} > 0.5$ & $I_{tot} + I_{uv} > 1$ (see [4] for details), it is found that TWs are visited for about 10% of the time.

The numerical simulations confirm the experimental observations [3]: the turbulent flows do transiently and recurrently resemble the TWs. These instances of coherence, however, occur for only 10% of the time at $Re = 2400$ in a constant-mass-flux, $5D$-long, periodic pipe. The TWs visited tend to be those with intermediate wall shear stresses which are located in the region of phase space populated by turbulent flows. Interestingly, TWs with low wall shear stresses are all found to be on a surface which separates initial conditions which uneventfully relaminarise and those which lead to a turbulent evolution.

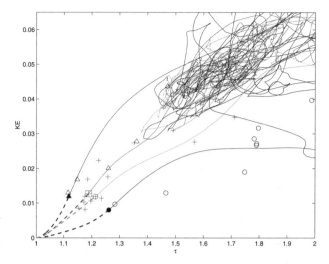

Fig. 2. The disturbance kinetic energy per unit mass, $\langle \frac{1}{2}(\mathbf{u} - \mathbf{u}_{lam})^2 \rangle$ in units of U^2, versus wall shear stress τ in units of $-8\rho U^2/Re$ starting at four lower branch TWs (the laminar state is represented by the point $(1,0)$). The solid line indicates the turbulent evolution for one sign of the eigenvalue perturbation for each TW and the thick dashed line traces out the uneventful relaminarisation for the other. All the TWs present are also plotted: $\triangle/+/\circ$ for 2/3/4-fold rotationally symmetric TWs.

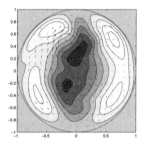 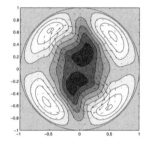

Fig. 3. Comparison plots of the instantaneous DNS flow (left) and a TW (right) (both with the laminar flow subtracted and shading as in figure 1) at a visit in a turbulent run indicated by I_{tot} and I_{uv}.

References

1. H. Faisst and B. Eckhardt. Phys. Rev. Lett. **91**, 224502 (2003)
2. H. Wedin and R. R. Kerswell: J. Fluid Mech. **508**, 333 (2004)
3. B. Hof, C.W.H. van Doorne, J. Westerweel, F.T.M. Nieustadt, H. Faisst, B. Eckhardt, H. Wedin, R.R. Kerswell and F. Waleffe: Science **305**, 1594 (2004)
4. R. R. Kerswell and O.R. Tutty: http://arXiv.org/physics/0611009 (2006)

Analysis of the Unsteady Flow around a Wall-Mounted Finite Cylinder at Re=200 000

O. Frederich, M. Luchtenburg, E. Wassen and F. Thiele

Berlin University of Technology, Institute of Fluid Mechanics and
Engineering Acoustics, Mueller-Breslau-Str. 8, 10623 Berlin, Germany
octavian.frederich@tu-berlin.de

The understanding of the unsteady flow pattern in a complex 3D separated flow is a main objective for the next decade in fluids engineering. Especially at relatively high Reynolds numbers, the extraction of the vortex shedding regime and the base of its formation is a challenge.

1 Scope and Used Methods

In the present work, the unsteady separated vortical flow around a wall-mounted finite circular cylinder is discussed as an example of a complex unsteady flow around a simple geometry. Although time-averaging is the basis for understanding the unsteadiness, it is not sufficient to describe the interaction of the main vortical flows [1]. Instead, a detailed analysis of the time series using POD [2], phase-averaging and also future structure tracing is used here. The temporal behaviour is predicted using different turbulence approaches and is compared to experiments [3] for validation.

The numerical investigations of the flow field are carried out using a Large-Eddy Simulation (LES) with the Smagorinsky subgrid-scale (SGS) model. In addition, a Detached-Eddy Simulation (DES) is applied to examine DES on a LES-capable grid and as alternative SGS model. The flow solver used is based on an implicit and conservative 3D finite-volume approximation of the Navier-Stokes equation with curvilinear coordinates.

The analysis of the flow is performed on the one hand using various visualisation techniques and on the other hand following the POD procedure for the flow field as well as DFT of local or global signals.

2 Selected Results

The simulations were carried out using a grid with 12.3 Mio grid points around a wall-mounted finite cylinder with an aspect ratio of 2 and a Reynolds number of $Re_D = 200\,000$. The time-averaged velocity field as ensemble of $40\,000$ time steps and the double correlations reveal a very good agreement between experiments and LES (fig. 1). The DES is problematic due to artifical eddy viscosity in the attached laminar boundary layer on the cylinder. Due to the congruence of LES and experimental results, the examination of the unsteady flow based on numerical results is also valid for the experiments.

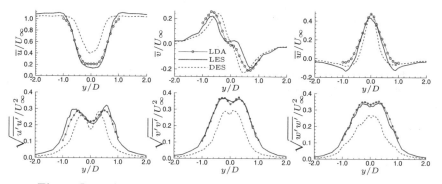

Fig. 1. Lateral velocity and RMS compared to LDA [3] in a wake line.

The POD amplitudes and the DFT of local signals reveal that the dominant frequency corresponds to a Strouhal number of 0.158 (fig 2). The ensemble for the POD consisted of 460 snapshots distributed equidistantly over 60 convective units D/U_∞. The first harmonic modes (1,2) capture 55% of the kinetic energy and the first six modes 85%, in which a strict decomposition of different frequencies is only found for the modes (1,2). The harmonic modes (1,2 and 5,6) become very similar to those of an infinite cylinder behind the recirculation region downstream of the cylinder (fig 3). This recirculation region [1] acts like an increase of the effective streamwise depth of the bluff body and the interaction of the harmonic modes in this region with the plate's boundary layer induces subharmonic modes (3,4). Single modes acting as change of the mean flow are not present, but are obtained if the ensemble of snapshots do not represent the second order statistics correctly. Accordingly, the evaluation of higher modes requires a basis of more snapshots.

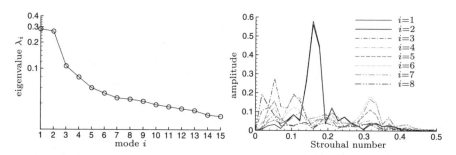

Fig. 2. POD eigenvalues and spectrum of Fourier coefficients, LES.

3 Conclusion and Outlook

The numerical and experimental analysis of the flow field are in very good agreement. Furthermore, the extraction of unsteady flow features provides an understanding of the interacting vortical flows. Ongoing work will focus on detailed insight of these unsteady flow phenomena using different methods.

The combination of numerical and experimental results as well as the comparison of different turbulence modelling approaches are intended to provide a combined database for the validation or development of numerical and experimental methods.

Acknowledgement

The work presented is supported by the German Research Foundation and the North German Cooperation for High Performance Computing.

References

1. Frederich, O., Wassen, E., Thiele, F. (2006) *Prediction of the flow around a short wall-mounted cylinder using LES and DES*, ICNAAM 2006, Wiley-VCH.
2. Lumley, J.L. (1970) *Stochastic Tools in Turbulence*, Academic Press, New York.
3. Richter, F. (2005) *Experimentelle Untersuchungen ... im Nachlauf eines Kreiszylinderstumpfes ...* . PhD thesis, University Rostock.

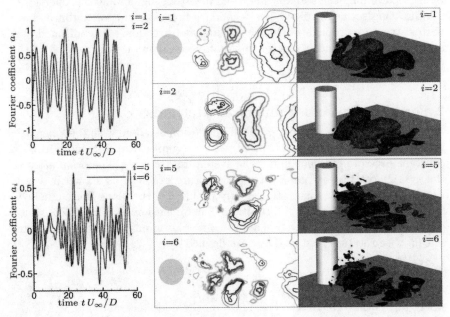

Fig. 3. First and second harmonic POD mode (1,2 and 5,6) represented by their Fourier coefficients, contour lines and iso-surfaces of the lateral v-velocity, LES.

Transition in plane Poiseuille flow with a stream-wise rotation

M. Nagata[1] and S. Masuda[2]

[1] Department of Aeronautics and Astronautics, Graduate School of Engineering, Kyoto University, Yoshida-honmachi, Sakyo-ku, Kyoto 606-8501 Japan
nagata@kuaero.kyoto-u.ac.jp
[2] shuichi_@kuaero.kyoto-u.ac.jp

1 Introduction

It is important to understand the stability of flows under a system rotation for both engineering and geophysical applications. Recently, tuebulent channel flow with a stream-wise system rotation has attracted attention theoretically[1] and experimentally[2]. The most striking feature among other turbulent properties of this flow is the generation of a cross flow in the span-wise direction. In the present short report we analyse the transition in this flow by first examining the stability and then investigating nonlinear properties of a bifurcating secondary flow. The secondary flow is found to exhibit a spiral vortex structure propagating in the stream-wise direction. It is confirmed that an anti-symmetric mean flow in the span-wise direction is generated in the secondary flow.

2 Formulation and Numerical Methods

We consider a viscous incompressible fluid motion of a fluid with the kinematic viscosity ν between two parallel plates with a gap width, $2d_*$, induced by a constant pressure gradient under the system rotation Ω_*. The orientation of the system rotation is parallel to the pressure gradient as shown in Fig.1. The basic flow with a quadratic velocity profile, *i.e.* the plane Poiseuille flow, $U_{B*}(z_*)$, is not affected by the rotation. The dimensionless parameters which control the motion in this system are the Reynolds number R and the constant stream-wise rotation rate Ω. They are defined by

$$R = \frac{U_{0*}d_*}{\nu}, \tag{1}$$

where U_{0*} is the value of the basic flow on the midplane, and

$$\Omega = \frac{2\Omega_* d_*^2}{\nu}. \tag{2}$$

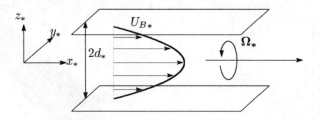

Fig. 1. The configuration of the model

.

We express the flow variables by means of the Fourier-Chebyshev expansions and use the Chebyshev collocation method to solve the eigenvalue problem for the linear stability analysis and the Newton-Raphson iterative method for finding a nonlinear equilibrium state which bifurcates from the laminar state.

3 Results and Discussion

3.1 Linear stability

It is found that the laminar flow becomes most unstable when perturbations are three-dimensional in the rotating case, i.e. the Squire theorem does not hold in this case. The critical Reynolds number R_c for the rotating case is far less than the well known value of 5772 for the non-rotating case with a span-wise independent perturbation (Squire mode), as can be seen in Fig.2.

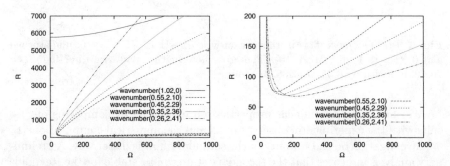

Fig. 2. The neutral curves on the $\Omega - R$ plane for various wavenumber pairs. (right: the blow-up near the origin)

3.2 nonlinear analysis

Nonlinear analysis shows that the bifurcation of a secondary flow with the stream- and the span-wise wavenumber pair $(\alpha, \beta) = (0.35, 2.36)$ occurs at $R_c = 69.3924$ when $\Omega = 140$. The secondary flow exhibits a three-dimensional

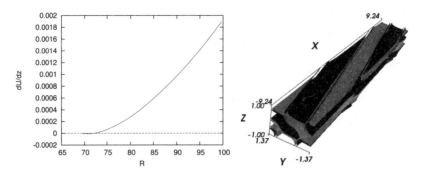

Fig. 3. The bifurcation diagram (left) and the stream-wise component of the vorticity with $\alpha = 0.35$ and $\beta = 2.36$ at $R = 70$ and $\Omega = 140$ (right).

spiral structure as shown in Fig.3. We find that the secondary flow distorts the symmetric mean flow in the stream-wise direction and generates an anti-symmetric cross flow in the span-wise direction as shown in Fig.4.

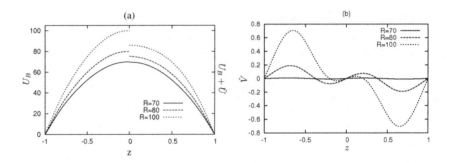

Fig. 4. (a): The mean flow in the stream-wise direction. $-1 \leq z \leq 0$: undisturbed and $0 \leq z \leq 1$: disturbed. (b): The cross flow in the span-wise direction. $R = 70, 80, 100$ with α=0.35, β=2.36, Ω=140.

We shall discuss nonlinear properties more in detail by comparing our results with the experimental observation[2] and the numerical computations[1].

Investigation on the stability of the secondary flow is under way. A preliminary analysis indicates that the secondary state can be stable for the Reynolds number slightly above its critical value.

References

1. M. Oberlack, W. Cabot and M. M. Rogers : *Procedings of the Summer Program*, 221 (1998)
2. I. Recktenwald, Ch. Brüker and W. Schröder : Advances in Turbulence X, *Proceedings of the Tenth European Turbulence Conference*, 561 (2004)

DNS of channel flows with pressure gradient

M. Marquillie, J.-P. Laval, and R. Dolganov

Laboratoire de Mécanique de Lille, CNRS UMR 8107, Blv Paul Langevin, 59655 Villeneuve d'Ascq, France. `matthieu.marquillie@univ-lille1.fr`

1 Introduction

The recent DNS of channel flow at high Reynolds number brings some new possibilities to study the complex dynamic of coherent structures near the wall. One of the easiest way to introduce an adverse pressure gradient is to use suction at one wall of a channel or to directly prescribe the adverse pressure gradient [1]. The advantage of this method is the possibility to use a numerical code adapted to plane boundary layer or plane channel flow. However the use of curved walls is more challenging. The numerical code used for the present study combine the advantage of the good accuracy of spectral resolution with fast integration procedure, for simulation over a smooth profile. This code was originally used to study the 2D and 3D instabilities of a boundary layers over a bump [2, 3]. The present study investigates the effect of pressure gradient on the turbulent structures from DNS data at $Re_\tau = \delta u_\tau/\nu \simeq 400$ (where δ is half the channel width and u_τ the friction velocity at the inlet). At this Reynolds number the flow slightly separates on the bump but not at the flat wall.

2 Description of the simulation

Rather to write the incompressible Navier-Stokes equations in curvilinear coordinates we merely transform the partial differential operators using a mapping which allows the physical domain to be transformed into a Cartesian one. For space discretization fourth-order central finite differences are used for the second derivatives in the streamwise x-direction. All first derivatives of the flow quantities appear explicitly in the time-advancing scheme and the first derivatives in x are discretized using eighth-order finite differences. Chebyshev-collocation is used in the wall-normal y-direction. The transverse direction z is assumed periodic and is discretized using a spectral Fourier expansion with N_z modes, the nonlinear coupling terms being computed using the conventional de-aliasing technique with $M > 3N_z/2$. For time-integration implicit second-order backward Euler differencing is used; the Cartesian part of the Laplacian is taken implicitly whereas an explicit second-order Adams-Bashforth scheme is used for the operators coming from the mapping as well as for the nonlinear convective terms. The three-dimensional system uncouples into N_z two-dimensional subsystems and the resulting 2D-Poisson equations are solved efficiently using the matrix-diagonalization technique. The inlet conditions was taken from a previous highly resolved LES of plane channel flow at the same Reynolds. The DNS grid ($1536 \times 257 \times 384$) leads to a resolution in wall units of ($\Delta x^+ = \Delta z^+ = 7.6$, $\Delta y^+_{max} = 8.2$) based on the maximum value of u_τ. This resolution corresponds to a maximum mesh size of 3.9η in the normal direction and 6.8η in the two other directions where $\eta = (\nu^3/\epsilon)^{1/4}$ is the local isotropic Kolmogorov scale. The simulation was integrated over 1.6 flow-through time after convergence of statistics.

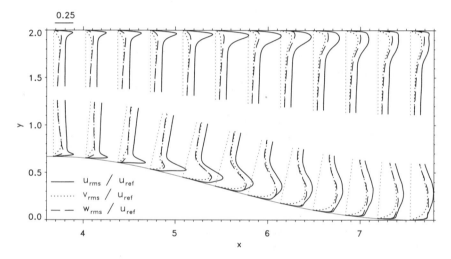

Fig. 1. Profiles of the rms velocity fluctuations normalized by the bulk velocity u_{ref} in the region of the bump

3 Results

Computing the probability of reverse flow, the region of transitory detachment on the bump do not exceed 10% of the bump height and the separation and reattachment points are located at $x = 4.23$ and $x = 5.4$ respectively. The complete balance of the Reynolds stresses has been computed. This allows a detailed comparison of the statistics of turbulence over the bump (with separation) and on the flat opposite wall (without separation). The maximum intensity of the rms of the streamwise velocity fluctuation (u_{rms}) is comparable in magnitude for the two sides unlike the spanwise and normal components which reach a maximum value of 50% lower at the upper wall as compared to the curved wall (see fig. 1). As reported by several authors for boundary layer flows with adverse pressure gradient [4], the profiles of turbulent normal stress exhibit a secondary maximum. On the downstream side of the bump, the first maximum of u_{rms} moves away from the wall indicating that the internal layer grows away from the wall. The coherent structures have also been compared and the same structures are detected on the opposite flat wall were no separation occurs (see fig. 2). The generation of intense coherent structures is stable in time and located slightly downstream at the flat upper wall. However, the typical size of the vortices seems comparable at both walls and their intensity slowly decrease in time.

Fig. 2. Isovalue of second invariant of the velocity gradient tensor $Q = \frac{1}{2}\left(\Omega^2 - S^2\right)$

4 Conclusion

This numerical code used for the present DNS allows us to perform simulations with smooth profiles at reasonable cost keeping the benefit of the efficiency and the accuracy of the spectral solver in spanwise and normal direction. The fact that this single configuration is able to account for on adverse pressure gradient flow with and without curvature makes an ideal test case for engineering. Moreover, the channel flow inlet conditions are well documented and easy to model for simulations with a turbulent model like Reynolds Averaged Navier Stokes or LES models. The statistics will be integrated to experimental database to study wall turbulence on adverse pressure gradient flows.

References

1. Y. Na and P. Moin, J. Fluid Mech. **374**, 379 (1998).
2. M. Marquillie and U. Ehrenstein, Computers & fluid **31**, 683 (2002).
3. M. Marquillie and U. Ehrenstein, J. Fluid Mech. **490**, 169 (2003).
4. D. R. Webster, D. B. Degraaf, and J. K. Eaton, J. Fluid Mech. **320**, 53 (1996).

An Investigation into the Evolution of Sub-Layer Streaks in Two- and Three-Dimensional Turbulent Boundary Layers

K.L. Kudar[1], P.W. Carpenter[1] and C. Davies[2]

[1] School of Engineering, University of Warwick, Coventry, CV4 7AL, UK.
[2] School of Mathematics, Cardiff University, Senghennyd Road, Cardiff, CF24 4AG, UK.

1 Introduction

Sub-layer streaks are known to be part of a quasi-cyclic process occurring in the near-wall region of turbulent boundary layers [1]. It is this process that is responsible for the high levels of drag that are generated. An understanding of the generation and evolution of the streaks will provide an insight into the process itself; hence the focus of the present study.

2 Numerical Procedure

To model the boundary-layer flow the highly efficient velocity-vorticity formulation of Davies & Carpenter [2] is used. A mean velocity profile is prescribed and the velocity and vorticity perturbations away from this mean profile are calculated. Initially, for two-dimensional flows, the mean velocity profile of Spalding [3] and Coles [4] is used. A line streamwise vorticity source, $(\omega_x)_{source}$, is used to approximate the quasi-streamwise vortices known to generate the streaks. It takes the form of a delta function in the streamwise (x) and wall-normal (z) directions and is sinusoidal in the spanwise direction (y) as follows:

$$(\omega_x)_{source} = -G\delta(x - x_f)\delta(z - z_f)exp(i\beta y) \tag{1}$$

where G is the strength of the source, x_f and z_f give the position of the source in the streamwise and wall-normal directions respectively and β is the spanwise wavenumber. The value of z_f is chosen so that the vorticity source generates the strongest streamwise velocity perturbation. A forcing time of τ^+ = 15 was initially derived from the experimental findings of [5]; however, it was subsequently found to be related to the theoretical period of the bursting cycle [1]. Finally, to investigate the evolution of the streaks in a swept three-dimensional boundary layer the mean velocity profiles derived by Degani et al. [6] were used.

3 Results and Discussion

The streaks generated (see Fig. 1) are found to be long in the streamwise direction and narrow in the spanwise and wall-normal directions. They also appear as alternating high- and low-speed streaks in the spanwise direction. The streaks produced display transient growth characteristics; the strength of the streaks is found to increase with time (even after the forcing has ceased) before reaching a maximum and then decaying. The strongest streak-response is produced at a distance of $z^+ \approx 13$ away from the wall and has a spanwise spacing of approximately $\lambda_{opt}^+ \approx 80$ (see Fig. 2). This is thought to be the spacing most likely to be seen in an experimental setting and also coincides with the range of mean streak-spacing found in many experimental investigations that have been carried out [7]. Varying the streamwise pressure gradient produces streaks that are stronger in adverse pressure gradients and weaker in favourable pressure gradients (see Fig. 2); this is consistent with the qualitative findings of Kline et al. [8].

Three-dimensional turbulent boundary layers are less efficient at generating shear stress than their two-dimensional counterparts; the near-wall structures are found to be weaker and more stable in three-dimensional flows [9]-[10]. When studying analogous structures (Klebanoff modes) in laminar three-dimensional boundary layers [11] it was found that the streak strength was much weaker in a swept boundary layer than in the corresponding two-dimensional boundary layer. This has therefore motivated us to investigate the effect of three-dimensionality on the sub-layer streaks and the results of this study will be presented.

References

1. P.W. Carpenter et al.: Phil. Trans. R. Soc. A, (in press) (2007).
2. C. Davies & P.W. Carpenter: J. Comp. Phys., **172**, p. 119 (2001).
3. D.B. Spalding: J. Applied Mech., **28**, p. 455 (1961).
4. D. Coles: J. Fluid Mech., **1**, p. 191 (1956).
5. M. Gad-el-Hak & F. Hussain: Phys. Fluids, **29**, p. 2124 (1986).
6. A.T. Degani et al.: J. Fluid Mech, **250**, p. 43 (1993).
7. M. Zacksenhouse et al.: Experiments in Fluids, **31**, p. 229 (2001).
8. S.J. Kline et al.: J. Fluid Mech., **30**, p. 741 (1967).
9. K.A. Flack: Experiments in Fluids, **23**, p. 335 (1997).
10. R.O. Kiesow & M.W. Plesniak: J. Fluid Mech., **484**, p. 1 (2003).
11. K.L. Kudar et al.: Proc. IUTAM Symp. on Laminar-Turbulent Trans., Bangalore, India, p. 167 (2004).

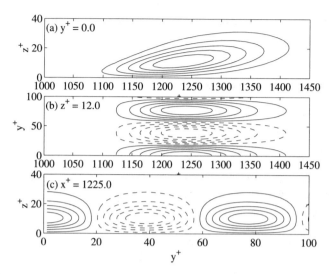

Fig. 1. Streamwise velocity perturbation contour plots of the optimum streaks generated when viewed (a) from the side, (b) from above and (c) into the streamwise direction. Solid and dashed lines indicate locally positive and negative velocity perturbations respectively.

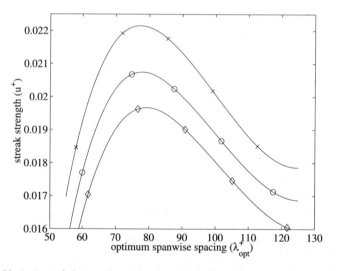

Fig. 2. Variation of the optimum spanwise spacing with strength of streaks generated in ($\times$) an adverse, ($\circ$) a zero and ($\diamond$) a favourable streamwise pressure gradient.

Wall-Shear Stress Assessment in Zero-Pressure Gradient Turbulent Flow using MPS^3

S. Große[1] and W. Schröder[1]

Institute of Aerodynamics, RWTH Aachen University, D-52062, Germany
s.grosse@aia.rwth-aachen.de

To determine the fluctuating wall-shear stress distribution is of utter importance in turbulence research. The mean wall-shear stress defines the character of wall-bounded flows and the dynamic part is a direct indicator of near-wall turbulent events. The micro-pillar shear-stress sensor MPS^3 described below measures the local shear stress indirectly by exploiting the linear relation between the velocity gradient in the viscous sublayer and the wall-shear stress. A detailed description of the MPS^3 sensor is given in [4, 5]. The pillars are manufactured from the elastomer PMDS. The pillar-tip deflection from a reference position at no velocity serves as a measure for the forces exerted on the structure and is detected using common optical equipment. A calibration of the sensor is performed in linear shear flow.

1 Wind Tunnel and Micro-Pillar Shear-Stress Sensor

The experiments were performed in the wind tunnel at the Laboratoire de Mécanique de Lille (LML), a detailed description of which is given in [2]. The wind tunnel of Göttingen type is temperature-stabilized ($\pm 0.2\ K$) and the test-section has a cross section of $2\ m \times 1\ m$. The position of the sensor is 19.6 m downstream of the leading edge of the flat plate. Measurements have been performed at freestream velocity U_∞ of 3, 5, 7, and 10 m/s. The Reynolds number based on the momentum thickness Re_Θ ranges from 7800 to 21000, the value based on the sensor position is $3.9 \cdot 10^6$ to $1.3 \cdot 10^7$.

Measurements using sensors with 10 micro-pillars aligned in the streamwise direction were performed. The sensor structure was flush mounted in the wind-tunnel wall. The micro-pillars have a height $L = 650\ \mu m$ and a mean diameter $D \approx 25\ \mu m$. Since the pillar has a minimum dimension in the wall-parallel directions, spatial averaging is reduced. Nonetheless, the sensor averages fluid fluctuations along its wall-normal extension and hence, one of the key parameters is the sensor height. For the range of Reynolds numbers in this study, the sensor height varies between 4.9 and 14.8 wall units and for

$Re_\Theta = 21000$, it is higher than the viscous sublayer, i.e., the linear dependence between sensor deflection and wall-shear stress is no longer valid.

The sensor is observed with a FastCam-X 1024 pci camera, operated at 50 Hz and 3000 Hz using 8192 samples. The data evaluation is performed using correlation routines with multi-passing. The error in determining the position of the pillar tips is ≈ 0.05 px, leading to an error of less than 1.5%.

2 Results

The results demonstrate the capability of the micro-pillar shear-stress sensor MPS^3 to accurately detect the wall-shear stress in a zero-pressure gradient turbulent boundary layer even at high Reynolds numbers.

The mean wall-shear stress (fig. 1(a)) shows reasonable agreement with data from hot-wire measurements in the wind-tunnel facility.

The intensity of the streamwise fluctuations (fig. 1(a)) is about 0.37 at $Re_\Theta = 7800$ and decreases to 0.31 with increasing Reynolds number. Note, this effect is rather due to the increased protrusion of the sensor into the boundary layer with increased Reynolds number showing good agreement with the findings [6] than due to spanwise spatial averaging, which is reported in [7, 10] to reduce the turbulence intensities measured with hot-wires.

The skewness of the fluctuations (fig. 1(b)) is $S_f \approx 1.25$ at the two lower Reynolds numbers and decreases slightly to $S_f = 0.9$ at higher Reynolds numbers. Values of $S_f \approx 1.0$ are reported in [1, 6] for hot-wires located at $y^+ \leq 4$ whereas in [3] S_f of $1.2 \div 1.3$ in the near-wall region and a vanishing S_f at values of $y^+ \geq 12$ are reported. The findings for the flatness F_f (fig. 1(b)) show a similar behavior. It reaches $F_f = 5.9$ at lower Reynolds numbers and decreases to a value of $F_f = 4.5$ at higher Reynolds numbers. Similarly high values are reported in [3] at $y^+ \leq 4$, whereas slightly lower values $4 \div 5$ in the very near-wall region were found in [6]. The decreasing skewness and flatness at higher Reynolds numbers result from an inadequate sensor length and an integration of fluctuations along the wall-normal direction up to higher values of y^+. Note, the flatness and skewness of the velocity fluctuations are reported to decay strongly with increasing y^+.

Power-density spectra [9] of the turbulent wall-shear stress fluctuations are given in fig. 1(c). Note, the sampling time was very short. The spectra are plotted in inner scaling (u_τ and ν) and are found to collapse on one universal curve, which is in good agreement with results from [8]. The pre-multiplied power-density spectra given in fig. 1(d) illustrate at least for the lowest Reynolds number the sensor to be able to capture the complete turbulence spectrum. At higher Reynolds numbers, the highest frequency structures can not be detected with the current setup due to the limited recording frequency and due to resonance of the sensor structure.

Acknowledgement. We would like to thank Prof. Stanislas from the LML for his support and for having the opportunity to perform the measurements in his facility.

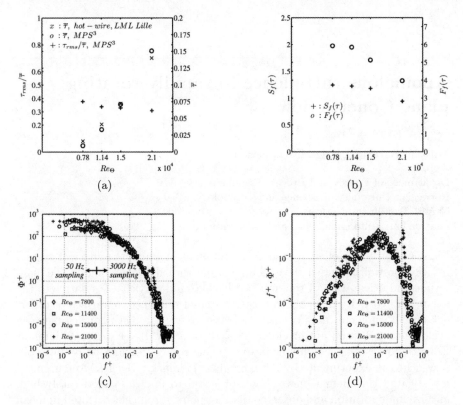

Fig. 1. (a) Mean wall-shear stress and RMS of wall-shear stress fluctuations. (b) Skewness and flatness of wall-shear stress fluctuations. (c) Power-density spectra of fluctuations. $\Phi^+ = \Delta T \cdot \frac{\Phi}{\overline{\tau} l_w^2} \cdot \frac{u_\tau^2}{\nu}$, $f^+ = \frac{f \cdot \nu}{u_\tau^2}$, ΔT is the sampling time. (d) Pre-multiplied power-density spectra of fluctuations.

References

1. Alfredsson, P.H., Johansson, A.V., Haritonidis, J., Eckelmann, H.: Phys. Fluids **31**, 1026–1033 (1988)
2. Carlier, J., Foucaut, J., Dupont, P., Stanislas, M.: In: Colloque AAAF (1999)
3. Fernholz, H., Finley, P.: Prog. Aerosp. Sci. **32, Nr. 4**, 245–311 (1996)
4. Große, S., Schröder, W.: submitted to Exp. Fluids (2006)
5. Große, S., Schröder, W., Brücker, C.: Meas. Sci. Technol. **17**, 2689–2697 (2006)
6. Khoo, B., Chew, Y., Li, G.: Exp. Fluids **22**, 327–335 (1997)
7. Ligrani, P., Bradshaw, P.: Exp. Fluids **5**, 407–417 (1987)
8. Madavan, N., Deutsch, S., Merkle, C.: J. Fluid Mech. **156**, 237–256 (1985)
9. Press, W., Teukolsky, S., Vetterling, W., Flannery, B.: Numerical Recipes in C (1992)
10. Ruedi, J.D., Nagib, H., Österlund, J., Monkewitz, P.: Exp. Fluids **36**, 393–398 (2004)

Anomalous turbulence in rapidly rotating plane Couette flow

Mustafa Barri and Helge I. Andersson

Department of Energy and Process Engineering,
Norwegian University of Science and Technology,
N-7491 Trondheim, Norway.
mustafa.barri@ntnu.no, helge.i.andersson@ntnu.no

Plane Couette flow offers a unique environment in which the effect of system rotation on wall-bounded shear flows can be studied. Depending on the magnitude and orientation of the imposed background vorticity Ω relative to the mean flow vorticity ω in the rotating frame-of-reference, a variety of different flow phenomena may occur. The ratio $S \equiv \Omega/\omega$ distinguishes between cyclonic ($S > 0$) and anti-cyclonic ($S < 0$) rotation. If $-1 < S < 0$, laminar Couette flow is *unstable* and counter-rotating roll cells occur, as recently observed in the experiments by Alfredsson & Tillmark.[1] Bech & Andersson[3,4] investigated turbulent Couette flow subjected to mild and moderately high anti-cyclonic rotation and observed distinct pairs of roll cells oriented in mean flow direction. Needless to say, these roll cells made a substantial contribution to the cross-sectional mixing and the wall-friction was higher than in the non-rotating case examined by Bech et al.[2]

The aim of the present study is to explore the turbulent flow field in a rapidly rotating Couette flow in the absence of roll cells. To this end we consider the shear-driven fluid motion between two infinite parallel planes separated a distance $2h$ in the y-direction. The turbulent flow is induced solely by the prescribed velocity difference $2U_w$ (in the x-direction) between the two planes. The Reynolds number is $Re \equiv U_w h/\nu = 1300$, i.e. well above 500 required for fully developed turbulence to persist.[1]

The relative importance of the imposed system rotation is given by the rotation number $Ro \equiv 2\Omega h/U_w$, which is a cross-sectional average of $-S$. While Bech & Andersson considered mild[3] and moderate[4] rotation with $Ro \in [0, 0.5]$, we will now focus on the high rotation number $Ro = 0.7$, which is believed to be representative for the range $0.5 < Ro < 1.0$. We therefore performed a direct numerical simulation of fully developed Couette flow, i.e. with periodic boundary conditions in the streamwise and spanwise directions. The size of the computational domain was $8\pi h \times 2h \times \frac{4}{3}\pi h$ and the first simulation reported here was made with a $64 \times 64 \times 64$ grid.

In accordance with our conjecture, rotational-induced roll cells, as observed for $Ro < 0.5$, did not occur at $Ro = 0.7$, as seen from the contour plots of the fluctuating velocity field in Fig. 1. The mean velocity variation $U(y)$ in Fig. 2a exhibits a substantial linear range which extends over 80% of the cross-section. Here, the slope $dU/dy \approx 2\Omega$ which makes $S = -1$. The extent of the Coriolis-dominated region is consistent with the criterion[5] that system rotation matters when $y > \delta_c$, where $\delta_c = u_\tau/\Omega$ is the Coriolis length scale. The tendency of a rotating flow to establish regions with zero vorticity in an inertial frame has been observed before both in Couette[4] and Poiseuille[5] flow. In the Couette flow, dU/dy cannot exceed U_w/h and $Ro = 1$ thus becomes an upper bound for which neutral stability $S = -1$ can be sustained.

(a) v-velocity, $y^+ = 176$

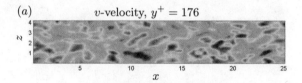

(b) v-velocity, $x^+ = 12.6$

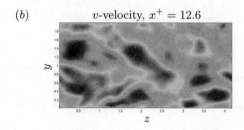

Fig. 1. Iso-contour plot of the instantaneous wall-normal velocity fluctuations. (a) (xz)-plane parallel with the walls; (b) Cross-sectional (yz)-plane.

The constancy of dU/dy in the Couette flow makes both contributions to the total mean shear stress $\mu dU/dy - \rho\overline{uv} = \tau_w \equiv \rho u_\tau^2$ constant in the core region, as seen in Fig. 2b. The rapidly rotating Couette flow therefore consists of a nearly homogeneous central region where rotational effects dominate and narrow regions adjacent to each wall where viscous effects prevail. The total production of turbulent kinetic energy can be expressed as:

$$\tilde{P} \equiv \frac{P}{U_w^3/h} = \left[\left(\frac{u_\tau}{U_w}\right)^2 - \frac{1}{Re}\frac{d\tilde{U}}{d\tilde{y}}\right]\frac{d\tilde{U}}{d\tilde{y}} \qquad (1)$$

where $\tilde{U} = U/U_w$ and $\tilde{y} = y/h$ are dimensionless quantities. From equation (1) the dimensionless production $\tilde{P}$ in the core region is found to be 2.35×10^{-3}, which is nearly three times higher than the production rate found for $Ro =$

$0.^2$ It is particularly noteworthy that the turbulence exhibits an abnormal anisotropy with $v > w > u$, in contrast to the conventional anisotropy $u > w > v$ found in non-rotating shear flows. These anomalies will be further explored by means of structural information based on, e.g., λ_2-structures and quadrant analysis.

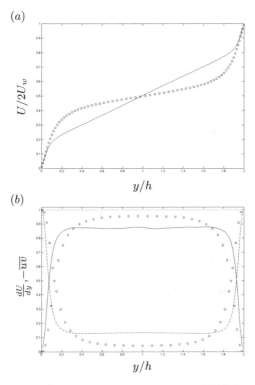

Fig. 2. Present results for $Ro = 0.7$ compared with DNS-data for $Ro = 0.0$ from Bech et al.^{2} (diamonds). (a) Mean velocity profile $U(y)$, (b) Turbulent(——) and viscous(- - -) shear stresses.

References

1. P.H. Alfredsson & N. Tillmark: *Proc. IUTAM Symposium*, Springer 2005, pp. 173–193.
2. K.H. Bech, N. Tillmark, P.H. Alfredsson & H.I. Andersson: J. Fluid Mech. **286**, 291–325 (1995)
3. K.H. Bech & H.I. Andersson: J. Fluid Mech. **317**, 195–214 (1996)
4. K.H. Bech & H.I. Andersson: J. Fluid Mech. **347**, 289–314 (1997)
5. K. Nakabayashi & O. Kitoh: J. Fluid Mech. **315**, 1–29 (1996)

Turbulent clustering of inertial particles and acceleration field

S. Goto,[1] J. C. Vassilicos[2] and H. Yoshimoto[1]

[1] Department of Mechanical Engineering and Science, Kyoto University, Japan.
goto@mech.kyoto-u.ac.jp
[2] Department of Aeronautics and Institute for Mathematical Sciences,
Imperial College London, UK.
j.c.vassilicos@imperial.ac.uk

1 Introduction

Inertial particles (in this article, we restrict ourselves to the case of small heavy particles) distribute in-homogeneously in space even in statistically homogeneous turbulence. This phenomenon is well-known as the preferential concentration of inertial particles. The preferential concentration in homogeneous isotropic turbulence is conventionally explained in terms of Kolmogorov-scale coherent eddies; that is, heavy particles form voids by being expelled from the coherent eddies (i.e. high vorticity regions) and cluster in the outer peripheries of these eddies (i.e. high strain-rate regions). However, it has been pointed out recently [1–3] that this approach does not fully explain the particle clustering in turbulence at high Reynolds numbers. More precisely, the clustering of inertial particles is a multi-scale phenomenon, and therefore cannot be described solely by the vorticity/strain fields which characterise the smallest-scale structures. Here, we propose a new approach to describe particle clustering: in terms of the clustering properties of the fluid acceleration field

$$a \equiv \partial u/\partial t + u \cdot \nabla u \qquad (1)$$

where $u(x,t)$ is the fluid velocity field.

2 Sweeping of acceleration field and clustering of its stagnation points

We propose the idea that particle clusters can be described in terms of clusters of the fluid acceleration field. In two-dimensional inverse-cascading turbulence, particle clusters directly reflect clusters of stagnation points of $a(x,t)$ when the Stokes time (see (2) below) τ_p is within the inertial time-scale range.

This suggestion is based on the following physical argument [2,4]. First, in developed turbulence, the acceleration field, and in particular its stagnation points, tend to be swept by the local velocity field (see figure 3a of Ref. [4]). In other words, the acceleration field hardly changes, on average, in the Lagrangian frame moving with the local fluid velocity. On the other hand, the velocity v_p of particles is well approximated by $u - \tau_p a$ when the Stokes time τ_p of particles is sufficiently shorter than a local time-scale of the turbulence. Then, at points where a vanishes, v_p tends to equal the fluid velocity u. This implies that inertial particles tend to move with stagnation points of the acceleration field for much longer than with points of any other non-zero value of the acceleration field. In this sense, acceleration-stagnation points are somehow sticky for particles. As the acceleration-stagnation points are found to cluster, the particle clusters mimic, as a result, the cluster of acceleration-stagnation points.

3 Numerical verification

In order to verify these ideas, we conduct direct-numerical simulations (DNS) of inertial particles in two- and three-dimensional homogeneous turbulence. In our DNS, the particle motions are simulated by solving

$$\frac{\mathrm{d}}{\mathrm{d}t} v_p(t) = \frac{1}{\tau_p} \left(u(x_p(t), t) - v_p(t) \right) \tag{2}$$

for each particle, where v_p and x_p are the velocity and position vectors of particles. Here, we have assumed that (i) particles (rigid spheres) are very much heavier than the background fluid, (ii) the particle diameter is sufficiently small for the surrounding flow to be approximated by a Stokes flow and (iii) gravity can be neglected. We further assume that the number density of particles is dilute enough for collisions between particles and feedback to fluid motion to be negligible.

In the two-dimensional simulation we impose a small-scale energy source and a large-scale energy sink. Then, as a result of the inverse energy cascade, the energy spectrum takes a $-5/3$ power law form in the inertial range. The particle distribution in such turbulence is surprisingly similar, over all length scales in the inertial range, to the spatial distribution of acceleration-stagnation points (see Fig. 8 of Ref. [2]). This observation strongly supports our proposed particle clustering mechanism.

In the three-dimensional case (which is based on conventional numerical turbulence forced at large scales), however, the cluster of acceleration-stagnation points is too sparse to totally describe the clusters of inertial particles. Hence, we identify points where only one well chosen component (based on the eigenvectors of the acceleration gradient tensor) of the acceleration a vanishes. These points are plotted together with inertial particles in Fig. 1.

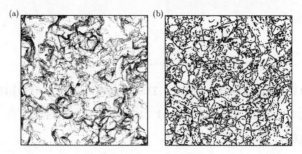

Fig. 1. (a) Spatial distribution of inertial particles ($\tau_p = 2\tau_\eta$ where τ_η is the Kolmogorov time) in three-dimensional homogeneous turbulence ($R_\lambda \approx 200$). The side length of the plots is 510η where η is the Kolmogorov length. Particles in a thin layer (5η thickness) are plotted. (b) Spatial distribution of zero-acceleration-component points in the thin layer.

Not only the cell-like structures of the inertial particle cluster are well captured by the cluster of these points, but also the locations of these two clusters fairly coincide with each other.

4 Remarks

An important advantage of our approach to particle clustering is that we can describe it without tracking the Lagrangian trajectories of particles. Furthermore, we expect this approach to be extendable beyond homogeneous isotropic turbulence to more general turbulent flows, because the above arguments only rely on the facts that the acceleration field clusters and that it is swept by the local fluid velocity field. These properties of the acceleration field are not expected to be unique to homogeneous isotropic turbulence.

Since, as mentioned in the introduction, the particle clustering in developed turbulence is a multi-scale phenomenon, the above results may imply that acceleration stagnation points have a multi-scale nature, although the statistics of acceleration are sometimes assumed to be described in terms of the smallest-scale structures only.

This research is supported by Grant-in-Aid for Young Scientists from the Ministry of Education, Culture, Sports, Science and Technology.

References

1. G. Boffetta, F. De Lillo, A. Gamba Phys. Fluids **16** L20–L23 (2004)
2. S. Goto, J. C. Vassilicos: Phys. Fluids **18** 115103 (2006)
3. H. Yoshimoto, S. Goto: J. Fluid Mech. in press (2007)
4. L. Chen, S. Goto, J. C. Vassilicos: J. Fluid Mech. **553** 143–154 (2006)

Effect of the Reynolds number and initial separation on multi-particle sets using Kinematic Simulations

A. Abou El-Azm Aly[1], F. Nicolleau[2], and A. ElMaihy[3]

[1] The University of Sheffield, Department of Mechanical Engineering, Mappin Street, S1 3JD, Sheffield, United Kingdom A.AboAzm@Sheffield.ac.uk
[2] F.Nicolleau@Sheffield.ac.uk
[3] Department of Mechanical Power and Energy, Military Technical College, Cairo, Egypt

We use Kinematic Simulation to study the dispersion of multi-particle clusters in isotropic turbulence. We study the evolution of the shape's characteristics of triangles and tetrahedrons. We also study the fractal dimension of line, surface and volume. We are particularly interested in the respective role of the Reynolds number and of the particle-cluster initial characteristic size. We generalise these studies to particles with inertia.

1 Kinematic Simulation

Kinematic Simulation (KS) is a Lagrangian model where a synthetic Eulerian velocity field $u(x,t)$ is assumed. It is then possible to track one, two or many particles and study their trajectories by integrating this velocity field. We use the KS developed in [1] based on [2] for incompressible isotropic turbulence.

2 Fractal dimension of set of particle

The evolution of the fractal dimension of material elements as a function of time is investigated using this KS. The fractal dimension of a surface is found to obey:

$$D = 2 + 0.044\frac{t}{\tau_\eta}$$

where $\tau_\eta = \frac{L}{u'}(Re_{KS})^{-\frac{1}{2}}$ is the Kolmogorov time scale and Re_{KS} the KS equivalent Reynolds number. The evolution of the fractal dimension is linear with time, independent of the surface characteristic size and the time needed for D to reach its maximum value is of the order of τ_η, that is inversely proportional to $\sqrt{Re_{KS}}$.

We also use KS to track particles released from a cube or a sphere and measure the fractal dimension of this set of particles as a function of time for different Reynolds numbers. The fractal dimension of the cube is found to decrease regularly towards 2. The cube's fractal dimension is found to be independent of the Reynolds number but function of the cube's initial size (S):

$$\frac{S(D-3)}{tu'} = 0.3 \left(\frac{tu'}{L} \right)^{-\frac{2}{3}}$$

See figure 1.

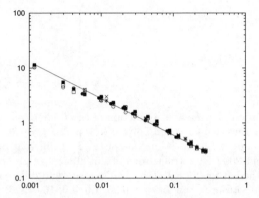

Fig. 1. Logarithmic plot of the evolution of $\frac{d(3-D)}{tu'}$ as a function of $t\frac{u'}{L}$. Cubical source, $L/\eta = 1000$: $+$ $s/L = 0.15$, $\times$ $s/L = 0.20$ and $*$ $s/L = 0.25$; spherical source, $L_1/\eta = 1000$: $\square$ $d/L = 0.15$, $\blacksquare$ $d/L = 0.20$ and $\odot$ $d/L = 0.25$; spherical source, $d/L = 0.15$: $\bullet$ $L_1/\eta = 36$ and $\triangle$ $L_1/\eta = 178$. The trend line is $y = 0.13x^{-\frac{2}{3}}$.

3 Triangles of particles

We study the evolution of three- and four-particle diffusion (triangle and tetrahedron) at large Reynolds numbers KS. We vary the Reynolds number and found that the geometrical characteristics of triangles and tetrahedrons as functions of time do not depend on the Reynolds number but only on the ratio Δ_0/L_1 that is on the portion of the inertial range that was contained within the triangle or tetrahedron at the initial time. We also study the effect of the modelling of the unsteadiness term in the KS.

Figure 2 illustrate this for the shape parameter $< I_2 >$ defined in [3].

The evolution of the triangle (respectively tetrahedron) depends on whether the value of the triangle size $< R^2 >^{1/2}$ (respectively tetrahedron volume

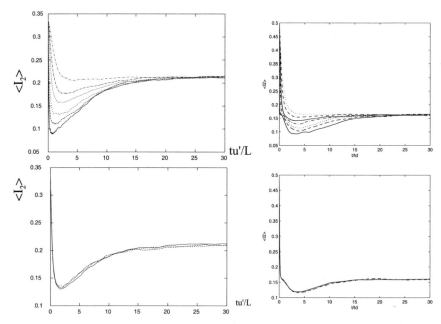

Fig. 2. Left: evolution of $< I_2 >$ as a function of tu'/L for $L/\eta =185$ and different initial tetrahedron sizes, namely from bottom to top $\Delta_0/\eta = 0.25, 1, 4, 16, 32$ and 64. Right: evolution of $< I_2 >$ as a function of tu'/L at different inertial subranges $L/\eta=185, 1000$ and 2000 for tetrahedrons of initial sizes of $\Delta_0/L =0.032$. Right: same cases as on the left for $St = 1$, Zero drift velocity and different triangle initial sizes, namely from bottom to top $\Delta_0/\eta = 0.25, 1.0, 4.0, 16.0, 32.0, 64.0$ and 92.5)

$< V >$) lies in the Richardson range. If it does, there is a non-trivial shape distortion. If $< R^2 >^{1/2}$ (respectively $< V >^{1/3}$) is larger than the largest length-scale, then the shape of the triangle (respectively tetrahedron) relaxes to an asymptotic value.

References

1. F. Nicolleau and A. ElMaihy, J. Fluid Mech. **517**, 229 (2004).
2. J. C. H. Fung, J. C. R. Hunt, N. A. Malik, and R. J. Perkins, J. Fluid Mech. **236**, 281 (1992).
3. A. Pumir, B. I. Shraiman, and M. Chertkov, Physical Review Letters **85**, 5324 (2000).

Lyapunov Exponents of Heavy Particles in Turbulent Flows

Jérémie Bec[1], Luca Biferale[2], Guido Boffetta[3], Massimo Cencini[4], Stefano Musacchio[5], and Federico Toschi[6]

[1] CNRS UMR6202, Observatoire de la Côte d'Azur, B.P. 4229, 06304 Nice Cedex 4, France
[2] Dept. of Physics and INFN, University of Rome "Tor Vergata", Via della Ricerca Scientifica 1, 00133 Roma, Italy
[3] Dept. of Physics and INFN, University of Torino, Via Pietro Giuria 1, 10125, Torino, Italy
[4] INFM-CNR, SMC Dipartimento di Fisica, Università di Roma "La Sapienza", Piazzale A. Moro 2, and ISC-CNR Via dei Taurini 19, I-00185 Roma, Italy
[5] Weizmann Institute of Science, Rehovot 76100, Israel
[6] CNR-IAC, Viale del Policlinico 137, I-00161 Roma, Italy, and INFN, Sezione di Ferrara, via G. Saragat 1, I-44100, Ferrara, Italy.

Impurities suspended in turbulent flows, being typically of finite size and heavier than the ambient fluid, posses inertia, which is responsible for the spontaneous generation of strong inhomogeneities in their spatial distribution.[1, 2] The intensity of particle clustering at small scales is related to the statistics of the Lyapunov exponents of particle trajectories.[3] The behavior of the Lyapunov exponents of inertial particles and the relation with particle clustering was recently investigated in random short-correlated flows.[4]

In this work we investigate the Lyapunov spectra of inertial particles, by means of high-resolution direct numerical simulations, varying their response time τ_s and the Reynolds number of the carrier turbulent flow. The Lagrangian equations for a small spherical particle with density much larger than the surrounding fluid, characterized by Stokes time τ_s, $\dot{x} = v$, $\dot{v} = -1/\tau_s [v - u(x(t), t)]$ are solved together with Navier–Stokes equations for the incompressible fluid velocity field $u(x, t)$

$$\partial_t u + u \cdot \nabla u = -\nabla p + \nu \Delta u + f, \quad \nabla \cdot u = 0. \tag{1}$$

The latter are integrated on cubic grid of size N^3 with $N = 128, 256, 512$ with periodic boundary conditions, by means of a fully de-aliased pseudo-spectral parallel code. In the simulations the Reynolds numbers based on Taylor's micro-scale are in the range $R_\lambda \in [65 : 185]$ (see Ref. [5] for further details) Particles inertia is measured in term of the Stokes number $St = \tau_s/\tau_\eta$ where τ_η is the Kolmogorov-scale turnover time.

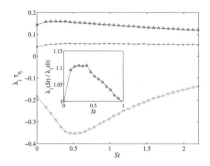 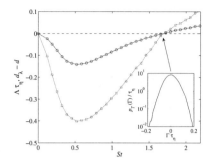

Fig. 1. *Left panel:* Lyapunov exponents λ_i for $i = 1$ (triangles), $i = 2$ (crosses) and $i = 3$ (circles) as a function of Stokes number. $R_\lambda = 185$. In the inset we show the relative growth of the first Lyapunov exponent $\lambda_1(St)/\lambda_1(0)$ occurring at small St. *Right panel:* Lyapunov dimension d_λ (squares) and volume growth rate $\Lambda = \sum_{i=1}^{3} \lambda_i$ (circles) as a function of Stokes number ($R_\lambda = 185$). Inset: PDF of the finite-time volume growth rate $\Gamma(T)$ for $T \approx 80\,\tau_\eta$.

The Lyapunov spectrum is obtained by following along each particle trajectory, the time evolution of infinitesimal volumes V^j, defined by j linear independent tangent vectors in the position-velocity phase space. Their growth rates $\sum_{i=1}^{j} \gamma_i(T) = (1/T) \ln \left[V^j(T)/V^j(0) \right]$ define the *finite-time Lyapunov exponents* $\gamma_i(T)$ (FTLE), which asymptotically converge to the *Lyapunov exponents* $\lambda_i = \lim_{T \to \infty} \gamma_i(T)$ (LE).

We find that for $St > 1$ the presence of inertia results in a generic reduction of chaoticity: the leading Lyapunov Exponent is smaller than the one of tracers. Remarkably, for $St < 1$ the opposite phenomenon is observed (See Fig. 1, left panel). This is due to the preferential concentration of particles in high-strain regions of the flow, which increases chaoticity. This phenomenon could not be predicted from analytical and numerical studies done in white-in-time random velocity fields, because such flows clearly possess no persistent structures.

Volume growth rate $\Lambda = \lambda_1 + \lambda_2 + \lambda_3$, which identically vanishes for fluid tracers, is negative for all Stokes numbers in the range $0 < St < 1.72$ (see Fig. 1, right panel), thus indicating the presence of clustering at small scale. The intensity of particle clustering can be measured by the Kaplan-Yorke dimension defined as $d_\lambda = J + \sum_{i=1}^{J} \lambda_i/|\lambda_{J+1}|$ where J is the largest integer such that $\sum_{i=1}^{J} \lambda_i > 0$. The minimum at $St \approx 0.5$ corresponds to maximal clustering. Notice that the fractal dimension observed is in close agreement with predictions made in the framework of white-in-time random flows [4]. For $St \approx 1.72$ the mean volume growth rate Λ vanishes, but finite-time growth rates $\Gamma = \gamma_1 + \gamma_2 + \gamma_3$ still experience large fluctuations. As a result, strong local inhomogeneities are present in the particle concentration also at larger values of St.

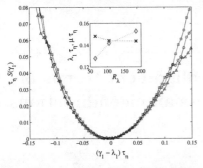 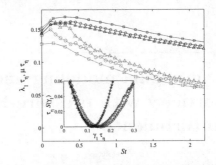

Fig. 2. *Left panel*: Cramér function $S(\gamma_1)$ for fluid tracers for for $R_\lambda = 65$ (squares), 105 (circles) and 185 (triangles). Inset: Lyapunov exponent λ_1 (crosses) and reduced variance $\mu = T\sigma^2$ (diamonds) as a function of R_λ.
Right panel: Mean value λ_1 (small symbols) and reduced variance μ (large symbols) of the FTLE as a function of the Stokes number, for $R_\lambda = 65$ (squares), $R_\lambda = 105$ (circles) and $R_\lambda = 185$ (triangle). Inset: Cramér function $S(\gamma_1)$ for $R_\lambda \approx 185$, and various values of the Stokes number: $St = 0$ (triangles), $St = 0.32$ (diamonds) and $St = 2.19$ (stars).

Intermittency affects the whole probability distribution function (PDF) of the largest finite-time Lyapunov exponents $\gamma_1(T)$. For T sufficiently large, the distribution of FTLE obeys a large-deviation principle, i.e. $p_T(\gamma_1) \propto \exp(-T\,S(\gamma_1))$ where $S(\gamma_1)$ is the Cramér (or rate) function. In Figure 2 we show the Reynolds number number dependence of the Cramér function, the Lyapunov esponent λ_1 and the reduced variance $\mu = T\langle(\gamma_1 - \lambda_1)^2\rangle$ of FTLE, which measures of the width of the Cramér function. Intermittency is responsible for non-dimensional scaling of λ and μ as a function of R_λ, and increases the probability of high stretching events, thus fattening the right tail of the distribution. These effects are found to persist also for large inertia, and act in the same direction as for fluid tracers.

Consistently with the behavior of the Lyapunov exponent, we find an enhancement of chaotic fluctuations for $St < 1$, signaled by the shift to higher value of the distribution of FTLE (See Fig 2 right panel). Conversely for $St > 1$ it shifts to lower values, and fluctuation become less probable.

References

1. J.K. Eaton and J.R. Fessler: Int. J. Multiphase Flow **20**, 169 (1994).
2. A. Pumir and G. Falkovich: Phys. Fluids **16**, L47 (2004).
3. J. Bec: Phys. Fluids **15**, L81 (2003).
4. K. Duncan, B. Mehlig, S. Ostlund, and M. Wilkinson: Phys. Rev. Lett. **95**, 240602 (2005).
5. J. Bec *et al.*: J. Fluid Mech. **550**, 349 (2006).

Lagrangian modeling and alignment trends of vorticity with pressure-Hessian eigendirections in turbulence

Laurent Chevillard[1] and Charles Meneveau[1]

Department of Mechanical Engineering and Center for Environmental and Applied Fluid Mechanics, The Johns Hopkins University, 3400 N. Charles Street, Baltimore, MD 21218, USA chevillard@jhu.edu, meneveau@jhu.edu

The local statistical and geometric structure of three-dimensional turbulent flow can be described by properties of the velocity gradient tensor $A_{ij} = \partial u_i / \partial x_j$. The Lagrangian evolution of this tensor is given by the gradient of the Navier-Stokes equations $\frac{d\mathbf{A}}{dt} = -\mathbf{A}^2 - \mathbf{P} + \nu \Delta \mathbf{A}$, where ν is the kinematic viscosity and $P_{ij} = \partial^2 p / \partial x_i \partial x_j$ is the pressure Hessian. Recently, based on prior works [1], a stochastic model has been developed for the Lagrangian time evolution of this tensor, in which the exact nonlinear self-stretching term $-\mathbf{A}^2$ which accounts for the development of well-known non-Gaussian statistics and geometric alignment trends [2] is kept exactly. The pressure Hessian $\mathbf{P}$ and viscous term $\nu \Delta \mathbf{A}$ entering in the Lagrangian evolution of the velocity gradient tensor $\mathbf{A}$ are accounted for by a closure that models the material deformation history of fluid elements. The resulting model can be written in the following non-dimensional way:

$$dA = \left(-\mathbf{A}^2 + \overbrace{\frac{\mathrm{Tr}(\mathbf{A}^2)}{\mathrm{Tr}(\mathbf{C}_\Gamma^{-1})} \mathbf{C}_\Gamma^{-1}}^{-\mathbf{P}} \underbrace{- \frac{\mathrm{Tr}(\mathbf{C}_\Gamma^{-1})}{3} \mathbf{A}}_{\nu \Delta \mathbf{A}} \right) dt + d\mathbf{W} \,, \tag{1}$$

where $\mathbf{P}$ is modeled using the recent material deformation as described by the (inverse) recent Cauchy-Green tensor $\mathbf{C}_\Gamma \equiv e^{\Gamma \mathbf{A}} e^{\Gamma \mathbf{A}^T}$ (T stands for matrix transpose). The viscous term is modeled as a friction term with a non-linear coefficient (see Ref. [2] for details). The forcing term $\mathbf{W}$ entering in Eq. (1) is chosen Gaussian and delta-correlated in time and describes the joint action of large-scale forcing and neigboring eddies at moderate Reynolds number. A model parameter $\Gamma = \tau/T$, defined as the ratio between a characteristic dissipative time scale and the integral time scale, has to be specified. For a moderate Reynolds number $\mathcal{R}_\lambda = 150$, where λ is the Taylor microscale, Γ has been estimated to be of the order of 0.1 [3] when τ is of the order of the Kolmogorov time scale τ_K . The resulting stochastic system reproduces many statistical and geometric trends observed in numerical and experimental 3D turbulent flows, including anomalous relative scaling and the preferential alignment of vorticity with the intermediate eigendirection of $\mathbf{S}$ (i.e. the symmetric part of $\mathbf{A}$) [3, 4].

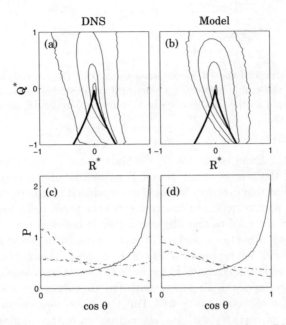

Fig. 1. Comparison of the joint PDFs and PDFs obtained from DNS ($\mathcal{R}_\lambda = 150$) and the model ($\tau/T = 0.1$) of (i) the invariants $Q = -\mathrm{Tr}(\mathbf{A}^2)/2$ and $R = -\mathrm{Tr}(\mathbf{A}^3)/3$ (Figs. a and b) and (ii) the cosine between vorticity and the eigendirections of the rate of Strain $\mathbf{S}$ associated to the smallest (dashed), intermediate (solid) and biggest (dot-dashed) eigenvalues (Figs. c and d).

As an example, we compare in Fig. 1(a-b) the joint probability density function (PDF) of the invariants R and Q obtained from Direct Numerical Simulations (DNS) and from the model. The agreement is excellent, in particular for the enstrophy-enstrophy production dominated region (upper left) and the dissipation-dissipation production dominated region (lower right). Some discrepancies occur in the two other quadrants (i.e. $R < 0, Q < 0$ and $R > 0, Q > 0$). In Fig. 1(c-d), we represent the PDF of the cosine of the angle between vorticity and eigendirections of the rate of strain $\mathbf{S}$. We can see that alignment of vorticity with eigendirections of $\mathbf{S}$ is also realistically reproduced, in particular the preferential alignment with the eigendirection associated to the intermediate eigenvalue.

We focus now on the precise topology and eigensystem of the resulting modeled pressure Hessian given in Eq. (1). Prior work for the inviscid limit

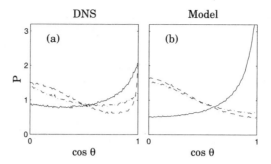

Fig. 2. Comparison of the PDFs obtained from DNS ($\mathcal{R}_\lambda = 150$) and the model ($\tau/T = 0.1$) of the cosine between vorticity and the eigendirections of the pressure Hessian associated to the smallest (dashed), intermediate (solid) and biggest (dot-dashed) eigenvalues.

described by the Euler equations [5] has shown that vorticity tends to be an eigenvector of the rate of strain tensor **S** as well as of the pressure Hessian **P**. We show that this property is exactly reproduced by the present model [3] because of the matrix exponential structure of the predicted pressure Hessian.

Then, DNS are used to test alignment trends between vorticity and pressure Hessian eigendirections, and results are compared to predictions from the Lagrangian model. We represent in Fig. 2 the PDFs of the cosine between vorticity and the eigendirections of **P** for both DNS and the model. We see that the model reproduces the alignment with the eigendirection associated with the intermediate eigenvalue very well. But the model describes only partially the orientation of vorticity with the eigendirection of the smallest eigenvalue, for which in DNS very strong alignment is found. This fact might explain why probabilities are underestimated in the lower-left quadrant of the joint distribution of R and Q, and overestimated in the upper-right quadrant. The analysis results provide, overall, additional support for the accuracy of the Lagrangain model proposed in [2] but highlight areas where improvements are required.

References

1. P. Vieillefosse, Physica A **125**, 150 (1984). B.J. Cantwell, Phys. Fluids A **4**, 782 (1992). S.S. Girimaji and S.B. Pope, Phys. Fluids A **2**, 242 (1990). M. Chertkov, A. Pumir, and B.I. Shraiman, Phys. Fluids **11**, 2394 (1999). E. Jeong and S. S. Girimaji, Theor. Comput. Fluid Dyn. **16**, 421 (2003).
2. L. Chevillard and C. Meneveau, Phys. Rev. Lett. **97**, 174501 (2006).
3. L. Chevillard, C. Meneveau, L. Biferale and F. Toschi, in preparation (2007).
4. L. Chevillard, C. Meneveau, `physics/0701274` (2007).
5. K. Ohkitani, Phys. Fluids A **5**, 2570 (1993).

Acceleration statistics of heavy particles in turbulent flows

A. S. Lanotte[1], J. Bec[2], L. Biferale[3], G. Boffetta[4], A. Celani[5], M. Cencini[6], S. Musacchio[7], and F. Toschi[8]

[1] CNR-ISAC and INFN, 73100 Lecce, Italy
[2] CNRS, Observatoire de la Côte d'Azur, B.P. 4229, 06304 Nice Cedex 4, France
[3] Dept. of Physics and INFN, Univ. "Tor Vergata", 00133 Roma, Italy
[4] Dept. of Physics and INFN, Univ. of Torino, 10125 Torino, Italy
[5] CNRS, INLN, 06560 Valbonne, France
[6] SMC-INFM c/o Dept. of Physics, Univ. "La Sapienza", and CNR-ISC, 00185 Roma, Italy
[7] Dep. of Complex Systems, Weizmann Institute of Science, Rehovot 76100, Israel
[8] CNR-IAC, 00161 Roma, Italy and INFN, 44100 Ferrara, Italy

The motion of small particles in a turbulent flow is encountered in many natural phenomena, the best example is perhaps that of water droplets in the turbulent air of an atmospherical cloud [1]. Depending on the particles size, concentration and the density contrast with respect to the carrying fluid, their dynamical modelization can be quite complex. Here we report the results from a Direct Numerical Simulations (DNS) study of high-resolution turbulent flows (up to 512^3 grid points), seeded with millions of particles much heavier than the carrier fluid [2]. To deal with the simplest yet non-trivial particle dynamics, we assume that both particle reaction on the flow and particle-particle interactions can be safely neglected. The Newton law describing very small particles dynamics is then simply : $\tau \ddot{\boldsymbol{X}} = \boldsymbol{u}(\boldsymbol{X}, t) - \dot{\boldsymbol{X}}$ where $\boldsymbol{u}(\boldsymbol{x}, t)$ is the fluid flow. The particle response time τ is proportional to the square of the particles size and to the density contrast with the fluid. The non-dimensional parameter relating the particle dynamics to the flow characteristic time scales is the Stokes number $St = \tau/\tau_\eta$, where τ_η is the Kolmogorov time scale associated to the smallest turbulent eddies.

We shall focus on the behaviour of particle acceleration at varying both the Reynolds Re_λ and the Stokes St numbers, i.e. the flow turbulence and the particle inertia, respectively.

For fluid tracers, it is known that trapping into vortex filaments [3, 4] is the main source of strong acceleration events. On the other hand, little is known about the acceleration statistics of heavy particles, whose spatial inhomogeneous distribution may play a crucial role. We start by considering, in Fig. 1, the plot of the autocorrelation function for one component of the acceleration

$C_{a_x}(t) = <a_x(t_0)a_x(t_0 + t)>$ at varying the particle inertia. As it is evident the gross features of the correlation functions do not seem to be strongly dependent on the Stokes value. In particular, we note that, similarly to what is observed for tracers, the behavior of $C_{|a_x|}(t)$ is characterized by two exponentials with different decay rates (see inset). The first one associated with the typical small scale decorrelation, and the second slower one associated with trapping in vortices, which are typically living for times longer than τ_η. For these observables, inertia is not statistically relevant.

A much better understanding of the role of inertia is achievable if we consider

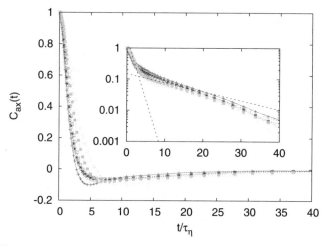

Fig. 1. The normalised autocorrelation functions of one component, $a_x(t)$ of the acceleration, for a subset of Stokes values ($St = 0.16, 0.37, 0.58, 1.01, 2.03, 3.31$ left to right) at $R_\lambda = 185$. Inset: the same but for the modulus of the chosen component. The two straight lines are $\sim exp(t/(1.1\tau_\eta))$ and $\sim exp(t/(15\tau_\eta))$.

the Probability Density Function (PDF) of the particle acceleration. For very small inertia, small St, the particle acceleration essentially coincides with the fluid acceleration; however, heavy particles are not homogeneously distributed in the flow and concentrate preferentially inside low vorticity regions [2]. As a result, the net effect of inertia is a drastic reduction of the root-mean-squared acceleration $a_{rms} = \langle \mathbf{a}^2 \rangle^{1/2}$, due essentially to preferential concentration. As shown in the inset of Fig. 2, the normalised a_{rms} drops off very fast already at small St values. We also notice that, at a given St, the normalised acceleration variance increases with the Reynolds number similarly to what has been observed for fluid tracers [4].

When the particle's response time is much larger than τ_η, larger St, another mechanism comes into play. A careful inspection of particle motion shows that the main effect of inertia is a low-pass filtering of fluid velocity differences, with a suppression of fast frequencies larger than the one typical of the particle motion, $1/\tau$. This is evident by looking, in the main body of Fig. 2, at the tails of the probability density function of the acceleration, which decrease when increasing the Stokes number. For very small ($St = 0.16$) and large ($St = 3.3$) inertia, the acceleration statistics are clear: preferential concentration of particles controls the PDF at small St, while filtering induced by the particle response time becomes dominant at larger St. Both effects are present for intermediate Stokes numbers. Future research will be devoted to better highlight the correlation between particles positions and regions of the flow with very large or very small acceleration values.

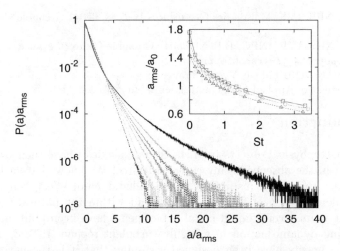

Fig. 2. Probability Density Function of the normalised acceleration a/a_{arms} for a subset of Stokes values ($St = 0, 0.16, 0.37, 0.58, 1.01, 2.03, 3.31$ from top to bottom) at $R_\lambda = 185$. Inset: the normalised acceleration variance a_{rms}/a_0 where $a_0 = (\epsilon^3/\nu)^{1/4}$, as a function of the Stokes number for $R_\lambda = 185$ (empty box); $R_\lambda = 105$ (circle); $R_\lambda = 65$ (triangle). Here, ϵ is the flow energy dissipation, and ν is the fluid viscosity (see [2] for details of the DNS).

References

1. R. A. Shaw, Annu. Rev. Fluid Mech. **35**, 183-227 (2003)
2. J. Bec, L. Biferale, G. Boffetta, A. Celani, M. Cencini, A. Lanotte, S. Musacchio, and F. Toschi: J. Fluid Mech. **550**, 349 (2006)
3. A. La Porta, G.A. Voth, A.M. Crawford, J. Alexander, and E. Bodenschatz, Nature **409**, 10171019 (2001)
4. L. Biferale, G. Boffetta, A. Celani, B.J. Devenish, A. Lanotte, and F. Toschi: Phys. Rev. Lett. **93**, 064502 (2004)

Influence of large scale flow fluctuations on the dynamo threshold

M. Peyrot[1,2], C. Fargant[1], F. Plunian[1], C. Normand[3], and A. Courvoisier[4]

[1] LGIT, CNRS, UJF, Maison des Géosciences B.P. 53, 38041 Grenoble Cedex 9, France
   LEGI, CNRS, UJF, INPG, B.P. 53, 38041 Grenoble Cedex 9, France
[2] Marine.peyrot@ujf-grenoble.fr
[3] CEA Saclay 91191 Gif sur Yvette, France
[4] Departement of Applied Mathematics, University of Leeds, UK

1 Introduction

Most of astrophysical objects (planets, stars, galaxies,) have their own magnetic field in the absence of any external source. We know that such a field results from a magnetic instability, the so-called dynamo effect, generated by the turbulent motion of the conducting fluid within the natural object. In the recent years several liquid metal experiments have been built, aiming at reproducing dynamo action in a highly turbulent regime [1, 2, 3]. Most of these experiments have been designed assuming that it is the mean (time-averaged) part of the flow $\bar{U}$ which generates the dynamo action. In fact, the flow has a highly turbulent nature and then presents a full spectrum of scales and frequencies. In particular the time fluctuation $\tilde{U}$ of scale comparable with the largest scale of $\bar{U}$ is suspected to modify the dynamo threshold. Knowing whether this large scale fluctuation can endanger or help for the dynamo instability is of high interest for the dynamo experiments [4].

In the present paper we consider two cases (i) and (ii) depending on the scale separation between L_B and $L_{\bar{U}}$, the largest scales of the generated magnetic field and of the mean flow. Case (i) corresponds to $L_B \approx L_{\bar{U}}$, the magnetic field is thus generated by a mean flow of comparable scale. Case (ii) corresponds to $L_B \gg L_{\bar{U}}$, the magnetic field is thus generated by a mean flow of much smaller scale. Both cases, (i)-without or (ii)-with scale separation (between B and $\bar{U}$), have been the object of several studies showing that the dynamo mechanisms are highly different [5].

Now, adding to the mean flow $\bar{U}$ a time-dependent fluctuation $\tilde{U}$ of same scale ($L_{\tilde{U}} = L_{\bar{U}}$) may drastically change the dynamo threshold. We want to compare the two types of dynamo mechanism (i) and (ii) under the influence of such a flow fluctuation. In particular is there one dynamo mechanism less sensitive than the other to such a fluctuation, and therefore more robust for a dynamo experiment ?

2 Model

For flow (i), we take a smooth Ponomarenko flow [6] defined in cylindrical coordinates (r, φ, z) by

$$\mathbf{U} = (0, r\Omega, V)h(r) \quad \text{with} \quad h(r) = \begin{cases} 1, & r < 1 \\ 0, & r > 1 \end{cases}, \tag{1}$$

corresponding to a helical flow in a cylindrical cavity infinite along the z-direction, the external part being at rest. Each velocity component, angular and vertical, is defined as the sum of a stationary and fluctuating part

$$\Omega = \left(\bar{R}_m + \tilde{R}_m f(t)\right)(1 - r), \quad V = \left(\bar{R}_m \bar{\Gamma} + \tilde{R}_m \tilde{\Gamma} f(t)\right)(1 - r^2) \tag{2}$$

where $\bar{R}_m$ and $\bar{\Gamma}$ (resp. $\tilde{R}_m$ and $\tilde{\Gamma}$) are the magnetic Reynolds number and a factor of helicity for the stationary (resp. fluctuating) part of the flow. In the rest of the paper, we consider a periodic fluctuation of frequency ω_f, $f(t) = \cos(\omega_f t)$.

For flow (ii) we take the Roberts flow [7] defined in cartesian coordinates (x, y, z) by

$$\mathbf{U} = \nabla \times (\psi \mathbf{e}) + V\mathbf{e}, \tag{3}$$

$$\psi = \left(\bar{R}_m + \tilde{R}_m f(t)\right)(\cos x - \cos y), \quad V = \left(\bar{R}_m \bar{\Gamma} + \tilde{R}_m \tilde{\Gamma} f(t)\right)(\cos x - \cos y)$$

where $\mathbf{e}$ is the unit vector in the z-direction. This flow corresponds to a periodic network of parallel vortices of alternated directions.

The magnetic field satisfies the induction equation

$$\frac{\partial \mathbf{B}}{\partial t} = \nabla \times (\mathbf{U} \times \mathbf{B}) + \nabla^2 \mathbf{B}. \tag{4}$$

where the dimensionless time t is given in units of diffusive time scale, implying that ω_f is also a dimensionless quantity. As the velocity field does not depend on z (neither on φ for case (i)), each magnetic mode in z (and φ for case (i)) is independent from the others. Therefore we look for monochromatic solutions of the form

(i) $\mathbf{B}(r, \varphi, z, t) = \exp i(m\varphi + kz) \, \mathbf{b}(r, t)$, (ii) $\mathbf{B}(x, y, z, t) = \exp(ikz) \, \mathbf{b}(x, y, t)$
$$\tag{5}$$

where k (resp. m) is the vertical (resp. azimuthal) wave number of the field. To solve (4), we use for case (i) a Galerkin method in r [8] and RK4 in time, and for case (ii) a Fourier decomposition in x and y (FFTW) and RK2 in time.

3 Results and conclusions

We plot on figure 1, the dynamo threshold $\bar{R}_m$ versus the fluctuation intensity $\rho = \tilde{R}_m / \bar{R}_m$ for $\bar{\Gamma} = 1$ and different values of $\tilde{\Gamma}$.

For the case (i), if the fluctuation has not an appropriate geometry (curves (c) to (g)), we find that the threshold is drastically increased even if ρ is fairly small. On the other hand, if it has an appropriate geometry, it can help for $\rho \geq 1$ (curves (a), (b) and (e)).

For the case (ii), similar results are obtained. The main difference is that in case (ii) if ρ is sufficiently large, the threshold decreases versus ρ, whatever the value of $\tilde{\Gamma}$.

Unfortunally for dynamo experiments, for both cases (i) and (ii), we did not find any subcritical dynamo state at small ρ.

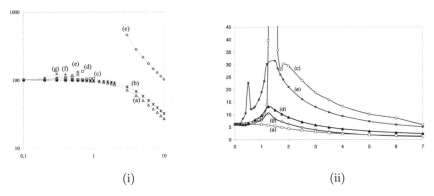

(i)   (ii)

Fig. 1. *Dynamo threshold $\bar{R}_m$ versus $\rho = \tilde{R}_m / \bar{R}_m$. For the case (i), $\bar{\Gamma} = 1$, $m = 1$, $k = -1$ and $\omega_f = 50$. The different curves correspond to $\tilde{\Gamma} = $ (a) 1; (b) 0.5; (c) 0; (d) -0.5; (e) -1. For the case (ii), $\bar{\Gamma} = 1$, $k = 1$ and $\omega_f = 0.1$. The different curves correspond to $\tilde{\Gamma} = $ (a) 1; (b) 0.5; (c) 0.25; (d) 2; (e) 4*

References

1. P. Cardin, D., D. Jault, H.-C. Nataf and J.-P. Masson, "Towards a rapidly rotating liquid sodium dynamo experiment", Magnetohydrodynamics **38**, 177-189 (2002)
2. P. Frick, V. Noskov, S. Denisov, S. Khripchenko, D. Sokoloff, R. Stepanov, A. Sukhanovsky, "Non-stationary screw flow in a toroidal channel: way to a laboratory dynamo experiment", Magnetohydrodynamics **38**, 143-162 (2002)
3. R. Monchaux, M Berhanu, M. Bourgoin, Ph Odier, M. Moulin, J.-F Pinton, R. Volk, S. Fauve, N. Mordant, F. Pétrélis, A. Chiffaudel, F. Daviaud, B. Dubrulle, C. Gasquet, L. Marié, and F. Ravelet. "Generation of magnetic fiel by a turbulent flow of liquid sodium", Phys. Rev. Lett. **98**, 044502 (2006)
4. R. Volk, F. Ravelet, R. Monchaux, M. Berhanu, A. Chiffaudel, F. Daviaud, Ph. Odier, J.-F. Pinton, S. Fauve, N. Mordant, and F. Pétrélis, "Transport of Magnetic Field by a Turbulent Flow of Liquid Sodium", Phys. Rev. Lett. **97**, 074501 (2006)
5. A.D. Gilbert, "Dynamo theory", In: Handbook of Mathematical Fluid Dynamics, **2** (ed. S. Friedlander, D. Serre), 355-441, Elsevier (2003)
6. Y.U. Ponomarenko, "Theory of the hydromagnetic generator", J. Appl. Mech. Tech. Phys. **6**, 755 (1973)
7. G. O. Roberts, "Dynamo Action of Fluid Motions with Two-Dimensional Periodicity" Phil. Tran. R. Soc. Lond. A **271**, 411 (1972)
8. L. Marié, C. Normand and F. Daviaud, "Galerkin analysis of kinematic dynamos in the von Kármán geometry", Phys. Fluids **18**, 017102 (2006)

The Magnetoelliptic Instability in the presence of Inertial Forces

K. Mizerski and K. Bajer

University of Warsaw, Inst. of Geophysics, ul. Pasteura 7, 02-093 Warszawa, Poland, krzysztof.mizerski@gmail.com

The elliptical instability, namely the problem of stability of a two-dimensional flow with elliptical streamlines with respect to three-dimensional perturbations is expected to play an important role as a secondary instability in the transition to turbulence [2]. We investigate the influence of uniform magnetic field, perpendicular to the plane of the flow, and of the inertial forces in a rotating frame (Coriolis force and angular acceleration) on the stability of the flow. Though the magnetic field and the inertial forces in this context have already been studied in isolation [5], [4], there has been no comprehensive investigation of their joint influence. Our study is motivated by possible geophysical and astrophysical applications since the results may be used to comment on the influence of tidal and rotational distortion of the Earth's (or another planet's or satelite's) mantle on the outer core's stability with possible dynamo implications (together with the stability analysis of the boundary layer at the core-mantle boundary [3], [6]) and the magnetoelliptic instability may prove to be an important factor in turbulent transition and therefore in understanding the energy balance in the Solar corona.

The system considered consists of an infinite two-dimensional vortex with elliptical streamlines in which the velocity field is defined as $\mathbf{u} = \gamma[-Ey, E^{-1}x, 0]$ with $\gamma > 0$ and $E \leq 1$ in a frame of reference rotating with angular velocity $\mathbf{\Omega} = \Omega\hat{\mathbf{e}}_z$ in the presence of a uniform magnetic field perpendicular to the plane of the flow $\mathbf{B}_0 = B_0\hat{\mathbf{e}}_z$. The fluid is inviscid and incompressible. Such configuration permits resonances between *inertial waves* which can grow exponentially [1]. Perturbations in the form of inertial waves are therefore introduced into the Navier-Stoke's and induction equations and by application of the Floquet theory we obtain ordinary matrix equations which then are solved numerically to obtain the growth rate dependence on the parameters characterizing the system such as the angle ϑ between the wave vector and the axis of the vortex (the z-axis), the eccentricity E, the Rossby number $Ro = \frac{\gamma(E+E^{-1})}{2\Omega}$ and a parameter measuring the magnetic field strength $\beta = \frac{B_0 k_0}{\gamma \sqrt{\mu \rho}}$, where μ and ρ are the magnetic permeability and the density of the fluid, and k_o^{-1} is the lengthscale of the perturbation.

Some analytical progress in solving the matrix equation is also possible through perturbation analysis in the parameter $\epsilon = \frac{1}{2} \left(E - E^{-1} \right)$ measuring the departure of the basic flow from axial symmetry with the Puiseux theory used to obtain the growth rates [4]. By the use of the geometrical optics approximation (WKB) this study may be easily transformed to the short-wavelength stability analysis of Riemann ellipsoids in the presence of the magnetic field to comment the possibility of the influence of the Earth mantle's tidal and rotational distortion on the stability of the outer core and throuh that on the Earth's dynamo.

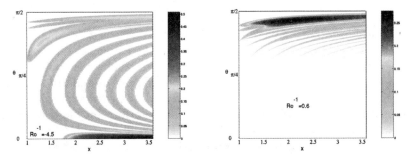

Fig. 1. The growth rates on $\vartheta - \delta$ surface, where $\delta = \frac{1}{2} \left(E + E^{-1} \right)$ and $2\delta\gamma$ is the vorticity of the basic flow, for the parameter $\beta = 4$ and two different values of the Rossby number $Ro^{-1} = -4.5$ on the left and $Ro^{-1} = 0.6$ on the right.

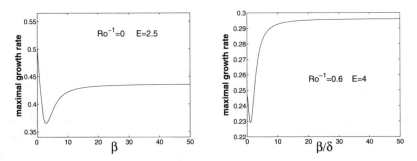

Fig. 2. The maximal growth rate as a function of the magnetic field strength.

We begin with presenting values of the growth rates as color-maps on Fig. 1. The values of the Rossby number where chosen such as to compare with [5] and to see the influence of the magnetic field on rotating elliptical vortex. The influence clearly exists and is very strong producing new possible directions for propagation of the unstable waves and strongly modyfying the already existing unstable regions of the $\vartheta - \delta$ plane. This can be also deduced by analyzing the resonances between the inertial modes. Strong influence is also observed in the horizontal instability, i.e. for waves propagating in the z direction, which, when present, dominates among other ustable modes. For $\beta = 0$ this instability type is present only for a bounded set of values of Ro^{-1}. When the magnetic field is turned on, the instability exists for all $Ro^{-1} < -\beta^2/4\delta^2$. Furthermore, Fig. 2 represents the dependence of the maximal growth rate on the magnetic field strength (parameter β) for two different values of the Rossby number. In qualitative agreement with asymptotic predictions for small eccentricity the value of the maximal growth rate tends to a constant value with increasing magnetic field (and the associated direction of propagation of the unstable wave (angle ϑ) tends to $\pi/2$). However, in both cases the growth rate dependence is not monotonic. To visualize the ustable mode, on Fig. 3, we also present snap-shots of the one period evolution of the projections of the streamlines on the XY plane.

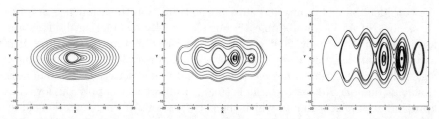

Fig. 3. The snap-shots of the one-period evolution of the projections on indicated planes of the streamlines taken (from left to right) at the times $t = 0, \pi, 2\pi$, for $E = 2.5$, $\beta = 4$ and without inertial forces.

We conclude by stating that the combined impact of the Coriolis and Lorentz forces on elliptical instability is qualitatively different from the effect they exert in isolation.

1. B.J. Bayly 1986 Three-dimensional instability of elliptical flow. *Phys. Rev. Lett.* **57** (17), 2160-2163.
2. B.J. Bayly, S.A. Orszag 1988 Instability mechanisms in shear flow transition. *Ann. Rev. Fluid Mech.* **20**, 359-391.
3. B. Desjardins, E. Dormy, E. Grenier 2001 Instability of Ekman-Hartmann boundary layers, with application to the fluid flow near the core-mantle boundary. *Phys. Earth. Planet. Int.* **124**, 283-294.
4. N.R. Lebovitz, E. Zweibel 2004 Magnetoelliptic instabilities. *The Astrophys. Journal* **609**, 301-312.
5. T. Miyazaki 1993 Elliptical instability in a stably stratified rotating fluid. *Phys. Fluids A* **5** (11), 2702-2709.
6. K.A. Mizerski, K. Bajer 2007 On the effect of mantle conductivity on the super-rotating jets near the liquid core surface. *Phys. Earth. Planet. Int.* **160**, 245-268.

Instability and transition to turbulence in a free shear layer affected by a parallel magnetic field

Anatoly Vorobev[1] and Oleg Zikanov[1]

Department of Mechanical Engineering, University of Michigan - Dearborn, 4901 Evergreen Road, Dearborn, MI 48128-1491, USA zikanov@umich.edu

1 Introduction

We consider the instability and subsequent transition to turbulence in a free shear layer of an incompressible viscous electrically conducting fluid. A uniform constant magnetic field is imposed in the direction of the flow. It is assumed that, as typical for laboratory and industrial flows of liquid metals, molten oxides, and other electrically conducting materials, the magnetic Reynolds number $Re_m \equiv U_0 L \sigma \mu_0$ is small. It allows us to apply the low-Re_m approximation, according to which the perturbations of the magnetic field induced by the fluid motion adjust instantaneously to the variations of the flow and can be neglected in comparison with the imposed field $\mathbf{B}$ in the expression for the Lorentz force. The governing equations reduce to conventional Navier-Stokes system with the additional Lorentz force term and the Poisson equation for the electric potential. The non-dimensional parameters are the Reynolds number and the magnetic interaction parameter $N \equiv \sigma B^2 L / \rho U_0$, which gives an estimate to the ratio between the Lorentz and inertia forces.

The main aspect of the low-Re_m interaction between the flow and the magnetic field is the suppression of flow gradients in the field direction. In our problem, the basic flow is not affected and can be assumed to have the form of a conventional viscous shear layer. The growing unstable perturbations are suppressed by the magnetic field. The rate of suppression of their energy (given by the rate of Joule dissipation) is proportional to the square of the cosine of the oblique angle θ between the directions of the wavenumber vector and the magnetic field. This introduces a new anisotropy into the problem. The strongest suppression is for the classical two-dimensional (purely spanwise) perturbations, while the suppression rate decreases with θ and becomes zero for purely streamwise modes. This have dramatic consequences for the instability as was first recognized in [1]. It was shown that, at sufficiently, strong magnetic fields, the first instability is to three-dimensional perturbations with $\theta \neq 0$.

We report the results of the extended analysis of the instability and the resulting transition to turbulence for a free shear layer with the basic profile $U(z) = \text{erf}(\sqrt{\pi}z/2)$. The detailed description of the results can be found in [2].

The linear stability analysis confirms the earlier predictions of [1] that the instability occurs due to three-dimensional disturbances if the applied magnetic field is sufficiently strong ($N > N_F = 0.106$). The three-dimensional eigenmodes have the form of horizontal rolls, which are not perpendicular to the direction of the basic flow but, rather, inclined to it at the oblique angle θ, which increases with the strength of the magnetic field and tends to $\pi/2$ as $N \to \infty$. The critical Reynolds numbers Re_c and the oblique angles θ_c of the neutral modes are given by $Re_c = Re_F N_F/N$ and $\theta_c = \cos^{-1}(N_F/N)$.

We investigated the fastest growing perturbations at supercritical Reynolds numbers. At N between 0.106 and 0.339, such perturbations are three-dimensional only in a finite range of the Reynolds numbers, becoming two-dimensional at higher Re. At $N > 0.339$, the fastest growing perturbations are always three-dimensional and have the form of oblique rolls. Interestingly, the strength of the magnetic field affects only the spatial orientation of these modes. The oblique angle increases with N, while the horizontal and vertical dimensions, and the spatial shape of the rolls remain unchanged.

We performed direct numerical simulations of the transition to turbulence resulting from the instability. Two types of the initial conditions were considered. In both cases, the initial velocity field consisted of the basic flow, the fastest growing eigenmode, and low-amplitude noise. The difference was in the composition of the eigenmodes, for which a single oblique wave or a combination of two symmetric oblique waves with positive and negative θ were taken. We have found that the flow evolution strongly depends on the form of the initial disturbance. Two symmetric waves result in stronger nonlinear growth, excitation of a wider range of length scales, stronger vorticity, and, generally, faster transition to more intense turbulence. This is accompanied by much stronger turbulent mixing, which, at moderate N, is comparable with the mixing in the non-magnetic case at the same Re.

We analyzed the evolution of the two-wave solution and found it to be principally different from the well studied scenarios of the classical non-magnetic Kelvin-Helmholtz instability. The first stage of the Kelvin-Helmholtz transition is always the development of two-dimensional billows. The three-dimensionality and turbulence appear as results of the secondary instability through pairing, spanwise waves in the braid region, etc. On the contrary, the two growing symmetric oblique waves generate the three-dimensionality immediately, in their direct interaction with each other, basic flow, and the magnetic field. We have seen that this interaction can lead to a fast transition to turbulence and efficient mixing.

Another important difference between the MHD and the non-magnetic cases is the state of the developed turbulent mixing layer. Not only the magnetic field adds the suppression by the Joule dissipation, it also introduces

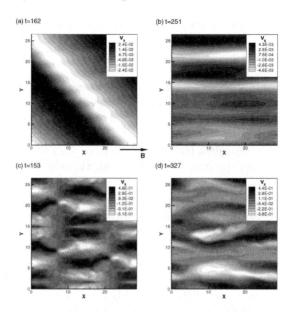

Fig. 1. Contours of streamwise velocity (in the $z = 0$ plane) are shown for solutions at $Re = 500$, $N = 0.5$, $k_{max} = 0.323$, and $\theta_{max} = 0.262\pi$ obtained with one wave ((a), (b)) and two waves ((c), (d)) in the initial conditions. The snapshots are taken at the stages of the break-up of the growing waves ((a), (c)) and decay of turbulence ((b), (d)).

anisotropy. The flow structures are elongated in the direction of the magnetic field. We saw that, as typical for the decaying turbulence, the anisotropy is quite strong even at moderate magnetic interaction parameters such as $N = 0.5$. An illustration of the transition to turbulence caused by one or two growing oblique modes is given in fig. 1

The work was initiated during the authors' participation in the 2005 MHD Summer Program at the Université Libre de Bruxelles, Belgium. The authors wish to thank the organizers, especially D. Carati and B. Knaepen, for their hospitality. The work was supported by the grant DE FG02 03 ER46062 from the U.S. Department of Energy. The authors have benefited from fruitful discussions with A. Thess, J. C. R. Hunt, K. Moffatt, and S. Molokov. Special acknowledgment is to P. Sarathy for his help in developing the parallel code for direct numerical simulations.

References

1. J. C. R. Hunt: Proc. Roy. Soc. Lond. A **293**, No. 1434, 342 (1966).
2. A. Vorobev and O. Zikanov: J. Fluid Mech. **574** 131 (2007).

Kinetic energy repartition in MHD turbulence

P. Burattini, M. Kinet, D. Carati, B. Knaepen

Physique Statistique et des Plasmas, Université Libre de Bruxelles, B-1050
Brussels, Belgium — paolo.burattini@ulb.ac.be

The forthcoming construction of the first nuclear fusion reactor (ITER) will require the development of cooling devices where magnetohydrodynamics (MHD) turbulence in liquid metals is likely to play a crucial role.

Here, homogeneous MHD turbulence is studied with direct numerical simulations (DNS), using a pseudo-spectral code [1]. An electrically-conducting, incompressible fluid is subject to a uniform magnetic field $\mathbf{B}$, and statistical stationarity is achieved by forcing the flow at large scales. The magnetic Reynolds number is $\ll 1$, so that the quasi-static approximation can be used [2], while the Taylor microscale Reynolds number (in the absence of magnetic field) is $R_\lambda \simeq 90$, λ being the Taylor microscale. Different values $(0,1,3,5)$ of the interaction parameter N are considered; N measures the intensity of the Lorenz force vis-à-vis inertia. The simulation domain is a cubic, triperiodic box, discretised by 256^3 Fourier modes, and the maximum resolution is $k_m = 128$.

Under the quasi-static approximation, the evolution equation for the turbulent energy spectrum $E(\mathbf{k})$, with $\mathbf{k}$ the wavenumber vector, is

$$F(\mathbf{k}) = T(\mathbf{k}) - 2\nu k^2 E(\mathbf{k}) - \frac{\sigma B_0^2}{\rho} \cos^2(\psi) E(\mathbf{k}), \qquad (1)$$

where F is the forcing term, T the nonlinear energy transfer, ν the kinematic viscosity, σ the electric conductivity, ρ the fluid density, B_0 the intensity of $\mathbf{B}$, and ψ the angle between $\mathbf{k}$ and $\mathbf{B}$. The last term on the right side is the Joule dissipation.

As observed in previous experimental and numerical works (e.g., [3, 4]), the magnetic force renders the velocity field anisotropic. In fact, the additional Joule dissipation attenuates the velocity gradients in the direction parallel to $\mathbf{B}$. The ratio of the one-dimensional (1D) energy spectra $\phi_\parallel(k_\parallel)$, $\phi_\perp(k_\perp)$ (defined as the 1D Fourier transforms of the velocity field in the directions parallel $\parallel$ and normal $\perp$ to $\mathbf{B}$), fig.1, indicates that the anisotropy is scale-dependent, and it is reduced for a range of length scales near $k \simeq \lambda$. Further, at small scales, the velocity component parallel to **B** is more attenuated than those in the normal direction. This latter observation seemingly contrasts with the conclusion based on the ratio of the three-dimensional (3D) spectra (calculated by averaging the energy spectrum of the normal and parallel velocity components over thin spherical shells) $E_\parallel(k)/E_\perp(k)$, fig.2. This ratio suggests that the small fluctuations of the velocity in the parallel direction may be more intense [5]. The discrepancy is rooted in the different averaging procedures involved in the 1D and 3D spectra. It is remarkable, however, that the anisotropy of the velocity 1D spectra — which can be directly measured in experiments — is closely reflected in that of a passive scalar transported by the flow (not shown).

More complete information regarding the distribution of the energy in Fourier space can be gained from the two-dimensional (2D) spectra $E\left(k_\perp, k_\parallel\right)$ [6]. These are obtained by averaging eq. (1) over thin spherical rings whose centre lies on the $k_\parallel$ axis. The 2D spectra, see fig.3 — where the distributions are divided by $\sin(\psi)$, to remove the spurious dependence on ψ, due to the spherical coordinate system — indicate that the Joule dissipation attenuates the turbulent energy in a region close to the $k_\parallel$ axis, compared to the case $N = 0$. Although such region approaches the shape of a cone, when N increases, fig.3(c), at small N, fig.3(b), the turbulent energy is still distributed rather evenly.

Figure 4 shows the 2D distributions of the nonlinear transfer $T\left(k_\perp, k_\parallel\right)$. (The hatched region near the origin has been excluded from the plot, to increase the dynamic range in the rest of the domain.) While, for $N = 0$, the contours of T are independent of ψ (apart from statistical fluctuations), for $N = 1$ and, more especially, $N = 3$, most of the transfer is located near the $k_\perp$ axis. The profiles suggest that — contrary to the case of isotropic turbulence, where there is only energy flux in the radial direction, from small to large k — in the presence of magnetic field there is also energy flux in the angular direction, from large to small ψ. This can be shown by examining the detailed transfers between different regions of the wavenumber space. Detailed contributions to T also indicate that the angular and radial transfers have a similar degree of locality.

References

1. B. Knaepen, P. Moin: Phys. Fluids **16**, 1255 (2004)
2. B. Knaepen, S. Kassinos, D. Carati: J. Fluid Mech. **513**, 199 (2004)
3. A. Alemany, R. Moreau, P. L. Sulem, U. Frisch: J. Mécanique **18**, 277 (1979)
4. O. Zikanov, A. Thess: J. Fluid Mech. **358**, 299 (1998)
5. A. Vorobev, O. Zikanov, P.A. Davidson et al: Phys. Fluids **17**, 125105 (2005)
6. R. Moreau: *Magnetohydrodynamics*, (Kluwer, Dordrecht 1990) pp 272-275

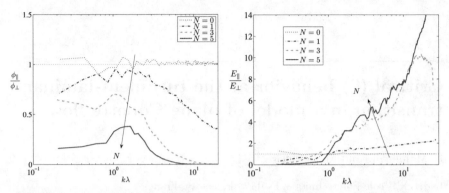

Fig. 1. Ratio of 1D spectra.   **Fig. 2.** Ratio of 3D spectra.

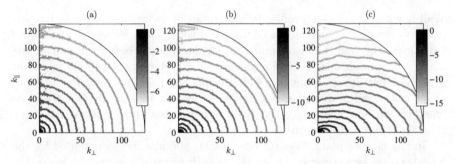

Fig. 3. Contours of $\log\left(E(k_\perp, k_\parallel)/\sin(\psi)\right)$. (a), $N = 0$; (b), $N = 1$; (c), $N = 3$.

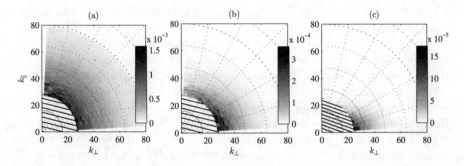

Fig. 4. Contours of $T(k_\perp, k_\parallel)/\sin(\psi)$. (a), $N = 0$; (b), $N = 1$; (c), $N = 3$.

Critical (?) behavior at the turbulent-laminar transition in a model of plane Couette flow

P. Manneville and M. Lagha

LadHyX, Ecole Polytechnique, F-91128 Palaiseau, France
paul.manneville@ladhyx.polytechique.fr

Wall flows with velocity profiles lacking inflexion points are prone to by-pass linear viscous instability modes (if any) and to experience direct transition to turbulence through the development of localized perturbations of mathematically finite amplitude, even if physically small. Non-normal linear transient growth of perturbation energy obviously plays a role, but only the size and shape of the initial disturbance can decide, usually in a highly sensitive way, whether the flow eventually leaves the laminar base flow's attraction basin (consult [1] for a discussion).

In the case of plane Couette flow (pCf), abundantly documented by the Saclay group and reviewed in [2, §6.3.4] with a broader setting in mind, the 'laminar → turbulent' transition at increasing Reynolds number[1] R takes place through the nucleation and growth of *turbulent spots*. For $R > R_g \simeq 325$, a large fraction of spots evolve into sustained turbulent flow, whereas turbulence is only transient as long as $R < R_g$. Since reaching the turbulent state may depend sensitively on the amplitude and shape of the disturbance, it is in fact preferable to determine the threshold R_g with less ambiguity by studying the 'turbulent → laminar' transition at decreasing R. Experimentally, when starting from a uniformly turbulent regime prepared at some initial high value $R_i \gg R_g$ and *quenching* the system to some final value $R_f < R_g$, one observes that turbulence breaks down into laminar flow after a long *turbulent transient*. Furthermore, by repeating the experiment with the same end points, one obtains a distribution of lifetimes in the form $\Pi(\tau_{tr} > \tau) \propto \exp(-\tau/\langle\tau\rangle)$ where $\Pi(\tau_{tr} > \tau)$ is the probability that a transient of duration $\tau_{tr} > \tau$ occurs, and $\langle\tau\rangle$ is a characteristic decay time function of $R_f = R$. Experiments further suggested $\langle\tau\rangle \propto (R_g - R)^{-1}$ [3], though this behavior was recently contested [4].

Here we study the 'turbulent → laminar' transition within the framework of a semi-realistic model of pCf derived from the Navier–Stokes equations by

---

[1] $R = U_p h/\nu$ where $2U_p$ is the relative speed of the plates driving the flow, $2h$ is the gap between the plates, and ν the kinematic viscosity of the fluid.

a cross-stream (coordinate y) Galerkin expansion truncated at lowest significant order. This model extends a previous one valid for stress-free boundary
conditions [5] to the no-slip case [6]. It is expressed as a set of partial differential equations for a few amplitudes functions of the in-plane coordinates x
(streamwise), z (spanwise), and time t, and from which the full velocity field
can be reconstructed. Experiments having shown that, in the low-R range of
the transitional regime, the most important features of the flow are well captured by fluctuations at the scale of the gap between the plates , we keep only
amplitudes U_0 and W_0 accounting for the *streamwise streaks* ($|U_0| \gg |W_0|$
almost everywhere); three supplementary amplitudes U_1, V_1, W_1 are used to
describe *streamwise vortices* generating the streaks and their interplay. Such
a low order truncation is further backed by qualitatively reasonable results
obtained in the corresponding stress-free model [7] but with the restriction
that turbulence here rather means *spatio-temporal chaos*.

As expected, the no-slip model is seen to improve over the stress-free
model, by better including viscous effects and mean-flow corrections within
turbulent regions, thus giving a more realistic value of R_g than obtained earlier but still underestimating the experimental value due to the neglect of
small-scale processes —energy transfer and dissipation— in the cross-stream
direction ($R_g^{\mathrm{mod}} \sim \frac{1}{2} R_g^{\mathrm{exp}}$, instead of $\frac{1}{5}$ in the stress-free case).

Simulations are performed using periodic boundary conditions and an
Adams–Bashforth pseudo-spectral scheme ($\delta x = \delta z = 0.25$, $\delta t = 0.01$). In this
preliminary work, a system of moderate size is considered ($L_x = L_z = 32$);
larger domains will be studied in the future. Series of *quench* experiments are
carried over by starting from $R_i = 200$ and quenching the system to variable
R_f. Results are presented in Fig. 1. Panel (a) leads to $R_g \simeq 173$ and (b) displays the distribution of lifetimes for several values of $R < R_g$. Taken together,
graphs in (c) and (d) describing the behavior of the inverse of the transients'
average lifetime with R suggest a critical behavior at $R = R_g$ obtained by
extrapolation as $R_g \approx 172.65$, rather than an indefinite exponential growth
(i.e. similar to the original analysis of experimental data [3] and not to their
reinterpretation [4]).

Extending the work to larger systems and longer times (i.e. closer to
threshold) appears essential to elucidate the nature of the transition to turbulence in pCf since, to us, transients seem best described by *spatio-temporal
intermittency* concepts and studied using the tools of *statistical mechanics*
rather than those of *dynamical systems theory* that were privileged in the
study of the companion case of the Poiseuille pipe flow (Ppf), the subject of
considerable attention recently [4,8,9].[2] In the case of Ppf, the 'laminar → turbulent' transition was studied by triggering perturbations in the laminar flow
and lifetimes of transient *turbulent puffs* were measured. Critical behavior at

---

[2] It is worthwhile to note that, when using (more physical) Reynolds numbers
based on the average shear rate, experimentally determined transitional values of
R become comparable in pCf and in Ppf (and even in plane Poiseuille flow) [2].

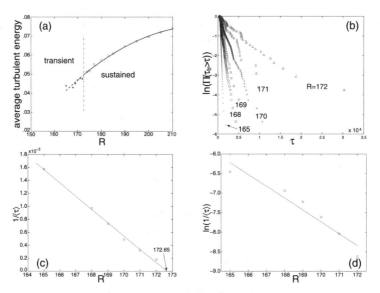

Fig. 1. (a) Average turbulent energy with R for sustained turbulence ($\square$, $R > R_g$) or during the turbulent part of a transient ($*$, $R < R_g$). (b) Distribution of transient durations for different values of $R_f = R$. (c) and (d) variations of $1/\langle\tau\rangle$ and its logarithm as functions of R.

$R_g \simeq 1800$ was observed [8,9], corresponding to a chaotic repellor becoming an attractor *via* a crisis. On the other hand, indefinite exponential growth observed in [4] was interpreted as resulting from the repelling properties of a persisting chaotic saddle. Discrepancies could be attributed to finite size effects (length of experimental apparatus or computational domain) [4,9] but the case is not fully settled and the availability of a 1D (streamwise) model of Ppf equivalent to our 2D model of pCf could be of interest.

References

1. T. Mullin & R. Kerswell, eds. *IUTAM Symposium on Laminar-Turbulent Transition and Finite Amplitude Solutions* (Springer, 2005).
2. P. Manneville: *Instabilities, Chaos and turbulence*, (Imperial College Press, London, 2004), §6.3.4.
3. S. Bottin & H. Chaté, Eur. Phys. J. B **6** (1998) 143–155.
4. B. Hof, J. Westerweel, T.M. Schneider & B. Eckhardt, Nature **443**, 59 (2006).
5. P. Manneville & F. Locher, C.R. Acad. Sci. Paris **328** Serie IIb (2000) 159–164.
6. M. Lagha & P. Manneville, "Modeling transitional plane Couette flow," submitted to Eur. J. Phys. B.
7. P. Manneville, Theor. Comput. Fluid Dynamics, **18** (2004) 169–181.
8. J. Peixinho & T. Mullin, Phys. Rev. Lett. **96**, 094501 (2006).
9. A.P. Willis & R.R. Kerswell, Phys. Rev. Lett. **98** (2007) 014501.

Interactions between finite-length streaks and breakdown to turbulence

Luca Brandt[1] and H. C. de Lange[2]

[1] Linné Flow Centre, KTH Mechanics, SE-100 44 Stockholm, Sweden
luca@mech.kth.se
[2] TUE Mechanical Engineering, 5600 MB Eindhoven, The Netherlands
h.c.d.lange@tue.nl

The occurrence and breakdown of streamwise velocity streaks have been identified as key elements in transition to turbulence and in the dynamics of
turbulence in wall-bounded shear flows. In the presence of high-levels of background disturbances, e.g. turbulent wall-bounded flows and bypass transition,
many streaks are forming randomly in time and space and interactions among
them are likely to occur. The aim of the present work is to investigate such
interactions and show when and how they can lead to turbulent production
or transition to turbulence. This study is motivated by observations of transition to turbulence in flat-plate boundary layers subject to high levels of
free-stream turbulence both from the direct numerical simulations [2] and the
experiments [3] performed in our groups in Stockholm and Eindhoven. These
previous studies have shown the importance of unsteadiness and interaction
among streaks. This is found to be an important triggering mechanism of the
streak breakdown: high-frequency oscillations are seen to form and grow in
those regions of strongest shear which are induced by streaks in relative motion. Here we present a model for the interaction between streaks, assumed
spanwise periodic but of finite length and small frequency, i.e. slowly moving
in the spanwise direction and show that the interaction is able to produce
breakdown without the need for additional random noise.

The presented simulation results are obtained using a spectral method
to solve the three-dimensional, time-dependent, incompressible Navier-Stokes
equations [2]. The periodic boundary conditions in streamwise direction are
combined with a spatially developing boundary layer by the implementation
of a "fringe region". In this region, the flow is forced to the prescribed inflow
velocity field, which in this case consists of a Blasius boundary layer profile
(zero-pressure gradient) with additional low-frequency steamwise streaks. The
streak considered are optimally growing perturbations, characterised by the
spanwise wavenumber β and the time frequency ω (low in order to obey the
boundary-layer approximation). The value of β which will be considered has
been chosen in order to reproduce the spacing observed at the breakdown

Fig. 1. Vortical structures characteristic of the symmetric breakdown at t=450. Dark (blue) surfaces represents low-streamwise velocity ($u - u_{blas} = -0.15$) while light gray (gold) surfaces high-speed regions ($u - u_{blas} = 0.15$). Gray (green) represents negative values of λ_2, used to identify vortical structures.

stages in the experiments [3]. The control parameters governing the subsequent evolution are therefore i) the time interval between the tail of the first streak and the head of the second one, in other words the initial distance between the two colliding structures; ii) the relative spanwise phase $\Delta\phi$, that is the relative position between the downstream low-speed region and the upstream high-speed region. It is observed that a longer initial distance would only require a larger amplification distance but lead to the same interaction only further downstream. Therefore, the effet of the latter parameter, i.e. the relative spanwise position, is mainly examined.

When varying the relative spanwise position three different scenarios are observed. For low values of $\Delta\phi$ the two subsequent streak structures join. The initial gap between both streaks is eventually bridged. For $\Delta\phi \approx \pi$ the upstream high-speed streak collides almost symmetrically with the downstream low-speed streak. For intermediate values of $\Delta\phi \approx \pi/2$ the interaction also leads to breakdown. In this case the flow field shows the asymmetry of the collision. A three-dimensional view of the vortical structures characteristic of the symmetric and asymmetric transition scenario are reported in figure 1 and 2 respectively. The interacting streaks are also shown. For the symmetric case, the typical vortical structures consists of Λ-structures pointing downstream and upstream, as observed in the case of the varicose instability of a steady streak [1] and from the simulations of transition induced by freestream turbulence [2]. In the asymmetric scenario, single quasi-streamwise vortices following the bending of the streaks are identified.

In the symmetric breakdown a frontal collision between the incoming high-speed fluid and the downstream low-speed streak is found. Conversely, in the asymmetric scenario, the upstream and downstream streaks slide along each other's side. Symmetric and asymmetric breakdown is also observed in the numerical simulations in [4] of a boundary layer exposed to high levels of free-stream turbulence. The latter authors show the appearance of upward and downward-moving fluid at the ends of a low-speed region leading to the formation of a hairpin structure (symmetric case) as well as the formation of one quasi-streamwise vortex in the asymmetric scenario.

Fig. 2. Vortical structures characteristic of the asymmetric breakdown, $\Delta\varphi = \pi/2$, at $t=450$. Same color coding as in figure 1.

Despite the different symmetry at the breakdown the detrimental interaction involves for both cases the tail of a low-speed region and the head of a high-speed streak. Further, the breakdown appears in both scenarios as an instability of three-dimensional shear layers formed between the two streaks. During the breakdown oscillations of the streaks in the direction normal to the shear surfaces are observed. In the symmetric case, wall-normal oscillations lead to splitting of the low-speed streak, whereas in the asymmetric case, the oscillations, oblique in the cross (y, z)-plane, induce a region of low-momentum fluid among high-speed fluid.

A parameter study of the interaction between streaks of different relative spanwise position and different obliqueness is also performed. The two different scenarios are observed in approximately equal ranges of the phase shift, where the asymmetric breakdown is slightly more likely to occur. This distribution does not change significantly for spanwise moving streaks. Threshold amplitudes for the breakdown in absence of any additional disturbance are investigated. It is observed that the local amplitudes at the interaction location are well below the critical amplitudes found for the secondary instability of steady streaks. Interestingly, the streak amplitude at which the symmetric breakdown is first observed is about 20% lower than that necessary for an asymmetric breakdown.

References

1. L. Brandt. Numerical studies of the instablity and breakdown of a boundary-layer low-speed streak. *Eur. J. Mech./B Fluids*, 26(1):64–82, 2007.
2. L. Brandt, P. Schlatter, and D. S. Henningson. Transition in boundary layers subject to free-stream turbulence. *J. Fluid Mech.*, 517:167–198, 2004.
3. J. Mans, E. C. Kadijk, H. C. de. Lange, and A. A. van. Steenhoven. Breakdown in boundary layer exposed to free-stream turbulence. *Exp. Fluids*, 39(6):1071–1083, 2005.
4. V. Ovchinnikov, U. Piomelli, and M. M. Choudhari. Numerical simulations of boundary-layer bypass transition due to high-amplitude free-stream turbulence. *J. Fluid Mech.*, 2006. under consideration.

Optimal secondary growth and transition in a plane channel flow

Carlo Cossu[1], Mattias P. Chevalier[2], and Dan S. Henningson[3]

[1] LadHyX, CNRS-École Polytechnique, F-91128 Palaiseau, France
[2] The Swedish Defence Research Agency (FOI), SE-164 90 Stockholm, Sweden
[3] Department of Mechanics, Royal Institute of Technology (KTH), S-10044 Stockholm, Sweden

We compute the linear 'secondary' optimal transient energy growth supported by an unsteady primary optimally growing basic flow in a plane channel at $R = 1500$ using a direct-adjoint technique. . The primary flow is generated by giving as initial condition the Poiseuille solution plus 'primary' optimal spanwise periodic vortices of finite amplitude A_I which then evolve into transiently growing streaks. The secondary optimal initial condition is given at $t = 0$, like the primary one, and is computed on the unsteady basic flow by using an adjoint technique. We find that the growth of the secondary perturbations can be larger than the primary optimal growth, as already observed in the case of frozen boundary layer streaks [2]. As long as the primary streaks are locally stable, the optimal secondary perturbations are uniform in the streamwise direction like the primary ones. For larger amplitudes of the primary vortices, inducing streaks that are locally unstable, the secondary optimal perturbations are sinuous. The relevance of these secondary perturbations to

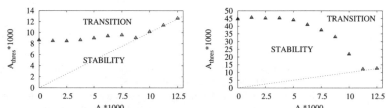

Fig. 1. Threshold values for transition of the initial perturbation amplitude The amplitude A_I of the primary vortices is also drawn (dotted line). Remark the different scales of the vertical axis in the two plots.

the transition process is then investigated by looking at the combinations of primary and secondary optimal perturbations that is most efficient in inducing transition; every such a combination of primary and secondary optimal perturbations satisfy the necessary condition for nonlinear optimality for transition ([1]). It is found (Figure 1) that a threshold level of the amplitude of the primary vortices exists, above which only exponentially small secondary perturbations are sufficient to induce transition; which corresponds to the well studied *streamwise-vortices* (SV) scenario. For values of A_I below this SV-threshold, finite amplitudes secondary perturbations are needed to induce transition and there is no clearly preferred combination of primary and secondary optimals in terms of minimum total amplitude threshold. A sketch of the threshold initial conditions and of their evolution at the first peak of rms-energy is given in Figure 2. The transition scenario followed with a primary

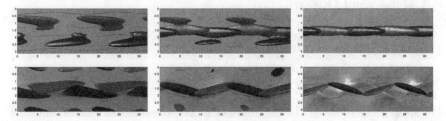

Fig. 2. Initial conditions (top row) nearest the threshold for transition for respectively $A_I = 0.01$ (left), $A_I = 0.0075$ (center) and $A_I = 0.0025$ (right). The corresponding solutions at the time of maximum deviation from the Poiseuille flow are reported in the bottom row. Only the bottom half of the channel is considered. In green is reported the surface where the streamwise velocity is 40% of its maximum value in the whole channel. In respectively blue and red are reported the surfaces where the streamwise vorticity is $\pm 55\%$ of its maximum value.

flow leading to stable streaks is reminiscent of the (STG) transition scenario proposed by Schoppa & Hussain [3]. Here we find that bended vortices can lead to transition even in the absence of primary vortices and therefore streaks with roughly the same amount of energy as in the SV scenario.

References

1. C. Cossu. An optimality condition on the minimum energy threshold in subcritical instabilities. *C. R. Mc.*, 333:331–336, 2005.
2. J. Hœpffner, L. Brandt, and D. S. Henningson. Transient growth on boundary layer streaks. *J. Fluid Mech.*, 537:91–100, 2005.
3. W. Schoppa and F. Hussain. Coherent structure generation in near-wall turbulence. *J. Fluid Mech.*, 453:57–108, 2002.

Nonlinear Disturbance Evolution and Transition to Turbulence in a Compressible Swirling Mixing Layer

S.B. Müller and L. Kleiser

Institute of Fluid Dynamics, ETH Zurich, 8092 Zurich, Switzerland
{se,kleiser}@ifd.mavt.ethz.ch

1 Introduction

We apply DNS to study the evolution of a compressible swirling mixing layer (i.e. a circular mixing layer with an additional swirl component which is present only in the shear layer) at $Ma = 0.8$ and $Re = 5000$. We investigate the nonlinear disturbance development and transition to turbulence of the layer perturbed at the inflow by viscous spatial instabilities. Different instability families as well as their nonlinear interactions are considered.

2 Numerical aspects

The conservative formulation of the compressible Navier-Stokes equations expressed in cylindrical coordinates and a mapping from Cartesian to cylindrical coordinates (r, θ, z) is employed. The governing equations are non-dimensionalized using the jet radius $r_j^* = D_j^*/2$, jet centerline inflow velocity W_j^*, density ρ_j^*, dynamic viscosity μ_j^* and temperature T_j^* as reference quantities. The convective as well as diffusive terms are discretized using sixth- to tenth-order (at interior points) compact central schemes. The centerline singularity of the governing equations is treated by a method proposed in [1]. Time integration is done by a low-storage explicit third–order Runge-Kutta method. The size of the computational domain is $L_r \times L_z = 10 \times 24$ and the resolution of our simulations is $N_r \times N_\theta \times N_z = 255 \times 150 \times 445$ grid points. The computational mesh is refined in the radial (r) and the axial (z) direction by coordinate transformations. At the inflow plane, Dirichlet conditions are applied on all conservative variables to precisely define the time-dependent inflow. In addition, a damping sponge zone is imposed to absorb any upstream-traveling acoustic disturbances. At the radial and outflow boundaries non-reflecting conditions (accounting for the curvilinear radial boundary) are employed and supplemented by sponge layers.

3 Results

The inflow forcing is based on the superposition of viscous spatial linear instabilities consisting of individual wavelike solutions of the form $\hat{q}(r)\cdot \exp\{i\cdot(\alpha z + n\theta - \omega t)\}$ where $\hat{q}$ is the complex eigenfunction, $\alpha = \alpha_r + i\alpha_i$ the complex streamwise wavenumber, n the azimuthal wavenumber and ω the circular frequency. For the base flow type under consideration, inviscid [2] and viscous [3] linear stability investigations have been performed. The maximum base-flow swirl velocity is set to $\bar{v}_{max} = 0.4$. Two distinct disturbance modes are obtained: a centrifugal and a shear instability. Three different inflow forcings are considered. The first case T1 consists of an axisymmetric $(n = 0)$ and two helical $n = 1$ shear instabilities with positive and negative circular frequency using a maximum axial-velocity total disturbance amplitude of 1.5%. The second case T2 is based on six centrifugal instabilities ($n = 14$ and $n = 15$) at distinct negative frequencies (amplitude 0.09%). The third case T3 is based on six shear instabilities ($n = 6$ and $n = 7$) at distinct positive frequencies (total amplitude 3%).

Figure 1 displays Schlieren flow visualizations using the instantaneous density-gradient magnitude. Simulation T1 (Figure 1 (a)) consists of an early developing region, subsequently followed by an elongated transitional zone and the eventual breakdown into turbulence. The dominant disturbances are axial shear instabilities. The simulation reveals a mutual competition among the disturbances imposed at the inflow. Centrifugal instabilities, which could be generated by weakly nonlinear interactions in the transitional zone, appear not to be dominant in this specific simulation. Simulation T2 (Figure 1 (b), employing centrifugal instabilities) shows an initial transitional region that is clearly confined to the outer swirling shear layer. Mushroom-like structures develop and grow in the radial direction. The subsequent turbulent mixing is intense but reaches the core stream only slowly. Thus the core flow remains laminar over a comparatively long downstream distance. The simulation does not appear to generate shear instabilities as were found in the other simulations. Results of simulation T3, shown in Figure 1 (c), reveal the characteristic long-wave nature of the most amplified (finite n) shear instabilities. A comparison of these flow fields with solutions consisting of the base flow and superimposed eigensolution (not shown here) helps understanding the interaction of the various disturbance modes and their nonlinear evolution. In [4] a massively swirling jet with a core rotating as a solid body was demonstrated to exhibit vortex breakdown.

References

[1] K. Mohseni and T. Colonius. Numerical treatment of polar coordinate singularities. *J. Comput. Phys.*, 157:787–795, 2000.

[2] G. Lu and S. K. Lele. Inviscid instability of compressible swirling mixing layers. *Phys. Fluids*, 11(2):450–461, 1999.

[3] S. B. Müller and L. Kleiser. Spatial linear stability analysis of compressible swirling jet flow. *Submitted*, 2006.

[4] S. B. Müller and L. Kleiser. LES of vortex breakdown in compressible swirling jet flow. *Submitted*, 2006.

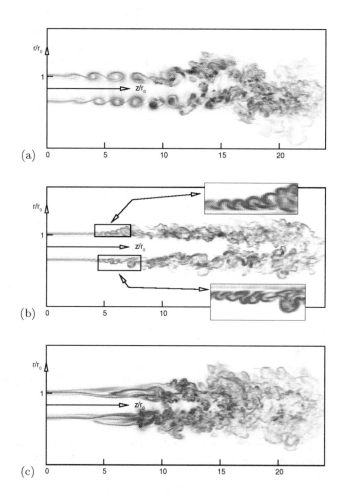

Fig. 1. Flow-field Schlieren visualization using $|\nabla\rho|$. (a) Kelvin-Helmholtz-type shear instabilities ($n = 1$ and $n = 0$); (b) Rayleigh-type centrifugal instabilities ($n = 14$ and $= 15$); (c) Kelvin-Helmholtz-type shear instabilities ($n = 6$ and $n = 7$).

Wake influence on boundary layers under severe adverse pressure gradients

Mark Phil Simens[1] and Javier Jiménez[1,2]

[1] School of Aeronautics Madrid, UPM, 28040 Madrid, Spain.
mark@torroja.dmt.upm.es, jimenez@torroja.dmt.upm.es
[2] Also at Center for Turbulence Research, Stanford University, CA 94305, USA

1 Introduction

The flow over blades in the low pressure part of a turbine is laminar, and it separates easily under the influence of an adverse pressure gradient. Reynolds numbers are normally moderate and the separation bubble is unstable, causing transition to turbulence. This normally causes reattachment of the separated boundary layer, which is important, as the flow should be attached at the trailing edge to prevent loss in efficiency. It is therefore desirable to effectively force the instability of the flow to cause transition, which in turbine applications, wakes have been shown to effectively do [2]. They are therefore important to increase the efficiency of turbines.

This subject has been studied experimentally [2], and numerically [6]. However a lot about the physics is still unknown. This is partly due to experimental difficulties and to the high computational demands necessary to simulate these flows. In this article we would like to contribute to the knowledge of initially laminar separated flow under the influence of wakes.

2 Numerical method and results

We study the flow using a numerical approach based on the equations in primitive-variable form, discretized on a staggered grid. A fractional-step method [1] is used to solve for the pressure and to conserve mass. Time integration is explicit for all terms, except for the viscous terms in the wall-normal direction being treated implicitly, using a third order Runge-Kutta scheme. The convective and viscous terms in the streamwise (x) and wall-normal (y)-directions, are discretized using fourth-order compact finite difference schemes, as described in [3]. The spanwise direction (z), is assumed periodic, and thus a pseudo-spectral method is used. The Poisson equation is discretized using a standard second-order discretization in the x and y directions.

Three-dimensional simulations were done of a boundary layer over a flat plate, with and without incoming wakes. The size of the numerical domain is $L_x \times L_z \times L_y \approx 20 \times 6 \times 3$ exit displacement thicknesses (δ^*) in the unforced case, and $40 \times 6 \times 13$ in the perturbed case. The equations are discretized using $N_x \times N_z \times N_y = 769 \times 512 \times 256$ points. The Reynolds number $Re_\theta = U_\infty(x)\theta(x)/\nu$, based on the momentum thickness $\theta(x)$ and on the free-stream velocity $U_\infty(x)$, ranges from $Re_\theta \approx 100$ at the inlet, to $Re_\theta \approx 2000$ (unforced) and $Re_\theta \approx 1000$ (forced) at the end of our numerical domain.

A very severe constant adverse pressure gradient ($\beta = (\delta^*/\tau_w)dP/dx \approx 150$ after separation in the unforced case) was induced by imposing a nearly constant suction velocity at the top boundary ($y = L_y$). There the u and w velocities are obtained imposing a zero-vorticity condition. The turbulent wake is assumed to be generated by moving rods with a diameter d. Its mean velocity profile has been approximated by $u_{wake}(\eta) = U_\infty C(x_w/d)^{-0.5} \exp(-1/4\eta^2)$ [4]. Here $x_w/d \approx 60$ is the distance to the flat plate. The wake moves past the inflow plane with a velocity $U_w = -0.83U_\infty^0$. The turbulent fluctuations in the wake have been neglected, as was also tested in [6]. The horizontal u and vertical v velocities at the inlet are given by a Hiemenz laminar profile plus the wake component, $u_H + u_{wake}\cos(U_w/U_\infty^0)$ and $v_H + u_{wake}\sin(U_w/U_\infty^0)$ while $w = 0$ always. An additional constant perturbation is added to the u component of the inflow velocity, in the spanwise direction, to introduce three-dimensionality in the flow. No-slip and impermeability conditions are imposed at the bottom wall on all three velocity components. The outflow velocities are obtained using the convective outflow boundary conditions, adjusted to conserve mass globally [5].

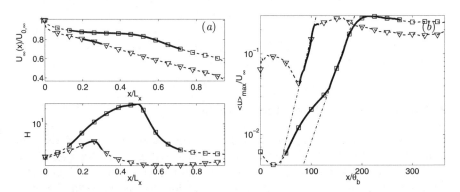

Fig. 1. For both figures: ▽: case with incoming wakes, □: the unperturbed case. —— indicates separated flow and ---- shows attached flow. (a) The U velocity at infinity and the shape factor $H = \delta^*/\theta$. (b) The u'/U_∞^0 indicating the amplification rate. ·—·—·: $0.001\exp(\alpha x/\theta_b)$ where θ_b is a bubble-related momentum thickness measured without incoming wakes. For the unforced case $\alpha = 0.042$ while with wakes, $\alpha = 0.06$.

In figure 1 results are shown for the case with and without incoming wakes. In both cases an instability is found, but without wakes a lower amplification rate is found than with wakes. This instability causes the bubble to attach. It remains so during the turbulent flow development, after the instability saturates. The adimensional wake frequency is $St \sim \mathcal{O}(0.01)$. As this is close to the most unstable Kelvin-Helmholtz frequency, it suggests that a shear layer instability is triggered. However the influence of wakes goes beyond triggering an instability. It also causes the separation point to move downstream compared with the initial unperturbed separation point. Transition to turbulence is found here, while in [6] it was almost absent. This is probably due to the constant three-dimensional perturbation imposed at the inlet.

The shape factor in figure 1(a) shows again the dramatic influence the wakes have, also decreasing the value downstream of the original separated zone. The potential flow U_∞ in figure 1(a) indicates that the wake reduces the height of the bubble, diminishing the interaction between it and the potential flow. In that case the imposed v velocity is seen to cause an almost linear adverse pressure gradient.

3 Conclusions

It has been shown that wakes have a twofold influence on the control of separation bubbles. First they cause the separation point to move downstream. Secondly they promote transition to turbulence. They reduce the height as well as the length of the separation bubble. As a consequence the shape factor is considerably reduced, which in practical applications will reduce losses and the risk of turbulent separation.

This work was financially supported by ITP S.A., the EU TMR network (HPRN-CT-2002-00300), the Airbus CAFEDA network, and CICYT contract DPI2003-03434.

References

1. J.K. Dukowicz and A.S. Dvinsky. Approximate factorization as a high order splitting for the implicit incompressible flow equations. *J. of Comp. Phys.*, 102:336–347, 1992.
2. P.H. Hodson and R.J. Howell. Bladerow interactions, transition, and high-lift aerofoils in low-pressure turbines. *Ann. Rev. Fluid Mech.*, 37:71–98, 2005.
3. S. Nagarajan, S.K. Lele, and J.H. Ferziger. A robust high-order compact method for large eddy simulation. *J. of Comp. Phys.*, 191:392–419, 2003.
4. H. Schlichting. *Boundary-Layer Theory.* Mc Graw-Hill Book Company, 1978.
5. M.P. Simens and J. Jiménez. Alternatives to Kelvin-Helmholtz instabilities to control separation bubbles. ASME Paper GT2006-90670, 2006.
6. J.G. Wissink, W. Rodi, and H.P. Hodson. The influence of disturbances carried by periodically incoming wakes on the separating flow around a turbine blade. *Int. J. of Heat and Fluid Flow*, 27:721–729, 2006.

Boundary layer structure in highly turbulent convection

R. du Puits[1], C. Resagk[2], and A. Thess[3]

[1] Dept. of Mechanical Engineering, Ilmenau University of Technology,
   POB 100 565, 98684 Ilmenau, Germany `ronald.dupuits@tu-ilmenau.de`
[2] Dept. of Mechanical Engineering, Ilmenau University of Technology,
   POB 100 565, 98684 Ilmenau, Germany `christian.resagk@tu-ilmenau.de`
[3] Dept. of Mechanical Engineering, Ilmenau University of Technology,
   POB 100 565, 98684 Ilmenau, Germany `thess@tu-ilmenau.de`

1 Introduction

Turbulent thermal convection is the most frequently occurring type of fluid flow in Nature. This kind of flows are responsible for the global motion of air in the earth atmosphere or water in the earth oceans as well as for the air exchange in rooms with high concentration of humans like passenger compartments open-plan offices or large auditoria. In spite of its widespread occurrence, the local properties of the temperature and velocity fields especially in the boundary layer between a solid surface and a surrounding fluid are still poorly understood. Presently a multitude of theoretical models exist to describe both fields, ranging from the model of the simple laminar boundary layer with linear profiles of mean horizontal velocity and mean temperature, over the turbulent boundary layer with a logarithmic profile of the mean velocity [4] to a power law relation for the temperature profile and the rms temperature fluctuations as predicted by Prandtl [2] and Priestley [3]. The purpose of the present work is to bridge this gap by performing local velocity measurements inside the boundary layers of turbulent Rayleigh–Bénard (RB) convection which have a higher spatial resolution than in previous experimental investigations.

2 Experimental setup

Our large-scale facility represents a classical RB system which is sketched in figure 1. We use air ($Pr = 0.7$) in a cylindrical enclosure with a diameter $D = 7.15$ m which is heated from below by an electrical heating system and cooled from above by a free-hanging water-cooled sandwiched aluminium plate with a weight of approximately 5 metric tons. An active compensation heating system prevents lateral heat losses and renders the sidewall effectively adiabatic. A detailed description of the Barrel of Ilmenau will be given elsewhere [1].

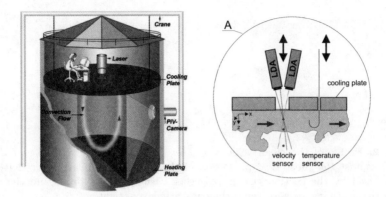

Fig. 1. Sketch of the large-scale experimental facility "Barrel of Ilmenau" (left) and the setup of the velocity and the temperature measurement at the centre of the cooling plate (right)

We performed local measurements of the two horizontal velocity components along the central axis of the box for distances up to 90 mm away from the cooling plate (see figure 1) using two single laser Doppler velocimetry probes. Time series of v_x and v_y over one hour are captured at different z–positions and the magnitude $v = (v_x^2 + v_y^2)^{1/2}$ as well as the angle $\phi = \arctan(v_y/v_x)$ of the velocity vector are computed as functions of time.

3 Results

Two series of experiments covering a range of Ra between $Ra \approx 10^9$ and $Ra \approx 10^{12}$ have been carried out in each of which only a single parameter was varied. In the first series of experiments the aspect ratio was maintained at $\Gamma \approx 1$ and the temperature difference between the heating and the cooling plate was changed, while in the second one the temperature difference was kept fixed and the aspect ratio was changed between $\Gamma \approx 1$ and $\Gamma \approx 10$.

We restrict our discussion to the case of constant aspect ratio $\Gamma = 1.13$ in which the one role structure of the global flow often called wind is well known from various high Rayleigh number experiments. It leads to a boundary layer at the cooling plate (and at the heating plate too) with similar characteristics according to free shear flows. Profiles of the mean horizontal velocity $v(z)$ as well their fluctuations were measured in the central axis of the experiment close to the cooling plate. One example of the measured profiles for the parameter set $Ra = 7.5 \times 10^{11}$, $\Gamma = 1.13$ and $\Delta T = 40K$ is plotted in figure 2. In

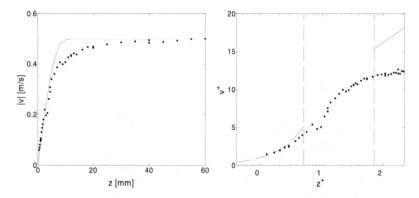

Fig. 2. Profile of the mean velocity for $Ra = 7.5 \times 10^{11}$, $\Gamma = 1.13$ and $\Delta T = 40K$: against the theoretical prediction of the laminar boundary layer (left) and normalized by the friction velocity together with the inner linear and the outer logarithmic scaling laws for the turbulent shear layer (right).

our talk we compare it with theoretical predictions of the mean velocity profile of laminar or turbulent boundary layers [4] and we discuss the deviations.

A theoretical explanation for these findings is still missing but possibly the activity of the cold (hot) plumes which arise from the boundary layer and interact with the global flow is still underestimated and leads to the change in the shape of those profiles. Presently we can not answer this question definitely but this problem will be addressed in our future work.

4 Acknowledgments

The authors wish to acknowledge the financial support of the Deutsche Forschungsgemeinschaft under the grant numbers TH 497 and of the Thueringer Ministerium fuer Wissenschaft, Forschung und Kunst for the work reported in this paper. Particularly we thank F. Busse and A. Tilgner for useful discussions and V. Mitschunas, K. Henschel and H. Hoppe for technical help.

References

1. du Puits, R., Busse, F.–H., Resagk, C., Tilgner, A., Thess, A.: J. Fluid Mech. **572**, 231–254 (2007).
2. Prandtl, L.: Beitr. z. Phys. Atmos. **19**, 188–202 (1932).
3. Priestley, C. H. B.: Austral. J. Phys. **7**, 176–201 (1954).
4. Schlichting, H., Gersten, K.: *Boundary layer theory*, (Springer, Berlin Heidelberg New York 2000)

Final states of decaying 2D turbulence in different geometries with no-slip walls

Kai Schneider[1] and Marie Farge[2]

[1] MSNM–CNRS & CMI, Université de Provence, Marseille, France
   `kschneid@dmi.univ-mrs.fr`
[2] LMD–CNRS, Ecole Normale Supérieure, Paris, France
   `farge@lmd.ens.fr`

Summary. Direct numerical simulations of two-dimensional decaying turbulence in domains of different geometries having no-slip walls are presented. Starting from random initial conditions the flow rapidly exhibits self-organization into coherent vortices. At the same time viscous boundary layers are formed on the walls, become unstable and produce new coherent vortices which are injected into the bulk flow. The computation uses a pseudo-spectral method with volume penalization to model the walls. Each flow is integrated until a quasi-final state is reached.

Two–dimensional turbulence in wall bounded domains has many applications in geophysical flows, *e.g.*, the prediction of coastal currents in oceanography, the transport and mixing of pollutants. Direct numerical simulations of 2D turbulence in circular and square domains can be found, *e.g.* in [2, 3, 5]. The aim of the present paper is to study the influence of the geometry of the domain on the flow dynamics and in particular on the long time behaviour of the flow. We consider different geometries, a circle, a square, a triangle and a torus.

The numerical technique we use here is based on a Fourier pseudo–spectral method with semi-implicit time discretization and adaptive time-stepping [4]. The Navier–Stokes equations are solved in a double periodic square domain of size $L = 2\pi$ using the vorticity–velocity formulation. The bounded domain is thus imbedded in the periodic domain and the no-slip boundary conditions on the wall $\partial\Omega$ are imposed using a volume penalisation method. A mathematical analysis of the method is given in [1], proving its convergence towards the Navier–Stokes equations with no–slip boundary conditions. Details on the code together with its numerical validation can be found in [4]. The governing equations in vorticity-velocity formulation are,

$$\partial_t \omega + \boldsymbol{u} \cdot \nabla \omega - \nu \nabla^2 \omega + \nabla \times (\frac{1}{\eta} \chi \boldsymbol{u}) = 0$$

where $\boldsymbol{u}$ is the divergence-free velocity field, *i.e.* $\nabla \cdot \boldsymbol{u} = 0$, $\omega = \nabla \times \boldsymbol{u}$ the vorticity, ν the kinematic viscosity and $\chi(\boldsymbol{x})$ a mask function which is 0 inside the fluid, *i.e.* $\boldsymbol{x} \in \Omega$, and 1 inside the solid wall. The penalisation parameter η is chosen to be sufficiently small ($\eta = 10^{-3}$) [4].

Starting with random initial conditions we compute the flow evolution in different geometries for initial Reynolds numbers of about 1000. Figure 1 shows the time evolution of energy, enstrophy and palinstrophy, and figure 2 the vorticity fields at early, intermediate and late times, for different flows in circular, square, triangular and toroidal geometries. More details and discussion of the results can be found in [6].

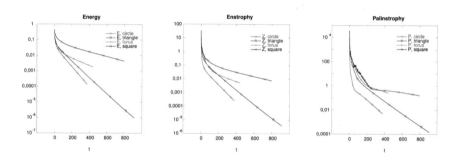

Fig. 1. Two-dimensional decaying turbulence in bounded domains. Time evolution until the final state is reached for energy (left), enstrophy (middle) and palinstrophy (right).

Acknowledgments :
We thankfully acknowledge financial support from the ANR project M2TFP and from the CEA-Euratom contract V.3258.001.

References

1. P. Angot, C.H. Bruneau and P. Fabrie. *Numer. Math.* **81**, 497–520, 1999.
2. H.J.H. Clercx, A.H. Nielsen, D.J. Torres and E.A. Coutsias. *Eur. J. Mech. B - Fluids* **20**, 557–576, 2001.
3. S. Li, D. Montgomery and B. Jones. *Theor. Comput. Fluid Dyn.* **9**, 167–181, 1997.
4. K. Schneider. *Comput. Fluids,* **34**, 1223–1238, 2005.
5. K. Schneider and M. Farge. *Phys. Rev. Lett.,* **95**, 244502, 2005.
6. K. Schneider and M. Farge. Decaying two–dimensional turbulence in bounded domains. *Preprint, 2007.*

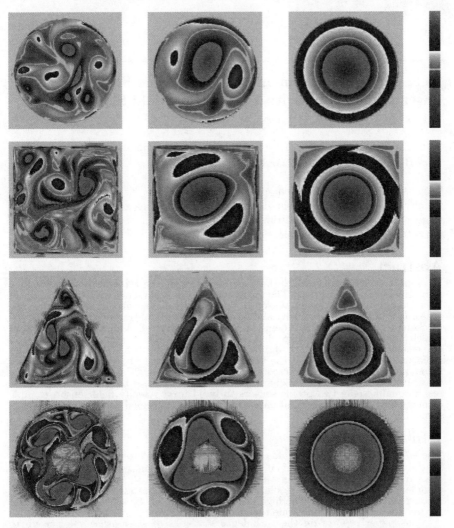

Fig. 2. Two-dimensional decaying turbulence in bounded domains. Vorticity fields at early (left), intermediate (middle) and late times (right). From top to bottom: circular, square, triangular and toroidal domains.

Near-Wall Measurements of Turbulence Statistics with Laser Doppler Velocity Profile and Field Sensors

Katsuaki Shirai, Christian Bayer, Andreas Voigt, Thorsten Pfister,
Lars Büttner and Jürgen Czarske

Technische Universität Dresden
Faculty of Electrical Engineering and Information Technology
Chair of Measurement and Testing Techniques
shirai@iee.et.tu-dresden.de

Abstract

We report on the near-wall turbulence statistics of the streamwise velocity in a fully developed channel flow. The measurement was conducted with two systems of laser Doppler velocity profile sensor at three different Reynolds numbers. Since the sensor provides both the velocity and position of individual tracer particles inside the measurement volume without being mechanically traversed, a high spatially resolved velocity measurement is achieved. The resulting turbulence statistics show a good agreement with available data of direct numerical simulations up to the fourth order moments. This demonstrates the velocity profile sensor to be one of the promising techniques for turbulent flow research with the advantage of a spatial resolution more than one order of magnitude higher than a conventional laser Doppler technique. The results qualifies the two profile sensors to build up a velocity field sensor, capable of measuring velocity distribution within a two-dimensional area without using a camera.

1 Introduction

The laser Doppler velocity profile sensor [1, 2] has been proposed to overcome the spatial averaging of conventional laser Doppler anemometry. The sensor resolves the particle's axial position as well as transverse velocity inside the measurement volume, hence has a spatial resolution at least one magnitude of order higher than a conventional technique. The high spatial resolution of the sensor is advantageous for the investigation of fine scale turbulence structures.

This paper reports on the turbulence statistics of streamwise velocity up to fourth order moment in the near-wall region of a fully developed turbulent channel flow. They were at three different Reynolds numbers with two independent velocity profile sensor systems. The statistics were compared with available direct numerical simulation (DNS) data.

2 Flow Apparatus and Velocity Profile Sensor

Flow measurements were carried out in the near-wall region of a fully developed air channel flow [3, 4]. A pair of glass plates was attached flush to the walls of the measurement section to have an optical access and not to disturb the flow (see Fig. 1). The insertion of the glass plate does not yield any serious problem as long as the sensor is calibrated through a glass plate with same material and thickness.

The principle of the laser Doppler velocity profile sensor is based on the use of two fringe systems in a single measurement volume [1, 2]. In contrast to a conventional laser Doppler technique, it employs two fan-like fringes, namely, diverging and converging fringe systems. The measured pair of Doppler frequencies provides the position of individual tracer particles through calibration curve, since the fringe

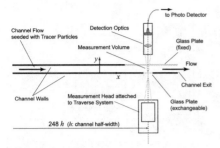

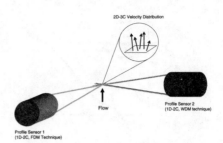

Fig. 1. Flow measurement configuration in the channel flow. The sensor has a spatial resolution in the direction of optical axis.

Fig. 2. Concept of velocity field sensor. The combination of two 1d2c sensors (velocity profile sensors) enables a 2d3c sensor.

space curves are unique functions of the coordinate along the optical axis. The fringe space at the local position finally gives the velocity of the particle. Because of the spatial resolution inside the measurement volume, high spatially resolved velocity measurement is achieved independent of the velocity magnitude. Two sensor systems realized with different techniques (WDM: wavelength- and FDM: frequency-division-multiplexing) were used for the measurements [1, 2, 4, 5, 6]. The nominal parameters of both the sensors are listed in Table 1.

Table 1. Characteristics of the sensors and the measurement conditions (L: working distance, l_x, l_z, l_y: measurement volume dimensions in x, z, y direction, σ_y: spatial resolution, σ_U/U: relative accuracy of velocity measurement, Sw_m: slot width of mean velocity, N_m : approximate data points in a slot for mean velocity, Sw_{rsf}: slot width for higher order moments (rms, skewness, flatness), N_{rsf} : approximate data points in a slot for the higher order moments)

sensor type	L [mm]	$l_x \times l_z \times l_y$ [μm]	σ_y [μm]	σ_U/U [%]	u_τ [m/s]	Re_τ	l_τ [μm]	Sw_m [μm]	N_m	Sw_{rsf} [μm]	N_{rsf}
WDM	80	100×100×500	1.5	0.06	0.26	420	60	30	400	140	1700
WDM	80	100×100×350	1.5	0.06	0.49	780	32	80	450	160	850
FDM	310	100×100×900	6	0.085	0.71	1100	23	50	650	120	1450

3 Results and Discussions

The measurements conditions are listed in Table 1. The coordinates x, y, z were taken for the streamwise, wall-normal, spanwise direction, respectively. The Reynolds number is based on the channel half-width h and the wall friction velocity u_τ determined from the streamwise wall-static pressure gradient.

The turbulence statistics of streamwise velocity are shown in Fig. 3 up to fourth order central moment. The horizontal axis was normalized with the viscous length scale l_τ ($=\nu/u_\tau$). For comparison, available DNS data at Re_τ=400 and 640 [7, 8] are shown in the same plots. The statistics were calculated using a constant-width slot-technique, where the statistics are calculated for the samples within each slot. Different slot widths were applied for mean and higher order moments (rms, skewness, flatness) taking account for the balance between the statistical convergence and the spatial resolution (see Table 1). The velocity bias was corrected using the weighting function based on the inverse of the instantaneous velocity [9].

The obtained statistics in Fig. 3 are well scaled with the wall variables especially close to the wall. The mean and rms velocity distributions in Fig. 3(a, b) coincide well for different Reynolds numbers and show good agreements with the DNS data [7, 8]. The third and fourth order moments (skewness and flatness factors) in Fig. 3(c, d) also show reasonably good agreements with the DNS data. Slight differences are observed for the higher order moments (rms, skewness and flatness factors) at different Reynolds numbers, which might indicate possible Reynolds number effects. A higher number of data points is required to clarify this point. The higher order moments show some scatter, which also indicate the necessity of a higher number of data for making further statements.

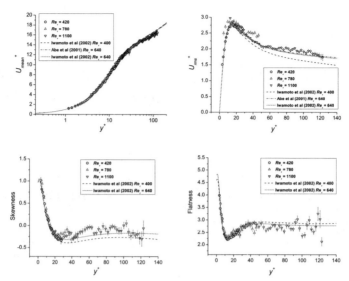

Fig. 3. Near-wall turbulence statistics at three different Reynolds numbers compared with available DNS data. (a) mean velocity (left top), (b) rms velocity (right top), (c) skewness factor (left bottom), (d) flatness factor (right bottom)

4 Outlook and Concluding Summary

The profile sensor is currently combined to build up a laser Doppler velocity field sensor (see Fig. 2), which enables the measurement of two-dimensional velocity field without using any camera. The field sensor resolves the out-of-plane velocity component in a two-dimensional area. The effect of spatial resolution on turbulence measurements will be systematically investigated in two orthogonal directions. The field sensor would be a powerful measurement technique for the investigation of small structures in turbulent flows.

In this paper, the application of a velocity profile sensor to the near-wall regions of a fully developed turbulent channel flow was reported. The flow measurements were conducted with two different sensor systems at three different Reynolds numbers. The turbulence statistics of the streamwise velocity up to fourth order moment was calculated with high spatial resolution. The obtained statistics show good agreements with the available DNS data. Taking higher number of samples with well controlled flow conditions would provide credible data up to higher order moments with full resolution of the sensors. Currently, the sensors are combined into a velocity field sensor capable of measuring 2d3c velocity distribution. As a conclusion, the laser Doppler velocity profile sensor has been demonstrated to be a promising technique for such an investigation of turbulent flows where a high spatial resolution is required.

Acknowledgments: Dr. G. Yamanaka, Dr. S. Becker, Mr. H. Lienhart and Prof. F. Durst are acknowledged for their supports on the flow measurements. The support from the Deutsche Forschungsgemeinschaft (DFG) is greatly appreciated (CZ55/18-1).

References

1. J. Czarske: Meas. Sci. Technol. **12** 52 (2001)
2. J. Czarske, L. Büttner, T. Razik, H. Müller: Meas. Sci. Technol. **13** 1979 (2002)
3. E.-S.Zanoun, F. Durst, H. Nagib: Phys. Fluids **15** 3079 (2003)
4. K. Shirai, T. Pfister, L. Büttner, J. Czarske, H. Müller, S. Becker, H. Lienhart, F. Durst: Exp. Fluids **40** 473 (2006)
5. J. Czarske: Meas. Sci. Technol. **17** R71 (2006)
6. T. Pfister, L. Büttner, K. Shirai, J. Czarske: Appl. Opt. **44** 2501 (2005)
7. H. Abe, H. Kawamura, Y. Matsuo: Trans. ASME J. Fluids Eng., **123** 382 (2001)
8. K. Iwamoto, Y. Suzuki, N. Kasagi: Int. J. Heat and Fluid Flow, **23** 678 (2002)
9. D.K. McLaughlin, W.G. Tiederman: Phys. Fluids **16** 2082 (1973)

Detached Eddy Simulation of Flows over Rough Surfaces

A. Silva Lopes and J.M.L.M. Palma

CEsA — Centro de Estudos de Energia Eólica e Escoamentos Atmosféricos
Faculdade de Engenharia da Universidade do Porto, Porto, Portugal
asl@fe.up.pt

Introduction

Many flows of practical interest are over surfaces with irregularities whose size is similar or larger than the viscous sublayer ($k^+ \gtrsim 5$) and much smaller than the boundary-layer thickness ($\delta/k \gtrsim 50$). In flows as common as the wind over grass, the size of the roughness elements is usually large enough ($k^+ \gtrsim 70$) for the momentum exchange to be due almost entirely to the pressure drag, which is then considerably larger than the viscous stresses in the same flow over a smooth surface. Clearly, the accurate prediction of the increased transfer of momentum is important in the study of such flows.

Since resolving all the details of the flow near each roughness element in large-eddy simulations (LES) is not feasible with currently available resources, the momentum transfer must be modelled. Most works use models based on an equilibrium boundary-layer [1, 2], which can be inaccurate if the flow is accelerating, decelerating or over curved surfaces. For instance, it was not possible to obtain grid convergence of the flow over an isolated hill [3], since the size of the separation in the lee side increased when the grid was refined.

The objective of this work is to evaluate the performance of two Detached Eddy Simulation (DES) models for flows over rough surfaces in the simulation of two boundary-layer flows: over a flat surface [4] and over two-dimensional sinusoidal waves [1]. DES is an hybrid model, based on the Spalart-Allmaras RANS model [5], where a length-scale is defined in such a way that it works as a LES away from solid surfaces, where the grid resolution is enough to resolve the turbulent motions. It will be used here with the two variants for flows over rough surfaces proposed in Aupoix and Spalart [6]. Since DES solves one additional transport equation for the eddy-viscosity, with terms accounting for the acceleration and deceleration of the flow, better results are expected than when using models based only on equilibrium. The goal is to determine the best variant of the model for flows with separation, to be used in future studies of atmospheric flows.

Methods

The filtered continuity and Navier-Stokes equations were discretised on a non-staggered grid, using the finite-volume technique. The DES approach uses the Spalart-Allmaras model and follows the formulation by Nikitin *et al.* [7]. The Boeing and ONERA variants of the model for flows over rough surfaces are presented in Aupoix and Spalart.

Results

Comparing the streamwise evolution of the friction coefficient using the DES model variants for flows over smooth and rough surfaces, we found that both Boeing and ONERA variants predicted the increased friction due to surface roughness (figure 1a). However, comparing with the experiments of Acharya *et al.* [4], we found that the DES underpredicted the friction even more than in the case of the smooth surface. DES underprediction of the friction in the case of smooth surfaces was already reported [7, 8] and is due to insufficient turbulence production in the transition between the RANS and LES zones (sometimes called "grey area"). We found that the error was larger in the case of the rough surface with $k^+ \approx 70$: 33 or 45%, respectively with the Boeing and ONERA variants, while it was 25% in the smooth surface (comparing DES with RANS). The growth rate of the momentum thickness was also underestimated (figure 1b), which was due to the lower friction and consequently lower momentum exchange. The observed errors are essentially due to the DES methodology, since using the Spalart-Allmaras RANS model the errors are smaller, especially with the ONERA variant, that underestimated the friction by less than 10%, while the Boeing variant overestimated it around 20%.

Conclusions and future work

Preliminary results of detached-eddy simulations (DES) of flows over rough surfaces were obtained. We found that DES correctly predicted the increased friction due to the surface roughness in a flow over a flat surface, but underestimated it, with an error even larger than in the case of a smooth surface (33 or 45%, depending on the variant considered, instead of 25%). This work will proceed with the simulation of the flow over rough two-dimensional sinusoidal waves, which has some separation. We hope to found better agreement with experiments in that case, since the large turbulent eddies generated by the separation can attenuate the insufficient production in the transition between the RANS and LES zones observed in the flat surface.

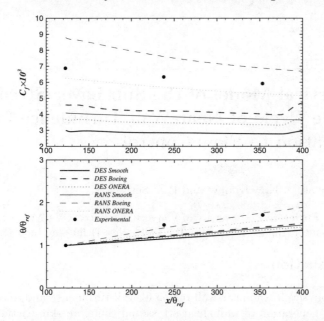

Fig. 1. Streamwise evolution of *(a)* friction coefficient and *(b)* momentum thickness in a flat plate boundary layer.

References

1. W. Gong, P.A. Taylor, and A. Dornbrack. Turbulent boundary-layer flow over fixed aerodynamically rough two-dimensional sinusoidal waves. *Journal of Fluid Mechanics*, 312:1–37, 1996.
2. I. Marusic, G.J. Kunkel, and F. Porté-Agel. Experimental study of wall boundary conditions for large-eddy simulation. *Journal of Fluid Mechanics*, 446:309–320, 2001.
3. A. Silva Lopes, J.M.L.M. Palma, and F.A. Castro. Simulation of the Askervein flow. Part 2: Large-eddy simulations. Submitted to Boundary-Layer Meteorology, 2007.
4. M. Acharya, J. Bornstein, and M.P. Escudier. Turbulent boundary layers on rough surfaces. *Experiments in Fluids*, 4:33–47, 1986.
5. P.R. Spalart and S.R. Allmaras. A one-equation turbulence model for aerodynamic flows. *La Recherche Aérospatiale*, 1:5–21, 1994.
6. B. Aupoix and P.R. Spalart. Extensions of the Spalart-Allmaras turbulence model to account for wall roughness. *International Journal of Heat and Fluid Flow*, 24:454–462, 2003.
7. N.V. Nikitin, F. Nicoud, B. Wasistho, K.D. Squires, and P.R. Spalart. An approach to wall modeling in large-eddy simulations. *Physics of Fluids*, 12:1629–1632, 2000.
8. A. Keating and U. Piomelli. A dynamic stochastic forcing method as a wall-layer model for large-eddy simulation. *Journal of Turbulence*, 7:1–24, 2006.

Theoretical Model of the Sub-Layer Streaks and the Cycle of Near-Wall Turbulence for Application to Flow Control

P.W. Carpenter[1], K.L. Kudar[1] and P.K. Sen[2]

[1] School of Engineering, University of Warwick, Coventry, CV4 7AL, UK.
[2] Department of Applied Mechanics, IIT Delhi, New Delhi - 16, India

1 Introduction

The flow physics in the near-wall region of wall turbulence and, most importantly, the generation of wall shear stress and therefore skin-friction drag, is known to be governed by a quasi-periodic cycle. Most authors agree that the near-wall region is characterized by sub-layer streaks of relatively high- and low-speed streamwise velocity that are generated through linear advection by streamwise vortices. Focussing on the temporal development of the sublayer streaks, these grow in strength, move away from the wall (ejection), undergo instability of some sort culminating in explosive growth (sometimes termed 'bursting'); this is followed by rapid movement of relatively high-speed fluid towards the wall (the sweep process) during which high levels of wall shear stress are generated. And so the cycle continues.

2 The Theoretical Model of the Near-Wall Cycle

Several previous authors have proposed theoretical models for the near-wall cycle [1-3]. Most of the previous theoretical models feature sublayer streaks of streamwise velocity generated by streamwise vortices through linear advection. The streaks then undergo some sort of nonlinear instability that leads to rapid, even explosive, growth. The vortices are then regenerated by some mechanism. However, these previous models do not appear to explain how the cycle period can be determined, nor what selects the spanwise streak spacing seen in experiments and numerical simulations alike. We wish to present a new model of the near-wall cycle which we introduced in a preliminary way recently [4]. It is illustrated schematically in Fig. 1. Unlike previous models it is based on a kind of dual cycle whereby streamwise vortices generate sublayer streaks more or less at the same time as other vortical elements generate quasi-plane

Tollmien-Schlichting-like waves. The waves and streaks then interact nonlinearly to generate oblique waves. This nonlinear interaction is characterized by a period of modulation and also leads to explosive (algebraic) growth. The streamwise vortices are regenerated by a nonlinear interaction between the plane and oblique waves. The strongest interaction occurs for oblique waves propagating at an angle of 65 degrees and we can expect to see such waves in experiments and numerical simulations. This value for the propagation angle is in close agreement with the results of a POD study by Sirovich et al. [5]. It also corresponds to the statistically averaged values of the spanwise streak spacing of around 100 wall units seen in the experiments of Kline et al. [6] and others. The corresponding period of modulation is identified with the period of the near-wall cycle. For a flat-plate boundary layer at $R = U_\infty \delta^*/\nu = 5000$ (where U_∞ is the freestream velocity, δ^* is the boundary-layer displacement thickness, and ν is the kinematic viscosity) the period of low-frequency modulation is about $36\nu/U_\tau^2$ (where U_τ is the friction velocity). This is close to the experimental value of the 'bursting' cycle measured by Klewicki et al. [7] at very high Reynolds number and of the same order of magnitude as the values of 72 measured by Kline et al. [6] and 60 obtained in numerical simulations [8].

It was commonly thought that Tollmien-Schlichting-like waves could not exist in turbulent boundary layers. However, Sen and Veeravalli [9] have definitely established, both experimentally and theoretically, that such waves do exist in turbulent channel flows and moreover can grow. Both plane and oblique waves were also shown by Sirovich et al. [5] to play an important role in near-wall turbulence.

3 Typical Result

To investigate the possibility of nonlinear interaction between sub-layer streaks and waves, we derived [9,10] an extended 3D Orr-Sommerfeld-type equation whereby the molecular viscosity was replaced by the sum of it and an anisotropic eddy viscosity. The undisturbed mean velocity profile was taken as the mean flat-plate boundary-layer profile. We then investigated the nonlinear interaction between a time-invariant sublayer streak and plane and oblique waves. This generated an inhomogeneous Orr-Sommerfeld-type system which was integrated numerically. The outcome depended on the amplitude of the streak velocity perturbation. For weak streaks the interaction generated a streamwise periodic variation in growth rate somewhat reminiscent of turbulent spots. For the stronger streaks more typical of the turbulent near-wall region a typical result is shown in Fig. 2 for a lightly damped 2D wave. We see an envelope curve $f_1(t)$ showing a cycle with initially strong algebraic growth followed by decay. The plain curve f_2 plots the individual 2D wave. If the 2D wave were growing, which is perfectly possible, the algebraic growth would be much more explosive.

References

1. J.M. Hamilton et al.: J. Fluid Mech., **287**, p. 317 (1995).
2. J. Jiménez & A. Pinelli: J. Fluid Mech., **389**, p. 335 (1999).
3. W. Schoppa & F. Hussain: J. Fluid Mech., **453**, p. 57 (2002).
4. P.W. Carpenter et al.: Phil. Trans. Roy. Soc. Lond. A (in press) (2007).
5. L. Sirovich et al.: Phys. Fluids A, **2**, p. 2217. (1986).
6. S.J. Kline et al.: J. Fluid Mech, **30**, p. 741 (1967).
7. J.C. Klewecki et al.: Phys. Fluids, **7**, 857 (1995).
8. J. Kim & P.R. Spalart : Phys. Fluids, **30**, p. 3326 (1987).
9. P.K. Sen & S.V. Veeravalli : Curr. Sci., **79**, p. 840 (1997).
10. P.K. Sen et al.: Sadhana, (in press) (2007).

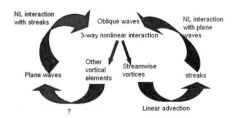

Fig. 1. Schematic of our theoretical model for the cycle of near-wall turbulence.

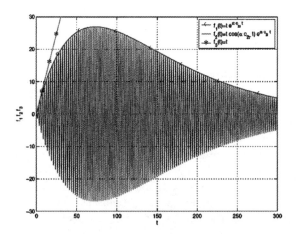

Fig. 2. Variation of the secular part of the total perturbation velocity due to the interaction between a sublayer streak and a damped 2D wave generating an algebraically growing 3D wave. R=5000.

Two-dimensional turbulence on a confined domain with no-slip walls

GertJan van Heijst[1] and Herman Clercx[1]

JM Burgers Centre for Fluid Dynamics and Department of Physics Eindhoven
University of Technology Eindhoven, the Netherlands
e-mail: G.J.F.v.heijst@tue.nl

The inverse energy cascade active in two-dimensional (2D) turbulence generally results in the emergence of increasingly larger vortex structures. In the
case of a turbulent flow in a confinement with no-slip walls, these vortices tend
to be domain-filling, at least when the initial energy spectrum satisfies certain
criteria. This tendency has been confirmed both by laboratory experiments
with contained rotating or stratified fluids and by 2D numerical flow simulations based on spectral methods. Close inspection of the results has revealed
the crucial role played by the no-slip boundaries: boundary layers associated
with the no-slip character of the walls are scraped off by neighbouring vortices,
leading to intense vorticity filaments that are subsequently advected into the
interior of the flow domain, as illustrated by the numerical simulation results
shown in Fig. 1. This mechanism affects the evolution of the 2D turbulent
flow considerably, even at relatively large distances from the walls, as is confirmed by the spectral characteristics of numerically simulated decaying flows
in such geometries. Near the no-slip walls, the spectrum revealed a -5/3 slope,
ranging from the wave number associated with the boundary layer thickness
to the smaller wave number associated with the larger vortex structures [1].
The spectra for the flow in the interior of the domain reveal a slope steeper
than -3, which is related to the enstrophy cascade. The role of the solid no-slip
boundaries as sources of vorticity has been investigated in more detail for the
case of forced flow by Wells et al. [2, 3], who considered the quasi-2D flow in
a rotating square container whose angular speed Ω was modulated in time
according to $\Omega(t) = \Omega_0(1 + A sin(ft))$, with A and f the amplitude and the
frequency of the modulation, respectively. With this specific type of forcing,
boundary layers are continuously created and subsequently removed from the
walls in the form of filamentary vorticity structures. Although the forcing of
the flow takes place at the edge of the domain, dipolar and single vortices
generated near the walls are observed to penetrate into the interior, thus effectively creating a 2D turbulent flow all over the domain - at least, when the
ratio $F = f/A\Omega_0$ is of the order unity or smaller. A dye-visualization of this

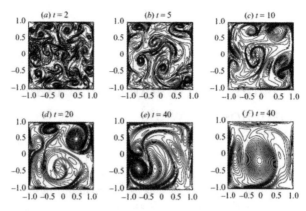

Fig. 1. Sequence of vorticity contour plots for decaying 2D turbulence on a square domain with no-slip boundaries, as simulated numerically. The flow was initialized by a slightly disturbed array of 10 x 10 vortices with alternating rotation sense, for Re = 2000, the Reynolds number Re being based on the initial total kinetic energy (taken from [5]).

process is shown in Fig. 2. For larger F-values, the wall-produced vortices fail to penetrate into the interior, and only the outer edge region of the domain is turbulent. The spectra measured and calculated in this combined experimen-

Fig. 2. Snapshots of the dye-visualized turbulent flow in a square tank rotating with a modulated rotation speed. The photographs were taken at $t/T = 0.175$ (a) and 0.325 (b), with $T = 2\pi/f = 200$s the modulation period. The dye was released at t = 0 in one of the corners, and $F = 0.52$ and $Re = 15,000$ (taken from [3]).

tal / numerical study are in agreement with earlier findings [1]. Fig. 3 shows the energy spectrum $E(k)$ calculated from experimental data for the case Re = 15,000 and $F = 1.57$ for two subsequent times in the flow evolution. For this purpose, data from a 2D fast Fourier transform of the kinetic energy were collapsed onto a 1D graph by computing the energy spectrum according to

$$E(k) = \int_{k_1}^{k_2} E(k_x, k_y) dk, \qquad (1)$$

with $k = (k_x^2 + k_y^2)^{1/2}$. To reduce the effect of the non-periodicity of the flow and hence the kinetic energy field, the area-averaged kinetic energy has been subtracted from the data, and a so-called Hanning window was applied before Fourier transforms of the kinetic energy data were computed. Obviously, the experimentally measured energy spectrum at large wave numbers behaves like $E(k) \sim k^{-5/3}$.

In a related line of approach, Molenaar [4] has investigated the behaviour

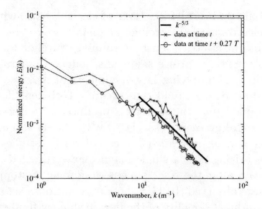

Fig. 3. Two measured energy spectra for the oscillating spin-up experiment in a rotating square tank with $Re = 15,000$ and $F = 1.57$ (taken from [3]).

of 2D turbulence on a square domain with no-slip walls for the case of random, but statistically-steady, uniform forcing. His numerical simulations have revealed that - depending on the forcing conditions - the forced flow may reach an 'organized' state of a single domain-filling cell. As in the decaying case, this organized flow cell is a result of the self-organizing tendency associated with the inverse energy cascade. Under certain forcing conditions, this central cell may be seriously eroded by wall-generated vorticity filaments, possibly leading to its complete destruction. After a while, a new organized state may be established, in the form of a new large vortical structure of the same or of opposite sign. As shown in Figure 4, the total angular momentum of the evolving flow then shows a remarkable flip-flop behaviour, with sign changes (see also [5, 6]). In order to investigate the specific role of the solid no-slip boundaries on the flow evolution, high-resolution numerical simulations were carried out in our group by Keetels [7] and Kramer [8], both focusing on vorticity transport and enstrophy production and on scalar transport properties.

In another line of approach, laboratory experiments have been performed

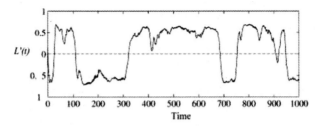

Fig. 4. Evolution of the normalized angular momentum $L'(t) = L(t)/L_{sb}(t)$, with Lsb the angular momentum of the fluid for the case that the actual turbulent motion with kinetic energy E(t) would have been converted into a pure rigid-body rotation (taken from [5]).

on quasi-2D turbulence in a shallow layer of electrolyte, in which the motion is generated by electromagnetic forcing: magnets placed underneath the fluid layer, with a uniform electrical current running through the fluid between two electrodes placed on opposite side walls, induce Lorentz forces that set the fluid in motion. By applying larger numbers of magnets, e.g. placed in a regular 10 x 10 array, one could thus create a turbulent flow. Although it is commonly assumed that the flow thus produced is quasi-2D, i.e. planar, although with vertical gradients associated with the no-slip condition at the tank bottom (see e.g. [9]), detailed Stereoscopic-PIV studies of elementary vortex structures arising near a single magnet have revealed that the flow is essentially 3D: even after the electromagnetic forcing is stopped, significant vertical motions are observed [10,11]. These observations cast doubt upon the assumption of two-dimensionality of the flow in shallow-fluid experiments, for decaying as well as forced flows.

The lecture aims at providing an overview of recent and ongoing work in our group concerning the behaviour of both forced and decaying quasi-2D flow in a no-slip confinement and will in particular highlight some of the aspects mentioned above.

References

1. H.J.H. Clercx and G.J.F. van Heijst - Energy spectra for decaying 2D turbulence in a bounded domain, Phys. Rev. Lett. **85**, 306-309 (2000).
2. H.J.H. Clercx, G.J.F. van Heijst, D. Molenaar and M.G. Wells - No-slip walls as vorticity sources in two-dimensional bounded turbulence, Dyn. Atmos. Oceans **40**, 3-21 (2005)
3. M.G. Wells, H.J.H. Clercx and G.J.F. van Heijst - Vortices in oscillating spin-up, J. Fluid Mech. **573**, 339-369 (2007)
4. D. Molenaar - Forced Navier-Stokes flows on a bounded two-dimensional domain, PhD thesis, Eindhoven University of Technology (2005)
5. D. Molenaar, II.J.II. Clercx and G.J.F. van Heijst - Angular momentum of forced 2D turbulence in a square no-slip domain, Physica D **196**, 329-340 (2004)

6. G.J.F. van Heijst, H.J.H. Clercx and D. Molenaar - The effects of solid bound-
    aries on confined two-dimensional turbulence. J. Fluid Mech. **554**, 411-431
    (2006)
7. G.H. Keetels - Statistical properties of 2D turbulence on a bounded domain.
    Proceedings ETC11, this issue (2007)
8. W. Kramer - The enstrophy cascade in bounded two-dimensional turbulence,
    Proceedings ETC11, this issue (2007)
9. J. Paret, D. Marteau, O. Paireau and P. Tabeling - Are flows electromagneti-
    cally forced in thin stratified layers two-dimensional? Phys. Fluids **9**, 3102-3104
    (1997)
10. R.A.D. Akkermans *et al.* - Stereoscopic-PIV study of a dipole in a shallow fluid
    layer. Proceedings ETC11, this issue (2007)
11. A.R. Cieslik *et al.* - Three-dimensional structures during a dipole-wall collision
    in shallow flows. Proceedings ETC11, this issue (2007)

Conformal invariance in two-dimensional turbulence

Guido Boffetta[1], Denis Bernard[2], Antonio Celani[3], and Gregory Falkovich[4]

[1] Dipartimento di Fisica Generale and INFN, Università di Torino,
   via Pietro Giuria 1, 10125 Torino, Italy boffetta@to.infn.it,
[2] Service de Physique Théorique de Saclay, CEA/CNRS,
   Orme des Merisiers, 91191 Gif-sur-Yvette Cedex, France
[3] CNRS, INLN, 1361 Route des Lucioles, 06560 Valbonne Sophia Antipolis, France
[4] Physics of Complex Systems, Weizmann Institute, Rehovot 76100, Israel

The existence of two cascades is the most distinguishing feature two-dimensional Navier Stokes equations. As predicted by Kraichnan [1], in a forced 2d flow an inverse cascade of kinetic energy to large scales and a direct cascade of enstrophy to small scales develop. The inverse cascade has been observed in many laboratory experiments [2] and in numerical simulations [3] with a statistical accuracy which has revealed the absence of intermittency corrections to the dimensional scaling [4]. In this contribution we discuss the recently discovered *conformal invariance* of the inverse cascade on the basis of very high resolution direct numerical simulations of forced 2D Navier-Stokes equations [5] and the extension to other two-dimensional turbulent models [6].

Conformal invariance is a powerful tool in equilibrium statistical mechanics at criticality [7]. In these systems, cluster boundaries were recently found to belong to a remarkable class of curves that can be mapped into Brownian walk (called Stochastic Loewner Equation or SLE curves) [8]. These curves are parameterized by the diffusivity κ of the associated Brownian motion which classify the different universality classes of critical phenomena. In this contribution we investigate the statistical properties of clusters of transported fields in two-dimensional turbulent systems, defined as the connected regions with the same sign of the field (for example, vorticity in the case of Navier-Stokes turbulence). As suggested by the example shown in Fig. 1, these are self-similar, fractal regions, suggesting the possibility to have conformal invariant boundaries. By a numerical implementation of the SLE we show that turbulent cluster boundaries are indeed conformal invariant curves statistically equivalent to boundaries in critical phenomena.

A large class of two-dimensional turbulent systems can be described by the equation for the transport of a scalar quantity $\theta(\mathbf{x}, t)$:

$$\partial_t \theta + \mathbf{v} \cdot \nabla \theta = \nu \nabla^2 \theta + f \tag{1}$$

in which the incompressible velocity $\mathbf{v}$ is determined by the stream function $\psi = \int d\mathbf{y}|\mathbf{x}-\mathbf{y}|^{\alpha-2}\theta(\mathbf{y},t)$ as $\mathbf{v} = \hat{\mathbf{z}} \times \nabla\psi$. α parameterizes among the different equations: $\alpha \to 2$ corresponds to Navier-Stokes (NS) equation (and θ becomes the vorticity), for $\alpha = 1$ (1) represents the Surface Quasi-Geostrophic model (SQG) [9], and for $\alpha = -2$ the Charney-Hasegawa-Mima model [10]. Dimensional arguments for the inverse cascade of energy predicts a scaling exponent for the scalar field $\delta_r\theta \sim r^h$ with $h = (2 - 2\alpha)/3$ which is indeed observed in direct numerical simulation [5, 6].

Fig. 1. Left: clusters (connected domains of positive vorticity) in a simulation of 2D turbulence in the inverse cascade. Negative vorticity regions are black. Right: Fractal dimensions as the slopes of the length-diameter dependencies in log-log coordinates for the cluster boundaries (red) and the outer boundaries (blue). Solid lines have the slopes predicted for SLE_6 curves, 7/4 and 4/3.

In the case of Navier-Stokes equations one has $h = -2/3$ therefore vorticity organizes on small scale structures, generating scale invariant objects on larger scales. Figure 1 shows a snapshot of positive vorticity clusters in an inverse cascade simulation at resolution 8192^2. The fractal dimension of the cluster boundaries is found to be close to $D = 7/4$ and the dimension of the external boundaries (i.e. without self-intersection) is close to $D^* = 4/3$. This numerical result suggests that vorticity clusters are statistically equivalent to cluster of critical percolation [7]. Therefore we conjecture that the boundary of vorticity cluster can be described, in the scaling limit, by SLE_κ curves with diffusivity $\kappa = 6$.

We have implemented a numerical procedure for evaluating the *driving function* associated to the vorticity boundaries. The outcome of the analysis, shown in Fig. 2, indicates that driving functions associated to vorticity boundaries have a diffusive behavior $\langle \xi(t)^2 \rangle \simeq \kappa t$ with diffusivity $\kappa = 6 \pm 0.3$ and thus zero-vorticity isolines are statistically indistinguishable from SLE_6 traces.

It is natural to address the issue of generality of this result in the realm of two-dimensional turbulent systems. We have therefore investigate the conformal invariance in the inverse cascade of another system of geophysical interest, the SQG model corresponding to $\alpha = 1$ in (1). In this case the scaling exponent for the scalar field θ is $h = 0$ which corresponds to iso-θ lines with fractal dimension $D = 3/2$ [6]. Taking into account the relation between fractal dimension and diffusivity for SLE curves, $D = 1 + \kappa/8$ [8], one expects for temperature isolines $\kappa = 4$. The direct verification of this hypothesis is shown in Fig. 2.

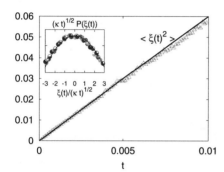 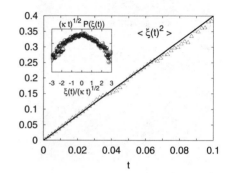

Fig. 2. Variance of the driving function as a function of time for NS (left) and SQG (right) turbulence. Lines represent the $\kappa = 6$ and $\kappa = 4$ diffusivities. In the inset it is shown the probability density function of $\xi(t)/\sqrt{\kappa t}$ at four different times compared with a Gaussian.

References

1. R.H. Kraichnan, Phys. Fluids **10**, 1417 (1967).
2. J. Paret and P. Tabeling, *Phys. Rev. Lett.*, **79**, 4162, (1997).
3. L. Smith and V. Yakhot, *Phys. Rev. Lett.*, **71**, 352, (1993).
4. G. Boffetta, A. Celani and M. Vergassola, *Phys. Rev. E*, **61**, R29 (2000).
5. D. Bernard, G. Boffetta, A. Celani, and G. Falkovich, *Nature Phys.*, **2**, 124 (2006)
6. D. Bernard, G. Boffetta, A. Celani, and G. Falkovich, *Phys. Rev. Lett.*, **98**, 024501 (2007)
7. J. Cardy, *Scaling and Renormalization in Statistical Physics* Cambridge Univ. Press, (1996).
8. J. Cardy, *Ann. Physics*, **318**, 81 (2005).
9. R. Salmon, *Geophysical Fluid Dynamics*, Oxford Univ. Press, (1998).
10. J. C. Charney, J. Atmos Sci. **28**, 1087 (1971); A. Hasegawa and K. Mima, Phys. Fluids **21**, 87 (1978).

Statistical properties of 2D turbulence on a bounded domain

G.H. Keetels[1], H.J.H. Clercx[1,2] and G.J.F van Heijst[1]

[1] J.M. Burgerscentre, Fluid Dynamics Laboratory, Eindhoven University of Technology, P.O. Box 513, 5600 MB Eindhoven, The Netherlands.
g.h.keetels@tue.nl
[2] Department of Applied Mathematics, University of Twente, PO Box 217, 7500 AE Enschede, The Netherlands

1 Introduction

The self-organization and dual cascade of two-dimensional unconfined, isotropic turbulence have been fascinating topics for almost 40 years. In this period the role of the boundary condition has mostly been disregarded, basically due to the lack of homogeneity and isotropy in the bounded case and the computational constraints in pursuing DNS of turbulence in complex geometries. It is however not straightforward to compare the computational results on a double periodic computational domain with experimental observations like, for example, forced quasi-two-dimensional turbulence in shallow fluid layers by Paret & Tabeling [7] and decaying turbulence in linearly stratified fluids by Maassen et al. [6]. Simulations of moderate Reynolds number flow (see ref [5]) using a standard Chebyshev pseudospectral scheme have revealed the crucial role of no-slip boundaries as sources of vorticity filaments, which significantly affect the evolution of forced and decaying 2D flows. In this study we use the volume-penalization method proposed by Angot et al. [1] with a Fourier pseudospectral scheme to achieve higher Reynolds numbers then previously possible with the standard Chebyshev pseudospectral scheme.

2 Results and discussion

We consider forced 2D turbulence on a square domain with no-slip boundaries. The forcing wavenumber is in the range $7 < k_f < 9$. The integral Reynolds number of the flow based on the total kinetic energy and half width of the domain in the steady state is approximately $Re = 25000$. It can be observed in Fig. 1a that the flow indeed evolves towards a statistically steady state in the total kinetic energy. Another remarkable observation is that a spontaneous production of angular momentum observed in previous computations

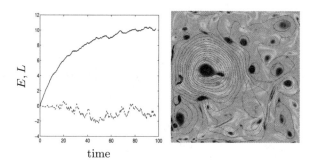

Fig. 1. *(a) Total kinetic energy (solid line) and angular momentum (dashed line) with respect to the center of the domain. (b) Snapshot of vorticity and stream function of a forced turbulence simulations with $7 < k_f < 9$. Resolution $N = 2048^2$ Fourier modes and penalization parameter $\epsilon = 10^{-8}$.*

[5] with a lower Reynolds number (around $Re = 3000$) is still present for the high Reynolds number achieved here. The stream function in Fig. 1b reveals that the kinetic energy is transferred towards larger scales. In most realizations one of the vortex structures is dominant. This vortex does not reach the size of the domain, which indicates that excitation of the lowest available modes is prevented due to viscous diffusion at the domain boundaries. This process can be seen as an arrest of the inverse energy cascade. For flows on double-periodic domains energy accumulation in the lowest-available modes or condensate formation [8] is usually prevented by appling additional friction terms. Here it is shown that the domain boundaries can act as a similar sink for the energy that arrives at the largest scales.

It is very well known that interactions (not necessary transfers) between different scales in two-dimensional turbulence on a double periodic domain can be very nonlocal, especially in the enstrophy inertial range. This is usually explained by strong elongation of small-scale vorticity filaments by the large scale energy containing flow structures. The dominant vortex in Fig 1b is present for many large-scale turnover times and moves only slowly through the domain. As a consequence, it can be expected that very long time averaging is required to reach isotropic statistics for the large-scale velocity separations. Fig. 2a demonstrates, on the other hand, that for the small-scale separations $k > k_f$ an averaging over about twenty large-scale turnover times is sufficient to reach isotropy. Thus the anisotropy of the large scales is not reflected to the small scales, eventhough strong nonlocal interactions drive the enstrophy inertial range.

To verify the presence of a direct enstrophy cascade the exact result of Bernard [3] in the enstrophy inertial range $\langle \Delta u \Delta \omega^2 \rangle = -2\eta r$ is used where η represents the enstrophy dissipation density, r the separation, u the velocity and ω

(a)   (b)

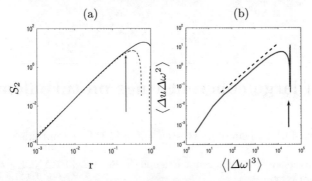

Fig. 2. *(a) Comparison of directly measured (solid) transverse second-order velocity structure function and transverse structure function computed with the longitudinal second-order structure function using the isotropic relation $S_2^{\perp}(r) = (1 + rd_r)S_2^{\parallel}(r)$. (b) Mixed third-order structure function $\langle \Delta u \Delta \omega^2 \rangle$ versus third-order vorticity structure function. Also a reference line with slope +1 (dashed) is shown. Velocity and vorticity separations are taken in the interior of the domain.*

the vorticity. In Fig. 2b the mixed third-order structure function demonstrates the presence of a direct enstrophy cascade in the center of the domain. Here, the concept of extended self-similarity of Benzi *et al.* [2] is applied relative to the third-order vorticity structure function. An important observation is that small-scale vorticity of the boundaries does not act as secondary injection scale as observed in decaying flows near the wall by Clercx & van Heijst [4]. This might be explained by a role-up of boundary layer vorticity into length-scales close to the forcing scale. These vortices move into the interior of the domain and feed a direct enstrophy cascade. To further analyze this process a similar study will be performed using a different forcing method proposed by Wells *et al.* [9]. By applying an oscillating background rotation they have shown that in every cycle vorticity produced at the no-slip boundaries moves into the interior and maintains the turbulence. The interesting aspect is that there are no volume forces present since all the vorticity is produced by the boundaries solely.

References

1. P. Angot *et al.*: Num. Math. **81**, (1999) pp. 497–520.
2. R. Benzi *et al.*: Phys. Rev. E **48**, (1993) pp. R29–R32.
3. D. Bernard: Phys. Rev. E **60**, (1999) pp 6184-6187.
4. H.J.H. Clercx and G.J.F van Heijst: Phys. Rev. Lett. **85**, (2000) pp. 306–309.
5. G.J.F. van Heijst *et al.*: J. Fluid Mech. **554**, (2006) pp. 411–431.
6. S.R. Maassen *et al.*: Phys. Fluids **14**, (2002) pp. 2150-2169.
7. J. Paret and P. Tabeling: Phys. Fluids **10**, (1998) pp. 3126–3136.
8. L.M. Smith and V. Yakhot: J. Fluid. Mech. **274**, (1994) pp. 115-138
9. M.G. Wells *et al.*: J. Fluid Mech. **573**, (2007) pp 339–369.

Effect of large coherent rings on turbulent field

J.F. Krawczynski,[1] L. Danaila,[1] B. Renou,[1] and P.E. Dimotakis[2]

[1] CORIA, University of Rouen, Saint Etienne du Rouvray, FRANCE -
   danaila@coria.fr
[2] California Institute of Technology, Pasadena, California, U.S.A. -
   dimotakis@caltech.edu

1 Introduction

In a partially stirred, forced-box-turbulence reactor (Fig. 1, left), fluid is injected through 16 (top/bottom) pairs of opposed jets that exits via porous top and bottom walls. A schematic of the flow is given on Fig. 1, right, that depicts trajectories of notional particles transported by the mean velocity field. This flow creates pairs of axisymmetric counter-current shear layers, in which large-scale (mean-field) vorticity essentially matches the large-scale strain rate. A simplified image of the mean flow is large-scale coherent rings rotating in alternate sense. The large-scale properties of the mean flow are quasi–2-D and lead to turbulence in local rotation at large scales. Azimuthal (toroidal) flow is also present along ring cores and the mean flow is better described in terms of large-scale spirals. We focus on the small-scale properties of the vorticity field that is injected at large scales, in pairs of coherent rings rotating in an alternating sense.

2 Results

The results discussed here derive from velocity PIV measurements in the jet planes. The (azimuthal) vorticity of the mean flow is represented on Fig. 2, left. This flow is characterized by large-scale sheared regions, in the interior of which the vorticity field is correlated. The ratio between the transport time in the reactor by the mean velocity field and the turbulence characteristic time is ~ 0.16, i.e., slow (quasi-frozen) turbulence is rapidly transported by the mean velocity field. The energy spectrum, $E(k)$, exhibits a $\sim k^{-3}$ region (not shown here) that is well-resolved in the PIV data and similar to that of turbulence in rotation (e.g., [1]). In the Kolmogorov-theory framework, two-point vorticity correlations decrease as $r^{-4/3}$ in the inertial range (e.g., [2]). In contrast, the second-order structure function of vorticity (any of the 3

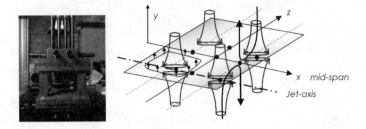

Fig. 1. Left: Experimental set-up. Right: Schematic representation of the flow.

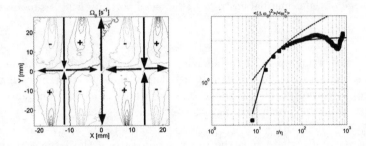

Fig. 2. Left: Large-scale vorticity Ω_θ $[s^{-1}]$. The signs correspond to the rotation sense. Right: Second-order (normalized) azimuthal vorticity structure functions ($\square$): $< (\Delta\omega_\theta)^2 > / < \omega_\theta^2 >$. Dashed line: $\sim \log(r)$. Solid line: $a + b/r$.

components) does not exhibit power-law behaviour [2]. In this flow, normalized second-order (azimuthal) vorticity structure functions are calculated as, $\langle (\Delta\omega_\theta)^2 \rangle (r) \equiv \langle (\omega_\theta(x+r) - \omega_\theta(x))^2 \rangle$ and depicted in Fig. 2 (right) as a function of r/η, with η the Kolmogorov scale (based on PIV estimates of the mean energy dissipation rate, with a maximum error of $\sim 10\%$). It is noteworthy that the azimuthal vorticity correlation length is large and equal to the scale over which the mean velocity field strongly varies in either the contraction (central) region, or the shear region (the rings). This behaviour is different from that usually encountered in 3-D turbulence, in which the vorticity correlation length is (maximum) 15 Kolmogorov scales. Figure 3 (left) reveals an instantaneous planar strain-rate field. The strong contraction region between two opposed jets is evident, with a large horizontal extent, reminiscent of pancakes. At the edge of these regions, fluid whose enstrophy has a strong azimuthal component is detrained (Fig. 3, right). Another high-enstrophy region is present in the central region between four pairs of opposed jets, where a secondary stagnation point is present. These structures are advected rapidly (to the exit), compared with their own life time, so vorticity stays correlated over relatively large scales.

Vorticity second-order structure functions do not exhibit a power-law behav-

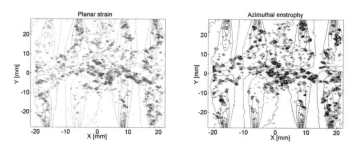

Fig. 3. Left: Instantaneous image of the planar strain-rate field. Right: Instantaneous image of the azimuthal enstrophy (the mean vorticity field is superposed as an aid).

iour in the inertial range. The behavior is $\sim \log(r)$ at intermediate scales. This was predicted by [3] for strictly 2-D turbulence. Further investigations of enstrophy variation over a range of separations is useful. The enstrophy transport equation (with no viscous effects) is, $\frac{D}{Dt}\langle(\Delta\omega_i)^2\rangle = \mathcal{F}(r)$, with $\mathcal{F}(r) = -2\frac{\partial}{\partial r_\alpha}\langle\omega_i\omega_\alpha^+ u_i^+ - \omega_i^+ \omega_\alpha u_i\rangle$. We approximate the space-time relation as $r\exp(St) = L$, where S is the mean strain rate of the large-scale velocity field in the rotation plane. Considering only the long-range vorticity in the forcing term, acting at scales larger than the ring size, leads to a logarithmic behaviour of the enstrophy structure function in the inertial range that corresponds to an energy spectrum $\sim k^{-3}$ [3]. Using the experimental information, we consider now the scale variation of the enstrophy forcing that behaves as $\mathcal{F} \simeq [1 + \log(r)]/r$, and after assuming isotropic relations between $\langle(\Delta\omega_i)^2\rangle$ and $\langle(\Delta\omega_\theta)^2\rangle$, we get $\langle(\Delta\omega_\theta)^2\rangle \propto a + b/r$, where a and b are constants, as confirmed by our experimental results over a significant range of scales (Fig. 2, right). In conclusion, considering the large-scale mean-vorticity effect over small-scale vorticity, one can predict the inertial-range behaviour of the second-order vorticity structure functions. This is important for understanding small-scale intermittency in 3-D turbulent flows and also provides evidence of nonlocal interactions in the enstrophy cascade.

This work was financially supported by ANR (Agence Nationale de la Recherche), under Grant 05-BLAN-0242-01.

References

1. Hossain M., 1994, Reduction of the dimensionality of turbulence due to a strong rotation, *Phys. Fluids*, Vol. 6, pp. 1077.
2. Antonia R.A., Ould–Rouis M., Zhu Y. and Anselmet F., 1998, Transport of turbulent vorticity increments, *Phys. Rev. E*, Vol. 57, pp. 5483.
3. Falkovich G. and Lebedev V., 1994, Nonlocal vorticity cascade in two dimensions, *Phys. Rev. E*, Vol. 49, pp. R1800.

Quasi-steady and unsteady Goertler vortices on concave wall: experiment and theory.

A. V. Boiko, A. V. Ivanov, Y. S. Kachanov, and D. A. Mischenko

Institute of Theoretical and Applied Mechanics of SB RAS,
630090, Novosibirsk, Russia. andi@itam.nsc.ru

1 Background and Goals of the Study

The Goertler instability is responsible for amplification of streamwise counter-rotating vortices, whose growth is able to lead to premature laminar-turbulent transition in boundary layers flows on concave walls [1,2]. The majority of previous experiments were devoted to the steady Goertler vortices, despite the unsteady ones are also observed in real transitional flows very often. Moreover, even for the steady Goertler vortices no good quantitative agreement between the experimental and theoretical linear-stability characteristics was obtained, especially for disturbance growth rates. These distinctions may be explained with a rather poor accuracy of experimental measurements at zero disturbance frequency; with a possible influence of nonlinearity, since the typical vortex amplitudes in experiments were typically about 10% [3]; and, at last, with the influence of near-field effects, i.e. by presence near the vortex source of a plenty of attenuating modes with identical spanwise wavenumbers [4].

The present work had the aim to overcome all these difficulties by means of: (i) tuning out from the exact zero frequency of Goertler vortices by means of working with quasi-steady perturbations rather than with steady ones, (ii) performing measurements at low disturbance amplitudes (hundredths of a percent) and (iii) minimization and careful estimation of the disturbance-source near-field by means of utilizing a special controlled disturbance source and performing special numerical computations for exact experimental conditions. Simultaneously, we had the aim to measure all linear-stability characteristics with respect to essentially unsteady Goertler vortices, to carry out linear theory calculations, and to perform a detailed direct comparison of the theoretical and experimental data.

2 Ranges of Parameters and Used Approaches

The hot-wire experiments were conducted in the low-turbulence wind tunnel T-324 of ITAM SB RAS in non-gradient 2D boundary layer developed over a concave wall with a radius of curvature 8.37 m. The Goertler number $G = (U_e\delta_1/\nu)(\delta_1/R)^{1/2}$ grew downstream inside the region of measurements from 10 to 17. The linear-stability characteristics were both measured and calculated for Goertler vortices with the spanwise wavelengths λ_z of 8, 12 and 24 mm $(\Lambda = (U_e\lambda_z/\nu)(\lambda_z/R)^{1/2} = 775, 274$ and 149. In the beginning of the measurement region the indicated values corresponded to non-dimensional spanwise wavenumbers $\beta\delta_1 = 0.354, 0.708$ and 1.063 and in the end of measurement region: $\beta\delta_1 = 0.496, 0.992$ and 1.488. The vortex frequencies 0.5, 2, 5, 8, 11, 14, 17 and 20 Hz were studied. The Goertler vortices of controlled frequencies and spanwise wavelengths were excited with a universal disturbance source of blowing-suction type. The theoretical part of the study included two approaches: (i) parallel approach based on the linear theory of Goertler and (ii) nonparallel approach based on parabolic linear stability equations. Both approaches were generalized to the case of non-stationary perturbations.

3 Most Important Results

The performed measurements and calculations have shown that at very low frequencies of excitation the perturbations formed in the boundary layer represent quasi-steady Goertler vortices of the first mode. The vortices are very long in this case and practically undistinguishable of classical steady Goertler vortices $(f=0)$. With the frequency growth, the vortices become shorter in streamwise direction and get inclined to the wall in (x,y)-plane forming more complex multilayer structure of partly overlapping vortices.

A very good quantitative agreement between all theoretical and experimental stability characteristics of the unstable vortices is found for all studied frequencies and spanwise wavenumbers, including the disturbance increments (see Fig. 1), the phase velocities (Fig. 2), and the eigenfunctions (Fig. 3). It is found (Figs. 1 and 2) that the linear non-parallel stability theory (LNST) provides better agreement with experiment than linear parallel stability theory (LPST). The calculations have shown that the neutral stability curve is subjected to only minor modifications as the Goertler vortex becomes slightly unsteady. Stability diagram for quasi-steady vortices $(f = 0, 5$ Hz $)$ shown in Fig. 4a, is practically undistinguishable from that for the stationary mode. However, in essentially non-stationary regimes the neutral curve is divided into isolated regions, the lowest of which (at G around 10 and less) corresponds to the first Goertler mode (see Fig. 4b). The position and shape of the neutral curve also correlate with the measurements: closed symbols correspond to the vortex growth, open ones - to decay, asterisks - to neutral behavior. A paradoxical result is observed for non-stationary Goertler vortices: in a certain range of wavelength parameter Λ the growth of Goertler number G leads to stabilization of the flow with respect to *unsteady* Goertler modes.

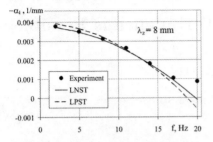

Fig. 1. Goertler vortex increments vs. frequency. Points - measurements, lines - calculations. $G=15$.

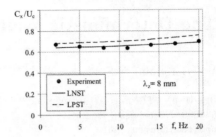

Fig. 2. Phase velocities of Goertler vortices vs. frequency. Points - measurements, lines - calculations. $G=15$.

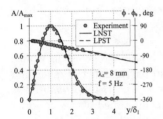

Fig. 3. Measured and calculated disturbance eigen functions.

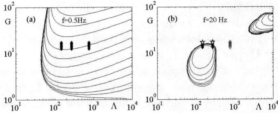

Fig. 4. Stability diagrams for quasi-steady (a) and unsteady (b) vortexes. Lines - calculations, points - experiment.

The work is supported by Russian Foundation for Basic Research (grant No 06-01-00519).

References

1. J.M. Floryan: On the Gortler instability of boundary layers. *Prog. Aerosp. Sci.*, **28**, 235–271 (1991)
2. W. S. Saric: Gortler vortices. *Ann. Rev. Fluid Mech.*, **26**, 379–409 (1994)
3. H. Bippes , H. Gortler: Dreidimensionale Storungen in der Grezchicht an einer konkaven Wand. *Acta Mech.*, **14**, 251–267 (1972)
4. J.P. Denier, S.O. Seddougui, P. Hall: On the receptivity problem for Gortler vortices: vortex motions induced by wall roughness. *Phil. Trans. Roy. Soc. London, Ser. A*, **335**, 51 (1991)

The Deterministic Wall Turbulence is Possible

V.I. Borodulin, Y.S. Kachanov, A.P. Roschektayev

Institute of Theoretical and Applied Mechanics of SB RAS,
630090, Novosibirsk, Russia
bo@itam.nsc.ru

Definitions. *Microscopic structure* of the flow is an instantaneous spatial velocity field of fluid particles. *Stochastic turbulence* is the turbulence, which microscopic structure *can not* be reproduced repeatedly (although the average structure – can be reproduced). *Deterministic (causal) turbulence* is the turbulence, which microscopic structure *can* be reproduced repeatedly from one realization to another at reproduction of the same initial conditions for perturbations incoming into the boundary layer.

Hypotheses. Let assume that there are some boundary layers in which: (*a*) laws of flow evolution are *fully deterministic* (in particular, the molecular chaos does not play any significant role in such flows) and (*b*) *stochastic* flow properties result entirely from *external perturbations* amplified by *instabilities*. We also assume that if all instabilities in these boundary layers are *convective* (what is general in 2D case) then the stochastic external perturbations may not have enough time to be amplified and the flow may remain deterministic even if it is turbulent. The experimental grounds for these hypotheses are summarized below.

Problem Setup and Experimental Technique. The experiments are devoted to a detailed investigation of "super-late" stages of laminar-turbulent transition in an adverse-pressure-gradient boundary layer. A post-transitional turbulent boundary layer was studied as well. Main goals of the experiments were the following: (*i*) by means of excitation of boundary-layer perturbations, which are random at small and moderate time scales but periodical at very large time scales, to produce a flow with a quick transition to turbulence and to study it at stages of the transition process completion and formation of a developed wall turbulent flow; (*ii*) to investigate a degree of the flow determinancy (repeatability) at stages where all main averaged flow characteristics correspond to the developed turbulent boundary layer; (*iii*) to examine characteristics of vortical structures, observed in post-transitional turbulent boundary layer and to compare the structures' properties with those found both at late stages of transition and in developed turbulent boundary layers; (*iv*) based on the obtained results to make a conclusion about the possibility of existence of the deterministic (model) wall turbulence and of its experimental realizability.

The experiments are performed in a self-similar boundary layer with Hartree parameter $\beta_H = -0.115$. The transition (up to its late non-linear stage) in this flow have been studied experimentally in [1]–[4].

The tests were carried out in a low-turbulence wind-tunnel T-324 of ITAM SB RAS. The measurements were done by means of a hot-wire anemometer in a broad spatial region starting with stages of small-amplitude TS waves, developing in the laminar boundary layer, and ending with the post-transitional turbulent flow. The universal disturbance source of instability waves [1] and the 'deterministic noise' method [5], [3] were used in these experiments to excite the flow. The transition was initiated by a mixture of a quasi-2D TS-wave and 3D broadband perturbations (a "noise" of TS-waves). The broadband perturbations were random within 20 periods of the fundamental TS-wave but then repeated periodically. During each repetition the flow passed the model several times. Thus, the "noise" was random from the viewpoint of the flow but it was deterministic (even periodic) from the viewpoint of data processing giving us the possibility to perform ensemble averaging of the hot-wire signals.

Main Results. It is found that the instability waves lead to transition with formation of characteristic vortical structures similar to those found in [4]. Farther downstream formation of post-transitional turbulent boundary layer is observed. At this stage ($560 < x < 650$ mm in present work) the mean velocity profiles, the profiles of rms velocity fluctuations, and the disturbance spectra correspond to those observed in developed turbulent boundary layers, see Fig. 1, 2. At the same time it turned out that the flow remains deterministic to a considerable degree. The coherence coefficient, defined as a ratio of the rms intensity of the ensemble-averaged (deterministic) and total (non-averaged) velocity fluctuations, remains very large (80 to 60%) even in the post-transitional flow. This circumstance gives us the possibility to obtain instantaneous velocity and vorticity fields in the (x, y, z, t)-space and to perform a computer-aided, quantitative visualization of the instantaneous flow structure in the post-transitional turbulent boundary layer. It is found that the structures resemble very much the vortical structures found earlier at later stages of transition [4], as well as the coherent vortical structures observed in developed turbulent boundary layers.

Conclusions. The experiments have shown that the produced post-transitional turbulent flow is mainly deterministic and can be simulated (modeled) experimentally. Such "model turbulence" represents one of possible microscopic realizations of the turbulent boundary layer (among infinite number of them), which possess all main average properties, characteristic of the wall turbulence. At the same time, the microscopic (instantaneous) structure of this turbulence is mainly deterministic and reproducible.

Thus, we can conclude that the deterministic wall turbulence does exist and can be used in turbulence research.

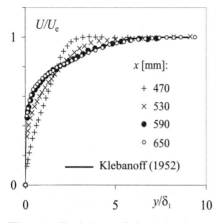 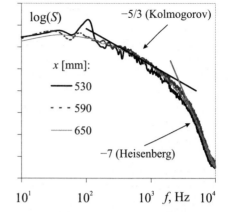

Fig. 1. Evolution of laminar mean-velocity profile to the turbulent one and comparison with classic experiment [6].

Fig. 2. Spectra of total velocity fluctuations and comparison with Kolmogorovs law at middle frequencies and Heisenbergs law at high frequencies.

References

1. V.I. Borodulin, Y.S. Kachanov, D.B. Koptsev: Experimental study of resonant interactions of instability waves in self-similar boundary layer with an adverse pressure gradient: I. Tuned resonances: *J. Turbulence* **3**(62) pp 1–38 (2002)
2. V.I. Borodulin, Y.S. Kachanov, D.B. Koptsev, A.P. Roschektayev: Experimental study of resonant interactions of instability waves in self-similar boundary layer with an adverse pressure gradient: II. Detuned resonances: *J. Turbulence* **3**(63) pp 1–22 (2002)
3. V.I. Borodulin, Y.S. Kachanov, D.B. Koptsev: Experimental study of resonant interactions of instability waves in self-similar boundary layer with an adverse pressure gradient: III. Broadband disturbances: *J. Turbulence* **3**(64) pp 1–19 (2002)
4. V.I. Borodulin, Y.S. Kachanov, A.P. Roschektayev: Turbulence production in an APG-boundary-layer transition induced by randomized perturbations: *J. Turbulence* **7**(8) pp 1–30 (2006)
5. F.N. Shaikh, M. Gaster: The non-linear evolution of modulated waves in a boundary layer: *J. Eng. Mathematics* **28** pp 55–71 (1994)
6. P.S. Klebanoff, Z.W. Diehl: Some features of artificially thickened fully developed turbulent boundary layers with zero pressure gradient. *NACA Report 1110* (1952)

The effect of free-stream turbulence on growth and breakdown of Tollmien-Schlichting waves

Philipp Schlatter[1], Rick de Lange[2], and Luca Brandt[1]

[1] Linné Flow Centre, KTH Mechanics, 100 44 Stockholm, Sweden
   pschlatt@mech.kth.se
[2] TUE Mechanical Engineering, 5600MB Eindhoven, The Netherlands

1 Introduction

In boundary layers with ambient free-stream turbulence (FST) intensities Tu of about 1% or less, laminar-turbulent transition is dominated by the classical scenario, *i.e.* exponential growth of Tollmien-Schlichting (TS) waves. As the TS-waves reach amplitudes of around 1% of the free-stream velocity U_∞, they experience resonant three-dimensional amplification and break down into a turbulent spot by the formation of Λ and hairpin vortices. However, in the presence of higher levels of FST, transition occurs more rapidly, bypassing the classical scenario. The new scenario, denoted bypass transition, is characterised by the formation of streamwise streaks of alternating high and low streamwise velocity, whose amplitude grows downstream and can reach values on the order of $0.1U_\infty$ prior to breakdown [1, 2]. It is yet unclear how the changeover from classical to bypass transition occurs, in particular, which role TS-waves play in the presence of FST. Boiko *et al.* [1] performed experiments with both FST and TS-waves and found that the TS growth rate within a streaky flow is reduced, as long as the TS-waves are at modest amplitudes. Similarly, the study by Cossu and Brandt [3] showed stabilisation of TS-waves by steady streaks in the Blasius boundary layer. In the presence of streaks, the unstable TS-waves evolve from 2D waves into spanwise modulated, less unstable waves, referred to as streaky TS-waves. In this contribution, the effect of FST on the behaviour of TS-waves is studied numerically, including modified growth rates, change in transition location, and transitional flow structures in comparison with both pure bypass and classical (*e.g.* K-type) transition.

2 Simulation Approach and Results

Numerical Method. The presented simulation results are obtained with a fully spectral method to solve the incompressible Navier-Stokes equations, similar to the recent DNS of bypass transition [2]. The fringe-region technique is

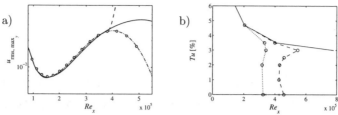

Fig. 1. a) Comparison of TS-waves obtained by LES and PSE showing the wall-normal maximum of u_{rms} for – – – TS wave with random 3D disturbances, ——— 2D nonlinearly-saturated TS-wave, —·— linear low-amplitude TS-wave (rescaled), $\circ$ PSE. b) Variation of the downstream position of the transition point $\gamma = 0.5$ as a function of the turbulence intensity Tu, ——— no TS-waves ($A_{TS} = 0\%$, pure bypass transition), – – – $A_{TS} = 0.76\%$, ········ $A_{TS} = 1.52\%$.

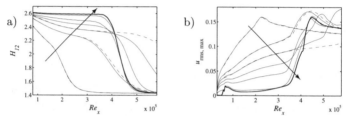

Fig. 2. Evolution of a) the shape factor H_{12} and b) the wall-normal maximum of the streamwise fluctuation u_{rms} for cases with FST and TS-waves, $A_{TS} = 0.76\%$. Arrows indicate decreasing turbulence intensity, $Tu = 4.7\%$, 3.5%, 3%, 2.5%, 2%, 1%, ——— only TS-wave forcing ($Tu = 0\%$), – – – only FST ($A_{TS} = 0\%$).

employed to allow simulations of a spatially developing boundary layer. The simulated region extends from $Re_x = 30000$ to $Re_x = 600000$. Large-eddy simulation (LES) is employed with the ADM-RT model [4], which was found to be well suited for spectral simulations of transitional flows.

The incoming FST is forced in the fringe region and is modelled by a superposition of modes of the continuous spectrum of the Orr-Sommerfeld/Squire operators, allowing the specification of both Tu and an integral length scale [2]. The TS-waves are introduced at $Re_x = 60000$ by volume forcing with $F = 120$. Small-amplitude steady, spanwise random noise is added to trigger K-type transition after branch II ($Re_x = 387000$). Figure 1a) displays the growth of the wall-normal maximum of the streamwise velocity fluctuations compared to results from the parabolic stability equations (PSE). Good agreement is obtained for the present LES.

Results. In figure 1b) the variation of the transition location (defined as the intermittency $\gamma = 0.5$) is shown for different turbulence intensities Tu and TS amplitudes A_{TS}. It is evident that the presence of TS-waves always leads to transition promotion, however this effect is weak for higher Tu. On the other hand, the streaks formed by the lower levels of FST can actually lead to transition delay compared to the case without FST, *e.g.* for $Tu = 3\%$ and

a)
$Tu = 4.7\%$

b)
$Tu = 3.0\%$

c)
$Tu = 2.0\%$

d)
$Tu = 0.0\%$

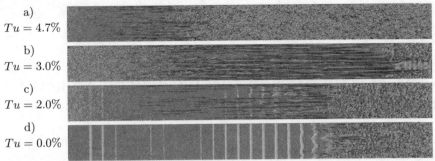

Fig. 3. Top view of the three-dimensional flow structures for $A_{\mathrm{TS}} = 0.76\%$ and the given Tu. Light isocontours represent the λ_2 vortex-identification criterion, darker colours are positive and negative streamwise disturbance velocity, respectively (in a) slightly higher contour levels than other cases). Flow from left to right.

$A_{\mathrm{TS}} = 0.76\%$ transition is delayed by about 20%. The effect of the FST on the boundary layer is illustrated in figure 2. The reduction of the shape factor H_{12} clearly shows the stabilising effect of the streaks [3], whereas the initial streak growth is mainly unaffected by the TS-waves (Fig. 2b). Instantaneous flow visualisations are presented in Fig. 3 clearly showing the appearance of streaks and TS-waves in the boundary layer. For the two highest Tu no TS-waves can be detected in the data, Figs. 3a) and b), whereas for $Tu = 2\%$ a distorted signature can be observed after branch I. The breakdown for that case appears to be a mixture between the roll-up of Λ vortices and intermittent turbulent spots due to the presence of the streaks.

3 Conclusions

The effect of ambient free-stream turbulence on the behaviour of TS-waves in a transitional boundary layer has been investigated by means of large-eddy simulation. The experimental results of Boiko *et al.* [1] showing a stabilisation of the TS-waves in the presence of FST has been verified. It is further shown that a higher turbulence intensity leads to transition completely dominated by the FST, whereas for lower Tu an interaction between TS-waves and boundary-layer streaks can be observed.

References

1. A. V. Boiko, K. J. A. Westin, B. G. B. Klingmann, V. V. Kozlov, and P. H. Alfredsson: *J. Fluid Mech.*, 281:219–245 (1994)
2. L. Brandt, P. Schlatter, and D. S. Henningson: *J. Fluid Mech.*, 517:167–198 (2004)
3. C. Cossu and L. Brandt: *Phys. Fluids*, 14(8):L57–L60 (2002)
4. P. Schlatter, S. Stolz, and L. Kleiser: *Int. J. Heat Fluid Flow*, 25:549–558 (2004)

Vortical structures in turbulent plane Couette flow

Anders Holstad[1], Helge I. Andersson[2] and Bjørnar Pettersen[1]

[1] Department of Marine Technology, Norwegian University of Science and Technology, N-7491 Trondheim, Norway. anders.holstad@ntnu.no bjornar.pettersen@ntnu.no

[2] Department of Energy and Process Engineering, Norwegian University of Science and Technology, N-7491 Trondheim, Norway. helge.i.andersson@ntnu.no

1 Introduction

Coherent structures in a turbulent plane Couette flow at low Reynolds number are studied by extracting vortical structures from data from a direct numerical simulation. The structures are extracted using the λ_2 vortex definition proposed by Jeong & Hussain [1] in which a vortex core is defined as a connected region of negative λ_2. The detection and ensemble-averaging of the structures are based on the scheme devised by Jeong et al. [2]. The resulting dominant near-wall structure of the plane Couette flow is then compared to that of the plane Poiseuille flow. The core region is also studied with the aim of identifying large-scale structures previously reported for the Couette flow, e.g. Komminaho et al. [3], Kitoh et al. [4] and Tsukahara et al. [5]. Information resulting from the λ_2 vortex definition will also be compared to information resulting from POD (proper orthogonal decomposition) analysis of the same flow field.

The flow considered is fully developed turbulent plane Couette flow at Reynolds number $Re = 1300$ ($Re = U_w h/\nu$, where U_w is half the velocity difference between the walls, h is half the channel height and ν is the kinematic viscosity). The computational domain has a streamwise length $L_x = 4\pi h$, spanwise length $L_y = \frac{4}{3}\pi h$ and wall-normal length $L_z = 2h$. Periodic boundary conditions are used in the streamwise (x) and spanwise (y) directions, and no-slip boundary conditions are used at the walls. The computational mesh is inhomogeneous in the wall-normal (z) direction.

2 Results and discussion

Figure 1 shows the distribution of dimensionless mean and root-mean-square values of $-\lambda_2$ for both the Couette flow (a) and the Poiseuille flow (b). The peak in the root-mean-square value of $-\lambda_2$ is located at $z^+ \approx 20$ for both flows, suggesting that the majority of the vortices will be found in the buffer region. This is in agreement with the findings of Jeong et al. [2] and Solbakken & Andersson [6]. However, in the Couette flow, the root-mean-square value exhibits only a modest reduction beyond its peak level and settles at a core level of about 50% of its maximum level. This is in contrast to the Poiseuille flow, in which the root-mean-square value tends to vanish at the center. This supports earlier findings, e.g. Komminaho et al. [3] and Kitoh et al. [4], indicating that outer-scale vortices play a more prominent role in the Couette flow than in the Poiseuille flow.

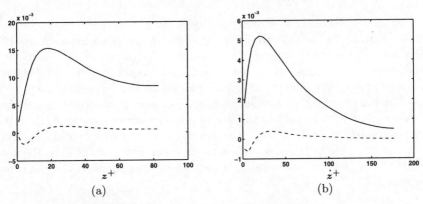

Fig. 1. Distribution of dimensionless mean (- - -) and root-mean-square (——) values of $-\lambda_2$. (a) Plane Couette flow at $Re_\tau = u_\tau h/\nu = 82$, where $u_\tau = (\tau_w/\rho)^{1/2}$ is the wall friction velocity. τ_w is the wall shear stress and ρ is the density of the fluid. (b) Plane Poiseuille flow at $Re_\tau = 180$.

POD analysis of the same flow field, Holstad et al. [7], confirms the prominent role played by outer-scale vortices in the Couette flow. Figure 2 (right) shows a vector plot of the velocities in the yz-plane of the second POD mode for the Couette flow. A pair of counter-rotating streamwise vortices filling the entire channel is easily seen. The wall-normal variation of the mode is shown in Figure 2 (left) indicating that the streamwise component carries appreciably more energy than the spanwise and wall-normal components.

An isosurface plot of λ_2 taken from an instantaneous Couette flow field (not shown here) shows that the dominant near-wall structures consist of elongated streamwise vortices such as those described by Robinson [8]. However, closer examination of these instantaneous λ_2-fields indicates that the vortical structures found in the buffer layer in the Couette flow extends further into the core region as compared to similar structures found in the Poiseuille flow. This observation can be explained by the consistently non-zero mean velocity gradient dU/dz in the Couette flow, causing increased interaction between the wall layer and the core region.

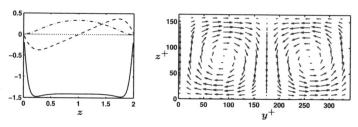

Fig. 2. The second POD mode with quantum numbers $(m, n, q) = (0, 1, 1)$ for the Couette flow. Left: Wall-normal variation of the streamwise (——), spanwise (- - -) and wall-normal (-·-) components. Right: Velocities in the yz-plane.

Ensemble-averaged λ_2-structures may be used as a generic model for near-wall vortices. For the Couette flow, a preliminary examination indicates that the dominant vortical structure in the region $10 < z^+ < 40$ is inclined $10°$ in the vertical (x, z)-plane and tilted $5°$ in the horizontal (x, y)-plane. This is in agreement with the findings of Jeong et al. [2] for the Poiseuille flow.

Work is underway to further investigate near-wall structures in the plane Couette flow with respect to the length and number of structures, as well as configuration, inclination and tilt angles of the ensemble-averaged structures. This analysis will be based on a Couette flow simulation at the same Reynolds number, but with a considerably better resolution and larger computational domain.

This work has received support from The Research Council of Norway (Programme for Supercomputing) through a grant of computing time.

References

1. J. Jeong & F. Hussain: Journal of Fluid Mechanics **285**, 69–94 (1995)
2. J. Jeong, F. Hussain, W. Schoppa & J. Kim: Journal of Fluid Mechanics **332**, 185–214 (1997)
3. J. Komminaho, A. Lundbladh & A. Johansson: Journal of Fluid Mechanics **320**, 259–285 (1996)
4. O. Kitoh, K. Nakabyashi & F. Nishimura: Journal of Fluid Mechanics **539**, 199–227 (2005)
5. T. Tsukahara, H. Kawamura & K. Shingai: Journal of Turbulence **7**, 1–16 (2006)
6. S. Solbakken & H.I. Andersson: Fluid Dynamics Research **37**, 203–230 (2005)
7. A. Holstad, P.S. Johansson, H.I. Andersson & B. Pettersen: On the influence of domain size on POD modes in turbulent plane Couette flow. In: *Direct and Large-Eddy Simulation VI*, ed by E. Lamballais, R. Friedrich, B.J. Geurts & O. Métais (Springer 2006) pp. 763–770
8. S. Robinson: Annual Review of Fluid Mechanics **23**, 601–639 (1991)

Wall Effects in Turbulent Rayleigh-Bénard Convection in a Long Rectangular Cell

M. Kaczorowski, A. Shishkin, and C. Wagner

German Aerospace Center (DLR) – Institute of Aerodynamics and Flow
Technology, Bunsenstr. 10, D-37073 Göttingen, Germany
e-mail: Matthias.Kaczorowski@dlr.de

1 Introduction

Numerical investigations of Rayleigh-Bénard convection are typically conducted in cylindrical containers or infinitely extended fluid layers. However, the shape of the container has a non-negligable influence on the properties of the flow. Therefore direct numerical simulations of turbulent Rayleigh-Bénard convection have been carried out in a rectangular geometry with two different configurations employing either solid walls or periodic boundaries in longitudinal direction.

2 Computational Setup

The incompressible three dimensional Navier-Stokes equations are solved in non-dimensional form using the finite volume method presented in [3]. Density variations of the fluid are accounted for by the Boussinesq approximation. The simulations are conducted in a rectangular cell of aspect ratios $W/H = 1$ and $L/H = 5$ of the cross section and the longitudinal direction respectively. The horizontal walls of the convection cell are isothermal with non-dimensional temperatures $\theta_b = +0.5$ and $\theta_t = -0.5$ at the lower hot and the upper cold wall, respectively, while the vertical walls are assumed to be adiabatic. Two different types of boundary conditions are considered, so that Rayleigh-Bénard convection is simulated with periodic boundary conditions as well as adiabatic solid walls in longitudinal direction.

3 Results

Figure 1 illustrates the impact of the boundary condition in longitudinal direction. While the flow field in the periodic cell is in steady state and 2-D,

3-D behaviour is observed in the closed box geometry where several plumes can be identified.

However, a similar scaling of Nusselt number is observed for the two different setups as can be seen from Fig. 2. For the 3-D regimes an effective exponent $\beta \approx 0.284$ is obtained for the scaling law $Nu \sim Ra^{\beta}$ which is in good agreement with Grossmann and Lohse's theoretical prediction [1] $\beta \approx 0.29$ for $1 \lesssim Pr \lesssim 7$.

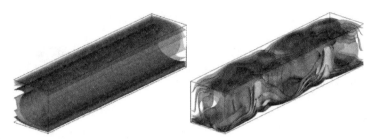

Fig. 1. Isotherms of the instantaneous flow field with periodic boundaries (left) and adiabatic walls (right) in longitudinal direction for $Ra = 3.5 \times 10^5$.

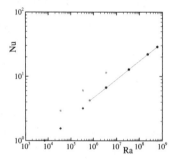 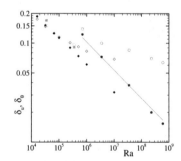

Fig. 2. Nusselt number Nu as a function of Rayleigh number Ra for Rayleigh-Bénard convection in a closed (■) and a periodic (●) rectangular cell; $Nu \sim Ra^{0.284}$ (—).

Fig. 3. Viscous (open) and thermal (closed symbols) boundary layer thicknesses of the closed (□) and the periodic (○) rectangular cell and an infinitely extended fluid layer [2] (◇); $\delta_\theta \sim Ra^{-0.284}$ (—).

Thermal and kinetic boundary layer thicknesses are plotted in Fig. 3 for an infinitely extended fluid layer [2], the periodic cell and the closed rectangular box. It is observed that all boundary layers show similar tendencies over the considered range of Rayleigh numbers. However, the thicknesses of the periodic rectangular cell differs significantly from those of the other setups.

Fig. 4 illustrates the impact of the horizontal walls on the distribution of energy across the range of scales. With increasing wall distance the equilib-

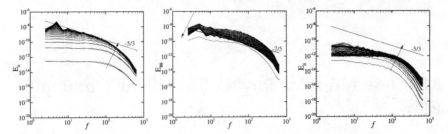

Fig. 4. Energy spectra of the u-velocity components (left) and temperature (right) for $Ra = 8.6 \times 10^8$ extracted from the centre of the periodic cell, $z/W = 0.5$ at various positions between the vertical walls. The right diagram shows the energy spectra of the w-velocity components close to the vertical walls. Arrows indicate increasing wall distance.

rium ranges of the velocity and temperature field are adjusted to follow the Kolmogorov and the Bolgiano law respectively. It can be noticed that all spectra have smaller exponents in the vicinity of the wall. However, the amount of energy held within the respective scales is decreasing for the temperature spectra, whereas it is increasing for the velocity spectra.

4 Conclusions

The comparison of the two rectangular cells shows that the periodic geometry delays the onset of convection, and hence the transition to a three-dimensional flow field. This is reflected by significantly thicker boundary layers and smaller heat transfer between top and bottom walls.

Analysis of the energy spectra at various wall distances midway between the vertical walls indicate that at small distances from the horizontal walls small scale fluctuations contain a large amount of energy, which is transferred into the large scales of the velocity field as the distance increases.

References

1. S. Grossmann and D. Lohse. Fluctuations in turbulent Rayleigh-Bénard convection: The role of plumes. *Phys. Fluids*, 16(12):4462 – 4472, Dezember 2004.
2. T. Hartlep, A. Tilgner, and F. H. Busse. Transition to turbulent convection in a fluid layer heated from below at moderate aspect ratio. *J. Fluid Mech.*, 554:309–322, 2005.
3. O. Shishkina and C. Wagner. A fourth order finite volume scheme for turbulent flow simulations in cylindrical domains. *Computers and Fluids*, 36(2):484–497, 2007.

Master-Mode Set for 3D Turbulent Channel Flow

Sergei I. Chernyshenko and Maxim E. Bondarenko

University of Southampton, Highfield, Southampton, SO17 1BJ, UK
chernysh@soton.ac.uk, M.Bondarenko@soton.ac.uk

Abstract. Master-mode set of the turbulent channel flow is calculated and stored in a database available online. The master-mode set is a subset of the modes used in direct numerical simulation. While being much smaller than the set of modes needed for fully resolved DNS the master-mode set contains full information about the developed turbulent flow. Master-mode set may be used directly as an approximate representation of the flow, while the remaining modes can be recovered with an additional calculation. The online database allows a remote user to upload a Fortran code for data analysis, compile and run it on the server with a direct access to the database.

The cost of direct numerical simulation (DNS) of a turbulent flow is high. For example, the DNS of a plane channel flow for $Re_\tau = 2003$ [1] required about 6×10^6 processor-hours. Rather than repeating such a calculation it is tempting to store the data for future analysis. However, a database of the full time history of a turbulent flow would be very large. For example, the above-mentioned DNS produced about 25 TB of raw data. DNS databases often contain only a limited number of snapshots of the flow field complemented with files with various statistics. Such a database cannot replace a full calculation, since there are always features of interest obtainable from DNS but not from an incomplete database. It is desirable to develop an approach for creating relatively small DNS databases nonetheless containing, may be in a nontrivial way, the full time history of the flow.

To reduce the size of the database one can try to store only a limited number of modes. A mode is a product of a time-dependent coefficient and a basis function of spatial coordinates. In DNS the solution is often represented as a sum of such terms. Proper orthogonal decomposition can reduce the number of modes required; however, the simplest option is to store the modes that were used in the DNS calculation. In any case, if the number of stored modes is less than the number of modes used in DNS, the information contained in the stored data can be incomplete.

Storing only the modes which form a master-mode set solves the problem. Master-mode set provides a full description of a developed turbulent flow while being much smaller than the set of modes needed for resolved DNS. The present work consists in calculating the master-mode set in 3D turbulent channel flow, establishing the properties of this set, and developing a tool for online access to the database of master modes.

Master-mode set contains full information about the flow in the following sense. If the sufficiently long time history of a master-mode set is known then the remaining modes can be restored by time-marching a DNS calculation with an arbitrary initial condition and replacing the amplitudes of the master-mode set of modes with their stored values at each time step. As time progresses, the amplitudes of the rest of the modes will converge to the values they had in the solution having the master-mode set that is stored. More formal definition follows. In DNS, at i-th time step the solution is represented as $\mathbf{u}_i(\mathbf{x}) = \sum_{n=1}^{S} \hat{u}_{in} \phi_n(\mathbf{x})$, where $\phi_n(\mathbf{x})$ is a given set of functions. Usually, this set of functions forms a full basis, as it is typical for Galerkin methods. In the particular numerical method used in the present work $\phi_n(\mathbf{x})$ is a product of the Chebyshev polynomial in the wall-normal direction and a Fourier mode in wall-parallel directions. Time marching is performed using the formula

$$\hat{\mathbf{u}}_{i+1,n} = D_n(\mathbf{u}_{i,1}, \mathbf{u}_{i,2}, \ldots, \mathbf{u}_{i,S}), \qquad n = 1, \ldots, S, \tag{1}$$

where functions D_n describe the particular numerical method. (We consider the case when the boundary conditions and body forces are independent of time so that functions D_n are independent of i.) Let $\hat{\mathbf{u}}_{in}, i = 1, 2, \ldots$ be a numerical solution so that it satisfies (1). For a given set $M = \{m_1, m_2, \ldots, m_K\}$ of the mode numbers consider a sequence $\mathbf{v}_i(\mathbf{x}) = \sum_{n=1}^{S} \hat{\mathbf{v}}_{i,n} \phi_n(\mathbf{x})$ such that $\hat{\mathbf{v}}_{i,m_j} = \hat{\mathbf{u}}_{i,m_j}$ for $j = 1, \ldots, K$, while for the rest of the modes $\hat{\mathbf{v}}_{i+1,n} = D_n(\mathbf{v}_{i,1}, \mathbf{v}_{i,2}, \ldots, \mathbf{v}_{i,K}), \qquad n = 1, \ldots, S$.

If $\|\mathbf{v}_i(\mathbf{x}) - \mathbf{u}_i(\mathbf{x})\| \to 0$ as $i \to \infty$ for any $\mathbf{v}_1(\mathbf{x})$ then the set M is a master-mode set of the numerical solution $\mathbf{u}_i(\mathbf{x})$.

In the more rigorous but narrower scope of 2D turbulence Olsen and Titi [2] calculated a master-mode set by running two DNS codes in parallel with different initial conditions, and replacing the amplitudes of a set of modes in one (slave) code with their values in the other (master) code at each time step. When this set is actually a master-mode set the solution of the slave code converges to the solution of the master code. So far extensions to 3D turbulence have been discussed only theoretically. In the present work the same technique is applied for 3D turbulent channel flow. Figure 1 shows that 2 per cent of the modes is sufficient for a very fast convergence. Comparisons demonstrated that the master-mode set alone gives a reasonable approximation of the velocity field. For two-point correlations recovering the full flow may be needed. A master-mode database can be used, for example, for conducting a search for rare events. A database search can be repeated with a criterion adjusted by the results of the previous search; without a database repeating the full

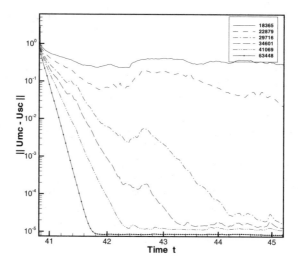

Fig. 1. Difference bewtween the slave and the master solutions as a function of time for different number of modes passed from master to slave code; $Re_\tau = 360$, box size is $6 \times 2 \times 3$, $128 \times 128 \times 160$ modes (longitudinal×wall-normal×transversal) were used.

DNS calculations for each search would be needed. Once the event is found the full set of modes at the time of the event can be restored.

The results obtained indicate that a master-mode database can resolve the contradiction between the advantages of a smaller database and the benefits of storing a full flow history. The first master-mode database (channel flow, $Re_\tau = 360$, box size $6 \times 2 \times 3$, with $128 \times 128 \times 160$ modes (longitudinal×wall-normal×transversal)) is available at http://www.dnsdata.afm.ses.soton.ac.uk as an experimental service. The user can upload a code and compile and run it on the server with a direct access to the database. Calculations of the master-mode set for a higher Re and a larger box are being performed; the results will be included into the database at a later time.

References

1. S. Hoyas and J. Jiménez: Scaling of the velocity fluctuations in turbulent channels up to Re_τ=2003. Physics of Fluids, **18**, 011702 (2006)
2. E. Olson and E.S. Titi: Determining modes for continuous data assimilation in 2-D turbulence. Journal of Statistical Physics, **113**, 799–840 (2003)

Dilute polymers in an oscillating grid turbulent flow

A. Liberzon[1,2], U. Reiter[2,3], M. Holzner[2], M. Guala[2,4], and W.Kinzelbach[2]

[1] Dept. Fluid Mechanics & Heat Transfer, Tel Aviv University,
    `alexlib@eng.tau.ac.il`
[2] Institute of Environmental Engineering, ETH Zurich, Switzerland,
    `holzner,kinzelbach@ifu.ethz.ch`
[3] Laboratory for Energy Systems Analysis, PSI, Switzerland,
    `ulrich.reiter@psi.ch`
[4] Swiss Federal Institute for Snow and Avalanche Research, Davos, Switzerland,
    `guala@slf.ch`

1 Introduction

The physical mechanisms underlying the phenomenon of turbulent flows modified by dilute solutions of polymers, discovered by Toms [10], remain poorly understood [4]. It is desirable to devise simple experiments in which the effects at large and small scales are easily separated and the effect associated with the boundaries (i.e. mechanism of turbulence agitation) is isolated from the effect in the turbulent bulk. Here we study the oscillating grid-induced turbulent flow with dilute polymers, in which focus on the *turbulent entrainment*, i.e. the propagation of the turbulent front into the quiescent fluid. This flow is characterized by two distinct scales (e.g. [9, 3, 12, 5]): it is driven by the large scale motion, but at the interface 'the work' is done by the small scales, e.g. [2]. We demonstrate in the following that *entrainment velocity* u_e in dilute polymer solutions is *higher* and the *entrainment rate*, u_e/u_{rms}, is also *larger*.

2 Experimental methods, results and discussion

We used the setup shown schematically in Fig. 1a as a part of the study of the entrainment interface, utilizing particle image velocimetry (PIV) [5], direct numerical simulations and three-dimensional particle tracking velocimetry (3D-PTV), [6, 7]. A description of the experimental methods is given in [5] and omitted here for the sake of brevity. We only recapitulate a short description of the experimental apparatus. The glass tank ($20 \times 20 \times 30$ cm) filled with water or aqueous dilute poly(ethylene oxide) solutions of Polyox WSR 301, $MW = 4 \times 10^6$ is fitted with a vertically oscillating grid (a fine woven screen of circular bars of $d = 1$ mm, mesh-size is $M = 4$ mm, $f = 6$ Hz. When the grid starts to oscillate, each grid bar generates regions of vorticity [12], that, after a short initial stage, grow beyond the mesh size M, and lead to a vortical, turbulent flow, propagating away from the grid [9, 3, 12]. The distance of the turbulent/non-turbulent interface from the grid is known to grow according to the power law $H \propto (Kt)^{1/2}$. In our apparatus, the front propagates according to $t^{1/2}$ for various experimental conditions [5, 6].

Fig. 1b illustrates the different propagation of turbulence (velocity field - arrows, vorticity component ω_z - contours) in flows of water (left) and 25 ppm dilute polymer solution (right). The thick dashed lines exemplifies the location of the interface, $H(t)$, identified by the threshold of vorticity [5]. The propagation of the turbulent front in polymer solution is *faster* during

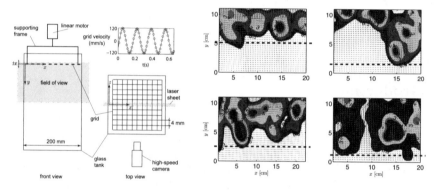

Fig. 1. a) Schematic view of the oscillating grid-stirred tank experimental setup. b) Vector fields and contours of the vorticity component in the water (left panel) and dilute polymer solution (right panel) at the initial stage (top panel) and later stage (bottom panel) of the turbulent front propagation. All the experimental parameters are identical.

the first stage (top panel), but not so at later times (bottom panel). The conditions of the experiment (i.e. the grid, the stroke and the frequency of oscillations) are the same for both cases. In order to understand differences, we varied two parameters of the experiment: a) *concentration* of the polymer solution increased to 50 wppm, and b) *stroke* of the grid for the concentration of 25 wppm reduced to $S = 6$ mm. The quantitative results are shown in the left panel of Fig. 2. The right panel of Fig. 2 shows the entrainment rate, defined as the ratio between the entrainment velocity, $u_e = dH/dt$, and local r.m.s. velocity, u_{rms}. The slopes of the plotted curves were estimated by regression analysis (as in [5]), using $H = k^{\frac{1}{2}}t^n$. For the experiments in water we obtain $n \sim 0.5$ (in agreement with [3, 5, 9, 12]) and a steeper slope $n \sim 0.8$ for the dilute polymer solution flows. Within the limits of the experimental uncertainty no notable change of the slope was observed when increasing the polymer concentration. As expected, the flow agitated by the *smaller stroke* yields *the same slope*, but a smaller value of the grid action, K. This confirms that we can in fact isolate the effect of polymers on the forcing from the effects in the bulk of the flow. Our results indicate that the propagation of the turbulent front in the dilute polymer solution flow is faster than the propagation speed in the water flow, presumably due to the effects of polymers *at the interface* and *in the bulk* of turbulent flow.

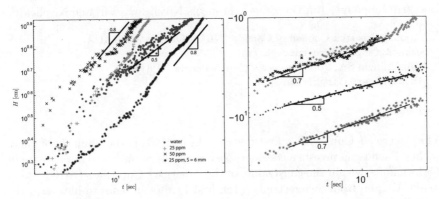

Fig. 2. Left: Distance of the front, $H(t)$. Right: entrainment rate, u_e/u_{rms}

References

1. O. Cadot, D. Bonn, S. Douady: Phys. Fluids **10**, 426 (1998)
2. S. Corrsin and A.L. Kistler: *The free-stream boundaries of turbulent flows* (NACA TR-1244 1954) pp 1033-1064
3. S.C. Dickinson and R.R. Long: Phys. Fluids **21** (10), 1698-1701 (1978)
4. A. Gyr, H.-W. Bewersdorff: *Drag reduction of turbulent flows by additives*, (Kluwer 1995)
5. M. Holzner, A. Liberzon, M. Guala, A. Tsinober, W. Kinzelbach: Exp. Fluids **41** 711-719 (2006)
6. M. Holzner, A. Liberzon, M. Guala, A. Tsinober, K. Hoyer, W. Kinzelbach: An experimental study on the propagation of a turbulent front generated by an oscillating grid In: *Proc. of THMT5*, ed by K. Hanjalić, Y. Nagano and S. Jakirlić(Durbrovnik, Croatia, 2006)
7. M. Holzner, A. Liberzon, N. Nikitin, A. Tsinober and W. Kinzelbach: Phys. Fluids submitted (2007)
8. A. Liberzon, M. Guala, B. Lüthi, and W. Kinzelbach A. Tsinober: Phys. Fluids **17**, 031707 (2005)
9. R.R. Long: Phys. Fluids **21** (10), 1887-1888 (1978)
10. B.A. Toms: Some observations on the flow of linear polymer solutions through straight tubes at large Reynolds numbers. In: *Proceedings of the 1st International Congress on Rheology*, vol II, (North-Holland, Amsterdam 1949) pp 135
11. M.P. Tulin, J. Wu: Phys. Fluids **20**(10), S109–S111 (1977)
12. S.I. Voropayev and H.J.S. Fernando: Phys. Fluids **8** (9), 2435-2440 (1996)

Transition Detection and Turbulence Measurements on Alinghi Yacht SUI-64 at Sea

Bernard Tanguay[1], Jim Bungener[2], François Nivelleau[2], Virginie Nivelleau[2]

[1] National Research Council of Canada (NRC), Ottawa, CANADA
bernard.tanguay@nrc-cnrc.gc.ga
[2] Team Alinghi SA, Port America's Cup, Valencia, SPAIN
jim.bungener@alinghi.com

Introduction

The America's Cup, with a first match held in 1851, is the pinnacle of yacht racing. Each team invests a considerable amount of resources to improve yacht performance, within the constraints of the America's Cup Class (ACC) Rules. Team Alinghi, from Switzerland, is the first European team to have won the Cup, in 2003. In the context of its preparations for the upcoming America's Cup match to be held in July 2007 in Valencia, the defender requested the expertise of the NRC's Aerodynamics Laboratory to examine boundary layer transition on the yacht's keel. The design of the keel of an ACC yacht is the best kept secret of this discipline. The keel is optimized to provide maximum side force and minimum drag, to allow the yacht to sail efficiently on the up-wind course. The tool of choice for the iterative design of the keel shape is CFD. However, this tool cannot yet reliably predict the onset of boundary layer transition. This information is critical since skin friction represents a significant fraction of the total hydrodynamic drag of the keel. In this context, it was imperative for Team Alinghi to obtain transition data in sailing conditions, in order to tune their CFD codes. The NRC was called in to design transition experiments that were conducted aboard Alinghi yacht SUI-64, in Valencia. Sister experiments were also performed on a large keel model at matching Reynolds numbers, in the NRC's 9m x 9m Wind Tunnel.

Experimental Set-up and Procedures

For the experiments at sea, it was essential to characterize the free-stream disturbances triggering instabilities in the boundary layers developing on the keel. A sting-mounted hot-film probe fastened to the keel leading edge was used to measure turbulence signals in the freestream. The turbulence probe was calibrated 'on-the-fly' against a reference velocity sensed by an adjacent Prandtl tube. The zones of boundary layer transition needed to be delimited for various sailing conditions, both on the wing section of the keel and on the bulb. To that effect, eight hot-film foil sensors were bonded to either element of the keel, at strategic locations (*Figure 1a*). These delicate experiments required precautionary measures due to the corrosive nature of sea water that could damage sensors or signal leads. All hot-film sensors were controlled by a CTA anemomener in high-power mode. The signals were low-pass filtered, amplified and simultaneously digitized using a PXI-based system located within the hull of the yacht. The custom-designed LabView software allowed the immediate appraisal of data quality.

Discussion of results

A typical set of transition signals is presented in *Figure 1b*, emphasizing the stream-wise evolution of the bulb boundary layer from a laminar state to turbulence. It is noteworthy that the signals provide two explicit indicators for identifying this change: *(1)* The DC level raises when the state changes from laminar to turbulent, due to the enhanced heat transfer from the hot-film into the fluid; *(2)* the turbulent signal reaches high fluctuation amplitudes. It is also remarkable that in the intermittent stage, each spike is characterized by a sharp rise corresponding to the arrival of the wavefront at the sensor location, followed by a smoother return to laminar flow at the trailing edge of the disturbance. Such a signature is typical of turbulent spots. A power spectrum of 'free-stream turbulence' measured at sea is shown in *Figure 1c*, emphasizing the large amplitudes of low-frequency motions. It must be noted that the turbulence probe responds to any stream-wise motion relative to the fluid, including that due to the pitching of the yacht on the waves. Such motions are at the low end of the frequency spectrum. The auto-correlation plot of *Figure 1d* confirms the dominance of the 'large-scale eddies' in the turbulence measured at sea. Transition experiments were also conducted at matching Reynolds numbers on a large keel model in the wind tunnel (*Figure 1e*). The turbulence intensity measured at sea was reproduced by means of a turbulence-generating net located upstream of the model. It was observed that the wind tunnel simulation yielded premature transition of the boundary layers. This was attributed to the higher energy content in the high-frequency, small-scale turbulent fluctuations (see *Figure 1c,d*). These disturbances are prone to trigger instabilities in the fore portions of the appendages.

Conclusions

The transition and turbulence experiments at sea were successful at providing Team Alinghi with precious information for tuning their CFD codes. In order to improve the correspondence of the wind tunnel simulation with the marine environment, it might be judicious, instead of matching the turbulence intensity, to select a turbulence-generating grid that will yield a spectrum whose high-frequency tail will match that of the marine spectrum of free-stream disturbances. This region of the spectrum is likely to have the largest influence on boundary layer stability.

Acknowledgements

The authors are thankful to the Alinghi crew whose dedication ensured the successful completion of the sea experiments. The preparations for both the sea and wind tunnel tests were masterfully conducted by NRC technical staff at the 9m x 9m Wind Tunnel. We thank V. Nguyen, G. Larose, M. Deslauriers and Y. Mébarki for their precious support during the wind tunnel tests.

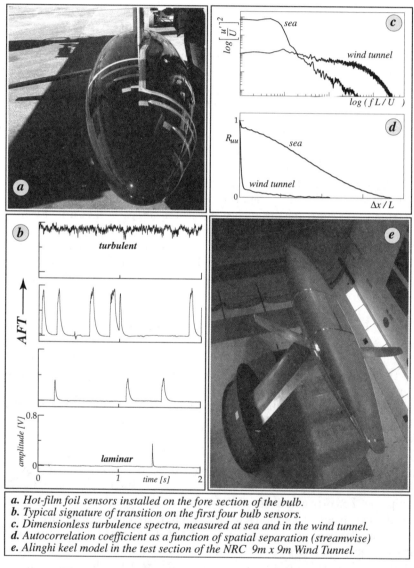

a. *Hot-film foil sensors installed on the fore section of the bulb.*
b. *Typical signature of transition on the first four bulb sensors.*
c. *Dimensionless turbulence spectra, measured at sea and in the wind tunnel.*
d. *Autocorrelation coefficient as a function of spatial separation (streamwise)*
e. *Alinghi keel model in the test section of the NRC 9m x 9m Wind Tunnel.*

Fig. 1. Experimental arrangements and selected results

Numerical Analysis of the Excited Jets Using Large Eddy Simulation - Parametric Study

Artur Tyliszczak and Andrzej Boguslawski

Institute of Thermal Machinery, Czestochowa University of Technology
Al. Armii Krajowej 21, 42-200 Czestochowa, Poland.
atyl@imc.pcz.czest.pl, abogus@imc.pcz.czest.pl

1 Introduction

Controlled flow of the jet was first reported by Crow and Champagne [1] who observed that application of suitable excitation (forcing) at the jet inlet intensifies mixing phenomena. Spectacular examples of this control technique are the bifurcating jets [4] occurring under particular excitation conditions. They may be characterized as jets which split downstream the potential core into two separate well defined streams. In this work excited jets are analyzed numerically using LES method, we study behavior of jet forced by helical and axial excitations with various excitation parameters. The main attention is devoted to determination of range of parameters causing the occurrence of the bifurcation. Experimental works [4] concerning the bifurcating jets showed that the frequency of excitation is a crucial parameter determining the flow field (bi-, trifurcating, blooming jets), however there is rather little knowledge about the importance of the amplitude of excitations and their relations to the initial momentum thickness, Reynolds number and turbulence intensity of the jet. We perform a series of computations varying mentioned parameters in the range allowing for reliable conclusions. Due to limited length of the paper we cannot show all data obtained, we present analysis of the influence of the the Strouhal number of excitation and forcing amplitude for varying inlet turbulence intensity.

2 Numerical method

The numerical code bases on the projection method to determine the pressure field. Spatial discretization is performed with 6th order compact difference scheme combined with the Fourier pseudospectral method. The time integration is performed with the low storage three step Runge-Kutta method. We use the filtered structure function model [3] for the subgrid stress tensor. The numerical code was extensively verified [5] and it reveals very good accuracy.

3 Boundary conditions

The dimensions of the computational domain are: $10 \times 10 \times 16$ jet diameters and the numerical mesh consist of $256 \times 256 \times 160$ uniformly spaced nodes. The inlet velocity profile applied in our computations is a superposition of the mean velocity corresponding to the hyperbolic tangent profile [2, 5], white noise disturbation and combination of axial and helical forcing. At the outflow we applied the so-called convective boundary condition. Fourier pseudospectral method applied in the directions perpendicular to the jet axis enforces to use periodic boundary conditions on the lateral boundaries. Performed analysis has shown that for the domain $10 \times 10 \times 16D$ this type of the boundary conditions practically do not affect the flow field in case of the natural jet. In case of bifurcating jet we observe that branches of the jet are attached to the periodic boundaries starting from $12D$ - this behavior is seen in the presented figures. However, this effect does not influence the center of the domain close to the jet inlet where the jet bifurcates ($\approx 5 - 6D$). Therefore one may assume that imposed periodicity has no influence on generation/destruction of the bifurcation phenomena.

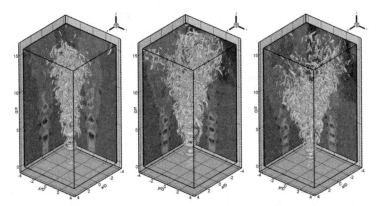

Fig. 1. 3D view of the isosurfaces of instantaneous Q parameter for computations with $St_a = 0.3, 0.4$ and 0.5. The side planes show axial velocity contours in crossections of the domain.

4 Results and conclusions

In this section we analyze the influence of two excitation parameters: the Strouhal number and the amplitude of the forcing. Figure 1 shows solutions obtained for $St_a = 0.3, 0.4, 0.5$ where one may see that the jet starts to bifurcate for $St_a = 0.4$, then increasing the Strouhal number to $St_a = 0.5$ the bifurcation angle is bigger and the bifurcation occurs closer to the jet inlet. For $St_a = 0.6$ and $St_a = 0.7$ (not shown) the bifurcation phenomena vanish but the spreading rate of the jet in bifurcating plane is considerably bigger than in

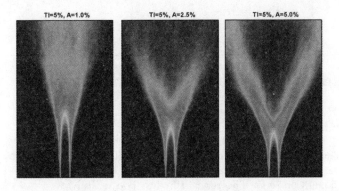

Fig. 2. Influence of the forcing amplitude.

perpendicular (bisecting) plane. Figure 2 shows statistically steady contours of the axial velocity obtained for three different values of the forcing amplitude $A = 1.0\%, 2.5\%, 5.0\%$ for the case when the turbulence intensity in the mixing layer was equal to 5%. Here one may observe that the jet bifurcates only for higher values of the amplitudes (center and right figures in Fig.2). However further study have shown that even for the small forcing amplitude $A = 1.0\%$ the jet can undergo bifurcation when the turbulence intensity (white noise perturbation in our study) is smaller - we observed bifurcation for values of turbulence intensity smaller than 2.5%.

The support for the research was provided within statutory funds BS-1-103-301/2004/P and the EU FAR-Wake Project AST4-CT-2005-012238. The authors are grateful to the TASK and Cyfronet Computing Centers in Poland for access to the computing resources.

References

1. Crow S.C. and Champagne F.H. Orderly structure in jet turbulence. *Journal of Fluid Mechanics*, 48:547–691, 1971.
2. da Silva C.B. and Metais O. Vortex control of bifurcating jets: A numerical study. *Physics of Fluids*, 14(11):3798–3819, 2002.
3. Ducros F., Comte P., and Lesieur M. Large-eddy simulation of transition to turbulence in a boundary layer developing spatially over a flat plate. *Journal of Fluid Mechanics*, 326:1–36, 1996.
4. Reynolds W.C., Parekh D.E., Juvet P.J.D., and Lee M.J.D. Bifurcating and blooming jets. *Annual Review of Fluid Mechanics*, 35:295–315, 2003.
5. Tyliszczak A. and Boguslawski A. LES of the jet in low Mach variable density conditions. In Geurts B.J. Metais O. Lamballais E., Friedrich R., editor, *Direct and Large Eddy Simulations VI*, ERCOFTAC, pages 575–582. Springer, 2006.

Local dissipation scales in turbulence

Jörg Schumacher[1], Katepalli R. Sreenivasan[2] and Victor Yakhot[3]

[1] Mechanical Engineering, Technische Universität Ilmenau, Germany
   `joerg.schumacher@tu-ilmenau.de`
[2] International Centre for Theoretical Physics, Trieste, Italy
   `krs@ictp.it`
[3] Mechanical and Aerospace Engineering, Boston University, USA
   `vy@bu.edu`

1 Motivation

The Kolmogorov theory of turbulence assumes the existence of a single mean dissipation length at the small-scale end of the turbulent cascade where the viscous dissipation term of the underlying Navier-Stokes equations becomes comparable to the nonlinear advection term. This scale is known as the Kolmogorov length η_K and is defined as $\eta_K = \nu^{3/4}/\langle \epsilon \rangle^{1/4}$ where ν is the kinematic viscosity and $\langle \epsilon \rangle$ the mean energy dissipation rate of the flow. The classical definition of the dissipation scale does not, however, capture the strongly intermittent nature of the energy dissipation field which is now a well-accepted fundamental building block of our understanding on turbulence. It therefore seems natural to include these fluctuations and to extend the notion of a single mean dissipation scale to that of a whole continuum of local dissipation scales, i.e. to a fluctuating random field η. The finest local dissipation scales will then be associated with the steepest velocity gradients, or alternatively, with the most intensive energy dissipation events in turbulence. These ideas were put forward within the multifractal formalism (see [1] for references).

In a recent theoretical approach which stayed close to the underlying Navier-Stokes equations, the distribution of local dissipation scales was directly calculated by one of the authors [2]. Here, we want to present high-resolution direct numerical simulations (DNS) of homogeneous isotropic turbulence, that confirm the theoretically predicted shape of the local dissipation scale distribution. A standard pseudospectral method is used to simulate homogeneous isotropic turbulence in a fully periodic box. However, grid resolutions used were significantly finer than the ones applied in standard cases. We enforce the spectral resolution criterion to values of $k_{max}\eta_K \geq 10$. Four different runs are conducted at Taylor microscale Reynolds numbers $R_\lambda = 10, 24, 42$ and 64 with grid resolutions of $N^3 = 512^3, 1024^3, 1024^3$ and 2048^3 points, respectively (for more details, see [1]).

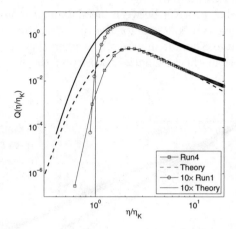

Fig. 1. Comparison of numerical and theoretical results for $Q(\eta, Re)$ for Runs 1 $(R_\lambda = 10)$ and 4 $(R_\lambda = 64)$. The data for Run 1 are shifted upwards in the diagram for a better visibility.

2 Results

Figure 1 shows the distribution $Q(\eta)$ of the local dissipation scales as determined from DNS for two Reynolds numbers. The distribution agrees qualitatively with the theoretical predictions [2]. First, one can observe that the range of excited sub-Kolomogorov scales increases with growing Reynolds number, which underlines that intermittent fluctuations from the inertial range sweep deeper into the viscous scale range. While the tail of the distribution for scales $\eta > \eta_K$ agrees well with the theory, deviations for the tail $\eta < \eta_K$ are detected.

An interesting connection of these results to the decay of the energy spectrum $E(k)$ for wavenumbers $k > \eta_K^{-1}$ should follow. The yet unanswered question is how the intermittency, which is detected in physical space, manifests itself in the decay of the energy spectrum in the viscous range. In order quantify the exponential decay of the energy spectra in the dissipation range we fit in Figure 2 the following dimensionless local slope to the data:

$$\frac{\mathrm{d}\log(\tilde{E}(k))}{\mathrm{d}\log(\tilde{k})} = \alpha - \beta\tilde{k} \quad \text{for } \tilde{k} \geq 1, \tag{1}$$

which transforms the exponential decay law of the spectrum into a linear function. Here, $\tilde{E}(k) = E(k)/(\nu^5\langle\epsilon\rangle)^{1/4}$ and $\tilde{k} = k\eta_K$. Our findings extend former results for the spectral tails [3, 4] to larger spectral resolutions and support the validity of (1). The lower panels of Figure 2 show both coefficients as functions of the Reynolds number. They are in the same range as in [3, 4].

Unfortunately, their asymptotic values for high Reynolds numbers cannot be deduced from data in this limited range.

This behavior at the crossover between inertial and viscous ranges of turbulence might may have consequences for the turbulent mixing of scalar concentration fields at large Schmidt numbers, i.e. when the scalar diffusivity is significantly smaller than the kinematic viscosity of the fluid. They will be discussed elsewhere.

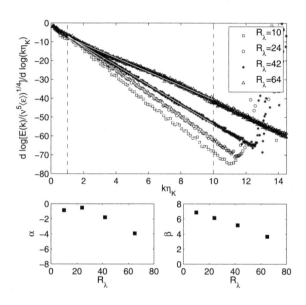

Fig. 2. Decay of the energy spectra $E(k)$ in the viscous scale range for different Reynolds numbers. Upper picture: Local slope of the spectrum as a function of the wavenumber. Least square fits for all data were performed between $1 \leq k\eta_K \leq 10$ (dashed lines). The fit results are indicated as gray lines for each data set. Lower pictures: Constants α (left) and β (right) as a function of the Reynolds number.

References

1. J. Schumacher, K. R. Sreenivasan, V. Yakhot: arXiv nlin.CD/0604072, submitted to New J. Phys. (2007).
2. V. Yakhot: Physica D **215**, 166 (2006).
3. S. Chen, G. Doolen, J. R. Herring, R. H. Kraichnan, S. A. Orszag, Z. S. She: Phys. Rev. Lett. **70**, 3051 (1993).
4. T. Ishihara, Y. Kaneda, M. Yokokawa, K. Itakura, A. Uno: J. Phys. Soc. Jpn. **74**, 1471 (2005).

Intermittency via Self-Similarity in New Variables

Mogens V. Melander

Dept. of Mathematics, SMU, Dallas TX 75275-0156, USA melander@smu.edu

Summary. We consider the classical subject of inertial range scaling in high Reynolds number isotropic turbulence driven into statistical equilibrium by steady large scale forcing. Using DNS data from a periodic box calculation, we look at the problem two ways. First, we use longitudinal velocity increments. Their statistics show intermittency, i.e., the pdf varies from Gaussian at the largest scale to highly non-Gaussian at the smallest scale. However, when we employ a new set of variables to look at the same data, we find self-similarity. The new variables are constructed in wave number space and the self-similarity expresses a particular symmetry. We discuss how the traditional variables (velocity increments) hide this symmetry.

Inertial range intermittency has been a roadblock to the advancement of turbulence theory since the deficiencies of Kolmorov's 1941 theory (K41) emerged some fifty years ago [1]. Intermittency occurs in the classical equilibrium setting where isotropic high Re turbulence is subject to steady large scale forcing. Inertial range statistics is generally believed to be universal, i.e, independent of both forcing and dissipation. That makes the inertial range an obvious candidate for self similarity. In fact, K41 calls for "statistical self-similarity" of velocity increments, which means that only the standard deviation of the pdf should depend on the length scale ℓ. Experimental and computational data conclusively contradict K41 [1]. The shape of the pdf varies with scale as illustrated in Fig. 1a where rescaling does not coalesce the different curves. It would then appear that the statistics are substantially different across scales with depressing implications for turbulence modeling. There is, however, another possibility. Maybe the usual velocity increments hide self-similarity. In fact, by using new variables, we do find symmetry (Fig. 1b) in the same 512^3 DNS dataset used to generate Fig. 1a. This paper is about that symmetry and how to generate Fig. 1b from DNS-data.

Let us start by describing what motivated Fig. 1b. A similar figure first emerged in intermittent shell model statistics [2]. Shell models are dynamical systems that attempt to describe turbulence with a very small set variables; in our case, just two real amplitudes per shell. Each shell represents an octave

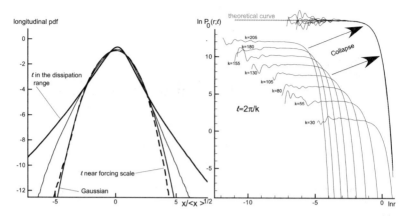

Fig. 1. 512^3 DNS dataset processed two ways: (a) pdf for $\delta v_\parallel(\ell)$; (b) The radial profiles $P_0(r;\ell)$ and their collapse. Dataset from LANL group Holm, Kurien and Taylor.

in wavenumber space, i.e., $2^n \le |\boldsymbol{k}| < 2^{n+1}$. In order to use the shell model as a guide for analyzing DNS data, we must work in Fourier space; the shell model has no analogs of the one dimensional velocity increments $\delta v_\parallel(\ell)$ and $\delta v_\perp(\ell)$. Correspondingly, we define scale three-dimensionally: $\ell = 2\pi/|\boldsymbol{k}|$.

Let $\hat{\boldsymbol{u}}(\boldsymbol{k})$ be the velocity in Fourier space. Using spherical coordinates in Fourier space, the incompressibility condition, $i\boldsymbol{k} \cdot \hat{\boldsymbol{u}}(\boldsymbol{k}) = 0$, is easy to satisfy. The complex helical waves decomposition [3] does that by writing $\hat{\boldsymbol{u}}(\boldsymbol{k}) = u^-(\boldsymbol{k})(\boldsymbol{b}(\boldsymbol{k}) - i\boldsymbol{a}(\boldsymbol{k})) + u^+(\boldsymbol{k})(\boldsymbol{b}(\boldsymbol{k}) + i\boldsymbol{a}(\boldsymbol{k}))$, where $\boldsymbol{a}(\boldsymbol{k})$ and $\boldsymbol{b}(\boldsymbol{k})$ are, respectively, azimuthal and longitudinal unit vectors on the sphere of constant $|\boldsymbol{k}|$. The amplitudes u^- and u^+ are easy to compute: $2u^\pm(\boldsymbol{k}) = \hat{\boldsymbol{u}}(\boldsymbol{k}) \cdot (\boldsymbol{b}(\boldsymbol{k}) \pm i\boldsymbol{a}(\boldsymbol{k}))$. They obey conjugate symmetry, e.g., $u^-(-\boldsymbol{k}) = u^{-*}(\boldsymbol{k})$ because the velocity is real in physical space. Using two real amplitudes, χ^- and χ^+, we satisfy conjugate symmetry identically: $u^\pm(\boldsymbol{k}) = \frac{1}{2}(\chi^\pm(\boldsymbol{k}) + \chi^\pm(-\boldsymbol{k})) + \frac{i}{2}(\chi^\pm(\boldsymbol{k}) - \chi^\pm(-\boldsymbol{k}))$. These real amplitudes are unconstrained and easy to compute: $\chi^\pm = Re(u^\pm) + Im(u^\pm)$. Navier Stokes equations can even be written as a dynamical system in terms of them.

In isotropic turbulence, all $\chi^\pm$ on a shell of constant $|\boldsymbol{k}|$ are statistically equivalent. So we can use $\chi^\pm$ as random variables in a statistical description. Each scale, $\ell = 2\pi/|\boldsymbol{k}|$, is thus described by the joint pdf: $J(x,y;\ell)dxdy \equiv \Pr\{x < \chi^-(\boldsymbol{k}) < x + dx, y < \chi^+(\boldsymbol{k}) < y + dy\}$. Switching to polar coordinates, the axisymmetric component of J is characterized by its radial profile $P_0(r;\ell)$. Fig. 1b shows $P_0(r;\ell)$ at various ℓ. The practical matter of obtaining $P_0(r;\ell)$ from the data involves two steps: (1)Using kernel density estimation, compute the pdf $\psi(\xi;\ell)$ for $\xi \equiv 0.5\ln\left((\chi^+)^2 + (\chi^-)^2\right)$; (2) Construct $P_0(r;\ell)$ from $\psi(\xi;\ell)$; the formula is $P_0(r;\ell) = \psi(\ln r;\ell)/(2\pi r^2)$.

We observe symmetry in Fig. 1b: shifts together with linear horizontal stretching coalesce the curves. That is affine symmetry in $\ln r$:

$$P_0\left(r;\ell\right) = C(\ell)f\left(\frac{\ln r - \mu\left(\ell\right)}{\sigma\left(\ell\right)}\right). \tag{1}$$

f is a similarity profile, and normalization determines $C\left(\ell\right)$. μ and σ reset the origin and unit on the $\ln r$-axis; that is akin to the common statistical practice of centering on the mean and scaling by the standard deviation. The difference here is that mean and standard deviation on the $\ln r$-axis do not exist, so μ and σ are set differently. Note, if $\sigma(\ell)$ is constant then (1) is K41-scaling in r.

In line with the literature, e.g., [1], we find that structure functions $S_p(\ell)$ are power laws in the inertial range:

$$S_p(\ell) \equiv \left\langle \left((\chi^+)^2 + (\chi^-)^2\right)^{p/2}\right\rangle = 2\pi \int_0^\infty r^{p+1}P_0\left(r;\ell\right)dr = C_p\left(\ell/\ell_o\right)^{\zeta_p} \tag{2}$$

We are now faced with an interesting theoretical question: what kind of statistics allow both (1) and (2)? If $\sigma(\ell)$ is constant, then there is no problem: K41-scaling takes care of everything without constraining f. If, on the other hand, $\sigma(\ell)$ is not constant, then (1) and (2) together impose severe constraints. By elevating (1) and (2) to hypotheses, we have undertaken a mathematical analysis of which f, μ, σ, C_p and ζ_p are possible [4]. The results are very pleasing; f, μ, σ, C_p and ζ_p are found up to a few undetermined constants. The normalized similarity profile f depends on only one parameter $0 < \beta < 3$. By analyzing the structure functions, we find that $\beta \approx 1.22$ for the present DNS data. As Fig. 1b shows, the data matches that theoretical curve well.

Figure 1 shows that the choice of statistical variables matters. Our variables reveal a simple symmetry (1), but the velocity increments $\delta v_\parallel(\ell)$ do not. By comparing the two types of variables, we get an idea why the velocity increments hide the symmetry. There are two problems. First, the velocity increments define the scale ℓ as a separation between two points in physical space; in Fourier space, that ignores two components of the wave vector. Consequently, there is improper scale separation in the Fourier sense. This point is well understood for energy spectra [5], but seems unappreciated in the context of intermittency. The second problem is that the symmetry involves a two dimensional pdf, not a projection thereof; for an analytic example see [6].

Acknowledgements: The author is grateful for the DNS data from D. Holm, S. Kurien and M. Taylor as well as for discussions with B. Fabijonas.

References

1. Frisch, U. 1995 *Turbulence*, Cambridge Univ. Press, Cambridge.
2. Melander, M.V. & Fabijonas, B 2006 eprint arXiv:physics/0603185
3. Lesieur, M. 1990 *Turbulence in Fluids*, Kluwer Acad. Press, Dordrecht.
4. Melander, M.V. 2007 Analysis of a Symmetry leading to an Inertial Range Similarity Theory for Isotropic Turbulence. eprint arXiv:physics/0702073
5. Hinze, J.O. 1987 *Turbulence* McGraw-Hill, New York.
6. Melander, M.V. & Fabijonas, B 2006 eprint arXiv:physics/0512198

Mixing of a passive scalar emitted from a random-in-time point source

Antonio Celani[1], Marco Martins Afonso[2], and Andrea Mazzino[3]

[1] CNRS, INLN, 1361 Route des Lucioles, 06560 Valbonne, France
[2] Department of Physics of Complex Systems, The Weizmann Institute of Science, Rehovot 76100, Israel   marcomar@fisica.unige.it
[3] Department of Physics - University of Genova & CNISM and INFN - Genova Section, via Dodecaneso 33, 16146 Genova, Italy

1 Introduction and classical homogeneous results

We study the stationary distribution of a passive scalar field – e.g. concentration of a pollutant or temperature in the absence of buoyancy – advected by a turbulent incompressible flow, in dimension $d > 1$. We focus on the Kraichnan advection model [1, 2], in which the velocity is assumed as a zero-mean white-in-time Gaussian random field, featuring stationarity, homogeneity and isotropy (and with power-law spatial increments of Hölder exponent ξ).

Well-known results are available in the classical framework, in which also the forcing term satisfies the aforementioned properties but is endowed with a spatial support on the large scale L. With these hypotheses, indeed, the stationary two-point equal-time scalar correlation function $C(\boldsymbol{x}, \boldsymbol{x}', t)$ depends on the absolute value of the relative separation ($\boldsymbol{r} \equiv \boldsymbol{x} - \boldsymbol{x}'$) only and, in the limit of vanishing diffusivity, shows two distinct well-defined behaviours [3]:

$$C(r) = \begin{cases} \alpha - \beta r^{2-\xi} & \text{for } r \ll L \text{ (inertial–convective range)} \\ \gamma r^{-(d+\xi-2)} & \text{for } r \gg L \text{ (equilibrium unforced regime)} . \end{cases} \quad (1)$$

The equilibrium regime corresponds to the interval where the problem is unforced, i.e. the correlation function is here a zero mode, while in the inertial–convective range the classical direct cascade process (from the injection scale L toward smaller and smaller scales, with constant scalar flux) takes place.

2 Extension to inhomogeneous forcing

The goal of our paper is to show what happens when the forcing term is not statistically homogeneous. In this case, C is expected to depend not only on the pair distance but, in general, on both the full vector r and the center-of-mass coordinate $z \equiv (x + x')/2$ [4]. The former homogeneous, classical results thus serve as a reference point to analyse the possibility of a small-scale restoration of homogeneity, simply by comparing them with the inhomogeneous behaviour we are going to derive.

The analysis is carried out by projecting onto bases invariant under translations and rotations, i.e. performing Fourier transform in z and SO(d) decomposition [5] in r, respectively. In Fourier space, the overall effect of inhomogeneity is represented by the appearance of a new "mass" term in the second-order differential equation which expresses the balance between advection and forcing in the inertial-convective range. The zero modes of the latter equation are thus no more power laws, as in the homogeneous case (1), but rather modified Bessel functions. In other words, a new lengthscale is introduced, for separations above (or of the order of) which inhomogeneities spoil the pure power-law behaviour. However, if the forcing has a discrete spectrum with a finite upper bound on the excited wavenumbers, it can be shown that homogeneity is well restored at small scales, because of the little weight that such "spoiling" modes play upon superposition. The same conclusion cannot be drawn if inhomogeneity is introduced at arbitrarily small scales, or in the presence of a forcing with generic continuum spectrum, because of the difficulty in performing the antitransformed.

3 The point-source problem

We shall thus focus on a special case, very simple but paradigmatic, as it can be considered the prototype of inhomogeneity: the problem of scalar emission (or absorption) from a point source, located e.g. in the origin, at a random rate, such as to maintain our initial hypotheses of statistical stationarity and Gaussianity with zero average and delta-correlation in time.

Back to the physical space, the final result in the isotropic sector is

$$C(r, z) \propto r^{d\xi/2 - 2d - \xi + 2} \left[1 + \text{const.} \times z^2 r^{-(2-\xi)} \right]^{1 - d(4-\xi)/2(2-\xi)} . \qquad (2)$$

Different regimes thus arise according to the values of r and z. In particular, for r sufficiently smaller than z, the asymptotic behaviour of (2) is

$$C(r, z) \sim z^{2 - d(4-\xi)/(2-\xi)} \left[1 + \text{const.} \times z^{-2} r^{2-\xi} \right] . \qquad (3)$$

The issue of the possible small-scale restoration of homogeneity can now be easily addressed simply by comparing (3) with (1). A clear parallelism appears with the inertial–convective-range homogeneous behaviour $r^{2-\xi}$ shown in (1), rather than with the one ($r^{-(d+\xi-2)}$) observed for separations larger than the forcing correlation length L, even if here the latter formally vanishes.

In the absence of large-scale forcing, this cascade-like mechanism can be explained as follows. The velocity field sweeps the scalar, initially concentrated where it was released, and generates structures which, for every point, are correlated on the scale corresponding to the distance of this point from the source; in other words, correlations between each point and the origin are created. In the centre-of-mass frame of reference, this means that in every point z a local cascade can then take place, starting from separations r of the order of (or sufficiently smaller than) z, which thus plays the role of a local forcing correlation length. This allows us to conclude that this case represents an interesting example of *persistence of inhomogeneity at small scales*.

Moreover, it can be shown that anisotropic contributions are always subdominant in the presence of a positive Corrsin integral, because anisotropy can only stem out from the interplay between inhomogeneity and finite-size effects. This interplay can easily be shown introducing a $2d$-dimensional formalism. Defining the hypervector $\vec{y} \equiv (\boldsymbol{r}, \boldsymbol{z})$, the evolution equation for the correlation function reads

$$\partial_t C + \vec{\partial}_{\vec{y}} \cdot \vec{J} \propto \delta(\vec{y}) , \qquad (4)$$

where $\vec{J}(\vec{y}, t) \equiv (\boldsymbol{J_r}, \boldsymbol{J_z})$ can be interpreted as a current expressed by $\vec{J} = \mathsf{K} \cdot \vec{\partial}_{\vec{y}} C$ and the $2d \times 2d$ matrix $\mathsf{K}(\vec{y})$ is made up of two non-zero $d \times d$ blocks in the upper-left and lower-right positions, respectively. The former block depends exclusively on $\boldsymbol{r}$ (and is the only one present in the homogeneous situation), so if the latter block were only dependent on $\boldsymbol{z}$ a complete decoupling between the two variables would take place and (4) would be invariant under independent rotations of $\boldsymbol{r}$ and $\boldsymbol{z}$. However, this is the case only if one takes into account separations much smaller than the velocity integral scale, as we did throughout the paper as a first approximation. If the two lengths become comparable, finite-size effects appear via the lower-right block, whose presence is due to inhomogeneity, and break the global isotropy. Back to the cascade process, equation (4) also provides the correct mathematical interpretation. It implies indeed that, away from the origin and in the steady state, $\partial_r \cdot \boldsymbol{J_r} = -\partial_z \cdot \boldsymbol{J_z}$. This suggests that the classical scalar flux toward smaller and smaller scales (in the variable r, corresponding to the left-hand side) is supplied in this case not by the large-scale forcing, as in the classical homogeneous situation described in Sect. 1, but rather by an analogous flux in the variable z toward larger and larger distances from the source (as implied by the minus sign on the right-hand side).

References

1. R.H. Kraichnan: Phys. Fluids **11**, 945 (1968)
2. R.H. Kraichnan: Phys. Rev. Lett. **72**, 1016 (1994)
3. G. Falkovich, K. Gawędzki, M. Vergassola: Rev. Mod. Phys. **73**, 913 (2001)
4. M. Martins Afonso, M. Sbragaglia: J. Turb. **6**, 10 (2005)
5. L. Biferale, I. Procaccia: Phys. Rep. **414**, 43 (2005)

The Coupled LES - Subgrid Stochastic Acceleration Model (LES-SSAM) of a High Reynolds Number Flows

V. Sabel'nikov[1], A. Chtab[2], and M. Gorokhovski[3]

[1] DEFA/EFCA ONERA, Palaiseau, France `Vladimir.Sabelnikov@onera.fr`
[2] CORIA-UMR CNRS 6614, University of Rouen, France `Anna.Chtab@coria.fr`
[3] LMFA UMR CNRS 5509, Ecole Centrale de Lyon, France `Mikhael.Gorokhovski@ec-lyon.fr`

Abstract. In the present work, in order to take into account the intermittency phenomena, we proposed a new approach to the LES of constant-density flows. In this approach, the main step is to reconstruct an approximation for unfiltered velocity field based on stochastic simulation of non-resolved (unknown) fluid particle acceleration. The last one is modeled with accounting for its significant dependency on the Reynolds number. In comparison with classical LES, the model reproduces effects which are recently observed by measurements in a high Reynolds number homogeneous turbulence.

Formulation of LES-SSAM. When the Navier-Stokes equations are filtered, one obtains two sets of equations, one is for filtered field :

$$\bar{a}_i \equiv \overline{\left(\frac{du_i}{dt}\right)} = -\frac{\partial \bar{P}}{\partial x_i} + \nu \Delta \bar{u}_i \tag{1}$$

$$\frac{\partial \bar{u}_k}{\partial x_k} = 0 \tag{2}$$

another is for residual one:

$$a'_i \equiv \left(\frac{du_i}{dt}\right)' = -\frac{\partial P'}{\partial x_i} + \nu \Delta u'_i \tag{3}$$

$$\frac{\partial u'_k}{\partial x_k} = 0 \tag{4}$$

Here $\bar{a}_i$ and a'_i represent the filtered total acceleration and the residual one, respectively. In order to close these equations, the subgrid scale modeling is required. While this is done usually by considering the residual velocity field u'_i, our main emphasis in this work is on modeling of the non-resolved total acceleration $a'_i = (\frac{du_i}{dt})'$. When the Reynolds number is high, this acceleration may have values largely superior compared to resolved accelerations.

Indeed, in accordance with Kolmogorov's scaling, the simple estimation gives: $(\bar{a}_k\bar{a}_k)/(a_i'a_i') \approx (\eta/\Delta)^{2/3}$ where Δ is the filter width, $\eta = L/Re_L^{3/4}$ is the Kolmogorov's scale, L is the integral scale of turbulence, $Re = \sigma_u L/\nu$ and σ_u denotes the rms velocity. We replace the exact Eq. (3) by the model equation:

$$\left(\frac{du_i}{dt}\right)' = -\frac{\partial \widehat{p}}{\partial x_i} + \widehat{a}_i' \tag{5}$$

in which acceleration $\widehat{a}_i'$ is simulated stochastically in accordance with main features of the acceleration field, recognized by experiments and DNS. The pressure $\widehat{p}$ is introduced in Eq. (5) in order to guarantee the velocity vector to be solenoidal. This equation, together with Eq. (1), allows to reconstruct a modeling approximation for the unfiltered velocity field u_i. Two assumptions are made. First, the equation (1) is considered in the framework of classical LES closure using the Smagorinsky model. Second, it is assumed that the reconstructed velocity $\widehat{u}_i = \bar{u}_i + \widehat{u}_i'$ is governed by the following model equation:

$$\frac{\partial \widehat{u}_i}{\partial t} + \widehat{u}_k \frac{\partial \widehat{u}_i}{\partial x_k} = -\frac{\partial \widehat{P}}{\partial x_i} + \frac{\partial}{\partial x_k} \nu_{tur} \left(\frac{\partial \widehat{u}_i}{\partial x_k} + \frac{\partial \widehat{u}_k}{\partial x_i}\right) + \widehat{a}_i' \tag{6}$$

where ν_{tur} is the Smagorinski subgrid model.

Stochastic model of unresolved acceleration The stochastic process for $\widehat{a}_i$ is based on the well-known from Kolmogorov-Oboukhov (1962) [1] estimation of the square of acceleration, which is conditionally averaged on the viscous dissipation rate :

$$\langle a_i a_j | \varepsilon \rangle = a_0' \varepsilon^{3/2}/\nu^{1/2} \delta_{ij} \tag{7}$$

where $\langle |\varepsilon \rangle$ denotes the conditional mean and a_0' is constant. Using the Oboukhov's log-normality conjecture on the stochastic ε field, the following stochastic equation for the norm of acceleration $\widehat{a} = |\widehat{a}_i'|$ can be obtained:

$$d\widehat{a} = -\widehat{a}\left(\ln\frac{\widehat{a}}{a_\eta} - \frac{3}{16}\sigma_\chi^2\right)T_\chi^{-1}dt + \frac{3}{4}\widehat{a}\sqrt{2\sigma_\chi^2 T_\chi^{-1}}dW(t) \tag{8}$$

where $dW(t)$ is the increment of standard Brownian process, i.e. $\langle dW \rangle = 0$, $\langle (dW)^2 \rangle = dt$, and $a_\eta = \langle \varepsilon \rangle^{3/4}/\nu^{1/4}$ is the Kolmogorov's acceleration. Using Ito transformation, this equation was derived from the Pope and Chen's [2] stochastic equation, proposed for in the case of homogeneous stationary turbulence. The parameters of stochastic equation (8) are introduced in specific way:

$$T_\chi = \frac{\nu_{tur}}{\Delta^2}; \quad \sigma_\chi^2 = A + \mu \ln Re_\Delta^{3/4}; \quad Re_\Delta = \frac{\nu_{tur}}{\nu} \tag{9}$$

where according to [3]: $A = -0.863$; $\mu \approx 0.25$. The acceleration components are then computed by introducing the unit vector $e(t)$ with random direction in time: $\widehat{a}_i' = \widehat{a}(t)e_i(t)$. The set of equations (6), (8), (9) represents the formulation of LES-SSAM approach.

Computational results and discussion In this work, the model (6), (8), (9) was applied for $3D$ box stationary homogeneous turbulence (32^3 grid points). The stationarity was provided by forcing scheme from [4]. The Reynolds number, the kinetic energy, the micro-scales of Kolmogorov and Taylor were close to those mentioned in Mordant *et al.* [5] ($Re_\lambda = 740$; $\sigma_v^2 = 1$ m/s). The computation reproduced the effects observed in experimental study [5]. While at integral times, the Lagrangian velocity increment was normally distributed, at small time lag this increment displayed a strong non-gaussianity. This is shown in the figure on the left-hand side: the distributions exhibit a growing central peak with stretched tails. This implies that the small amplitude events are alternating with events of large acceleration (middle part of the figure). This effects have been not observed by classical LES (no stretched tails, maximal accelerations of order of Kolmogorov's one, $a \approx 500$ m/s^2). The measurement in [5] showed also that autocorrelation $\langle a_i(t)a_i(t+\tau)\rangle$ decreased very rapidly with progressing of τ. However the time scale in decreasing of autocorrelation of the norm of acceleration $\langle |a_i(t)||a_i(t+\tau)|\rangle$ was of order of integral time $T_L \approx L\sigma_u$. Such effect of the acceleration norm "memory" is obtained numerically and demonstrated on the right-hand side of the figure.

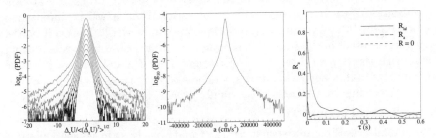

Fig. 1. Left-hand side: distributions of the velocity increment $\Delta_\tau U_i = U_i(t+\tau) - U_i(t)$ at different time lags $\tau = 0$; 0.15; 0.3; 0.6; 1.2; 2.5; 4.9; 9.8; 20 *et* 39 ms; middle part: distribution of acceleration; right-hand side: autocorrelations of acceleration $R_{a_i}(\tau) = \frac{\langle a_i(t)a_i(t+\tau)\rangle}{\sigma_{a_i}^2}$ and of its norm $R_{|a|}(\tau) = \frac{\langle |a(t)|\cdot|a(t+\tau)|\rangle}{\sigma_{|a|}^2}$.

References

1. A.S. Monin, A.M. Yaglom: *Statistical fluid mechanics : mechanics of turbulence*, (The MIT press 1981) pp 900
2. S.B. Pope, Y.L. Chen: Phys. Fluids A **2**, 1437 (1990)
3. A.G. Lamorgese, S.B. Pope, P.K. Yeung , B.L.Sawford: submitted to JFM
4. V. Eswaran, S.B. Pope: Computers and Fluids **16**, 257 (1988)
5. N. Mordant, E. Leveque, J.-F. Pinton: New Journal of Physics **6**, 116 (2004)

Stereo-PIV of sinuous and varicose breakdown

J.Mans[1], M. Brouwers, and H. C. de Lange

TUE Mechanical Engineering, P.O. Box 513, 5600 MB Eindhoven, The
Netherlands h.c.d.lange@tue.nl

Many features of the bypass transition process in flat boundary layers exposed
to free-stream turbulence have been examined by previous experimental and
numerical studies. However, it is not yet clear how the flow breaks down
into turbulent spots through secondary instabilities on isolated streaks in the
flow field. In this study the natural, no external triggering is used, breakdown process in a flat plate boundary layer exposed to strong free-stream
turbulence (Tu=6%) is experimentally investigated in a water channel using a combined PIV-visualization technique and Stereo-PIV. The evolution of
structures present in the transitional boundary layer is observed while moving
along with the flow in downstream direction. This enables one to determine
and analyze events appearing during the breakdown process. This paper pin-
points the events by which the flow finally breaks down into turbulence.

Experimental setup

The optical accessible measuring section (made of glass) in which a plate is
positioned, is 2.7 m long, 0.57 m wide and 0.45 m high. The flat plate is
exposed to free-stream turbulence generated by a static turbulence grid. The
main flow velocity in the set-up is set to around 0.12 m/s, which results in a
turbulence intensity of 6% at the leading edge. The estimated integral length
scale, Taylor microscale and Kolmogorov scale at the leading edge are respectively 25 mm, 5 mm and 0.7 mm. More information on the set-up can be
found in Mans et al. [1].

Sinuous and Varicose breakdown

The sinuous secondary instability measured with the combined PIV-visualization
technique is shown in figure 1. The figure shows the typical anti-symmetric
spanwise oscillation of the instability. The disturbance field presents the low
and high speed streaks in the boundary layer and the sinuous oscillation of
the unstable low speed streak. From the analysis of a sequence of disturbance
fields it is found that a discontinuity (a patch of low speed fluid) appears
in the adjoining high speed streak. This discontinuity seems to trigger the
sinuous instability. Figure 2 shows the varicose instability and its character-

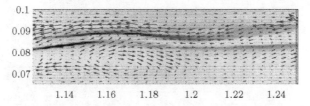

Fig. 1. The sinuous instability, $U_\infty = 0.125$ m/s and $1.52 \times 10^5 < Re_x < 1.56 \times 10^5$. The vector plot presents the disturbance field while the gray-scale indicates the dye intensity, here black states high dye intensity and white low intensity.

istic symmetric spanwise oscillation. The analysis of the varicose instability revealed that it is initiated by a frontal collision between an upstream high speed streak and a down stream low speed streak.

The evolution of three-dimensional structures

The Stereo-PIV results in figure 3 reveal the three-dimensional structures of a similar sinuous instability (bottom image) and varicose instability (top image). The bottom image presents only a half period from the sinuous oscillation, approximately corresponding to the streamwise domain $1.18 < x < 1.26$ in figure 1. The image shows the presence of two streamwise vortex tubes positioned in the high shear region between the streaks. The first tube is located in the region with the angular points $393, 19; 393, 24$ to $410, 22; 410, 26$, while the second tube is located just after the maximum amplitude of the sinuous oscillation at $411, 18; 411, 23$ to $425, 18; 425, 22$. It is clear that the second vortex tube obtains a rotation opposite to the first tube. The rotation in the horizontal plane reveals that the first tube is inclined away from the wall, while the second tube is inclined towards the wall. As a consequence of the interaction between both tubes the typical sinuous spanwise motion is induced in the overlap region, $412, 32$. All these features correspond well with the results from Brandt *et al.* [2].The inclination angle of the second tube increases under the influence of the main shear until at a certain point it is pushed over after which the instability breaks down into three-dimensional structures. The

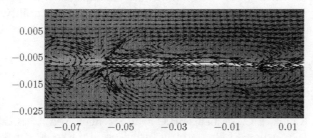

Fig. 2. The varicose instability, $U_\infty = 0.125$ m/s and $1.45 \times 10^5 < Re_x < 1.53 \times 10^5$. The vector plot presents the disturbance field while the gray-scale indicates the dye intensity, here white states high dye intensity and black low intensity.

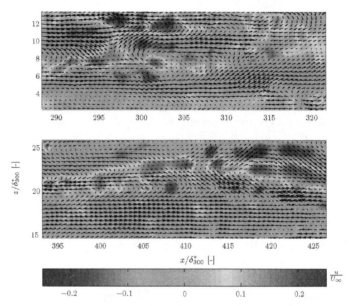

Fig. 3. Three-dimensional structures present in a varicose instability (top image) and a sinuous instability (bottom image), $U_\infty = 0.125$ m/s. Velocity vectors represent the disturbance components u' and w'. Background coloring refers to the amplitude of v'.

wave packet of the sinuous instability consist generally of around 4 periods and it possesses a wavelength and amplitude of respectively $\simeq 28\,\delta_{300}^*$ and 0.22. The amplitude is made dimensionless using the wavelength.

The Stereo-PIV results of the varicose instability show that two vortex tubes are positioned symmetrically around the unstable low speed streak. Both tubes are inclined away from the plate en form a λ-like structure. The preceding velocity data reveals that this λ-like structure results from a frontal collision between a high and a low speed streak. With the evolution of the flow in streamwise direction the collision process will result in a train of λ-like structures which finally breakdown into turbulence. The wave packet of the varicose instability consists, likewise as the sinuous instability, of around 4 periods and the wavelength and amplitude are respectively $\simeq 19\,\delta_{300}^*$ and 0.59.

References

1. Mans, J., Kadijk, E.C., de Lange, H.C., van Steenhoven, A.A.: Breakdown in a boundary layer exposed to free-stream turbulence. In: Exp. Fluids **39(6)** (2005) 1071–1083
2. Brandt, L., Schlatter, P., Henningson, D.S.: Transition in boundary layers subject to free-stream turbulence. In: J. Fluid Mech. **517** (2004) 167–198

Direct Numerical Simulation of Turbulent Taylor-Couette Flow with High Reynolds Number

Wenqi He, Mamoru Tanahashi, and Toshio Miyauchi

Dept. of Mechanical and Aerospace Engineering, Tokyo Institute of Tech., 2-12-1 Ookayama, Meguro-ku, Tokyo 152-8550, Japan. mtanahas@mes.titech.ac.jp

1 Introduction

It is well known that Taylor-Couette flow changes its flow pattern with the increase of Reynolds number [1]. Quasi-periodicity in the azimuthal direction can be observed in weakly turbulent state and disappears for higher Reynolds number cases [2]. Reynolds number dependence of the mean torque at the inner cylinder shows a transition at $Re \approx 10000$ [3]. However, turbulence structures in each state have not yet been clarified in detail. In this study, direct numerical simulation (DNS) has been conducted to investigate turbulence transition process and fine scale strcutures in Taylor-Couette flow.

2 Numerical methods

In this study, DNS of Taylor-Couette flow with a fixed outer cylinder and a rotating inner cylinder has been conducted by using Fourier-Chebyshev spectral methods. The incompressible Navier-Stokes equations are normalized by gap width (D) and the rotating speed of inner cylinder (U_{in}). The radius ratio is set to be 0.8. The periodic boundary conditions are used in azimuthal and axial directions. The calculation domain in the axial direction (H) is selected to be $4D$ and $5D$. The maximum Reynolds number $(Re = DU_{in}/\nu)$ calculated in the present study is 12000 where $97 \times 1280 \times 512$ grid points are used. The calculations were continued until the statistically steady state from an impulse start. The energy spectra show that the computational grids in all cases can resolve velocity fluctuation in high wavenumber adequately and distinct inertial subrange is observed for high Reynolds number cases. The numerical results of mean torque of the inner cylinder agree with Wendt's empirical formula [4].

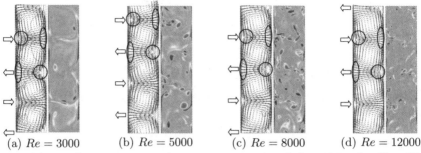

(a) $Re = 3000$ (b) $Re = 5000$ (c) $Re = 8000$ (d) $Re = 12000$

Fig. 1. Mean velocity vectors and instantaneous azimuthal vorticity for $H = 4D$ (left arrows: outflow boundaries, right arrows: inflow boundaries).

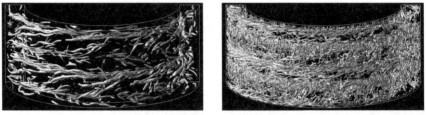

Fig. 2. Iso-surfaces of the second invariant of velocity gradient tensor for $H = 4D$ ($Q = 1.0$) (left: $Re = 5000$, right: $Re = 12000$).

3 Large scale Taylor vortices and fine scale eddies

Figure 1 shows the mean velocity fields and the instantaneous azimuthal vorticity on a typical $r - z$ cross section. The mean velocity fields show that large scale Taylor vortices which are formed for low Reynolds number cases still exist even for the highest Reynolds number case. Between Taylor vortices, outflow boundaries and inflow boundaries are formed. Near the walls, there are ejection and sweep regions which are denoted by circles and ellipses respectively. For lower Reynolds number cases, fine scale eddies are mainly formed near outflow boundaries due to strong shear layers induced by ejections. With the increase of Reynolds number, fine scale eddies also appear near the inflow boundaries. Especially, lots of fine scale eddies are observed in ejection regions. Figure 2 shows the iso-surfaces of the second invariant of velocity gradient tensor (Q). With the increase of Reynolds number, the number of fine scale structures increases. For $Re > 10000$, fine scale eddies parallel to the axial direction are formed in sweep regions, which can be observed only for high Reynolds number cases and is unexpected structure.

Fine scale eddies in turbulent Taylor-Couette flow have been identified by the method developed in our previous studies [5][6]. It has been shown that fully developed turbulence consists of a universal fine scale structure which is called coherent fine scale eddy (CFSE). Figure 3 shows the probability density functions (PDFs) of the diameter (D_e) and maximum azimuthal velocity ($v_{\theta,max}$) of fine scale eddies which are normalized by local Kolmogorov scale (η) and local Kolmogorov velocity (u_k) respectively. The most expected

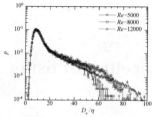 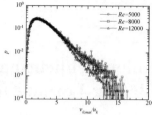

Fig. 3. PDFs of the normalized diameter (left) and maximum azimuthal velocity (right) of CFSEs for $H = 4D$.

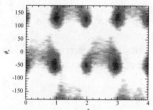 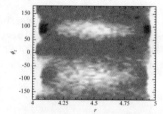

Fig. 4. JPDF of the inclination angle and axial position for $Re = 12000$.

Fig. 5. JPDF of the tilting angle and radial position for $Re = 12000$.

diameter and maximum azimuthal velocity are 8η and $1.7u_k$ for the highest Reynolds number. These results coincide with those of other turbulence reported in previous studies [5][6].

The statistical characteristics of inclination angle (ϕ_r) and tilting angle (ϕ_z) have been analyzed to investigate spatial distribution of CFSEs. ϕ_r is the angle between the tangential diretion of the wall and vorticity vector at the CFSE center on $r - \theta$ plane, and ϕ_z is the angle between tangential diretion and the vorticity vector on $r - z$ plane. Figure 4 shows the joint probability density function (JPDF) of ϕ_r and axial position. The peaks of JPDF is at about -45 and 135 degree near outflow and inflow boundaries, which corresponds to the fact that CFSEs have specific angles with respect to the azimuthal direction mentioned above. Figure 5 shows the JPDF of ϕ_z and radial position. Near the walls, tilting angle is close to 90 degree for high Reynolds number. These eddies correspond to CFSEs parallel to the axial direction as shown in Fig. 2.

References

1. M. Gorman, H.L. Swinney: J. Fluid Mech. **117** (1982)
2. Y. Takeda: J. Fluid Mech. **389** (1999)
3. D.P. Lathrop, J. Fineberg, H.L. Swinney: Phys. Rev. Lett. **68** (1992)
4. F. Wendt: Archive of Applied Mechanics **4** (1933) pp 577–595
5. M. Tanahashi, S.-J. Kang, T. Miyamoto, T. Miyauchi: Int. J. Heat and Fluid Flow **25** (2004)
6. M. Tanahashi, S. Iwase, T. Miyauchi: J. Turbulence **2** 6 (2001)

Tollmien-Schlichting wave cancellation using an oscillating Lorentz force

T. Albrecht[1], H. Metzkes[1], G. Mutschke[2], R. Grundmann[1], and G. Gerbeth[2]

[1] Institute for Aerospace Engineering, Technische Universität Dresden, 01062 Dresden, Germany thomas.albrecht@tu-dresden.de
[2] MHD Department, Institute of Safety Research, Forschungszentrum Dresden, P.O. Box 51 01 19, 01314 Dresden, Germany {mutschke,gerbeth}@fzd.de

Introduction

Given a flat plate boundary layer in a low free stream turbulence environment, the transition from laminar to turbulent flow is initiated by the amplification of small-amplitude, two-dimensional wave-like velocity fluctuations, the so-called Tollmien-Schlichting (TS) waves. Since the turbulent wall friction can exceed the laminar one by more than an order of magnitude, damping of these waves, hence delaying transition, reduces drag. Various methods of actuation have been proposed, where modifying the mean velocity profile by wall suction or blowing is probably the most common one. Although successfully tested even during in-flight experiments, mean velocity profile modification demands a comparably large power input which degrades its efficiency. In contrast, Milling [1] reported experimental results of wave cancellation by superposition of a counter-phase wave using vibrating wires, requiring only a fraction of control power input. Following this principle, several numerical and experimental works appeared since then, e.g. [2, 3]. If the fluid under consideration is low conductive, such as seawater with an electrical conductivity $\sigma \approx 5\,\mathrm{S/m}$, a Lorentz force is able to control the flow [4]. Driven by an external electric field, almost arbitrary time signals may be generated. An application to wave cancellation is obvious, but, to our knowledge, has not been published before. In the present study, we investigate this by 2d/3d direct numerical simulation.

Methods

Fig. 1 shows the actuator design and its resulting Lorentz force. Being inhomogeneous in spanwise direction with strong peaks appearing above the electrode's and magnet's edges, the force has significant wall-normal and spanwise components near the actuator's ends, but is mostly streamwise oriented in the middle. Its decay in wall-normal direction is approximately exponential and scales with the actuator's stripe size a. To produce a counter-phased wave, the

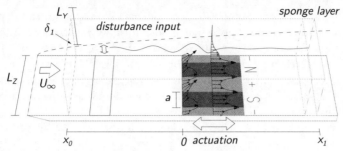

Fig. 1. Computational domain, actuator design and resulting Lorentz force

force oscillates sinusoidally in time using the TS wave frequency. The modified Hartmann number $Z = (j_0 M_0 a^2)/(8\pi\rho U_\infty \nu)$, describing the ratio of Lorentz and viscous forces, serves as a non-dimensional force amplitude, where j_0, M_0, and ρ denote the applied current density, the magnetization of the permanent magnets, and the fluid density, respectively. The Reynolds number, based on the inflow displacement thickness δ_1, is 585. Extending over 800×50 units, the computational domain is decomposed into 1105 spectral elements of polynomial degree 7. Boundary conditions include a Blasius velocity profile at the inflow, no-slip condition $\boldsymbol{u} = 0$ at the wall, and outflow condition $(\boldsymbol{n} \cdot \nabla)\boldsymbol{u} = 0$ at both free-stream and downstream boundary, where a sponge region technique prevents non-physical reflections. For 3d calculations, a Fourier ansatz with 32 modes models the periodic spanwise direction. Artificial disturbances are introduced near the inflow boundary by means of an oscillating body force of fixed frequency $F^+ = (2\pi f\nu/U_\infty^2) * 10^4 = 1.07$, from which TS waves arise and propagate downstream. The local wave amplitude is then determined by finding the maximum root mean square value $\hat{u}_{rms}$ of the streamwise velocity component over the wall-normal direction at given downstream position x.

Results

Plotted in Fig. 2 (left), the uncontrolled TS wave ($Z = 0$, dotted line) grows until it reaches branch II of the neutral stability curve at $x = 381$ (local Re = 1193), which is in agreement to linear stability theory (LST). Beyond this point, it decreases again. During many 2d calculations using a spanwise averaged force, we adjusted amplitude, penetration depth, and phase φ of the actuation to minimize the remaining TS wave amplitude. Results are shown by the dashed line: downstream of the actuation at $0 \le x \le 18$, the amplitude gradually reduces by more than an order of magnitude. For $x > 100$, it oscillates at a level $\hat{u}_{rms} \approx 10^{-4}$. When applying the more realistic, inhomogeneous force (3d case, solid lines) at the same conditions a, Z, φ, due to the four-peak structure, the TS wave is no longer purely two-dimensional after actuation, but modulated in spanwise direction. To determine its amplitude and two-dimensionality, we extracted the $\hat{u}_{rms}$-value from the peak plane $z = 0$ (named peak amplitude in the following) and performed spanwise FFT's at these $\hat{u}_{rms}$ (x,y)-positions, respectively. For clarity, only the first three non-zero modes are shown. Similar to 2d, the mean wave amplitude (mode 0) reduces during actuation, but higher modes initially rise from zero to almost the same level as mode 0, thus indicating a highly three-dimensional TS wave, and actually *increasing* the peak amplitude temporarily. For $x > 30$, however, all modes decrease, and the higher modes quickly settle down around 10^{-7}. Finally, at $x \approx 180$, the TS wave can be considered two-dimensional again, further evolving as predicted by LST.

Measured at $x = 179$, the 2d calculation shows a TS wave reduction, defined as $1 - \hat{u}_{rms}/\hat{u}_{rms,Z=0}$, by 97%, which is only slightly degraded when using the 3d force distribution. This maximum is found at $a = 5.5$ and $Z = 0.57$ for both 2d and 3d. At different penetration depths, however, the actuator performs similarly well as shown in Fig. 2 (right), if the force amplitude Z is adjusted accordingly. This allows for a reduction $> 90\%$ within a range $4.5 \le a \le 7.5$.

We thank Prof. G. E. Karniadakis for the donation of the SEM code. Computational facilities were provided by ZIH, TU Dresden. Financial support by Deutsche Forschungsgemeinschaft (SFB 609 TP C1/C5) is gratefully acknowledged.

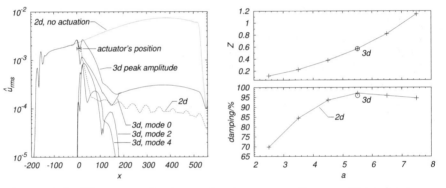

Fig. 2. Left: TS wave amplitudes vs. downstream coordinate x. Right: maximum TS wave damping for various a and corresponding force amplitude Z.

References

1. R. W. Milling. Tollmien-Schlichting wave cancellation. *Phys. Fluids*, 24(5):979 – 981, 1981.
2. H. W. Liepmann and D. M. Nosenchuck. Active control of laminar-turbulent transition. *J. Fluid Mech.*, 118:201–204, 1982.
3. V. V. Kozlov and V. Y. Levchenko. Laminar-turbulent transition control by localized disturbances. In *Turbulence Managment and Relaminarisation*. Springer Verlag, 1987. IUTAM-Symposium, Bangalore, India, 1987.
4. T. Albrecht, R. Grundmann, G. Mutschke, and G. Gerbeth. On the stability of the boundary layer subject to a wall-parallel Lorentz force. *Phys. Fluids*, 18:098103, 2006.

Temporal dynamics of small perturbations for a 2D growing wake

Scarsoglio S.[1], Tordella D.[1], and Criminale W.O.[2]

[1] Dipartimento di Ingegneria Aeronautica e Spaziale, Politecnico di Torino, Torino 10129, Italy stefania.scarsoglio@polito.it, daniela.tordella@polito.it
[2] Department of Applied Mathematics, University of Washington, Box 352420, Seattle, WA 98195-2420, USA lascala@amath.washington.edu

1 Introduction

A general three-dimensional initial-value perturbation problem is presented to study the linear stability of a two-dimensional growing wake. The base flow has been obtained by approximating it with an expansion solution for the longitudinal velocity component that considers the lateral entrainment process [1]. By imposing arbitrary three-dimensional perturbations in terms of the vorticity, the temporal behaviour, including both the early time transient as well as the long time asymptotics, is considered [2], [3], [4]. The approach has been to first perform a Fourier transform of the governing viscous disturbance equations and then resolve them numerically by the method of lines. The base model is combined with a change of coordinate [5]. Base flow configurations corresponding to a R of 35, 50, 100 and various physical inputs are examined. In the case of longitudinal disturbances, a comparison with recent spatio-temporal multiscale Orr-Sommerfeld analysis [6], [7] is presented.

2 The initial-value problem

The base flow is viscous and incompressible. To define it, the longitudinal component of an approximated Navier-Stokes expansion for the two-dimensional steady bluff body wake [1], [8] has been used. The x coordinate is parallel to the free stream velocity, the y coordinate is normal. The coordinate x_0 plays the role of parameter of the system together with the Reynolds number. The analytical expression for the wake profile is $U(y; x_0, R) = 1 - a(R)x_0^{-1/2}e^{-(Ry^2)/(4x_0)}$, where $a(R)$ depends on the Reynolds number [8]. By changing x_0, the base flow profile locally approximates the behaviour of the actual wake generated by the body. The equations are

$$\nabla^2 \widetilde{v} = \widetilde{\Gamma} \tag{1}$$

$$\frac{\partial \widetilde{\Gamma}}{\partial t} + U\frac{\partial \widetilde{\Gamma}}{\partial x} - \frac{\partial \widetilde{v}}{\partial x}\frac{d^2 U}{dy^2} = \frac{1}{R}\nabla^2 \widetilde{\Gamma} \tag{2}$$

$$\frac{\partial \widetilde{\omega}_y}{\partial t} + U\frac{\partial \widetilde{\omega}_y}{\partial x} + \frac{\partial \widetilde{v}}{\partial z}\frac{dU}{dy} = \frac{1}{R}\nabla^2 \widetilde{\omega}_y \tag{3}$$

where $\widetilde{\omega}_y$ is the transversal component of the perturbation vorticity, while $\widetilde{\Gamma}$ is defined as $\widetilde{\Gamma} = \dfrac{\partial \widetilde{\omega}_z}{\partial x} - \dfrac{\partial \widetilde{\omega}_x}{\partial z}$. All physical quantities are normalized with respect to the free stream velocity, the spatial scale of the flow D and the density. By introducing the moving coordinate transform $\xi = x - U_0 t$ [5] and performing a Fourier decomposition of the dependent variables in terms of ξ and z, the governing equations become

$$\frac{\partial^2 \hat{v}}{\partial y^2} - (k^2 - \alpha_i^2 + 2i\alpha_r\alpha_i)\hat{v} = \hat{\Gamma} \tag{4}$$

$$\frac{\partial \hat{\Gamma}}{\partial t} = -ikcos(\phi)(U - U_0)\hat{\Gamma} + ikcos(\phi)\frac{d^2 U}{dy^2}\hat{v}$$

$$+ \alpha_i(U - U_0)\hat{\Gamma} - \alpha_i\frac{d^2 U}{dy^2}\hat{v} + \frac{1}{R}[\frac{\partial^2 \hat{\Gamma}}{\partial y^2} - (k^2 - \alpha_i^2 + 2i\alpha_r\alpha_i)\hat{\Gamma}] \tag{5}$$

$$\frac{\partial \hat{\omega}_y}{\partial t} = -ikcos(\phi)(U - U_0)\hat{\omega}_y - iksin(\phi)\frac{dU}{dy}\hat{v}$$

$$+ \alpha_i(U - U_0)\hat{\omega}_y + \frac{1}{R}[\frac{\partial^2 \hat{\omega}_y}{\partial y^2} - (k^2 - \alpha_i^2 + 2i\alpha_r\alpha_i)\hat{\omega}_y] \tag{6}$$

where $\hat{f}(y, t; \alpha, \gamma) = \displaystyle\int_{-\infty}^{+\infty} \int_{-\infty}^{+\infty} \widetilde{f}(\xi, y, z, t)e^{i\alpha\xi + i\gamma z}d\xi dz$ is the Fourier transform of a general dependent variable, $\phi = tan^{-1}(\gamma/\alpha_r)$ is the perturbation angle of obliquity, $k = \sqrt{\alpha_r^2 + \gamma^2}$ is the polar wavenumber and $\alpha_r = kcos(\phi)$, $\gamma = ksin(\phi)$ are the wavenumbers in ξ and z directions respectively. We choose periodic and bounded initial conditions:

CASE I (symmetric initial condition): $\hat{v}(0, y) = e^{-y^2}cos(\beta y)$, $\hat{\omega}_y(0, y) = 0$

CASE II (asymmetric initial condition): $\hat{v}(0, y) = e^{-y^2}sin(\beta y)$, $\hat{\omega}_y(0, y) = 0$

3 Results and Conclusions

The amplification factor G is defined as the normalized energy density [3], namely $G(t; k, \phi) = E(t; k, \phi)/E(t = 0; k, \phi)$. It effectively measures the growth of the energy at time t, for a given initial condition at $t = 0$ (fig. 1). By defining the temporal growth rate [4] as $r = log|E(t)|/(2t)$ ($E(t)$ is the total perturbation energy) and the angular frequency f as the temporal derivative of disturbance phase, we can evaluate the initial stages of exponential growth and, in the case of 2D disturbances, compare them with the normal mode theory results [6] (fig. 2).

Figure 1 yields three differing examples of early transient periods. Case (a) shows that a growing wave becomes damped, increasing the obliquity angle beyond $\pi/4$. Case (b) corresponds to dispersion relation values far from the saddle point and shows that spatially damped/amplified waves can be temporally amplified/damped. Case (c) demonstrates that perturbations normal to the base flow are stable. Figure 2 presents the comparison between the initial value problem and the Orr-Sommerfeld problem. The results are parameterized with respect to the position x_0 through the polar wavenumber $k = k(x_0)$. Equations are integrated in time beyond the transient until the temporal growth rate asymptotes to a constant value. We observed a very good agreement with the stability characteristics given by the Orr-Sommerfeld theory for both the symmetric and asymmetric arbitrary disturbances considered.

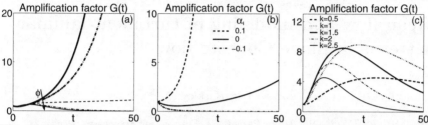

Fig. 1. The amplification factor G as a function of time. **(a)**: $R = 100$, $k = 1.2$, $\alpha_i = -0.1$, $\beta = 1$, $x_0 = 10.15$, $\phi = 0, \pi/8, \pi/4, (3/8)\pi, \pi/2$, symmetric perturbation (case I). **(b)**: $R = 50$, $k = 0.3$, $\beta = 1$, $\phi = 0$, $x_0 = 5.20$, $\alpha_i = -0.1, 0, 0.1$, symmetric perturbation (case I). **(c)**: $R = 100$, $\alpha_i = -0.01$, $\beta = 1$, $\phi = \pi/2$, $x_0 = 7.40$, $k = 0.5, 1, 1.5, 2, 2.5$, symmetric perturbation (case I).

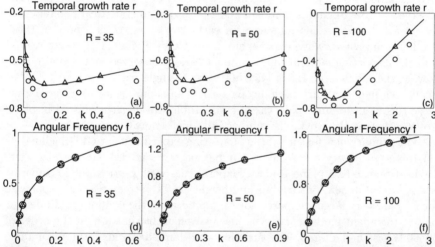

Fig. 2. $\beta = 1$, $\phi = 0$. **(a, b, c)** Temporal growth rate and **(d, e, f)** angular frequency. Comparison between present results (triangles: symmetric perturbation, case I; circles: asymmetric perturbation, case II) and normal mode analysis by Tordella, Scarsoglio and Belan, 2006 Phys. Fluids (solid lines). The wavenumber $\alpha = \alpha_r(x_0) + i\alpha_i(x_0)$, $\alpha_r(x_0) = k(x_0)$ is the most unstable wavenumber in any section of the near-parallel wake (dominant saddle point in the local dispersion relation). The wake sections considered are in the interval $3D \leq x_o \leq 50D$.

References

1. D. Tordella, M. Belan: Phys. Fluids **15**, 7 (2003)
2. P. N. Blossy, W.O. Criminale, L.S. Fisher: J. Fluid Mech. submitted, (2006)
3. D.G. Lasseigne, et al.: J. Fluid Mech. **381**, (1999)
4. W. O. Criminale, et al.: J. Fluid Mech. **339**, (1997)
5. W.O. Criminale, P.G. Drazin: Stud. in Applied Math. **83**, (1990)
6. D. Tordella, S. Scarsoglio and M. Belan: Phys. Fluids **18**, 5 (2006)
7. M. Belan, D. Tordella: J. Fluid Mech. **552** (2006)
8. M. Belan, D. Tordella: Zamm **82**, 4 (2002)

Mean flow and modeling of turbulent-laminar patterns in plane Couette flow

Laurette S. Tuckerman[1] and Dwight Barkley[2]

[1] PMMH (ESPCI, CNRS, Univ. Paris 6, 7), France laurette@pmmh.espci.fr
[2] University of Warwick, United Kingdom barkley@maths.warwick.ac.uk

The fields and phenomena of pattern formation and of turbulence have generally remained well-separated, the first restricted to regular periodic patterns observed at low Reynolds numbers and the second to statistical behavior observed at high Reynolds numbers. In early 2000s, an overlap betwen these two fields appeared, when experimentalists at GIT-Saclay [1, 2, 3] discovered that the coexisting laminar and turbulent regions observed in Taylor-Couette flow in the 1960s and 1980s [4, 5, 6] were a single wavelength of a statistically regular, spatially periodic pattern. The same group discovered that these turbulent-laminar patterns of comparable wavelengths existed in plane Couette flow at the Reynolds numbers at which transition is observed.

The primary new feature of the pioneering experiments at GIT-Saclay that led to the discovery of turbulent-laminar patterns is very simple: size. The Taylor-Couette and plane Couette experiments were carried out in apparati whose lateral dimensions (radius and height for TC; streamwise and spanwise for PC) are very large compared to the gap between the confining cylinders or plates. turbulent-laminar patterns have wavelengths that are on the order of 30–60 times the half-gap. The turbulent and laminar bands form at an angle of 20°–40° to the streamwise direction, at Reynolds numbers (based on half the imposed velocity difference and half the gap) between 300 and 400.

We have computed these turbulent-laminar patterns in plane Couette flow via direct numerical simulation of the 3D time-dependent Navier-Stokes equations. Our simulation uses a spectral element/Fourier code [7] in a rectangular periodic domain which is *tilted* in the direction of the expected pattern as shown in figure 1 (left), economizing on the length in the direction in which the pattern is homogeneous. More specifically, we impose periodic boundary conditions in the lateral directions x, z and rigid boundary conditions $\mathbf{u}(y = \pm 1) = \pm(\mathbf{e}_x \cos \theta + \mathbf{e}_z \sin \theta)$ in the cross-channel direction y. Our domain is of size $L_x \times L_y \times L_z = 10 \times 2 \times 40$ or $10 \times 2 \times 120$, with $O(10^6)$ modes or gridpoints. Our computations show a rich variety of turbulent-laminar patterns as the angle θ and Reynolds number are varied, including spatio-temporal intermittency, branching and travelling states, and localized states analogous to spots [8, 9] some of which are shown in figure 1 (right).

We have conducted a quantitative analysis [10] of the mean flow $\mathbf{U}(y, z) \equiv \langle \mathbf{u} \rangle$ averaged over $T = 2000$ and $L_x = 10$ (the direction in which the pattern is statistically homogeneous), which obeys the averaged Navier-Stokes equations

$$0 = -\left(\mathbf{U} \cdot \nabla\right)\mathbf{U} - \mathbf{f} - \nabla P + \frac{1}{Re}\mathbf{\Delta U} \tag{1}$$

where $\mathbf{f} = (f^U, f^V, f^W) \equiv -\langle (\tilde{\mathbf{u}} \cdot \nabla)\,\tilde{\mathbf{u}} \rangle$ is the Reynolds stress force. We find that the mean flow in the quasi-laminar region of the pattern is not the linear profile of laminar plane Couette flow, but instead represents a non-trivial balance between the viscous and advective forces. Surprisingly, both $\mathbf{U}$ and $\mathbf{f}$ are almost exactly centrosymmetric, i.e. they obey $\mathbf{g}(y, z) = -\mathbf{g}(-y, -z)$, as shown in figure 2. Even more surprisingly, we find that the z-dependent components of $\mathbf{U}$ and $\mathbf{f}$ are almost exactly trigonometric, i.e. that both can be represented by only three functions of y as:

$$\mathbf{g}(x, y, z) = \mathbf{g}_0(y) + \mathbf{g}_c(y)\cos(kz) + \mathbf{g}_s(y)\sin(kz) \tag{2}$$

Substituting the form (2) for $\mathbf{U}$ and $\mathbf{f}$ into (1) leads to a system of 6 ODEs in the 12 functions of y representing the mean flow and the Reynolds stress force. Turbulence modelling would close this system by providing a relation between $\mathbf{f}$ and $\mathbf{U}$. The primary difficulty encountered is that turbulence models are usually formulated and calibrated for high Reynolds numbers, not those near transition, and also present difficulties near walls [11]. Work on this is currently underway, based on the $k - \omega$ model [12].

References

1. A. Prigent, G. Grégoire, H. Chaté, O. Dauchot & W. van Saarloos: Phys. Rev. Lett. **89**, 014501 (2002).
2. A. Prigent, G. Grégoire, H. Chaté & O. Dauchot: Physica D **174**, 100 (2003).
3. S. Bottin, F. Daviaud, P. Manneville & O. Dauchot: Europhys. Lett. **43**, 171–176 (1998).
4. D. Coles: J. Fluid Mech. **21**, 385 (1965).
5. D. Coles & C.W. van Atta: AIAA J. **4**, 1969 (1966).
6. C.D. Andereck, S.S. Liu, & H.L. Swinney: J. Fluid Mech. **164**, 155 (1986).
7. R.D. Henderson & G.E. Karniadakis: J. Comput. Phys. **122**, 191 (1995).
8. D. Barkley & L.S. Tuckerman: Phys. Rev. Lett. **94**, 014502 (2005).
9. D. Barkley & L.S. Tuckerman: Turbulent-laminar patterns in plane Couette flow. In *IUTAM Symposium on Laminar-Turbulent Transition and Finite Amplitude Solutions*, ed. by T. Mullin & R. Kerswell (Springer, Dordecht, 2005), pp. 107–127.
10. D. Barkley & L.S. Tuckerman: J. Fluid Mech., to appear.
11. P.A. Durbin & B.A. Pettersson Reif: *Statistical Theory and Modeling for Turbulent Flows* (Wiley 2001).
12. D.C. Wilcox: AIAA Journal **26**, 1414 (1988).

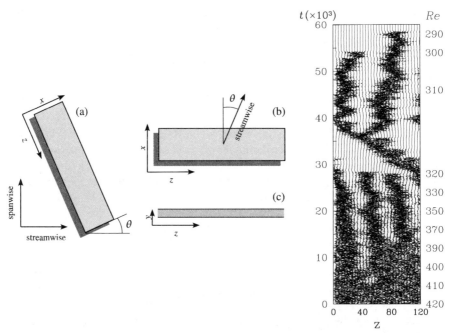

Fig. 1. Left: Computational domain oriented at angle θ to the streamwise-spanwise directions. The z direction is aligned to the pattern wavevector while the x direction is perpendicular to the pattern wavevector. (a) Domain oriented with streamwise velocity horizontal. (b) Domain oriented with z horizontal. (c) View between plates. Right: Timeseries of $w(x = 0, y = 0, z = z_i)$, $z_i = L_z i/N_z$ for $L_z = 120$ and $\theta = 24°$. The Reynolds number (indicated on the right) is lowered in steps over a time $0 \leq t \leq 60,000$. Uniform turbulence at $Re = 420$ is succeeded by formation of three bands at $Re \approx 390$. Two bands disappear almost simultaneously at $Re = 320$. The remaining band moves left, periodically emitting turbulent spurs, of which one finally becomes a second turbulent band. Single band at $Re = 300$ is a localized state, succeeded by laminar Couette flow at $Re = 290$.

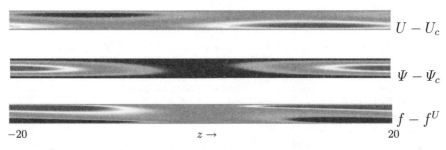

Fig. 2. Mean flow, averaged over x and t, of turbulent-laminar pattern is centro-symmetric in the (y, z) plane. Turbulent region centered around $z = 0$, laminar region around $z = \pm 20$. Top: $U - U_c$ is deviation from laminar Couette profile, here $y \cos \theta$. Middle: $\Psi - \Psi_c$ is deviation of (V, W) streamfunction from that of Couette profile. Bottom: Reynolds stress force.

Streamwise velocity fields in fully developed turbulent pipe and channel flows obtained experimentally

Jason P. Monty, James A. Stewart, Rob C. Williams, and Min S. Chong

Walter Bassett Aerodynamics Laboratory, Mechanical and Manufacturing
Engineering, University of Melbourne, Victoria 3010 Australia.
montyjp@unimelb.edu.au

It is well-known that the large-scale behaviour of turbulent flow over a solid
boundary is not characterised by completely random motions, but rather a
complex interaction of organised vortical structures. Over the past 50 years,
there have been a number of studies suggesting the vortical structures are
characteristically hairpin- or 'Λ'- shaped; see Adrian *et al.* [1] for example.
The behaviour of these hairpin vortices is of great interest since they appear to be the dominant structure in the logarithmic region of the flow. More
recently, it has been proposed that these hairpins interact with each other,
organising themselves into 'packets', that is, long streamwise-aligned chains
of hairpins [3]. The individual 'Λ' vortices lean approximately 45° to the wall
and their legs entrain the fluid such that the flow outside of the vortex structure experiences a relatively higher streamwise velocity, whereas flow between
their legs is retarded relative to the mean flow. Hutchins *et al.* [2], have experimentally shown that very long streaks of low momentum fluid characterise
the logarithmic region of a boundary layer. These streaks may be interpreted
as signatures of packets of eddies. Until recently, the largest structures were
generally accepted to be around two to three boundary layer heights long,
however, the results of [2] & [3] suggest lengths upwards of 10 heights.

This study was undertaken primarily to further the understanding of the
large-scale structures in pipes and channels. Innovative rakes of hot-wire
probes were built to make simultaneous velocity measurements at a given
wall-distance. These are shown scematically in figure 1. For pipe flow, the
array consisted of 15 custom-made hot-wire probes with a pitot-static tube
pair included on the ring for calibration. The hot-wires were spaced $8.4mm$
apart so that the entire array spanned $2.34R$, where $R = 49.4mm$ is the pipe
radius. In the channel flow case, hot-wire probes were constructed by weaving
sewing needles to a piece of standard electronic stripboard with $2.5mm$ hole
spacing. The resultant array contained 10 hot-wire probes and had a total
width of $0.9h$, where $h = 50mm$ is the channel half-height.

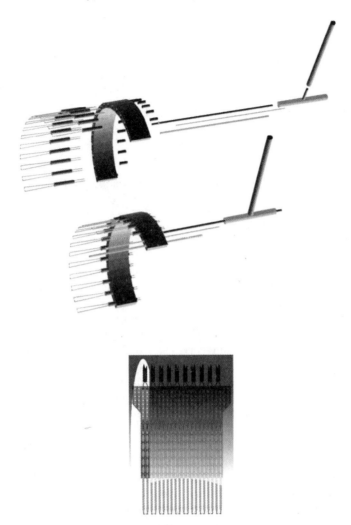

Fig. 1. Upper: schematic of the azimuthally spaced hot-wire rake for pipe flow; two views are shown to illustrate the detachable probe tip design. Lower: channel flow hot-wire array with shading to highlight the important features.

By taking each velocity record and subtracting the spatial and temporal mean, velocity fluctuation fields were produced as shown in figures 2 & 3. The figures show streaks of low momentum (white) fluid in both pipe and channel flow data indicating large-scale structures up to and, in some cases, greater than $20h$ or $20R$ in length. It is also interesting to note that the low speed regions meander around the planes considerably. This is especially interesting in the pipe flow case, where meandering means the structures causing the low speed regions are precessing around the pipe. In some cases we have

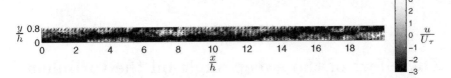

Fig. 2. Channel flow velocity fluctuatin field at $y = 0.14h$ with $Re_\tau = 3200$.

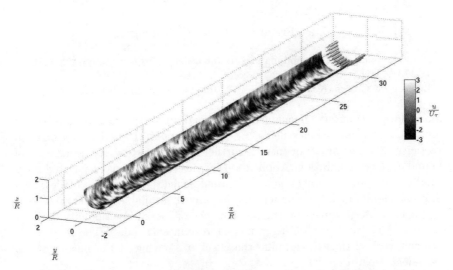

Fig. 3. Pipe flow velocity fluctuation field at $y = 0.15R$ with $Re_\tau = 3000$.

observed low speed regions moving sharply across the entire arry, suggesting the overlying structures have rotated 180°.

The large amount of data gathered from these rakes provides many other unique opportunities to study the flow structure. Energy spectra, two-point correlations and auto-correlations, for example, may all be calculated from the measurements. The result of all of these analyses is a new, far greater insight into the large-scale behaviour of these two important cases of turbulent wall-bounded fluid flows.

References

1. R. J. Adrian, C. D. Meinhart, and C. D. Tomkins. Vortex organization in the outer region of the turbulent boundary layer. *J. Fluid Mech.*, 422:1–54, 2000.
2. N. Hutchins, B. Ganapathisubramani, and I. Marusic. Dominant spanwise fourier modes, and the existence of very large scale coherence in turbulent boundary layers. In *Proc. 15th Aust. Fluid Mech. Conf.* University of Sydney, Aust., 2004.
3. K. C. Kim and R. J. Adrian. Very large-scale motion in the outer layer. *Phys. Fluids*, 11(2), 1999.

The effect of the sweep angle on the turbulent separation bubble on a flate plate

Astrid H. Herbst[1,2], Luca Brandt[2] and Dan S. Henningson[2]

[1] Aerodynamics and Thermodynamics, Bombardier Transportation Sweden AB, 72 173 Västerås, Sweden herbsta@mech.kth.se
[2] Linné Flow Centre, KTH Mechanics, SE-100 44 Stockholm, Sweden

Many flows encountered in engineering applications are characterised by three-dimensional mean velocity fields, *i.e.* they are characterised by mean axial vorticity. A wide variety of them is further complicated by the occurrence of boundary-layer separation. Applications of three-dimensional separated flows can be found among others on airfoils, road vehicles and turbine blades. Separation is still difficult to predict by the presently available turbulence models and it has a deep impact on the performance or lost of functionality of many devices. This justifies the increasing interest in such flows. The aim of the present work is thus to examine the effect of skewing of the mean flow on turbulent separation.

The pressure-induced separation of a turbulent boundary layer evolving over a semi-infinite swept flat plate is considered in the present work and studied by means of Direct Numerical Simulation (DNS). The flow under consideration is therefore spanwise invariant: the mean velocity field is characterised by three velocity components but these can be expressed only as a function of two spatial independent variables. Therefore the spanwise derivative of any flow quantity is zero. The present configuration is thus an obvious first step in the generalisation from a two-dimensional and coplanar mean flow to the fully three-dimensional case. The spanwise invariant mean flow can still be seen as two-dimensional but it is not coplanar. The few related studies found in the literature all consider three-dimensional separation triggered by the geometrical configuration, a sharp edge [1, 4], and to the best of our knowledge no studies on turbulent three-dimensional pressure-induced separation are available. Recently, the effect of the sweep angle on the disturbance amplification and the onset of transition on a laminar separation bubble on a flat plate was studied [3].

The simulations of the turbulent boundary layer exposed to an adverse to favourable pressure gradient have been performed using a spectral code developed at KTH Mechanics.units long, 50 units high and 24 units wide (all length are made non-dimensional with the inflow boundary-layer displacement

(a)

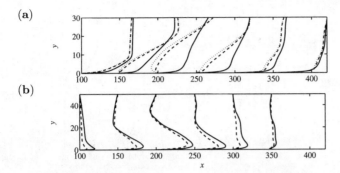

(b)

Fig. 1. Mean velocity profiles in the coordinate system aligned with the free-stream streamline for the different sweep angles under consideration (*solid line* $\alpha_0 = 45°$, *dashed line* $\alpha_0 = 11°$, *dotted line* $\alpha_0 = 0°$) at different streamwise positions. (**a**) streamline component $x + 70 * \hat{U}/C$, (**b**) spanwise component $x + 140 * \hat{W}/C$.

thickness). A resolution of 480 modes in streamwise direction, 193 modes in wall-normal direction and 64 modes in spanwise direction is used, which gives a total of 6 million grid points. The description of the free-stream boundary condition used to establish the bubble can be found in [2]. Two sweep angles will be considered here, $\alpha_0 = 11°$ and $\alpha_0 = 45°$, where α_0 denotes the angle between the incoming free-stream velocity and the direction of the imposed pressure gradient measured in the region of zero pressure gradient upstream of the bubble.

Some of the results obtained by averaging in time and in the spanwise direction are reported in the figures. In figure 1(a) the mean velocity profiles in the direction of the local free-stream streamline are shown for the three cases under consideration. When increasing the sweep, the profiles appear fuller and no counterflow is observed. This is explained by the fact that the angle with respect to the direction of the pressure gradient is increasing and therefore the negative velocity in the streamwise direction is not longer yielding a dominant projection in the streamline direction. The cross-flow mean velocity is reported in figure 1(b). By comparison with Fig. 1(a), it can be observed that the thickness of the cross-flow layer is larger than that of the reverse flow region and of the viscous boundary layer relative to the streamline component $\hat{U}$.

The turbulent kinetic energy normalised by U_0C_0 and C_0^2 is shown in figures 2(a) and (b) respectively, where U_0 and C_0 are the inflow free-stream velocities in the direction of the imposed pressure gradient and of the outer streamline. The increase in the turbulent fluctuations in the free-shear layer limiting the size of the bubble is evident in both plots. The normalisation with U_0C_0 seems to be more appropriate in the outer part of the boundary layer, whereas the scaling with the total velocity C_0 seems to capture the near-wall dynamics for the attached flow better. An increased turbulent activity is observed close to the wall in the separated region for the case of largest sweep, yielding a double-peaked structure of the kinetic energy profiles. Such

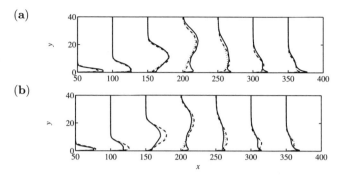

Fig. 2. Wall-normal profiles of turbulent kinetic energy for the different sweep angles under consideration (*solid line* $\alpha_0 = 45°$, *dashed line* $\alpha_0 = 11°$, *dotted line* $\alpha_0 = 0°$) at different streamwise positions **(a)** $x + 2000 * k/C_0U_0$ **(b)** $x + 2000 * k/C_0^2$.

an increase close to the wall can be explained by the fact that with increasing sweep the mean velocity component parallel to the separation line W can provide near-wall shear increasing turbulent production.

It is found that the location of separation is not changed by the introduction of sweep for the angles examined here. The reduction of the separated region is in agreement with previous studies on spanwise invariant three-dimensional separation, where deviations from the sweep-independence principle are observed for angles $\alpha > 40°$. As observed in previous studies, it is not possible to find an universal scaling for the mean-flow quantities and statistics of the turbulent fluctuations valid at all locations. However, it can be deduced from the data that, in general, the velocity C of the external streamline seems to provide the best scaling for the flow quantities at separation and inside the separated region, whereas the streamwise velocity U yields a better data collapse in the reattachment region. The latter can be explained by the fact that U is more directly related to the external pressure distribution. Comparison with the simulations of three-dimensional boundary layers subject to weak adverse pressure gradient indicates that the strength of the pressure gradient, the strain rate and the occurrence of separation have a deeper impact on the turbulence structure than the skewing of the mean flow.

References

1. HANCOCK, P. E. *Exp. Fluids* **27**, 53 (1999).
2. HERBST, A. H. & HENNINGSON, D. S. *Flow Turb. Combust.* **76**, 1 (2006).
3. HETSCH, T. & RIST, U.: The effect of sweep on laminar separation bubbles. In: *Sixth IUTAM Symposium on Laminar-Turbulent Transition* 2006 (ed. R. Govindarajan), pp. 395–400. Springer, Dordrecht, The Netherlands.
4. KALTENBACH, H.-J. *Eur. J. Mech. B/Fluids* **23**, 501 (2004).

Effects of the Streamwise Computational Domain Size on DNS of a Turbulent Channel Flow at High Reynolds Number

Hiroyuki Abe[1], Hiroshi Kawamura[2], Sadayoshi Toh[3] and Tomoaki Itano[4]

[1] Japan Aerospace Exploration Agency, 7-44-1 Jindaiji-higashi, Chofu, Tokyo 182-8522, Japan habe@chofu.jaxa.jp
[2] Tokyo University of Science, 2641 Yamazaki, Noda, Chiba 278-8510, Japan
[3] Kyoto University, Kitashirakawa, Sakyo-ku, Kyoto, Kyoto 606-8502, Japan
[4] Kansai University, 3-3-35 Yamate-cho, Suita, Osaka 564-8680, Japan

1 Introduction

Owing to the rapid increase in the computer power, high Reynolds-number (Re) turbulence phenomena are now being increasingly investigated by performing direct numerical simulations (DNSs). In a turbulent channel flow, several DNSs have been performed over $Re_\tau = u_\tau \delta / \nu = 1000$ (u_τ is the friction velocity, ν the kinematic viscosity and δ the channel half-width) with large computational domains so as to capture large-scale structures in the outer layer as accurately as possible (e.g. [1–3]). However, these DNSs are practically expensive to perform so that another strategy, i.e. use of a small computational domain, may be required for the high Re DNS.

The effects of the computational domain size on the DNS of a turbulent channel flow were discussed at low Re in terms of the smallest domain which sustains turbulence, where the key element is to accommodate the quasi-streamwise vortices and the near-wall streaks [4, 5]. At high Re, one may consider another key element such as the large-scale structures in the outer layer so that the minimal flow unit found in the low Re DNSs [4, 5] may not produce proper turbulence statistics in the high Re DNS. However, this issue is not quantitatively pursued so far, which is indispensable in performing the DNS and also the large eddy simulation (LES) at high Re with the lowest computational cost. In the present study, we investigate the effects of the streamwise computational domain size on the DNS of a turbulent channel flow at $Re_\tau = 1020$. In particular, we focus on how large the streamwise domain is required to obtain turbulence statistics comparable to those obtained from the DNS with a large domain.

2 Numerical Methodology

DNSs with five different domains have been performed in a turbulent channel flow at $Re_\tau = 1020$ by shortening the streamwise domain size for the DNS

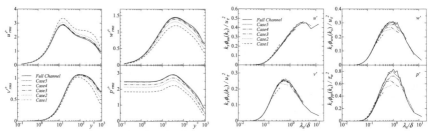

Fig. 1. Rms values of u', v', w', p' normalized by wall units.

Fig. 2. Streamwise pre-multiplied spectra of u', v', w', p' at $y/\delta = 0.5$.

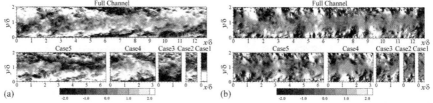

Fig. 3. Contours of the instantaneous u' and p' in the $x - y$ plane: (a) u'^+; (b) p'^+.

of Abe et al. [1] (Full Channel). The domain size, number of grid points, and spatial resolution for Full Channel are $L_x \times L_y \times L_z = 12.8\delta \times 2\delta \times 6.4\delta$ ($L_x^+ \times L_y^+ \times L_z^+ = 13056 \times 2048 \times 6528$), $N_x \times N_y \times N_z = 2048 \times 448 \times 1536$ and $\Delta x^+ = 6.38$, $\Delta y^+ = 0.15 \sim 7.32$, $\Delta z^+ = 4.25$ (x, y, z denote the streamwise, wall-normal and spanwise directions, respectively). The present computational parameters are the same as those of Full Channel except for the domain size and the number of grid point in the x direction, i.e. $L_x = 0.4\delta$, 0.8δ, 1.6δ, 3.2δ, 6.4δ and $N_x = 64$, 128, 256, 512, 1024 for Cases1, 2, 3, 4, 5, respectively. Note that a superscript + denotes the normalization by wall units. Details of the present numerical methodology are given in Abe et al. [1].

3 Results and Discussion

The root-mean-square (rms) values of the streamwise, wall-normal, and spanwise velocity fluctuations, u', v', w', and the pressure fluctuations, p', are normalized by wall units and are given in Fig. 1. As expected, Case1 whose L_x is comparable to the minimal flow unit [4,5] does not show the same distributions as Full Channel. The magnitude of u'^+_{rms} increases with decreasing L_x, while those of v'^+_{rms}, w'^+_{rms}, p'^+_{rms} decrease with decreasing L_x. The distributions of Cases4 and 5 are almost identical with those of Full Channel.

To examine the effects of L_x in detail, the streamwise pre-multiplied spectra of u', v', w', p' at $y/\delta = 0.5$ are shown in Fig. 2, where large-scale u' structures appear clearly (e.g. [6]). Agreement between Case5 and Full Channel is excellent except for the long-wavelength behavior. Even Case4 gives a relatively good approximation to Full Channel, where the u' spectrum captures the energy up to a peak found in Full channel at the streamwise wavelength, $\lambda_x/\delta \approx 3$, and the peak energy in the v', w', p' spectra is well reproduced.

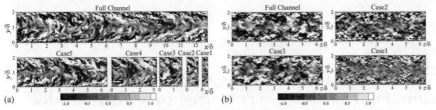

Fig. 4. Contours of the instantaneous w'^+: (a) $x - y$ plane; (b) $y - z$ plane.

This behavior is indeed associated with large-scale shear layers, i.e. backs [7], which play an important role in transferring the momentum in the outer layer as shown in Fig. 3 where contours of the instantaneous u' in the $x - y$ plane are given together with the instantaneous p'. When L_x is short as Case1, the instantaneous u' does not evolve spatially and the instantaneous p' shows small-scale structures only. With increasing L_x, the instantaneous u' tends to evolve spatially and exhibit the large-scale shear layers involving large-scale p' structures. One may notice that the streamwise size of the large-scale shear layers is likely to be about 3δ, which corresponds to the peak wavelength in the u' spectrum shown in Fig. 2. To keep one unit of the large-scale shear layers in the computational domain, L_x must be larger than 3δ. Among the five cases, such cases are Cases4 and 5. This is the reason why Cases4 and 5 give relatively good approximations to Full Channel in the rms values and the spectra. Interestingly, the effects of L_x appear significantly in the spanwise behavior of w' (see Fig. 4), although L_z stays the same as Full Channel for all the cases examined. This is because the large-scale shear layers are the three dimensional structures as the internal shear layers.

4 Conclusions

The effects of L_x were examined by performing the DNSs of a turbulent channel flow at $Re_\tau = 1020$ with the five different domains. Unlike the low Re DNSs, the present DNS with L_x comparable to the minimal flow unit [4, 5] does not produce turbulence statistics properly, whereas those with $L_x \geq 3.2\delta$ give relatively good approximations to Full Channel in the rms values and the spectra. The key element is to keep one unit of the large-scale shear layers in the computational domain, which involve the large-scale p' structures. The present finding is of great importance for saving the computational cost when one performs the DNS and also the LES at high Re.

References

1. H. Abe, H. Kawamura, Y. Matsuo: Int. J. Heat and Fluid Flow **25**, 404 (2004)
2. K. Iwamoto, N. Kasagi, Y. Suzuki: In: *Proc. of the 6th Symp. on Smart Control of Turbulence*, (2005) pp 327-333
3. S. Hoyas, J. Jiménez: Phys. Fluids **18**, 011702 (2006)
4. J. Jiménez, P. Moin: J. Fluid Mech. **225**, 221 (1991)
5. J. Jiménez, A. Pinelli: J. Fluid Mech. **389**, 335 (1999)
6. J. C. del Álamo, J. Jiménez: Phys. Fluids **15**, L41 (2003)
7. S. K. Robinson: Annu. Rev. Fluid Mech. **23**, 601 (1991)

On the use of Taylor's hypothesis in constructing long structures in wall-bounded turbulent flow

David JC Dennis and Timothy B Nickels

Engineering Department, University of Cambridge, Trumpington Street, Cambridge. djcd2@cam.ac.uk

Summary. Taylor's hypothesis of frozen flow has frequently been used to convert temporal experimental measurements into a spatial domain. This technique has led to the 'discovery' of long meandering structures in the log-region of a turbulent boundary layer. There is some contention over whether Taylor's approximation is valid over large distances. This paper presents an experiment that compares velocity fields constructed using Taylor's approximation with those obtained from particle image velocimetry (PIV), i.e. true spatial data. It was found that Taylor's approximation was largely valid over the spatial range ($\approx \delta$) of the PIV images.

1 Introduction

Taylor's approximation of frozen flow is commonly used to infer spatial information from temporal information in turbulent flows, e.g. in [1]. The hypothesis states that, "if the velocity of the airstream which carries the eddies is very much greater than the turbulent velocity, one may assume that the sequence of changes in u at the fixed point are simply due to the passage of an unchanging pattern of turbulent motion over the point" [2]. In order to test this hypothesis in a turbulent boundary layer we have used high speed PIV and compared the true spatial structure to that inferred from Taylor's hypothesis.

2 Experimental procedure

The experiments were conducted in a water tunnel of dimensions 0.9m $\times$ 0.5m $\times$ 8m length at a velocity of 0.4m/s and at streamwise station of 6m giving a Reynolds number of $Re_\delta \approx 48000$. The experiments involved using a measurement plane parallel to the wall to obtain a velocity field in the streamwise-spanwise plane at a height above the wall at which long structures have previously been found ($y/\delta \approx 0.15$). One spanwise line of vectors is treated as if it were the readings from a rake of hot-wires. These results are used, together with Taylor's hypothesis, to generate a pseudo-spatial image of the structures. This constructed velocity field can then be compared to full spatial field from the PIV images in order to test Taylor's hypothesis

3 Results

Figure 1 shows a contour plot of the streamwise fluctuation, hence dark areas show regions of reduced streamwise momentum. The field constructed using Taylor's approximation compares well with that of the spatial field. This is observed throughout all the data collected, implying that Taylor's approximation is valid for lengths of at least 1δ.

One argument against the use of Taylor's approximation to identify long structures is that it elongates the structures far beyond their genuine length. Figure 1 shows the velocity field in which there appears the end of a some kind of structure (i.e. the end of the dark area). Taylor's hypothesis does not seem to overestimate the length of the structure in this case.

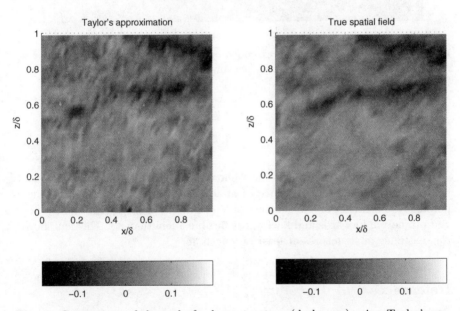

Fig. 1. Comparison of the end of a long structure (dark area) using Taylor's approximation and the real PIV correlation. (Shading scale is fluctuating streamwise velocity in m/s.)

As it has been shown that Taylor's approximation is valid for lengths of at least 1δ it is possible to use a true spatial field (of size δ) and project backwards and forwards in time a distance of δ using Taylor's approximation to construct a velocity field of 3δ in the streamwise direction. This can be assumed to be valid because Taylor's approximation has only been used to project over the length that has already been shown to be correct (i.e. 1δ). An example of this is shown in Fig. 2. A slow-moving structure can be seen to span the whole 3δ.

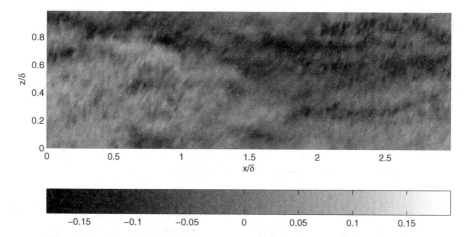

Fig. 2. True spatial field (centre) extended by 1δ either side using Taylor's approximation. (Shading scale is fluctuating streamwise velocity in m/s.)

4 Conclusion

The results of this experiment have shown that Taylor's approximation of frozen flow is valid for a distance of at least 1δ. Since this is the case it is possible to construct a velocity field of length 3δ by using a true spatial field and projecting forwards and backwards in time a length of 1δ. This has shown the existence of structures at least of length 3δ.

References

1. N. Hutchins and I. Marusic. Evidence of very long meandering features in the logarithmic region of the turbulent boundary layers. Under consideration for publication in J. Fluid Mechanics.
2. G.I. Taylor. The spectrum of turbulence. In *Proc. Roy. Soc. Lond.*, volume 164 of *A*, pages 476–490, 1938.

Flow Development in Boundary Layers with Pressure Gradient

Kapil A. Chauhan[1a], Hassan M. Nagib[1b], and Peter A. Monkewitz[2]

[1] Illinois Institute of Technology, 10 W. 32nd St., Chicago, IL 60616, USA
  [a]chaukap@iit.edu, [b]nagib@iit.edu
[2] Swiss Federal Institute of Technology, CH-1015, Lausanne, Switzerland
  peter.monkewitz@epfl.ch

The development of flat plate turbulent boundary layers (TBLs) with streamwise pressure gradient is studied with respect to its mean flow parameters. The approach is based on composite logarithmic mean velocity profiles, similar to one presented recently [1] for zero pressure gradient (ZPG) TBLs. The inner function used is of the form given by Musker[2] and an exponential wake function is utilized in the outer part[3]. In the region far away from the wall, U^+_{musker} develops into the logarithmic law of the overlap region and hence the outer profile is given by the well known form as,

$$U^+ = \frac{1}{\kappa} \ln(y^+) + B + \frac{2\Pi}{\kappa} W_{exp}(y/\delta), \qquad y^+ >> 1 \tag{1}$$

In all the experimental data considered here, the skin-friction was measured directly. The inner normalized mean velocity profiles are utilized to fit the composite form while minimizing least-square errors for $y^+ > 50$ by optimizing κ, B, Π and δ. The variation of κ and B in the overlap region for boundary layers with pressure gradient has been recently documented [4]. In the present continued effort, the deviation of mean velocity profiles from equilibrium ZPG behavior is further studied by analyzing recent and past experiments[4, 5, 6, 7, 8, 9] under strong adverse and favorable pressure gradients. The classical pressure gradient parameter β for these experiments lies between -3 and 20.

Figure 1(a) shows the variation of a turbulent time scale with a mean flow time scale estimated by δ/u_τ and $(-dU_\infty/dx)^{-1}$, respectively, for experiments in the NDF[4]. The ratio of these time scales, γ, represents the slope of lines starting from the origin for various pressure-gradient flows. We note that for a particular pressure gradient, flows with different freestream velocities but same x location are represented along lines of constant γ. Also shown in the figure are the limiting cases of $\gamma \rightarrow \pm\infty$ for separation or re-laminarization. The variation of $\gamma(x)$ with normalized streamwise distance, $(x-x_i)/(x_f-x_i)$, is plotted in Fig. 1(b). Under favorable pressure gradient (FPG), γ decreases, while it increases under adverse pressure gradient (APG). This implies that

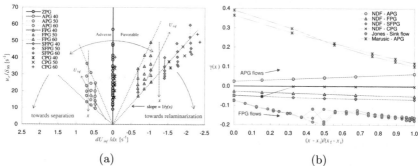

(a)   (b)

Fig. 1. Turbulent and mean flow time scales in flat plate turbulent boundary layers

the presence of pressure gradient not only affects the structure of the mean flow but also that of the turbulence. Higher γ under APG represents a flow condition with increased turbulent activity and vice versa for FPG. As noted in Fig. 1(a), γ is only a function of streamwise distance and not of freestream velocity for a fixed tunnel geometry. This is clear from the collapse of γ at a particular x location for different U_∞ in Fig. 1(b). Fully developed ZPG flows exhibit self-similar asymptotically constant balance of these times scales with increasing Re[10]. Therefore it is speculated that analogous to ZPG flows, a constant ratio of time scales can be an important parameter to describe and evaluate equilibrium turbulent flows with pressure gradient. Figure 1(a) and 1(b) suggest that an equilibrium condition translates to a constant γ. However, neither the present data nor any of the past experiments had conditions with constant γ to confirm this conjecture.

Figure 2(a) shows the variation of the log-law parameters with γ through their product, κB. Unlike ZPG, where κB is a constant, an overall trend of decreasing κB is clearly seen with increasing γ. Relative to ZPG, κB is higher for FPG flows, while it is lower for APG flows. The global behavior of κB with γ is consistent for flows near separation[7], flows near re-laminarization[6], and ZPG flows; although, there is local scatter in the data. Hence, one may assume that a flow with constant γ will have constant κB, i.e., a characteristic of equilibrium flows. Figure 2(b) exhibits the variation of κB with the additive constant B, which helps to understand their individual behavior. We find a remarkable coherence of data from different experiments with varying pressure gradients. This indeed confirms that κ and B are inter-dependent and well-correlated by the imposed pressure gradient, and are not just arbitrarily chosen constants to fit the overlap region. For the range of B shown in the figure, κ is found to vary between 0.2 to 1.1! The behavior of κB with B, is also consistent with constants proposed for Pipe, Channel and ZPG TBLs.

Figure 3 shows the variation of Coles wake parameter Π with normalized Reynolds number. Correlated with changes in the overlap region, the outer part of pressure-gradient boundary layers also deviates from the ZPG behavior. Consistent with the classical theory we find that, Π increases under APG

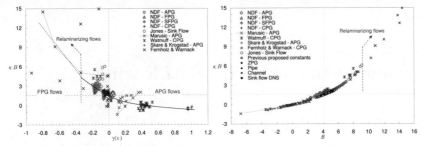

Fig. 2. Variation of log-law parameters κ and B.

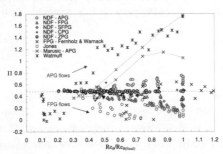

Fig. 3. Variation of wake parameter Π with Re_θ for various pressure gradients.

and decreases under FPG. For strong FPG flows, negative Π values are observed indicating an absence of outer flow, while for strong APG flows near separation, Π values greater than one are achieved. The changes in Π are correlated with behavior of γ implying that the large scale turbulence in the outer part is most affected by the presence of the pressure gradient. We also find that the wake function $\mathcal{W}_{exp}(\eta)$ derived from ZPG flows describes the outer part of PG data equally well.

References

1. Chauhan, K. and Nagib, H. (2006), In *IUTAM Symposium on Computational Physics and New Perspectives in Turbulence, Nagoya, Japan, Sep 11-14*.
2. Musker, A. (1979), AIAA J., vol. 17, pp. 655-657.
3. Chauhan, K., Nagib, H. and Monkewitz, P. (2007), AIAA Paper 2007-532.
4. Chauhan, K., Nagib, H., and Monkewitz, P. (2005), In *iTi Conference on Turbulence, Bad Zwischenahn, Germany, Sep 25-28*, Springer.
5. Marusic, I. and Perry, A. (1995), J. Fluid Mech., vol. 298, pp. 389-407.
6. Warnack, D. and Fernholz, H. (1998), J. Fluid Mech., vol. 359, pp. 357-381.
7. Skare, P. and Krogstad, P. (1994), J. Fluid Mech., vol. 272, pp. 319-348.
8. Watmuff, J. (1998), Data from AGARD-AR-345.
9. Jones, M. (1998), PhD thesis, University of Melbourne, Australia.
10. Nagib, H., Chauhan, K. and Monkewitz, P. (2007), Phil. Trans. Royal Soc. A, vol. 365, pp. 755-770.

Dynamic behaviour of a HALE wing

W.F.J. Olsman, R.R. Trieling, A. Hirschberg and G.J.F. van Heijst

Fluid Dynamics Laboratory, Department of Applied Physics, Eindhoven University of Technology, P.O. Box 513, 5600 MB Eindhoven, The Netherlands
w.f.j.olsman@tue.nl, r.r.trieling@tue.nl, a.hirschberg@tue.nl,
g.j.f.v.heijst@tue.nl

1 Introduction

Within the European project VortexCell2050 a new design High-Altitude Long-Endurance aircraft wing is being developed [2]. From a structural point of view it is desirable to make relatively thick wings. However, thick wings promote flow separation. In order to prevent flow separation or massive vortex shedding, the newly designed airfoil will be equipped with a cavity (vortex cell), in which a vortex is trapped using active flow control.

The goal of the present research is to estimate the unsteady forces due to flow-induced vibrations on an airfoil with a vortex cell without active flow control. The unsteady forces are calculated using a commercial discrete vortex method [1] under prescribed vibration conditions.

2 Estimation of forces

As a first approach the airfoil is modelled by a flat plate without thickness oscillating normal to the flow direction. The dynamic behaviour of an oscillating flat plate in a uniform flow can be solved analytically using linear theory [3]. The numerical results are compared with the results obtain with linear theory.

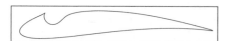

Fig. 1. Airfoil with small cavity.

Fig. 2. Airfoil with huge cavity, the base airfoil is NACA0016.

Numerical simulations were performed on two airfoils with different cavities, both airfoils are shown in Fig. 1 and 2. For all the results presented here the ratio of vertical velocity v of the airfoil to the freestream velocity U is 0.1 and the mean angle of attack is zero. The results of these numerical simulations are presented in Fig. 3 and 4. The figures show the nondimensional lift force L/L_0 for different Strouhal numbers $Sr = \frac{c\omega}{2U}$, where L_0 is the quasi-steady lift force of a flat plate, c is the chord of the plate and ω is the frequency of oscillation in rad/s. On the x-axis and y-axis the imaginary and real part of L/L_0 is shown respectively. The modulus of a point on the curve gives the amplitude of the force whereas the argument of that point gives the phase shift with respect to the vertical velocity of the airfoil or plate.

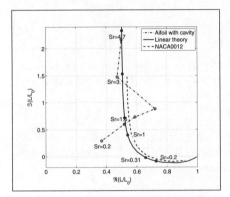
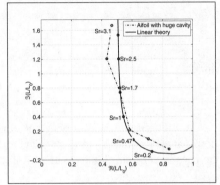

Fig. 3. Nondimensional lift force for different values of Sr for an airfoil with a small cavity (dash-dotted line), linear theory for a flat plate (solid line) and NACA0012 (dashed line).

Fig. 4. Nondimensional lift force for different values of Sr for an airfoil with a large cavity (dash-dotted line) and linear theory for a flat plate (solid line).

The solid lines in Fig. 3 and 4 show the results of linearised theory for a flat plate. The dash-dotted lines show the results for an airfoil with a relatively small cavity and a large cavity in Fig. 3 and 4 respectively. In Fig. 3 the dashed line shows the numerical results of a NACA0012 profile, it is observed that this airfoil shows similar dynamical behaviour as the flat plate. For the airfoil with the small cavity there are significant deviations from the linearized theory both in amplitude and phase of the force.

Figure 5 shows the lift coefficient c_l as a function of nondimensional time t for $Sr = 0.16$. A periodic build up of vorticity behind the cavity is observed, and the liftcoefficient rises. At about t=43.0 the vorticity behind the cavity starts to move downstream and induces a vortex of opposite strength at the trailing edge. Now both vortices form a vortex dipole of opposite strength and the lift coefficient drops very quickly. After this the entire process of the build up and subsequent shedding of a vortex dipole structure start all over again. The time scale of the periodic shedding of a dipole vortex structure is larger than the time scale of the oscillation of the airfoil i.e. it takes several oscillation cycles for the vorticity to build up and be shed.

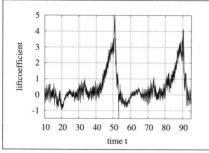

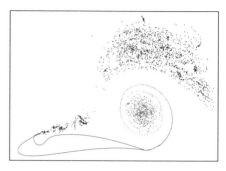

Fig. 5. Nondimensional lift force as a function of nondimensional time for an airfoil with a small cavity.

Fig. 6. Snapshot of the vortex blob distribution at $t = 52.4$ for an airfoil with a small cavity.

At high Strouhal numbers the forces on the airfoil or plate are mainly dominated by inertia forces and the dynamic behaviour of the airfoils with cavities is very similar to that of a flat plate without thickness.

3 Conclusions

From the results it can be concluded that a cavity in the airfoil can cause a significant change in the amplitude and phase of the forces on an airfoil with a cavity with respect to a flat plate without thickness or a symmetric airfoil, like a NACA0012 profile. Consequently the flutter behavior of an airfoil with a cavity can be very different with respect to classic airfoils. Experiments to measure the unsteady forces on an airfoil with a cavity are currently in preparation at the Eindhoven University of Technology in The Netherlands.

References

1. http://www.cmhands.com, Virtual Oscillation–2D, College Master Hands, Inc.
2. http://www.vortexcell2050.org
3. Fung, Y.C. *An Introduction to the Theory of Aeroelasticity*, Dover Phoenix Editions, Toronto, Canada (1969).

POD analysis of large-scale structures through DNS of turbulent plane Couette flow

T. Tsukahara[1], K. Iwamoto[1,2] and H. Kawamura[1]

[1] Dept. of Mech. Eng., Tokyo Univ. of Science,
   2641 Yamazaki, Noda, Chiba 278-8510, Japan `tsuka@rs.noda.tus.ac.jp`
[2] Present address: Dept. of Mech. Sys. Eng., Tokyo Univ. of A & T,
   2-24-16 Nakacho, Koganei, Tokyo 184-8588, Japan `iwamotok@cc.tuat.ac.jp`

1 Introduction

Turbulent plane Couette flow has been extensively investigated, since this is one of the canonical flow cases. In addition, its monotonic velocity profile gives a significantly different character to the flow in the core region, as compared to a pressure-driven channel flow. In the Couette flow, it has been revealed that the central part contains large-scale low and high speed regions (large-scale structure, LSS) extending over a very long streamwise distance. Recently, authors' group [1] carried out the DNS at the moderate Reynolds numbers with applying the large computational domain. We reported that the finite-length LSS is captured when the streamwise domain size is larger than 64δ.

In this paper, the turbulent plane Couette flow is analyzed by means of the proper orthogonal decomposition (POD) in order to extract three-dimensional spatial modes from a flow field obtained from DNS. Moehlis *et al.* [2] and Holstad *et al.* [3] used POD to extract spatial modes from DNS of the Couette flow at low Reynolds numbers with relatively small domains. The present objective is to examine POD modes at low and moderate Reynolds numbers through the large-scale DNS, with emphasis on their Reynolds-number dependence.

2 Numerical condition

The flow is fully developed turbulent plane Couette flow as shown in Fig. 1. Calculations are performed at the Reynolds numbers $Re_w = 4U_w\delta/\nu = 3000$ and 8600. Details of the numerical scheme for the flow field can be found in the previous paper [1]. Instantaneous velocity field can be decomposed as a linear superposition of the eigenfunctions $\phi^{\mathbf{k}}$, where each eigenfunction is specified with a triplet $\mathbf{k} = (k_x, k_z, q)$. Note that an eigenvalue $\lambda^{\mathbf{k}}$ represents the energy in each POD mode of $\phi^{\mathbf{k}}$, so that the eigenfunctions can be sorted according to their contributions to the turbulent kinetic energy.

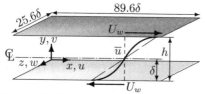

Fig. 1. Configuration of plane Couette flow. The periodic boundary condition is imposed in the horizontal directions.

Table 1. Eigenvalues of the POD modes.

	$Re_w = 3000$		$Re_w = 8600$	
Index	(k_x, k_z, q)	$E^k_\%$	(k_x, k_z, q)	$E^k_\%$
1	$(1, 5, 1)$	1.54	$(1, 6, 1)$	2.45
2	$(0, 6, 1)$	1.49	$(1, 5, 1)$	2.19
3	$(1, 7, 1)$	1.17	$(2, 5, 1)$	1.87
4	$(1, 6, 1)$	1.09	$(0, 5, 1)$	1.11
5	$(0, 7, 1)$	1.04	$(1, 7, 1)$	0.81

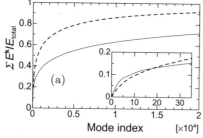

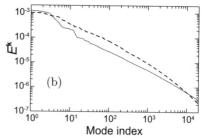

Fig. 2. (a) Cumulative energy summation *versus* a number of POD modes. (b) Energy content $E^k (= \lambda^k)$, non-dimensionalized by U_w and δ, of each single mode.

3 Result

A listing of the first fifth modes is presented in Table 1 with a comparison to the different Reynolds number simulation with the same box size. Here, $E^k_\% = (E^k / E_{total})$ is the percentage of average total energy contained in the (k_x, k_z, q) mode. As shown in Table 1, the energetic POD modes are characterized by the low wavenumbers of $k_x = 0$–2 and $k_z = 5$–7. Especially, the first and second POD modes for $Re_w = 8600$ represent large energy fractions. Their spatial wavelengths of 90δ and 4.2–5.1δ in x and z directions are in good agreement with the scales of the finite-length LSS observed in the previous study with the two-point velocity correlation (refer Fig. 3 in [1]).

The cumulative energy sums for each Reynolds number are compared in Fig. 2 (a). The contributions of the first several modes are much higher than those of the following modes. We also observe that the convergence of the cumulative contribution of POD modes towards the 100% is relatively slow in the case of the higher Reynolds number. However, if emphasis is placed on the energy sum of the 1–15 dominant POD modes, the value for $Re_w = 8600$ is larger than that for the lower Re_w. It is due to the significant contribution of the first three modes for $Re_w = 8600$. In Fig. 2 (b), it is interesting to note that the energies of the first three POD modes are slightly increased (or unchanged) with the increasing Re_w. Therefore the first conclusion drawn from the present result is that the energy of the dominant POD modes (associated with the LSS) is of constant, at least in the range of the present Reynolds numbers.

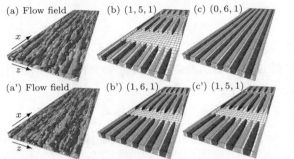

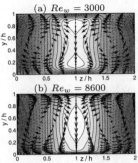

Fig. 3. Iso-surface of u' of instantaneous flow field (a, a'), and of u' from reconstruction of the first (b, b') and the second POD modes (c, c'). Light gray, positive; gray, negative. (a–c) $Re_w = 3000$, (a'–c') 8600.

Fig. 4. Flow field of the energetic POD mode. Vector, v' and w'; contour, u' (negative contour is dashed).

For $Re_w = 8600$, the large-scale staggered pattern similar to the energetic modes is observed in an actual flow realization (see Figs. 3 (a'–c')). On the other hand, the second mode for $Re_w = 3000$ is characterized by $(0, 6, 1)$ without streamwise dependence. This elongated structure exceeds the domain, and its energy contribution is as much as the first mode. Here, if you look at the longitudinal two-point correlation coefficient $R_{uu}(\Delta x)$ (not shown here, please refer to [1]), it gradually decreases to zero at $Re_w = 3000$, while the $R_{uu}(45\delta)$ becomes negative at $Re_w = 8600$. This tendency is consistent with the experiment study [4]. This observation indicates the Reynolds-number dependence of the dominant structure. In the (z, y) plane, Fig. 4 shows the vectors of the most energetic mode and the counter-rotating streamwise vortices. The significant Reynolds-number dependence of the u' distribution is found: its magnitude for $Re_w = 3000$ is largest at $y^+ = 15$–20, while the peak for the higher Re_w is located at the channel center. In consequence, the present result implies that the dominant modes at $Re_w = 3000$ are essentially different from the modes of LSS at $Re_w = 8600$ due to the low-Reynolds-number effect.

This work was conducted in Research Center for the Holistic Computational Science (Holcs) supported by MEXT. The first author (T.T.) is granted by JSPS Fellowship (18-81). The computations were performed with the supercomputer at Information Synergy Center of Tohoku University.

References

1. T. Tsukahara, H. Kawamura, K. Shingai: *J. Turbulence* **7**, No. 19 (2006)
2. J. Moehlis, T.R. Smith, P. Holmes, H. Faisst: *Phys. Fluids* **14**, 2493 (2002)
3. A. Holstad, P.S. Johansson, H.I. Andersson, B. Pettersen: In *Proc. Sixth Int. ERCOFTAC Workshop on Direct and Large-Eddy Simulation* (edited by E. Lamballais, R. Friedrich, B.J. Geurts, O. Metais), Springer, pp 172–173 (2005)
4. N. Tillmark: PhD Thesis, Royal Institute of Technology, Stockholm (1995)

Analysis of a bursting vortex using continuous and orthogonal wavelets

Jori E. Ruppert-Felsot[1], Marie Farge[2], and Philippe Petitjeans[3]

[1] Laboratoire de Météorologie Dynamique du CNRS (UMR 8539), Ecole Normale Supérieure, 24, rue Lhomond - 75231 Paris Cedex 5 - France `jori@lmd.ens.fr`
[2] Laboratoire de Météorologie Dynamique du CNRS, Ecole Normale Supérieure, 24, rue Lhomond - 75231 Paris Cedex 5 - France `farge@lmd.ens.fr`
[3] Laboratoire de Physique et Mécanique des Milieux Hétérogènes, UMR CNRS 7636, Ecole Supérieure de Physique et de Chimie Industrielles, 10, rue Vauquelin - 75231 Paris Cedex 5 - France `phil@pmmh.espci.fr`

We study the time evolution of the quasi-periodic bursting of a laboratory produced vortex using orthogonal and continuous wavelets.

1 Laboratory Experiment

The vortex is produced in laminar channel flow. The vortex is both stretched by axial suction and strained by the channel flow and eventually breaks down, resulting in a burst that leads to the production of turbulence. A new vortex is formed after each burst, and the cycle repeats quasi-periodically. The current bursting vortex has been the subject of previous studies of the buildup of the turbulence due to the bursting [1, 2, 3], however previous measurements were not well resolved simultaneously in time and space.

We measure the velocity field in a plane perpendicular to the vortex by particle image velocimetry and calculate the vorticity component perpendicular to the plane, shown in Fig. 1 (a). The current measurements are sufficiently well resolved in time and space to allow us to study the transient buildup.

2 Why Wavelets?

The vortex under study is a quasi-stationary coherent structure which moves in space before bursting. After bursting the evolution of the remaining pieces which have been spread in space is highly nonlinear. As a result the measured signal is inhomogeneous and non-stationary. It is therefore more natural to analyze this flow using a spatially localized set of basis functions rather than a

Fourier basis. Wavelets consist of translations and dilations of a compact function localized in both physical and spectral space. A wavelet basis is a better choice to analyze signals that contain features well localized in physical space and non-stationary in time [4]. Indeed, it has been found in simulation [5] and laboratory experiment [6] that the dynamics of turbulent flows are dominated by the contribution of a relatively small fraction of wavelet coefficients corresponding to the coherent structures.

3 The Orthogonal and Continuous Wavelet Transforms

The orthogonal wavelet transform (OWT) permits a signal to be decomposed into independent contributions, which can be separately reconstructed, possibly after filtering out some coefficients. To insure orthogonality the transform should be performed with discrete values of translation and dilation corresponding to a dyadic grid in wavelet space. A loss of translation invariance results which makes it difficult to read the OWT coefficients. In contrast, the continuous wavelet transform (CWT) permits the dilations and translations to vary continuously, making the coefficients in wavelet space easy to read and interpret [4]. The CWT unfolds signals in space and scale (and possibly direction), allowing one to study how energy is distributed in space and scale by reading the modulus of complex-valued wavelet cofficients. However this also results in a redundancy of the wavelet coefficients and in a correlation between neighboring coefficients which hinders interpretation.

4 Results

We use the OWT to separate the measured vorticity field into a coherent and an incoherent component [shown in Fig. 1 (b) and (c)], following reference [5]. The coherent field retains the dynamical and statistical properties of the total field, such as the evolution of the non-Gaussian PDF and large-scale

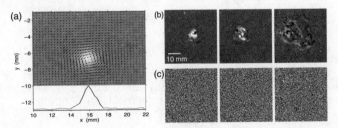

Fig. 1. (a) Close-up of a vortex prior to bursting and a 1D cut of the vorticity field along its center. The velocity field is superimposed on the vorticity field. The largest velocity (vorticity) value corresponds to 0.37 m/s (200 s^{-1}). (b) Time evolution of the coherent and (c) incoherent fields during bursting at 0.33 second intervals

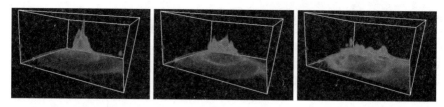

Fig. 2. Isosurfaces of the modulus of the CWT coefficients of three snapshots of the coherent vorticity field during bursting [corresponding to Fig. 1 (b)]. The hortional axes correspond to physical space and vertical axis corresponds to the scale of the transform, with smaller scales at the top

energy spectrum. It is efficiently captured by a small percentage of the large amplitude wavelet coefficients. In contrast, the incoherent field, corresponding to the remaining small amplitude wavelet coefficients, is uncorrelated and featureless with quasi-Gaussian statistics.

We calculate the CWT of the coherent vorticity field, shown in Fig. 2, using a complex-valued Morlet wavelet. The square modulus of the coefficients is thus the local enstrophy density in space and scale. We use the coefficients of the CWT to calculate the evolution of the local intermittency factor, i.e. the deviation from the mean energy spectrum at each location in space [4]. We can thus identify which locations actively contribute to the nonlinear cascade in the inertial range, and which locations are dominated by viscous dissipation.

5 Conclusion

Orthogonal wavelets were used to separate the flow field into a coherent component, capturing the nonlinear dynamics and statistics of the bursting, and an incoherent component void of structure and with quasi-Gaussian statistics. The CWT has allowed us to visualize the wavelet coefficients and track the time evolution of the coherent enstrophy in space and scale. This gives us better insight to interpret the nonlinear cascade of turbulent flows. Each transform has its advantages, thus we recommend that a mixture of the two analyses should be used, each one complementing the other.

References

1. Cuypers Y, A. Maurel, and P. Petitjeans (2003) Phys. Rev. Lett. 91:194502
2. Cuypers Y, Maurel A, and Petitjeans P (2004) J. Turb. 5:N30
3. Cuypers Y, Maurel A, and Petitjeans P (2006) J. Turb. 7:N7
4. Farge M (1992) Ann. Rev. Fluid Mech. 24:395–457
5. Farge M, Schneider K, and Kevlahan N (1999) Phys. Fluids 11:2187–2201
6. Ruppert-Felsot JE, Praud O, Sharon E, and Swinney HL (2005) Phys. Rev. E 72:016311

Flow structure in a bi-axially rotating sphere: a compact turbulence generator

S. Kida, K. Nakayama and S. Goto

Department of Mechanical Engineering and Science, Kyoto University
Yoshida-Honmachi, Sakyo, Kyoto 606-8501, JAPAN
kida@mech.kyoto-u.ac.jp

The flow structure in a precessing sphere is investigated experimentally and numerically. The state diagram, classifying the steady, periodic, aperiodic states, is constructed. In a case of steady state the sphere is shown to be filled with infinitely many streamline tori.

1 Bi-axially rotating sphere

We consider the motion of an incompressible viscous fluid in a precessing sphere, of which the spin angular velocity Ω_s and the precession angular

velocity Ω_p are constant in time and perpendicular to each other (see Fig. 1). The governing equations and boundary condition for the fluid velocity have only two non-dimensional parameters, i.e. the Reynolds number $Re = a^2\Omega_s/\nu$ and the precession rate $\Gamma = \Omega_p/\Omega_s$, where a is the sphere radius and ν is the kinematic viscosity of fluid. The flow characteristics are therefore completely determined by these two parameters apart from the initial condition.

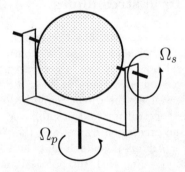

Fig. 1. Bi-axially rotating sphere.

2 State diagram

The flow state may be classified most conveniently into three groups by the temporal behaviour, namely, the steady, periodic, and aperiodic states. The global behaviour of the flow state is shown in Fig. 2, where symbols ○ and ■ denote the laminar (steady or periodic) state and the turbulent (aperiodic) state, respectively. This diagram is obtained by laboratory experiment using a spinning sphere, filled with water, of radius 5cm on a turntable. The velocity field on a rectangular laser sheet, 17mm apart from the sphere centre and

perpendicular to the spin axis, is measured by PIV using video camera fixed on the turntable. Two-time correlation function of the velocity is calculated and averaged over the space. The above three temporal states are classified by the functional form of this correlation function. As anticipated from the fact that the flow in a mono-axially rotating sphere tends to a solid-body rotation, the flow is stable in the regions where $\Gamma \ll 1$ and $\Gamma \gg 1$. The lowest Reynolds number at which the flow can be time-dependent is about 3,200 at the precession rate of about 0.1.

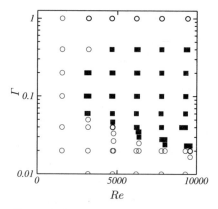

Fig. 2. State diagram. ○, Laminar (steady or periodic); ■, Turbulent.

3 Toral streamlines

It is well-known that the flow in a rotating sphere with constant angular velocity without precession eventually tends to a solid-body rotation. A non-trivial flow structure is generated in a precessing sphere even with a small precession rate. Here, we show a flow field composed of many streamline tori observed by the direct numerical simulation of the Navier-Stokes equation. The spectral method is employed in the spherical polar coordinate system in the precession frame, and the time integration is made using the second-order Adams-Bashforth and the Crank-Nicolson schemes. The velocity field is represented with the poloidal and troidal functions which are expanded in terms of 'Jacobi' polynomials and the spherical harmonics in the radial and two angular directions, respectively. The details of the numerical method will be reported elsewhere.

In Fig. 3 we show the flow structure at $Re = 10$ and $\Gamma = 0.1$. At this relatively small Reynolds number and precession rate the flow is steady and close to the solid-body rotation around the spin (x) axis. Each of the two tori shown in Fig. 3(a) is the full orbit of a single streamline, which is nearly axi-symmetric around the spin axis. Note that these streamlines are slightly twisted on the tori and that the torsion is too weak to distinguish the indi-

(a)   (b)

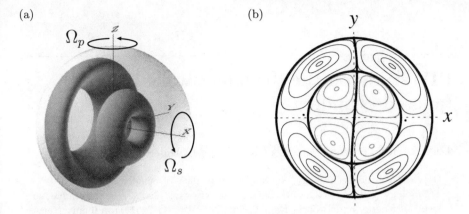

Fig. 3. Structure of a steady flow. (a) Two typical toral streamlines. (b) Cross-section of streamlines on the (x, y)-plane. $Re = 10$. $\Gamma = 0.1$.

vidual lines. The whole sphere is filled with infinitely many such streamline tori.

The cross-sections, on the (x, y) plane, of several of such streamline tori are depicted with thin solid curves in Fig. 3(b). The two coordinate axes are drawn with dashed lines. The corresponding closed curves above and blow the x axis are the cross-sections of the common streamline tori. The whole sphere is divided into four regions by three thick curves. In each region the cross points of the streamlines move either clockwise (e.g. in the inner region of the first quadrant) or counter-clockwise (e.g. in the outer region of the first quadrant).

4 Discussion

A bi-axially rotating sphere (or a precessing sphere) is a simple and fundamental fluid system which creates a variety of flow states by changing the spin and precession angular velocities. We have shown that the turbulent state can be easily generated in laboratory by a highly controlled way. This encourages us to consider it as a standard turbulence generator to study various problems in turbulence, such as mixing. According to our preliminary numerical results, very complicated fluid motions appear in this system probably due to some complicated interactions between the effects of two rotations, the viscosity and the spherical geometry. It may provide us with various fundamental fluid phenomena relevant to the geophysics, such as the geodynamo problem. An extensive numerical and experimental study is under way to make more elaborate state diagram, to examine the flow structure in more detail, to find the stability boundary of steady flow, and so on.

PIV study on the turbulent wake behind tapered cylinders

Jan Visscher[1], Bjørnar Pettersen[1] and Helge I. Andersson[2]

[1] Dept. of Marine Technology, Norwegian University of Science and
Technology (NTNU); NO-7491 Trondheim, Norway; `jan.h.visscher@ntnu.no`
[2] Dept. of Energy and Process Engineering, NTNU; NO-7491 Trondheim, Norway

1 Introduction

A constant uniform incoming flow to a cylinder creates a time-dependent three-dimensional wake. More complex three-dimensional wakes develope behind cylinders in a shear current (nonuniform inflow profiles). This is also the case when we have a constant uniform inflow to a cylinder with linearly changing diameter. The characteristic parameter of these tapered cylinders is the taper ratio, defined as $R_T=l/(d_2 - d_1)$, with d_1 and d_2 being the diameter of the smaller and the larger end, respectively. Tapered cylinders can be found in numerous applications, like marine offshore structures, chimneys, broadcasting towers etc. While the flow around tapered cylinders is subject to several recent numerical investigations [1], there is little actual experimental data available. Piccirillo and Van Atta [2] used hot-wire anemometry and smoke-wire visualization in laminar flow with Reynolds numbers from $Re_2=87$ to $Re_2=179$, based on the large diameter d_2 and the uniform inflow velocity U. Higher Reynolds number regimes have been studied with hot-wire anemometry by Hsiao et al. [3] ranging from $Re_m= 4 \times 10^3$ to $Re_m= 14 \times 10^3$, based on the mean diameter d_m.

2 Experimental setup

This study is part of a joint project which involves experimental and computational analysis of cylinder wake flow and covers multiple geometries and Reynolds numbers. In the present study we use Particle Image Velocimetry (PIV) to measure a large area of the flow field in the cylinder wake with image rates up to 100 Hz. Two cylinder models (A1, B1) of the same length $l = 600$mm with different diameters and aspect ratios (based on the mean diameter, $a = l : d_m$) are used (see Table 1). The taper ratio R_T is fixed to 75:1 to match with existing results. Furthermore, a straight cylinder model (S) is being tested for comparison and check for end effects. All models are equipped with thin circular endplates ($D = 3d_2$) to eliminate disturbances caused by free ends, an effect which has been reported by several authors. An entirely uniform inflow velocity was achieved by towing the models and the measurement system through a still water basin.

3 Results

The results presented here are taken from experiments with the A1, B1 and S cylinder models at Reynolds numbers between $Re_m = 1.15 \times 10^3$ and 18.4×10^3. The data processing is still being done and more results will be shown in the oral presentation. The results obtained from tapered models show the characteristic features for this kind of flow, i.e. oblique vortex shedding, vortex dislocations and cellular shedding behaviour. It becomes obvious in the distribution of the local Strouhal number ($St_{local} = f_s \cdot d_{local}/U$, where f_s is the local shedding frequency) along the cylinder span (Fig. 2): Within shedding cells, St_{local} changes linearly, while cell boundaries appear as abrupt jumps. Correspondingly, the straight cylinder shows a flat St_{local} distribution. Fig. 3 shows the evolving vortices through the cross-flow velocity component sampled along a line parallel to the cylinder axis located 2 d_m downstream of it, as shown in Fig. 1. Repetitive vortex splits are visible which form multiple cells with constant shedding frequencies inside. Additionally, a remarkable change in the shedding behaviour is recorded: The first half of the record (t<17s) reveals two splits (three cells along the span), while the second half (18<t<30s) shows only one split, located in the middle (two cells). However, Hsiao et al. [3] found three shedding cells for this Reynolds number range on cylinders with slightly smaller aspect ratios (10:1). For comparison Fig. 4 displays the flow behind the straight model: Slight obliqueness and spanwise non-uniformities are visible while no vortex dislocations take place. The results obtained so far show the advantage of PIV compared to hot-wire anemometry and encourage further investigation. The focus lies on long timeseries to reveal possible periodicities in shedding schemes and on high-resolved analysis of single vortex split events.

References

1. Narasimhamurthy, V.D., Schwertfirm, F., Andersson, H.I., and Pettersen, B., "Simulation of unsteady flow past tapered circular cylinders using an immersed boundary method", *ECCOMAS CFD06*, (2006).
2. Piccirillo, P. S. and Van Atta, C. W., "Experimental-study of vortex shedding behind linearly tapered cylinders at low Reynolds-numbers", *J. Fluid Mech.*, 246 (1993) 163–195.
3. Hsiao, F., and Chiang, C., "Experimental study of cellular shedding vortices behind a tapered circular cylinder", *Expt. Th. and F. Sc.*, 17 (1998) 179–188.

Table 1. Model sizes and test parameters

Model	Taper ratio	d_1	d_m	d_2	$a = l : d_m$	min Re_m	max Re_m
	[-]	[mm]	[mm]	[mm]	[-]	[-]	[-]
A1	75:1	42	46	50	13:1	2.3×10^3	18.4×10^3
B1	75:1	19	23	27	26:1	1.15×10^3	9.2×10^3
S	∞	42	46	50	26:1	1.15×10^3	9.2×10^3

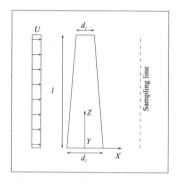

Fig. 1. Coordinate system

Fig. 2. Local Strouhal number distribution

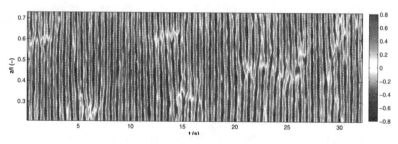

Fig. 3. Time evolution of the normalized cross-flow velocity component v/U for $\mathrm{Re}_m = 18.4 \times 10^3$, tapered cylinder model A1

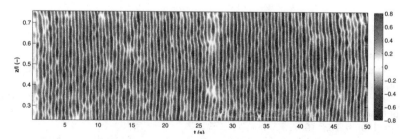

Fig. 4. Time evolution of the normalized cross-flow velocity component v/U for $\mathrm{Re}_m = 2.3 \times 10^3$, straight cylinder model S

Turbulence of drag-reducing polymer solutions

R. Piva[1], C.M. Casciola[1] and E. De Angelis[2]

[1] Dipartimento di Meccanica e Aeronautica, Università di Roma *La Sapienza*,
Via Eudossiana 18, 00184 Roma, Italy
renzo.piva@uniroma1.it, carlomassimo.casciola@uniroma1.it
[2] DIEM, *Alma Mater Studiorum*-Università di Bologna,
via Fontanelle 40, 47100 Forlì
e.deangelis@unibo.it

A small amount of long chain polymers dissolved in an otherwise Newtonian flow is known to deeply modify the structure of turbulence. The most striking effect is perhaps the reduction of drag in wall bounded flows, where the two coefficients in the Prandtl-Kármán friction law, $f^{-1/2} = \hat{A} \log \left(Re_b f^{1/2} \right) - \hat{B}$, which implicitly relates the Fanning friction factor $f = 2 \left(u_*/U_b \right)^2$ to the bulk Reynolds number $Re = U_b D/\nu$, increase from the Newtonian values of 4. and 0.4 to 19. and 32.4, respectively, when the drag-reduction asymptote is reached. A corresponding alteration is found in the mean velocity profile, where the slope the log-law of the wall passes from 2.5 to 11.7 [1].

At equilibrium the chains take the form of random coils. The dimension of the coil, which depend not only on chemistry and degree of polymerization N, but on the solvent also, is characterized by its gyration radius R_0 ($\propto N^{1/2}$ according to Kuhn) which typically takes values order of 100 nm. The contour length of the chain $R_c \propto N$, is several thousands of nanometers (a value taken from [1] is 6×10^3 nm). As an example, in a pipe with diameter $D = 25$ mm operated at a friction Reynolds number $Re_* = u_* D/\nu = 0.7 \times 10^3$, well above the onset for drag reduction, the viscous length is $\ell_* \simeq 18. \times 10^{-3}$ mm $= 18. \times 10^3$ nm, i.e. the polymers are considerably smaller than the smallest scale of turbulence with effective size somewhere between R_c and R_0. Clearly the figures can vary a lot in different conditions.

The parameter which couples the dynamics of the polymer chains and the turbulence is in fact the ratio of two times scales, namely the Deborah (or Weissemberg) number $De_* = \tau_p/\tau_*$. Here $\tau_* = \nu/u_*^2$ is the friction time scale and τ_p is the principal relaxation time of the chain, the estimated time needed to recover equilibrium after the external strain is removed. As an order of magnitude, the principal relaxation time follows from Zimm law, $\tau_p = .325\mu_s R_0^3/(k_B T)$. Taking water as solvent ($T = 25^0 C$, $\nu_s = 1. \times 10^{-6}$ m^2/s), in the example above we have $\tau_p = 0.3 \times 10^{-3}$ s, which gives $De_* = 2.7$.

The discrepancy of scales rules out the possibility that the details of the polymer could be significant for its interaction with turbulence. In terms of time scales, a polymer chain presents a spectrum of relaxation times. The nth time scale can be estimated as $\tau_n \simeq n^{-3/2}\tau_{\text{p}}$, which for the second one yields $\tau_2/\tau_* \simeq 0.35De_*$. We see that the internal dynamics of the chain is unlikely to get substantially coupled to the turbulence. This opens the way for an hyper-simplified description of the polymer, as a system with a single internal degree of freedom. We end up with the dumbbell model consisting of two mass-less beads, acted upon by friction in the relative motion with respect to the carrier fluid, connected through an elastic force of Brownian origin. Given the huge size of the chains, their diffusivity in the solvent is negligible. Hence to a first approximation the dumbbells are transported along the trajectories of the supporting flow, where they experience the local relative velocity difference at the two beads which stretches them by friction – the friction force is proportional to the relative velocity. Given the sub-viscous length, the velocity difference of the flow at the two beads location can be approximated in terms of the local velocity gradient $\nabla \mathbf{u}$ and the vector separation of the two beads $\mathbf{R}$, leading to the equation

$$\frac{D\mathbf{R}}{Dt} = -\frac{1}{De_*}(\mathbf{R} - \boldsymbol{\xi}) + \mathbf{R} \cdot \nabla \mathbf{u},$$

where the dumbbell time scale entering the Deborah number is the ratio of friction coefficient to elastic constant, $\tau_{\text{p}} = \zeta/H$. In the equation above $\boldsymbol{\xi}$ is a Gaussian delta-correlated random process of unit variance introduced to mimic the thermal noise. The end-to-end vector $\mathbf{R}$ is normalized by the equilibrium gyration radius and everything else is expressed in viscous units. The time derivative is understood along the trajectories of the carrier flow. At zero strain, the restoring force shrinks any overextended dumbbell to its equilibrium length. During this phase mechanical energy is dissipated by friction, now $\propto \dot{\mathbf{R}}$. As we will see shortly, this extra-dissipation is crucial in turbulence where the velocity gradient fluctuates and the dumbbell takes a time order τ_{p} to adjust to the environment. In the meanwhile the dumbbell is swept away, carrying its elastic energy on the way to be dissipated.

Clearly, the reaction force exerted on the carrier fluid exactly equals the elastic force. This originates an additional stress in the carrier fluid,

$$\mathbf{T}_{\text{p}} = [n]\frac{HR_0^2}{\rho u_*^2}\left\langle\!\left\langle \mathbf{R} \otimes \mathbf{R} - \frac{\mathbf{I}}{3}\right\rangle\!\right\rangle,$$

where $[n]$ is the number density of dumbbells and double-angular brackets denote averaging with respect to processes at dumbbell scale. The coefficient in front is rearranged as η_{p}/Re_*, with $\eta_{\text{p}} = (\mu_0 - \mu_{\text{s}})/\mu_{\text{s}}$ the ratio of the difference between viscosity of the solution at zero shear stress and solvent viscosity over solvent viscosity.

1 The model for dilute polymer solutions

Averaging the force balance on the dumbbell leads to the evolution equation for the conformation tensor $\mathcal{R} = \langle\langle \mathbf{R} \otimes \mathbf{R} \rangle\rangle$,

$$\frac{D\mathcal{R}}{Dt} = \mathcal{R} \cdot \nabla\mathbf{u} + \nabla\mathbf{u}^{\mathrm{T}} \cdot \mathcal{R} - \frac{1}{De_*}(\mathcal{R} - \mathbf{I}) \ .$$

The model is completed with the momentum equation for an incompressible flow ($\nabla \cdot \mathbf{u} = 0$) augmented with the extra-stress

$$\frac{D\mathbf{u}}{Dt} = -\nabla p + \frac{1}{Re_*}\nabla^2\mathbf{u} + \nabla \cdot \mathbf{T}_{\mathrm{p}} \ .$$

This system of equations admits a conserved quantity, given as the sum of the kinetic energy of the carrier fluid and the elastic (free) energy of the population of dumbbells, E_{p}, proportional to the trace of the conformation tensor $E_{\mathrm{T}} = 1/2u^2 + \eta_{\mathrm{p}}/Re_* tr(\mathcal{R})$. The evolution equation for E_{p} reads

$$\frac{DE_{\mathrm{p}}}{Dt} = \Pi_{\mathrm{p}} - e_{\mathrm{p}} \ ,$$

where the dissipation term – sum of average and fluctuation – is proportional to the energy, $e_{\mathrm{p}} = -2/De_* E_{\mathrm{p}}$, and $\Pi_{\mathrm{p}} = 2\eta_{\mathrm{p}}/Re_* tr(\mathcal{R} \cdot \nabla\mathbf{u})$ is the energy transfer towards the polymer microstructure. The evolution equation for the kinetic energy $E_{\mathrm{k}} = 1/2u^2$,

$$\frac{DE_{\mathrm{k}}}{Dt} = -\nabla \cdot (p\mathbf{u}) + \frac{1}{Re_*}\nabla \cdot \left[(\nabla\mathbf{u} + \nabla\mathbf{u}^T) \cdot \mathbf{u}\right] - \Pi_{\mathrm{p}} - e_{\mathrm{N}} \ ,$$

where e_{N} is the standard Newtonian dissipation, shows that the polymers gain energy at the expenses of the carrier fluid.

On average, the energy drive of the pressure drop balances the total dissipation rate $\langle e_{\mathrm{T}} \rangle$, sum of the two aforementioned components. At fixed pressure drop, drag reduction amounts to an increased throughput Q and is measured by the fractional flow rate enhancement compared to Newtonian, $S = (Q_{\mathrm{p}} - Q_{\mathrm{N}})/Q_{\mathrm{N}}$.

2 Gross features of drag-reducing flows

The left panel of figure 1 shows several mean profiles from direct numerical simulations of the equations above, see [2, 3, 4, 5] for a description of different algorithms. The configuration is a channel flow between planar walls. The fluctuations are periodic in the streamwise, x, and span-wise, z, coordinates. The numerics consists of a Fourier × Chebyshev × Fourier spectral method. The various cases, at the same friction Reynolds number $Re_* = u_* h/\nu$, where h is the channel half-width, differ for the Deborah number. From the relationship

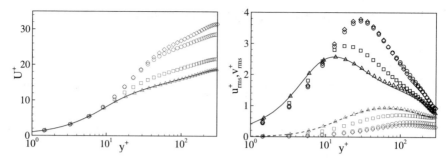

Fig. 1. Left panel: Mean velocity profiles in viscous variables. In the different cases the friction Reynolds is the same, $Re_* = 300$. The solid line gives the Newtonian data. The polymeric data are: $De_* = 18$, $S = 0.004$ (triangles), $De_* = 36$, $S = 0.15$ (squares), $De_* = 72$, $S = .52$ (circles), $De_* = 90.$, $S = .64$ (diamonds). Right panel: Fluctuation intensities in viscous variables, streamwise $u_{rms}^+ = \sqrt{(u'/u_*)^2}$ (heavier symbols) and wall-normal $v_{rms}^+ = \sqrt{(v'/u_*)^2}$ (lighter symbols). The values of De_* are coded as in the left panel.

$$Re_* = \sqrt{h^2/(\nu\tau_p)}\sqrt{De_*}\,,$$

given polymer parameters and Reynolds number, De_* is inversely proportional to pipe diameter or channel height squared. A decrease in D produces a strong flow enhancement, see figure 2.a of [1]. In figure 1 we see this affect in terms of growing De_*. As De_* is increased we observe at first (mild drag-reduction) the outward shift of the logarithmic portion of the profile. At higher De_* a change in the slope of the log-law is distinctly detected. The right panel of the same figure provides the fluctuation velocities. The streamwise component increases with the flow enhancement, whereas the wall normal decreases, consistently with experiments which show outward shift and increased value of the peak u_{rms}^+. The depletion of the Reynolds stresses with drag reduction is apparent,

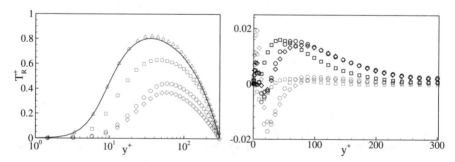

Fig. 2. Left panel: Reynolds shear stress, $-T_R^+ = \langle u'v'\rangle/u_*^2$. Symbols as in fig. 1. Right panel: Contributions of the polymers to the turbulent kinetic energy balance, symbols as in fig. 1, $-\pi_p$ (heavier symbols) and $d\langle\mathbf{T}_p' \cdot \mathbf{u}'\rangle/dy$ (lighter symbols).

left panel of figure 2.

The contribution of the polymers to the turbulent kinetic energy balance,

$$-\langle u'v'\rangle\frac{dU}{dy} - \frac{d\langle q'^2 v'\rangle}{2dy} - \frac{d\langle p'v'\rangle}{dy} + \frac{1}{Re_*}\frac{d^2\langle q'^2\rangle}{2dy^2} - \epsilon_N - \pi_p + \frac{d\langle \mathbf{T}'_p \cdot \mathbf{u}'\rangle}{dy} = 0\,,$$

where q'^2 is the squared modulus of the fluctuation velocity and $\pi_p = \langle \mathbf{T}'_p : \nabla\mathbf{u}'\rangle$, is discussed in the right panel of figure 2. It is splitted in two components, $-\pi_p$ represents the energy drain in favour of the miscrostructure, $\langle \mathbf{T}'_p \cdot \mathbf{u}'\rangle$ contributes to the spatial flux in the wall normal direction, $\phi_s = \langle q'^2 v'\rangle/2 + \langle p'v'\rangle - \langle \mathbf{T}'_p \cdot \mathbf{u}'\rangle - d\langle q'^2\rangle/(2Re_* dy)$. The energy available at a certain location, due to local production and divergence of the spatial flux, is partly dissipated by ordinary viscosity, ϵ_N, and partly moved to the polymers, π_P.

3 Small scale dynamics and energy containing scales

One of the basic effects of polymers in turbulence is to alter the dissipation mechanisms in the system. The basic concepts were introduced in § 1 and reiterated in § 2. Here we emphasize the effect polymers have on the small scales in the simplest context of a homogeneous and isotropic flow, see the experiments on decaying turbulence in [6]. The issue is typically dealt with the classical Kolmogorov equation, extended here to polymeric flows [7, 8],

$$\nabla_r \cdot \langle \delta q'^2 \delta\mathbf{u}' + 2\mathbf{T}'^*_p \cdot \delta\mathbf{u}'\rangle = -4\epsilon_T + 2\langle \delta\mathbf{u}' \cdot \delta\mathbf{f}'\rangle + 2\nu\nabla_r^2\langle \delta q'^2\rangle$$

where derivatives are taken with respect to the separation vector $\mathbf{r}$ between the two points among which the fluctuating velocity difference $\delta\mathbf{u}'$ is evaluated, $\delta q'^2 = \delta\mathbf{u}' \cdot \delta\mathbf{u}'$ and $\mathbf{T}'^*_p$ is the half-sum of the extra-stress at the two points. After the equation is integrated in $\mathbf{r}$-space, in Newtonian turbulence one ends

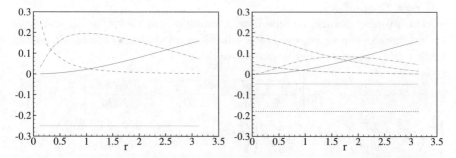

Fig. 3. Scale-energy budget for the Newtonian (left), and a viscoelastic case De = .54 (right). Φ_c/r (dash-dotted), Φ_p/r (dash-double-dotted), ϵ_N (dotted), ϵ_P (dashed), ϵ_T (heavy line), velocity-force correlation (solid), viscous correction (long-dashed).

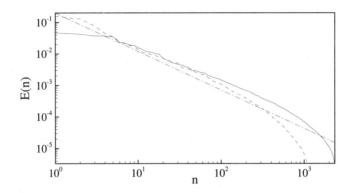

Fig. 4. Karhunen-Loewe expansion. Newtonian, $E_N(n)$ (solid line), and viscoelastic, $E_{VE}(n)$ (dashed line), energy content vs mode numbrr n. The dash-dotted line is Kolmogorov prediction.

up with the four-fifths law of the inertial range, $\langle \delta u_\parallel'^3 \rangle = -4/5 \epsilon_N \, r$. Typical results for a polymeric flow are show in figure 3, where $\Phi_c = \langle \delta q'^2 \delta u_r' \rangle$ and Φ_p is the flux to the polymers (second term on the l.h.s of the equation). The dominance of Φ_p and the depletion of Φ_c in the small scale range is apparent. Note the reduction in ϵ_N, with $\epsilon_T \simeq \epsilon_p$.

Clearly, drag-reduction cannot be understood properly without the assessment of the dissipative dynamics. At any rate the striking aspect of polymers is their ability to modify the larger scales of the system. The large scale are obviously non-universal, and must be dealt with expansions in terms of empirical modes, a posteriori determined by the geometry and the statistics of the field (Karhunen-Loewe decomposition, or after Lumley, proper orthogonal decomposition, POD). For a channel flow figure 4 [9] shows the energy distribution among the modes for a mildly drag-reducing and a Newtonian flow. We clearly detect the increase in the energy content of the largest modes (n small) and the depletion of the smallest ones (n large), as compared to the reference Newtonian case.

For a planar channel or pipe flow, a simultaneous view of small and large scale dynamics can be achieved by a suitable generalization of the scale-energy budget [10], here extended to polymeric flows,

$$\frac{\nabla_r \cdot \langle \delta q'^2 \delta \mathbf{u}' \rangle}{4} + \left[\frac{\langle \delta u' \delta v' \rangle}{2} \frac{dU}{dy} + \frac{\partial \langle v'^* \delta q'^2 \rangle}{4 \partial y} + \frac{\partial \langle \delta p' \delta v' \rangle}{2 \rho \partial y} \right] =$$

$$\left[\frac{\nu}{2} \left(\nabla_r^2 + \frac{\partial^2}{8 \partial y^2} \right) \langle \delta q'^2 \rangle - \epsilon_N^* \right] + \left[\left(\nabla_r \cdot \langle \mathbf{T}'^*_p \cdot \delta \mathbf{u}' \rangle + \frac{\partial \langle \hat{\mathbf{y}} \cdot \delta \mathbf{T}'_p \cdot \mathbf{u}' \rangle}{4 \partial y} \right) - \epsilon_p^* \right],$$

where the asterisk is defined as before, $\hat{\mathbf{y}}$ is the wall-normal unit vector, u', v' and p' are streamwise, wall-normal velocity fluctuations and pressure fluctuation, respectively. This simplified form holds when the two points lay at

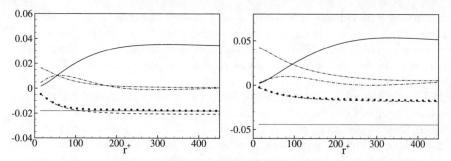

Fig. 5. Scale by scale budget at $y^+ = 150$ for the $De_* = 36$. (left panel) and $De_* = 72$. (right panel). P_e (solid), V_e (dashed), I (dash-dotted). In addition, for the viscoelastic case, G_e (dash-double-dotted) and E_p (dotted).

the same wall-normal distance. The different terms are integrated in r-space over square domains of length $2r$ on planes parallel to the walls and normalized by the area. After integration certain terms are grouped together. We end up with: An inertial transfer term I issuing from the first term on the l.h.s; An effective production P_e – from the square brackets on the l.h.s – with the meaning of scale-energy source at scale r and wall-normal location y due to "Reynolds stress", wall-normal and pressure transport; A Newtonian viscosity related term V_e, from the first bracket on the r.h.s; An overall polymer transport G_e, from the round bracket in the last term on the r.h.s; A scale-energy flux E_p towards the microstructure related to $\epsilon_p^* = \langle \mathbf{T}_p' : \nabla \mathbf{u}' \rangle^*$. Examples of scale-energy budget for the channel flow are provided in the left and right panel of figure 5. In the Newtonian case (not shown) we definitely find a cross-over scale below which the inertial transfer I becomes the leading term, overwhelming the effective production P_e [10]. An interesting feature emerges in the polymer case: The cross-over is between production (solid) and polymer transport (dash-double-dotted). Below the cross-over scale the leading term is the polymer transport, instead of being the inertial transfer as in the Newtonian case. The larger scales are dominated by production effects, the smaller scales are controlled by the polymers. Across the cross-over, energy production and polymer transfer interact directly, a clear indication of the effect of polymers on the structure of turbulence in the large scale range of the system, presumably at the origin of the strong alteration of the energy containing scales we have commented on in figure 4. Comparing the two cases of figure 5, we see how the effect increases with Deborah number.

4 Comments and Remarks

As shown in homogeneous isotropic turbulence, the energy flux intercepted by the polymers is accumulated in the microstructure as elastic energy and dissipated by the relative friction between polymers and solvent. In typical conditions, no evidence of transfer of such energy towards smaller scales has been found, thus indicating that the dissipation in the microstructure is local in wave-number space. This opens the possibility to model the effect of the polymers on the velocity field as a viscous-like process occurring on the whole range of scales where the polymers are acting.

Clearly, near a solid wall the physics is more complex. The classical mean velocity profile is substantially modified in the so-called elastic layer, where the logarithmic slope increases to a limiting value which corresponds to the maximum drag reduction asymptote. As the amount of drag reduction increases, the region occupied by the elastic layer prevails up to filling the entire region. Several phenomenological models (e.g. those of Lumley [11], De Gennes [12], L'vov-Procaccia [13], see also [14]) have been proposed to describe the sequence and the rational of the events taking place in such conditions.

A generalized form of scale-energy budget has been proposed for a deeper evaluation of the alteration of turbulence. In fact, the polymers must interfere directly with the production of turbulent kinetic energy so as to modify the large scales of the flow. This leads to the notion of a range of scales where the draining of energy by the polymers is prevailing. The existence of such elastic range is crucial for the establishment of the altered state of turbulence which provides the observed drag-reduction phenomenology.

References

1. P. S. Virk. *AIChE Journal*, 21:625–655, 1975.
2. R. Sureshkumar, A.N. Beris, R.A. Handler. *Phys. Fluids*, 9:743–755, 1997.
3. E. De Angelis, C.M. Casciola, R. Piva. *Computers & Fluids*, 31:495–507, 2002.
4. G. Boffetta, A. Celani, A. Mazzino. *Phys. Rev. E*, 71(3):036307/1–5, 2005.
5. C.D. Dimitropoulos, Y. Dubief, E.S.G. Shaqfeh, P. Moin.   *Phys. Fluids*, 17:011705/1–4, 2005.
6. E. Van Doom, C.M. White, K.R. Sreenivasan. *Phys. Fluids*, 11:2387–2393, 1999.
7. E. De Angelis, C.M. Casciola, R. Benzi, R. Piva. *J. Fluid Mech.*, 531:1–10, 2005.
8. C.M. Casciola, E. De Angelis. *J. Fluid Mech.*, 2007. To appear.
9. E. De Angelis, C.M. Casciola, V. L'vov, I. Procaccia, R. Piva. *Phys. Rev. E*, 67:056312/1–11, 2003.
10. N. Marati, C.M. Casciola, R. Piva. *J. Fluid Mech.*, 521:191–215, 2004.
11. J.L. Lumley. *J. Pol. Sci., Macrom. Rew.*, 7:263–290, 1963
12. M. Tabor, P.G. De Gennes. *Europhys. Lett.*, 2:519–522, 1986
13. V.S. L'Vov, A. Pomyalov, I. Procaccia, V. Tiberkevich.   *Phys. Rev. Lett.*, 92:244503–1, 2004.
14. K.R. Sreenivasan, C.M. White. *J. Fluid Mech.*, 409:149–164, 2000.

Stereoscopic-PIV study of a dipole in a shallow fluid layer

R.A.D. Akkermans[1], A.R. Cieslik[1], L.P.J. Kamp[1], H.J.H. Clercx[1,2], and G.J.F. van Heijst[1]

[1] Dept. of Applied Physics & J.M. Burgerscentrum, Eindhoven University of Technology, P.O. Box 513, 5600 MB Eindhoven, The Netherlands, e-mail: `R.A.D.Akkermans@tue.nl`

[2] Dept. of Applied Mathematics & J.M. Burgerscentrum, University of Twente, P.O. Box 217, 7500 AE Enschede, The Netherlands

The large-scale dynamics of the Earth's atmosphere and oceans can be considered as quasi-two-dimensional (Q2D). This motivates laboratory experiments in shallow fluid layers aimed at studying the dynamics of vortices and 2D turbulence. The geometrical confinement, when the vertical dimension $\mathcal{H}$ is significantly smaller than the horizontal length scale $\mathcal{L}$, is then a commonly used argument for the flow to behave in a Q2D fashion. When these experiments are performed, a no-slip boundary condition applies to the bottom of the tank and a stress-free condition at the free surface. The resulting shear in the vertical direction leads to a 3D structure of the velocity field and will set up a secondary circulation in the case of a monopolar vortex [1]. Beside this secondary circulation, the emergence of the so-called frontal circulation is seen in the dipolar vortex case [2].

In this contribution, we demonstrate that in general the assumption of Q2D flow in a shallow fluid layer is questionable. For this purpose, we study one of the elementary vortex structures of 2D turbulence, the dipolar vortex, in a shallow fluid layer. In this investigation both experimental and numerical techniques are used. Stereoscopic-PIV is used to measure instantaneous (three-component) velocity fields at several heights inside the fluid layer. Beside this, 3D numerical simulations are performed which provide the full 3D velocity and vorticity field. The flow is characterized by the horizontal Re-number $(= \mathcal{U}\mathcal{L}/\nu)$ and vertical Reynolds number Re_α $(= \mathcal{U}/\alpha\mathcal{L})$. The latter is based on the bottom friction coefficient α, defined as $\nu(\pi/2\mathcal{H})^2$. The characteristic velocity $\mathcal{U}$ is defined as the maximum horizontal velocity of the dipole.

The experimental setup consists of a 52×52 cm^2 square tank with a magnet below the bottom and two electrodes on opposite sides of the tank. A salt solution serves as the conducting fluid enabling electromagnetic forcing. The forcing protocol constitutes of a 1 s pulse of constant current. The fluid layer is thin (up to 12 mm), thus 3D motions are assumed to be suppressed by the shallowness. The settings of the numerical simulations are identical to the experimental ones. Results of the forcing and free evolution phase are presented.

We start with the flow field during the forcing period. An instantaneous velocity field is shown in Fig. 1(a) at time $t = 0.96$ s, for total fluid height of 9.3 mm (measurement plane 5 mm above the bottom). As the total forcing

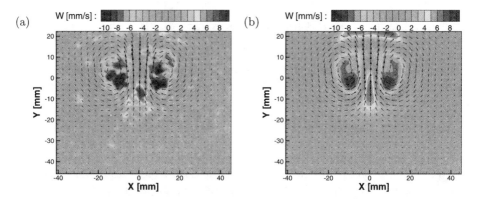

Fig. 1. Instantaneous velocity field of dipolar vortex during the forcing phase ($t = 0.96$ s). Vectors represent in-plane velocities and colors the out-of-plane velocity. (a) Experimental result.(b) Numerical result.

time is 1 s and time equals zero at the onset of forcing, this is at the late stage of the forcing period. As can be seen in Fig. 1(a), two well-defined regions of negative velocities are present inside the vortices and upward motion at the tail of the dipole. During the entire forcing phase, a buildup of downward motion is seen inside the vortices. As a comparison, we show in Fig. 1(b) a numerical result for the same time as the experimental result. A striking quantitative resemblance is seen with the experimental result, although the upward motion at the frontal side is not clearly visible in the experimental result of Fig. 1(a). The origin of the negative velocities inside the vortex cores cannot be explained solely by the acting Lorentz force in the vertical direction since this is directed oppositely with respect to the plane $X = 0$.

In Fig. 2 we see an experimental (a) and numerical result (b) of the dipole during the free-evolution, 1 s after the forcing stopped. A good qualitative resemblance is seen. In front of the dipole we see a recirculation zone, the frontal circulation, upward motion close to the dipole and downward further away. The maximum vertical velocity is 25% of the horizontal velocity. This makes the assumption of 2D behavior of shallow fluid layers indeed questionable. Inside the vortex cores upward motion is seen similar to the secondary circulation in the monopolar vortex [1], although the driving mechanism is quite different. During the forcing phase there is a partial magneto-hydrostatic bal-

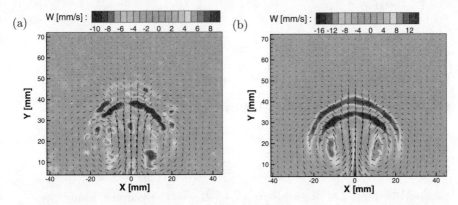

Fig. 2. Instantaneous velocity field of dipolar vortex during the evolution phase ($t = 2.00$ s). (a) Experimental result.(b) Numerical result.

ance between the vertical pressure gradient and the vertical component of the Lorentz force. Departure from this equilibrium (viz. after the forcing stopped) leads to a vertical motion observed which cannot be explained solely by the effect of the no-slip boundary condition. This was confirmed by numerical simulations with both stress-free surface and bottom boundary conditions, which even questions quasi-two-dimensionality of flows in some stratified two-layer fluids. As time progresses we see an oscillating motion of the vertical velocity component inside the two vortex cores, which we identify as inertial oscillations. Initially two well-defined vorticity patches are seen, at later time instances this vertical component of the vorticity becomes more fragmented.

Our findings question the assumption of two-dimensionality in laboratory experiments based on geometric confinement, for decaying as well as forced Q2D turbulence. For this purpose, we have identified regions of significant vertical motion during the forcing and evolution phase. We show a good qualitative agreement between experiments and numerical simulations. A regime diagram is constructed which enables to identify regions in the (Re, Re_α) parameter space where the flow might be considered as Q2D.

The authors wish to thank the Foundation for Fundamental Research on Matter (Stichting voor Fundamenteel Onderzoek der Materie, FOM) for financial support.

References

1. M.P. Satijn, A.W. Cense, R. Verzicco, H.J.H. Clercx, G.J.F. van Heijst: Phys. Fluids.**13**, 1932 (2001)
2. D. Sous, N. Bonneton, J. Sommeria: Phys. Fluids.**16**, 2886 (2004)

Three-dimensional structures during a dipole-wall collision in shallow flows

A.R. Cieslik[1], R.A.D. Akkermans[1], L.P.J. Kamp[1], H.J.H. Clercx[1], and G.J.F. van Heijst[1]

Dept. of Physics & J.M. Burgerscentrum, Eindhoven University of Technology, P.O. Box 513, 5600 MB Eindhoven, The Netherlands, a.r.cieslik@tue.nl

Turbulence in real laboratory experiments or in the atmosphere and oceans is always three-dimensional (3D). Nevertheless, in certain cases the flow might be approximated by a quasi-two-dimensional (Q2D) model. An example of enormous practical importance is the behaviour of the large-scale flows in the Earth's atmosphere and oceans where two-dimensionality is enforced by the presence of rotation of the Earth, the density stratification, and the geometrical confinement of the flow.

Aiming at a better understanding of these geophysical flows, numerous investigators have designed experimental set-ups in order to study Q2D turbulence. Shallow fluid layers are a typical example, see e.g. Ref. [2].

In previous studies it was assumed that shallow flows are Q2D, which means that 3D motions are suppressed by geometrical confinement. It is, however, important to realize that this type of flow is essentially 3D due to the no-slip condition at the bottom and interactions of the flow with lateral walls. Therefore, in the present contribution we investigate the influence of no-slip walls on the distinction between 3D and Q2D dynamics by direct measurements supported by numerical simulations.

Our experimental set-up is similar to that of Rivera and coworkers [2] and consists in our case of a 52 cm × 52 cm square tank with (for our purpose) a single magnet below the bottom and two electrodes on opposite sides of the tank. A salt solution serves as a conducting fluid. The thickness of the fluid layer is 10.4 mm. The interaction of the magnetic field with the electric current creates a dipole, which propagates and collides with the side wall. Initially the dipole has the size of $2L \approx 5$ cm and the propagation velocity of $U \approx 2.5$ cm/s, which results in the Reynolds number $Re = UL/\nu \approx 625$. In order to obtain quantitative data a Stereoscopic Particle Image Velocimetry technique has been developed, which enables the measurement of the three components of the velocity field in a plane [3]. Velocity fields at different heights have been obtained: close to the bottom, in the middle of the fluid layer and near the free surface.

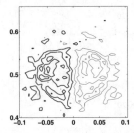 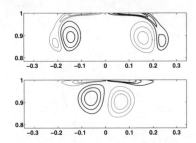

Fig. 1. Left: Initial state, three contours of dimensionless vorticity: 100, 250 and 400; Right: Collision, three vorticity contours: 30, 70 and 110; Thicker contours indicate positive and thinner negative values. A part of computational domain is displayed.

Here numerical 2D simulations with a Chebyshev pseudo-spectral code [1] serve as a framework for interpreting our results. A horizontal velocity field obtained from experiments (just after switching off the current) is placed in the computational domain $[-1,\ 1] \times [-1,\ 1]$, see Fig. 1. The initial velocity field and distances were non-dimensionalized by the rms value of the velocity, $U \approx 2$ mm/s (based on the full flow domain), and the half of the domain size $L = 25$ cm, respectively. The Reynolds number of our computation was $Re = 500$ which results from this scaling. The resolution of our computation was 256×256 modes.

An experimentally obtained dipole has strong inhomogeneities in the vorticity field. After releasing it in the code the dipole reorganizes according to diffusion and advection in the plane of motion, forms a compact structure, collides with the no-slip wall and rebounds, see Fig. 1.

Unlike in the 2D numerical simulation, the initially disordered dipole enhances its inhomogeneities when released in the real 3D experiment, see Fig. 2. This results in a formation of oppositely signed vorticity in the cores of the dipole due to 3D dynamics of the flow. As an effect the vorticity of a core becomes shielded from the frontal part and no substantial rebound was observed. Such a scenario differs dramatically from the one known from 2D flows.

Furthermore, we show the sequence of events that is leading to the creation of 3D motion close to the bottom during the collision process with the lateral side wall, see Fig. 3. While in the middle layer the dipolar structure is well developed, close to the bottom the flow is more complicated with a downward motion in the front. The dipole above is much faster than the structure that is being created close to the bottom and collides earlier with the side wall. Consequently, when a dipole collides with the side wall it will result in a production of a subsequent downward and adverse flow close to the wall and near the bottom of the container: 3D small-scale motion, see the right panel of Fig. 3.

Our results of the dipole-wall collision clearly show significant discrepancies between the 2D flow and those from a real 3D experiment. This research

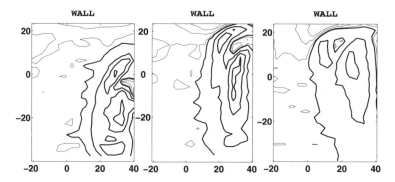

Fig. 2. Sequence of events during collision in the middle layer. Vorticity contours are displayed: -0.5, -0.25, -0.05, 0.05, 0.5, 1, 1.5 s^{-1}. Thick and thin contours as in Fig. 1. Values on axes are in mm.

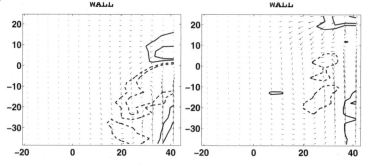

Fig. 3. Sequence of events during collision in the bottom layer. Vertical velocity contours are displayed: -1, -0.5, 0.5 and 1 mm/s; Values on axes are in mm. Solid contours indicate negative and dotted positive values.

is a first step towards a better comprehension of a more complicated case as continuously forced Q2D turbulence in shallow flows. In our view this simple experiment can shed some light on the future interpretation of kinetic energy spectra, especially those measured close to lateral boundaries. At this stage of research, conclusions from the dipole experiment suggest that the measured 2D kinetic energy spectra might be seriously contaminated by small-scale 3D turbulence due to the presence of lateral side walls.

References

1. H.J.H. Clercx. *J. Comput. Phys.*, 137:186, 1997.
2. M.K. Rivera, W.B. Daniel, S.Y. Chen, and R.E. Ecke. *Phys. Rev. Lett.*, 90:104502, 2003.
3. L.J.A. Van Bokhoven, R.A.D. Akkermans, H.J.H. Clercx, and G.J.F. Van Heijst. *Exp. Fluids*, 2007. Under Revision.

The enstrophy cascade in bounded two-dimensional turbulence

W. Kramer[1,2], H.J.H Clercx[1,2,3], and G.J.F. van Heijst[1,2]

[1] Fluid Dynamics Laboratory, Department of Applied Physics, Eindhoven University of Technology, P.O. Box 513, 5600 MB Eindhoven, The Netherlands
w.kramer@tue.nl
[2] J.M. Burgers Centre, Research School for Fluid Dynamics, The Netherlands
[3] Department of Applied Mathematics, University of Twente, The Netherlands

In two-dimensional turbulence there is both a cascade of energy from the forcing scale to larger scales and a cascade of enstrophy towards the smaller scales. The inverse energy cascade, related to $-5/3$ power-law slope in the energy spectrum, causes the formation of a domain-sized structure. The shape of these structures depends strongly on the applied boundary conditions. Clercx and van Heijst found that for decaying 2D turbulence the energy spectrum, measured near no-slip walls, revealed a $-5/3$ slope ranging from the wave number related to the boundary layer thickness up to smaller wave numbers [1]. In the interior a slope steeper than -3 was observed, which is related to the enstrophy cascade. This suggests that the no-slip wall acts as a source of small-scale vortices, that are injected into the flow. In a laboratory experiments with a rotating fluid and numerical simulations Wells, *et al.* found that when the background rotation was slightly modulated the viscous boundary layers are continuously created and subsequently detach from the wall. They form small-scale vortices and vorticity filaments [2] and the spectra from the simulations are consistent with [1]. In forced two-dimensional turbulence the wall-generated vorticity structures interact with the domain-size structure, which can lead to the total collapse of the central structure [3].

To investigate the special role of the no-slip boundaries in forced two-dimensional turbulence, we have set up a simulation on a periodic channel for an aspect ratio of four enclosed by no-slip walls. The flow is governed by the two-dimensional vorticity equation,

$$\frac{\partial \omega}{\partial t} + \mathbf{u} \cdot \nabla \omega = \nu \nabla^2 \omega + F_\omega, \tag{1}$$

with $\mathbf{u}$ to two-dimenesional velocity field, ω the scalar vorticity field, ν the kinematic viscosity and F_ω an external forcing. A pseudo-spectral method based on a expansion in Fourier and Chebyshev polynomials is used to resolve the flow. The fluid, initially in rest, is forced using a standard volume forcing

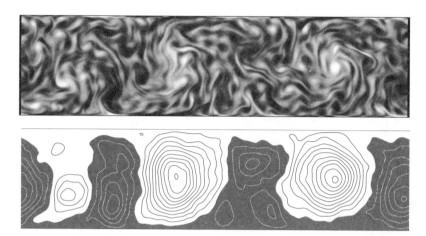

Fig. 1. Vorticity field (upper panel) and streamfunction (lower panel) of forced two-dimensional turbulence on a rectangular periodic channel with no-slip walls.

that is restricted to a small range of wave numbers. The forcing scale is about 1/8th of the domain width ensuring a broad direct enstrophy cascade range. In several steps the forcing amplitude is increased and the viscosity decreased to obtain a Reynolds number of Re $= UL/\nu \approx 20,000$. The r.m.s. velocity is used as the typical velocity and half the channel height as typical length scale. The maximum number of active Fourier and Chebyshev modes is 2048 and 512, respectively.

In the statistically stationary state the flow field consists of a number of domain-sized circulation cells for a large aspect ratio channel (Fig. 1). These circulation cells are not necessarily related to domain-sized vortices, but can be caused by a clustering of like-signed vortices. The no-slip wall is actively producing small-scale vorticity structures and are advected in to the domain by the circulation cells. These small-scale structures hinder the formation of a well structered domain-size vortex.

At lines parallel to the no-slip walls we have computed the one-dimensional spectra and structure functions (Fig. 2). For instance, the p-th order vorticity structure function is given by

$$S_\omega^p(r) = \langle (\omega(x+r, y_0) - \omega(x, y_0))^p \rangle, \qquad (2)$$

where the separation r is taken along the line $y = y_0$. Averaging is performed over ten integral-scale eddy turnover times. In this period there is only a slight increase of total kinetic energy by about ten percent. In the interior the spectra reveal a slope steeper than -3 ranging from the forcing scale towards the Kolmogorov scale. For the spectra measured near the wall we do not observe a $-5/3$ slope, as was found by Clercx and van Heijst for decaying turbulence [1] and Wells $i.e.$ for forced turbulence [2]. The energy input by the forcing mechanism is dominating the spectra and hence the effect of vorticity injection by the no-slip walls is less pronounced.

In addition to the spectra we have investigated the scaling of the velocity structure functions. The third-order structure functions reveals a clear power law scaling range both in the interior and near the wall, where the exponent is equal to three as is expected for the enstrophy cascade range [4]. Extended self-similarity predicts a power-law slope of $p/3$ for the scaling of the pth-order structure function relative to third-order structure function. We observe a clear $p/3$ scaling for structure functions up to order eight, indicating that there is no intermittency in the enstrophy cascade range in the interior (Fig. 2b). However, near the wall the a deviation from this scaling is observed for large scales. The observed intermittency most likely stems from the active creation of small-scale vorticity at the no-slip wall. The same conclustion can be made when comparing the second-order structure functions of the vorticity field and those of a passive tracer. For large separations it can be expected that structure functions are indpendent from the separation, which is indicates the most chaotic uncorrelated state. This behavior of the structure function is observed for the passive tracer. In the vorticity structure function a slight dependence on the separation is found near the wall, which points to the presence of some intermittency.

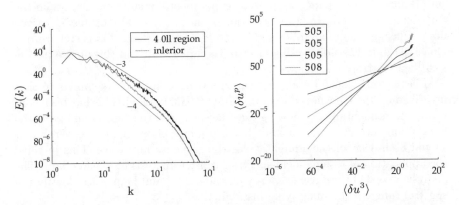

Fig. 2. (a) Longitudinal energy spectrum in the interior and near the wall. (b) Power law scaling of velocity structure functions reveals extended self-similarity

References

1. H. J. H. Clercx and G. J. F. van Heijst: Phys. Rev. Lett., **85** 306 (2000)
2. M. G. Wells, H. J. H. Clercx and G. J. F. van Heijst: J. Fluid Mech, to be published (2007)
3. D. Molenaar, H. J. H. Clercx and G. J. F. van Heijst: Physica D, **196** 329 (2004)
4. A. Babiano, B. Dubrelle and P. Frick, Phys. Rev. E **52** 3719 (1995)

Spectra of quasi-2D turbulence in plasma and fluid during spectral condensation

M.G. Shats[1], H. Xia[2], and H. Punzmann[3]

[1] Research School of Physical Sciences and Engineering,
The Australian National University, Canberra, ACT 0200
Australia, Michael.Shats@anu.edu.au
[2] Hua.Xia@anu.edu.au
[3] Horst.Punzmann@anu.edu.au

Studies of the turbulence spectra modifications in toroidal plasma have revealed interesting similarities between turbulence evolution in plasma during low-to-high confinement transitions and spectral condensation of two-dimensional fluid turbulence [1]. Both show self-organization of turbulence into a largest-scale vortex (in a bound 2D fluid system), or into a zonal flow (in plasma). In this paper we present experimental results on changes in the turbulence spectra during the formation of the largest-scale turbulence-driven flows. Plasma results are obtained in the H-1 heliac (helical axis stellarator). Quasi-2D fluid is studied using electromagnetically driven turbulence in thin stratified electrolyte layers.

Plasma turbulence in our experiments is in many aspects similar to that described by the Charney-Hasegawa-Mima (CHM) model. This has been confirmed experimentally by showing that the spectra of the density and potential fluctuations are very similar and that the phase between them is zero, as it should be in case of the adiabatic electron response in a wave. This justified the application of a single-field spectral transfer analysis to plasma data. The analysis revealed the inverse energy cascade [2, 3] and helped to identify the spectral range of the underlying instability.

In this paper we analyze the shape of the spectra of electrostatic potential fluctuations in low (L) confinement mode characterized by a high level of turbulence and in high (H) confinement mode, where the turbulence level is substantially reduced. Due to obvious difficulties in visualizing turbulence in the high-temperature plasma, the wave number spectra are rarely available. Frequency spectra can, in principle, carry information about the wave number spectra. For example, in the laboratory frame of reference frequencies of the fluctuations are Doppler shifted due to the presence of the $E \times B$ drift: $\omega_{lab} = \omega_{plasma} + k_\theta V_{E \times B}$. The $E \times B$ drift often dominates over the phase velocity in the plasma frame. In this case the fluctuation frequencies in the lab frame are proportional to poloidal wave numbers. Since in the broadband turbulence the wave number spectra are usually isotropic, $k_\theta \approx k_r$, one can assume that $k \approx \sqrt{2}k_\theta \propto \omega$. The $E \times B$ Doppler shift plays in such cases a role of the wave number spectrograph. We have experimentally confirmed linear $k_\theta - f$ relationship, such that the frequency spectra are at least indicative of the shape of k_θ spectra.

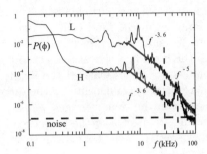

Fig. 1. : Power spectra of the fluctuations in the floating potential, $\tilde{V}_f$ in L and H modes at $\rho = 0.85$.

Fig. 1 shows spectra of the electrostatic potential fluctuations measured at the transport barrier region in plasma in L and H modes. The H-mode spectrum shows strong spectrally broadened zero-mean-frequency zonal flow at $f < 0.3 kHz$ which has poloidal and toroidal wave numbers of $m = n = 0$ [7]. The H-mode spectrum also shows a distinct peak at $f \approx 50$ kHz. This peak is believed to be coinciding with the spectral range of the unstable drift wave driving turbulence in H-mode [3]. It thus marks the high-k (high-frequency) end of the (inverse) energy cascade interval.

Spectra in both regimes show the power-law scalings which are consistent with those theoretically predicted and observed in numerical simulations for the inertial ranges in CHM-type turbulence [4, 5, 6]. In the low confinement mode, the potential enstrophy inertial range shows $E(k) \propto k^{-5}$ scaling, while for lower wave numbers $E(k) \propto k^{-3.6}$ is observed. The $k^{-3.6}$ power law is in agreement with the predicted $k^{-11/3}$ scaling in the inverse energy cascade range. The interface between these two ranges coincides with the unstable spectral range k_{UR} deduced from the spectral transfer analysis [2, 3]. In the high confinement mode turbulence levels are reduced, and a strong mean-zero-frequency zonal flow develops in the plasma. In addition to this, broadband turbulence at $k < k_{UR}$ is observed, again showing $E(k) \propto k^{-11/3}$, similarly to L-mode. These results constitute the first experimental confirmation of the spectra scalings in the Hasegawa-Mima-type turbulence in plasma.

To generate the spectral condensate in a fluid we used the experimental procedure similar to that reported in [8, 9]. Turbulence is generated in stratified thin layers of NaCl solution of different concentrations. An electric current driven through the fluid cell interacts with a 30×30 matrix of permanent magnetic dipoles placed below the bottom of the cell. The size of the boundary is varied between 100 mm and 300 mm such that the actual number of the forcing vortices varies between 100 and 900. The flow is visualized using small (50um) particles suspended in the electrolyte and illuminated using a laser sheet. Images of the tracer particles are captured using high-resolution still camera (12.8 Mpixel) shooting at 3fps and standard video camera at 25 fps. The velocity field is computed using particle image velocimetry algorithm.

When the linear damping in the fluid cell is sufficiently low, the inverse energy cascade leads to the shift of the maximum of the energy spectrum to small wave numbers. If the boundary size exceeds the integral (resistive) scale of 2D turbulence, steady-state is achieved. If the resistive scale exceeds the boundary, spectral energy starts condensing at the largest scale, forming a monopole vortex, as shown in the time averaged velocity fields of Fig. 2.

Changes in the turbulence spectra in the presence of the condensate and without it are somewhat similar to the changes in spectra of the plasma turbulence during L to H transitions.

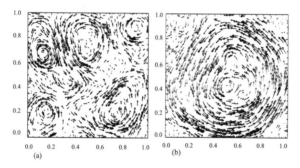

Fig. 2. : Measured velocity fields during the spectral condensate development: (a) early stage, and (b) final steady-state.

References

1. M.G. Shats, H. Xia, H. Punzmann: Phys. Rev. E **71**, 046409 (2005)
2. H. Xia and M. Shats: Phys. Rev. Lett. **91** 155001 (2003)
3. H. Xia and M. Shats: Phys. Plasmas **11** 561 (2004)
4. A. Hasegawa, C. G. Maclennan, Y. Kodama: Phys. Fluids **22**, 2122 (1979)
5. N. Kukharkin, S.A. Orszag, V. Yakhot: Phys. Rev. Lett. **75** 2496 (1995)
6. T. Watanabe, H. Fujisaka, T.Iwayama: Phys. Rev. E **55**, 5575 (1997)
7. H. Xia, M. Shats and H. Punzmann, Phys. Rev. Lett. **97** 255003 (2006).
8. J. Paret and P. Tabeling: Phys. Fluids **10**, 3126 (1998)
9. J. Paret and P. Tabeling: Phys. Rev. Lett. **79**, 4162 (1997)

Heat Transfer during Growth of a Boiling Bubble on the Wall in Turbulent Channel Flow

J.G.M. Kuerten, C.W.M. van der Geld and B.P.M. van Esch

Department of Mechanical Engineering, Technische Universiteit Eindhoven, P.O.Box 513, NL–5600 MB Eindhoven, The Netherlands j.g.m.kuerten@tue.nl

1 Introduction

Heat transfer to a growing boiling bubble at a plane wall is complicated because of a microlayer between vapor and solid, convection and diffusion and the interaction of flow and heat transfer in the wall. However, typical bubble growth histories [1] and heat flow rate histories [2] are known from experiments. This allows to decouple the growth of a vapor bubble from its effect on temperatures and to consider the bubble as a distribution of heat sinks. One of the important issues to be addressed is the extent to which turbulent fluid velocity fluctuations affect bubble growth. Here, this is investigated by means of DNS of turbulent flow and heat in the presence of a bubble. The geometry is a plane channel, where on both walls a heat flux is supplied. The temperature is split into a periodic part and a part which varies linearly with streamwise coordinate and matches the energy supplied at the walls.

2 Numerical method

The Navier-Stokes equation and the heat equation for the temperature of the fluid and of the wall are solved by a spectral method, consisting of a Fourier-Galerkin method in the two periodic directions and a Chebyshev collocation method in the wall-normal direction. Integration in time is second-order accurate. A vapor bubble is created at a position on the wall where the temperature first exceeds a preset value. From that time onwards the bubble starts growing and subtracts energy from both fluid and wall, in a ratio that is prescribed (2:1) as measured in a dedicated set-up in our laboratory [2]. Also the bubble radius history is based on measurements. The bubble is simulated by heat sources of equal strength distributed over a line in y-direction, that ends where the top of the bubble at that time is situated. In the wall, equal heat sinks at three neighboring collocation points in y-direction are employed.

3 Results and analysis

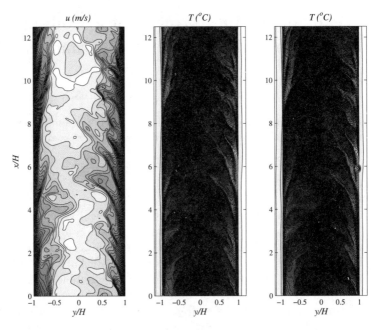

Fig. 1. Velocity and temperature fields at bubble inception in a cross-section of the channel flow (left). Temperature field at bubble detachment (right), with cooling near the bubble clearly visible in the center of the channel wall on the right.

The results show that during the time of growth of a bubble the velocity and temperature near the bubble site hardly change, see Fig.1, and that the velocity length scales exceed those of the temperature, because of the Prandtl number exceeding 1. Moreover, where velocity is low, temperature is high. Temperature variations at the bubble site due to convection are negligible, and diffusion would control heat transfer to the bubble interface if it would be stationary. Figure 2 shows that the liquid near the bubble is cooled by heat consumed for evaporation, whereas further away temperature is constant during bubble growth. Diffusion of heat to the bubble is barely visible in this (y-)direction, since penetration depth due to diffusion is too small. The isotherms in the fluid are shifted because of the growth of the bubble. The isotherms in the wall are shifted because of heat extraction for $y/H < 1.02$ and because of diffusion of heat further away in the wall. Temporal changes in temperature due to diffusion of heat are controlled by the Fourier number, $Fo \overset{\text{def}}{=} \alpha t/L$. Here, L is a typical length scale, with α the heat diffusion constant, $\lambda/(\rho c_p)$, and t time. Both α and λ are about 20 times as larger in the wall than in the fluid. The diffusion penetration depth at certain time for

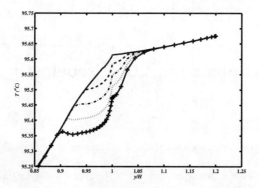

Fig. 2. History of temperatures at the line normal to the wall at the bubble center (at $y = H$). Solid line is at bubble inception, bottom line at bubble detachment, other lines at equidistant time steps in between.

a given isotropic heat sink located at the wall can be estimated to be about $3\sqrt{4\alpha t}$, and diffusion is therefore clearly visible in the wall. Note that in our simulation of a bubble, twice as much heat is taken from the fluid than from the wall. The temporal variation of temperatures near the bubble site and near the wall is relatively small since the bubble was initiated at a spot of high temperature that corresponds to a low velocity. If the thermal properties of the fluid and the wall would be selected differently, such that the α-ratio would change, similar observations could be made. However, the ratio 2 of heat taken from the fluid to that taken from the wall would probably need to be selected differently as well, and no data exist to support the choice. If the wall thickness would be decreased to about the diffusion penetration depth in the wall, a further spread in flow direction of isotherm changes would be found.

The main conclusion is that turbulent fluctuations hardly affect heat transfer to a boiling bubble when bubble-bubble interaction at neighboring sites can be ignored. This observation does not depend on the Reynolds number, but turbulent stresses might affect the forces involved in bubble detachment.

References

1. C.W.M. van der Geld, W.G.J. van Helden, P.G.M.T. Boot: On the effect of the temperature boundary condition on single bubble detachment in flow boiling. In: *Convective flow boiling*, ed by J.C. Chen, 149–154, 1996.
2. M. Kovačević, C.W.M. van der Geld: Single bubble growth in saturated flow boiling on a constant wall temperature surface in uniform approaching flow, International Topical Team Workshop on Two-Phase Systems for Ground and Space Applications, Brussels, September, 2006.

Elastic Turbulence in 2D Viscoelastic Flows

S. Berti[1], A. Bistagnino[2], G. Boffetta[2], A. Celani[3], and S. Musacchio[4]

[1] Department of Mathematics and Statistics, University of Helsinki, P. O. Box 4, 00014 Helsinki, Finland
[2] Dipartimento di Fisica Generale and INFN, Università degli Studi di Torino, Via Pietro Giuria 1, 10125, Torino, Italy
[3] INLN-CNRS, 1361 Route des Lucioles, Sophia Antipolis, 06560 Valbonne, France
[4] Physics of Complex Systems, Weizmann Institute of Science, 76100, Israel

Contact address: `stefano.berti@helsinki.fi`

1 Introduction

It is well known that the addition of small amounts of long chain polymers produces dramatic effects on flowing fluids. A remarkable effect, recently observed experimentally, is the onset of "elastic turbulence" in the limit of very low Reynolds number Re, provided elasticity is high enough [1]. In this regime, the polymer solution displays features typical of turbulent flows. Consequently, elastic turbulence has been proposed as an efficient technique for mixing in very low Re flows as, e. g., microchannel flows [2]. Despite its wide technological interest, elastic turbulence is still poorly understood from a theoretical point of view. Recent predictions are based on simplified versions of viscoelastic models and on the analogy with MHD equations [3].

In this study we investigate the phenomenology of elastic turbulence by means of direct numerical simulations (DNS). The remarkable agreement with experimental results suggests the possibility to understand elastic turbulence on the basis of known viscoelastic models.

2 Viscoelastic Model

The dynamics of the dilute polymer solution is described by the linear viscoelastic Oldroyd-B model:

$$\partial_t \boldsymbol{u} + (\boldsymbol{u} \cdot \boldsymbol{\nabla})\boldsymbol{u} = -\boldsymbol{\nabla}p + \nu\Delta\boldsymbol{u} + \frac{2\eta\,\nu}{\tau}\boldsymbol{\nabla} \cdot \boldsymbol{\sigma} + \boldsymbol{f} \tag{1}$$

$$\partial_t \boldsymbol{\sigma} + (\boldsymbol{u} \cdot \boldsymbol{\nabla})\boldsymbol{\sigma} - (\boldsymbol{\nabla}\boldsymbol{u})^T \cdot \boldsymbol{\sigma} + \boldsymbol{\sigma} \cdot (\boldsymbol{\nabla}\boldsymbol{u}) \qquad \frac{2(\boldsymbol{\sigma} - 1)}{\tau} \tag{2}$$

where $\boldsymbol{u}$ is an incompressible, two-dimensional, velocity field; the matrix of velocity gradients is defined as $(\boldsymbol{\nabla u})_{ij} = \partial_i u_j$ and $\boldsymbol{f}$ is an external forcing; the symmetric positive definite matrix $\boldsymbol{\sigma}$ is the conformation tensor of polymer molecules and $\mathbf{1}$ the unit tensor. The (slowest) polymer relaxation time is denoted by τ; ν is the kinematic viscosity of the solvent and η is the zero-shear contribution of polymers to the total viscosity $\nu_T = \nu(1 + \eta)$ of the solution.

3 Numerical results

The equations of motion are integrated, for a Kolmogorov flow configuration corresponding to $\boldsymbol{f} = (F\cos(y/L), 0)$, by means of a pseudo-spectral method implemented on a grid of resolution 512^2 with periodic boundary conditions; the Weissenberg number Wi, controlling elastic instabilities, is varied in a wide range and Re is kept fixed at a small value.

It is known that DNS of viscoelastic models are limited by instabilities associated with the loss of positiveness of the conformation tensor [4], which are particularly important at high elasticity and restrict the possibility to investigate the elastic turbulent regime by direct integration of Eqs. (1)-(2). To overcome this problem we developed an algorithm based on a Cholesky decomposition of the conformation matrix ensuring symmetry and positive definiteness [5]. To further control numerical instabilities, the simulations have been performed adding a small diffusivity κ for polymers to Eq. (2), the corresponding Schmidt number $Sc \equiv \nu/\kappa$ being always greater than 5×10^2.

The polymer solution flow is found to display features of a strongly non-linear state such as irregular temporal behaviour and spatial disorder (see Fig. 1).

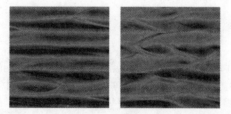

Fig. 1. Vorticity field from DNS, for Wi slightly above the elastic instability threshold (left panel) and for a larger value of Wi (right panel).

At moderate values of Wi, a secondary flow develops in the form of thin filaments superimposed to the basic flow. These small-scale structures are elastic waves reminiscent of Alfvén waves propagating in a plasma. The possibility to observe elastic waves in polymer solutions was theoretically predicted within a simplified uniaxial model [3]. At higher values of elasticity the flow develops active modes at all the scales and finally reaches a turbulent-like state characterized by a power law spectrum of velocity fluctuations slightly steeper than k^{-3}, in excellent agreement with laboratory observations [1].

Since in this elastic turbulence regime the flow is smooth, a suitable characterization of mixing is given by the Lagrangian Lyapunov exponent λ_L. The behaviour of λ_L as a function of Wi is shown in Fig. 2; in the inset of this figure we also plot the Cramer function $G(\gamma)$ which is defined from the probability distribution of Lyapunov exponent fluctuations $P_t(\gamma) \sim \exp{(-tG(\gamma))}$.

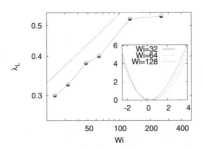

Fig. 2. Lagrangian Lyapunov exponent at varying Weissenberg number ($Re = 1$); the straight dashed line has slope 0.31. Inset: Cramer function $G(\gamma)$.

We observe a growth approximately following a power law $\lambda_L \sim Wi^{0.3}$, accompanied by larger and larger fluctuations γ. In particular, the distribution of γ becomes asymmetric with a larger relative probability of positive fluctuations, in close analogy to the usual behaviour of Newtonian fluids at growing Re.

4 Conclusions

In conclusion, we numerically reproduced the basic phenomenology of elastic turbulence in a 2D configuration. It would be very interesting to extend our work to a 3D setup, in order to better compare with experiments and gain more insight on the basic physical mechanisms underlying the phenomenon.

References

1. A. Groisman and V. Steinberg: Nature **405**, 53 (2000)
2. A. Groisman and V. Steinberg: Nature **410**, 905 (2001)
3. A. Fouxon and V. Lebedev: Phys. Fluids **15**, 2060 (2003)
4. R. Sureshkumar and A.N. Beris: J. Non-Newtonian Fluid Mech. **60**, 53 (1995)
5. T. Vaithianathan and L.R. Collins: J. Comp. Phys. **187**, 1 (2003)

Experimental verification of a theoretical model for the influence of particle inertia and gravity on decaying turbulence in a particle-laden flow

G.Ooms, C.Poelma, M.J.B.M.Pourquie and J.Westerweel[1] and Pietro Poesio Italy[2]

[1] Delft University of Technology, The Netherlands g.ooms@tudelft.nl
[2] Brecia University, Italy Pietro.Poesio@ing.unibs.it

1 Introduction

Recently L'vov et al. (2003) developed a one-fluid theoretical model for a homogeneous, isotropic turbulent flow with particles, paying particular attention to the two-way coupling effect. It is based on a modified form of the Navier-Stokes equations with a wavenumber-dependent effective density of the particle-laden flow and an additional damping term representing the fluid-particle friction. In experiments the effect of gravity due to the difference in density between the particles and the carrier fluid, is present. It causes a production of turbulence due to the net vertical movement of the particles with respect to the fluid. This effect is not included in the model of L'vov et al. So for a proper comparison with experiments it was necessary to extend the theoretical model by including also the gravity effect. Such an extension has recently been made by Ooms and Poesio (2005). In the present paper a comparison is given between predictions made with this extended model and experimental data (given in the literature) for a decaying turbulent particle-laden flow derived from measurements in a water tunnel or wind tunnel.

2 Brief review of the theoretical model

Ooms and Poesio extended the budget equation derived by L'vov et al. for the spectral density of turbulent kinetic energy of the suspension $E_s(t, k)$, with the energy production term $G(t, k)$ representing the turbulence generation due to particle settling

$$\frac{\partial E_s(t, k)}{2 \partial t} + [\gamma_0(k) + \gamma_p(k)] E(t, k) = W(t, k) + R(t, k) + G(t, k). \quad (1)$$

The left-hand side of this equation includes, apart from the time-dependent term, two damping terms: $\gamma_0(k)E_s(t,k)$ caused by the effective viscosity of the suspension and $\gamma_p(k)E_s(t,k)$ caused by the fluid-particle friction. The right-hand side includes the source of turbulent energy $W(t,k)$ due to a possible stirring force (localized in the energy-containing interval of the spectrum), $R(t,k)$ the energy redistribution term due to the interaction between turbulence eddies and (as mentioned) the turbulent energy production term due to particle settling. When integrated over k, $E_s(t,k)$ yields the total turbulent kinetic energy of the effective fluid averaged over all directions in real space.

Using some closure relations, applying an initial condition and a boundary condition (in wave number space) and formulating the problem in a dimensionless way it becomes possible to solve equation (1) numerically for the case of a decaying ($W(t,k) = 0$) turbulent suspension. The details are given in the publication by Ooms and Poesio. They found the following dimensionless groups: the particle volume fraction ψ, the particle mass fraction ϕ, the dimensionless particle response time δ, the fluid Reynolds number Re_f, the Froude number Fr, and the ratio Λ/L. The dimensionless particle response time is defined as $\delta = \tau_p/\tau_L$ with τ_p the particle response time and τ_L the integral timescale of turbulence. The fluid Reynolds number is given by $Re_f = Lu_L/\nu$ with L the integral length scale, u_L the integral velocity scale and ν the kinematic viscosity of the fluid. The Froude number is equal to $Fr = (\rho_p u_L{}^2)/(\Delta\rho\, g\, L)$ in which ρ_p is the mass density of the particles, $\Delta\rho$ the difference in mass density between the particles and the carrier fluid and g the acceleration due to gravity. Λ/L is the ratio of the length scale Λ of the turbulence generated by the settling particles and the integral length scale L of the turbulence generated by the grid in a water tunnel or wind tunnel.

3 Comparison with experiments

We compared model predictions for the turbulent kinetic energy u'^2 with experimental results given in the literature. In our calculations we started with the case of decaying turbulence without particles. To that purpose we estimated the turbulent Reynolds number Re_f at the start of the decay process from the data given in the literature and computed the turbulence decay. To be able to compare model predictions with the experimental results we calculated the distance x travelled by the turbulent flow downstream of the grid in the water tunnel or wind tunnel from the time t in the model by means of the following expression $x = Ut$, in which U is the average velocity in the tunnel. Thereafter we estimated the values of the other parameters at the start of the decay process and calculated the turbulence decay for the case with particles. By comparison with the result for the case without particles the influence of the particles on the turbulence decay process could be studied. This calculated influence was finally compared with the experimentally determined particle influence. The calculations for a turbulent fluid with particles were carried out in two steps; first without the gravity effect and then with the gravity effect included. In this way we were able to study in particular the importance of turbulence production due to settling of the particles on the decay process.

As an example we compare in figure 1 model predictions with experiments carried out by Poelma et al. (2006) in a water tunnel with glass particles with a mass fraction of $\phi = 0.0065$. From the case with particles but without gravity effect it can be seen, that according to the model the inertia effect is not negligible. It causes a significant additional damping of the turbulence. However, the gravity effect nearly balances the inertia effect and the decay rate becomes similar to that for the case without particles. The agreement with experiments is reasonable. This holds also for the comparison with the other experiments reported in the literature.

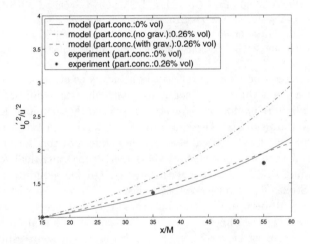

Fig. 1. Normalized inverse value of the turbulent kinetic energy as a function of the (dimensionless) distance behind the grid. u'^2 has been normalized by its value $u_0'^2$ at a certain distance $(x/M = 15)$ behind the grid. Comparison between experimental data of Poelma et al. and model predictions for glass particles in water. The parameters at the start of the calculation have the following values $\psi = 0.0026$, $\phi = 0.0065$, $\delta = 0.0044$, $Re_f = 800$, $Fr = 0.0017$ and $\Lambda/L = 0.2$.

References

1. V.S. L'vov, G. Ooms and A. Pomyalov: Phys. Review E **67**(4), no: 046314 part 2 (2003)
2. G. Ooms, and P. Poesio: Phys. Fluids **17**, no:125101 (2005)
3. C. Poelma, J. Westerweel and G. Ooms: Submitted to Journal of Fluid Mechanics.

Stationary states, Fluctuation-Dissipation Theorem and effective temperature in a turbulent von Karman flow

R. Monchaux[1], P. Diribarne[1], P-H. Chavanis[2], P. Diribarne[1], A. Chiffaudel[1], F. Daviaud[1], and B. Dubrulle[1]

[1] Service de Physique de l'État Condensé, DSM, CEA Saclay, CNRS URA 2464, 91191 Gif-sur-Yvette, France romain.monchaux@cea.fr

[2] Laboratoire de Physique Théorique (UMR 5152), Université Paul Sabatier, 118, route de Narbonne 31062 Toulouse, France chavanis@irsamc.ups-tlse.fr

A yet unanswered question in statistical physics is whether stationary out-of-equilibrium systems share any resemblance with classical equilibrium systems. A good paradigm to explore this question is offered by turbulent flows. Incompressible flows subject to statistically stationary forcing generally reach a kind of equilibrium (in the statistical sense), independent of the initial conditions. Description of turbulence with tools borrowed from statistical mechanics is a long-standing dream, starting with Onsager. In 2D, equilibrium states of the Navier-Stokes equations have been classified through statistical mechanics principle by Robert and his collaborators [6; 1]. More recent advances have been done for 3D axisymmetric flows (an intermediate situation between 2D and 3D) by Leprovost *et al.* [2]. In the following, we present results obtained within this framework for a von Kármán flow.

1 Axisymmetric steady states

We consider the Euler limit of a turbulent flow, in which both forcing and dissipation are removed. In this limit, an axisymmetric flow is characterized by a number of global conserved quantities like energy and helicity. In addition, owing to the symmetry, there is conservation of angular momentum along a velocity line resulting in a Liouville theorem and additional global conserved quantities as Casimirs of angular momentum. This allows the definition of a mixing entropy, and derivation of Gibbs states of the problem through a procedure of maximization of the mixing entropy under constraints of conservation of the global quantities. From the Gibbs state, one can derives general identities characterizing the steady states [2; 4]: $\sigma = F(\Psi)$, $\xi - \frac{FF'}{r^2} = G(\Psi)$, with $\xi = \omega_\theta/r$, where F and G are arbitrary functions linked with conservation laws of the system, σ is the angular momentum, ψ the stream function and ω_θ the azimuthal vorticity. Furthermore, $r^{-1}\partial_r\left(r^{-1}\partial_r\psi\right)+r^{-2}\partial_z^2\psi = -\xi$. We have verified the existence of similar relations for steady states in an experimental von Kármán flow for different impellers and a wide range of Reynolds number (from 100 to 314000). Details of the experimental setup are given in Ref. [5; 3]. A representative result is provided on figure (1). One sees that while data on the whole vessel display significant scatter, preventing the outcome of a well-defined F, the data far from boundaries and impellers gather onto a cubic-shaped function fitted by a two parameters cubic $F(\Psi) = p_1\Psi + p_3\Psi^3$. This fit is then used to obtain G. Furthermore, we found that they depend

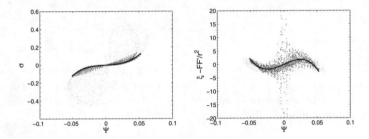

Fig. 1. F (left) and G (right) at $Re = 5 \times 10^5$. The light gray dots corresponds to the whole flow, the dark gray crosses to 50% of the flow: $r/R \in [-0.6; 0.6]$, $z/R \in [-0.4; 0.4]$ and the solid line to a cubic fit.

on the impellers and on Reynolds number: with other impellers F (resp. G) tend to be linear (resp. equal to zero) as Reynolds number is increased [4]. This can be interpreted as a signature of a Beltramization of the flow (*ie* a depletion of nonlinearities).

2 Fluctuations

In the Beltrami limit, the Gibbs states of the Euler equation can be used to derive two relations between steady solutions and their fluctuations:

$$\overline{(\mu\sigma)^2} - \overline{\mu\sigma}^2 = -\frac{\delta\overline{\mu\sigma}}{\delta\overline{\xi}} = \frac{\mu^2}{\beta}r^2, \tag{1}$$

$$\overline{\xi^2} - \overline{\xi}^2 = -\frac{\delta\overline{\xi}}{\delta\mu\overline{\sigma}} = \frac{\beta}{\mu^2}\frac{1}{r^2}.$$

where β^{-1} is an effective temperature and μ is a vortical susceptibility (the Lagrange parameter associated to the energy and helicity respectively). We have tested these two relations for fluctuations in our experiment. Results are provided on figure (2). Once more the relations are satisfied in the core of the flow. Temperature and vortical susceptibility can be derived from the fit parameters. The study of the dependance of these thermodynamic quantities with Reynolds number and impellers is under way [3].

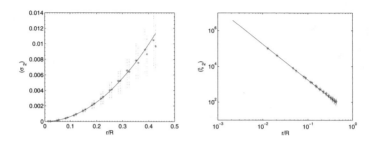

Fig. 2. Fluctuation relation at $Re = 5 \times 10^5$ with the same window in r and z as on figure 1. Left: relation for $\sigma_2 = \overline{\sigma^2} - \overline{\sigma}^2$. Right: relation for $\xi_2 = \overline{\xi^2} - \overline{\xi}^2$. The small dots are the experimental data for all values of z/R, the stars are the averages over z at given r, the line is a $(r/R)^2$ (resp. $(R/r)^2$) fit.

References

[1] P. H. Chavanis. Generalized thermodynamics and Fokker-Planck equations: Applications to stellar dynamics and two-dimensional turbulence. *Phys. Rev. E*, 68:036108, 2003.

[2] N. Leprovost, B. Dubrulle, and P. H. Chavanis. Dynamics and thermodynamics of axisymmetric flows: Theory. *Phys. Rev. E*, 73:046308, 2006.

[3] R. Monchaux, P. Diribarne, B. Dubrulle, A. Chiffaudel, and F. Daviaud. Fluctuation theorem and effective temperature in turbulent flow. *submitted to Phys. Rev. Lett.*, 2007.

[4] R. Monchaux, F. Ravelet, B. Dubrulle, A. Chiffaudel, and F. Daviaud. On the properties of turbulent axisymmetric flows. *Phys. Rev. Lett.*, 96:124502, 2006.

[5] F. Ravelet, L. Marié, A. Chiffaudel, and F. Daviaud. Multistability and Memory Effect in a Highly Turbulent Flow: Experimental Evidence for a Global Bifurcation. *Phys. Rev. Lett.*, 93:164501, 2004.

[6] J. Sommeria and R. Robert. Statistical equilibrium states for two-dimensional flows. *J. Fluid Mech.*, 229:291, August 1991.

The Influence of External Turbulence on a Wall-Bounded Jet

C. Poelma[1], F. Beati[2], J. Westerweel[1] and J.C.R. Hunt[1]

[1] Laboratory for Aero and Hydrodynamics, Delft University of Technology;
Leeghwaterstraat 21, 2628 CA Delft (The Netherlands) c.poelma@tudelft.nl

[2] Department of Mechanics and Aeronautics - University of Rome "La Sapienza";
Via Eudossiana, 18 - I-00184 Roma (Italy)

1 Introduction

The effect of external turbulence on flows near a wall is of considerable interest for the understanding and modeling of many industrial processes. Additionally, it is an interesting case for the validation of turbulence models: besides the 'internal' time and length scales of the wall-bounded jet, the external turbulence introduces a complete new set of scales. Finally, many fundamental questions remain regarding the interactions between a shear layer and inhomogeneous turbulence [1]. In this study, the effect of strong external turbulence on a wall-bounded jet is studied experimentally. This work is an extension of an earlier study by Tsai et al. [2].

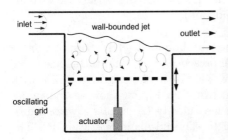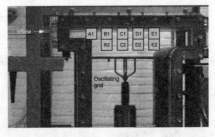

Fig. 1. Schematic representation (*left*) and photograph (*right*) of the experimental facility. The locations of the measurement stations are denoted by the labeled boxes (A1-E2).

The experimental facility (see Fig.1) consists of a box filled with water. Near the top of this facility, a turbulent channel flow enters the box, to form a wall-bounded jet. Opposite of the entrance is a larger channel where the jet exits. The Reynolds number of the inlet flow, based on the channel half-height ($h/2 = 20mm$) and mean centerline velocity ($\overline{U_c} = 0.11m/s$) is 2200. Strong external turbulence with nearly zero mean flow is generated by means of an oscillating grid (centered at $y=5h$).

Velocity measurements are obtained by means of high-speed Particle Image Velocimetry. This technique was chosen because it can give insight in the spatial structure of the jet interface. Additionally, the data can be used to calculate both spatial *and* temporal second-order statistics, such as correlation functions with sufficient resolution.

2 Results & Discussion

Measurements are obtained at a number of stations in a vertical plane covering a large part of the jet region, indicated in Fig. 1. Experiments were performed first without the external turbulence and subsequently with oscillating frequencies $G=1Hz$ and $G=3Hz$. Initial results indicate that the external turbulence has a significant effect on the wall-bounded jet; an example is the increased spreading rate of the jet. This can be explained by the fact that the initially localized shear region is perturbed by the strong external vortices. Associated with this is the fact that the mean velocity profile rapidly approaches a self-similar form, see Fig. 2. In this figure, the non-dimensional mean horizontal velocity profiles at 4 downstream locations (B, C, D and E, as denoted in the graphs) are reported. As can be seen in the right-hand side graph, the data collapses on a single curve for the case with oscillating grid ($G = 3Hz$). In contrast, the velocity profile is still evolving in the case without external turbulence ($G = 0Hz$, left-hand side).

The flow in the inner region of the wall-bounded jet (corresponding to $y/y' \approx 0.5$) changes dramatically in the presence of external turbulence. The turbulence level increases from 2% for the case without grid to 6.7% with the grid oscillating at 3Hz. This is illustrated in Fig. 3, showing the values of $\langle u'_x u'_x \rangle$, normalized using the centerline velocity U_m, along a vertical line (x=150 mm, x/h = 3.4, corresponding to the mid-plane of station B). The dashed line indicates a fit to the region that is dominated by the grid-generated turbulence. The spatial decay of the latter is proportional to $(y - y_g)^{-1}$, with y_g the location of the grid. As can be seen in the figure, at the location of the shear layer, the grid-generated fluctuations can be estimated to be of the same order of magnitude as the fluctuations of the shear layer itself. Thus, the shear layer is not expected to 'shield' the wall-bounded jet. This is in agreement with the observed change in turbulence in the inner region.

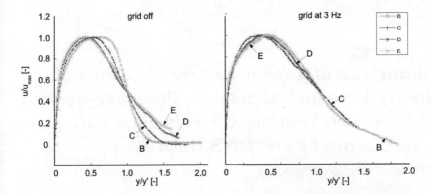

Fig. 2. Non-dimensional mean streamwise velocity without external grid *(left)*, and with oscillating grid (3Hz) *(right)*. The velocities are normalized using the maximum velocity of each profile, while the vertical position is scaled with $y' = y(u = u_{max}/2)$.

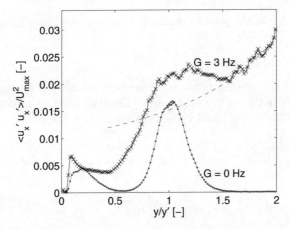

Fig. 3. The influence of the grid on the variance of the streamwise velocity, normalized with the maximum (centerline) velocity. Note that the centerline velocity did not change significantly for the purpose of this scaling.

References

1. J.C.R. Hunt, I. Eames and J. Westerweel "Mechanics of inhomogeneous turbulence and interfacial layers", Journal of Fluid Mechanics (2006) vol. 554 pp. 499-519
2. Y.S. Tsai, J.C.R. Hunt, F.T.M. Nieuwstadt, J. Westerweel and B.P.N. Gunasekaran: "Effect of strong external turbulence on a wall jet boundary layer", Flow, Turbulence & Combustion (submitted)

Comparison of Tensor Representations of Velocity-Pressure-Gradient, Pressure-Strain, and Pressure-Velocity Correlations with Plane Channel Flow DNS Data

11. ETC, jun 25–28, 2007, Porto [PRT]

G.A. Gerolymos[1], D. Sénéchal[1], I. Vallet[1], and B.A. Younis[2]

[1] Université Pierre-et-Marie-Curie (Paris VI), 4 place Jussieu, 75005 Paris, France; geg@ccr.jussieu.fr, dsenecha@ccr.jussieu.fr, vallet@ccr.jussieu.fr
[2] University of California, Davis, CA 95616, USA; bayounis@ucdavis.edu

1 Scope

The purpose of this paper is to examine various tensor representations for correlations between pressure and velocity, such as pressure-strain ϕ_{ij}, pressure-diffusion d_{ij}^p and velocity-pressure-gradient Π_{ij}

$$\phi_{ij} = \overline{p' \left(\frac{\partial u_i'}{\partial x_j} + \frac{\partial u_j'}{\partial x_i} \right)} \quad ; \quad d_{ij}^p = -\frac{\partial}{\partial x_\ell} \left(\overline{p' u_i'} \delta_{j\ell} + \overline{p' u_j'} \delta_{i\ell} \right) \tag{1}$$

$$\Pi_{ij} = -\left(\overline{u_i' \frac{\partial p'}{\partial x_j} + u_j' \frac{\partial p'}{\partial x_i}} \right) \quad ; \quad \Pi_{ij} \equiv d_{ij}^p + \phi_{ij} \tag{2}$$

which are used in Reynolds-stress models (RSMs). It is well known that redistribution (ϕ_{ij}) has a major influence on the Reynolds-stresses transport budgets [7], and pressure-diffusion (d_{ij}^p) appears to have a strong influence in 3-D flows with high helicity (secondary flows) [10]. Most models treat these terms separately, following Rotta [8, 9]. However, recent work [13] models directly Π_{ij}, following Chou [1].

2 DNS

There are several available plane-channel-flow DNS data for the above terms [4, 6], but the available data are mainly concerned with the y-wise (normal-to-the-wall) distributions of the various terms. An approach more appropriate for the evaluation of tensor representations was implemented in the DNS computations used in the present paper [3]. Kim [5] has developed the now standard Green's function approach of splitting DNS-computed pressure fluctuations to rapid p'_r and slow parts p'_s [1], which can be used [5] to separate the rapid and slow parts of the correlations (Eqs. 1, 2). The integrodifferential solution of the Poisson equations for the slow and rapid terms, is expressed as the sum of a volume-integral and a surface integral [1], but is awkward to use because the surface-integral contains p' at the surface [5]. In the present work [3], the volume integrals $p'_{(r\mathfrak{V})}$ and $p'_{(s\mathfrak{V})}$ are computed, and their difference from the Kim [5] complete solutions, yields the terms associated with the surface integrals $p' = p'_r + p'_s = p'_{(r\mathfrak{V})} + p'_{(rw)} + p'_{(s\mathfrak{V})} + p'_{(sw)}$. The various terms of the present p'-splitting can be used to compose the appropriate correlations [3], *eg*

$$\phi_{ij} = [\underbrace{\overline{p'_{(r\mathfrak{V})}S'_{d_{ij}}}}_{\phi_{ij}^{(r\mathfrak{V})}} + \underbrace{\overline{p'_{(rw)}S'_{d_{ij}}}}_{\phi_{ij}^{(rw)}} + \underbrace{\overline{p'_{(s\mathfrak{V})}S'_{d_{ij}}}}_{\phi_{ij}^{(s\mathfrak{V})}} + \underbrace{\overline{p'_{(sw)}S'_{d_{ij}}}}_{\phi_{ij}^{(sw)}}] \tag{3}$$

for ϕ_{ij} in the limit $[M \longrightarrow 0; \rho, \ \mu \cong \text{const}]$. The last term (surface integral) is obviously associated with reflection from the wall (also known as wall blocage), and this approach allows to separate the *"homogeneous"* (volume-integrals) part, from the *"inhomogeneous"* (echo-terms) part. This processing of DNS data is important in separating the contributions of various terms, which are modelled separately [2].

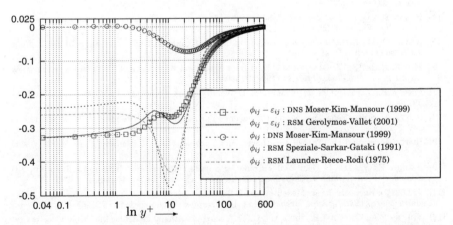

Fig. 1. *A priori* compraison of ϕ_{ij} closures with DNS data [6] for $Re_\tau = 590$ incompressible plane channel flow.

3 Tensor Representations

Tensor representations for various terms are reviewed and completed. These include the well-known representations for ϕ_{ij} [14], and recent representations for d_{ij}^p [11, 12] and for the velocity-pressure-gradient Π_{ij} [13]. These representations are extended in the present work to include the unit-vector pointing in the turbulence inhomogeneity direction [2], resulting in extended tensor representations.

4 Results

The above tensor-representations are compared with DNS data, for plane channel flows, both from existing databases (Fig. 1), and from the present computations. Particular care is taken to evaluate independently various integrals, and in this way to assess separately the "homogeneous" part, and the wall-blockage effect.

References

[1] CHOU P.Y.: On Velocity Correlations and the Solutions of the Equations of Turbulent Fluctuations, *Quart. Appl. Math.* **3** (1945) 38–54.

[2] GEROLYMOS G.A., SAURET E., VALLET I.: Contribution to the Single-Point-Closure Reynolds-Stress Modelling of Inhomogeneous Flow, *Theor. Comp. Fluid Dyn.* **17** (2004) 407–431.

[3] GEROLYMOS G.A., SÉNÉCHAL D., VALLET I.: Pressure Fluctuations in Compressible Channel Flow, AIAA Paper 2007–3863, 37. AIAA Fluid Dynamics Conference, 25–28 jun 2007, Miami [FL, USA] (2007).

[4] HOYAS S., JIMÉNEZ J.: Scaling of the Velocity Fluctuations in Turbulent Channels up to $Re_\tau = 2003$, *Phys. Fluids* **18** (2006) 011702·1–4.

[5] KIM J.: On the Structure of Pressure Fluctuations in Simulated Turbulent Channel-Flow, *J. Fluid Mech.* **205** (1989) 421–451.

[6] MOSER R.D., KIM J., MANSOUR N.N.: Direct Numerical Simulation of Turbulent Channel Flow up to $Re_\tau = 590$, *Phys. Fluids* **11** (1999) 943–945.

[7] POPE S.B.: *Turbulent Flows* (Cambridge University Press, Cambridge [UK] 2000), ISBN 0–521–59125–2.

[8] ROTTA J.: Statistische Theorie nichthomogener Turbulenz — 1. Mitteilung, *Z. Phys.* **129** (1951) 547–572.

[9] ROTTA J.: Statistische Theorie nichthomogener Turbulenz — 2. Mitteilung, *Z. Phys.* **131** (1951) 51–77.

[10] SAURET E., VALLET I.: Near-Wall Turbulent Pressure Diffusion Modelling and Influence in 3-D Secondary Flows, *ASME J. Fluids Eng.* **129**, (in print; accepted for publication, sep 2006).

[11] VALLET I.: Reynolds-Stress Modelling of 3-D Secondary Flows with Emphasis on Turbulent Diffusion Closure, *ASME J. Appl. Mech.* **74**, (in print; accepted for publication, oct 2006).

[12] YOUNIS B.A., GATSKI T.B., SPEZIALE C.G.: Towards a Rational Model for the Triple Velocity Correlations of Turbulence, *Proc. Royal Society Lon. A* **456** (2000) 909–920.

[13] YOUNIS B.A., SMITH G.F., FEIGENBAUM H.P.: On Modelling the Fluctuating Pressure-Velocity Correlations using Tensor Representation Theory, *(preprint)* .

[14] YOUNIS B.A., SPEZIALE C.G., BERGER S.A.: Accounting for Effects of a System Rotation on the Pressure-Strain Correlation, *AIAA J.* **36** (1998) 1746–1748.

Near Wall Measurements in Rough Surface Turbulent Boundary Layers

Brian Brzek[1], Raúl Bayoán Cal[2], Gunnar Johansson[3] and Luciano Castillo[4]

[1] Rensselaer Polytechnic Institute, Troy, NY `brzekb@rpi.edu`
[2] The Johns Hopkins University, Baltimore, MD `bayoan.cal@jhu.edu`
[3] Chalmers University of Technology, Gothenburg, Sweden `gujo@chalmers.se`
[4] Rensselaer Polytechnic Institute, Troy, NY `castil2@rpi.edu`

Near wall measurements using 2-D LDA have been performed for the mean velocity and Reynolds stresses over smooth and rough surfaces at 11 streamwise measurement locations. This allows for the calculation of the skin friction coefficient using the Navier-Stokes equations with an accuracy of 3-5%. It was found that the destruction of the viscous stress near the wall with increasing roughness led to reduced Reynolds number dependence in the flow field. In addition, it will be shown that a significant reduction in the near wall peak of the streamwise Reynolds stress was seen with increased roughness.

1 Experimental Setup

An experiment has been performed in the L2 wind tunnel facility in the department of Applied Mechanics at Chalmers University of Technology to study the effects of roughness in the wall region and its relation to Reynolds number dependence. A flat aluminum plate mounted vertically in the wind-tunnel was used to create the boundary layer and had dimensions of 2.5 m long, 1.25 m wide and 5 mm thick. The roughness for this particular experiment is a 24 grit sand paper as well as a 60 grit sand paper. Three different wind tunnel speeds of 5, 10 and 20m/s provide a wide range of the roughness parameter, $k^+ = ku_\tau/\nu = 0 - 108$ (i.e., hydraulically smooth to fully rough) and range in Reynolds number from $1,500 < Re_\theta = U_\infty\theta/\nu < 10,500$. The roughness height, k, in this definition is the ten point height, $k = 1.552$ mm which is an average of the 5 largest peaks and five deepest valleys. These new sets of experiments are compared to the smooth wall measurements of Castillo & Johansson (1) for the same inlet conditions.

2 Results

2.1 Skin Friction Coefficient
The values of c_f were determined from the integrated boundary layer equation (Eq. 1) as well as the momentum integral equation, $c_f/2 = d\theta/dx$ and are shown in Fig. 1(a). This experiment utilized 11 streamwise measurement

locations so that the required x-dependence could be realized, giving accuracy in c_f within 3 and 5% for smooth and rough surfaces respectively.

$$\frac{\tau_w}{\rho} = \nu \frac{\partial \overline{U}}{\partial y} - \overline{uv} - \int_0^y \frac{\partial \overline{U}^2}{\partial x} dy' + \overline{U} \int_0^y \frac{\partial \overline{U}}{\partial x} dy' - \int_0^y \frac{\partial \overline{u^2}}{\partial x} dy' + \int_0^y \frac{\partial \overline{v^2}}{\partial x} dy' + U_\infty \frac{dU_\infty}{dx} y$$

Notice, that there is a significant increase in skin friction due to roughness as would be expected. For the three cases in the transitionally rough regime, $5 < k^+ < 70$, the skin friction shows a dependence on roughness and Reynolds number. However, for the case at 20 m/s, (i.e., in the fully rough regime, $106 < k^+ < 108$) the skin friction is Reynolds number invariant, indicating that the effect of viscosity is not present due to roughness and the skin friction is composed of form drag.

2.2 The Role of Viscosity in Rough Surfaces

In rough surfaces, the structure of the various layers of a boundary layer break down to various degrees depending on the roughness height and Reynolds number. This can be seen in Fig. 1(b) where the various regions of a smooth boundary layer and corresponding roughness regimes are seen. Notice that the roughness height, k, is taken in relation to the virtual origin, thus, $k = \epsilon$ in this case. For example, in the hydraulically smooth regime, $k^+ < 5$, roughness extends into the viscous dominated region of the inner flow. As k^+ increases into the transitionally rough regime, $5 < k^+ < 70$, the roughness elements protrude into the buffer layer and into the lower edge of the outer layer. For flows in the fully rough regime, $k^+ > 70$, the roughness elements completely destroy the viscous region near the wall and the flow ceases to be Reynolds number dependent as seen in Fig. 1(a).

In order to determine the extent at which the rough surface changes the role of viscosity in the inner region, the viscous stress is examined in Fig. 1(c). This analysis utilizes smooth and rough surface data at approximately a fixed Reynolds number of $\delta^+ = \delta u_\tau / \nu \approx 1000$ so that the different regions of the boundary layer can be compared. In addition, a profile at a slightly lower Reynolds number is used where measurements were performed in the viscous sublayer. Notice, that for the smooth surface profile, the viscous stress is completely dominant inside $(y + \epsilon)^+ \approx 10$. For the rough profile at 5 m/s over the 24 grit surface, (i.e., $k^+ = 25$, $\epsilon^+ = \epsilon u_\tau / \nu = 13$), and 10 m/s over the 60 grit surface, (i.e., $k^+ = 15$, $\epsilon^+ = 16$), the roughness elements extend into the buffer region of the flow. This reduces the influence of viscosity, since the maximum contribution of the viscous stress is approximately 30% for both cases. For the profile at 10 m/s over the 24 grit sand paper (i.e., $k^+ = 54$, $\epsilon^+ = 36$), the roughness elements extend through the buffer layer and into the Mesolayer. Thus, the viscous stress region is nearly destroyed where the contribution of the viscous stress is approximately 7%. This reduction in the viscous stress leads to less Reynolds number dependence in the skin friction coefficient as was examined in Fig. 1(a).

The effect of viscosity is also important in the streamwise Reynolds stress, where the high peak near the wall occurs in the viscous dominated region, (i.e., $y^+ \approx 15$). This peak has been shown to be the result of viscous generated vortices lifting off from the wall. Thus, as the roughness parameter increases, it destroys the viscous region and the high peak near the wall is reduced as seen in Fig. 1(d). Once the fully rough regime has been reached, the high peak ceases to exist and the profiles become flat as seen in the investigation of Ligrani & Moffat (2). Consequently, isotropy is promoted near the wall for rough surfaces.

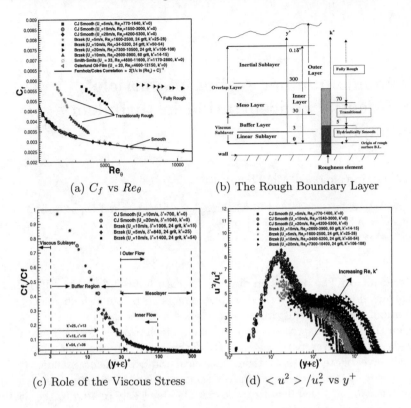

(a) C_f vs Re_θ (b) The Rough Boundary Layer

(c) Role of the Viscous Stress   (d) $< u^2 > / u_\tau^2$ vs y^+

Fig. 1. *Skin Friction and Mean Profiles*

3 Conclusion

These 2-D LDA measurements have allowed us to investigate the near wall behavior in rough surface turbulent boundary layers. Using 11 streamwise measurements, the Navier-Stokes equations could be used to determine the skin friction within 5% accuracy for rough surfaces. The role of the viscous term proved to be important in the transitionally rough regime where the roughness elements increasingly protrude into the viscous regions of the boundary layer. This results in a reduction of the Reynolds number dependence on the skin friction and other roughness parameters. Furthermore, the viscous generated peak in the streamwise Reynolds stress decreases with increasing roughness height.

References

[1] Castillo, L., and Johansson, G., "The effects of the upstream conditions on a low Reynolds number turbulent boundary layer with zero pressure gradient", Journal of Turbulence, **3**, 031, 2002.

[2] Ligrani, P., and Moffat, R., "Structure of transitionally rough and fully rough turbulent boundary layers," J. Fluid Mech., 162, pp. 69-98, 1986.

Reynolds number scaling of particle preferential concentration in turbulent channel flow

Cristian Marchioli and Alfredo Soldati[†]

Centro Interdipartimentale di Fluidodinamica e Idraulica and Dipartimento di Energetica e Macchine, University of Udine, 33100 Udine, Italy
marchioli@uniud.it, soldati@uniud.it

Introduction The dispersion of particles with finite inertia in wall-bounded turbulent flows is characterized by phenomena such as non-homogeneous distribution, large-scale clustering and preferential concentration in the near-wall region [1]. These macroscopic phenomena, which are due to the inertial bias between the denser particles and the lighter surrounding fluid, have been widely investigated in theoretical, numerical and experimental studies (see [1] for a review). From these studies, it appears that the degree of preferential concentration at given flow Reynolds number is closely related to the particle Stokes number, defined as:

$$St = \frac{\tau_p}{\tau_f} = \frac{\rho_p d_p^2}{18\mu} \cdot \frac{u_\tau^2}{\nu} \tag{1}$$

where τ_p is the particle relaxation time, ρ_p is the particle density, d_p is the particle diameter, μ is the fluid dynamic viscosity, $\tau_f = \nu/u_\tau^2$ is the characteristic time-scale of the flow, u_τ is the shear velocity[1] and ν is the fluid kinematic viscosity. How particle preferential concentration scales with the Reynolds number of the flow is still an open question. Numerical investigations on the Reynolds number scaling properties have been performed for the case of heavy particle dispersion in a synthetic turbulent advecting field [2] and for the case of homogeneous isotropic turbulence [3]. The focus of the present study is to investigate the same effect for the case of wall-bounded turbulent flows trying to emphasize possible scaling properties.

Numerical Methodology We used an MPI parallel pseudo-spectral Direct Numerical Simulation code [1, 4] to compute the turbulent Poiseuille flow

---

[†] Also affiliated with Department of Fluid Mechanics, CISM, 33100 Udine, Italy.

[1] The shear velocity u_τ is defined here as $u_\tau = \sqrt{\tau_w/\rho}$, where τ_w is the wall shear stress and ρ is the gas density

$St^l = St\|_{Re^l_\tau}$	τ_p (s)	d_p^+	d_p (μm)	$V_s^+ = g^+ \cdot St$	$Re_p^+ = V_s^+ \cdot d_p^+ / \nu^+$
0.2	$0.227 \cdot 10^{-3}$	0.068	9.1	0.019	0.00139
1	$1.133 \cdot 10^{-3}$	0.153	20.4	0.094	0.01444
5	$5.660 \cdot 10^{-3}$	0.342	45.6	0.472	0.16127

$St^h = St\|_{Re^h_\tau}$	τ_p (s)	d_p^+	d_p (μm)	$V_s^+ = g^+ \cdot St$	$Re_p^+ = V_s^+ \cdot d_p^+ / \nu^+$
1	$0.283 \cdot 10^{-3}$	0.153	10.2	0.094	0.01444
5	$1.415 \cdot 10^{-3}$	0.342	22.8	0.472	0.16127
25	$70.75 \cdot 10^{-3}$	0.765	51.0	2.360	1.80505

Table 1. Particle simulation parameters.

of air (incompressible and Newtonian) in a two-dimensional channel at two different shear Reynolds numbers [2]: $Re_\tau^l = 150.$ and $Re_\tau^h = 300$. The reference geometry consists of two infinite vertical flat parallel walls with periodic boundary conditions in the streamwise (x) and spanwise (y) directions and no-slip conditions at the walls. The dimensions of the computational domain are $1885 \times 942 \times 300$ wall units (obtained using ν and u_τ) in x, y and z, discretized with $128 \times 128 \times 129$ grid nodes for the Re_τ^l simulations, and $3770 \times 1885 \times 600$ wall units with $256 \times 256 \times 257$ grid points for the Re_τ^h simulations.

Since we are interested in characterizing the collective behavior of particles, we chose a simplified numerical setting in which particles are pointwise, rigid, elastically rebounding spheres that have no effect onto the turbulent field and do not collide with each other. The Lagrangian equation of particle motion includes only the effects of particle inertia and nonlinear Stokes drag. Initially, particle number concentration is uniform in the computational box and particle position is chosen randomly.

Results and Discussion For the simulations presented here, large samples of $O(10^5)$ particles, characterized by different Stokes numbers (see Table 1), were considered for each simulation setting. The characteristic time scale of the flow changes depending on the specific value of the shear velocity, namely on the specific value of the shear Reynolds number. Elaborating, we find:

$$\frac{St^h}{St^l} = \frac{\tau_f^l}{\tau_f^h} = \left(\frac{u_\tau^h}{u_\tau^l}\right)^2 = \left(\frac{Re_\tau^h}{Re_\tau^l}\right)^2. \tag{2}$$

Thus, if the shear velocity is the proper scaling parameter to quantify the Reynolds number effect on particle preferential concentration then comparisons should be possible between particle Stokes numbers that match the two different Reynolds numbers according to Eq. 2. Namely $St^l = 0.2$ and $St^h = 1$ (Fig. 1a), $St^l = 1$ and $St^h = 5$ (Fig. 1b), $St^l = 5$ and $St^h = 25$ (Fig. 1c). Also the averaging time span must be chosen to match the following:

[2] The shear Reynolds number is defined here as $Re_\tau = u_\tau h / \nu$, where h is the channel half-height

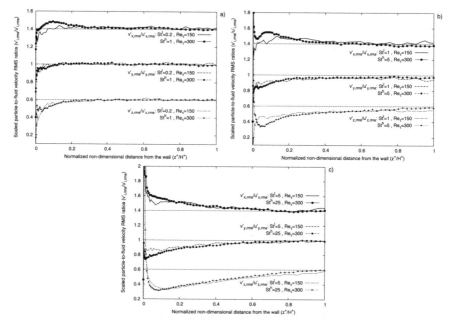

Fig. 1. Scaled particle-to-fluid velocity rms ratios at low Reynolds number, $(v'_{i,rms}/u'_{i,rms})|_{St^l,Re^l_\tau}$, and at high Reynolds number, $(v'_{i,rms}/u'_{i,rms})|_{St^h,Re^h_\tau}$.

$t^+|_{Re^h_\tau} = \left(Re^h_\tau/Re^l_\tau\right)^2 \cdot t^+|_{Re^l_\tau}$. Figure 1 compares *vis-à-vis* the ratios between the particle and the fluid velocity rms components, $v'_{i,rms}/u'_{i,rms}$, normalized to wall variables using the shear velocity as the scaling parameter. For ease of reading, $v'_{i,rms}/u'_{i,rms}$ profiles for the streamwise and wall-normal rms components are shifted by a factor of 0.4, up and down respectively. The profiles (though a bit ragged) overlap quite well even in the near-wall region, thus confirming i) that particle time scale normalized to wall variables (the shear fluid velocity being the scaling parameter) may be used as the representative particle Stokes number, and ii) that universality of statistics may persist to Reynolds numbers much higher than those considered here [3, 5].

Support from HPC Europa Transnational Access Program is gratefully acknowledged.

References

1. A. Soldati: Z. Angew. Math. Mech., **85**, 683-699 (2005).
2. P. Olla: Phys. Fluids, **14**, 4266-4277 (2002).
3. L.R. Collins, A. Keswani: New J. Phys., **6**, 1-17 (2004).
4. C. Marchioli: Parallelization of a DNS code for dispersed turbulent channel flow, http://www.sara.nl/projects/projects_04_01_466/466_report.pdf (2006).
5. L.M. Portela, P. Cota, R.V.A. Oliemans: Powd. Tech., **12**, 149-157 (2002).

Dissipative structure in multi mode stretched-spiral vortex

Kiyosi Horiuti and Takeharu Fujisawa

Dept. Mechano-Aerospace Eng., Tokyo Institute of Technology, 2-12-1 O-okayama, I1-64, Meguro-ku, Tokyo 152-8552, Japan, khoriuti@mes.titech.ac.jp

1 Introduction

It is well known that the dissipation field is highly intermittent [1]. As for an example, we show in figure 1 the PDFs of the rate of dissipation, ε normalized by its average $< \varepsilon >$, obtained using the DNS data for decaying homogeneous isotropic turbulence. The long tails of large ε values are discernible. DNS is conducted using three different grid resolutions with $k_{\max}\eta \approx 1, 2$ and 4 ($\eta = (\nu^3/ < \varepsilon >)^{1/4}$) at $R_\lambda \approx 77$. Notable feature is that the tails of PDFs extend further to larger values as the grid is refined. In this study, we investigate on the structures which generate this intermittent dissipation field in homogeneous isotropic turbulence.

Turbulent structures are overall classified as the vortex sheet and tube. Figure 2 shows the gray scale isocontours of the eigenvalue, $[A_{ij}]_+ (= [S_{ik}\Omega_{kj}+ S_{jk}\Omega_{ki}]_+)$ and the velocity vectors at an early stage ($t = 1.05$). In this study,

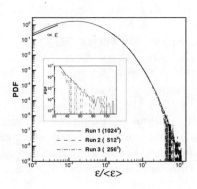 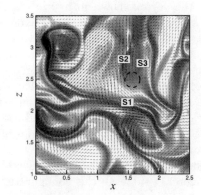

Fig. 1. PDFs of $\varepsilon/ < \varepsilon >$ from Runs with $k_{\max}\eta \approx 1, 2, 4$.

Fig. 2. Distribution of velocity vectors and gray-scale isocontours of $[A_{ij}]_+$.

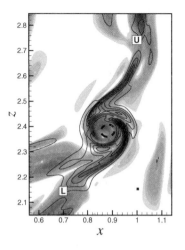
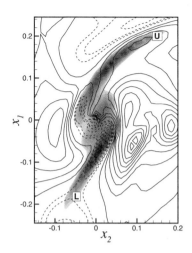

Fig. 3. Isocontours of $[A_{ij}]_+, \varepsilon, p$.   **Fig. 4.** Isocontours of $[A_{ij}]_+$ and D.

the sheet is identified using the eigenvalue $[A_{ij}]_+$ [2]. It can be seen that the field is filled with many vortex sheets. These sheets evolve into the tubes.

2 Dissipation field associated with spiral vortex

In figure 2, the stagnation flow can be seen along the sheet marked as S1. This stagnation flow strains and stretches the sheet marked as S2 and the recirculating flow is generated in the region indicated by the circle drawn using the heavy dashed line through an interaction with another sheet marked as S3. The low pressure region asssociated with this recirculating flow concentrates and the vortex shown in figure 3 is formed at $t = 1.95$. The vortex sheets are spiralling around the core region of tube shown using the heavy dashed line. The sheets are stretched to very thin thickness, and the spiral turns are formed. Its structure is similar to the stretched-spiral vortex (SSV) [3]. In figure 3, the solid lines display the dissipation rate. Intense dissipation takes place along these sheets. Note that this SSV is formed not by a rolling-up of a single sheet. In this SSV, the arrangement of the vorticity vectors on the lower sheet (marked as L) and the upper sheet (U) is asymmetric (Mode 2). SSV consists of the three modes of configurations regarding the vorticity vector alignment on the two sheets. Although the vorticity vectors on both L and U are initially perpendicular to those along the recirculating flow as in figure 2 (Mode 3), this configuration is transformed into either Mode 2 or another symmetric mode (Mode 1) with lapse of time.

This inter-mode transformation is primarily attributed to an occurrence of negative strain-rate eigenvalue σ_s, which is caused primarily by the action of the pressure Hessian on the basis of the principal strain eigenvectors, $\widetilde{\Pi}_{ss}$.

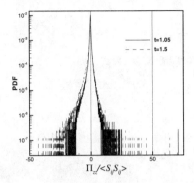
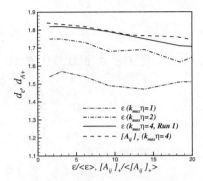

Fig. 5. PDFs of the $\widetilde{\varPi}_{ss}$ term normalized by $< S_{ij}Sij >$.

Fig. 6. Threshold value dependence of fractal dimensions of ε and $[A_{ij}]_+$.

Figure 5 shows the PDFs of the $\widetilde{\varPi}_{ss}$ term. At an early stage ($t = 1.05$), $\widetilde{\varPi}_{ss}$ is skewed to positive values. As a result, the positive σ_s is converted to negative value, and the reorientation of vorticity vectors takes place on the sheets due to negative vortex-stretching term, $\sigma_s \omega_s^2$. Later at $t = 1.5$, as $\widetilde{\varPi}_{ss}$ becomes skewed to negative value, an occurrence of this conversion ceases and the vorticity in the converted direction grows.

Formation of the spiral turns is attributed to the differential rotation incurred by the vortex tube and sheets. Figure 4 shows the strength of differential rotation, $D(= r\partial/\partial r(u_\theta/r))$, in the polar coordinate around the tube axis at $t = 1.85$. The D term is large along the vortex sheet, which induces stretching and thinning of the sheets. Subsequently, intense energy cascade and dissipation take place along the stretched sheets due to thinning of the sheet (its thickness $\approx 2.0 \, \eta$ in Run with $k_{\max}\eta = 4$. Therefore, the fine scale dissipation field critically depends on $k_{\max}\eta$.

Figure 6 shows the fractal dimension of ε, d_ε, estimated using the box-counting of the most intense sets defined by thresholding of ε [4]. As $k_{\max}\eta$ is increased, d_ε increases and ≈ 1.8 in Run with $k_{\max}\eta = 4$ with small dependence on the threshold value, suggesting structures in the form of (wrinkled) sheets. For comparison, the fractal dimension of $[A_{ij}]_+$, $d_{[A_{ij}]_+}$, is included. Closeness of $d_{[A_{ij}]_+}$ and d_ε indicates that fractal properties of the sheet and dissipation field are similar.

References

1. K.R. Sreenivasan: Turbulence and Combustion **72**, 115–131, (2004).
2. K. Horiuti, Y. Takagi: Phys. Fluids **17**, 121703, (2005).
3. T.S. Lundgren: Phys. Fluids **25**, 2193–2203, (1982).
4. F. Moisy, J. Jiménez: J. Fluid Mech. **513**, 111–133.

Modeling of a turbulent vortex ring

F. Kaplanski[1] and Y. Rudi[1]

Tallinn University of Technology, Akadeemia tee 23A, Tallinn 12618, Estonia
feliks.kaplanski@ttu.ee

1 Introduction

In a gross sense the overall mean vortex ring flow in a turbulent regime to behave like a very viscous flow with constant Reynolds number [1]. We use this consideration to develop our model. Our objectives in this paper are twofold. First, we analyse the capability of the earlier solution[2] to reproduce the behavior of the real rings at the post-formation stage for a relatively high Reynolds numbers. Second, we present attempt to model a turbulent vortex ring flow by using an effective 'eddy' viscosity [3].

2 Solution

We restrict our attention on the finding of the solution for the problem of an axisymmetric vortex ring. Following the paper [2], it is possible to find such solution for the azimutal component of voricity ζ by ignoring the nonlinear terms. For generality, we will employ the arbitrary power law for the time in the length and vorticity scales:

$$\zeta_0 \sim t^\lambda, \ell \sim t^b. \tag{1}$$

Introducing the dimensionless variables

$$\sigma = \frac{r}{\ell}, \eta = \frac{x}{\ell}, \tau = \frac{R_0}{\ell}, \Phi = \frac{\psi}{\zeta \ell^3}, \omega = \frac{\zeta}{\zeta_0}, \tag{2}$$

we obtain

$$-\frac{\lambda \ell^2}{t\nu}\omega - \frac{\ell \dot{\ell}}{\nu}\left(\sigma\frac{\partial \omega}{\partial \sigma} + \eta\frac{\partial \omega}{\partial \eta} + \tau\frac{\partial \omega}{\partial \tau}\right) - \left(\frac{\partial^2 \omega}{\partial x^2} + \frac{\partial^2 \omega}{\partial r^2} + \frac{1}{r}\frac{\partial \omega}{\partial r} - \frac{\omega}{r^2}\right)$$

$$= Re\left(\frac{\partial}{\partial \sigma}\left(\omega\frac{1}{\sigma}\frac{\partial \Phi}{\partial \eta}\right) + \frac{\partial}{\partial \eta}\left(-\omega\frac{1}{\sigma}\frac{\partial \Phi}{\partial \sigma}\right)\right), \tag{3}$$

where $Re = \zeta_0 \ell^2 / \nu$. When

$$\frac{\lambda}{b} = -3 \tag{4}$$

and $\nu = \ell \dot{\ell}$, an approximate solution of Eq.(3) is given by

$$\omega = \exp\left(-\frac{\sigma^2 + \eta^2 + \tau^2}{2}\right) I_1(\sigma\tau). \tag{5}$$

where I_1 denotes the first-order modified Bessel function of the first kind. The condition (4) maintains both the invariance of the vorticity impulse $M = I/\rho$ and existing of the solution (5). The laminar regime is characterized by length scale $\ell = \sqrt{2\nu t}$ with constant viscosity ν. For this case

$$\zeta_0 = \frac{2M}{(4\pi\nu)^{3/2} R_0 t^{3/2}}, \quad Re = \frac{M}{2(\pi\nu)^{3/2} R_0 t^{1/2}}. \tag{6}$$

For description of the 'turbulent' regime, we use turbulent eddy viscosity instead of the kinematic viscosity $\nu = \ell \dot{\ell}$, with a length scale $\ell = \alpha M^{1/4} t^{1/4}$, where α is empirical constant, which shoud be defined from the comparison with experimental data. Since behavior of the turbulent flow manifests self-similarity, i.e the overall motion is completely determined by just one parameter, M, we ignore dependence on R_0 [3]. This yelds

$$\omega = \sigma \exp\left(-\frac{\sigma^2 + \eta^2}{2}\right), \zeta_0 = \frac{1}{(2\pi)^{3/2} \alpha^4 t}, Re = \frac{4}{(2\pi)^{3/2} \alpha^4}. \tag{7}$$

3 Results

The evolution of the vortex ring propogation velocities in the initial and final stages are well approximated by the Saffman[4] and Rott and Cantwell[5] formulae, respectively. The similar property have the translational velocity U, which is obtained on the basis of the solution (5)[2] and have asymptotic those are identical with Saffman and Rott-Cantwell formulae. Figure 1 shows that U is also in agreement with the velocity profile described by Saffman additional formula [3] with $k = 14.4$ and $k'=7.8$. According to the experimental data [6] the evolution of the vortex ring propogation velocities in the Reynolds - number range $830 < Re < 1650$ are well approximated by this formula. Entrainment diagrams calculated on the basis of Eqs. (5) and (7) show (Fig.2) that as the the pattern for the laminar regime indicates the presence of the bifurcation (appearance of the new critical point- saddle) for some values of τ and Re_0 [2], the pattern for the turbulent regime contains only one saddle on the axis of symmetry and does not exhibits the structures corresponding to the combination of a wake and a vortex ring. Although this difference in the predicted behaviors is due to the assumption that for the describing of the turbulent ring can be used the ring model described by Phillip's solution, the result is in agreement with the experimental data[7].

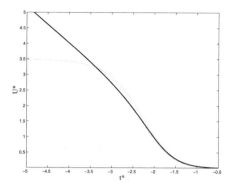

Fig. 1. Evolution of the translational velocity. The thick solid line corresponds to the normalized velocity U^*, the thin solid line refers to Saffman formula and dashed line corresponds to the value of $k'=9.2[8]$.

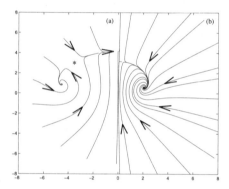

Fig. 2. Typical entrainment diagrams based on the solution(5)(a) and on the solution(7)(b).

References

1. B.J. Cantwell: *Introduction to Symmetry Analysis*, (Cambridge University Press Cambridge, 2002) pp 1–612
2. F. Kaplanski, Y. Rudi: Phys. Fluids **17**, 087101 (2005).
3. B. Lugovtsov: Arch. Mech. **28**, 759 (1976).
4. P. G. Saffman: Stud. Appl. Math. **49**, 371 (1970).
5. N. Rott, B. Cantwell: Phys. Fluids A **5**, 1443 (1993).
6. A. Weigand, M. Gharib: Exps. Fluids **22**, 447 (1997)
7. A. Glezer, D. Coles: J. Fluid Mech. **211**, 243 (1990).
8. Y. Fukumoto, H. K. Moffatt: J. Fluid Mech. **417**, 1 (2001).

Experimental study of longitudinal horizontal roll vortices in a convective flow in rectangular box

Batalov V., Frick P., and Sukhanovsky A.

Institute of Continuous Media Mechanics, 614013 Perm, Korolyov Str, 1, Russia

Contact address: frick@icmm.ru, san@icmm.ru

Horizontal roll vortices in atmospheric boundary layer have been observed by aircraft, radar and surface stations. Usually rolls are nearly aligned with the mean wind over sea surface. Recent observations showed existence of roll vortices in the boundary layer of tropical cyclones. These organized coherent structures may strongly affect the heat and mass air-sea exchange. Theoretical studies and numerical simulations showed that origin and dynamics of roll vortices are determined by hydrodynamical and convective instabilities in the boundary layer. Most of the experimental studies concern buoyancy-driven convective longitudinal rolls, which arise in plane Poiseuille flow, or rolls in inclined convective layers. Previously, we have observed longitudinal roll structures in a boundary layer above the localized heat source in rotating layer. These rolls play an essential role in the formation of large scale cyclonic vortex.

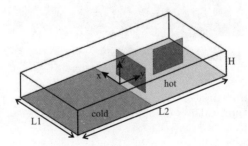

Fig. 1. Scheme of the model

Below, we present results of detailed study of longitudinal rolls in convective flow in a free upper surface layer of oil. Experiments were carried out in a rectangular box ($L1 = 10cm, L2 = 20cm, H = 8cm$), the bottom of which consists of two heat exchangers (Fig.1). The depth of the layer h was changed from 0.8 cm up to 7 cm. The convective flow develops due to temperature difference on heat exchangers $\triangle T$, $T_1 = T_r - \triangle T/2$, $T_2 = T_r + \triangle T/2$, where T_r is temperature of the air in the room, which was kept constant during experiments. Measurements of the temperature were done by thermocouples and velocity fields were measured by PIV system "Polis". Examples of vertical cross-sections where measurements were done are shown in Fig.1.The structure and dynamics of the horizontal rolls in a convective flow over heated plate were studied.

The large scale mean flow in the plane XOZ is determined by the aspect ratio $\gamma = h/L2$ and the temperature difference ($\triangle T$). At small aspect ratio and low values of $\triangle T$ the flow is concentrated near the border between the cold and the hot heat exchangers, forming a roll parallel to the axis OY. The increase of $\triangle T$ led to growth of this vortex and formation of mean advective flow, which finally occupies the whole box.

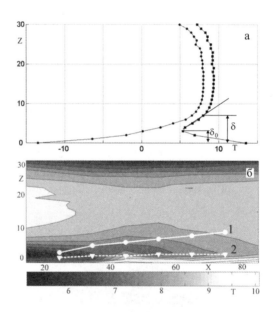

Fig. 2. a - temperature profiles for $x = -20mm$(circles) and $x = 15mm$ (squares); b - averaged temperature field over the hot plate (plane XOZ) $\triangle T = 26.8K, h = 30mm$

The formation of secondary flows strongly depends on temperature distribution. Results of temperature measurements are shown in Fig.2. Temperature distribution over cold heat exchanger showed very weak dependence on x (Fig.2a, left curve). Cold flow in the lower layer over hot heat exchanger formed boundary layer with depth δ_0 characterized by unstable temperature gradient. A depth of the layer with quasi-linear growth of temperature is δ.

Potentially unstable temperature distribution led to formation of longitudinal rolls aligned with the mean flow in the convective boundary layer over the hot plate (Fig.3). The size of rolls grows with the coordinate x together with the thickness of the temperature boundary layer. It is important to mention that growth of the rolls happened mostly by stretching their upper part. Size of the rolls is shown by curve 1 (Fig.3b) and the location of maximum vorticity (center of roll) is shown by curve 2. The center of rolls is determined by value of δ_0 and the size of rolls by δ. We have studied the characteristics of rolls for different H and $\triangle T$. The growth of $\triangle T$ led to increase of the mean flow velocity and decrease of the depth of temperature boundary layer. Longitudinal rolls begin to form farther from the boundary between heaters (fig.3c). A very weak dependence of the roll size and intensity on H was observed. Experiments for fixed H show that increasing $\triangle T$ led to decreasing of the roll size and growth of their intensity.

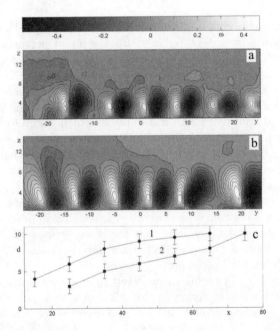

Fig. 3. Averaged vorticity field ω_x (plane ZY) for $x = 35mm$ (a) and $x = 65mm$ (b), $\triangle T = 26.8K, h = 30mm$; c - dependence of rolls size on x for $\triangle T = 10.7K(1)$ and $\triangle T = 26.8K(2)$

This study was supported by grant RFBR-Ural 07-01-96007

Nonlinear Evolution of Disturbed Vortex Rings

Yuji Hattori[1] and Yasuhide Fukumoto[2]

[1] Faculty of Engineering, Kyushu Institute of Technology, Kitakyushu, Japan
    hattori@mns.kyutech.ac.jp
[2] Graduate School of Mathematics, Kyushu University, Fukuoka, Japan
    yasuhide@math.kyushu-u.ac.jp

1 Introduction

Evolution of a vortex ring is one of the important problems in vortex dynamics not only in its own right but also from a broad perspective of vortical structures which have curvature. Recently we have discovered a new linear instability of a vortex ring, which is a direct effect of curvature, by normal-mode analysis[1] and local stability analysis[2]. It is an $O(\epsilon)$ curvature effect, with ϵ being a ratio of the core radius to the ring radius of the vortex ring, in contrast to the well-known Widnall instability[3] which is an $O(\epsilon^2)$ effect. Since ϵ is a small parameter, the maximum growth rate of this *curvature instability* is larger than that of the Widnall instability in a large region of parameter space. Taking account of viscous effects perturbatively, however, this region shrinks for finite Reynolds number.

Having explored the new mechanism of linear instability, next question should be: what is the role of the curvature instability in practical situations? Our previous studies[1, 2] assume that the disturbance be infinitesimal and are essentially limited for inviscid case and for particular types of vorticity distribution. In reality both nonlinearity and viscosity would have significant effects; the vorticity distribution observed in the experiments are likely to depend on the particular situation. In this paper we study these effects by two methods: weakly nonlinear analysis and direct numerical simulation.

2 Weakly Nonlinear Analysis

First, weakly nonlinear analysis is applied to a disturbed vortex ring. Linear instability and nonlinear effects compete when the non-dimensionalized amplitude of unstable modes is $O(\epsilon^{1/2})$. The resulting amplitude equations are

$$\frac{\mathrm{d}A_\pm}{\mathrm{d}t} = aB_\pm + i\left(c_1\left|A_\pm\right|^2 + c_2\left|B_\pm\right|^2\right)$$

$$+ c_3 \left| A_\mp \right|^2 + c_4 \left| B_\mp \right|^2 + d_1 C_\pm + d_2 C_\mp \Big) A_\pm + i e_1 A_\mp B_\pm \overline{B_\mp},$$

$$\frac{dB_\pm}{dt} = b A_\pm + i \left(c_5 \left| A_\pm \right|^2 + c_6 \left| B_\pm \right|^2 \right.$$

$$\left. + c_7 \left| A_\mp \right|^2 + c_8 \left| B_\mp \right|^2 + d_3 C_\pm + d_4 C_\mp \right) B_\pm + i e_2 B_\mp A_\pm \overline{A_\mp},$$

$$\frac{dC_\pm}{dt} = A_\pm \overline{B_\pm} + \overline{A_\pm} B_\pm.$$

Here $A_\pm$ and $B_\pm$ are the complex amplitudes of the Kelvin waves whose azimuthal wavenumbers are m and $m + 1$, respectively, and the subscript denotes chirality of the modes; $C_\pm$ is the real amplitude of mean-flow correction. The coefficients a, b, c_i, d_i, e_i are evaluated by inner-product formulation of the stability problem. The above form of the equations can be derived by considering symmetry breaking of the flow: $O(2) \times SO(2) \Rightarrow O(2)$[4]. There are five invariants, of which three are independent

$$E = E_+ + E_-, \quad E_\pm = K_A \left| A_\pm \right|^2 + K_B \left| B_\pm \right|^2 + K_C C_\pm,$$

$$F = \left| A_+ \right|^2 + \left| A_- \right|^2 - a \left(C_+ + C_- \right),$$

$$G = \left| B_+ \right|^2 + \left| B_- \right|^2 - b \left(C_+ + C_- \right),$$

$$H_\pm = e_2 \left| A_\pm \right|^2 + e_1 \left| B_\pm \right|^2 - \left(a e_2 + b e_1 \right) C_\pm.$$

In Fig. 1 we show examples of numerical simulation of the above amplitude equations. It shows that asymmetry between $+$ and $-$ modes in the initial conditions leads to chaotic behaviour, while periodic motions are observed for symmetric initial conditions or reduced degrees of freedom (e.g. $A_- = B_- = C_- = 0$). The amplitudes of the modes saturate because of nonlinear effects. The maximum amplitudes are greater for asymmetric cases than for symmetric cases.

3 Direct Numerical Simulation

Next, direct numerical simulation is performed by solving the three-dimensional Navier-Stokes equations. Three different configurations are considered: vortex ring in (i) a periodic box; (ii) a periodic cylinder; and (iii) a torus.

In the third case the inner wall of the torus rotates in poloidal direction so that Kelvin's vortex ring is formed inside the torus. This configuration is free from critical layer which, if present, makes Kelvin waves decay exponentially so that the curvature instability is difficult to be observed in the first and second cases. A novel numerical method with high precision is developed for the third case. Figure 2 shows time evolution of energy of the modes corresponding to curvature instability. The mode energies are seen to grow exponentially with the rates close to the linear stability analysis (shown by solid lines). Thus the curvature instability is captured by direct numerical simulation for the first time. The results of further investigation of both the curvature and Widnall instabilities will be presented in the talk.

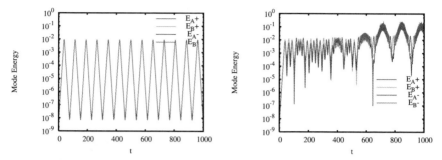

Fig. 1. Time evolution of mode energies. Amplitude equations. (Left) symmetric initial conditions, (right) asymmetric initial conditions.

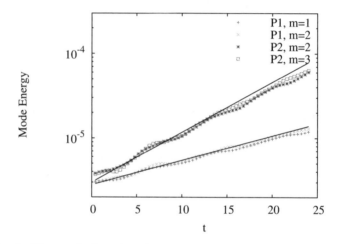

Fig. 2. Time evolution of mode energies. Direct numerical simulation.

References

1. Fukumoto, Y. and Hattori, Y.: Curvature instability of a vortex ring. *J. Fluid Mech.* **526**, 77 (2005).
2. Hattori, Y. and Fukumoto, Y.: Short-Wavelength Stability Analysis of Thin Vortex Rings. *Phys. Fluids* **15**, 3151 (2003).
3. Widnall, S. E. and Tsai, C.-Y.: The instability of the thin vortex ring of constant vorticity. *Phil. Trans. R. Soc. Lond. A* **287**, 273–305 (1977).
4. Knobloch, E., Mahalov, A. & Marsden, J. E.: Normal forms for three-dimensional parametric instabilities in ideal hydrodynamics. *Physica D* **73**, 49 (1994).

LES of turbulent mixing in a confined coaxial jet with 0.8 velocity ratio

P.M. Areal[1,3] and J.M.L.M. Palma[1,2]

[1] CEsA — Centro de Estudos de Energia Eólica e Escoamentos Atmosféricos
[2] Faculdade de Engenharia da Universidade do Porto
[3] Instituto Superior de Engenharia do Porto
   paa@isep.ipp.pt

Introduction

We present large-eddy simulations of turbulent and confined coaxial jets, where the inner jet contains a fully mixed solution of water and a fluorescent dye, Rhodamine B. This is an important engineering flow, common in injection systems, which is also the model for confined mixing in more complex geometries. Accurate prediction of such flows is required. The objective of this work is to demonstrate that an accurate simulation can be obtained in pipe flows using a standard non-orthogonal code and to analyze the computational results obtained, validated with the experimental results, to gain insight into the dynamics of confined coaxial jets.

Model

The geometry was chosen to match the experiments of [3], where the diameter and velocity ratios of the outer $(_o)$ and inner $(_i)$ jets were $\beta = D_o/D_i = 2.33$ and $r_u = u_o/u_i = 0.8$, the Reynolds number was equal to 35000. The simulations were carried out using a fractional-step based, 2nd order, LES code in a non-orthogonal grid [4].

Solution of the passive scalar transport equation, whose solution is bounded by the initial field maximum and minimum, required a special treatment of the convective terms, to preserve boundedness. Two schemes were used: a second order TVD scheme, from [6] (TVD1) and a bounded second order scheme similar to [1] (TVD2).

To account for the effects of the filtered small scales on the large resolved ones, a Lagrangian dynamic subgrid model was used for the velocity field and extended to the scalar field, following the method outlined in [2] that uses a dynamic procedure to calculate a spatially variable subgrid scale Schmidt number.

The inlet section was set at the first measurement station, located 0.14 inner tube diameter (D_i) after the pipe outlet. Planes (slices) captured in precursor pipe and annulus simulations were fed at the inlet of a precursor full run (RUN1), along with a scalar flat profile. Slices containing velocity and scalar field were collected and rescaled to fit the experimental mean profile, following the method of [5] (RUN2).

Results

Figures 1 and 2 shows the statistics available in the experimental results (circles) and compares them with LES results.

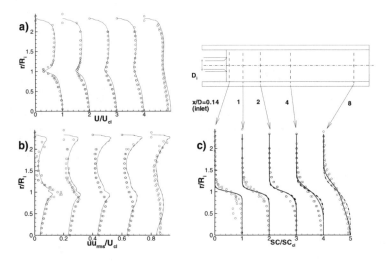

Fig. 1. Statistics at x/D=0.14 (inlet), 1, 2, 4 and 8, made dimensionless by the mean centerline values and with appropriate offsets

The axial velocity profiles, both U (mean, figure 1a) and uu_{rms} (fluctuation, figure 1b) show that the models and numerics used are capable of accurately replicate the velocity field. Good agreement is also demonstrated for the mean passive scalar in figure 1c.

The passive scalar turbulent statistics, however, show significant differences in the reproduction of both $scsc_{rms}$ and usc (respectively, the fluctuation, figure 2a and the turbulent scalar flux, figure 2b). The inability to reproduce $scsc_{rms}$ and usc remained similar both without a passive scalar subgrid scale model and with the dynamic scalar subgrid scale model. Using fully developed profiles at the inlet (RUN1) or rescaled profiles (RUN2) was irrelevant, as was moderate grid refinement. It seems that the decisive factor is the discretization of the convective term and the three line types presented

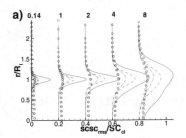 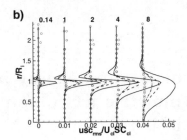

Fig. 2. Statistics at x/D=0.14 (inlet), 1, 2, 4 and 8, made dimensionless by the mean centerline values and with appropriate offsets

in figure 2 refer to three distinct schemes: upwind scheme (dash-dot line), second order TVD (TVD1 - dashed line) and bounded scheme (TVD2 - solid line).

Even though only first order accurate, the upwind method is the one closest to the two experimental quantities. However, the TVD scheme (TVD1) was the preferred method, since it is second order accurate (reverts to upwind at only 6 % of the control volumes). The bounded scheme, TVD2, would seem to be more accurate, as it only reverts to upwind in 1.5 % of the control volumes. However, it is much worse than TVD1 in reproducing $scsc_{rms}$ and usc. As an explanation, we suggest that TVD2 cannot remove all numerical wiggles that, even though below the bounding values, contribute to the turbulent scalar flux over evaluation, comparing with TVD1.

References

[1] M. Herrmann, G. Blanquart, and V. Raman. Flux corrected finite-volume scheme for preserving scalar boundedness in large-eddy simulations. Technical report, Center for Turbulence Research, University of Stanford, 2004.

[2] A. J. Keating. *Large-Eddy simulation of heat transfer in turbulent channel flow and in the turbulent flow downstream of a backward-facing step.* PhD thesis, University of Queensland, 2004.

[3] M. M. C. L. Lima. *Simultaneous Measurements of Velocity and Concentration by Laser Doppler Anemometry and Laser Induced Fluorescence (In Portuguese).* PhD thesis, University of Porto, 2000.

[4] A. Silva Lopes and J.M.L.M. Palma. Simulations of isotropic turbulence using a non-orthogonal grid system. *Journal of Computational Physics*, 175(2):713–738, 2002.

[5] C. D. Pierce. *Progress-variable approach for large-eddy simulation of turbulent combustion.* PhD thesis, University of Stanford, 2001.

[6] P. Wesseling. *Principles of Computational Fluid Dynamics*, volume 29 of *Springer series in computational mathematics.* Springer-Verlag, 2001.

Turbulence and energy balance in an axisymmetric jet computed by Large Eddy Simulation

Christophe Bogey[1] and Christophe Bailly[2]

[1] LMFA, UMR CNRS 5509, Ecole Centrale de Lyon, 69134 Ecully, France
`christophe.bogey@ec-lyon.fr`
[2] Same address `christophe.bailly@ec-lyon.fr`

1 Introduction

An axisymmetric jet at Mach number M = 0.9 is computed by Large Eddy Simulation (LES) using high-order schemes with the aim of investigating the jet self-similarity region. The Reynolds number based on the inflow diameter and velocity is $\mathrm{Re}_D = 1.1 \times 10^4$, corresponding to that of the jet studied experimentally by Panchapakesan & Lumley [1] (hereafter referred to as P&L). In the jet self-preserving region, second- and third-order moments of velocity fluctuations, and pressure-velocity correlations are evaluated directly. The terms involved in the budgets for the turbulent kinetic energy and for the energy components are also all calculated explicitly. In this way, the original feature of this work is that the LES provides some turbulent quantities, such as the dissipation or the pressure diffusion in the kinetic energy budget, which cannot be easily measured, and are therefore usually neglected, estimated using turbulence modellings, or as the closing balance [1, 2, 3, 4, 5].

2 Parameters

The simulation is performed on a Cartesian grid using schemes displaying low dissipation and low dispersion [6]. All the scales discretized at least by four points per wavelength are neither distorted nor dissipated by the numerical algorithm. A selective filtering is applied to remove grid-to-grid oscillations without affecting the resolved scales, and to take into account the effects of the subgrid energy-dissipating scales. This LES approach based on explicit filtering has been previously used by the authors for jets at various Reynolds numbers [7, 8]. The effects of the filtering have been investigated from the turbulent kinetic energy budget. It has been in particular shown that this approach does not decrease artificially the flow Reynolds number [9, 10].

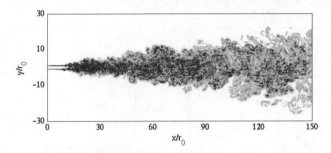

Fig. 1. Snapshot of the vorticity field in the $x - y$ plane at $z = 0$.

The present simulation is carried out using a grid of 44 million nodes, extending up to 150 radii in the axial direction, as illustrated by the snapshot of vorticity in Figure 1. For the convergence of statistics, the computation time is more than 3000 CPU hours using a Nec SX5.

3 Results

The centerline profiles of turbulence intensities u'/u_c and v'/u_c are presented in Figure 2(a). Their values are nearly constant for $x \geq 110r_0$, indicating that the jet self-similarity region is reached, in good agreement with the results of P&L. In what follows, the turbulence properties are therefore calculated for $x \geq 110r_0$. Example of terms which are usually not available experimentally are given in figure 2(b) showing the pressure-velocity correlations $<p'u'>$ and $<p'v'>$ across the jet. These terms are directly calculated, whereas in [4] for instance they are estimated from the energy dissipation curves assuming isotropic or axisymmetric turbulence.

The budget of the turbulent kinetic energy in the jet self-similarity region is presented in figure 3(a). The main terms correspond to mean convection, production, dissipation, turbulence diffusion and pressure diffusion, the dissipation term being the sum of the viscous dissipation and of the filtering dissipation. The LES results are compared with the experimental results of P&L for a jet at the same Reynolds number. There is a very good agreement for the four curves associated with convection, production, dissipation and turbulence diffusion. The pressure-diffusion term calculated by LES, albeit not negligible, is found to be small with respect to the other energy terms.

The budgets is also computed for the different energy components. The budget for $<v'v'>/2$ is for instance shown in figure 3(b). The pressure-diffusion term and the pressure-strain rate term (dashdot lines) are in particular provided separately, whereas in experiments only the sum, namely the pressure-gradient term, is classically obtained. Such results will help to clarify the role of different terms in redistributing the turbulent kinetic energy within the jet.

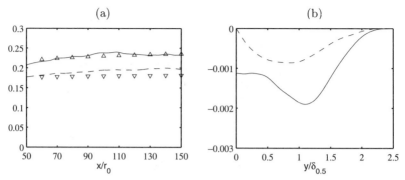

Fig. 2. (a) Turbulence intensities along the jet centerline, u'_{rms}/u_c: ——— LES, $\triangle$ P&L, and v'_{rms}/u_c: – – – LES, $\triangledown$ P&L. (b) Pressure-velocity correlations obtained across the jet by LES: ——— $<p'u'>/(\rho_c u_c^3)$, – – – $<p'v'>/(\rho_c u_c^3)$. The curves are averaged over $110r_0 \leq x \leq 140r_0$ (ρ_c and u_c: centerline mean density and axial velocity, $\delta_{0.5}$: jet half-width).

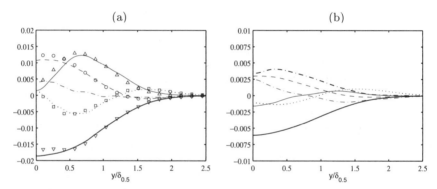

Fig. 3. Budgets across the jet for: (a) turbulent kinetic energy, (b) component energy $< v'v' > /2$: mean convection (– – – LES, o P&L), production (——— LES, $\triangle$ P&L), dissipation (——— LES, $\triangledown$ P&L), turbulence diffusion (······ LES, $\square$ P&L), pressure diffusion (– · – · LES) and pressure strain-rate (– · – · LES). The curves are normalized by $\rho_c u_c^3 \delta_{0.5}$, and averaged over $110r_0 \leq x \leq 140r_0$.

References

1. N.R. Panchapakesan, J.L. Lumley: J. Fluid Mech. **246**, 197 (1993)
2. S. Sami: J. Fluid Mech. **29**(1), 81 (1967)
3. I. Wygnanski, H. Fiedler: J. Fluid Mech. **38**(3), 577 (1969)
4. H.J. Hussein, S.P. Capp, W.K. George: J. Fluid Mech. **258**, 31 (1994)
5. T.H. Weisgraber, D. Liepmann: Experiments in Fluids **24**, 210 (1998)
6. C. Bogey, C. Bailly: J. Comput. Phys. **194**(1), 194 (2004)
7. C. Bogey, C. Bailly: Theoret. and Comput. Fluid Dynamics **20**(1), 23 (2006)
8. C. Bogey, C. Bailly: Phys. Fluids **18**, 065101 (2006)
9. C. Bogey, C. Bailly: AIAA Journal **43**(2), 437 (2005)
10. C. Bogey, C. Bailly: Int. J. Heat and Fluid Flow **27**, 603 (2006)

Shear-Improved Smagorinsky Model

F. Toschi[1], E. Lévêque[2], L. Shao[3] and J.-P. Bertoglio[3]

[1] IAC-CNR, Viale del Policlinico 137, I-00161, Roma, Italy and INFN, Sezione di Ferrara, Via G. Saragat 1, I-44100 Ferrara, Italy. toschi@iac.cnr.it
[2] Laboratoire de Physique, CNRS, Université de Lyon, École normale supérieure de Lyon, France.
[3] Laboratoire de Mécanique des Fluides et Acoustique, CNRS, Université de Lyon, École centrale de Lyon, Université Lyon I, INSA de Lyon, France.

We present results validating the recently proposed Shear-Improved Smagorinsky Model (SISM) for Large Eddy Simulation (LES) of non-homogeneous turbulent flows. The SISM is rooted in the phenomenology of strongly sheared turbulent flows, e.g. of relevance close to solid boundaries. The SISM is a simple modification over the standard Smagorinsky model (SM) [2] and consists in subtracting the averaged shear: $\nu_T = (C_s\Delta)^2(|\overline{S}| - |\langle\overline{S}\rangle|)$. This is enough to correct for the overdissipative character of SM near wall boundaries.

In last recent years, several studies have addressed the question of how the presence of a strong shear affects the statistical properties of turbulence [3,5]. Here, we will only sketch the basic features of the model. For a statistically homogeneous turbulent shear flow, the subgrid-scale (SGS) energy budget, that can be derived analytically from the Navier-Stokes equations, indicates that the mean SGS energy flux should involve two separate contributions of order $\Delta^2\langle|\overline{S'}|^3\rangle$ and $\Delta^2|\langle\overline{S}\rangle|\langle|\overline{S'}|^2\rangle$ respectively. The resolved rate-of-strain $\overline{S}$ is here decomposed into a statistical mean (a priori ensemble-averaged) and a fluctuating part: $\overline{S} = \langle\overline{S}\rangle + \overline{S'}$. Accordingly, the *shear-improved eddy-viscosity* reads:

$$\nu_T(\mathbf{x}, t) = (C_s\Delta)^2 \cdot \left(|\overline{S}(\mathbf{x}, t)| - |\langle\overline{S}(\mathbf{x}, t)\rangle|\right) \tag{1}$$

with corresponding SGS energy flux: $F_{\text{sgs}} = (C_s\Delta)^2 \left(\langle|\overline{S}|^3\rangle - |\langle\overline{S}\rangle|\langle|\overline{S}|^2\rangle\right)$. Straightforwardly, this flux vanishes if the resolved turbulence disappears, i.e, $\overline{S'} = 0$. Furthermore,

- where $|\overline{S'}| \gg |\langle\overline{S}\rangle|$, the contribution of order $\Delta^2\langle|\overline{S'}|^3\rangle$ dominates the mean SGS energy flux and turbulence is locally homogeneous and isotropic. In that case, our model yields the Smagorinsky estimate $F_{\text{sgs}} \simeq (C_s\Delta)^2\langle|\overline{S'}|^3\rangle$ in agreement analytical developments.

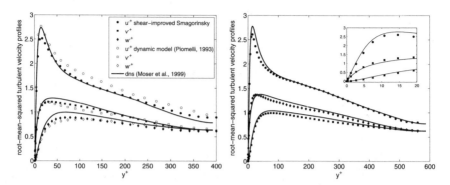

Fig. 1. Left: Turbulent intensity profiles at $Re_\tau = 395$ (this corresponds to a grid $64 \times 65 \times 64$) in comparison with DNS data and LES data obtained with the dynamic Smagorinsky model. Right: Turbulent intensity profiles at $Re_\tau = 590$ (with a grid $96 \times 97 \times 96$). The inset focuses on near-wall behavior in comparison with DNS data.

- where $|\langle \overline{S} \rangle| \gg |\overline{S'}|$, eddies of size comparable to the grid-scale have no time to adjust dynamically and are rapidly distorted by the mean shear. There, the SGS energy flux is driven by the contribution of order $\Delta^2 |\langle \overline{S} \rangle| \langle |\overline{S'}|^2 \rangle$. By assuming that $\langle |\overline{S}|^3 \rangle \approx \langle |\overline{S}|^2 \rangle^{3/2}$, one consistently obtains $F_{\mathrm{sgs}} \simeq 1/2 \, (C_s \Delta)^2 |\langle \overline{S} \rangle| \, \langle |\overline{S'}|^2 \rangle$.

Comparisons with the dynamic Smagorinsky model [1] and direct numerical simulations (DNS) are shown to be very satisfactory for mean velocity, turbulent kinetic energy (Fig. 1) and Reynolds stress profiles (Fig. 2).

Another appealing feature of the SISM, as compared to the SM, is its capability to automatically switch on (or off). In Fig. 3, the time development of the Reynolds number $Re_\tau(t)$ is displayed : at $t = 0$ the flow is laminar

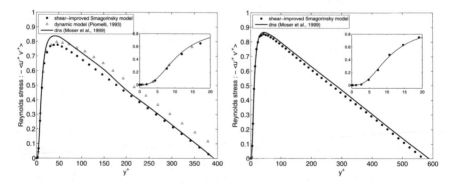

Fig. 2. Left: Reynolds stress at $Re_\tau = 395$ (computed from the resolved velocity). Right: The Reynolds stress at $Re_\tau = 590$. The insets focus on the near-wall behavior.

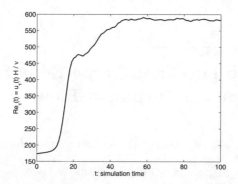

Fig. 3. The time development of the Reynolds number $Re_\tau(t)$ based on the plane-averaged friction velocity $u_\tau(t)$.

(plus a small random perturbation); as time goes on turbulence grows in the channel (drag crisis); the model becomes more and more dissipative as the average velocity profile develops; asymptotically $Re_\tau = 590$ is reached (simulation performed on a $96 \times 97 \times 96$ grid).

First tests of the SISM for turbulent plane-channel flows indicate that the model possesses a good predictive capability (almost equivalent to the dynamic Smagorinsky model) with a computational cost and a manageability comparable to the original Smagorinsky model. Preliminary tests in more complex geometry were also performed, see [4]. A challenge for the future will be to validate and implement appropriate generalization of the model to deal with generic non stationary flows.

References

1. M. Germano. Turbulence: the filtering approach. *J. Fluid Mech.*, 238(325-36), 1992.
2. J. Smagorinsky. General circulation experiments with the primitive equations. i. the basic experiment. *Mon. Weather Rev.*, 91:99, 1963.
3. F. Toschi, G. Amati, S. Succi, R. Benzi, and R. Piva. Intermittency and structure functions in channel flow turbulence. *Physical Review Letters*, 82(25):5044–5047, 1999.
4. F. Toschi, H. Kobayashi, U. Piomelli, and G. Iaccarino. Backward-facing step calculations using the shear improved smagorinsky model. CTR - Proceedings of the 2006 Summer Program.
5. F. Toschi, E. Lévêque, and G. Ruiz-Chavarria. Shear effects in nonhomogeneous turbulence. *Phys. Rev. Lett.*, 85(7):1436, 2000.

Hybrid Two Level and Large-Eddy Simulation of Wall Bounded Turbulent Flows

Ayse Gul Gungor, Martin Sanchez-Rocha and Suresh Menon

School of Aerospace Engineering, Georgia Institute of Technology, Atlanta, Georgia 30332-0150 suresh.menon@ae.gatech.edu

A novel approach to model high-Re wall-bounded flows is developed based on coupling the outer flow large-eddy simulation (LES) approach to a near-wall two-level simulation (TLS) approach. Here, the near-wall small-scale field is simulated rather than modeled. Simulations of turbulent channel flows, boundary layer and flow around a surface mounted hill demonstrate the ability of this model for capturing the turbulent flow behavior using very coarse grids.

1 Introduction

LES of high-Re wall bounded flows is too expensive due to the need to accurately predict dynamically important scales in the near-wall region, and many methods have been developed in the past to reduce the near-wall resolution requirement [1]. In this paper, we extend the recently developed multi-scale decomposition approach called TLS[2, 3, 4] as a near-wall model, and combine it with conventional localized dynamic kinetic energy LES (LDKM-LES) approach [5] in the far field. The TLS approach decomposes the velocity field into large-scale (LS) and small-scale (SS) components [3, 4] and both the LS and SS are explicitly simulated and coupled together. A key attribute of the TLS approach is that the SS velocity fluctuations are not suppressed, and especially near the wall this is an unique capability unlike other near-wall LES models [1]. In this approach, TLS is used only in the near-wall region, while traditional LES is used in the rest of the domain. A consistent formulation of this TLS-LES approach was demonstrated earlier [6].

2 Results and Discussion

Fig. 1(a) shows a very fine TLS mesh embedded inside the coarser LES mesh. The TLS field evolves between each LES time-step and the coupling between the two-scales occurs when then energy content in the TLS field approximately

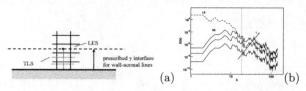

Fig. 1. (a) Illustration of wall-normal discretization for the hybrid TLS-LES model; (b) Evolution of SS energy spectrum.

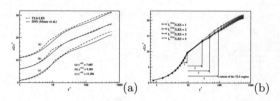

Fig. 2. Sensitivity of the TLS-LES approach for channel flow (obtained using a $32 \times 40 \times 32$ grid) to (a) the first LES location; (b) the extension of the TLS region.

matches with the LES resolved spectra (Fig. 1(b)) [6]. Figures 2(a) and (b) show results obtained by keeping the LES and TLS grids constant but with different overlap region. Changes are observed as the TLS region is extended from one to three LES cells, but very little difference is seen when it is extended to five LES cells. These results are consistent with the earlier observation [4] that we only need to resolve the very near-wall energetic region using TLS .

A spatially evolving turbulent flow over a boundary layer at $Re_\theta = 1400$ is also simulated. The inflow conditions were generated using a re-scaling procedure. Mean flow statistics predicted by LDKM-LES and TLS-LES (using a much coarser grid) are shown in Figs. 3(a) and (b). For TLS-LES the first LES point is at $\Delta y_1^+ = 5$ and the TLS model is used up to $y^+ = 25$. Although no grid optimization has been conducted so far it can be seen that the TLS-LES captures the near-wall statistics quite reasonably using a substantial coarser wall-normal grid spacing than the original LDKM-LES. Additionally, TLS-LES is able to reproduce the characteristic structures generated in the near wall zone that controls the generation of turbulent kinetic energy.

Finally, a flow over an 3D hill is investigated with the TLS region represented by five LES cells extending up to $y^+ \approx 40$. Results are compared with LDKM-LES and data in Figs. 4)(a) and (b). Although further optimization is needed, the TLS-LES approach in the near-wall region appears to provide consistent results when compared to the much higher resolution LDKM-LES.

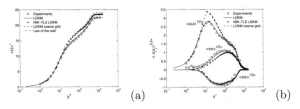

(a)   (b)

Fig. 3. Turbulent boundary layer obtained from LDKM-LES on $201 \times 91 \times 121$ grid with $\Delta y_1^+ = 0.5$, LDKM-LES using $101 \times 45 \times 61$ grid with $\Delta y_1^+ = 0.5$, and TLS-LES using $101 \times 45 \times 61$ grid with $\Delta y_1^+ = 5.0$ (a) Mean velocity; (b) Rms velocities.

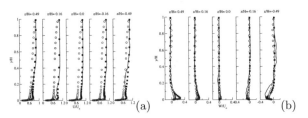

(a)   (b)

Fig. 4. Comparison of the mean flow quantities obtained by LDKM-LES (solid lines) using a $192 \times 184 \times 144$ grid with $\Delta y_1^+ = 4$ and TLS-LES (filled symbols) using a $48 \times 46 \times 36$ grid with $\Delta y_1^+ = 10$ (a) U/U_0; (b) W/W_0.

3 Conclusion

A novel approach to simulate the high-Re number flows has been developed based on coupling a near-wall simulation using TLS with a more conventional LES approach in the far field. A coupled system of TLS and LES equations are implemented to simulate both separated and separating-reattaching flows without any model changes. Results obtained so far suggest that the hybrid TLS-LES approach has the potential for capturing the near-wall dynamics by using very coarse grids. Future studies will further optimize this approach.

This work is supported by the Office of Naval Research. Computational time has been provided by DOD HPC at NAVOCEANO and ARL.

References

1. Piomelli, U. and Balaras, E. *Annual Rev. of Fluid Mech.* **34**, 349–374 (2002).
2. Kemenov, K. and Menon, S. In *Adv. in Turbulence IX*, 203–206. CIMNE, (2002).
3. Kemenov, K. and Menon, S. *J. of Computational Physics* **220**, 290–311 (2006).
4. Kemenov, K. and Menon, S. *J. of Computational Physics* **222**, 673–701 (2007).
5. Kim, W.-W. and Menon, S. *International J. of Numerical Fluid Mechanics* **31**, 983–1017 (1999).
6. Gungor, A. G. and Menon, S. *AIAA paper 2006-3538* (2006).

Finite dimensional models for perturbed self-similar turbulent flows

Stephen L. Woodruff[1] and Robert Rubinstein[2]

[1] Florida State University, Tallahassee FL, USA woodruff@caps.fsu.edu
[2] NASA Langley Research Center, USA r.rubinstein@larc.nasa.gov

Two-equation turbulence modeling poses a significant theoretical question: given that a steady state of homogeneous isotropic turbulence is characterized by its energy flux and integral scale, can a time-dependent state of homogeneous isotropic turbulence be described by closed evolution equations for these two parameters alone?

It seems evident that answering this question requires dynamic equations for turbulence evolution; because it is strictly a steady-state theory, the Kolmogorov theory itself cannot be sufficient. In previous work [1], we derived such equations for small, slow perturbations of steady state turbulence as *compatibility conditions* for construction of a perturbation series, much like the Hilbert expansion in kinetic theory. These equations are less dynamic equations than constraints that force the evolution to remain in a space of nearly steady states. The viewpoint is related to that of Yoshizawa [4], according to which the two equation model is a condition permitting reformulation of the equations in terms of any two equivalent variables.

Since the analysis was based on the classical Heisenberg closure model, it is natural to ask whether similar results could be found for other closures. Similarly, we can ask for a characterization of closure models that admit an abridged description by a two-equation model. Another question is whether the analysis can be extended to weak perturbations of general self-similar states ('equilibrium' states, in the unsatisfactory terminology of the modeling literature), and even ultimately to inhomogeneous problems like perturbed boundary layers. This work is a first step toward answering these questions.

The spectral evolution equation is $\dot{E}(\kappa, t) = P(\kappa, t) - \partial \mathcal{F}/\partial \kappa - 2\nu\kappa^2 E(\kappa, t)$. where $\mathcal{F}$ is the energy flux. In a steady state, the production $P(\kappa) = P_0(\kappa; \bar{P}, L_P)$ is characterized by a total rate of energy input $\bar{P}$ and a forcing scale L_P, and the energy spectrum $E(\kappa) = E_0(\kappa; \epsilon, L)$ is characterized by the energy flux ϵ and its integral scale L. A steady state of course requires $\epsilon = \bar{P}$ and $L = L_P$.

We now let the parameters $\bar{P}$, L_P, ϵ, and L be functions of a slow time variable τ as a result of a perturbed production spectrum $P(\kappa, t) =$

$P_0(\kappa; \bar{P}(\tau), L_P(\tau)) + \delta \cdot P_1(\kappa; \bar{P}(\tau), L_P(\tau))$. Then the energy spectrum admits the corresponding expansion
$E(\kappa, t) = E_0(\kappa; \bar{P}(\tau), L_P(\tau)) + \delta \cdot E_1(\kappa; \bar{P}(\tau), L_P(\tau); \epsilon(\tau), L(\tau))$. Standard multiple-scale perturbation methods lead to an inhomogeneous linear integral equation,

$$\frac{\partial}{\partial \kappa} \mathcal{L}[E_1(\kappa, \tau)] = -\frac{\partial}{\partial \tau} E_0(\kappa; \epsilon_0(\tau), L_0(\tau)) + P_1(\kappa, \tau) - D_1(\kappa, \tau).$$

where $\mathcal{L}$ is the energy transfer linearized about the steady state and $D_1 = 2\nu \int_0^\infty d\kappa \; \kappa^2 E_1$. Compatibility conditions to solve this equation are of the form

$$\int_0^\infty d\kappa \; \Psi_i(\kappa) \frac{\partial}{\partial \tau} E_0(\kappa, \epsilon(\tau), L(\tau)) = \int_0^\infty d\kappa \; \Psi_i(\kappa) \left[P_1(\kappa, \tau) - D_1(\kappa, \tau) \right] \quad (1)$$

where the Ψ_i are solutions of the homogeneous adjoint equation $\mathcal{L}^\dagger[\partial \Psi / \partial \kappa] = 0$ and the index i ranges over the number of such solutions. One solution is always $\Psi_1 = $ constant; the corresponding compatibility equation is simply the energy balance obtained by integrating the spectral evolution equation over all wavenumbers.

Straightforward algebraic manipulations express the compatibility equations as equations for $\partial \epsilon / \partial \tau$ and $\partial L / \partial \tau$. This calculation was completed [1] for the classical Heisenberg closure, for which a second compatibility condition was found. The analysis left open the question of whether the number of compatibility equations and number of slowly varying scalars are necessarily equal: the number of equations equals the dimension of the null-space of $\mathcal{L}^\dagger$, but because $\mathcal{L}$ is not self-adjoint, there is no simple general way to compute it.

It is therefore useful to carry out the analysis for different closure models. Two closely related models are the Kovasznay and Ellison models, which we write in the common form $\mathcal{F} = \eta(\kappa)\kappa E(\kappa)$. For the Kovasznay model $\eta = \sqrt{\kappa^3 E}$ and $\mathcal{L}^\dagger[\Psi] = \kappa^{5/2} E_0(\kappa)\Psi'(\kappa)$ where the prime denotes differentiation, and E_0 is the steady state spectrum. The only null solution is a constant. In this case, only one compatibility condition exists.

For the Ellison model, $\eta(\kappa)^2 = \int_0^\kappa d\mu \; \mu^2 E(\mu)$ and

$$\mathcal{L}^\dagger[\Psi] = \eta_0 \kappa \Psi'(\kappa) + \frac{1}{2}\kappa^2 \int_\kappa^\infty dp \; \frac{p E_0(p)}{\eta_0(p)} \Psi'(p)$$

where η_0 is the integral strain rate evaluated on the steady spectrum E_0. The null solution is

$$\Psi'(\kappa) = \frac{C}{\kappa \eta_0} \exp\left[\int d\kappa \; \frac{\kappa^2 E_0(\kappa)}{2\eta_0(\kappa)^2}\right] \quad (2)$$

As for the Heisenberg model, two compatibility equations for slowly varying spectral evolution can be found: one is the energy balance from constant Ψ, and the second arises from the non-constant solution in which $C \neq 0$ in Eq. (2). In an inertial range, this solution has the form $\Psi \sim \kappa^{-1/6}$.

We can extend the analysis to perturbations about time-dependent reference states. As a model problem, consider decaying turbulence in which the large scales are of the form Ak^2, with constant A. The similarity solution is $E(\kappa, t) = A^{3/5}t^{-4/5}\phi(x)$ with $x = A^{1/2}\kappa^{5/2}t$. The existence of such solutions is a geometric consequence of scaling invariances of the Navier-Stokes equations [3]. For the Heisenberg model, the similarity solution for decay is defined by ϕ satisfying

$$-\frac{4}{5}\phi + x\phi' = \frac{2}{5}x^{3/5}\frac{d}{dx}\left[\int_0^x y^{-3/5}\phi(y)dy \int_x^\infty z^{-6/5}\phi(z)^{1/2}dz\right] \quad (3)$$

Suppose that we allow breaking of the scaling invariance responsible for the self-similar solution [3] through slow variation $A = A(\tau)$ and $B = B(\tau)$ where, as before, τ denotes a slow time variable. Evolution is then through a series of locally self-similar states, but the evolution as a whole is not self-similar.

The slow variation *ansatz* $E(\kappa, t) = A(\tau)^{3/5}t^{-4/5}\phi(B(\tau)^{1/2}\kappa^{5/2}t)$ does not lead to a solution of the spectral evolution equation unless A and B are constants. Assume therefore that the energy spectrum contains a correction

$$E(\kappa, t) = A(\tau)^{3/5}t^{-4/5}\phi(B(\tau)^{1/2}\kappa^{5/2}t) + \varepsilon E_1(\kappa, t)$$

Then

$$\frac{3}{5}\frac{\partial A}{\partial \tau}\phi(x) + \frac{1}{2}A^{1/2}\phi'(x)B^{-1/2}\frac{\partial B}{\partial \tau}x = \mathcal{L}[E_1] \quad (4)$$

where $\mathcal{L}$ is the Heisenberg transfer linearized about the similarity solution Eq. (3).

Compatibility conditions to solve Eq. (4) are again determined by solutions of the adjoint equation $\mathcal{L}^\dagger[\partial\Psi/\partial\kappa] = 0$. These conditions are expressed by integration over the similarity variable x; this effectively separates the time dependence of the similarity solution from the slow time dependence of the variables A and B.

The principal conclusion is that these derivations require a dynamic description of turbulence evolution; at this point, we invoke simple classical closures only because of their analytical simplicity. Reduced order models cannot be derived from Kolmogorov steady-state arguments in principle. Making precisely such arguments caused previous attempts to derive models to fail.

References

1. S. L. Woodruff and R. Rubinstein, J. Fluid Mech. **565**, 95 (2006).
2. R. Rubinstein and T. Clark, Theor. Comput. Fluid Dyn., **17**, 249 (2004).
3. M. Oberlack, J. Fluid Mech. **379**, 1 (1999).
4. A. Yoshizawa, Phys. Fluids **30**. 628 (1987).

Turbulent flow of viscoelastic shear-thinning liquids through a rectangular duct

M P Escudier, A K Nickson and R J Poole

University of Liverpool, Department of Engineering, Liverpool, L69 3GH UK
m.p.escudier@liv.ac.uk, k.nickson@liv.ac.uk, robpoole@liv.ac.uk

1 Introduction

Large reductions in turbulent frictional drag occur when high molecular-weight polymers, surfactants etc are added to a Newtonian solvent [1]. Recent advances in numerical modelling, especially Direct Numerical Simulations [2, 3, 4, 5] have enhanced our understanding of how the additives interact with and modify the turbulence and reduce the frictional drag. The purpose of this work is to provide for comparison purposes a more comprehensive experimental database for higher polymer concentrations than has been available hitherto for planar duct flow. Selected results are presented here.

2 Experimental arrangement and instrumentation

Complete details of the new experimental facility used will be given elsewhere. The duct consists of seven $1.2m$ long modules with a rectangular internal cross section of height $h = 25mm$ and width $w = 298$ mm (hydraulic diameter $D_H = 46$ mm). Reynolds numbers are defined as $Re = \rho D_H U_B / \mu_W$ where for CMC and XG μ_W, was determined from τ_W and the flow curves and U_B is the bulk velocity. For water $Re = 18900$, CMC $Re = 13500$, XG $Re = 13000$. A perspex test section, length $100mm$, is located $6m$ (130 D_H) from the inlet. We used a 2D LDA system (measuring volume length 200 μm, diameter 20 μm) to measure the transverse variation of the mean and fluctuating velocities. In the test section we use a unique open-slot technique, inspired by Poggi et al [6], which allows unimpeded access of the laser beams to the flow, simultaneous and coincident measurement of u, u' and v', and hence the determination of the Reynolds shear stress $\overline{uv}$, without the need for a complex optical arrangement. w' was measured separately.

3 Results and discussion

Mean velocity profiles. The excellent quality of the data is apparent from Figure 1(a) which shows the mean velocity distributions plotted in u^+ vs log y^+ form. For CMC and XG, $y^+ = \rho u_\tau y / \mu_W$. For water the data closely follow the standard log law for $y^+ > 30$ and extend down to $y^+ \approx 10$. For CMC there is an upshift $\Delta u^+ \approx 3$ corresponding to the modest drag reduction (DR = 28%). For the high DR (67%) XG the data follow $u^+ = y^+$ from $y^+ \approx 1.5$ to ca 15 and then lie close to Virk's ultimate profile, $u^+ = 11.7 \ln y^+$ - 17. The sub-layer data for XG confirm that the surface slit has no discernible influence on the flow. In each case, the smallest y-value is $0.5mm$

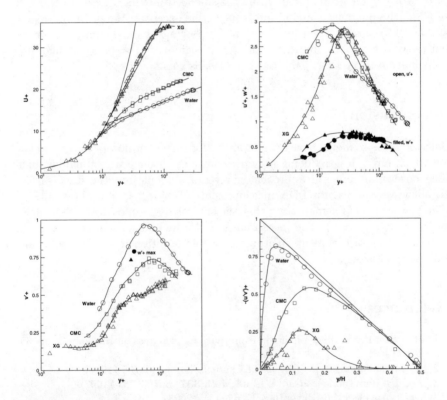

Fig. 1. Normalised mean velocity, Reynolds normal and shear stresses in wall units (a) u^+, (b) u'^+ and w'^+, (c) v'^+ and (d) $\overline{uv}^+$

Normal-stress profiles. The u'^+ vs y^+ data in Figure 1(b) (open symbols) show that the peak values in all three cases are roughly the same but the peak moves to higher y^+ values with increasing drag reduction and at the same time the distribution narrows. The lateral (z-direction) fluctuation data

w'^+ in Figure 1(b) again show little difference in the peak values which are about 25% of those for u'^+. The v'^+ data (Figure 1(c)) show slightly more complex behaviour with a much greater spread of peak values, from 0.95 for water to about 0.55 for XG. What the data also reveal is that the degree of anisotropy increases with drag reduction. All fluctuations are reduced (normalised with bulk velocity) with the greatest reduction in u': for water, u'/v' ≈ 3.1, for CMC 4.2 and for XG 4.8 (for peak values) while $u'/w' \approx 4$ for all fluids.

Reynolds shear stress. For any fully developed flow, the variation of total shear stress τ_T must follow the diagonal straight line in Figure 1(d). For water, the difference $\tau_T + \rho\overline{uv}$ must equal the viscous contribution $\mu\partial u/\partial y$. For polymers, as u' and, even more, v' are suppressed, the difference $\tau_T + \rho\overline{uv}$ - $\mu\partial u/\partial y$ has to be made up by the so-called polymer stress. In the case of CMC the polymer contribution is negligible except in the near-wall region (y/H < 0.2) but for XG it is clear that the polymer stress dominates over the entire cross section. Only in the near-wall region (y/H $\approx$ 0.15) is there any contribution to τ_T from -$\rho\overline{uv}$, but even this is small.

4 Conclusions

Selected measurements have been reported for turbulent flow of water and two shear-thinning, drag-reducing polymer solutions through a rectangular duct. The novel experimental approach involves a slit cut into the duct surface to allow easy access for LDA measurements of u, u', v' and, above all, $\overline{uv}$. The measurements demonstrate that the slit has negligible effect on the near-wall flow (to $y^+ \approx 2$). The measurements also include w' and so permit the determination of turbulent kinetic energy k in addition to the Reynolds and polymer stresses.

References

1. B. Toms: Proceedings of the first international congress on rheology **2**, pp135–141 (1948)
2. R. Sureshkumar *et al* : Physics of Fluids **9**, 33 pp743–755 (1997)
3. J. M. J. den Toonder *et al*: J. Fluid Mech **337**, pp193–231 (1997)
4. E. De Angelis *et al*: Phys. Rev. E. **67**, (5), (2003)
5. T. Min *et al*: J. Fluid Mech **486**, pp213–238 (2003)
6. D. Poggi *et al*: Expts in Fluids **32**, (3), pp336–375 (2003)

Experimental studies of liquid-liquid dispersion in a turbulent shear flow

Florent Ravelet, René Delfos, and Jerry Westerweel

Laboratory for Aero and Hydrodynamics, Leeghwaterstraat 21, 2628 CA Delft, The Netherlands. florent.ravelet@ensta.org

1 Introduction

Liquid-liquid dispersions are encountered for instance in extraction or chemical engineering when contact between two liquid phases is needed. Without surfactants, when two immiscible fluids are mechanically agitated, a dispersed state resulting from a dynamical equilibrium between break-up and coalescence of drops can be reached. Their modelling in turbulent flows is still limited [1].

Experimental setup

We built up an experiment to study liquid-liquid dispersions in a turbulent Taylor-Couette flow, produced between two counterrotating coaxial cylinders, of radii $r_i = 100$ mm and $r_o = 110$ mm (gap ratio $\eta = r_i/r_o = 0.909$). The useful length is $L = 185$ mm. There is a space of 10 mm between the cylinders bottom ends and a free surface on the top. The cylinders rotation rates ω_i and ω_o can be set independently. We use the set of parameters defined in [2]: a mean Reynolds number Re based on the shear and on the gap; and a "Rotation number" Ro which is zero in case of perfect counterrotation ($r_i\omega_i = -r_o\omega_o$). At maximal rotation frequency, $Re \simeq 5.6 \times 10^4$ for pure water.

For the two-phases studies, we use a low-viscous oil (Shell Macron 110), with $\mu = 3 \times 10^{-3}$ Pa.s, $\rho = 800$ kg.m^{-3}. The acqueous phase is either pure water or an NaI solution with a refractive index matched to that of oil (1.445 at 20°C). The viscosity, density and interfacial tension for the NaI are $\mu = 2 \times 10^{-3}$ Pa.s, $\rho = 1550$ kg.m^{-3} and $\sigma \simeq 15$ mN.m^{-1}. The viscosity ratio is close to unity and the density ratio is close to 2.

We give here some preliminary results on thresholds for dispersion, and on the increase of wall shear stress, based on global measurements of the torque T, using an HBM T20WN rotating torquemeter.

2 Results and discussion

Turbulent Taylor-Couette flow in exact counterrotation

Though the Taylor-Couette flow has been widely studied, few experimental results and theories are available for $Ro = 0$ [2]. We first report torque measurements on pure fluids in Fig. 1. The torque due to the bottom part has been removed by studying different filling levels. Esser and Grossmann [3] proposed a formula for the first instability threshold. Here it gives $Re_c(\eta, Ro) = 338$. Moreover, Dubrulle and Hersant [4] proposed a formula for the dimensionless torque $G = T/(\rho \nu^2 L)$ in the turbulent regime which is also consistent with our data (Fig. 1b).

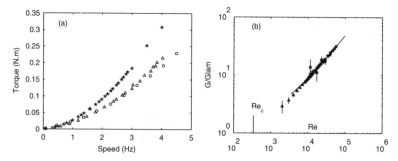

Fig. 1. (a): Torque *vs.* rotation frequency in exact counterrotation for Oil ($\triangle$), Water ($\circ$) and NaI ($\star$). (b): Dimensionless torque G normalised by laminar torque $G_{lam} = 2\pi\eta/(1-\eta)^2 Re$ *vs.* Re. Solid line is a fit of the form $G = a\, Re^2/(ln(b\, Re^2))^{(3/2)}$ [4], with $a = 16.5$ and $b = 7 \times 10^{-5}$.

Increase of dissipation in presence of turbulent dispersions

At rest, two immiscible fluids are separated into two layers. Increasing the speed, the fluids gradually get dispersed one into another. The torque shows up a fast increase around $f = 3\,\mathrm{Hz}$ (Fig. 2a), corresponding to an homogeneous dispersion. When thereafter decreasing the speed, there seems to be hysteresis in the system. In fact the time scales involved in the dispersion and separation processes are very long, the latter being even longer (Fig. 2b-c). For this volume fraction of oil and water, the two-layer system is the stable state below 2.5 Hz and the fully dispersed is stable above 3 Hz. Complicated behaviours including periodic oscillations (with a period of 20 min) have been observed in between and requires further investigation.

The torque per unit mass in a dispersed state at 4 Hz is maximum for a volume fraction of the acqueous phase around 33% and is 2 times that of pure Oil (Fig. 3). Similar behaviours have been evidenced recently in liquid-liquid turbulent pipe flow experiments [5].

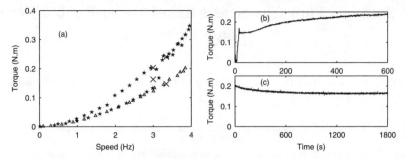

Fig. 2. (a): Torque vs. rotation frequency for pure Oil (△) and 33% Water in Oil (⋆). Speed is increased by steps of 0.2 Hz every 60 s, up to 4 Hz and then decreased back to 0. (×) are values taken from experiments (b) and (c). (b): From a two-layer system at rest, the speed is set to 3.5 Hz. (c): The speed is then decreased to 3 Hz.

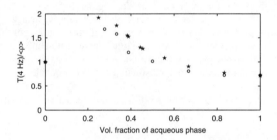

Fig. 3. Massic torque at 4 Hz *vs.* volume fraction of Water (○) and NaI (⋆).

Perspectives

Further investigations on the scaling for the counterrotating monophasic flow will be performed, on a wider Reynolds number range. We have already optimised the calibration procedure for Stereoscopic PIV measurements in a new setup of better optical quality. Further work is needed to understand what are the relevant parameters for dispersion to arise (Froude number, Weber number) We will then address the question of drop size distribution using Light Induced Fluorescence. First results have shown the presence of multiple drops.

References

1. L. M. Portela, R. V. A. Oliemans: Flow Turbulence Combust. **77**, 381 (2006)
2. B. Dubrulle, O. Dauchot, F. Daviaud et al: Phys. Fluids **17**, 095103 (2005)
3. A. Esser, S. Grossmann: Phys. Fluids **8**, 1814 (1996)
4. B. Dubrulle, F. Hersant: Eur. Phys. J. B **26**, 379 (2002)
5. K.Piela, R.Delfos, G.Ooms et al: Int. J. Multiphase Flows **32**,1087 (2006)

Study on Flow Characteristics of Micro-Bubble Two-Phase Flow

Zensaku Kawara[1], Hiromasa Yanagisawa[1], Tomoaki Kunugi[1], Akimi Serizawa[2]

[1] Kyoto University, Department of Nuclear Engineering, Yoshida, Sakyo-ku,
606-8501, Kyoto, Japan
[2] Daikin Industries, Ltd., Kanaoka, Kita-ku, 591-8511,Sakai, Japan
kawara@nucleng.kyoto-u.ac.jp

Abstract. Experimental study was carried out on micro-bubble containing bubbly two-phase flow in a circular pipe. Flow resistance in a pipe reduces relative to single-phase flow, especially in the transient region from laminar to turbulent flow. Local velocity profile shows increase of liquid velocity near the wall, and overall velocity profile is planarized with it. Visual observation with high-speed imaging system shows that comparatively large bubbles with slip velocity tend to be observed more frequently near the wall, and this suggests the existence of the interaction between coherent turbulent structure and micro-bubbles.

Keywords: micro-bubble, two-phase flow, drag reduction, velocity profile, visualization

1. Introduction

Recently micro-bubble has become to be strongly interested in wide area of science and technology fields. Serizawa et al.[1] reported that flow resistance of micro-bubble containing milky bubbly flow in a pipe was significantly reduced compared with that of single-phase flow. Though reduction rate of flow resistance possibly depend on many parameters such as Reynolds numbers, void fractions and bubble size etc., we have not enough knowledge on this phenomena to clarify mechanism of reduction of resistance. For this point of view, experimental work are carried out on upward water flow in a pipe with air micro-bubbles. Measurements are taken of pressure drop and velocity profile of liquid phase. Flow is also observed using by a high-speed video camera with a high-powered microscope.

2. Experimental Setup and Procedure

Test section is a 4m long transparent acrylic resin circular pipe with 20mm in inner diameter, which is set up vertically. Working fluid is water and air. Micro-bubble generator is set at 500mm upstream of the inlet of test section. Micro-bubbles are averagely 40 micron in diameter generated by pressurized dissolution method and/or

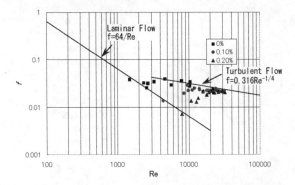

Fig.1 The comparison between friction factor and Reynolds number.

cavitation method.   Pressure drop is measured by using a manometer between pressure taps located at 2000, 2500, 3000mm downstream of the inlet. Local velocity profile is measured at 3500mm downstream by using a small-sized Pitot tube anemometer. For direct observation on micro-bubbles, high-speed imaging system is set up near the wall at 4000mm downstream of the inlet. This system consists of a high-speed video camera (Vision Research Co., Phantom 7.1, 4700 frame per second, exposure time 67 micro second) and a high-powered microscope. Visualized video images on micro-bubbles are recorded at several positions near the wall. Averaged void fraction on the cross-section of flow channel was estimated from total mass of two-phase mixture mass which was sampled at the exit of the test section.

3.   Results and Discussions

Fig.1 shows the relationship between friction coefficient of pipe f and Reynolds number Re.   Measurements are conducted for void fraction α =0, 0.10%, 0.20%. Friction coefficient of pipe f and Reynolds number Re are defined as follows:

$$\frac{\Delta P_f}{L} = f\frac{\rho V^2}{2D}, \qquad Re = \frac{DV}{\nu} \qquad (1)$$

Friction coefficient for single-phase flow indicates well-known transition from laminar flow to turbulent flow at Re=2300. On the other hand, friction coefficient of micro bubble flow significantly decrease with increase of void fraction, as though the transition point from laminar flow to turbulent flow moved to higher Reynolds region. Fig.2 shows the local velocity profile according to velocity defect law. y^+ and u^+ are defined with following equations:

$$y^+ = \frac{ru_\tau}{\nu}, \qquad u^+ = \frac{u}{u_\tau} \qquad (2)$$

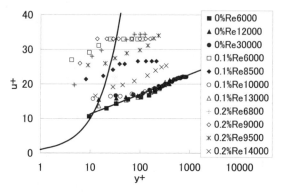

Fig.2 Universal velocity profile of various Reynolds numbers and void fractions. The result indicates that the existence of micro-bubbles leads to the increase of the velocity near the wall.

Velocity profiles for micro-bubble containing two-phase flow shift to upper value u^+ according to the reduction of friction. This is considered because of the decrease of friction velocity. Main flow region has planarized profile like plug flow.

Fig.3 shows examples of high-speed video image focused at 50 and 400 microns from the wall. Void fraction is 0.10% and Re=5500, or the data point is on the laminar flow line in Fig.1. Near the wall, comparably large (about 100 micron in diameter) bubbles, which have velocity slip relative to small bubbles less 50 micron in diameter, are observed with high frequency. This suggests the possibility of the interaction between coherent turbulent structure and micro-bubbles, and this interaction is considered to concern mechanism of the reduction of flow friction. .

Reference

1. Serizawa, A., Inui, T., Yahiro, T. Kawara, Z.: 3rd Euro-Japan Two-Phase Flow Group Meeting, Certosa di Pontignano, 21-27 September (2003).

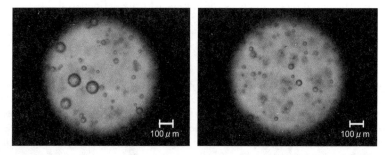

Fig.3 Visualized images near the wall : The left hand image recorded by focusing to the plane 50μm apart from the wall shows the existence of rather large bubbles near the wall. On the other hand, comparatively small number of large bubble were observed on the inner region (the right hand image recorded on the plane apart from 400μm.)

Velocity statistics in microbubble-laden turbulent boundary layer

B. Jacob[1], A. Olivieri[1], M. Miozzi[1], E. F. Campana[1] and R. Piva[2]

[1] INSEAN, Via di Vallerano 139, 00128 Rome, Italy (b.jacob@insean.it)
[2] DMA, University *La Sapienza*, Via Eudossiana 18, 00184 Rome, Italy

1 Introduction

This paper deals with the effects induced on the dynamics of turbulent boundary layers by the presence of a small quantity of microbubbles. The interaction between air bubbles and liquid turbulent flows is a problem essential for many industrial and environmental processes, and has therefore already been studied extensively. One of the most intriguing experimental findings concerns the ability of air bubbles to reduce the frictional drag of turbulent wall-bounded flows, an aspect of particular interest for ship transportation. The first evidence of this phenomenon was obtained by McCormick and Bhattacharyya [1]. Many successive experiments, carried out in order to assess the dependence of the drag reduction on the bubble dimensions and air concentrations, report reductions up to 80%, see e.g. [2]. However, because of the scatter in the results due to the large number of additional factors affecting the phenomenon, (e.g the bubble distribution in the boundary layer or the initial seeding conditions), no firm conclusion could be drawn. More recently, the problem has been tackled also by means of numerical simulations, e.g. [3, 4]. In the direct numerical simulation of a microbubble-laden turbulent boundary layer by Ferrante and Elghobashi, for instance, the bubbles were treated as rigid point-like spheres, and drag reductions up to 20% were obtained, depending on bubble concentration. The fundamental mechanism at the origin of the phenomenon was identified in the local positive divergence of the velocity field, induced by bubble-concentration inhomogeneities. As a consequence, the mean wall-normal velocity increases with respect to the Newtonian case, and the streamwise vortices are displaced away from the wall. The aim of this contribution is to examine how the features of a turbulent boundary layer over a flat plate are affected by the introduction of very small concentrations of microbubbles with dimensions comparable to the Kolmogorov scale. To this purpose, we have carried out an experimental investigation in conditions very similar to those adopted in the simulation described in [4].

2 Experimental setup

The experiment has been carried out in the recirculating water channel at INSEAN. A canonical zero-pressure gradient turbulent boundary layer grows on the lower surface of a 4 m long flat plate placed inside the test section. The free-stream velocity U_∞ is 0.75 m/s and the measurement location is 300 cm downstream the leading edge, yielding a Reynolds numbers Re_θ equal to 3000. An initially uniform distribution of microbubbles (mean diameter $\approx$ 100 μm) is produced at the leading edge of the plate by an array of wires connected to a power source, and located a few centimeters below the plate. The bubble Weber number $We = \rho u^2 d/\sigma$ is, in these conditions, significantly smaller than 1, indicating that the bubbles can be considered as rigid. Measurements of the streamwise and wall-normal components of the velocity field are obtained by means of standard PIV techniques. An ensemble of more than 2000 digital images, for both the Newtonian and the bubbly flows, is collected. Each image, corresponding to a physical region of 19.7 mm x 24.6 mm, is processed with a feature-tracking software [5] which resolves both the seeding and the bubble displacements, by allowing a full deformation of the interrogation area.

3 Results

We focus on the differences in the velocity statistics of the Newtonian case and a bubble-laden flow with a mean projected void fraction $C_s \approx 1\%$. For both cases, the streamwise mean velocity profiles are shown in Figure 1(left). The velocity $\overline{U}$ and the wall-normal distance y are non-dimensionalized by means of the corresponding inner scales of the Newtonian case, $u_\tau^* \approx 0.03\,m/s$ and $l^* = \nu/u_\tau^* \approx 0.03\,mm$. In presence of microbubbles, a substantial decrease of the mean velocity in the proximity of the wall is observed. This suggests that a reduction of the wall shear stress occurs, even if no direct measurement of the velocity gradient is available. A *rough* estimate of the friction velocity for the bubbly flow can be obtained by collapsing the near-wall data on the $\overline{U}(y)^+ = y^+$ law, yielding a figure of $u_\tau^b \approx 0.026\,m/s$. The observed increase in the *rms* value of the streamwise velocity fluctuations observed in the bubbly case is also a typical feature of drag-reducing flows, where the vertical fluctuations are only marginally affected. In Figure 1 (left) the distribution of the Reynolds stress within the boundary layer, $-\overline{uv}^+(y^+)$ is shown for both the Newtonian and the bubbly cases. The key point of the figure is the outwards shift of the distribution introduced by the bubbles, which results in the suppression of the Reynolds stress close to the wall. We note that this result is fully consistent with the plots presented in [4], and that the diminished value of the frictional drag in the bubbly situation may be interpreted in view of the relation between the skin friction coefficient C_f and the distribution of the weighted Reynolds stress, following the idea of Fukagata *et al.* [6] : $C_f \propto \int_0^\delta (1 - y)(-\overline{uv})dy$. Finally, the inset in Figure 1 (right) shows how the reduction of the mean

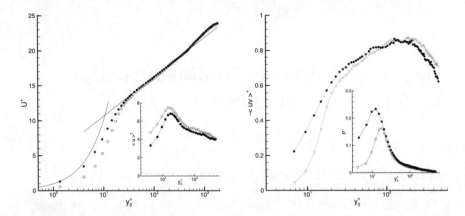

Fig. 1. (Left) Streamwise mean velocity profiles for the Newtonian and the bubble-laden cases (filled and open symbols, respectively). Inset: streamwise velocity fluctuations profiles (same symbols). (Right) Reynolds stress profiles for both the Newtonian (filled symbols) and the bubble-laden case (open symbols). Inset: production of turbulent kinetic energy(same symbols).

velocity gradient at the wall and the displacement of the Reynolds stress distribution in the bubble-laden flow affect the production of turbulent kinetic energy $\mathcal{P}^+ = (\partial \overline{U}^+/\partial y^+)(-\overline{uv}^+)$. Together with a substantial reduction of the kinetic energy production with respect to the Newtonian case, a displacement towards larger wall-normal distances of its peak value is again observed.

In conclusion, the experimental results presented so far, though restricted to very limited conditions (boundary layer flow, rigid bubbles, small concentrations), seem to provide very interesting results, in accordance with the findings of a direct numerical simulation performed with similar parameters. The explanation of the drag reduction provided in [4] in terms of a volumetric effect due to the bubbles inducing a displacement of the vortex structures away from the wall, seems to be plausible. However, completely different physical phenomena may be expected in other conditions (e.g. channel flow or deformable bubbles) which need further and deeper investigation.

References

1. M. E. McCormick and R. Bhattacharyya: Nav. Eng. J. **85** (1973).
2. H. Clark III and S. Deutsch, Phys. Fluids A. **3** (1991).
3. J. Xu, M.R. Maxey and G. Karniadakis: J. Fluid Mech. **468** (2002).
4. A. Ferrante and S. Elghobashi: J. Fluid Mech **503** (2004).
5. M. Miozzi, In: *Proceedings of the 12th International Symposium on Applications of Laser Techniques to Fluid Mechanics*, Lisbon, Portugal (2004).
6. K. Fukagata, K. Iwamoto and N. Kasagi:Phys. Fluids **14** (2002).

Drag reduction by non-Brownian rodlike particles in a channel flow

E. De Angelis[1] and E.S.C. Ching[2]

[1] II Facolta' di Ingegneria, Universita' di Bologna, I-47100 Forli', Italy
e.deangelis@unibo.it
[2] Department of Physics, The Chinese University of Hong Kong, Shatin, Hong Kong ching@phy.cuhk.edu.hk

Aim of the present contribution is the study of drag reduction (DR) in rodlike polymers by means of Direct Numerical Simulations. Even if in the past a number of investigators have tried to explain DR via the role of elasticity in flexible polymers, it is interesting to point out that non-Brownian fibres, that have no elasticity, can achieve significant reduction of the friction coefficient. The results we will discuss are obtained by using a model, already discussed in [3], which is an appropriate limit of the rodlike polymer ones. As customary in dilute solutions, the total stress tensor is obtained from the superposition to the Newtonian contribution, with dynamic viscosity μ, of a part due to the polymer molecules, T_{ij}^p. Specifically, in this approach each polymer is represented by a neutrally boyant axisymmetric particle whose configuration is given in terms of the vector n_i which accounts for the orientation. In the original model, at each point the evolution of an ensemble of particles forced by the fluid dynamic field and by a Brownian noise, is considered and the moments of the orientation vector are introduced $\mathcal{R}_{ij} = \langle n_i n_j \rangle$ and $\mathcal{R}_{ijkl} = \langle n_i n_j n_k n_l \rangle$ hence the field equation for the covariance matrix reads, see [1],

$$\frac{\partial \mathcal{R}_{ij}}{\partial t} + u_k \frac{\partial \mathcal{R}_{ij}}{\partial x_k} = K_{ir}\mathcal{R}_{rj} + \mathcal{R}_{ir}K_{jr} - 2E_{kl}\mathcal{R}_{ijkl} - 6\gamma_B \left(\mathcal{R}_{ij} - \frac{\delta_{ij}}{3} \right) \quad (1)$$

where K_{ij} and E_{ij} are the velocity gradient and its simmetric part, respectively, and γ_B is the Brownian rotational diffusion. For the stress tensor instead, again following [1], we have

$$T_{ij}^p = \mu_p \left[E_{kl}\mathcal{R}_{ijkl} + 6\gamma_B \left(\mathcal{R}_{ij} - \frac{\delta_{ij}}{3} \right) \right], \quad (2)$$

where μ_p is the rodlike polymers contribution to viscosity depending on the number density, which is proportional to the zero shear viscosity. In the present work we considered a case where the Brownian diffusion term can be neglected, i.e. $\gamma_B \to 0$. Let's note that in this limit the polymer model depends only on the parameter μ_p and that in the simulations the simple closure hypotesis, $\mathcal{R}_{ijkl} = \mathcal{R}_{ij}\mathcal{R}_{kl}$, was used. Considering the previous closure and by neglecting Brownian effects, the dynamics of the divergence-free velocity is described by a slightly modified form of the Navier-Stokes equation and an equation for the conformation tensor, namely

$$\frac{\partial u_i}{\partial t} + u_k \frac{\partial u_i}{\partial x_k} = -\frac{\partial p}{\partial x_i} + \frac{1}{Re}\frac{\partial^2 u_i}{\partial x_j \partial x_j} + \frac{\eta_p}{Re}\frac{\partial}{\partial x_j}\left(E_{kl}\mathcal{R}_{ij}\mathcal{R}_{kl}\right) \quad (3)$$

$$\frac{\partial \mathcal{R}_{ij}}{\partial t} + u_k \frac{\partial \mathcal{R}_{ij}}{\partial x_k} = K_{ir}\mathcal{R}_{rj} + \mathcal{R}_{ir}K_{jr} - 2E_{kl}\mathcal{R}_{ij}\mathcal{R}_{kl} . \quad (4)$$

Only two nondimensional numbers appear: the Reynolds number $Re = U_0 L_0/\nu$ and the relative viscosity $\eta_p = \mu_p/\mu$. For the present DNS data,

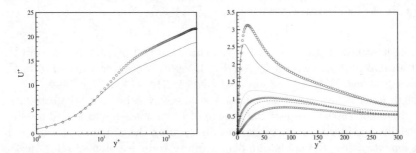

Fig. 1. Left: Mean velocity profile for the Newtonian (solid line) and polymer (symbols) simulations. Right: Root-mean square velocities, u'_{rms} solid line and circles, v'_{rms} dashed line and squares, w'_{rms} dotted line and triangles, Newtonian and rods, lines and symbols respectively.

the dimensions of the domain are $2\pi h \times 2h \times 1.2\pi h$, where h is half the channel height. The numerical method is standard pseudo-spectral with Fourier expansion in the directions parallel to wall and Chebyshev in the wall-normal one. The grid used is $128 \times 193 \times 64$. The simulations have been performed at a nominal Reynolds number of 10000, for both the cases, Newtonian and rodlike polymers. For the polymers simulation the value of η_p was fixed to 25. The flows have been forced with the same pressure drop, so the resulting Reynolds number based in the friction velocity is the same and equal to $Re_\tau = 300$. In this framework drag reduction correspond to an increased flow rate, which, when the velocity profile is plotted in viscous units results in an increased constant in the logarithmic region (see Fig. 1, left). Quantitatively drag reduction can be measured as the relative decrease of the friction factor

$$f = \frac{\tau_w}{\frac{1}{2}\rho \overline{U}^2} \quad (5)$$

which is evaluated as 23.5%. As customary for drag reducing flows, the turbulent intensities change with respect to a corresponding Newtonian simulation,

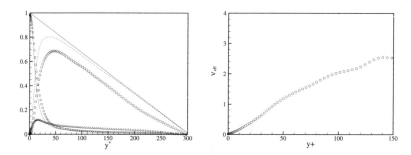

Fig. 2. Left: Momentum balances, Reynolds stress, dashed and circles, viscous contribution, solid and squares, Newtonian and rods, lines and symbols respectively, stress due to the rodlike particles, triangles. Right: Mean value of the effective viscosity for the rodlike particles simulations.

i.e. the fluctuations increase in the streamwise direction and decrease in the wall normal and spanwise ones, see Fig.1, right. Moreover a decrease in the Reynolds stress, W^+, is also observed (see Fig.2).

Though such a flow has been already studied by other investigators [5], a full discussion of the effect of the change of Reynolds number is still missing, especially considering some recent results arguing the presence of a *drag enhancement* at low Reynolds number in dilute rodlike polymer solutions [6]. Preliminary results in this direction, here not reported, do not show such a behaviour. As a final remark, it is worth pointing out that the present simulations confirm the idea that the drag reduction can be obtained introducing only viscous effects [4]. In this framework it is actually quite instructive analising the effective viscosity profile defined as $\nu_{eff}(y) = T_{12}^p/dU/dy$ and shown in Fig.2 right, which suggests that drag reduction can be obtained allowing for a linearly variable viscosity [7].

References

1. M. Doi, S.F. Edwards, Clarendon Press-Oxford, (1986).
2. E. De Angelis, C.M. Casciola, R. Piva, Computers & Fluids, **31**, (2005).
3. R. Benzi, E.S.C. Ching, T.S. Lo, V.S. L'vov, I. Procaccia, Phys. Rev. E, **72** (3) 016305, (2005).
4. R. Benzi, E. De Angelis, V.S. L'vov, I. Procaccia, V. Tiberkevich, J. Fluid Mech. **551**, 185-195, (2006).
5. J.S. Paschkewitz, Y. Dubief, C.D. Dimitropoulos, E.S.G. Shaqfeh, P. Moi,. J. Fluid Mech. **518**, 281-317, (2004).
6. Y. Amarouchene, D. Bonn, H. Kellay, T.S. Lo, V.S. Lvov, I. Procaccia, J. Fluid Mech. submitted.
7. E. De Angelis, C.M. Casciola, V.S. L'vov, A. Pomyalov, I. Procaccia, V. Tiberkevich, Phys. Rev. E, 70, 055301(R), (2004).

Use of dual plane PIV to assess scale-by-scale energy budgets in wall turbulence

N Saikrishnan[1], EK Longmire[1], I Marusic[1], N Marati[2], CM Casciola[2], R Piva[2]

[1] Dept. of Aerospace Engg. & Mechanics, Univ. of Minnesota neela@aem.umn.edu
[2] Dipartimento di Meccanica e Aeronautica, Università di Roma *La Sapienza*

Introduction

Turbulence in wall-bounded flows can be characterized by variations in physical space or in scale space. In physical space, the near wall region can be classified into a viscous sub-layer, the buffer region, the logarithmic region and the outer region. In the space of scales, turbulent energy is produced at the large scales and transferred to smaller scales, finally dissipating in the form of heat in a mechanism commonly known as the Richardson cascade. These are seemingly parallel approaches to describing turbulence, and in order to describe the scale-dependent dynamics in the presence of spatial inhomogeneities, a more general approach combining these two ideas is required.

For homogeneous shear flow, the generalized Kàrmàn-Howarth equation can be used in anisotropic conditions. By extending this equation to inhomogeneous flows, the relation between spatial fluxes and the energy cascade can be studied in detail. This modified equation is averaged in the space of scales on two-dimensional square domains in wall-parallel planes to yield

$$T_r(r, Y_c) + \Pi_e(r, Y_c) = E_e(r, Y_c) \tag{1}$$

where r represents the scale and Y_c represents the wall-normal coordinate. In other words, this expression implies that the transfer across scales plus the effective production must equal the effective dissipation at every scale and wall-normal location. Each of the terms shown in the above equation is a combination of individual terms in the derived equation. The derivation of this equation is described in detail in [1]. This relation will hold through the various regions of the boundary layer; however the relative importance of each term varies with the location. While the production and transfer terms are most important in the buffer layer, in the outer region, the dissipation and transfer are the most important terms. The log layer is nominally an equilibrium layer, where local production and local dissipation match each other and thus ensures a constant flux of energy across scales.

Results from DNS and PIV Data

Previously, data from a Direct Numerical Simulation (DNS) of a channel flow at a friction Reynolds number $Re_\tau = 180$ were analyzed at representative locations within the boundary layer to evaluate the terms of Eqn. 1 [1]. The present work extends the earlier study to larger Reynolds number flows. Data were examined from a DNS of a channel flow at $Re_\tau = 934$, which had a computational domain of size $8\pi \times 3\pi$ in the streamwise and spanwise directions, with a resolution of 11.46×5.73 viscous wall units respectively [2]. A second dataset was obtained experimentally in a zero pressure gradient boundary layer flow at $Re_\tau = 1160$ using dual plane Particle Image Velocimetry (PIV) [3]. The dual plane setup allows for the determination of the full velocity gradient tensor in a plane. The data were acquired in streamwise-spanwise planes in the logarithmic region ($Y_c^+ = 110$) with fields of size $1.1\delta \times 1.1\delta$ and velocities were resolved to 25 viscous wall units.

From Fig. 1(a), it is observed that at small scales, there is very little production and the transfer is the dominant term, thus reducing the scale dynamics to the classical Richardson cascade. As the scale increases, the production term increases, which is consistent with the idea of production of energy at the larger scales. Simultaneously, the transfer term reaches a maximum at the transfer scale $l_t^+ = 50$, and then starts to reduce. At the crossover scale $l_c^+ = 80$, the transfer and production terms are equal, and at scales larger than this, the production becomes more dominant as compared to the transfer. Although similar trends of production and transfer are observed with the PIV data (Fig. 1(b)), the value of $l_c^+ = 60$ is lower compared to the DNS value. Further, the rate of reduction of transfer with increasing scale is much less, which would have implications in the energy balance. This is clearly seen in the inset, where a mismatch between the terms of Eqn. 1 is seen for the PIV data, while a very good balance is seen for the DNS data.

This mismatch is likely due to errors in calculating the dissipation term, which is a result of the limited spatial resolution of the experiments. The resolution is constrained by the size of the interrogation window in the PIV processing. A study by [4] showed the effect of averaging and reduced spatial resolution in the computation of velocity and velocity gradients using PIV in a turbulent boundary layer. With this knowledge, additional PIV experiments were conducted at $Re_\tau = 1100$ and $Y_c^+ = 100$ with a smaller field of view and finer spatial resolution. These high resolution data were acquired in streamwise-spanwise planes in fields of size $0.45\delta \times 0.45\delta$, and velocities were resolved to 10 viscous wall units.

As can be seen from Fig. 1(c), the better resolution provides a more accurate value of the dissipation term at this wall-normal location, and thus the balance between the terms of Eqn. 1 is much better. It must be noted that the production term does not change substantially compared with Fig. 1(b) since it does not involve local gradients of fluctuating quantities. On the other hand, the transfer term is computed more accurately and hence the value of $l_c^+ = 80$ matches well with the value predicted by the DNS.

Conclusions and Future work

The balance between the various terms of the turbulent kinetic energy budget was verified for DNS and PIV data and it was shown that high resolution PIV experiments are required to accurately resolve the terms. The results from the DNS and high resolution PIV data matched each other very well, both in terms of trends observed and values of scales computed. [1] argued on the basis of the classical equilibrium theory that $l_c \simeq \kappa Y_c$, where κ is the Kármán constant. However, the value of l_c computed from the DNS and PIV data at a higher Reynolds number are larger than these predicted values. This suggests a dependence of this value on the Reynolds number of the flow. Further experiments at different wall normal locations and higher Reynolds numbers will enable a better understanding of these trends.

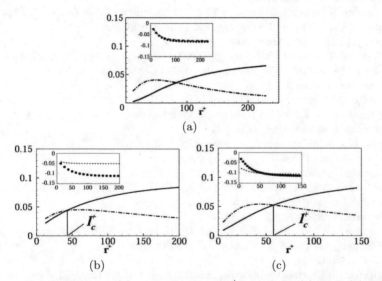

Fig. 1. Detailed balance (1) in the log-layer $Y_c^+ = 110$ for (a) DNS (b) PIV (c) High resolution PIV. The solid line is $-\Pi_e$ and the dash-dotted line is $-T_r$. (Inset) The sum $(T_r + \Pi_e)$ is represented by filled symbols and E_e is given by the dashed line. All terms are normalized by $(u^*)^4/\nu$.

References

[1] Marati N, Casciola CM, Piva R 2004 *J. Fluid Mech.* **521**
[2] Del Álamo JC, Jiménez J, Zandonade P, Moser RD 2004 *J. Fluid Mech.* **500**
[3] Ganapathisubramani B, Longmire EK, Marusic I 2006 *Phys. Fluids* **18**
[4] Saikrishnan N, Marusic I, Longmire EK 2006 *Exp. Fluids* **41**

Highly Time- and Space-Resolved Experiments on a High Reynolds Number Turbulent Boundary Layer

Murat Tutkun[1], Peter B. V. Johansson[2], William K. George[3],
Jim Kostas[4], Sebastien Coudert[5], Jean-Marc Foucaut[6], Michel Stanislas[7],
Carine Fourment[8], and Joël Delville[9]

[1] Turbulence Research Laboratory (TRL), Department of Applied Mechanics,
   Chalmers University of Technology, SE-41296 Gothenburg, Sweden
   murat.tutkun@chalmers.se
[2] Turbulence Research Laboratory (TRL), Department of Applied Mechanics,
   Chalmers University of Technology, SE-41296 Gothenburg, Sweden
   peter.johansson@chalmers.se
[3] Turbulence Research Laboratory (TRL), Department of Applied Mechanics,
   Chalmers University of Technology, SE-41296 Gothenburg, Sweden
   wkgeorge@chalmers.se
[4] Laboratoire de Mécanique de Lille (LML), UMR CNRS 8107, Bv Paul Langevin,
   Cité Scientifique F-59655 Villeneuve d'Ascq France
   dimitrios.kostas@univ-lille1.fr
[5] Laboratoire de Mécanique de Lille (LML), UMR CNRS 8107, Bv Paul Langevin,
   Cité Scientifique F-59655 Villeneuve d'Ascq France
   sebastien.coudert@univ-lille1.fr
[6] Laboratoire de Mécanique de Lille (LML), UMR CNRS 8107, Bv Paul Langevin,
   Cité Scientifique F-59655 Villeneuve d'Ascq France
   jean-marc.foucaut@ec-lille.fr
[7] Laboratoire de Mécanique de Lille (LML), UMR CNRS 8107, Bv Paul Langevin,
   Cité Scientifique F-59655 Villeneuve d'Ascq France
   stanislas@univ-lille1.fr
[8] Laboratoire d'Etudes Aérodynamiques (LEA), UMR CNRS 6609, 43 route de
   l'Aérodrome, F-86036 Poitiers Cedex France
   carine.fourment@lea.univ-poitiers.fr
[9] Laboratoire d'Etudes Aérodynamiques (LEA), UMR CNRS 6609, 43 route de
   l'Aérodrome, F-86036 Poitiers Cedex France
   joel.delville@lea.univ-poitiers.fr

1 Introduction and experimental setup

High Reynolds number flat plate turbulent boundary layer experiments have
been performed in the Laboratoire de Mécanique de Lille (LML) wind tunnel,
the test section of which is 20 m long and has a 1×2 m^2 cross section. The

(a) LML wind tunnel.   (b) LEA hot-wire rake.

Fig. 1. The wind tunnel and the hot-wire rake.

whole test section can be seen in figure 1(a). Two different Reynolds numbers based on momentum thickness, Re_θ, were tested, 10 000 and 20 000, corresponding to freestream velocities of 5 m/s and 10 m/s respectively. The boundary layer thickness at the measurement location was about 30 cm.

A hot-wire rake, figure 1(b), made of 143 single wire probes distributed in a plane normal to the flow (wall normal-spanwise (YZ) plane) was used for all measurements in order to provide both spatial and temporal information simultaneously. This rake was built by LEA. The probes were distributed on an array of 11×13 so that 11 probes were placed logarithmically in the wall normal direction on each vertical comb and 13 combs were staggered in the spanwise direction. The first row of probes was placed 0.3 mm away from the wall, corresponding to y^+ =7.5 at Re_θ= 20 000. The last probe on each vertical comb was located at y=306.9 mm, or near δ_{99}. The width of the rake was 280 mm in the spanwise direction, and the vertical combs were distributed symmetrically around z=0. The 144-channel TRL hot-wire anemometry system with data acquisition was used during the experiments. The sampling frequency of the hot-wire measurements for both cases was the same at 30 kHz. 2200 blocks of data, each 6 seconds long, were collected for each case.

Spatial information on the flow was obtained by the two stereo PIV systems (SPIV) recording a 30×30 cm^2 (YZ) plane located 1 cm upstream of the hot-wire rake plane as it is seen in figure 2. These two systems used a BMI 2×150 mJ dual cavity Yag Laser and 4 Lavision Image Intense PIV cameras with a CCD of 1376×1040 pixels and a sampling rate of 4 velocity fields/s. A

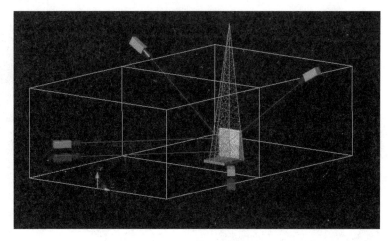

Fig. 2. Schematic of the synchronized systems.

third stereo PIV system was used to record a streamwise-wall normal (XY) plane on the plane of symmetry of the hot-wire rake at z=0. It used a BMI 2×150 mJ dual cavity Yag Laser and 2 Lavision Flowmaster PIV cameras with a CCD of 1280×1024 and a sampling rate of 4 velocity fields/s. These three SPIV systems were synchronized together and with the hot-wire rake (HWR) data acquisition system. The first 600 blocks of 2200 blocks of the data were acquired by synchronized HWR and SPIV systems.

In a second step, one high repetition rate SPIV system was used in the streamwise-spanwise (XZ) plane to obtain spatial and temporal information in the near-wall region. This system was based on a Quantronix dual cavity 2×20 mJ YLF laser and two Phantom V9 cameras. PIV planes were parallel to the wall at y^+ =100 and y^+ =50 for Re_θ=20 000 and Re_θ=10 000 respectively. The sampling rates were 3000 velocity fields/s for the high Reynolds number case, and 1500 velocity fields/s for the low Reynolds number case. A total of 1000 blocks of HWR data were synchronously recorded with the high repetition SPIV.

The data of the experiment is still under analysis. Final results will include extensive space-time correlations, as well as instantaneous snapshots and profiles.

Acknowledgments

This work has been performed under the WALLTURB project. WALLTURB (A European synergy for the assessment of wall turbulence) is funded by the CEC under the 6th framework program (CONTRACT No: AST4-CT-2005-516008).

Fully mapped energy spectra in a high Reynolds number turbulent boundary layer

Nicholas Hutchins, Ivan Marusic, and M. S. Chong

Walter Bassett Aerodynamics Laboratory, Mechanical and Manufacturing Engineering, University of Melbourne, Victoria 3010 Australia. nhu@unimelb.edu.au

Analysis of the surface formed from the pre-multiplied energy spectra of streamwise velocity fluctuations $k_x\phi_{uu}/U_\tau^2$ (as calculated at numerous wall-normal stations across the turbulent boundary layer) has provided much recent insight regarding the structure and energy content of the logarithmic region [1]. The manner in which this surface is formed is illustrated in figure 1. The axis system x, y and z refer to the streamwise, spanwise and wall-

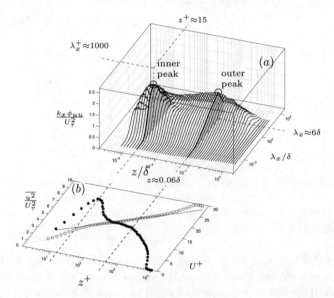

Fig. 1. (a) Premultiplied energy spectra of streamwise velocity fluctuation $k_x\phi_{uu}/U_\tau^2$ for a turbulent boundary layer at $Re_\tau \approx 7300$ as a three-dimensional surface plot for all wall-normal locations; (b) corresponding (○) mean velocity and (●) broadband turbulence intensity profiles. Dashed lines denote inner and outer spectral peaks.

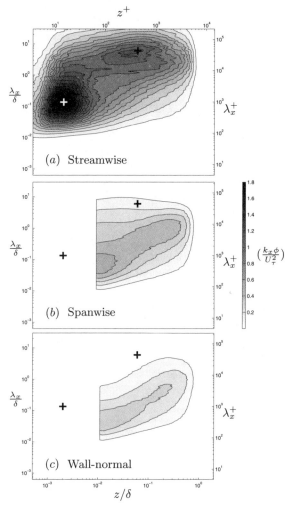

Fig. 2. Contours of pre-multiplied energy spectra for (a) streamwise $k_x\phi_{uu}/U_\tau^2$ (b) spanwise $k_x\phi_{vv}/U_\tau^2$ (c) wall-normal $k_x\phi_{ww}/U_\tau^2$ fluctuations. Contours are from 0.1 to 1.8 in steps of 0.1 (see gray scale). (+) symbols denote approximate locations of inner ($z^+ \approx 15, \lambda_x^+ \approx 1000$) and outer ($z/\delta \approx 0.06, \lambda_x/\delta \approx 6$) peaks in u spectra.

normal directions, with u, v and w denoting respective fluctuating velocity components. At adequate Reynolds number the surface in figure 1(a) assumes a bimodal appearance, comprised of an inner peak fixed in viscous wall-units and due to the near-wall cycle, along with an outer peak, scaling on boundary layer thickness and due to a much larger class of structure inhabiting the log region, termed the 'superstructure' [1]. As Reynolds number increases, the scale separation between these two peaks increases, in addition to which the outer peak becomes increasingly comparable in energy to the inner. Here we build on this picture, presenting similar energy surfaces for the wall-normal

(w) and spanwise (v) components. The novelty of these results lies not just in the manner of presentation and interpretation, but also in the excellent spatial and temporal resolution afforded by the high Reynolds number boundary layer facility at Melbourne. Details of this facility are given in [4]. Data are obtained using constant-temperature hot-wire anemometry (single-normal wires for u and $\times$-wires for v and w; nominal wirelength $l^+ = lU_\tau/\nu = 22$). The size of the $\times$-wire precludes the possibility of measuring v and w for $z^+ \lesssim 70$.

Figure 2 shows contour maps of pre-multiplied energy spectra for all three velocity components at $Re_\tau = 7300$. These maps are iso-contours of surfaces constructed from individual spectra (such as those shown in figure 1). A full description of the bimodal appearance in the u spectra of plot (a), along with structural origins, is given in [1, 2]. For now, we just highlight the locations of the inner and outer peaks in all plots (using + symbols). Immediately noteworthy in the v and w spectra of plots (b) and (c) is the inclined ridge of energy where length-scale (λ) is proportional to distance from the wall (z). This is indicative of attached eddies [6, 5] (this inclined behaviour is also evident in the u spectra of a). However, an important distinction between the v and w components occurs in the near-wall region where the wall-normal fluctuations lack a large-scale energetic contribution, in contrast to the spanwise fluctuations which exhibit near-wall energy at large 'superstructure' type length-scales ($\lambda_x \sim 6\delta$). Such behaviour is consistent with the notion of attached eddies (where the wall-normal fluctuations will lack a large-scale component at the wall due to blocking or image vortices[5]). It is also consistent with the notion that the superstructure is associated with very large counter-rotating roll-modes [2]. These results are suspected to hold for very high Reynolds numbers. Measurements made in the log region of the atmospheric surface layer have also reported ϕ_{ww} as scaling only with inner variables (z, U_τ) [3].

References

[1] N. Hutchins and I. Marusic. Evidence of very long meandering features in the logarithmic region of turbulent boundary layers. *J. Fluid Mech.*, 2007. in press.

[2] N. Hutchins and I. Marusic. Large-scale influences in near-wall turbulence. *Proc. R. Soc. Lond.*, 365:647–664, 2007.

[3] G. J. Kunkel and I. Marusic. Study of the near-wall-turbulent region of the high-Reynolds-number boundary layer using an atmospheric flow. *J. Fluid Mech.*, 548:375–402, 2006.

[4] T. B. Nickels, I. Marusic, S. Hafez, and M. S. Chong. Evidence of the k_1^{-1} law in a high-Reynolds-number turbulent boundary layer. *Phys. Rev. Letters*, 95:074501, 2005.

[5] A. E. Perry and M. S. Chong. On the mechanism of wall turbulence. *J. Fluid Mech.*, 119:173–217, 1982.

[6] A. A. Townsend. *The Structure of Turbulent Shear Flow*. Cambridge University Press, 1956.

Torque scaling in Taylor-Couette flow

Bruno Eckhardt[1], Siegfried Grossmann[1] and Detlef Lohse[2]

[1] Fachbereich Physik, Philipps-Universität Marburg, D-35032 Marburg, Germany
`bruno.eckhardt@physik.uni-marburg.de`
[2] Department of Applied Physics, University of Twente, 7500 AE Enschede, The
Netherlands

Building on a quantitative analogy between Rayleigh-Bénard flow and Taylor-
Couette flow we describe the angular momentum current and the difference
between turbulent and laminar dissipation in terms of the transverse motions.
The torque derived within this model is in quantitative agreement with ob-
servations by Lewis and Swinney. The model also predicts relations between
Reynolds number and torque at different radius ratios.

1 Introduction

The turbulence that develops in fluids put under stress by external heating or
shearing modifies the mean profiles and influences the transport of heat and
momentum. In the case of Rayleigh-Benard (RB) flows where experiments
have now covered Rayleigh numbers up to 10^{17} and a wide range of Prandtl
numbers, two of the authors have put forward a phenomenological theory for
the global heat transport that is in extremely good agreement with observations [1]. Exploiting analogies between Rayleigh-Benard and Taylor-Couette
(TC) flows (as well as other shear flows, [2]) we have have derived predictions
for TC flow and compared them with the experimental data of Swinney et al.
as well as of Wendt [3]. A more detailed account is given in [4].

2 Analogies and Results

At the heart of the Rayleigh-Bénard theory are two exact relations for the
total energy dissipation and the heat flux. The analogy to shear flows is based
on the identification of the corresponding quantities.

The total heat flux in Rayleigh-Bénard flow in Oberbeck-Boussinesq approximation can be obtained from the equation for the temperature field by
averaging over planes A parallel to the bottom and top plates (and time t).
With z the coordinate normal to the plates, we find

$$J_\theta = \langle u_z\theta \rangle_{A,t} - \kappa \partial_z \langle \theta \rangle_{A,t} \ . \tag{1}$$

The dimensionless Nusselt number is $N_\theta = J_\theta/\kappa\Delta d^{-1}$. The various physical parameters are the height d, the temperature difference Δ between the bottom and top plates, and the thermal conductivity κ. The temperature θ denotes the deviation from a reference temperature and u_z is the velocity field component in the direction of gravity, normal to the plates. By derivation from the equations of motion, this current is independent of the height z at which it is calculated.

The kinetic energy in the wind is dissipated by fluid viscosity ν. Multiplying the Navier-Stokes equations with the velocity $\mathbf{u}$ and integrating over the volume, the energy balance between dissipation and input reads

$$\epsilon_w = \nu^3 d^{-4} Pr^{-2} Ra(Nu_\theta - 1) \ . \tag{2}$$

Here, $Pr = \nu/\kappa$ is the Prandtl number and $Ra = \alpha_p g d^3 \Delta/(\kappa\nu)$ the Rayleigh number. α_p denotes the isobaric thermal expansion coefficient.

Similiar expressions exist for TC flow [2, 4]. The physical quantity which corresponds to the temperature field in RB flow is the azimuthal velocity u_φ, since u_φ is subject to an externally applied profile. Thus take the azimuthal (φ) component of the Navier-Stokes equation and average it over time and cylinder surfaces between the two rotating cylinders, i. e., perpendicular to the profile direction. Then

$$J_\omega = r^3 \left(\langle u_r\omega \rangle_{A,t} - \nu \partial_r \langle \omega \rangle_{A,t} \right) = \frac{d}{\kappa(\omega_1 - \omega_2) r_a^3} N_\omega \ . \tag{3}$$

is independent of the radius and hence the appropriate conserved current [4]. Here, $\omega = u_\varphi/r$ is the angular velocity, and ν the kinematic viscosity of the liquid in the gap. The other quantities are the arithmetic mean of the radius ratios $r_a = (r_1 + r_2)/2$, the gap width d, and $\kappa = \nu r_1^2 r_2^2/r_a^4$ is the equivalent of the thermal diffusivity. The second equality in eq. (3) serves to define the dimensionless analog N_ω to the Nusselt number also for TC flow.

For the energy dissipation one has to take into account that in TC flow there is already some dissipation in the laminar case. Since we are interested in the changes that occur when turbulence sets in, we have to consider not the full dissipation but the increase above the laminar level. Therefore, we take the difference in dissipation between the case with wind and the case without,

$$\epsilon_w = \epsilon - \epsilon_{lam} = r_a^{-1} d^{-1} (\omega_1 - \omega_2) J_{\omega,lam}(Nu_\omega - 1). \tag{4}$$

The fluid is driven by rotating the cylinder walls with angular velocities ω_1 and ω_2, respectively, and it responds with an angular velocity transport $Nu_\omega = J_\omega/J_{\omega,lam}$ and by setting up a wind ("Taylor vortices") with amplitude U_w that can be characterized by a Reynolds number $Re_w = U_w d/\nu$. In analogy to RB flow, the aim then is to predict Nu_ω and Re_w and their dependencies on the angular velocities ω_1, and ω_2, the radius ratio η, the aspect ratio, or other parameters.

From here on the analysis and modelling proceeds very much in parallel to the case of RB-flows. It gives predictions for the tranverse velocities, the torque and the gap width dependence. In order to demonstrate the quality of the fit, the figure shows a comparison between the experimentally obtained torque and the model values.

The model also leads to a prediction for the variation of the torque with radius ratio η (see [4] for details). Specifically, with $\hat{R}_1$ and $\hat{N}_\omega$ the reference values for Reynolds number and current for $\eta = 1$, the values for other η are obtained from

$$(\hat{R}_1, \hat{N}^\omega) \rightarrow \left(\frac{(1+\eta)/2}{\sqrt{\sigma}} \hat{R}_1, \sigma \hat{N}^\omega \right) . \tag{5}$$

This mapping implies that a pair of $(\hat{R}_1, \hat{N}^\omega)$ values moves towards smaller R_1 and larger N^ω as η decreases from 1 downwards.

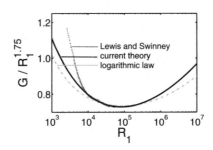

Fig. 1. Torque, rescaled by $R_1^{1.75}$, vs. Reynolds number R_1 for rotating inner cylinder and outer cylinder at rest. The full curve is the present modelling, the circles (partly on top of full curve) the experimental data of Lewis and Swinney, and the dashed line the results from a fit to a logarithmic profile. Differences for small Re are due to the dominance of vortices.

References

1. Grossmann S and Lohse D (2000) J Fluid Mech 407:27-56;
   – (2004) Phys. Fluids 16:4462-4472.
2. Eckhardt B, Grossmann S, and Lohse D (2006),
   Nonlinear Phenom Complex Sys 9:109–114.
3. Lewis G S and Swinney H S (1999) Phys Rev E 59:5457-5467.
   Wendt F (1933) Ingenieurs-Archiv 4:577-595.
4. Eckhardt B, Grossmann S, and Lohse D (2007) Torque scaling in turbulent Taylor-Couette flow between independently rotating cylinders. J Fluid Mech, in press.

Turbulent Shear Flows on a Sparse Grid

F. De Lillo and B. Eckhardt

Fachbereich Physik, Philipps-Universität Marburg, D-35032 Marburg (Germany)
filippo.delillo@physik.uni-marburg.de

The picture of a turbulent cascade suggests a reduction of the full Navier-Stokes dynamics to interactions on a limited number of shells or modes in wave number space. Instead of modelling small scales by means of an eddy diffusivity, dynamics at all scales can be represented on a sparse spectral grid, which gets coarser towards large wave numbers, while preserving the structure of nonlinear interactions. In this note we extend previous efforts in homogeneous turbulence to the case of wall bounded flows.

1 Introduction

Direct numerical simulations of turbulent flows are limited by the number of modes that can be included, with the current record being 4096^3[1]. To access higher Reynolds numbers, modelling assumptions are necessary. While methods such as LES can be useful to model high Re flows, the lack of resolution near the wall requires specific wall models [2]. We discuss here an extension of earlier efforts where the velocity field is represented on a subset Ω of the modes only [3, 4]. We show here that with such a model one can reproduce many features of plane Couette flow at a Reynolds number of 2000 with about 15% of the modes needed for a full DNS [5]. Within this scheme the total number of modes required to describe a turbulent flow does not increase like $Re^{9/4}$ but only as $Re^{3/4} \ln(Re)$. We therefore expect in the near future to increase the number of modes and study flows at Reynolds numbers an order of magnitude higher.

2 Wave Number Reduction

The geometry we consider is that of plane Couette flow. The velocity field can be decomposed in Fourier series in downstream and spanwise directions, and Chebyshev polynomials in the wall normal direction. In wall normal direction all modes are kept so as to not loose resolution there. In planes parallel to the surface only a reduced number of Fourier modes is retained: starting from a set Ω centered around $\mathbf{k} = \mathbf{0}$, a wave vector $\mathbf{k}$ is accepted only if there is a $\mathbf{K} \in \Omega$ such that $\mathbf{k} = 2^n \mathbf{K}$. In the computation of convolutions, modes that do not fulfill this condition are rejected. With this definition of the convolution the structure of the nonlinear terms in the Navier-Stokes equation and the conservation laws for the momentum current are preserved. This is in contrast to LES, where the modelling of the Reynolds stresses changes the structure of the interaction between modes.

In order to test this method, we use a code for fully resolved direct numerical simulations of plane Couette flow and implement the mode reduction after each time step. This allows us to continuously scan from the full resolution to a reduced one. The reduction is parametrized by the size n_c of Ω, which contains all modes with $|k_i| < 2\pi n_c / L_i (i = x, y)$.

3 Results

The results for different n_c were compared with fully resolved DNS. We find that the wall stress in the reduced simulations is slightly higher. However, for $n_c \geq 16$, the velocity profiles are compatible with a logarithmic variation (in the usual wall units) $u^+(y^+) = 1/\kappa \, \log(y^+) + C_l$. Interestingly, the value of the von Karman constant $\kappa = 0.4$ does not seem to be affected by the resolution. However, the additive constant C_l varies with the level of mode reduction (see Fig.1) in a way that can be traced back to the change in wall stress. The latter, at the lowest resolution shown, is about 12% larger than for the reference DNS.

We analysed the statistics of velocity fluctuations by computing the structure functions at different distances from the wall. In particular, we calculated the flatness $F(u; \delta x)$ of the longitudinal fluctuations of the streamwise component from the relative second and fourth order structure functions. For small δx this flatness decreases as a function of n_c, approaching the Gaussian value, in agreement with previous results for homogeneous flows [4]. However, for large δx, the flatness is close to the values found in the full simulation for all runs with a sparse grid. Thus, the mode reduction seems to supress small scale intermittency (as noted before) but preserves the effects of large scale dynamical features of the flow.

4 Conclusions

We propose an application to shear flows of the sparse-spectral-grid approach. While showing the expected deviations for small scale statistics, our results show a good agreement with a fully resolved DNS run, at large scales and away from the wall, where the usual law of the wall is reproduced, up to a

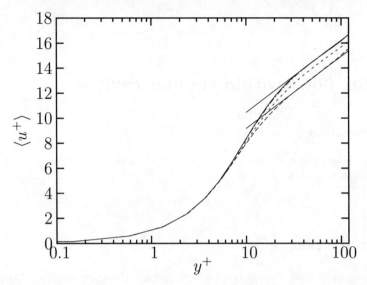

Fig. 1. Streamwise velocity profiles, in wall units, for a full DNS at a resolution of $256 \times 128 \times 65$ grid points (full line), and three runs with mode reduction applied in the horizontal directions. The size of the core shell Ω (see text) is $n_c = 16(- - -)$, $24(\text{- - -})$ and $32(- \cdot -,$ hidden by the DNS line). The friction Reynolds number of the full simulation is $\text{Re}_* = 120$. For each run the quantities are normalised with the respective wall stresses. The straight lines are the law of the wall, with $\kappa = 0.4$ and $C_l = 3.5$ and 4.7 respectively

shift due to an increased wall-shear stress. The ability to reproduce many of the results expected for shear flows with a sparse and numerically efficient grid without adjustable parameters (similar to the grid reduction scheme in [6]) holds great potential for extension to higher Reynolds numbers.

We thank the Deutsche Forschungsgemeinschaft for support.

References

1. Y. Kaneda, T. Ishihara, M. Yokokawa, K. Itakura and A. Uno, Phys. Fluids **15**, L21 (2003).
2. S.B. Pope, New Journal of Physics **6**, 35 (2004).
3. M. Meneguzzi, H. Politano, A. Pouquet and M. Zolver, J. Comp. Phys. **132**, 32 (1996).
4. S. Grossmann, D. Lohse and A. Reeh, Phys. Rev. Lett. **77**, 5369 (1996).
5. F. De Lillo and B. Eckhardt, submitted to Phys. Rev. E (2007).
6. J.C. Bowman, B.A. Shadwick and P.J. Morrison, Phys. Rev. Lett. **83**, 5491 (1999).

Vortex flows within circular cavities

Ralph Savelsberg & Ian P. Castro[1]

School of Engineering Sciences, University of Southampton, Southampton, UK
i.castro@soton.ac.uk

1 Introduction

Subsonic cavity flows have been widely studied. However, nearly all the published work has been in the context of rectangular cavities (e.g. [1]-[3]), although more recently circular cavity flows have been studied both experimentally and computationally, in the context of looking for ways to reduce the drag of aircraft wings (e.g. [4]-[5]). We report here the results and implications of initial experiments on the nominally two-dimensional flow within circular cavities of large spanwise aspect ratio [1] ($\alpha = 8.6$). The cavity ('cell') flow was driven by the separated shear layer above an opening (angle ϕ) in the wall of the circular cell; Fig.1 shows a sketch of the arrangement and a photograph of the (newly designed and built) rig itself. Particular interests in the present work include the characteristics of the vortex flow within the cell and the way in which these are influenced by the imposed axial pressure gradient, the

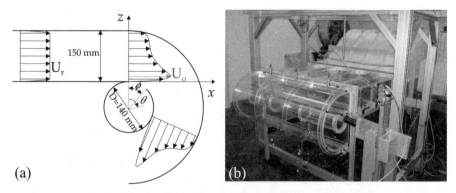

(a)   (b)

Fig. 1. (a): sketch of the geometry; (b) the rig itself.

---

[1] the ratio of the cavity's span to its inner diameter

nature of the boundary layer at the cavity lip ($x = 0$ in Fig.1a) and the cavity opening angle, ϕ (see Fig.1a). For the measurements presented here $\phi = 40°$.

2 Characteristics of the vortex

As sketched in Fig.1a, the velocity profile far upstream of the cavity, outside the boundary layers, is uniform, with a velocity U_r. The flow in the outer cylinder is largely irrotational. Hence, there the angular velocity is eventually proportional to $1/r$. Consequently, U_o, the velocity above the shear layer, is higher than U_r and the static pressure p_{cusp} is lower. Fig.2a shows the non-

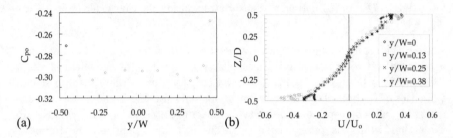

Fig. 2. Spanwise pressure distribution (a) and velocity profiles (b) at four different spanwise positions along $\theta = 0$ - the vertical diameter through the separation point

dimensionalised static pressure ($C_{po} = (p_s - p_{cusp})/\frac{1}{2}\rho U_o^2$) measured with pressure tappings mounted near the bottom of the cell ($\theta = 185°$). The pressure is practically constant for $-0.2 < y/W < 0.2$. Fig.2b shows mean velocity profiles inside the cavity measured (using LDA) along the vertical diameter ($\theta = 0, -0.5 \leq z/D \leq 0.5$) for four different spanwise positions. Where the pressure is constant, outside of the bounding shear layers, the profiles are very

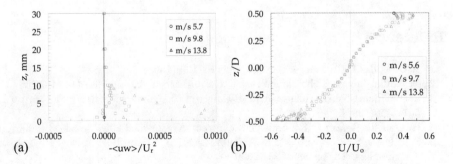

Fig. 3. (a): Shear stress in the upstream boundary layer; (b) velocity profiles in the cavity, for three main stream velocities

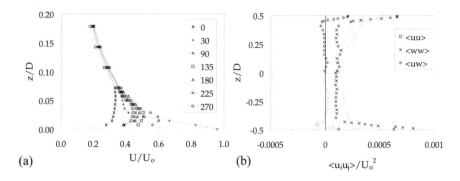

Fig. 4. (a) Boundary layer profiles at different angular positions θ, with the wall located at $z/D = 0$; (b) normal and shear stress inside the cavity, for $U_r = 9.7$ m/s

similar. This might suggest that, at least within this region, the mean flow is essentially two-dimensional. Figure 3a shows profiles of the shear stress in the upstream boundary layer (at $x = -54$ cm) for three different main stream velocities. The boundary layer is turbulent only for the highest velocity. Remarkably, the corresponding velocity profiles measured in the cavity, in Fig.3b, show that while the peak velocities in the cell do depend on the Reynolds number of the upstream flow, the non-dimensionalised rotation velocity in the core is practically the same in all three cases, even though there must be an overall balance between the maximum stress in the shear layer and the total wall-boundary stress. Fig.4a shows how the flow inside the cavity adjacent to the mixing layer and cylinder wall, respectively, accelerates and then gradually slows down as it approaches the top of the cell. Fig.3b shows that a solid-body-rotation-type of flow is *only* apparent over the central 40% or so of the cell diameter. However, Fig.4b shows that the shear stress is closely zero over a much larger part of the cell, despite fairly high turbulence intensities. For a two-dimensional flow this seems contradictory and suggests that three-dimensional effects are *not* wholly insignificant.

References

1. R.D. Mills: J. Roy. Aero. Soc. **69**, pp 116–120 (1965)
2. M. Nallasamy & K. Krishna Prasad: J. Fluid Mech. **79**, pp 391–414 (1977)
3. T.P. Chiang, W.H. Sheu & R.R. Hwang: Int. J. Numer. Meth. Fluids **26**, pp 557–579 (1998)
4. R. Donelli: Report D 3.2 under EU contract 'Vortex Cell 2050', pp 1-33 (2006)
5. S.A. Isaev, S.V. Guvernyuk, M.A. Zubin & Yu. S. Prigorodov: J. Eng. Phys. & Thermophys. **73**, pp 337–344 (2000)

New Criteria for the Eduction of Three-dimensional Turbulent Structures

Lionel Larchevêque[1] and Michèle Larchevêque[2]

[1] IUSTI, Université Aix-Marseille I, UMR CNRS 6595, 5 rue Fermi, F-13453 Marseille cedex 13, France (lionel.larcheveque@polytech.univ-mrs.fr)
[2] Institut d'Alembert, Université Paris VI, UMR CNRS 7190, boîte 162, 4 place Jussieu, F-75252 Paris cedex 5, France (larchevq@lmm.jussieu.fr)

Summary. A new family of criteria designed to geometrically educe coherent structures is proposed and validated using analytic and DNS three-dimensional flowfields.

1 Motivation and mathematical description

In the recent years a large number of methods have been proposed to educe vortical structures with application to the visualization and the study of turbulent coherent structures (see [1] for an analysis of the classical Δ, Q and λ_2 criteria). It is worth noting that most of these methods, although resulting in distinct definitions of a vortex core for three-dimensional flowfields, boil down for incompressible two-dimensional flows to the single Weiss criterion [2]. On the contrary, the two-dimensional criterion defined by Herbert *et al.* [3], based on the curvature of isovorticity lines, yields more precise dynamical results and educes much thinner structures. The purpose of the present work is to extend the former two-dimensional definition of [3] to a three-dimensional space.

Information of the local maxima and curvature of isovorticity surfaces can be fully recovered by analyzing the Hessian matrix of the vorticity norm $\mathbf{H}(\|\omega\|)$. For instance, regions with three negative eigenvalues λ_i are related to absolute maxima of vorticity. In the vicinity of such points, the isovorticity surfaces can be locally described by ellipsoids. Consequently, the negative eigenvalue condition is associated with vortical structures of pancake-shaped core. This condition can also be recast into a function of the three invariants I_i of the Hessian matrix and then reads ($I_1 < 0$, $I_2 > 0$, $I_3 < 0$), the criterion thus defined being hereafter denoted H_ω-p.

Next, the analytical Burgers vortex and Burgers vortex sheet three-dimensional flowfields are analyzed in order to refine the study of the invariants of on a physical basis. By enforcing $I_1 < 0$, the classical definition of the vorticity core $r^* = 2\sqrt{\alpha/\nu}$ of the Burgers' vortex is recovered whatever the $\mathrm{Re}_\Gamma = \Gamma/(4\pi\nu)$ is, as seen in Fig. 1. The same result also holds for the

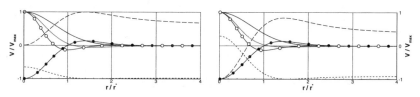

Fig. 1. Profiles of $\|\omega\|$ (———), Δ (······), Q (– –), λ_2 (— —), $I_1^{\mathbf{H}(\omega)}$ (–●–) and $I_2^{\mathbf{H}(\omega)}$ (–○–) for Burgers vortices of shear- (left) and rotation- (right) dominated core.

vortex sheet. It is therefore assumed that $I_1 < 0$ is a necessary condition in the present definition of a structure core. Note that Q and λ_2 exclude the whole Burgers' vortex sheet as well as, for low Re_Γ value, the Burgers' vortex.

Apart of the pancake shape, structures can be mainly classified into tubes or sheets possibly modeled in space as hyperboloids corresponding to one positive and two negative eigenvalues. In the invariant space, this condition reads $I_3 \geq 0$ providing that $I_1 < 0$. The discrimination between tubes and sheet is then carried out on the basis of their slenderness by means of the I_2 invariant. A positive value of I_2 ensures that the hyperbolic positive eigenvalue is low enough compared with the two elliptic positive ones. It is consequently associated with tubular regions. The $\mathbf{H}_\omega$-t criterion related to vortex tube therefore reads $(I_1 < 0, I_2 > 0, I_3 \geq 0)$. For Burgers' vortex, this criterion educes the region $r \leq r^\star < r^*$, with $r^\star$ being the I_2 cancellation radius that also corresponds to the inflexion point in the $\|\omega\|$ profile.

Lastly, sheets are characterized in the invariant space by the criterion $(I_1, I_2 < 0, I_3 \geq 0)$. However another subdivision between sheets located inside or outside of a vortex core is desirable [4] since it is generally acknowledged that vortex sheets surrounding vortex cores play an important role in the dissipation of k. Such a splitting appears to be difficult to obtain if based solely on the the three invariants of $\mathbf{H}_\omega$. It can nonetheless be achieved with help of an auxiliary condition on Q so as to discriminate between shear-dominated and rotation-dominated regions. The two criteria, labeled $\mathbf{H}_\omega$-rs for the rotational sheet regions and $\mathbf{H}_\omega$-ss for the shear sheet regions, are then respectively defined by $(I_1, I_2 < 0, I_3 \geq 0, Q > 0)$ and $(I_1, I_2 < 0, I_3 \geq 0, Q \leq 0)$.

2 Test cases

The H_ω criteria have been validated using synthetic three-dimensional flowfields. For instance, when applied to the synthetic ABC flow, these criteria are able to discriminate between pancake-like structures, tube-like structures surrounded by tubular sheets and "planar" sheets developing between the tubes while other classical criteria yield the eduction of a large lattice of structures.

The criteria have also been tested using snapshot fields issued from two DNS computations of freely decaying turbulence, either isotropic or vertically

	Δ	Q	λ_2	$I_1^{H(\omega)}$	$\cup\, H_\omega$-	H_ω-p	H_ω-t	H_ω-rs	H_ω-ss
Volume	65.8	41.7	41.4	53.1	44.4	6.5	12.1	13.3	12.5
ϵ	54.9	31.2	29.8	59.0	48.3	7.1	13.4	10.4	17.4
ϵ/volume	0.835	0.749	0.720	1.111	1.088	1.092	1.112	0.782	1.390

Table 1. Percentages of total volume and dissipation rate corresponding to volumes educed from a flowfield of $Re_\lambda \simeq 50$ isotropic turbulence using various criteria.

stratified[5] with $Re_\lambda \simeq 50$. Table 1 summarizes the volume ratio and ϵ contents in the isotropic case for the regions educed using various criteria. It is seen from it that regions educed using the H_ω criteria exhibit larger ϵ density than their Δ, Q or λ_2 counterparts. For instance, the ratio of ϵ density between Q regions where H_ω criteria are true compared to Q regions where they are not is equal to about 1.4. This is due to the fact that about 80% of the H_ω-p pancakes and 70% of the H_ω-t tubes are located within the Q regions. Lastly, the two last columns of Tab. 1 demonstrate that the splitting of the sheets according to the shear/rotation balance is meaningful in term of the dissipative behavior of the structures.

When applied to the stratified flow, The H_ω criteria reveal the strongly anisotropic and differentiated structure of the flow by highlighting stratified thin shear-dominated sheets as well as a high density of large pancakes surrounded by horizontally-aligned tubes.

3 Conclusion

A new family of geometric criteria designed to educe turbulent structures has been defined and has demonstrated promising results in highlighting regions of high dissipation rates. It is worth noting that these criteria , being geometrical in nature, are free from compressibility-related limitations and can be easily extended to fields other than $\|\omega\|$ such as passive scalars or their gradients.

References

1. J. Jeong and F. Hussain. On the identification of a vortex. *J. Fluid Mech.*, 285:69–94, 1995.
2. J. Weiss. The dynamics of enstrophy transfer in two-dimensional hydrodynamics. *Physica D*, 48(2-3):273–294, March 1991.
3. V. Herbert, M. Larchevêque, and C. Staquet. Identification des structures organisées en écoulement bidimensionnel. *C. R. Acad. Sci. Paris Sér. II b*, 323(8):519–526, 1996.
4. K. Horiuti and Y. Takagi. Identification method for vortex sheet structures in turbulent flows. *Ph. Fluids*, 17:121703, 2005.
5. L. Liechtenstein, F. S. Godeferd, and C. Cambon. Nonlinear formation of structures in rotating stratified turbulence. *Journal of Turbulence*, 6:1–18, 2005.

Structure of a tornado-like vortex

Koji Sassa[1], Kensuke Yamashita and Saki Takemura

Kochi Univ., Akebonocho 2-5-1, Kochi 780-8520 Japan `sassa@kochi-u.ac.jp`

1 Introduction

Tornadoes cause probably the most violent wind on earth. Though many researchers have observed and investigated the tornadoes, their structure has not been cleared yet because of their lifetime is very short and their scale is very small compared with the other meteorological phenomena. Therefore, many tornado simulators have been designed to measure the velocity field around the tornado in detail[1]. But, most of them generated a tornado in a closed chamber and could not move it as actual tornadoes running on fields.

We developed an original tornado simulator to realize a tornado running on a plane. It can simulate the updraft and the rotation of mesocyclone. The structure of the resultant vortex was measured in detail and found to have velocity profile similar to that of actual tornado.

2 Experiment

The tornado simulator is composed of an axial fan to generate updraft and a rotating porous disk to simulate rotation of mesocyclone. Various swirl ratio, $Sr = \Omega R/U$, can be set by controlling the speed of the fan and the rotation rate of the disk. In the present measurement, the driving condition in which the strong tornado-like vortex was observed stably was selected; the updraft velocity, $U = 270$ mm/s and the rotation rate, $\Omega = 33.5$ rad/s, respectively. The horizontal plane at various height was illuminated by an Argon gas laser sheet and filmed by a hi-speed camera at 1000 fps. The instantaneous velocity field was calculated from the resultant movies by using a dynamic PIV method. The mean velocity and vorticity fields were obtained by averaging 1000 instantaneous velocity fields. Hotwire measurements were also made by using an X-probe to obtain the temporal evolution of tornado velocity.

3 Result and Discussion

The resultant mean velocity vectors and contour maps of turbulence intensity
are shown in Fig. 1. Converging flow toward the center axis of the vortex is
clear observed at $Z = 4$ mm. The layer becomes higher, the rotating flow
becomes stronger than the converging flow gradually, and the rotating flow
dominates velocity fields in $Z > 14$ mm. Then, 3-D boundary layer like Ekman
boundary layer can be defined in $Z < 14$ mm. The radius of the vortex
that defined by the point where the maximum azimuthal velocity is observed
becomes larger at higher layer. The turbulence intensity is high around the
vortex axis at $Z = 4$ mm and near $r = 20$ mm where the maximum azymuthal
velocity is observed.

Circulation in the core region of the vortex is almost constant, $\Gamma =$
$0.031 mm^2/s$ except in the boundary layer. The velocity profile of the vortex is similar to that of the actual tornado[3]. The vortex Reynolds number
and the swirl ratio at $Z = 30$ mm were $Re = 320$ and $Sr = 0.9$, respectively.
The vortex Reynolds number almost equals to that of the actual tornado[4],
if we use eddy diffusivity, $\epsilon = 50 m^2/s$, as kinematic viscosity. Though the
large swirl ratio corresponds to the vallue of the multiple-vortex tornado[2],
the multiple-vortex was not observed due to relatively high intensity of background turbulence in the present experiment.

Velocity in the vortex fluctuated quite intermittently due to the rotational
swing of the vortex axis and the instability of the vortex itself. The energy
spectra of its fluctuation are shown in Fig. 2. Inertial subrange is observed in

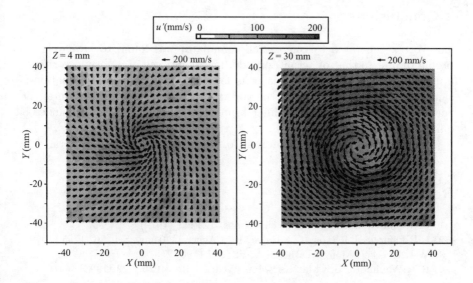

Fig. 1. Contour map of turbulence intensity around the tornado at Z=4mm(left)
and 30mm(right)

the lower frequancy range in both heights. But, the energy gap around $f = 3$ Hz was cleary observed in the spectra at $Z = 30$ mm. Such energy gap shows that the time scales of the swing motion of the vortex axis and the fluctuation of the vortex itself are apart remarkably from each other.

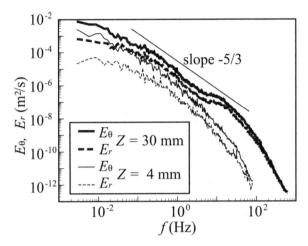

Fig. 2. Energy Spectra of velocity fluctuation

4 Conclusions

The vortex generated by our tornado simulator was found to have velocity profile similar to that of actual tornado. It is composed of a normal vortex in upper layer and 3-D boundary layer in lower layer. Turbulence fluctuation is caused by the swing motion of the vortex axis and by the fluctuation of the vortex itself. Thier scales are differ from each other and then turbulence is quite intermittent.

Authors are grateful to Photoron Co. who lent us a hispeed camera.

References

1. C.R.Church, J.T.Snow: J. Rech. Atmos. **12**, 111 (1979)
2. C.R.Church, J.T.Snow, G.L.Baker, E.M.Agee: J. Atmos. Sci. **36**, 1755 (1979)
3. W.H.Hoecker: Mon. Weather Rev. **88**, 167 (1960)
4. R.M.Wakimoto, B.E.Marther: Mon. Weather Rev. **120**, 522 (1992)

Self-similar structure formation process in thermal turbulence

Hiroki Yatou, Takeshi Ogasawara, Takeshi Matsumoto, and Sadayoshi Toh

Division of Physics and Astronomy, Graduate School of Science, Kyoto University
Kitashirakawa Oiwakecho Sakyo-ku, Kyoto 606-8502, Japan
TEL : +81-75-753-3805, sanmaya@kyoryu.scphys.kyoto-u.ac.jp

1 Introduction

Successive generation of small-scale structures via nonlinearity is a central issue in turbulence research. Much understanding can be obtained if an elementary process of the generation is extracted. We here show that such extraction can be done for 2D thermal convection turbulence in which many aspects (e.g., the inertial range and the temperature-variance cascade [1]) are common to 3D Navier–Stokes turbulence. In this convection system, the Rayleigh–Taylor instability plays a triggering role in the small-scale structure generation[2]. We focus on how thin temperature interface between hot and cold regions evolves through a chain of the instabilities. Initially the sheets are still and perturbed only in a large scale. The evolution can be considered as an elementary process of the generation of small-scale activities in this convection system. The 2D thermal convection turbulence is described by the 2D Boussinesq approximation equations:

$$(\partial_t + \boldsymbol{u} \cdot \nabla)\boldsymbol{u} = -\nabla p/\rho_0 + \nu \triangle \boldsymbol{u} + \alpha g T \boldsymbol{e}_y, \quad \nabla \cdot \boldsymbol{u} = 0, \tag{1}$$

$$(\partial_t + \boldsymbol{u} \cdot \nabla)T = \kappa \triangle T. \tag{2}$$

Here, $\boldsymbol{u}$, T, p and ρ_0 represent the velocity, the temperature, the pressure and the (constant) density respectively. And ν, α, g and κ are the kinematic viscosity, the thermal expansion coefficient, the gravity acceleration and the thermal diffusion coefficient, respectively. The Prandtl number ν/κ is unity. The unit vector in the buoyancy term is $\boldsymbol{e}_y = (0, 1)$. Notice that no mean temperature gradient is imposed and that the boundary condition is doubly periodic. The numerical method is a standard, fully-dealiased spectral method with a 4th order Runge-Kutta method. In practice the hyper viscosity/diffusivity (8th order Laplacian) is used to have a wider inertial range. We solve these equations with the initial field having thin temperature interfaces between hot and cold regions (Fig.1:left figure). The resolution is from 2048^2 to 16384^2. We write a program taking advantage of the symmetry of the flow, resulting in a fast and memory-saving simulation.

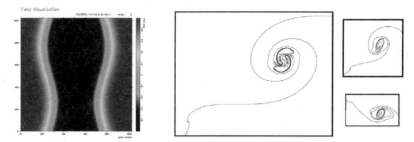

Fig. 1. Left: initial temperature field (blue:cold, red:hot). The whole domain $(2\pi \times 2\pi)$ is shown. Right: self-similar structures (temperature interfaces) at t=7.66 (biggest panel, domain: $0.23\pi \times 0.15\pi$), t=7.86 (medium panel, domain: $0.078\pi \times 0.068\pi$) and t=8.1(smallest panel, domain: $0.068\pi \times 0.029\pi$) respectively.

2 Results

In the beginning, the temperature interfaces become thinner and thinner. The important feature in the 2D convection system is that the high temperature gradient region is the same as the high vorticity region (the temperature gradient is the source term of the vorticity equation here). Hence, the thin temperature interfaces are actually thin vortex sheet at the same time. Soon the interfaces begin to roll up due to the Rayleigh–Taylor instability. The rolled-up part becomes then a vortex blob where the interface is entrained in a spiral shape. The rather straight part of the interface between the spiral parts gets unstable and rolls up in the sequel. The small-scale structure formation which we follow is this successive generation of the spiral parts in the interface.

It should be noted that the temperature interface studied here has a certain similarity to the slender vortex filaments observed in the 3D Navier–Stokes turbulence. The temperature interface is characterized by the large temperature gradient. If we use the variable $\chi = (\partial T/\partial y, -\partial T/\partial x)$, the equation for χ is written as

$$(\partial_t + \boldsymbol{u} \cdot \nabla)\chi = (\chi \cdot \nabla)\boldsymbol{u} + \kappa\Delta\chi. \tag{3}$$

Obviously this equation has the same structure as the 3D Navier–Stokes vorticity equations. We expect that knowledge of the 2D convection system will shed some light to the 3D Navier–Stokes turbulence.

The temperature interface is identified by numerically following a material line which is initially placed on the maximum of the temperature gradient. In the right of Fig.1, we show three instances of the birth of a new spiral. For example in the biggest panel, the newly born spiral is seen as the kink seen in the left bottom. The rest two panels have the same kind of kinks close to the big spirals. It is suggested that this formation process is self-similar.

In order to quantify this, we measure the distance between the kink and the center of the nearest big spiral in each instance. This distance is used as the length scale of the small-scale structure formation. For the time scale of the formation, we measure the time intervals between the successive appearance of the kinks. The relation between the length scale l and the associated time scale τ_l is shown in Fig.2. The data seem to follow the scaling law $\tau_l \propto l^{0.7}$. The length scales shown are within the inertial range except the smallest value in this simulation (resolution 2048^2). Do we have an explanation for the exponent 0.7? So far we do not. We list two observations related to this issue. It is known that the basic scaling property of this 2D convection turbulence (if it is a statistically steady state) is so-called Bolgiano–Obukhov scaling (see, for example, [1]) where the time scale of the spatial scale r follows $t_r \propto r^{2/5}$. The observed exponent 0.7 is different from the Bolgiano–Obukhov scaling exponent. The next attempt is the following. Since our simulated system is a transient process, we may not use the Bolgiano–Obukhov scaling argument. Nevertheless the energy spectrum $E(k)$ and the temperature variance spectrum $S(k)$ have a clean power-law behavior at each given instant (figures not shown). The exponents of the power laws change in time (the spectra approaches to the Bolgiano–Obukhov scalings from below). From the spectra the time scale t_k associated to the wavenumber k can be estimated dimensionally as $t_k \propto (\alpha g)^{-1} E(k)^{1/2} S(k)^{-1/2}$. When we put the transient scalings of $E(k)$ and $S(k)$ in this relation, the result is $t_k \propto k^{-n}$ where $0.25 \leq n \leq 0.32$. This is again different from the value 0.7.

With higher resolutions (8192^2 and 16384^2) at a later time we find that the process of the small-scale structure formation enters a different stage with a certain self-similarity. Comparison between the two resolutions suggests that the difference is not caused by the viscosity or diffusivity. This second stage shall be discussed in the conference.

Acknowledgements: the numerical calculations were carried out on SX8 at YITP in Kyoto University. This work is supported by the Grant-in-Aid for the 21st Century COE "Center for Diversity and Universality in Physics" from the Ministry of Education, Culture, Sports, Science and Technology (MEXT) of Japan.

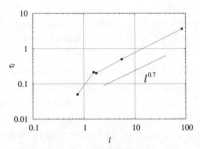

Fig. 2. Scaling law of the small scale generation process. Here l is the length scale of the structure and τ_l is the time interval between two successive structure formation.

References

1. S. Toh and E. Suzuki: Phys. Rev. Lett., **73**, 1501 (1994)
2. M. Chertkov: Phys. Rev. Lett., **91**, 115001 (2003)

Vortex Dynamics in the Reattaching Flow of Separation Bubbles with Variable Aspect Ratio

S. Courtine, A. Spohn[1], and J.P. Bonnet

[1] Laboratoire d'Etudes Arodynamiques
UMR CNRS 6609, Universit de Poitiers, ENSMA
Tlport 2 - 1, avenue Clment Ader BP 40109
86961 FUTUROSCOPE FRANCE
spohn@lea.ensma.fr

In many flow applications arise separation bubbles characterized by separating and reattaching flow. The unsteady behaviour of the turbulent reattaching flow increases wall shear stress and produces cyclic structural and thermal loading. Although this phenomenon is frequently observed there remains considerable confusion about the vortical structures involved and their impact on the turbulent flow in the reattachment region. According to Kiya & Sasaki (1985) [4], Muti Lin & Pauley (1996) [5] among others, large-scale spanwise vortices dominate the unsteadiness of separation bubbles. In contrast Watmuff (1999) [6] found three-dimensional vortex loops to play a key role in the turbulent reattaching flow. Very recently Burgmann *et al.* (2006) [1] observed C-shaped vortices in the reattachment region which undergo rapid transformations in a vortex burst. In order to further clarify the physical mechanisms involved we investigate in the present study the vortex dynamics in the reattachment region in more detail.

We used a half model with a rounded front part and variable width to create separation bubbles with different aspect ratios under controlled conditions. Figure 1 shows the model with height H and width L in the test section of a water channel. For visualization purpose the flow can be observed through all side walls and from behind in the upstream direction. The aspect ratio $\Lambda = L/H$ of the frontal cross-section varied between 2.2 and 8.8. The Reynolds number $\Re e = U_\infty H/\nu$ (U_∞ uniform upstream velocity in the test section, ν the kinematic viscosity of water) varies between $\mathcal{O}(10^3)$ and $\mathcal{O}(10^4)$. Time-resolved flow visualisations and PIV measurements are used to analyse the spatio-temporal evolution of the flow in the reattachment region.

Figure 2 compares flow visualisations obtained with the electrolytic precipitation technique and PIV measurements at the position of mean reattachment in a vertical plane perpendicular to the mean flow. In both cases the flow is observed in the upstream direction. For $\Lambda = 2.2$ both techniques indicate the existence of central periodic eruptions of fluid. The time-resolved

PIV measurements show in addition the correlation between these periodic ejections and the vertical displacement of a pair of streamwise counter rotating vortices. These vortices emerge from the foci situated at both ends of the separation line. Around these foci fluid is continuously pumped from outside towards the interior of the separation bubble thus increasing the quantity of fluid periodically ejected. In contrast for $\Lambda = 8.8$ several pairs of counter rotating vortices indicate the formation of secondary instabilities which lead to spanwise periodic deformations of the primary Kelvin-Helmholtz vortices. The view from above in figure 3 shows clearly the existence of these spanwise Kelvin-Helmholtz vortices for both aspect ratios. However only for $\Lambda = 8.8$ the rapid formation of a spanwise periodicity in the tracer distribution indicates the presence of such secondary instabilities. Figure 4 illustrates the spatio-temporal evolution the primary Kelvin-Helmholtz instability for $\Lambda = 2.2$. While in the front part of the bubble the shear-flow instability grows only slowly, the central eruptions in the reattachment region lead to a rapid increase in growth rate. Compared to a free shear layer the spatio-temporal evolution of the wall bounded shear layer looks more like a highly nonlinear bursting process as demonstrated by the steeper growth rate of the momentum thickness θ. Clearly both, the neighbourhood of a wall and the lateral pumping of fluid towards the plane of symmetry, greatly influence the vortex dynamics in the reattachment region and thus condition the modelling of turbulence further downstream.

References

1. S. Burgmann, C. Brücker and W. Schröder. Scanning PIV measurements of a laminar separation bubble. *Experiments in Fluids*, **41**, pp. 319-326 (2006)
2. S. Courtine. Etude exprimentale des dcollements provoqus par une paroi courbe : topologie et volution spatio-temporelle. *Ph.D. Thesis*, Poitiers University (2006)
3. C.M. Ho and P. Huerre. Perturbed free shear layers. *Annual Review of Fluid Mechanics*, **16**, pp. 365-424 (1984)
4. M. Kiya and K. Sasaki. Structure of large-scale vortices and unsteady reverse flow in the reattaching zone of a turbulent separation bubble. *Journal of Fluid Mechanics*, **154**, pp. 463-491 (1985)
5. J.C. Muti Lin and L.L. Pauley. Low Reynolds number separation on an airfoil. *AIAA Journal*, **24**-8, pp. 1570-1577 (1996)
6. J.H. Watmuff. Evolution of a wave packet into vortex loops in a laminar separation bubble. *Journal of Fluid Mechanics*, **397**, pp. 119-169 (1999)

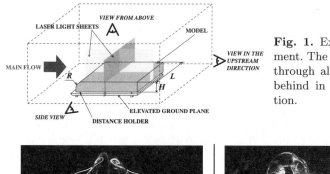

Fig. 1. Experimental arrangement. The flow can be observed through all sidewalls and from behind in the upstream direction.

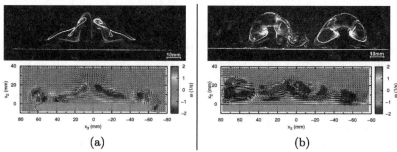

Fig. 2. Comparison of flow visualization and instantaneous velocity and vorticity fields at the mean reattachment location at $\Re e = 1520$ for two different aspect ratios. View from behind in the upstream direction. (a) $\Lambda = 2.2$; (b) $\Lambda = 8.8$.

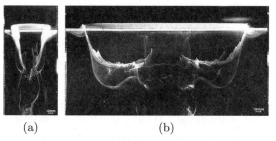

Fig. 3. Comparison of the spatial evolution of the separating and reattaching flow at $\Re e = 1520$ for two different aspect ratios. View from above. (a) $\Lambda = 2.2$; (b) $\Lambda = 8.8$.

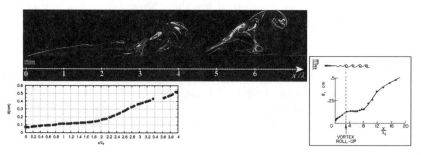

Fig. 4. Spatial evolution of the shedded vortices and measured growth rate of the corresponding momentum thickness ($\Re e = 1520$, $\Lambda = 2.2$). For comparison the corresponding much slower evolution for a free shear layer is indicated on the right (Ho & Huerre (1984) [3])

LES of Compressible Inert and Reacting Turbulent Shear Flows

Rainer Friedrich, Somnath Ghosh, and Inga Mahle

Fachgebiet Stroemungsmechanik,Technische Universitaet Muenchen,
Boltzmannstr. 15, 85748, Garching, Germany. `r.friedrich@lrz.tu-muenchen.de`

Abstract Compressible turbulent flows are an important element of high-speed flight. However, they are not completely understood. Especially interactions between compressible turbulence and combustion lead to challenging, yet unsolved problems. Large-eddy simulation (LES) represents a powerful tool which allows to mimic such flows in remarkable detail and to analyse underlying physical mechanisms, sometimes even those which cannot be accessed by experiment. The paper provides a short description of this tool and applications to complex turbulent flow through a supersonic nozzle and in a reacting mixing layer. The analysis focuses on effects of compressibility, acceleration and heat release on the transport characteristics of the streamwise Reynolds stress component only, but emphasises the need to model these effects based on the full Reynolds stress equations.

1 Introduction

In vehicles which fly at high speed, compressible turbulent flow plays a significant role in controlling drag, thrust and flight mechanics. A complete understanding of the flow dynamics is therefore a prerequisite for proper turbulence modelling and an improved design of these systems. The two most powerful tools to predict such flows are direct numerical (DNS) and large-eddy simulation (LES). While DNS provides a complete description of turbulent flows, LES computes only their large scales and models their small ones. LES is thus intermediate between DNS and statistical turbulence modelling and has the potential to predict those statistical quantities reliably which are mainly controlled by large turbulent scales.

Large-eddy simulation of compressible turbulent flows needs modelling of many unclosed nonlinear terms. The number of these terms increases when chemical reactions take place. Techniques are therefore of great value which circumvent the modelling of each individual unknown term in the governing LES equations. Such techniques are of pure mathematical nature and make no use of the flow physics in modelling effects of unresolved scales. Sagaut [1] uses the terminology 'structural modelling' in describing such models for incompressible flow and groups them into several categories. Domaradzki and Adams [2] discuss methods which model the unresolved primitive variables and compute the unclosed terms in a secondary step. We will focus here on the approximate deconvolution method (ADM) presented in [3] and extensions thereof. Starting from a generic one-dimensional transport equation and computing a low-wavenumber solution of this equation as in LES, means solving the following equation:

$$\frac{\partial \bar{u}^L}{\partial t} + \overline{\frac{\partial f(u)}{\partial x}}^L = 0, \quad or, \quad \frac{\partial \bar{u}^L}{\partial t} + G * \frac{\partial f(u)}{\partial x} = 0 \qquad (1)$$

in which f(u) is a nonlinear function of the unfiltered variable u(x,t). According to Leonard [4], a filtered variable is defined as

$$\bar{u}^L = G * u = \int G(x - x', \Delta) u(x') dx' \qquad (2)$$

where $G(x - x', \Delta)$ is a low-pass filter kernel and Δ the filter width. Note that the filtered (or LES) variable is a continuous function of the independent variables x and t, which attenuates scales smaller than the filter width Δ. Besides filtering, LES also implies an implicit projection of $\bar{u}^L$ onto a mesh of size $\Delta x \sim O(\Delta)$. If a Fourier spectral method is employed to solve (1), the LES projection is equivalent to an implicit sharp spectral filter with the cutoff wavenumber $k_c = \frac{\pi}{\Delta}$ [2]. The use of finite-difference methods entails smooth filters based on Padé approximants [5] and differentiation errors, which affect only the smallest resolved scales, but have the positive side-effect of keeping aliasing errors low. Filtered and unfiltered variables are related via the traditional decomposition $u = \bar{u}^L + u'$ where u' is called the subgrid scale component. We refer to [2] for a discussion of the fact that u' receives contributions from the subgrid and the resolved scales. Now, since $f(u)$ is nonlinear, equation (1) is unclosed. Stolz and Adams [3] close this equation by using a defiltered field u^* which replaces u in the nonlinear term and is given by:

$$u = u^*, u^*(x,t) = Q_N * \bar{u}^L = Q_N * G * u, Q_N = \sum_{\nu=0}^{N} (I - G)^\nu \approx G^{-1} \qquad (3)$$

Q_N is the approximate inverse of G, obtained from a van Cittert [6] series which is truncated at $N = 5$. The resulting effective cutoff wavenumber is $k_c = \frac{2\pi}{(3\Delta x)}$. G is a 2nd order Padé filter involving a parameter α that is linked to the cutoff wavenumber k_c. In a later formulation a relaxation term has been added to the RHS of the closed eq. (1) which improves the energy transfer between the resolved and unresolved scales and can even be applied to capture shocks [7]. Mathew et al. [8] have shown that ADM, as presented in [3] , can be implemented as single-step explicit filtering procedure using a $Q_N * G$ filter with $\alpha = 0.25$ and $N = 6$. Application of a $(Q_N * G)^2$ filter to the approximate unfiltered field at every time step even acts in a similar way as the relaxation term introduced in [7]. Higher-order closure has been proposed and tested in [9], by using the definition of u^* given in (3). This provides the following LES equation:

$$G * \left[\frac{\partial u^*}{\partial t} + \frac{\partial f(u^*)}{\partial x} \right] = G * \frac{\partial \big(f(Q_N * G * u^*) - f(u^*) \big)}{\partial x} \qquad (4)$$

Integration of (4) means evaluation of nonlinearities for u^*, and for $Q_N * G * u^*$. The LES results shown below have been obtained using explicit filtering of flow variables with a $(Q_N * G)^2$ filter at every simulation time step, rather than applying a technique as given in (4), because of its high computational costs. In section 2, G is a 2nd order filter with $\alpha = 0.25$ and in section 3, G is of 4th order with $\alpha = 0.5$.

2 Wall-bounded shear flow

We now turn to the prediction of turbulent supersonic flow in a nozzle of circular cross-section via LES. Our motivation is to see in which way acceleration and compressibility affect the flow variables and the transport characteristics of the streamwise Reynolds stress component, as one element of the full Reynolds stress tensor. We solve the compressible Navier-Stokes equations in a special pressure-velocity-entropy form [10] on non-orthogonal curvilinear coordinates using 6th order compact central schemes for spatial discretization. A 3rd order low-storage Runge-Kutta scheme advances the flow field in time. The flow entering the nozzle is fully developed turbulent pipe flow driven by a homogeneous body force which compensates the wall shear stress and allows the use of streamwise periodic boundary conditions. Table 1 gives an overview over flow parameters and computational data for DNS and LES of this flow. M is a global Mach number based on the speed of sound at

Case	M	Re_τ	N_x	N_θ	N_r	Δx^+	$r\Delta\theta^+_{max}$	Δr^+_{min}	Δr^+_{max}
DNSM1.5	1.5	245	256	128	91	9.5	12.0	1.3	3.73
LESM1.5	1.5	245	64	64	50	38.0	21.2	2.5	6.79

Table 1. Flow and computational parameters for DNS/LES of fully developed pipe flow

constant wall temperature and the bulk velocity, $u_{av} = 2 \int_0^1 \langle u_x \rangle \frac{r}{R} d\frac{r}{R}$. The friction Reynolds number, Re_τ, is based on friction velocity $u_\tau = \sqrt{\tau_w/\rho_w}$, pipe radius and kinematic viscosity $\nu_w(T_w)$. Re_τ is a result of the computation. The mesh sizes are non-dimensionalised by u_τ and ν_w. From Table 1 we conclude that the LES grid is an order of magnitude coarser than the DNS grid. In general we want LES grids to be much coarser than DNS grids. In the present LES, however, no further grid coarsening is possible, since the turbulent Reynolds number is low and the near wall flow needs sufficient resolution to capture the small energy-carrying structures. A comparison of DNS and LES data, shown in [11], reveals excellent agreement. We are therefore confident that the present LES of supersonic nozzle flow provides reliable data. The computational nozzle domain is $L = 10R$ long, with R the upstream pipe radius. The aperture angle of the nozzle is 2.5° and the ratio of maximum nozzle to pipe radius is 1.58, generating an acceleration of the core flow from a Mach number of 1.7 to 2.7. The mean pressure gradient averaged along the nozzle is $\frac{\delta^*}{\tau_w}\frac{\partial <p>}{\partial x} = -1.2$. The temperature is kept constant along the nozzle wall. The nozzle domain is discretized with $64 \times 64 \times 50$ points in streamwise, circumferential and radial directions. The periodic pipe flow and the nozzle flow are coupled in the LES using MPI routines. The concept of characteristics is used to set inviscid inflow conditions [12] for the nozzle flow. For

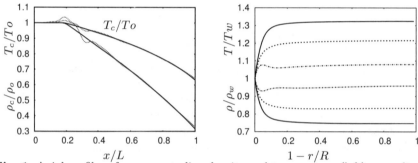

Fig. 1. Axial profiles of mean centerline density and temperature (left) normalized with upstream values. Solid lines represent isentropic flow. Right: Radial mean density and temperature profiles, normalized with wall values at $x/L = 0.0$ (—), 0.45 (- - -), 0.8 (-.-)

an understanding of compressibility and acceleration effects it is necessary to discuss the behaviour of mean flow variables first. Figure 1 shows profiles of mean temperature and density along the nozzle centreline in comparison to results obtained from the isentropic streamtube equations. The mean core flow behaves surprisingly close to accelerated isentropic flow. In fully developed pipe flow ($x/L = 0.0$) mean density and temperature are closely linked in radial direction, since the radial pressure gradient is negligibly small. The heat generated by dissipation in the wall layer strongly increases the mean temperature and leads to a heat flux out of the pipe. The mean density in turn drops from its high wall value to a low core value and thus reduces the pressure-strain correlations. This effect of compressibility has been explained for turbulent channel flow in [13] and holds for pipe flow as well. The described direct coupling between temperature and density in radial direction persists in the present accelerated flow (Figure 1, (right), $x/L > 0.0$). While in the nozzle core adiabatic cooling due to acceleration compensates dissipative heating, it is less pronounced in the near-wall region. The Reynolds stresses are damped through flow acceleration. The question here is, which role mean dilatation and hence compressibility, mean shear and extra rates of strain play in this flow. To see this, we write down the streamwise Reynolds stress transport equation using cylindrical (x, θ, r)-coordinates and splitting the 'kinetic' production term into contributions involving mean shear, extra rates of strain and mean dilatation:

$$\bar{\rho}\frac{\widetilde{Du''_x u''_x}}{Dt} = \underbrace{-2\overline{\rho u''_x u''_r}\frac{\partial \tilde{u}_x}{\partial r}}_{shear} \quad \underbrace{-2\overline{\rho u''_x u''_x}\frac{\partial \widetilde{u}^d_x}{\partial x}}_{extra\ rates\ of\ strain}$$

$$\underbrace{-\frac{2}{3}\overline{\rho u''_x u''_x}\frac{\partial \tilde{u}_l}{\partial x_l}}_{mean\ dilatation} \quad \underbrace{-\overline{u''_x}\frac{\partial \bar{p}}{\partial x}}_{enthalpic\ production}$$

$$\underbrace{+\overline{p'\Big(\frac{\partial u'_x}{\partial x} - \frac{1}{3}d'\Big)}}_{deviatoric\ pressure\text{-}strain\ corr.} \quad \underbrace{+\frac{1}{3}\overline{p'd'}}_{pressure\text{-}dil.} \quad +... \quad (5)$$

In eq. (5) we have used deviatoric strain rates and dilatation fluctuations as defined by:

$$\frac{\partial \widetilde{u}^d_x}{\partial x} = \frac{\partial \widetilde{u}_x}{\partial x} - \frac{1}{3}\frac{\partial \tilde{u}_l}{\partial x_l}, \quad d' = \frac{\partial u'_l}{\partial x_l} \quad (6)$$

The tilde and the double dash denote Favre averages and fluctuations. The single dash represents a Reynolds fluctuation. Besides 'kinetic' production the RHS of (8) contains enthalpic production as well. The pressure-strain correlation has been split into its deviatoric and isotropic parts. Figure 2 displays the streamwise components of the Reynolds stress and the pressure-strain correlation. The dramatic decrease of both tensor components as the flow

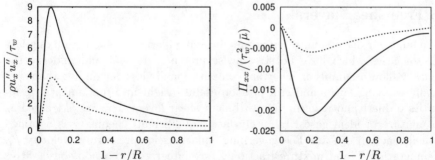

Fig. 2. Radial profiles of streamwise Reynolds stress (left) and pressure-strain correlation (right) at stations $x/L = 0.0$ (—); 0.45 (- - -)

accelerates is obvious. Since pressure-dilatation is so small, the full pressure-strain correlation is plotted here. Further investigations are needed to explain whether the rapid pressure-strain decay is due to mean density variation alone

(Fig. 1) or is also affected by mean dilatation. Figure 3 provides insight into the development of kinetic and enthalpic production during acceleration. In fully developed pipe flow, enthalpic production is negligibly small and kinetic production by mean shear is the only turbulence producing mechanism. This mechanism is reduced by a factor of two in the buffer layer of the nozzle (Fig. 3 right). Production by extra rate of strain and mean dilatation further reduce the total kinetic production term. Enthalpic production has only a small positive contribution. The terms in eq. (5) which are not written down due to

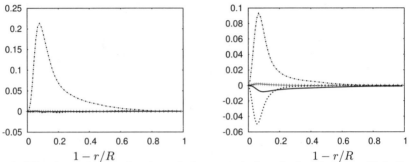

Fig. 3. Kinetic production by shear (-.-), extra strain rate (- - -), mean dilatation (—) and enthalpic production (+) at $x/L = 0.0$ (left) and $x/L = 0.45$ (right)

lack of space are: dissipation (D), viscous and turbulent diffusion (VTD) and mass-flux variation by shear stress. Like the pressure-dilatation correlation the compressible dissipation rate represents intrinsic effects of compressibility. It turns out that this term is small as well at this Mach number compared to the solenoidal dissipation rate which decreases strongly due to acceleration. For a more detailed discussion of acceleration and compressibility effects on the Reynolds stress transport based on DNS and LES data, we refer to [14].

3 Free shear layers

We limit our discussion to temporally evolving inert and reacting mixing layers of gases which have no streamwise pressure gradient and underlie the parallel-flow assumption. They are computationally less demanding than spatially evolving layers and still provide useful insight into fundamental properties of inert and reacting free turbulent shear layers. The gas is treated as a mixture of ideal gases. Infinitely fast (equilibrium) chemistry is assumed when reaction sets in. The governing equations for pressure p, velocity u_i, temperature T and mixture fraction Z are integrated in a non-conservative form using compact central finite-difference schemes of 6th order in space and a low-storage Runge-Kutta scheme of 3rd order in time. The transport coefficients of the mixture are computed with the help of the programme EGlib [15] which is integrated into our compressible Navier-Stokes solver. EGlib uses numerically efficient and accurate formulae derived from kinetic theory of gases and involving the local temperature, pressure and species composition. Soret and Dufour effects are neglected. Polynomial expressions are applied to compute the species enthalpies and specific heats. The Schmidt number has a fixed value of 0.7. The heat flux comprises a flux by conduction and diffusion. An infinitely fast H_2/O_2 one-step chemical reaction produces an infinitely thin flame sheet which separates oxidizer and fuel. On both sides of the flame, species profiles are linear functions of the mixture fraction Z, but their slopes are discontinuous at the flame front, according to the Burke-Schumann flame structure. Inert and reacting temporal mixing layers have been simulated directly and as LES for different Mach numbers. We report on low-Mach-number cases with $M_c = 0.15$, the flow and numerical parameters of which are described below in Table 2.

Gas1:Gas2	Case	M_c	Re_{δ_ω}	Da	$\frac{\rho_1}{\rho_2}$	$\frac{T_1}{T_2}$	$\frac{L_{x1}}{\delta_\omega}$	$\frac{L_{x2}}{\delta_\omega}$	$\frac{L_{x3}}{\delta_\omega}$	N_{x1}	N_{x2}	N_{x3}
$O_2 : N_2$	DNS	0.15	11540	0	1.14	1	129	32	97	768	192	576
$O_2 : N_2$	LES	0.15	11540	0	1.14	1	129	32	97	192	48	144
$O_2/N_2 : H_2/N_2$	DNS	0.15	23810	∞	1	1.92	345	86	172	768	192	432
$O_2/N_2 : H_2/N_2$	LES	0.15	23810	∞	1	1.92	345	86	172	192	48	108

Table 2. Flow and computational parameters for DNS/LES of inert and reacting temporal mixing layers

Convective Mach and Reynolds numbers are defined using the constant velocity difference Δu across the mixing layer and reference values, $c_0 = \frac{(c_1+c_2)}{2}$, $\rho_0 = \frac{(\rho_1+\rho_2)}{2}$, $\mu_0 = \frac{(\mu_1+\mu_2)}{2}$ viz. $M_c = \frac{\Delta u}{2c_0}$, $Re_{\delta_\omega} = \frac{\rho_0 \Delta u \delta_\omega}{\mu_0}$. δ_ω is the vorticity thickness. The Damkoehler number, Da, which is the ratio of an eddy-turnover time and the time of reaction is infinite in the reacting cases where a mixture of O_2/N_2 with the mass fraction $Y_{O_2} = 0.23$ in the upper stream and a mixture of H_2/N_2 with $Y_{H_2} = 0.0675$ in the lower stream mix and react in sheets where $Z_s = 0.3$.

When large-eddy simulations are performed, explicit filtering of the flow variables is applied as described in section 1 and only the heat release term in the temperature and pressure equations needs separate modelling based on the conditionally filtered scalar dissipation rate. For more details we refer to [16].

3.1 Mean flow variables and Reynolds stresses

The following results refer to flow states in which the mixing layers have reached self-similar states with constant momentum-thickness growth rates. A key aspect in understanding how combustion modifies free shear flow is the decrease in density due to heat release. Figure 4 contains profiles of mean density and streamwise velocity. Statistical averaging has been performed in the two homogeneous directions and by taking samples out of the self-similar state.

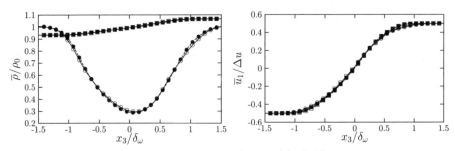

Fig. 4. Mean density normalized with $\rho_0 = (\rho_1 + \rho_2)/2$ (left) and mean streamwise velocity (right)for temporally developing shear layers. Self-similar state at $M_c = 0.15$. $\square$: DNS inert, $\blacksquare$: LES inert, $\circ$: DNS reacting, $\bullet$: LES reacting

The most striking feature is the strong reduction in density around the re-action zone. The mean streamwise velocity does not show as dramatic changes due to heat release as the mean density. Concerning the Reynolds stresses, we

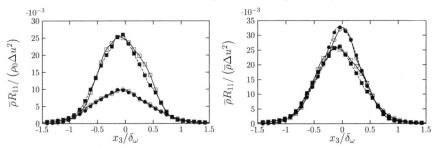

Fig. 5. Streamwise Reynolds stress,normalized by $\rho_0 \Delta u^2$ (left) and by $\bar{\rho} \Delta u^2$ (right) for inert/reacting shear layers. Self-simlar state at $M_c = 0.15$.Symbols as in Fig. 4

observe a strong reduction due to heat release in all components. In Figure 5, the streamwise Reynolds stress $\overline{\rho u_1'' u_1''} = \bar{\rho} R_{11}$ is plotted in two different normalizations which show that the reduction is mainly a mean density effect. The overshoot of the Favre-averaged stress R_{11} in the core of the shear layer is a Reynolds number effect. The Reynolds numbers of the reacting cases are by about a factor of 2 larger than those of the inert cases (Table 2).

3.2 Reynolds stress budget

The balance of the streamwise Reynolds stress reads:

$$\bar{\rho}\frac{DR_{11}}{Dt} = -\overline{\rho u_1'' u_3''}\frac{\partial \tilde{u}_1}{\partial x_3} - \frac{\partial\left(\overline{\rho u_1''^2 u_3''/2} - \overline{u_1' \tau_{13}'}\right)}{\partial x_3} + \overline{u_1''\left(\frac{\partial \overline{\tau_{13}}}{\partial x_3} - \frac{\partial \bar{p}}{\partial x_1}\right)}$$

$$+ \overline{p'\left(\frac{\partial u_1'}{\partial x_1} - 1/3d'\right)} + 1/3\overline{p'd'} - \overline{\tau_{1j}\frac{\partial u_1'}{\partial x_j}} \qquad (7)$$

Two important terms in this equation are kinetic production and pressure-strain correlation. They are displayed in Figure 6. There are no extra rates of strain in this flow and hence no mean dilatation terms that produce turbulence. The pressure-strain correlation contains an isotropic part which is an order of magnitude smaller than the complete term and hence not plotted. We note strong reductions in the production term due to heat release. Similar effects are observed in the pressure-strain correlation which transfers energy to the other two components. The behaviour of the pressure-strain correlation is primarily responsible for the reduction of the shear layer growth rate. An increase in convective Mach number leads to a further damping of the turbulence activity, but its underlying mechanism is of acoustic rather than entropic nature [17].

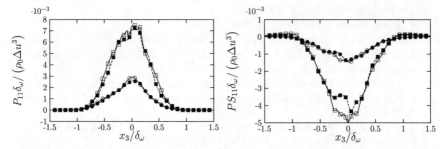

Fig. 6. Production (left) and pressure-strain correlation (right) in the streamwise Reynolds stress budget, normalized by $\rho_0 \Delta u^3 / \delta_\omega$ for inert and reacting shear layers. Self-similar state at $M_c = 0.15$. Symbols as in Fig. 4

4 Conclusions

Large-eddy simulations of non-reacting turbulent compressible flow in nozzles and shear layers show that explicit filtering with composite filters forms a convenient and reliable tool to predict these complex flows. When reaction takes place, further modelling of the filtered heat release terms is needed. Comparisons between DNS and LES data for low-speed inert and reacting shear layers reveal excellent agreement. The two flow cases considered demonstrate that effects of compressibility, flow acceleration and heat release all dampen the turbulence activity and modify the turbulence anisotropy, but via different mechanisms. Statistical turbulence modelling has therefore to be based on sophisticated second-order closure.

Acknowledgements
The financial support of this research by BMBF(03FRA1AC) and the Bavarian Research Network (FORTVER) is gratefully acknowledged.

References

1. P. Sagaut, *Large eddy simulation for incompressible flows*. 3rd Edition.Springer (2006)
2. J.A. Domaradzki, N.A. Adams, Direct modelling of subgrid scales of turbulence in large-eddy simulation.*J. Turbulence* 3, 024 (2003)
3. S. Stolz, N.A. Adams, An approximate deconvolution procedure for large-eddy simulation. *Phys. Fluids* 11, 1699-1701 (1999)
4. A. Leonard, Energy cascade in large-eddy simulations of turbulent fluid flows. *Adv. Geophys.* A 18, 237-248 (1974)
5. S.K. Lele, Compact finite difference schemes with spectral-like resolution. *J. Comp. Phys.* 103, 16-42 (1992)
6. P.H. van Cittert, *Z. Physik* 69, 298 (1931)
7. S. Stolz, N.A. Adams, L. Kleiser, The approximate deconvolution model for LES of compressible flows and its application to shock-turbulent-boundary-layer interaction. *Phys. Fluids* 13, 2985-3001 (2001)
8. J. Mathew, R. Lechner, H. Foysi, J. Sesterhenn, R. Friedrich, An explicit filtering method for large-eddy simulation of compressible flows. *Phys. Fluids* 15, 2279-2289 (2003)
9. J. Mathew, H. Foysi, R. Friedrich, A new approach to LES based on explicit filtering. *Int. J. Heat Fluid Flow*, 27 (2006) 594-602
10. J. Sesterhenn, A characteristic-type formulation of the Navier-Stokes equations for high-order upwind schemes. *Computers and Fluids*, 30:37-67 (2001)
11. S. Ghosh, J. Sesterhenn, R. Friedrich, DNS and LES of compressible turbulent pipe flow with isothermal wall.- In: *Direct and Large-Eddy Simulation VI*, Springer, (2006), pp. 721-728
12. T.J. Poinsot and S.K. Lele, Boundary conditions for direct simulations of compressible viscous flows, *J. Comp. Phys.*, 101:104-129 (1992)
13. H. Foysi, S. Sarkar, R. Friedrich, Compressibility effects and turbulence scalings in supersonic channel flow. *J. Fluid Mech.* 509, 207-216 (2004)
14. S. Ghosh, J. Sesterhenn, R. Friedrich, Supersonic turbulent flow in axisymmetric nozzles and diffusers.- In: *Proceedings of the 5th Int. Symp. on Turbulence and Shear Flow Phenomena*, Munich, August 27-29, 2007
15. A. Ern, V. Giovangigli, Fast and accurate multi-component transport property evaluation. *J. Comp. Phys.* 120,105-116 (1995)
16. I. Mahle, J.-P. Mellado, J. Sesterhenn, R. Friedrich, LES of turbulent low Mach number shear layers with active scalars using explicit filtering. - To appear in: *Progress in Turbulence* II, M. Oberlack et al. (Eds.), Springer, 2007.
17. I. Mahle, DNS and LES of inert and reacting compressible turbulent shear layers. *Ph.D. thesis*, TU Muenchen, 2007.

Coriolis induced compressibility effects in rotating shear layers

Bernard J. Geurts[1,2], Darryl D. Holm[3,4], and Arkadiusz K. Kuczaj[1]

[1] Multiscale Modeling and Simulation, NACM, J.M. Burgers Center, Faculty EEMCS, University of Twente, P.O. Box 217, 7500 AE Enschede, the Netherlands: b.j.geurts@utwente.nl
[2] Anisotropic Turbulence, Fluid Dynamics Laboratory, Department of Applied Physics, Eindhoven University of Technology, P.O. Box 513, 5300 MB Eindhoven, the Netherlands
[3] Mathematics Department, Imperial College London, SW7 2AZ, London, UK
[4] CCS2, Los Alamos National Laboratory, Los Alamos, NM 87545, USA

Rotation about a fixed axis introduces a competition between two- and three-dimensional tendencies in a turbulent flow [1, 2, 3]. At strong rotation rates, this competition expresses itself, e.g., by suppressing fluid motion along the axis of rotation. This yields a reduced mixing-efficiency in a rotating frame of reference that is particularly relevant in the context of atmospheric and oceanic flows. The exchange of gases between atmosphere and oceans, the transport of heat and the spreading of pollutants or large-scale plankton-populations, are all significantly affected by rotation in large-scale environmental flows.

The canonical problem of flow in a horizontal temporal mixing layer, subjected to rotation about a vertical axis is investigated with the use of direct numerical simulation. The compressible mixing layer is considered, composed of two counter-flowing horizontal slabs of fluid [4]. Periodic boundary conditions are adopted in the horizontal directions and free-slip conditions apply in the vertical far-field. High order finite volume discretization and explicit, compact storage, four-stage Runge-Kutta time-stepping are used.

Qualitatively, in the non-rotating case the flow shows vigorous mixing in the vertical direction leading to a complex three-dimensional flow. A snapshot of the vertical vorticity component in the turbulent regime is shown in Fig. 1(a). Adding rotation to this flow may completely alter the transitional and developed flow. As a result of rotation, the flow breaks up into vertically oriented cells that collectively undergo an oscillatory motion (cf. Fig. 1(b)). The frequency of this oscillation increases with rotation rate.

The effect of rotation was investigated at a Reynolds number $Re = 50$, based on the initial momentum-thickness of the shear layer, and a low convective Mach number of $M = 0.2$. The strength of the rotation is characterized

Fig. 1. Snapshot of the vertical component of vorticity in the developed regime at $t = 90$ for (a) the non-rotating case $Ro = \infty$ and (b) rotating flow at $Ro = 10$. Positive and negative values are shown in different colors.

by the Rossby number $Ro = u_r/(2\Omega_r \ell_r)$ where u_r, Ω_r and ℓ_r are reference velocity, rotation-rate and length-scale used in the non-dimensionalization. At high Rossby numbers Ro, i.e., low rotation rates, the flow develops very similarly to incompressible flow. However, at sufficiently low Rossby numbers $Ro < 1$, rotation was found to induce significant compressibility effects. These give rise to rapidly varying small scale flow-features. By comparing results obtained at a variety of time-steps, we verified that the explicit Runge-Kutta time-stepping properly captures these rapid variations. The spatial resolution was found to be adequate at 256^3 grid-cells for the purpose of studying the decay of the kinetic energy. Results at 256^3 grid-cells were compared with lower resolutions at 192^3 and 128^3, confirming the accuracy.

Rotation has an indirect influence on the decay-rate of kinetic energy $E = \langle \mathbf{u} \cdot \mathbf{u} \rangle / 2$. Here $\langle \cdot \rangle$ denotes averaging over the flow-domain. The evolution of the kinetic energy may be written as

$$\frac{dE}{dt} = D - W \quad ; \quad D = \langle p\nabla \cdot \mathbf{u} \rangle \quad ; \quad W = \langle \mathbf{S} : \mathbf{S}/(2Re) \rangle \tag{1}$$

where p denotes pressure, $\mathbf{u}$ the velocity field, $\mathbf{S}$ the rate of strain tensor and Re the Reynolds number. The viscous dissipation is represented in W while the compressible pressure/velocity-divergence is given by D. In (1) there is no direct effect of rotation. Rather, rotation effects may be distinguished in the dynamics of D and W. By investigating the various contributions to dD/dt and dW/dt we may identify the basic mechanisms that govern the kinetic energy dynamics, particularly the role of the Coriolis forces. The dominant mechanisms at various Ro in the dynamics of the viscous dissipation and the pressure/velocity-divergence can be extracted from the DNS-data.

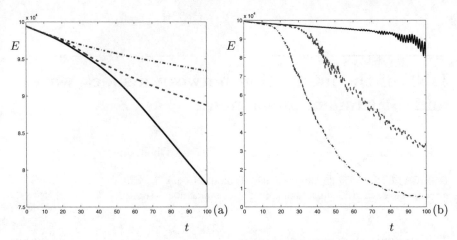

Fig. 2. Decay of kinetic energy at low (a) and high (b) rotation rates. In (a) $Ro = \infty$ (solid), $Ro = 10$ (dashed), $Ro = 5$ (dash-dotted); in (b) $Ro = 1$ (solid), $Ro = 0.5$ (dashed) and $Ro = 0.2$ (dash-dotted).

The marked differences in the decay of kinetic energy in slowly and in rapidly rotating shear layers are shown in Fig. 2. At high Ro an increase in the rotation-rate yields a decrease in the decay rate (cf. Fig. 2(a)). This reduction in decay-rate and the strict monotonicity of dE/dt are also found in incompressible flow. For sufficiently rapid rotation (in this flow $Ro \lesssim 1$; Fig. 2(b)) explicit compressibility effects become dominant. The behavior of the decay-rate is completely reversed and an increase in rotation-rate yields a strong increase in the 'average decay-rate'. Moreover, the kinetic energy is no longer a monotonously decaying function of time. Next to the strictly dissipative viscous contributions, the pressure/velocity-divergence correlations are found to become dynamically significant at low Ro. The observed rapid fluctuations are a purely compressible effect. A complete analysis is topic of current investigations.

A.K.K. gratefully acknowledges financial support from the Dutch Foundation for Fundamental Research of Matter (FOM) and from the Turbulence Working Group at Los Alamos National Laboratory - USA.

References

1. Mansour, NN, Cambon, C, Speziale, CG (1992) in *Studies in Turbulence* (ed. Gatski, TB, Sarkar, S, Speziale, CG). Springer.
2. Hossain, M (1994) Phys. Fluids 6:1077.
3. Smith, LM, Waleffe, F (1999) Phys. Fluids. 6:1608.
4. Geurts, BJ, Holm, DD (2006) J. of Turbulence. 7 1 - 33.

DNS of the interaction between a shock wave and a turbulent shear flow

Matthieu Crespo, Stéphane Jamme, and Patrick Chassaing

ENSICA, 1 Place Emile Blouin 31056 Toulouse Cedex 5, France
mcrespo@ensica.fr, jamme@ensica.fr, chassain@ensica.fr

1 Numerical considerations

We solve the three-dimensional Navier-Stokes equations in non-dimensional conservative form using a finite difference approach. The inviscid part is resolved using a fifth-order Weighted Essentially Non-Oscillatory scheme (WENO). Viscous terms are computed using a sixth-order accurate compact scheme. A third-order Runge Kutta algorithm is used to advance in time.

Equations are solved on a cubic domain of size 2π in the three directions and a grid of $176 \times 128 \times 128$ points is used. The mean flow is aligned with x. Periodic conditions are specified in the z direction, and non-reflecting boundary conditions with a sponge layer are used for the top and bottom boundaries along y as well as for the outflow where the flow is subsonic. At each time step, velocity, pressure, temperature, and density fields are specified at the inflow. These fields are superpositions of a supersonic mean flow and turbulent fluctuations in velocity, pressure, temperature, and density. The mean velocity at the inflow varies linearly across streamlines while the mean pressure is uniform. The mean temperature and density vary such as the mean Mach number is uniform : $\overline{U}_1(y) = U_0 + S(y - y_{\min})$, $\overline{V}_1 = \overline{W}_1 = 0$, $\overline{P}_1(y) = 1/(\gamma M_r^2)$, $\overline{T}_1(y) = M_r^2 U_1^2/M_1^2$, where the overbar denotes the conventional Reynolds average and the subscript 1 indicates the upstream state. A fluctuating field is then superposed onto the mean upstream flow and advected through the shock. This field comes from a preliminary calculation. The anisotropy of the turbulent velocity field used in the inflow plane is either of axisymmetric type or typical of a turbulent shear flow.

2 Results

Several simulations were conducted with the following values of the reference parameters: $Re_r = \frac{\rho_r^* u_r^* L_r^*}{\mu_r^*} = 94$, $M_r = \frac{u_r^*}{c_r^*} = 0.1$, $Pr = 0.7$, where f_r^*

refers to a dimensional reference variable. The mean Mach number is fixed to $M_1 = 1.5$, and the turbulence parameters in the inflow plane are the following :
$Re_\lambda = Re_r \frac{\lambda u_{1rms}}{\bar{\nu}} = 47$ and $M_t = \frac{q}{\bar{c}} = \frac{\sqrt{\widetilde{u_i' u_i'}}}{\bar{c}} = 0.173$. Table 1 summarizes the characteristics of the different runs. They differ by the nature of the mean flow (sheared or not), by the anisotropy of the upstream turbulent flow and by the amount of $\widetilde{u''^2}$ immediately upstream of the shock wave.

Table 1. Characteristics of the different runs (i = inflow ; b.s. = just before shock)

Run	S	$\left(\frac{q^2}{2}\right)_i$	$(\widetilde{u''^2})_{b.s.}$	$\left(\frac{\widetilde{u''^2}}{q^2}\right)_{b.s.}$	$\left(\frac{\widetilde{v''^2}}{q^2}\right)_{b.s.}$	$\left(\frac{\widetilde{w''^2}}{q^2}\right)_{b.s.}$	$\left(\frac{\widetilde{u''v''}}{q^2}\right)_{b.s.}$
RunSI	1.5	1.5	1.04	0.42	0.28	0.31	-0.14
RunSA1	1.5	1.5	1.10	0.44	0.27	0.30	-0.12
RunSA2	1.5	1.5	1.11	0.40	0.29	0.31	-0.17
RunI	0	2	1.00	0.35	0.33	0.32	0.005
RunA1	0	1.7	1.04	0.43	0.28	0.30	0.007

Previous works lead to the conclusion that the amplification of the kinetic energy behind the shock wave is strongly dependent of the upstream anisotropic state, and that it is clearly determined by the amount of the longitudinal normal Reynolds stress $\widetilde{u''^2}$ upstream of the shock (see e.g. Jamme et al. [1]). The mean flow was uniform without shear stress is these studies.

In the present work, we first compare three runs (RunSI, RunSA1 and RunSA2) where a mean shear has been introduced. The anisotropy of the turbulence is slightly different just before the shock for these three cases. The near-field amplification of $q^2/2$ behind the shock wave is found to depend on the amount of the correlation $\widetilde{u''T''}$ immediately upstream of the shock. This correlation is positive in the three cases, but its value is not the same. The more $\widetilde{u''T''}$ is high upstream, the less $q^2/2$ is amplified behind the shock.

In order to get rid of the effect linked to the amount of $\widetilde{u''^2}$, and trying to isolate the influence of the nature of the anisotropy of the incident turbulent flow itself, we conducted two more runs (RunI and RunA1) in which the amount of $\widetilde{u''^2}$ is the same as in RunSI just before the shock, but not the values of the other components of the Reynolds stress tensor. Figure 1 shows that the axisymmetric case displays a greater near-field amplification of $\widetilde{u''^2}$ than the isotropic case, whereas the opposite is true for the sheared case. Concerning the near-field behaviour of $q^2/2$, both the axisymmetric and sheared cases show a greater amplification than the isotropic case. Mahesh et al. [2] observed a decrease of $q^2/2$ across a $M_1 = 1.2$ shock for a sheared case, and they attributed this trend to the fact that $\widetilde{u''T''} > 0$ before the shock, which is known to inhibit the amplification of the kinetic energy. In the present case (RunSI), we have $\widetilde{u''T''} > 0$ upstream, but $q^2/2$ is still more amplified in the

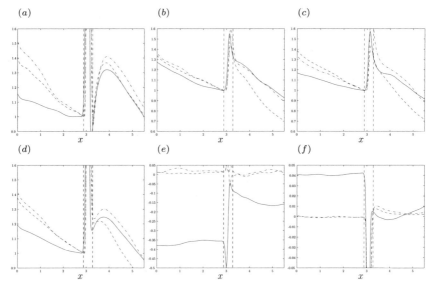

Fig. 1. Streamwise evolution of different turbulent statistics across the shock: (a) $\widetilde{u''^2}$; (b) $\widetilde{v''^2}$; (c) $\widetilde{w''^2}$; (d) $q^2/2$; (e) $\widetilde{u''v''}$; (f) $\widetilde{u''T''}$. In (a), (b), (c), (d), curves are normalized by their value immediately upstream of the shock wave. (——) RunSI ; (– – –) RunI ; (– · –) RunA1.

near field compared to the isotropic situation. This difference with Mahesh *et al.* [2] may be a consequence of the shock strength ($M_1 = 1.5$ in our case instead of $M_1 = 1.2$).

The behaviour of $\widetilde{u''v''}$ is found to be same as the one observed by Mahesh *et al.* [2] : we notice a decrease of the magnitude of $\widetilde{u''v''}$ across the shock wave.

Moreover, a clear influence of the shear stress can be seen on the streamwise component of the vorticity (not shown here). An increase of $\overline{\omega'^2_1}$ in the near field behind the shock is indeed observed for the three cases, but this tendency is much more pronounced for the sheared case. The vortex stretching by the mean flow is found to be responsible for this increase of $\overline{\omega'^2_1}$, which means that this term is enhanced in the sheared case.

References

1. S. Jamme, M. Crespo, and P. Chassaing. Direct numerical simulation of the interaction between a shock wave and anisotropic turbulence. In *35th AIAA Fluid Dynamics Conference and Exhibit*, Toronto, Canada, AIAA Paper 2005-4886, 2005.
2. K. Mahesh, P. Moin, and S.K. Lele. The interaction of a shock wave with a turbulent shear flow. Technical Report TF-69, June 1996.

Effects of compressibility and heat release on the turbulent mixing layer boundaries

Inga Mahle[1], Joseph Mathew[2] and Rainer Friedrich[1]

[1] Fachgebiet Strömungsmechanik, Technische Universität München, 85748 Garching, Germany inga.mahle@aer.mw.tum.de
[2] Department of Aerospace Engineering, Indian Institute of Science, Bangalore 560012, India

1 Introduction

Turbulent shear layers grow as fluid in the irrotational, nonturbulent regions acquires vorticity and becomes a part of the turbulent region. Over the past few decades this entrainment process had come to be described, most often, as comprising an engulfment of large packets of irrotational fluid from the surrounding non-turbulent region followed by disintegration and acquisition of vorticity well within the turbulent region. Recently, the DNS study of round jets by Mathew & Basu [1] revealed that the entrainment process may instead, more frequently, be a fast, small-scale process occurring very close to the turbulent-nonturbulent interface. Subsequently, some aspects of this study have been confirmed with PIV measurements in a laboratory jet which showed that engulfed fluid volume as well as its growth rate are very small compared to that of the turbulent regions and its growth rate, respectively [2]. Mathew & Basu [1] had argued that large scale quantities would suffice to predict entrainment rates even when the process is small scale if there is an equilibrium across scales—a situation obtained in the usual self-preserving canonical shear layers, and which has suggested the engulfment, or, the large-scale process view. They argued further that the entrainment flux estimated on large scales could equal that summed from the actual small scale processes when the interface is fractal. Then, the smaller velocities of the small-scale entraining motions occur over the larger area of the convoluted interface to provide the same entrainment flux. It is then of interest to examine geometrical properties of turbulent-nonturbulent interfaces of flows with changes in entrainment rates. In this study, we investigate compressible plane mixing layers with and without heat release because it is well-known that entrainment rates are reduced with increasing compressibility and heat release.

2 Preparation and analysis of DNS data

Direct numerical simulations (DNS) of compressible mixing layers were performed. The convective Mach number $Ma_c = \Delta u / (c_1 + c_2)$ is based on the velocity difference between the streams Δu and their sonic speeds c_1 and c_2. Flows at three values of Ma_c (0.15, 0.7, 1.1), with and without reaction (and heat release) were studied. The upper stream is pure oxygen and the lower stream is pure nitrogen for the inert cases. Initially, the temperature and pressure were uniform. For the reacting cases the single step reaction between hydrogen and oxygen carried in nitrogen streams was taken. Solutions are periodic in the streamwise and spanwise directions. The code is based on a pressure-velocity-entropy formulation and has been used for several previous studies [3].

After initial transients, there is a useful interval of self-similar evolution when the mixing layer growth rate is roughly constant. All analyses are over this self-similar stage. The Reynolds number Re_ω, defined with the vorticity thickness, δ_ω, exceeds 10000 in all cases. During the initial evolution at the lower Mach number, spanwise vortices undergo pairing, but later there is merely a continuous coalescence into a single large spanwise roller. At the higher Mach number, the mixing layer has little 2-dimensional organization until the eventual formation of the single roller.

Structural changes can be observed in the scalar fields shown in Figures 1 at instants during the self-similar stage of each simulation. For both inert and reacting flows, the dominant scale of the indentations of the mixing layer boundary increases with Ma_c, but the bending of the interface on the largest scales is greater at the smallest Mach number. In each case, with heat release, dominant length scales are still larger. Figure 2 shows a surface at the turbu-

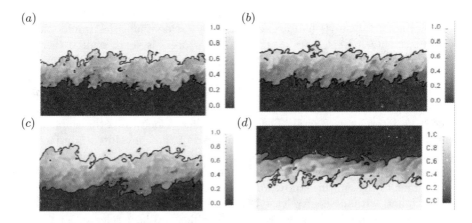

Fig. 1. Scalar fields of non-reacting and reacting flows. (*a*) $Ma_c = 0.15$, (*b*) $Ma_c = 0.7$, (*c*) $Ma_c = 1.1$, (*d*) $Ma_c = 1.1$ (reacting).

lence interface defined by iso-level of scalar concentration (inert: oxygen mass fraction $Y_{O2} = 0.95$, reacting: mixture fraction $z = 0.05$). Close examination reveals that the surface has small, evenly distributed protrusions into the non-turbulent region at lower Mach numbers (0.15, 0.7). Simulations at $Ma_c = 1.1$ indicate that these outward protrusions are thicker with a few smaller scale, deeper protrusions into the turbulent region. The dominant scales are significantly larger with heat release. These changes in dominant interface scales are consistent with the reduced entrainment at higher Mach numbers when the process is small scale and occurs at the interface.

A quantitative characterisation of the turbulence interface is obtained by measuring its fractal dimension. We find the dimension D_2 of sections such as those in Fig. 1 and obtain the surface dimension $D = D_2 + 1$. Each chosen section is covered by squares of side $r_j = j\Delta x$ $(j = 1, 2, \ldots)$ and the number of squares $N(r)$ which include a segment of the interface contour are counted. The dimension is $D_2 = \log N/\log(1/r)$. The values of D are 2.35, 2.31 and 2.37 (inert flow), 2.30, 2.20 and 2.26 (reacting flow) at $Ma_c = 0.15, 0.7$ and 1.1, respectively, which agree well with measurements in other flows [4]. For both inert and reacting flows, the dimension falls when Ma_c increases from 0.15 to 0.7 and then rises again. The fall is consistent with a decrease in entrainment and shear layer growth rate. The increase at 1.1 may be due to rapid changes in scale in the infrequent indentations, but the generally smoother boundary is nonetheless consistent with the observed further decrease in growth rate and entrainment by processes at the turbulence boundaries. This smoothening of the boundary could also account for the observed tapering off of the fall in growth rate at $Ma_c \approx 1.1$.

References

1. J. Mathew & A. J. Basu, *Phys. Fluids*, **14(7)**, 2065 (2002)
2. J. Westerweel, C. Fukishima, J. M. Pedersen & J. C. R. Hunt, *Phys. Rev. Lett.*, **95**, 174501 (2005)
3. J. Sesterhenn, *Computers & Fluids*, **30**, 37 (2001)
4. R. R. Prasad & K. R. Sreenivasan, *Phys. Fluids A*, **2(5)**, 792 (1990)

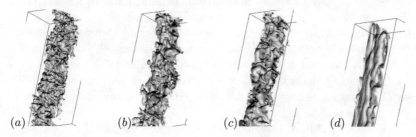

Fig. 2. Turbulence boundary defined by concentration. (*a*) $Ma_c = 0.15$, (*b*) $Ma_c = 1.1$. (*c*) $Ma_c = 0.15$, reacting. (*d*) $Ma_c = 1.1$, reacting.

Dual-Time PIV Investigation of the Sound Producing region of the Controlled and Uncontrolled High-Speed Jet

Jeremy Pinier and Mark Glauser

Syracuse University, Syracuse, NY 13244, USA jpinier@gmail.com

With the aim of being capable of controlling the sound created by high-speed jets or engine exhausts, a preliminary experiment, using a near-field azimuthal array of 15 pressure sensors and a far-field array of 6 microphones, pointed out the region of highest sound production in the turbulent flow where a dual-time particle image velocimetry (DT-PIV) investigation is now being carried out. This measurement technique enables access to time-derivatives of the velocity field which is a quantity of great interest for developing prediction capabilities crucial for closed-loop flow control. Jointly, a flow control device using synthetic jets is developed and PIV measurements are taken with and without the control to assess the effectiveness of such devices for aeroacoustic flow control.

1 Experimental setup and velocity fields

The experiment is carried out in the Syracuse University Skytop Anechoic Chamber which is equipped with a high-speed jet with a nozzle exit diameter 5.08 cm capable of transonic speeds. In this experiment the jet is run at a Mach number of 0.6 for reasons of running time capability. The Reynolds number of this highly turbulent flow is 690,000. The use of multiple PIV systems was proposed and shown by Kähler and Kompenhans [2] to develop a tool capable of measuring time derivatives of the velocity (i.e. acceleration) as well as spatial derivatives (velocity gradient tensors and vorticity vectors). Two stereoscopic PIV systems are used here to measure all 3 components of velocity in an identical cross-flow plane perpendicular to the jet axis. Using finite differencing, acceleration is computed for each pair of velocity field obtained. The process is reiterated for downstream positions ranging from 3 to 10 jet diameters. The experiment having taken more time than initially planned because of technical challenges the setup presents and the fine tuning as described earlier, the data has just been gathered and initial statistical analysis of the data from only one downstream position (x/D=8) is shown in this paper. Extensive results will be shown during the actual presentation. Figure 1 shows a color contour of the mean axial velocity (u_x) and the sum of the variances of all 3 components of velocity with a typical maximum average level of fluctuations in the shear layer. The smoothness of the data is a result of the large number of samples used to compute the statistics. Figure 2 shows instantaneous velocity fields taken by both systems with a delay of Δt=25.3 μs. The resulting out-of-plane component of the acceleration is shown in figure 3 where local maxima are noticeable in the core region of the jet. These instantaneous events could be the source of strong shear and could perhaps be effective at radiating sound, which is the subject of future work with this database.

2 Flow control devices for noise reduction

The aeroacoustic process that creates sound being subtle, a similarly subtle flow control strategy needs to be sought. The jet flow needs to be controlled without being denatured, the main feature of which would be diminished: thrust. Jet flow control devices have therefore been designed in this perspective. A circular array of synthetic jet actuators with exit velocities on the order of 50 m/s was designed and integrated to the jet nozzle. The synthetic jets are slightly inclined towards the jet to only manipulate the developing shear layer as it is convected downstream. The aim is, through closed-loop control, using downstream real time information, to perform the subtle changes in the near-field pressure around the jet that would inhibit sound propagation. Hall et al. [1] found that azimuthal mode 0 computed from near-field pressure was the mode that correlated best with the far-field sound and that at the same time the presence of azimuthal mode 1 inhibited the propagation of sound to the far-field. Using this information, a closed-loop approach will be devised using the synthetic jet actuators. Jointly, PIV measurements will be made in the region of highest sound production which was found to be from x/D=6 to x/D=10 to study, if noticeable, the effect of the flow control on the overall flow.

3 Dynamical system for flow control

Simple closed-loop control algorithms based on the proper orthogonal decomposition (POD) have been implemented and have shown successful [3]. In order to develop more sophisticated control algorithms, a low-order model of the jet flow is needed. In this aim, the experimental data collected is used to "train" a low order dynamical system, the form of which is taken from the Navier-Stokes equations. To identify the linear and quadratic coefficients, the moments method is used and an evaluation of the derivative of the POD expansion coefficient is needed. This double PIV experiment was performed to be able to capture directly time derivatives of the flow quantities.

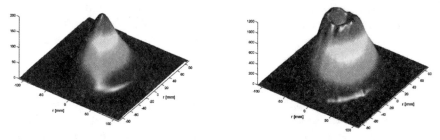

Fig. 1. Mean axial velocity u_x (left) and sum of the variances of all 3 components of velocity (right), x/D=8.

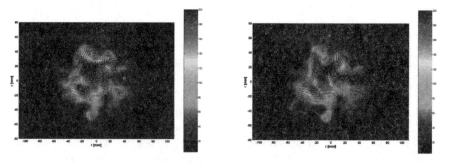

Fig. 2. Instantaneous velocity fields from systems 1 (top) and 2 (bottom), out-of-plane velocity in color contours, Δt=25μs, M=0.6, x/D=8.

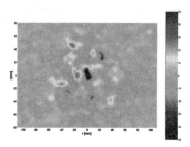

Fig. 3. Instantaneous out-of-plane acceleration field resulting from the finite difference of the fields shown below, M=0.6, x/D=8, t=265.

References

1. Hall, J., J. Pinier, A. Hall, and M. Glauser, 'Two-point Correlations of the Near and Far-field Pressure in a Transonic Jet'.
2. Kaehler, C. J. and J. Kompenhans: 2000, 'Fundamentals of multiple plane stereo particle image velocimetry'. *Exp. Fluids (Suppl)* **29**(7), S70–S77.
3. Pinier, J. T., A. M. Ausseur, M. N. Glauser, and H. Higuchi: 2007, 'Proportional closed-loop control of flow separation'. *AIAA Journal* **45**(1), 181 – 190.

Investigation of the behavior of noise sources in heated jets

Peter Moore[1] and Harmen Slot[1] and Bendiks Jan Boersma[1]

Laboratory for Aero and Hydrodynamics, Delft University of Technology, Leeghwaterstraat 21, 2628 CA Delft, The Netherlands. p.moore@wbmt.tudelft.nl

Abstract

In this paper we show the results of three simulated jets of Mach number 0.8, Reynolds number $4,000$ where the jet inflow to environment temperature ratio is 0.85, 1.0 and 1.1 respectively. We demonstrate that the obtained noise levels are in agreement with published experimental data. We then compare the sound sources of each of the jets with the framework of Lighthill's acoustic analogy. In particular we compare the relative strengths of quadrupole and entropy sources for each case.

1 Introduction

In the high Reynolds number experiments of [8] it has been found that the effect of heating a jet above room temperature is to increase the noise levels for Mach numbers lower than 0.7 and to decrease the noise levels for higher Mach numbers. Cooling gives the reverse temperature dependence. This effect has also been investigated numerically with LES [1], with some, but not complete success. One conclusion of [1] was that heating and cooling alters the noise emissions of the small scales of the jet, which are unresolved by LES. If this effect is also observed in low Reynolds number jets then this problem is a good candidate for investigation with Direct Numerical Simulation (DNS).

2 Simulation

The numerical method used for the DNS has been previously described [2]. A validation of the far field extension method is described in [6]. A description of the three completed simulations is given in [5]. In the DNS of the jets, both the turbulent flow field and acoustic field are determined with the acoustic far field being calculated through the use of the Ffowcs Williams Hawkings analogy [9]. Figure 1 represents the simulated jet with temperature ratio 1.0 through iso-contours of vorticity. The inflow of the jet is initially laminar (middle left of the figure), until a short distance downstream where the jet quickly transitions to turbulent flow. The jet then exhibits expected linear growth. The 2-d slices of acoustic pressure in the figure show sound waves emanating from the point of transition to turbulence.

3 Results

Figure 2 gives overall sound pressure levels of the jets. The neutral jet is compared to the experimental data of Stromberg [7] at a radius of 30 jet diameters (which had a Reynolds number $3,600$, $Ma = 0.83$). At the radius of 60 jet diameters the far field sound pressure levels of each simulated jet are shown. Even at this low Reynolds number, the hot jet exhibits lower sound pressure levels than the neutral jet, while the cold jet has higher sound pressure levels, except in the direction of peak radiation where the mach numbers coincide.

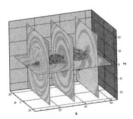

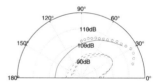

Fig. 1. Iso contours of positive and negative vorticity (red/blue) with four cross-sections of the acoustic near field.

Fig. 2. Directivity of SPL for hot (red dashed), cold (blue dashed) and neutral (black dashed) jets and comparison to Stromberg data (blue circles).

$$ p_{ac} = \frac{1}{4\pi c_0^2} \int \frac{(y - y_0)^2}{r^3} \left(\frac{\partial^2 (pv^2)}{\partial t^2} + \frac{\partial^2 (p - c_0^2 \rho)}{\partial t^2} \right) dV \qquad (1) $$

Equation (1) is the integrated form of the portion of Lighthill's Acoustic Analogy that radiates in the y direction to the far field. In figure 3 we plot the instantaneous radial variation of the portion of the quadrupole term (first term inside integral of equation (1)) averaged over all angles at that radius and in figure 4 the same for the entropy term (second term inside integral), at a downstream point. The cold jet has both stronger quadrupole and stronger entropic sources while the hot jet has weaker.

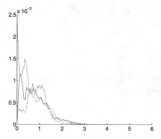

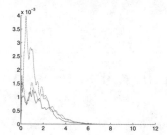

Fig. 3. Radial variation of average of Lighthill quadrupole term. Blue is cold jet, black is neutral and red is hot jet.

Fig. 4. Radial variation of average of Lighthill entropy term. Blue is cold jet, black is neutral and red is hot jet.

4 Conclusion

The sound pressure level temperature dependence from the high Reynolds number experiments of Tanna [8] is shown to occur in our simulations at Reynolds number of 4,000. Investigation of the noise sources reveals that a likely cause of the temperature dependence of sound pressure levels is the different jet spreading rates of each of the jets.

References

1. D.J. Bodony and S.K. Lele. Jet noise prediction of cold and hot subsonic jets using large-eddy simulation. *AIAA Paper*, 3022, 2004.
2. B.J. Boersma. A staggered compact finite difference approach for the compressible navier-stokes equations. *J. of Comp. Phys*, 208:675–690, 2005.
3. P. Di Francescantonio. A new boundary integral formulation for the prediction of sound radiation. *J. Sound and Vibration*, 202, No. 4:491–509, 1997.
4. M.J. Lighthill. On sound generated aerodynamically, part 1: General theory. *Proc. Roy. Soc. Lon.*, A 211:564–587, 1952.
5. P. Moore and B. Boersma. Investigation of the noise from cold and heated subsonic jets. *12th AIAA/CEAS Aeroacoustics Conference*, 2500, 2006.
6. P. Moore and B.J. Boersma. Use of surface integral methods in the computation of the acoustic far field of a turbulent jet. *Proceedings of Direct and Large Eddy Simulation 6*, 2005.
7. J.L. Stromberg, D.K. Mclaughlin, and T.R.. Troutt. Flow field and acoustic properties of a mach number 0.9 jet at a low reynolds number. *Journal of Sound and Vibration*, 72:159–176, 1980.
8. H.K. Tanna. An experimental study of jet noise i: Turbulent mixing noise. *Journal of Sound and Vibration*, 50, No. 3:405–428, 1977.
9. J.E.F. Williams and D.L. Hawkings. Sound generated by turbulence and surfaces in arbitrary motion. *Proc Roy. Soc. Lon.*, A264:321–342, 1969.

Dynamics of spheres in turbulent channel flow

P. H. Mortensen[1], H. I. Andersson[1], J. J. J. Gillissen[2] and B. J. Boersma[2]

[1] Department of Energy and Process Engineering, Norwegian University of Science and Technology, 7491 Norway. paal.h.mortensen@ntnu.no, helge.i.andersson@ntnu.no

[2] Laboratory for Aero and Hydrodynamics, TU-Delft, 2628 CA Delft, The Netherlands. j.j.j.gillissen@wbmt.tudelft.nl, b.j.boersma@wbmt.tudelft.nl

1 Introduction and methodology

The dynamics of small inertia particles suspended in turbulent flow fields has been a subject for research through several decades. These so-called dispersed multiphase flows arise in several engineering and environmental situations. Though several experimental and numerical studies have been performed in this field, there are still open questions on the physics of particle dynamics in turbulent flow fields.

In the present paper, the spin dynamics of mono-sized spherical particles suspended in a turbulent channel flow is investigated by direct numerical simulation. It is assumed that the particles are small, and the flow field at the particle positions can be considered locally Stokesian. Further, the particles only feel the effects of the Stokes drag and the hydrodynamic torque [1]. Also, the coupling between the fluid and the particles is one-way, i.e. particles only feel the effect of the flow field. The particle response times, based upon the translational and the rotational motions, equal 30 and 9 in wall units, respectively. The turbulent flow field was simulated until its statistics became steady. Thereafter, 10^6 particles were released randomly in the channel geometry.

2 Results and discussion

It is a well-known fact that there is a net transport of inertia particles towards the wall regions in a turbulent channel flow [2], [3]. Also, a few wall units away from the walls, the particles tend to concentrate in regions of large strain and low enstrophy, i.e. low-speed streaks. These trends are verified in figure 1, which shows the concentration of particles in the wall-normal direction and the distribution of particles in the plane $z^+ \approx 6$, respectively. Figure 2 shows the time development of the density function of the particle spin components.

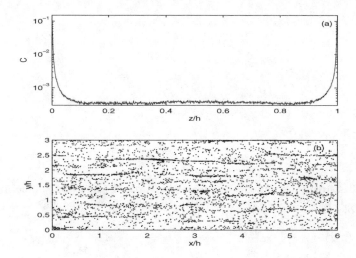

Fig. 1. Instantaneously distribution of particles at $t^+ = 5040$. a) Concentration C of particles versus wall-normal position, b) Particle distribution in plane $z^+ \approx 6$.

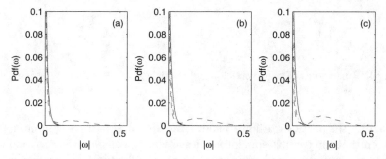

Fig. 2. Time development of particle spin; (—) ω_x, (- -) ω_y, (-.-) ω_z. a) $t^+ = 2889$, b) $t^+ = 3960$, c) $t^+ = 5040$.

It is noted that the particles have probability of larger spin about the spanwise axis (ω_y) than compared to spin about the streamwise (ω_x) and wall-normal (ω_z) axis, respectively. This is due to the mean vorticity of the flow, which is dominating in the spanwise direction. However, as time proceeds, it can be seen that the probability of large particle spin about the streamwise and spanwise axis increases, while the opposite is true for the spin about the wall-normal axis. This is probably due to the accumulation of particles in the viscous sublayer. In this region, the streamwise and spanwise fluid vorticity is considerably larger than its wall-normal component. This is confirmed by figure 3, which shows the instantaneous contours of streamwise and wall-normal vorticity in the plane $z^+ = 0.5$. This near-wall vorticity difference is

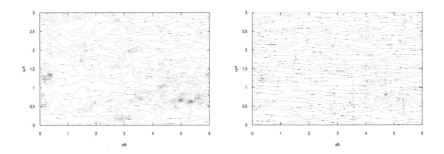

Fig. 3. Instantaneous contours of streamwise vorticity (left) and wall-normal vorticity (right) at $t^+ = 5040$. Contour levels are the same in both figures.

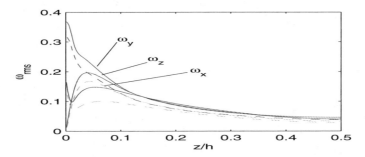

Fig. 4. Rms values of fluid vorticity and particle spin; (—) fluid, (- -) particles.

also felt by the particles. Figure 4 shows the fluid and particle spin intensities. It is seen that the particle spin intensities are lagging behind the corresponding fluid vorticity components. Close to the wall, the fluctuations in spanwise and wall-normal spin of the particles exceed the wall-normal spin component, thus supporting the increasing probability of larger wall-normal and streamwise spin with increasing concentration of particles in the viscous sublayer.

References

1. H. Brenner: Chem. Eng. Sci. **18**, 1 (1963)
2. J.K. Eaton, J.R. Fessler: Int. J. Multiphase Flow. **20**, 169 (1994)
3. C. Marchioli, A. Soldati: J. Fluid Mech. **468**, 283 (2002)

Stereoscopic PIV measurements in electromagnetically forced rotating turbulence

L.J.A. van Bokhoven, H.J.H. Clercx, G.J.F. van Heijst, and R.R. Trieling

J.M. Burgerscentre and Fluid Dynamics Laboratory,
Department of Applied Physics, Eindhoven University of Technology,
P.O. Box 513, 5600 MB Eindhoven, The Netherlands
Contact address: l.j.a.v.bokhoven@tue.nl

Rotating turbulence plays an important role in fields as diverse as engineering (e.g. turbomachinery), geophysics, and astrophysics. In rotating flows, the dynamics and structure of turbulence are affected by the Coriolis force, which tends to 'two-dimensionalise' the flow.

When a Coriolis force is suddenly imposed on initially isotropic turbulence, the following three main effects are shown. First, the energy cascade is partly inhibited, which is linked to a strongly reduced dissipation rate (such an effect can be mimicked by an empirical modification of the dissipation equation [2]). Second, because the dispersion relations for inertial waves are anisotropic, the initial isotropy is broken. This breaking of isotropy is reflected by an incomplete transition from 3D to 2D structure. Third, an asymmetry appears between cyclonic and anticyclonic fluctuating vertical (along the rotation axis) vorticity.

The effects mentioned above are intimately connected and result from both linear and nonlinear effects which interplay in a subtle way to drive the dynamics of rotating turbulence. Here, we focus on the third effect. As firstly pointed out by Bartello et al. [1], the dominance of cyclonic vorticity can be quantified by the vertical vorticity skewness $S_{\omega_3} \equiv \langle \omega_3^3 \rangle / \langle \omega_3^2 \rangle^{3/2}$, with the vorticity $\omega_i \equiv \epsilon_{ijk} \partial_j u_k$ and the brackets $\langle \cdot \rangle$ denoting spatial averaging, since the third order vorticity correlation $\langle \omega_3^3 \rangle$ can distinguish by its sign cyclonic prevalence ($\omega_3 > 0$) from anticyclonic prevalence ($\omega_3 < 0$). Bartello et al. [1] found a clear growth of S_{ω_3} using Large Eddy Simulation (LES) with hyperviscosity. This growth of S_{ω_3} has been confirmed by recent laboratory experiments performed by Morize et al. [3]. Furthermore, Van Bokhoven et al. [4] have found by means of Direct Numerical Simulation (DNS) that a lower initial Taylor Reynolds number and/or lower initial Rossby number — both implying more linearity — results in lower final values of S_{ω_3}.

In an attempt to extend the studies addressing the asymmetry between cyclonic and anticyclonic vorticity, a novel laboratory experiment has been set up (Fig. 1). In a square fluid container (dimensions $L_x = L_y = 500$ mm,

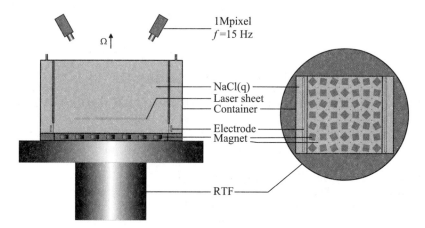

Fig. 1. Schematic drawing of the experimental setup.

$L_z = 300$ mm) subjected to rotation, highly three-dimensional turbulence is generated using electromagnetic forcing. This non-intrusive forcing allows one to stir only the bottom part of the bulk fluid, with typical Taylor-based Reynolds numbers Re_λ of the order of 150–200. (The Taylor-based Reynolds number is here defined as $Re_\lambda \equiv \langle u_i^2 \rangle^{1/2} \lambda / \nu$, with $\lambda \equiv \sqrt{15 \langle u_i^2 \rangle / W^{3/2}}$ the Taylor microscale, ν the kinematic viscosity, and $W \equiv \langle \omega_i^2 \rangle / 2$ the global enstrophy.) Consequently, the turbulence will decay with increasing vertical height H, where H is measured perpendicularly from the bottom of the container, and rotation will become more and more important.

The decay with vertical height is thus to some extent analogous to the decay in time in the above-mentioned numerical and laboratory experiments. To quantify the vertical decay of the turbulence a triangulation-based stereoscopic PIV technique [5] is applied to a horizontal plane at a vertical height H. This experimental procedure is illustrated in the left part of Fig. 1. The result is a three-component velocity field at the corresponding measurement plane. In what follows, we demonstrate how the stereoscopic PIV technique can be used to determine important characteristics of the turbulence at a certain height H.

To investigate the reproducibility, the stationarity, and the isotropy of the turbulence, we have performed two long-time measurements of the horizontal velocity field for identical conditions, viz. $H = 0.18 L_z$, a constant forcing from time $t = 0$ s, and no background rotation. These two measurements are denoted by the Roman numerals I and II, respectively. Figure 2(a) shows the vertical root-mean-square fluctuating velocity $\langle u_3 \rangle$ as a function of the time t for measurements I and II. Clearly, under the given circumstances the turbulent flow is highly reproducible and a stationary state is already established after a very short initialisation period (typically less than 15 s in this case). We have observed similar behaviour for $\langle u_1 \rangle$ and $\langle u_2 \rangle$ (not shown).

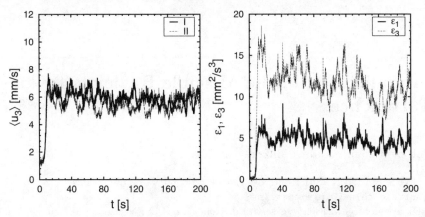

Fig. 2. (a) Time evolution of the vertical root-mean-square velocity $\langle u_3 \rangle$ for two cases with identical conditions. (b) Time evolution of the the dissipation rates ϵ_1 and ϵ_3 for the horizontal and vertical direction, respectively, for case I.

It thus seems that the turbulence at $H = 0.16 L_z$ is isotropic at the large scales. Comparison of the horizontal and vertical dissipation rates however, reveals that these rates differ almost by a factor of two [see Figure 2(b)], which may indicate anisotropy at the smallest scales. The increased vertical dissipation reflects the vertical decay that is inherent to the present experimental setup where the forcing is restricted to the bottom part of the fluid.

Our ultimate goal is to characterise the turbulent decay with height for various forcing rates and rotation rates in terms of global quantities, such as the kinetic energy, the global enstrophy, the dissipation rate, and the vertical vorticity skewness S_{ω_3}, but also in terms of spatial in-plane velocity correlation functions, temporal correlation functions, and probability distribution functions of both the velocity components and the vertical vorticity.

The authors wish to thank the Foundation for Fundamental Research of Matter (Stichting voor Fundamenteel Onderzoek der Materie, FOM) for financial support.

References

1. Bartello P, Métais O, Lesieur M (1994), J. Fluid Mech. 273:1–29.
2. Cambon C, Mansour NN, Godeferd FS (1997), J. Fluid Mech. 337:303–332.
3. Morize C, Moisy F, Rabaud M (2005), Phys. Fluids 17:095105.
4. Van Bokhoven LJA, Cambon C, Liechtenstein L, Godeferd FS, Clercx HJH (2007), Proc. Euromech Colloquium 477, Twente, The Netherlands, 21–23 June, 2006.
5. Van Bokhoven LJA, Akkermans RAD, Clercx HJH, Van Heijst GJF (2007), under revision for Exp. Fluids.

Intermittency in the Miscible Rayleigh-Taylor Turbulence

Antonio Celani[1], Andrea Mazzino[2] and Lara Vozella[2]

[1] CNRS and INLN 1361 Route des Lucioles, 06560 Valbonne, France
   antonio.celani@inln.cnrs.fr
[2] Dipartimento di Fisica, Università di Genova and INFN-Sezione di Genova, Via
   Dodecaneso 33, I-16146 Genova, Italy
   mazzino@fisica.unige.it, vozella@fisica.unige.it

The Rayleigh-Taylor (RT) instability is a well-know fluid-mixing mechanism occuring, generally, when a light fluid is accelerated into a heavy fluid [1, 2]. RT instability plays a crucial role in many field of science and technology (e.g. astrophysics [3]) At the final stage of RT instability, in the mixing zone, the fluid motions are turbulent. At this stage, there is a general consensus that the width of the mixing layer growing quadratically in time. On the other hand the value of prefactor and its possible universality is still a much debated issue (see, e.g., [5]). Despite the long history of RT turbulence approximation, a consistent phenomenological theory has been presented only very recently by Chertkov for two miscible fluids under the Boussinesq approximation [4]. About the 2D case, inside the mixing zone the small-scale statistics follow Bolgiano scaling. Our work is the first attempt to compare numerical results (by means of direct numerical simulations) with such phenomenological theory.

1 Simulations and Statistics

The equations ruling the fluid evolution in the 2D Boussinesq approximation are:

$$\partial_t T + \boldsymbol{v} \cdot \nabla T = \kappa \Delta T , \tag{1}$$

$$\partial_t \omega + \boldsymbol{v} \cdot \nabla \omega = \nu \Delta \omega - \beta \nabla T \times \boldsymbol{g} , \tag{2}$$

T being the temperature field, $\omega = \nabla \times \boldsymbol{v}$ the vorticity, $\boldsymbol{g}$ the gravitational acceleration, β the thermal expansion coefficient, κ the molecular diffusivity and ν the viscosity. The integration of both equations here is performed by a standard 2/3-dealiased pseudospectral method on a doubly periodic domain of horizontal/vertical aspect ratio 1 : 4. The resolution is 128×4096 collocation points. The time evolution is implemented by a standard second-order Runge-Kutta scheme. The integration starts from an initial condition corresponding to a zero velocity field and to a step function for the temperature (the system is at rest with the colder fluid placed above the hotter one).

Given that the system is intrinsically nonstationary, averages to compute statistical observables (e.g. structure functions) are performed over different realizations (about 40 in the present study). The latter are produced by generating initial interfaces with sinusoidal waves of equal amplitude and random phases [5].

2 Results

A snapshot of the temperature field obtained in our simulations, is shown in Fig. 1. Cold and warm regions of the fluid are coded in black and white, respectively. This one is a typical picture of the temperature field during the turbulent stage.

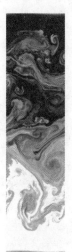

Fig. 1. Two-color coded image of the temperature field. Black (white) regions represents cold (warm) fluid.

Consistenly with previuos findings we observed a mixing layer growth behaving quadratically in time. This quadratic law has a simple physical meaning in term of gravitational fall and rise of thermal plumes. As a result, the low-order statistics of temperature and velocity follow Bolgiano scaling. Instead there are strong corrections (spatio-temporal intermittency) from fourth-order structure functions of temperature while this is not the case for the velocity field, which shows a close-to-Gaussian probability density function. The anomalous exponents are quantitatively in agreement with previous results obtained in 2D driven Boussinesq convection studied in [6].

Finally the behaviour of time-dependent adimensional variables such as Nusselt and Reynolds numbers as a function of Rayleigh number follows Kraichnan scaling associated with the elusive "ultimate state of thermal convection" (see, e.g., [7] and references therein).

3 Conclusions

2D RT turbulence is well described by the phenomenological theory by Chertkov, at least for low-order observables as spectra. However, strong intermittency effects appear for higher-order temperature statistics. The values of such corrections suggest that 2D RT turbulence corresponds to the case driven by a linear temperature profile with a mean gradient that adiabatically decreases in time. Moreover RT turbulence provides a natural framework where heat transport takes place exclusively by bulk mechanisms [7] and thus provides a physically realizable example of the Kraichnan scaling regime, inviting further experimental and numerical effort in this direction.

References

1. Lord Rayleigh: Proc. London Math. Soc. **14**, 170 (1883)
2. G.T. Taylor: Proc. R. London, Ser A **201**, 192 (1950)
3. M. Zingale, S.E. Woosley, J.B. Bell, M.S. Day and C.A. Rendleman: J. Phys., Conference Series **16**, 405 (2005)
4. M. Chertkov: Phys. Rev. Lett. **91**, 115001 (2003)
5. T.T. Clark: Phys. Fluids **15**, 2413 (2003)
6. A. Celani, A. Mazzino, T. Matsumoto and M. Vergassola: Phys. Rev. Lett. **88**, 054503 (2002)
7. D. Lohse, and F. Toschi: Phys. Rev. Lett. **90**, 034502 (2003)

Highly-resolved Simulation of Flow Over a Three-dimensional Hill

Ning Li and Michael A. Leschziner

Aeronautics Dept., Imperial College, South Kensington, London, SW7 2BY, UK.
n.li@imperial.ac.uk

1 Introduction

The flow over a relatively tall three-dimensional, hill-shaped obstacle in a duct contains most generic features encountered in almost all types of three-dimensional separation from continuous, curved surfaces. Because of the smoothly varying geometry, this separation process is very different from that provoked by a sharp edge. First and foremost, the separation process is intermittent, spotty and varies substantially in space and time over the curved surface. In some cases, a closed bubble-type recirculation zone may form; in other cases, the separation features a variety of complex topological features, such as curved detachment and attachment lines and nodes, focal points and saddles. Large vortical structures are intermittently or periodically shed from the surfaces, resulting in complex wake structure downstream of the obstacles. Such flows are also observed to be sensitive to turbulence in the upstream flow, especially when separation is preceded by a thick turbulent boundary layer, and in the free stream above the boundary layer. All these features pose major challenges to both statistical modelling and time-resolved simulation.

The configuration considered here is shown in Fig. 1. This has been the subject of an ongoing, broad research effort, with a particular focus on the predictive capabilities of zonal LES-RANS (hybrid) schemes (Tessicini et al. [1]). It is also a flow that had been observed

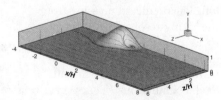

Fig. 1. Flow over a three-dimensional hill.

to be especially poorly predicted by statistical models, even at second-moment-closure level (e.g. Wang et al. [2]), an observation that has greatly motivated studies of the flow with LES. The Reynolds number, based on hill height and free-stream velocity, is 130000 - a high value for which the viscous wall-layer is of the order of only 0.1% of the hill height. A rich range of experimental data is available, obtained by Simpson et al. [3] and Byun & Simpson [4] with elaborate LDA techniques.

Large eddy simulations for this geometry have so far been performed, not only by the present authors (e.g. Persson et al. [5]), on relatively coarse, non-wall-resolving grids with hybrid RANS-LES schemes. These simulations have yielded encouraging results, in terms of flow topology and statistical quantities, but pose uncertainties in respect of refined near-wall processes, which conceivably influence the details of the separation process.

Against this background, a wall-resolving simulation has been performed on a 36.7-million-node mesh, which fully resolves the near-wall region above the bottom surface. The hill is subject to a thick incoming boundary layer, extending roughly to 50% of the hill height, which

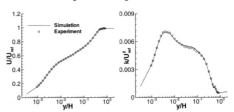

Fig. 2. The U and k profiles at inlet.

contains fully-established and influential turbulence. This is accounted for through the imposition at the computational inlet plane of time- and space-resolved turbulent fields from a precursor simulation in which a forcing method is used to enforce the correct mean-velocity and normal-stress profiles given by the experiment, as shown in Fig. 2. This is an important pre-requisite for placing full confidence in the simulation, especially if turbulence budgets are to be extracted from the simulation, as is currently being done.

2 Results

Presented in Fig. 3 is the pressure distribution above the hill surface on the geometric hill-centre plane, a sensitive indicator of the separation process in the leeward side of the hill. The predicted pressure from the highly-resolved LES (refers to as 'fine-grid' in figures to follow) matches the experimental data better than those wall-

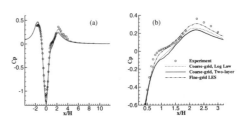

Fig. 3. Pressure coefficients along the hill surface at the centre-plane.

model simulations ('coarse-grid'), as reported in Tessicini et al. [1].

Fig. 4 shows the velocity-vector field across the centre plane, with the recirculation highlighted by the zero-velocity locus. The separation predicted by the high-resolution simulation is at $x/H = 1.25$, slightly too far downstream relative to the experimental value of 0.96. Curiously, a closer examination of the field reveals an extremely thin layer of reverse flow nestling over the leeward-side of the hill and reaching up to $x/H = 0.3$. A possible explanation for this feature, assuming it is not real, is that the integration period of $40H/U_{ref}$ over which statistical data are assembled is insufficiently long to account for (very-)low-frequency shedding motions revealed by POD studies of

Tessicini et al. [1]. On the other hand, the streakline pattern in the plane closest to the hill surface ($y^+ < 2$), given in Fig. 5, does show a very well-defined
and symmetric separation line at about $x/H = 0.3$. If this feature is indeed
physical, neither the LDV measurements nor the wall-model simulations have
had the resolution necessary to capture it.

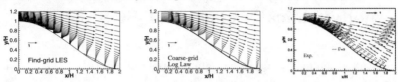

Fig. 4. Velocity field across the centre-plane in the leeward side of the hill.

A comparison of three hill-
surface topology maps is given
in Fig. 6. The figure includes
the experimental pattern, and
one zonal-model (coarse-grid)
result, both compared to the
pattern constructed from the
velocity field on the 15th wall-
normal grid plane of the fine-
grid simulation, so that they

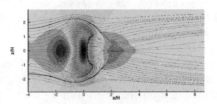

Fig. 5. Streakline pattern predicted by the high-
resolution LES at a near-wall plane.

are comparable in terms of wall-normal distance. All fields present a focal
point at about $x/H = 1.3$ and $z/H = 0.7$, from which the reverse flow in the
separation zone detaches from the hill surface. However, the high-resolution
LES reveals an additional secondary vortex around $x/H = 1.0$ and $z/H = 0.2$.
Related POD studies and budgets will be reported in a fuller paper to follow.

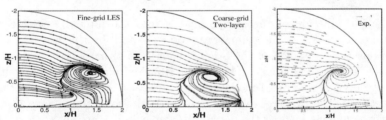

Fig. 6. Flow topology maps on the leeward-side of the hill.

References

1. F. Tessicini, N. Li, M.A. Leschziner: Int. J. Heat Fluid Flow (to appear) (2007)
2. C. Wang, Y.J. Yang, M.A. Leschziner: Int. J. Heat Fluid Flow **25** 499–512
   (2004)
3. R.L. Simpson, C.H. Long, G. Byun: Int. J. Heat Fluid Flow **23** 582–591 (2002)
4. G. Byun, R.L. Simpson: AIAA Paper, 2005-0113 (2005)
5. T. Persson, M. Liefvendahl, R.E. Bensow, C. Fureby: J. Turbulence **7**(4) 1–17
   (2006)

A new class of symmetry preserving and thermodynamically consistent SGS models

Dina Razafindralandy[†], Aziz Hamdouni[‡], Martin Oberlack[†]

[†]Fluid Dynamics Group, Technische Universität Darmstadt, Germany,
   drazafin@fdy.tu-darmstadt.de,
   oberlack@fdy.tu-darmstadt.de,
[‡]LEPTAB, Université de La Rochelle, France,
   ahamdoun@univ-lr.fr.

The Navier–Stokes equations admit symmetry properties, such as the two-dimensional material indifference [10], the invariance under the generalized Galilean transformation, under constant rotations or under certain scaling transformations [4]. These properties are fundamental for the understanding of fluid mechanics and, in particular, of turbulence. For example, they can be used for the derivation of conservation laws (Nœther's theorem [5]). Oberlack used the symmetry method to obtain turbulent scaling laws from the Navier–Stokes equations [7]. Further, symmetries enable to calculate analytical (self-similar) solutions ([2]). Some self-similar solutions are particularly important because they represent interesting physical solutions (for example, a vortex solution [3]) or an asymptotic behavior of the flow [1].

The introduction of turbulence models into the filtered Navier–Stokes equations may destroy the symmetry properties and, by this way, the physical properties contained in the original equations of motion. In addition, it has been shown by Oberlack in [6] and by Razafindralandy and Hamdouni in [8] that only a few subgrid models in the literature are compatible with the symmetries of the Navier–Stokes equations. At the same time, many existing turbulence models, such as the popular dynamic model, are not conform with the second law of thermodynamics because they may induce negative dissipation. Presently, we propose to build a new class of subgrid models which preserve the symmetries of the equations and which, moreover, are consistent with the second law of thermodynamics.

In order to have models preserving the translational and rotational symmetry properties of the Navier–Stokes equations, the subgrid stress tensor $\boldsymbol{\tau}_s$ is taken to be a function of the filtered strain rate tensor $\overline{\boldsymbol{S}}$, the subgrid-scale energy q and the dissipation rate ε. Tensor invariance theory leads to the following form of $\boldsymbol{\tau}_s$:

$$\tau_s^d = A(q, \varepsilon, \chi, \zeta) \, \overline{S} + \frac{1}{\sqrt{\chi}} \, B(q, \varepsilon, \chi, \zeta) \, \mathrm{Adj}^d \, \overline{S} \tag{1}$$

where the superscript d represents the deviatoric operator and $\chi = \mathrm{tr} \, \overline{S}^2$ and $\zeta = \det \overline{S}$ are the invariants of $\overline{S}$. The tensor $\mathrm{Adj} \, \overline{S}$ is the comatrix of $\overline{S}$, defined by:

$$(\mathrm{Adj} \, \overline{S}) \overline{S} = (\det \overline{S}) I_d, \tag{2}$$

I_d being the identity matrix. A and B are arbitrary scalar functions. The preservation of the scaling properties of the Navier–Stokes equations and a dimension analysis lead finally to

$$\tau_s^d = \frac{q^2}{\varepsilon} \left(A_1(v) \, \overline{S} + \frac{1}{\sqrt{\chi}} B_1(v) \, \mathrm{Adj}^d \, \overline{S} \right) \tag{3}$$

where $v = \chi^{-3/2} \zeta$. A_1 and B_1 are arbitrary, dimensionless scalar functions. Relation (3) defines a class of symmetry preserving subgrid models. In what follows, we refine this class to thermodynamically consistent models.

The Newtonian viscous constraint τ can be written in a "potential" form:

$$\tau = \frac{\partial \psi}{\partial S}, \tag{4}$$

where ψ is the "potential" defined by $\psi = \nu \, \mathrm{tr} \, S^2$. This potential form is important because the convexity and positivity of ψ ensure the positivity of the molecular dissipation, and then the conformity with the second law of thermodynamics. Since τ_s can be considered as a (subgrid-scale) constraint, we assume that it can also be written in a potential form:

$$\tau_s = \frac{\partial \psi_s}{\partial \overline{S}}. \tag{5}$$

Relations (3) and (5) induce that

$$\tau_s^d = \frac{q^2}{\varepsilon} \left[\left(2g(v) - 3v\dot{g}(v) \right) \overline{S} + \frac{1}{\sqrt{\chi}} \, \dot{g}(v) \, \mathrm{Adj}^d \, \overline{S} \right]. \tag{6}$$

where g is an arbitrary, scalar and non-dimensional function and $\dot{g}$ its derivative. With model (6), it is straightforward, using (2), to check that the total dissipation

$$\Phi = \rho \, \mathrm{tr}[(\overline{\tau} + \tau_s) \overline{S}]$$

is positive when g is such that

$$\nu + \frac{q^2}{\varepsilon} \, g(v) \geq 0. \tag{7}$$

Note that the positivity of the total dissipation guarantees the stability of the model ([8]).

In summary, relations (6) and (7) represent a class of subgrid-scale models which are compatible with the symmetry properties of the Navier–Stokes equations and are conform with the second law of thermodynamics. These models ameliorate those proposed in [9, 8].

References

1. G.I. Barenblatt. *Scaling, self-similarity and intermediate asymptotics.* Cambridge University Press, 1996.
2. W. Fushchych and R. Popowych. Symmetry reduction and exact solutions of the Navier-Stokes equations I. *Journal of Nonlinear Mathematical Physics*, 1(1):75–113, 1994.
3. V. Grassi, R.A. Leo, G. Soliani, and P. Tempesta. Vorticies and invariant surfaces generated by symmetries for the 3D Navier-Stokes equation. *Physica A*, 286:79–108, 2000.
4. N.H. Ibragimov. *CRC handbook of Lie group analysis of differential equations. Vol 1, 2, 3.* CRC Press.
5. E. Nœther. Invariante Variationsprobleme. In *Königliche Gesellschaft der Wissenschaften*, pages 235–257, 1918.
6. M. Oberlack. Invariant modeling in large-eddy simulation of turbulence. In *Annual Research Briefs*. Stanford University, 1997.
7. M. Oberlack. A unified approach for symmetries in plane parallel turbulent shear flows. *Journal of Fluid Mechanics*, 427:299–328, 2001.
8. D. Razafindralandy and A. Hamdouni. Consequences of symmetries on the analysis and construction of turbulence models. *Symmetry, Integrability and Geometry: Methods and Applications*, 2: Paper 052, 2006.
9. D. Razafindralandy, A. Hamdouni, and C. Béghein. A class of subgrid-scale models preserving the symmetry group of Navier-Stokes equations. *Communications in Nonlinear Science and Numerical Simulation*, 12(3):243–253, 2007.
10. C. Speziale. Some interesting properties of two-dimensional turbulence. *Physics of Fluids*, 24:1425–1427, 1981.

A turbulent-energy based mesh refinement procedure for Large Eddy Simulation

Naudin A., Vervisch L. and Domingo P.

INSA - Rouen, UMR-CNRS-6614-CORIA, Avenue de l'Université, BP 8, 76801 Saint Etienne du Rouvray Cedex, France
naudin@coria.fr vervisch@coria.fr domingo@coria.fr

Abstract

Large Eddy Simulation (LES) relies on four strongly coupled basic ingredients, the Sub-Grid Scale (SGS) closure, the time and space discretization methods, the boundary conditions and the mesh. The objective of this paper is to test a practical strategy to optimize LES grids according to the level of unresolved SGS energy. This approach was first discussed by Pope [1] and Klein [2].

The flow issuing from a swirl burner is simulated with three meshes, two having 0.7M cells and one with 6M cells. It is shown that the 0.7M mesh cells, which has been optimized from a SGS-energy based procedure, provides results of the same quality than those obtained with the 6M cells.

Flow configuration and numerics

The test-case is a swirling flow that was thoroughly studied experimentally by Schneider et al. [3]. The experimental device consists of a movable block type swirler, followed by a single annulus injecting the swirling flow with a coflow of air in a section at atmospheric pressure, where the measurements are performed. The presented case is simulated without combustion, with a swirl number of the order of 0.75, and a moderate turbulence intensity ($Re =$ 10,000).

LES is conducted using the AVBP solver developed by CERFACS [4]. The fully compressible Navier Stokes equations are solved using a third order explicit time integration and a second order space discretization based upon a cell-vertex approach for unstructured grids. The Navier Stokes Characteristic Boundary Conditions, specifically developed for compressible flow simulations, are prescribed at all boundaries [5]. The WALE eddy-viscosity model is used [6].

Mesh optimization from SGS energy

The distribution of grid points may be optimized according to two strategies. On the one hand, a local adaptive mesh refinement (AMR [7]) consists in generating multilevel refined grids overlaid on a coarsely resolved base-level grid, thus increasing the cost of each timestep while increasing the number of levels. On the second hand, as exposed in [2] and shown in the current work, a global mesh refinement may be employed based on quality assessment of the LES. This latter approach provides an adapted grid of an equivalent computational load (i.e. the number of grid points is constant).

All ingredients being fixed (i.e. numerical methods, boundary conditions and SGS modeling), the accuracy of LES is expected to increase when, $k_{SGS}(\underline{x},t)$, the amount of energy that is contained within the subgrid is reduced. In the asymptotic limit where $k_{SGS}(\underline{x},t)$ vanishes, Direct Numerical Simulation (DNS) is recovered. The balance between the energy of the LES resolved motions $K(\underline{x},t)$ and $k_{SGS}(\underline{x},t)$ may thus be used to start assessing the quality of a simulation. To this end, the number $M(\underline{x},t)$ characterizing the 'turbulence resolution' was introduced by Pope [1] and defined as:

$$M(\underline{x},t) = \frac{k_{SGS}(\underline{x},t)}{K(\underline{x},t) + k_{SGS}(\underline{x},t)} \tag{1}$$

$M = 1$ in the limit of Reynolds Average Navier Stokes (RANS) calculations and $M = 0$ in DNS. From this quantity, the concept of adaptive LES may be introduced in which a value of turbulence-resolution tolerance ϵ_M is specified ($\epsilon_M = 0.2$ corresponds to a resolution of 80%, which may be proposed as the threshold of an acceptable quality [2]). In this adaptive grid generation procedure, the LES mesh is refined so that the criterion $M(\underline{x},t) \leq \epsilon_M$ is verified.

The simulation is performed on a first grid of 0.7M cells, over a time T of the order of one flow-through time. The quantities useful to approximate $M(\underline{x},T)$ are estimated according to:

$$K(\underline{x},T) \approx 0.5 \left(\left\langle \widetilde{\mathbf{u}}^2 \right\rangle_T - \langle \widetilde{\mathbf{u}} \rangle_T^2 \right) \tag{2}$$

$$k_{SGS} \approx 0.5 \langle \tau_{ii} \rangle_T \tag{3}$$

where $\langle \widetilde{\mathbf{u}} \rangle_T$ denotes the time average of the resolved velocity vector $\widetilde{\mathbf{u}}$, which is cumulated over the duration T, and τ_{ij} is the SGS stress tensor expressed with the WALE closure. From this information, the grid is globally refined with the criterion $\epsilon_M = 0.2$, i.e. grid points are added where $M(\underline{x},t) \geq \epsilon_M$ and points may also be removed in locations where $M(\underline{x},t) < \epsilon_M \leq \epsilon_M$. The simulation is then pursued till full convergence of new time averaged values (i.e. at least twice longer than T).

Comparisons for the mean axial velocity and the turbulent kinetic energy measured in the experiment and computed on three different grids (6M cells

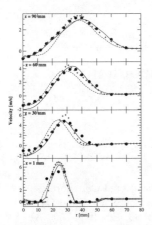 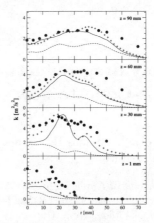

Fig. 1. Measurements vs. LES: axial velocity *(left)*, turbulent kinetic energy *(right)*. Measurements (•), 0.7M optimized mesh (—), 0.7M regular mesh (- -), 6M high resolution mesh (+).

and 0.7M cells before and after refinement) are shown in fig. 1. For the three grids, the major physical phenomena, as the precessing vortex core, were successfully captured, with matching frequencies. The refined 0.7M cells mesh provided results close to the 6M cells mesh, but for a computational effort that was reduced by a factor 15 (simulations performed on 32 processors). In the critical downstream region ($z > 60$mm), the grid independency of the flow is verified, showing that the 'turbulence resolution' used for the refinement procedure is an efficient and practical indicator for both assessing and refining the quality of an LES.

References

1. S B. Pope, *Ten questions concerning the large-eddy simulation of turbulent flows*, N. J. Physics, 6, 2004.
2. Klein, M., *An Attempt to Assess the Quality of Large Eddy Simulations in the Context of Implicit Filtering*, Flow, Turb. and Comb., 75-1:131–147, 2005.
3. C. Schneider, A. Dreizler, and J. Janicka, *Fluid dynamical analysis of atmospheric reacting and isothermal swirling flows*, Flow Turbulence and Combustion, 74(1):103–127, 2005.
4. Y. Sommerer, D. Galley, T. Poinsot, S. Ducruix, and S. Veynante, *LES of flashback and extinction in a swirled burner*, J. of Turb., 5(1), 2004.
5. T. Poinsot and S. K. Lele. *Boundary conditions for direct simulations of compressible viscous flows*, J. Comput. Phys., 1(101):104–129, 1992.
6. F. Nicoud and F. Ducros. *Subgrid-scale stress modelling based on the square of the velocity gradient tensor*, Flow, Turb. and Comb., 62:183–200, 1999.
7. Berger, M.J. and LeVeque, R.J., *An Adaptive Cartesian Mesh Algorithm for the Euler Equations in Arbitrary Geometries*, AIAA Paper, 1930:1989, 1989.

Bubbly drag reduction in turbulent Taylor-Couette flow

Detlef Lohse[1], Thomas H. van den Berg[1], Dennis P. M. van Gils[1], and Daniel P. Lathrop[2]

[1] Department of Applied Physics, IMPACT, International Collaboration for Turbulence Research (ICTR), and J.M. Burgers Center for Fluid Dynamics, Physics of Fluids group, University of Twente, P. O. Box 217, 7500 AE Enschede, Netherlands d.lohse@utwente.nl

[2] Institute for Research in Electronics and Applied Physics, University of Maryland, Collegepark, MD USA dpl@complex.umd.edu

1 Smooth wall case

In Taylor-Couette flow the total energy dissipation rate and therefore the drag can be determined by measuring the torque on the system [1]. We do so for Reynolds numbers between $Re = 7 \cdot 10^4$ and $Re = 10^6$ after having injected (i) small bubbles ($R = 1mm$) up to a volume concentration of $\alpha = 5\%$ and (ii) buoyant particles ($\rho_p/\rho_l = 0.14$) of comparable volume concentration [2]. In case (i) we observe a crossover from little drag reduction at smaller Re to strong drag reduction up to 20% at $Re = 10^6$. The crossover occurs at a bubble Weber number of about 1. In case (ii) we observe at most little drag reduction throughout. These results suggest that bubbly drag reduction is due to bubble deformation, as also suggested by numerical simulations [3].

2 Rough wall case

We can also localize where in the flow the relevant bubble deformations take place: In Taylor-Couette flow with rough walls the overall drag is drastically enhanced [4], as no smooth boundary layers can develop. Therefore we inject bubbles into turbulent Taylor-Couette flow with rough walls (with a Reynolds number up to $4 \cdot 10^5$), finding an enhancement of the dimensionless drag (see figure 1) as compared to the case without bubbles. The results demonstrate that bubbly drag reduction is a pure boundary layer effect [5].

References

1. D. P. Lathrop, J. Fineberg, and H. S. Swinney, Phys. Rev. Lett. **68**, 1515 (1992).

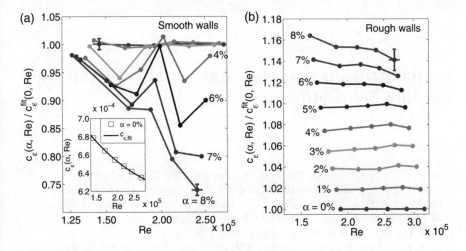

Fig. 1. Compensated drag coefficient $c_\epsilon(\alpha, Re)/c_\epsilon^{fit}(0, Re)$ vs. Reynolds number for increasing void fractions α for the smooth wall case (a) and the rough wall case (b). While in the smooth wall case (a) the drag decreases up to 25%, in the rough wall case (b) the bubble injection leads to a drag enhancement.

2. T. H. van den Berg, S. Luther, D. P. Lathrop, and D. Lohse, Phys. Rev. Lett. **94**, 044501 (2005).
3. J. C. Lu, A. Fernandez, and G. Tryggvason, Phys. Fluids **17**, 095102 (2005).
4. T. H. van den Berg, C. Doering, D. Lohse, and D. Lathrop, Phys. Rev. E **68**, 036307 (2003).
5. T. H. van den Berg, D. P. M. van Gils, D. P. Lathrop, and D. Lohse, Phys. Rev. Lett. **98**, 084501 (2007).

Concentration and segregation of particles and bubbles by turbulence

E. Calzavarini[1], M. Cencini[2], D. Lohse[1] and F. Toschi[3]

[1] Dept. of Applied Physics, University of Twente, 7500 AE Enschede, The Netherlands. e.calzavarini@tnw.utwente.nl, lohse@tnw.utwente.nl

[2] INFM-CNR, SMC Dept. of Physics, University of Rome La Sapienza, Piazzale A. Moro 2, 00185 Roma, Italy and CNR-ISC, Via dei Taurini 19, 00185 Roma, Italy. massimo.cencini@roma1.infn.it

[3] CNR-IAC, Viale del Policlinico 137, I-00161 Roma, Italy and INFN, Sezione di Ferrara, Via G. Saragat 1, I-44100 Ferrara, Italy. toschi@iac.cnr.it

1 The Problem

Understanding the spatial distribution of finite-size massive particles, such as heavy impurities, dust, droplets, neutrally buoyant particles or bubbles suspended in incompressible, turbulent flows is a relevant issue in industrial engineering and atmospheric physics. In a turbulent flow vortices act as centrifuges ejecting particles heavier than the fluid and entrapping lighter ones [1, 2]. This phenomenon produces on one side clusterization (also dubbed *preferential concentration*) on the other segregation (*de-mixing*) of particle species differing in size and densities.

Stated in a rather simplified form, i.e., assuming spherical, not-deformable particles smaller than the smallest scale of turbulence and gravity negligible, the equation of motion for a particle is [3]:

$$\ddot{\mathbf{x}} = \beta \left(\partial_t \mathbf{u} + (\mathbf{u} \cdot \partial)\mathbf{u} \right) - \left(\dot{\mathbf{x}} - \mathbf{u} \right) / \tau. \tag{1}$$

Here $\mathbf{u} = \mathbf{u}(\mathbf{x}(t), t)$ is the fluid velocity field described by the incompressible Navier-Stokes (NS) equation, while the parameters β and τ account for the physical properties of the particle. Specifically, β is a dimensionless number connected to the ratio between the particle density (ρ_p) and the fluid one (ρ_f), defined as $\beta \equiv 3\rho_f/(\rho_f + 2\rho_p)$. τ instead is the typical particle response-time, that is $\tau \equiv a^2/(3\beta\nu)$, with, a, the particle radius and, ν, the fluid kinematic viscosity. Equation (1) coupled to NS can be considered an accurate physical model as long as the particle suspension is dilute, namely it is almost collisionless and it does not exert feedback on the fluid, that is to say, it is passively advected by the flow.

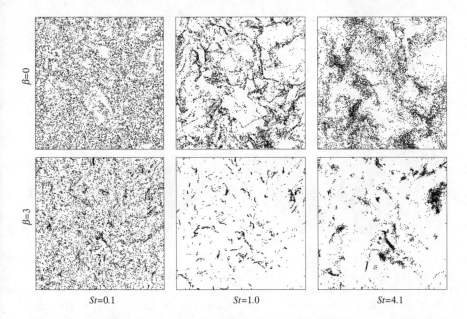

β=0

β=3

St=0.1 St=1.0 St=4.1

Fig. 1. Slices $320 \times 320 \times 8$ in size from a 512^3 DNS. Both very heavy particles, $\beta = 0$ (top), and bubbles, $\beta = 3$ (bottom) at different Stokes numbers, $St = 0.1, 1, 4.1$ (left to right) are reported. The underlying fluid flow field is the same in all cases. All particles were injected homogeneously into the fluid domain roughly one large-eddy-turnover-time before the snapshots.

2 A numerical study

We address the problem numerically. Here we present results from a series of direct numerical simulations (DNS) where passive suspensions of particles of variable density and size are tracked in a homogenous isotropic turbulent flow. We track up to ~500 sets of particles, corresponding to couples of values in the parameter-space β-St, where $St \equiv \tau/\tau_\eta$ stands for the Stokes number and τ_η is the dissipative time-scale. The total number of particles per type ranges between 10^5-10^6. Numerics are performed at different resolutions, 128^3 and 512^3, corresponding respectively to $Re_\lambda = 65$-185, and extended in time for few large-eddy-turnover times. As shown in fig.1, non-homogeneities in the particle/bubble distributions, their dependence on the Stokes number and de-mixing between different species are already evident from plain visualizations.

Correlation dimension and concentration conditioned to flow topology

To gain better insight into the small-scale features of clustering, we study the probability, $P_2(r)$, that the distance between two particles is less than r. In the small-distance limit such probability has a power law behavior, $P_2(r) \sim r^{D_2}$.

The exponent D_2, called correlation dimension of the spatial distribution, can be used as an estimator for the dimension of the set on which particles accumulates. Whether particle distribute *locally* uniformly D_2 equals the spatial dimension 3. If instead $D_2 < 3$ particles accumulate onto a fractal set.

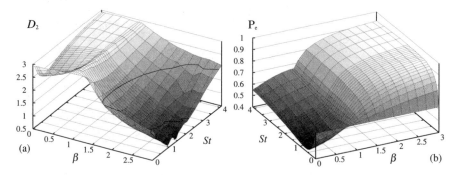

Fig. 2. (a) The correlation dimension D_2 as a function of the density parameter β and the Stokes number. Isolines are drawn at $D_2 = 1, 2$. (b) The probability P_e to find a particle in elliptic regions of the flow versus β and St. Note that for fluid tracers ($St = 0$ particles) it is here $P_e \simeq 0.6$, i.e., as expected elliptic regions in a turbulent flow extend over larger volumes than strain regions.

Both heavy and light particles at small St numbers concentrate on fractal sets, see fig.2(a), the minimum of D_2 being around $St \simeq 1$. Heavy particles always have D_2 above 2, light particles instead reach D_2 values around 1. Indeed, the extremely strong agglomeration occurring for light particles produces decimation of statistics and hence noisy D_2 for $\beta \gtrsim 2$. We conclude that at small-scales filament-like clusters are expected for light particles while heavy particles agglomerate on surface-shaped regions.

Segregation is addressed by looking at particle concentrations conditioned to the local topology of the flow field. Vortical (also called *elliptic*) regions of the flow are defined as the positions where the eigenvalues of the local strain matrix $(\partial_i u_j)$ have imaginary parts [4]. In fig.2(b) we report the probability (P_e) to find a particle, of given β-St value, in an elliptic region of the flow. Heavy particles are lacking in vortical regions, while extremely light particles concentrate almost completely in elliptic regions.

References

1. M. R. Maxey, J. Fluid Mech. **174**, 441-465 (1987).
2. K. D. Squires, J. K. Eaton, Phys. Fluids A **3**, 1169-1178 (1991).
3. M. R. Maxey, J. J. Riley, Phys. Fluids **26**, 883-889 (1983).
4. J. Bec *et al.* J. Fluid Mech. **550**, 349-358 (2006).

Clustering of heavy particles in turbulent flows

J. Bec[1], L. Biferale[2], M. Cencini[3], A. Lanotte[4], S. Musacchio[5], and
F. Toschi[6]

[1] CNRS UMR6202, OCA, Nice, France bec@obs-nice.fr
[2] University of Rome Tor Vergata, Rome, Italy
[3] CNR-ISC, Rome, Italy
[4] CNR-ISAC, Lecce, Italy
[5] The Weizmann Institute of Science, Rehovot, Israel
[6] CNR-IAC, Rome, Italy

Dust, droplets and other finite-size impurities with a large mass density suspended in incompressible flows are commonly encountered in many natural phenomena and industrial processes. The most salient feature of such suspensions is the presence of strong inhomogeneities in the spatial distribution of particles. This phenomenon, dubbed preferential concentration (see, e.g., [1]) can affect the probability to find particles close to each other and thus have influence on their possibility to collide or to have biological, chemical and gravitational interactions. The statistical description of such preferential concentrations remains largely an open question, with applications, for instance, in the growth of raindrops in sub-tropical clouds or in the formation of planetesimals in the early Solar System.

We present results of three-dimensional direct numerical simulations of heavy particles transported by an incompressible, homogeneous and isotropic turbulent fluid flow with, presently, a maximal resolution of 512^3 (corresponding to a Taylor-microscale Reynolds number $\mathrm{Re}_\lambda \approx 185$). Details on the simulation are given in [2]. The suspensions considered are very diluted, so that particle-to-particle hydrodynamic interactions and retroaction of the particles on the fluid can be disregarded. The particle motion is then integrated by a Lagrangian method, allowing a description in the full position-velocity phase space where their dynamics takes place. The particles are assumed to be much heavier than the fluid and much smaller than the smallest active scale of the flow, that is the Kolmogorov scale η. In these asymptotics, they are well approximated by point-like particles and they interact with the flow through a viscous drag, so that the trajectory $\boldsymbol{X}$ of such a particle obeys the Newton equation $\ddot{\boldsymbol{X}} = (1/\tau)(\boldsymbol{u}(\boldsymbol{X},t) - \dot{\boldsymbol{X}})$. The response time τ of the particles is usually non-dimensionalized by a characteristic time of the fluid flow (typically, the eddy turnover time $\tau_\eta = \varepsilon^{-1/3}\eta^{2/3}$ associated to the Kolmogorov scale) to define the Stokes number $\mathrm{S}_\eta = \tau/\tau_\eta$ that measures the inertia of the particles. The model we consider hence depends on only two parameters: the Stokes number of the particles and the Reynolds number of the carrier flow. This allows for a systematic investigation.

After relaxation of transients, the phase-space density of particles becomes singular with its support on a dynamically evolving fractal set. This attractor is characterized by non-trivial multiscaling properties at scales much smaller than η. Multifractality in phase space implies also multiscaling of the coarse-grained spatial distribution of the mass of particles. The scaling exponents of the latter are accurately determined from the simulations and are related to the spectrum of dimensions of the particle distribution (see [3]). Figure 1 represents the exponent of the second-order moment (equal to the correlation dimension $\mathcal{D}_2$) for various Reynolds numbers. The collapse of the curves evidences the fact that intensity of clustering in the dissipative range is almost independent of the Reynolds number and that the Kolmogorov time scale used to define S_η is there the relevant time scale of the flow.

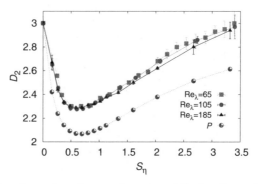

Fig. 1. Correlation dimension $\mathcal{D}_2$ as a function of S_η for three different values of Re_λ as labeled. The bullets represents the probability $\mathcal{P}$ to find particles in non-hyperbolic (rotating) regions of the flow (shown for $\mathrm{Re}_\lambda = 185$ and multiplied by an arbitrary factor for plotting purposes).

For larger length scales $r \gg \eta$ inside the inertial range of turbulence, the particle distribution is not scale invariant anymore. Dimensional analysis suggests that deviations from a uniform distribution can be measured in terms of the local Stokes number $S_r = \tau \varepsilon^{1/3}/r^{2/3}$, defined as the ratio between the response time of the particle and the eddy turnover time of the turbulent flow at the scale r considered. We show that this number does not fully characterizes the particle distribution but that deviations from uniformity depend only on a scale-dependent contraction rate $\gamma_{r,\tau}$ (see Fig. 2 Left) whose scaling $\propto \tau/r^{5/3}$ can be related to the scaling of pressure increments in the inertial range (see [3]). Figure 2 (Left) actually displays quasi-Lagrangian statistics, that is the statistics of particle number density in a ball centered on a given reference particle trajectory.

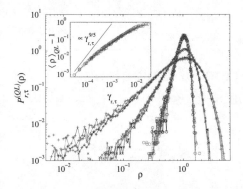 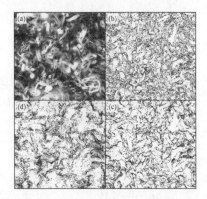

Fig. 2. Left: Probability distribution function of the coarse-grained mass in the inertial range for various values of τ and r associated to three different values of the non-dimensional contraction rate $\gamma_{r,\tau}$. Inset: deviation from unity $\langle \rho_r \rangle_{QL} - 1$ of the first-order quasi-Lagrangian moment of the coarse-grained particle mass density for scales r within the inertial range. For comparison, the behavior $\propto \gamma_{r,\tau}^{9/5}$ obtained when assuming point clusters of particles is shown as a solid line. Right: (a) Modulus of the pressure gradient, which gives the main contribution to fluid acceleration, on a slice $512 \times 512 \times 4$. Dark and light code low and high intensity, respectively. Particle positions in the same slice shown for (b) $S_\eta = 0.16$, (c) $S_\eta = 0.80$ and (d) $S_\eta = 3.30$. Note the presence of voids with sizes much larger than the dissipative scale.

Another important observation is that the particle distribution presents large voids where the mass is orders of magnitude below its average. Such regions are typically correlated with the vortical structures of the flow; this confirms the classical phenomenological pictures that in turbulent flow, eddies act as small centrifuges and eject heavy particles leading to their concentration in the strain-dominated regions. This behavior is evidenced from our numerical simulations as seen in Fig. 2 (Right). These voids, which are spanning all scales of the turbulent flow, have a signature on the coarse-grained mass probability distribution: they are responsible for an algebraic behavior at small mass densities. As qualitatively illustrated from Fig. 2 (Right) and also reported in [4], these voids are related to the non-uniform distribution of the points where the acceleration of the fluid is large. Understanding quantitatively the correlations between the particle positions and the spatial distribution of acceleration in the flow is still an open question

References

1. J.K. Eaton and J.R. Fessler: Int. J. Multiphase Flow **20**, 169 (1994)
2. J. Bec, L. Biferale, G. Boffetta et al: J. Fluid Mech. **550**, 349 (2006)
3. J. Bec, L. Biferale, M. Cencini et al: Phys. Rev. Lett. in press (2007)
4. L. Chen, S. Goto, and J. Vassilicos: J. Fluid Mech. **553**, 143 (2006)

Experimental investigation of turbulent transport of material particles

Nauman Qureshi[1], Mickaël Bourgoin[1], Christophe Baudet[1], Alain Cartellier[1], and Yves Gagne[1]

[1] L.E.G.I. - U.M.R. 5519 C.N.R.S./I.N.P.G./U.J.F.,
1025 rue de la Piscine, 38041 - Grenoble, France ; `mickael.bourgoin@hmg.inpg.fr`

Particle laden turbulent flows play an important role in several situations such as industrial systems or atmospheric dispersion of pollutants for instance. When the particles are neutrally buoyant and small (typically comparable in size with the dissipation scale of the surrounding turbulence) they behave as tracers for fluid particles. However, in many practical situations, the particles are heavier and/or larger. Their dynamics is then affected by inertial effects and it deviates from fluid particles dynamics[1, 2, 3]. The precise role of size and density of the particles in the modification of their dynamics with respect to fluid tracers, remains largely an open question.

Here, we report measurements of Lagrangian velocity and acceleration statistics of material particles transported in a grid generated windtunnel turbulent flow, with a Reynolds number (based on Taylor microscale) of $R_\lambda \sim 200$. The dissipation scale η is 200 μm and the energy injection scale L is 2.5 cm. As a first step, we only explore particles finite size effects. To decouple the role of size and density of the particles, we consider neutrally buoyant particles, which are soap bubbles inflated with helium and which diameter can be adjusted from 1.5 mm to 6 mm which corresponds to inertial range scales. The Lagrangian measurements are obtained with an acoustic Doppler velocimetry technique (figure 1a): from the instantaneous Doppler frequency shift of acoustic waves scattered by a particle in a turbulent flow, we measure the velocity of the particle [4]. The instantaneous frequency is determined with a parametric maximum of likelyhood algorithm [5]. The particles can be tracked over a period covering several dissipation time scales, corresponding to a significant fraction of the integral time scale of the flow.

In order to investigate the influence of particle size on its Lagrangian dynamics, we first consider how the Lagrangian velocity autocorrelation function is affected when we change the bubbles diameter. Figure 1b represents for instance the Lagrangian velocity autocorrelation function R_L for 6 mm particles. Note that we only show a relatively short time lags range, for which we have enough sufficiently long Lagrangian trajectories to ensure a good statis-

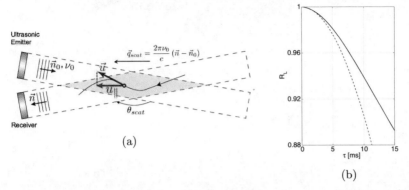

Fig. 1. (a) Principle of acoustical Doppler velocimetry : an ultrasonic plane wave is generated by an emitter and scattered by the particle. The Doppler shift of the scattered wave is directly related to the longitudinal component $u_{//}$ of the particle velocity. (b) Lagrangian velocity autocorrelation function for 6 mm bubbles. The dashed line represents the parabolic fit $1 - \frac{\tau^2}{\tau_\lambda^2}$ around $\tau = 0$ used to determine the Lagrangian microscale τ_λ.

tical convergence. From the curvature at $\tau = 0$ we can estimate an equivalent Lagrangian Taylor time scale $\tau_\lambda(D)$ associated to the Lagrangian dynamics of a particle of diameter D. Figure 2a shows a clear dependence of τ_λ on particle size. We note that as the particle size decreases, τ_λ appears to approach an asymptotic value (which we can estimate here around 25 ms) which corresponds to the intrinsic Lagrangian microscale of the turbulent flow, as smaller particles approach fluid tracers. The increase of the microscale of the Lagrangian dynamics of the particles as their size increases suggests a longer response time of larger particles to the turbulence forcing. This is consistent with the intuitive phenomenology, that large particles do not feel velocity gradients at scales smaller than their size, and therefore, they must filter in some way the turbulent energy cascade at some small scale related to their size. To test further this scenario, we analyse the Lagrangian velocity correlation function in the frame of two times stochastic models [6]. In this description, the autocorrelation function is given by a double exponential law :

$$R_L(\tau) = \frac{\tau_\lambda^2 e^{-2\tau\tau_D/\tau_\lambda^2} - 2\tau_D^2 e^{-\tau/\tau_D}}{\tau_\lambda^2 - 2\tau_D^2} \tag{1}$$

where τ_D is a small time scale characterizing the cut-off of the particles Lagrangian energy spectrum. For fluid particle tracers, for instance, τ_D is directly related to the viscous dissipation time τ_η. For particles with finite diameter D in the inertial range, in the scenario described above where the fluid turbulent energy is low-pass filtered by the particle at a scale corresponding to its diameter D, the corresponding cut-off time scale can be estimated in the framework of K41 phenomenology as $\tau_D \sim \epsilon^{-1/3}D^{2/3}$, where ϵ is the en-

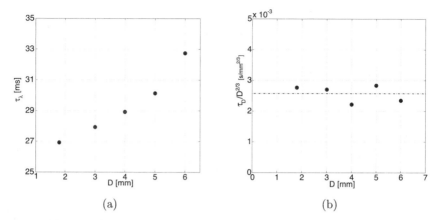

(a)   (b)

Fig. 2. (a) Time microscale τ_λ as a function of the particles diameter D. (b) Compensated cut-off time scale $\tau_D/D^{2/3}$ as function of D.

ergy dissipation rate. From our measurements, we determine the cut-off time scale τ_D as a function of the particle diameter by fitting the autocorrelation function with expression (1). We haven't represented the fit on figure 1b because it is almost undistinguishable from the measured correlation function. Figure 2b shows the compensated cut-off timescale $\tau_D/D^{2/3}$ measured for different particle diameter. In spite of some scattering, the agreement with the $D^{2/3}$ prediction is good and consistent with the small scales cut-off scenario. Other diagnosis, based for instance on measurements of the acceleration variance of the particles as a function of their diameter also confirm this idea. Further investigations will explore the role of particles density.

References

1. Maxey, M.R., Riley, J.J.: Equation of motion for a small rigid sphere in a nonuniform flow. Physics of Fluids **26**(4) (1983) 883–889
2. Aliseda, A., Cartellier, A., Hainaux, F., Lasheras, J.: Effect of preferential concentration on the settling velocity of heavy particles in homogeneous isotropic turbulence. Journal of Fluid Mechanics **468** (2002) 77–105
3. Ayyalasomayajula, S., Gylfason, A., Collins, L., Bodenschatz, E., Warhaft, Z.: Lagrangian measurements of inertial particle accelerations in grid generated wind tunnel turbulence. Physical Review Letters **97**(144507) (2006)
4. Poulain, C., Mazellier, N., Gervais, P., Gagne, Y., Baudet, C.: Spectral vorticity and lagrangian velocity measurements in turbulent jets. Flow, Turbulence and Combustion **72** (2004) 245–271
5. Mordant, N., Metz, P., Michel, O., Pinton, J.F.: Measurement of lagrangian velocity in fully developed turbulence. Physical Review Letters **87**(21) (2001) 214501
6. Sawford, B.L.: Reynolds number effects in lagrangian stochastic models of turbulent dispersion. Physics of Fluids A **3**(6) (1991) 1577–1586

Shear Effect on Lagrangian Acceleration in High-Reynolds Number Turbulence

Yoshiyuki Tsuji[1]

Department of Energy Engineering and Science, Nagoya University, Japan
c42406a@nucc.cc.nagoya-u.ac.jp

The motion of fluid particles as they are pushed along the trajectories by fluctuating pressure gradient is fundamental to transport and mixing in turbulence. It is essential in cloud formation, atmospheric transport, chemical reaction process, and in combustion system. In principle, fluid particle trajectories are easily measured by seeding a turbulent flow with small tracer particles and following their motions with an imaging system. But, in practice, this can be a very challenging task because we must fully resolve particle motions which take place on the order of Kolmogorov time scale. This kind of measurements was recently achieved by La Porta et al.[1] with using a specially designed particle tracking system in high energy physics. In a usual notation, acceleration vector is given by N.S. equation as follows; $\mathbf{a} = D\mathbf{u}/Dt = -\nabla(p/\rho) + \nu\nabla^2\mathbf{u}$. This means that acceleration is decomposed into the contribution from pressure gradient and viscous force while the fluid density is constant. In a fully developed turbulence, the viscous damping term is small compared with the pressure gradient term, therefore, the acceleration is closely related to the pressure gradient. In this study, lagrangian acceleration is evaluated by measuring the instantaneous pressure fluctuations.

We have developed the accurate pressure measurement technique. A standard Pito-static tube, 0.5mm outside diameter and 0.05mm in thickness is attached to the sensor (see Fig.1(a)). Four static pressure holes (0.15mm in diameter) are spaced 90 degrees apart. The sensor is a usual 1/8 inch condenser microphone. It is noted that microphone can catch a very small amplitude pressure fluctuation up to 70 kHz. Several points are taken into account for accurate measurements. For instance, HR (Helmholz resonance) and standing waves in static tube should be removed. Flow attach angle and special resolution are another important factors[2]. Measured pressure statistics were compared with those of DNS. Probability density function measured in nearly isotropic condition is skewed on the negative side and it departs from Gaussian considerably, which agrees quantitatively with DNS result. As predicted by Kolmogorov, pressure spectrum indicates the $-7/3$ power-law in the inertial range, but it can be realized for $R_\lambda \geq 600$. The pressure measurements in the boundary layer up to $R_\theta = 15000$ was also performed for the first time. Pressure intensity distribution, spectral shape, correlation between wall pressure and pressure in the boundary layer, and the velocity pressure correlation were measured. The results were carefully compared with DNS [3]. These results encourage us to evaluate the acceleration by means of pressure gradient measurements.

Pressure gradient measurement was performed by using two pressure probes. The pressure difference measured at a distance Δy(or Δx) becomes pressure gradient dp/dy(or dp/dx) as far as Δy (or Δx) is sufficiently small. Pressure gradient distribution has a stretched exponential shape as shown in Fig.1(b), in which the tails extend much further than a Gaussian distribution. This indicates that acceleration is an extremely intermittent variable. When the flatness of velocity gradient is compared with that of pressure, it is clearly understood that the acceleration field is more intermittent. In the acceleration spectra, the expected power-law exponent $-1/3$ is well observed in the present measurements. It is noted that the acceleration spectrum is hard to obtain as far as pursuing the particle trajectories. Following the Kolmogorov's idea, acceleration is scaled by the energy dissipation rate and kinematic viscosity as $\langle a_i a_j \rangle = a_0 \varepsilon^{3/2} \nu^{-1/2} \delta_{ij}$, where $a_i = (1/\rho)\partial p/\partial x_i$. The constant a_0 is expected to be universal. But the recent DNS (HIT) and La Porta's experiment do not show that a_0 is not constant but a function of Reynolds number. In a shear flow (Mixing layer), a_0 is much smaller than those of HIT. As R_λ increases, a_0 increases and approaches the values of HIT. Present result indicates that the local isotropy is realized at $R_\lambda \simeq 2000$ in the inertial range. But more detailed discussions are necessary for the shear effect (large scale anisotropy) on the acceleration.

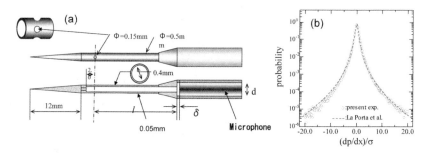

Fig. 1. (a) pressure probe used in this measurement. (b)probability density function of acceleration measured at $R_\lambda = 700$.

[1] A. La Porta, et al., Nature, vol.409, 1017 (2001)
[2] Y. Tsuji and T. Ishihara, Phys. Rev. E, vol.68, 026309 (2003).
[3] Y. Tsuji, J. Fransson, H. Alfredsson, A. Johansson, accepted for publication in JFM (2007).

The Fluid Mechanics of Gravitational Structure Formation

Carl H. Gibson

University of California San Diego, La Jolla, CA 92093-0411, USA
cgibson@ucsd.edu

The standard model for gravitational structure formation in astrophysics, astronomy, and cosmology is questioned. Cold dark matter (CDM) hierarchical clustering cosmology neglects particle collisions, viscosity, turbulence and diffusion and makes predictions in conflict with observations. According to CDMHC cosmology, the non-baryonic dark matter NBDM forms small clumps during the plasma epoch after the big bang that "cluster" into larger clumps. Growing "CDM halo clusters" collect the baryonic matter (H and He) by gravity so that after 300 Myr of "dark ages" huge, explosive (Population III) first stars appear and then galaxies and galaxy clusters. Contrary to CDMHC cosmology, "hydro-gravitational-dynamics" HGD cosmology shows the diffusive NBDM material cannot clump and the clumps cannot cluster. The big bang results from an exothermic turbulent instability at Planck scales (10^{-35} m). Big bang stresses cause inflation of space, which produces fossil density turbulence remnants that trigger gravitational instability at protosupercluster masses (10^{46} kg) in the H-He plasma. These fragment along plasma turbulence vortex lines to form protogalaxy masses (10^{42} kg) just before the transition to gas. The gas has $\times 10^{-13}$ smaller viscosity, so it fragments at earth-mass and globular star cluster masses (10^{25} and 10^{36} kg) to form the baryonic dark matter (BDM). Observations from the Hubble Space Telescope show protogalaxies (PGs) in linear clusters reflecting their likely fragmentation on vortex lines. From the BDM planets, these PGs gently form small stars in globular clusters ≤ 1 Myr after the big bang without the dark ages and superstars of CDM cosmology.

1 Hydro-Gravitational-Dynamics Theory

The hydro-gravitational-dynamics theory of gravitational structure formation [1, 2, 3, 4] covers a wide range of length scales from the big bang Planck scale $L_P = 1.62 \times 10^{-35}$ m of quantum gravitational instability to the present horizon scale $L_H \approx 10^{26}$ m. The Planck temperature $T_P = [c^5 h G^{-1} k^{-2}]^{1/2} =$

1.40×10^{32} K is so large that turbulence and turbulent mixing are needed to produce entropy and make the process irreversible.

Only Planck particles and Planck anti-particles can exist at such temperatures, plus their spinning combinations (Planck-Kerr particles), so the viscosity is low. Planck-Kerr particles are the smallest possible Kerr (spinning) black holes. They represent the big bang equivalent of positronium particles formed from electrons and positrons during the pair production process that occurs at 10^9 K supernova temperatures. Prograde accretion of Planck particles by Planck-Kerr particles can release up to 42% of the Planck particle rest mass $m_P = [chG^{-1}]^{1/2}$, resulting in the highly efficient exothermic production of turbulent Planck gas [3, 4]. Large negative turbulent Reynolds stresses $\tau_P = [c^{13}h^{-3}G^{-3}]^{1/2} = 2.1 \times 10^{121}$ m^{-1}s^{-2} rapidly stretch space until the turbulent fireball cools to the strong force freeze-out temperature $T_{SF} \approx 10^{28}$ K so that quarks and gluons can form. Besides Planck particles, the smallest possible Schwarzschild (non-spinning) black hole, only magnetic monopole particles are possible in the big bang temperature range.

Gluon viscosity damps the big bang turbulence at a Reynolds number $\approx 10^6$ and increases negative stresses and the rate of expansion of space. Turbulence and viscous stresses combine with false vacuum energy in the stress energy tensor of Einstein's equations to produce an exponential expansion of space (inflation) by a factor of about 10^{25} in the time range $t = 10^{-35} - 10^{-33}$ s [5]. Fossil temperature turbulence patterns produced by the big bang and preserved by nucleosynthesis and cosmic microwave background temperature anisotropies indicate a similar large value (10^5) for the big bang turbulence Reynolds number [6]. Only small, transitional Reynolds numbers $c^2t/\nu \approx 10^2$ are permitted by photon kinematic viscosity $\nu = 4 \times 10^{26}$ m^{2}s^{-1} at the time ($t = 10^{12}$ s) of first structure [2].

Gluon, neutrino, and photon viscosities dominated momentum transport and prevented turbulence during the electroweak, nucleosynthesis, and energy dominated epochs before $t = 10^{11}$ s, and also the formation of gravitational structures. Soon after this beginning of the matter dominated epoch the neutrinos ceased scattering on electrons and became super diffusive. Neutrinos were produced in great quantities at the 10^{-12} s electroweak transition that may still exist as part, or most, of the non-baryonic dark matter that dominates the mass of the universe. Momentum transport became dominated by photon viscosity, with the possibility of weak turbulence [2, 7].

The conservation of momentum equations for a fluid subject to viscous, magnetic and other forces is

$$\frac{\partial v}{\partial t} = -\nabla B + v \times \omega + F_g + F_\nu + F_m + F_{etc.} \tag{1}$$

where $B = p/\rho + v^2/2$ is the Bernoulli group, p is pressure, ρ is density, v is velocity, t is time, ω is vorticity, F_m is magnetic force, and $F_{etc.}$ are miscellaneous other forces. In the early universe, ∇B, F_m. and $F_{etc.}$ are small compared to the inertial-vortex force $\boldsymbol{v} \times \boldsymbol{\omega}$ that causes turbulence and the viscous force $\boldsymbol{F_\nu}$ and gravitational force $\boldsymbol{F_g}$.

When viscous and turbulence forces as well as diffusion are taken into account, the HGD cosmological criterion for gravitational instability at scale L becomes

$$L_H \geq L \geq L_{SX_{max}} = max[L_{SV}, L_{ST}, L_{SD}] \tag{2}$$

where L_H and the Schwarz scales L_{SV}, L_{ST}, L_{SD} define the competing regimes of fluid motion [8].

The initial stages of gravitational instability are very gentle, driven by either positive or negative density variations $\delta\rho$. All forces other than $\boldsymbol{F_g}$ on the right hand side of the momentum equations vanish. Pressure support cannot prevent gravitational instability because any forces from enthalpy p/ρ gradients are perfectly balanced by gradients of kinetic energy $v^2/2$ as the fluid starts to move ($\nabla B = 0$). Because the universe is uniformly expanding with rate-of-strain $\gamma \approx t^{-1}$, gravitational condensations on density maxima are inhibited by the expansion but gravitational fragmentations at density minima are enhanced. The first gravitational structures occurred by fragmentation at density minima when the horizon scale L_H increased to exceed the plasma Schwarz scales, which were all nearly equal in the low Reynolds number hot plasma epoch, $L_{ST} \approx L_{SV} \approx L_{SD}$.

2 Conclusions

The standard CDMHC cosmology, the Jeans 1902 criterion for gravitational structure formation, and the "collisionless" concepts of galactic dynamics and frictionless tidal tail formation are in fundamental conflict with modern fluid mechanics and make predictions that are increasingly in conflict with observations. They must be abandoned.

The Jeans theory and CDMHC cosmology have been modified and extended to include important effects of quantum gravity, viscosity, turbulence, fossil turbulence and diffusion on gravitational structure formation and cosmology in a new paradigm termed hydro-gravitational-dynamics (HGD) [1, 2, 3, 4, 7, 8].

References

[1] Gibson, C. H. 1996, Appl. Mech. Rev., 49, 299, astro-ph/9904260
[2] Gibson, C. H. 2000, J. Fluids Eng., 122, 830, astro-ph/0003352.
[3] Gibson, C. H. 2004, Flow, Turbulence and Combustion, 72, 161179
[4] Gibson, C. H. 2005, Combust. Sci. and Tech., 177, 1049-1071
[5] Guth, A. H. 1997, The Inflationary Universe, Addison-Wesley, NY.
[6] Bershadskii, A. 2006, Physics Letters A, 360, 210-216
[7] Gibson, C. H. 2006, astro-ph/0606073
[8] Gibson, C. H. & Schild, R. E. 2007, astro-ph/0701474

DNS of structural vacillation in the transition to geostrophic turbulence

W.-G. Früh[1], P. Maubert[2], P. L. Read[3], and A. Randriamampianina[2]

[1] Heriot Watt University, Edinburgh EH14 4AS, UK. mailto:w.g.fruh@hw.ac.uk
[2] IRPHE, UMR 6594 CNRS, Marseille, France
[3] Atmospheric, Oceanic & Planetary Physics, University of Oxford, UK

Summary. The onset of small-scale fluctuations around a steady convection pattern in a rotating baroclinic annulus filled with air is investigated using Direct Numerical Simulations (DNS). In previous laboratory experiments of baroclinic waves, such fluctuations have been associated with *Structural Vacillation* which is regarded as the first step in the transition to fully-developed geostrophic turbulence. Here we present an analysis which focusses on the small-scale features.

1 Introduction

The differentially-heated, rotating cylindrical annulus has proved a fruitful means of studying fully-developed, nonlinear baroclinic instability in the laboratory, e.g. [1]. Transitions within the regular wave regime follow canonical bifurcations to low-dimensional chaos, but disordered flow appears to emerge via a different mechanism involving small-scale secondary instabilities. Not only is the transition to geostrophic turbulence less well understood than those within regular waves, but also the classification and terminology for the weakly turbulent flows is rather vague. Various terms applied include Structural Vacillation or Tilted-Trough Vacillation which both refer to fluctuations by which the wave pattern appears to change its orientation or structure in a roughly periodic fashion. These fluctuations have been explained by the growth of higher order radial mode baroclinic waves, by barotropic instabilities, or by small-scale secondary instabilities or eddies, which lead to erratic modulations of the large-scale pattern. Further within this 'transition zone', the gradual and progressive breakdown of the wave pattern leads ultimately to a form of stably-stratified 'geostrophic turbulence' [4].

2 The numerical model and the case discussed

The model [3] is that of air between two vertical coaxial cylinders of inner radius $a = 34.8$mm and outer radius $b = 60.2$mm, held at constant temper-

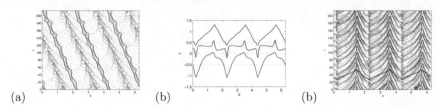

(a)   (b)   (b)

Fig. 1. (a) Space-time plot along an azimuthal ring near the warm outer wall Contour lines show the temperature with a contour interval of 0.1. (b)Time-averaged temperature profiles at the three radial positions from the temperature profiles shifted to a co-rotating frame. (c) Temperature residuals near the outer wall in a space-time plot with temperature contours 0.025. The time-averaged profile is superimposed.

atures, here $T_b = 308$K and $T_a = 278$K, and two horizontal insulating rigid lids separated by a distance d. The cavity rotates around the central axis, here $\Omega = 52$rad/s. The model equations are solved using a pseudo-spectral collocation-Chebyshev in the radial and vertical, and Fourier method in the azimuthal direction with varying model resolutions, here $108 \times 108 \times 128$. The time integration, based on a combination of Adams-Bashforth and Backward Differentiation Formula schemes is semi-implicit and second order accurate.

3 Results

This flow, and the transition to it, is presented elsewhere [5]. This paper focusses on the small-scale structures by an alternative method to filter out the large-scale flow, applied to azimuthal temperature profiles at mid-height and three radial positions, at mid-radius and around 15% from each wall. As the solution was characterised by an almost steady wave drifting along the channel on which small-scale fluctuations were evident (Fig. 1a for the profiles near the outer wall), the results could be transformed to a frame moving with the wave structure. From this, the time-averaged spatial structure could be obtained (Fig. 1b for all radial positions), which could then be subtracted from the standing wave to obtain time series of the residual temperature profiles (Fig. 1c). The space-time plots of the residuals highlight that small perturbations are emitted fairly regularly (but not periodically) near the warm outer wall from the cold jet which brings cold air towards that wall, and that these perturbations travel, initially fairly rapidly, in the azimuthal direction but then slow down as they approach the hot jet, which originates near this wall and takes fluid towards the inner wall. The case is similar near the cold inner wall with perturbations originating from the incoming hot jet. In the centre of the gap, however, the fluctuations appear to be localised to the jets.

Spatially averaged power spectra from the temperature residuals at the three profiles are shown in Figure 2. The main frequency (~ 0.03) is that of

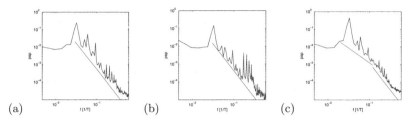

Fig. 2. Power spectra from the time series of the temperature spectra, spatially averaged over each profile:. (a) near the inner wall, (b) at mid-radius, and (c) near the outer wall. The frequency is in units of the inverse of the reference time scale, i.e. 2Ω. All three show a line $\sim f^{-3}$; (c) also shows a line $\sim f^{-5/3}$ at lower frequencies.

the emission of the perturbations. The spectrum falls off at all radial positions, with some distinctive peaks over the general decay. While the decay at higher frequencies is largely consistent with a f^{-3} law at all postions, a slight flattening in the frequency range between that of the main perturbation and about 0.1 can be observed near the outer wall. The behaviour of the spectrum in that range appears to be closer to a $f^{-5/3}$ law.

The spectral evidence suggests that the flow investigated here consists of a fairly steady large-scale convection pattern in the form of three pairs of hot and cold radial jets. From those jets, smaller perturbations are emitted at relatively regular intervals where a jet approaches a wall. The overall flow responds in a cascade of faster (and smaller) fluctuations which appear consistent with two-dimensional, quasi-geostrophic turbulence over a wide range of frequencies similar to results of DNS of geostrophic turbulence. Recently, similar energy spectra were observed for geostrophic turbulence in a square box [2, 6].

References

1. W.-G. Früh and P.L. Read. Wave interactions and the transition to chaos of baroclinic waves in a thermally driven rotating annulus. *Phil. Trans. R. Soc. Lond. (A)*, 355:101–153, 1997.
2. E. Lindborg and K. Alvelius. The kinetic energy spectrum of the two-dimensional enstrophy turbulence cascade. *Phys. Fluids*, 12(5):945–947, 2000.
3. A. Randriamampianina, W.-G. Früh, P.L. Read, and P. Maubert. Direct numerical simulations of bifurcations in an air-filled rotating baroclinic annulus. *J. Fluid Mech.*, 561:359–389, 2006.
4. P.L. Read. Transition to geostrophic turbulence in the laboratory, and as a paradigm in atmospheres and oceans. *Surveys in Geophys.*, 22:265–317, 2001.
5. P.L. Read, P. Maubert, A. Randriamampianina, and W.-G. Früh. DNS of transitions towards structural vacillation in an air-filled, rotating, baroclinic annulus. *Phys. Fluids*, submitted, 2007.
6. M.L. Waite and P. Bartello. The transition from geostrophic to stratified turbulence. *J. Fluid Mech.*, 568:89–108, 2006.

Stereo-PIV measurements in turbulent rotating convection

R.P.J. Kunnen[1], B.J. Geurts[1,2], and H.J.H. Clercx[1,2]

[1] Dept. of Physics & J.M. Burgerscentrum, Eindhoven University of Technology, P.O. Box 513, 5600 MB Eindhoven, The Netherlands, R.P.J.Kunnen@tue.nl
[2] Dept. of Applied Mathematics & J.M. Burgerscentrum, University of Twente, P.O. Box 217, 7500 AE Enschede, The Netherlands

Convection and rotation are prevalent influences on many geophysical flows, such as the flows in the atmosphere and in the oceans. Also in turbo-machinery and chemical process engineering such situations arise. A convenient model for studying these effects is provided by rotating Rayleigh–Bénard convection (RRBC) in a cylindrical container: a fluid enclosed by a rotating cylinder is heated from below and cooled from above.

RRBC is geverned by several dimensionless parameters: the Rayleigh number $Ra = g\alpha\Delta T H^3/(\nu\kappa)$, the Taylor number $Ta = (2\Omega H^2/\nu)^2$, the Prandtl number $\sigma = \nu/\kappa$, and, the geometry aspect-ratio $\Gamma = D/H$. Here D and H are diameter and height of the cylinder, g is the gravitational acceleration, ΔT the temperature difference between the plates, Ω the rotation rate antiparallel to gravity, and α, ν, κ are thermal expansion coefficient, kinematic viscosity and thermal diffusivity of the fluid, respectively.

Here we present results from an experimental study of RRBC. Stereoscopic particle image velocimetry (SPIV) is applied [1]. This measurement technique provides the three velocity components as measured in a planar cross-section of the flow domain. The setup is displayed schematically in Fig. 1. A closed plexiglas cylinder of height and diameter $H = D = 23$ cm ($\Gamma = 1$) is filled with water seeded with small polyamid tracer particles. A laser light sheet illuminates a horizontal cross-section of the cylinder at half-height. Two cameras are mounted above the cylinder at off-axis angles, to record the particle motions in the light sheet plane. All parts, including cooling and heating units for the cylinder, are placed on a rotating table. The experimental set-up allows to monitor the flow in a window of roughly 9×12 cm.

In the measurements constant $Ra = 1.1 \times 10^9$ and $\sigma = 6.4$ have been used. The measurements were conducted in two different rotation regimes: the first at very small rotation-rates ($0 \leq \Omega \lesssim 0.03$ rad/s, $0 \leq Ta \lesssim 2 \times 10^7$), and the second at larger rotation rates (up to $Ta = 2.2 \times 10^{10}$).

At low rotation-rates the entire domain is filled by a so-called large-scale circulation (LSC). Hot plumes rise on one side of the cylinder, while cold

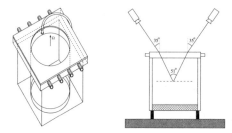

Fig. 1. Sketches of the experimental set-up and the positioning of the cameras. A square box filled with water surrounds the cylinder to ease optical access. The water circulation chamber on top (with temperature sensor) is used to cool the cylinder from above, while still being transparent.

plumes sink on the other. The LSC displays dynamics on widely varying time scales, from an azimuthal oscillation with period of order one minute to sudden rotations and cessations at time intervals of several hours to a day [2]. The azimuthal oscillation is investigated as follows. In the interrogation plane the LSC is organized into a few large parts of the flow domain with either positive or negative vertical velocity w, cf. the velocity snapshot in Fig. 2(a). The centroid of w in either the $w > 0$ or $w < 0$ regions is computed. The angle ϕ between the positive x-axis and the line connecting these centroids is taken as the orientation of the LSC. The centroids and the connecting line are also indicated in Fig. 2(a). The time history and autocorrelation of the orientation ϕ is shown in Fig. 2(b). The peaks in the autocorrelation directly provide the oscillation period $\tau_0 = (1.4 \pm 0.1) \times 10^2$ s, which, nondimensionalized as $H^2/(\kappa\tau_0) = (2.6 \pm 0.2) \times 10^3$, matches well with the result $H^2/(\kappa\tau_0) = 0.084 Ra^{0.50 \pm 0.01} = 2.8 \times 10^3$ of [2]. At very small but nonzero rotation rates the LSC displays an anticyclonic precession, this motion is currently under study.

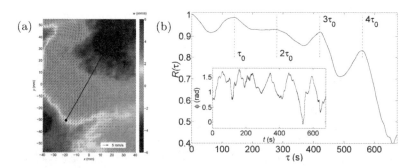

Fig. 2. (a) Example velocity snapshot. Arrows represent the horizontal component, while the greyscale is for the vertical component. The centroids and the orientation line are also included. (b) Autocorrelation $R(\tau)$ of the LSC orientation ϕ. Inset: time history of ϕ.

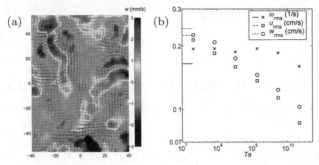

Fig. 3. (a) Velocity snapshot at $Ta = 2.2 \times 10^{10}$. (b) Root-mean square velocity (u_{rms}, w_{rms}) and vorticity (ω_{rms}) as a function of Ta. The line segments on the vertical axis indicate the rms values at $Ta = 0$.

The regime of strong rotation covers the Taylor numbers in the range $10^7 \lesssim Ta \lesssim 2 \times 10^{10}$. From velocity snapshots it can be concluded that horizontal length scales decrease as rotation is increased. The convective plumes now contain considerable vorticity. Fig. 3(a) displays a velocity snapshot at the highest $Ta = 2.2 \times 10^{10}$ used. The attenuating effect of rotation on the turbulence intensities can be observed in Fig. 3(b). The horizontal and vertical rms velocity fluctuations are plotted as a function of Ta, along with the rms value of the vertical vorticity component. These rms values are averaged over the measurement area and in time. Both horizontal and vertical rms velocities become smaller under rotation, with a power-law drop-off that scales as $Ta^{-0.13\pm0.01}$. The rms vorticity is nearly constant at a higher value than at $Ta = 0$. Only at very large Ta this value shows a decrease.

In this experiment the LSC of nonrotating convection was characterized independently, based on SPIV measurements. The oscillation frequency matched well with previous results. The addition of a very small background rotation causes anticyclonic rotation of the LSC; this is currently under study. At higher rotation rates the LSC has fragmented into small, but intense vortical regions. We observed a decrease of the turbulence intensities as well as a decreased horizontal length scale. Further investigations will include statistics of the vortical plumes and properties of the convective turbulence, such as structure functions.

The authors wish to thank the Foundation for Fundamental Research of Matter (Stichting voor Fundamenteel Onderzoek der Materie, FOM) for financial support.

References

1. M. Raffel, C. Willert, J. Kompenhans: *Particle Image Velocimetry*, (Springer, Berlin 1998) pp 174–184
2. H.-D. Xi, Q. Zhou, K.-Q. Xia: Phys. Rev. E **73**, 056312 (2006)

Lagrangian statistics in rotating turbulence through Particle Tracking experiments

L. Del Castello, H.J.H. Clercx, R.R. Trieling, and L.J.A. van Bokhoven

J.M. Burgercentrum and Fluid Dynamics Laboratory,
Department of Applied Physics, Eindhoven University of Technology,
P.O. Box 513, 5600 MB Eindhoven, The Netherlands
Contact address: l.delcastello@tue.nl

Turbulent dispersion is one of the most important subjects in turbulence research, from a fundamental point of view as well as for its numerous applications: dispersion models are used to describe diffusion of every kind of particle in fluid flows, as contaminants in the atmosphere and in marine environments, plankton in oceans, droplets in controlled combustion problems, sand grains in coastal environments.

The background rotation of the Earth starts to play a role in the flow dynamics when the relative importance of the nonlinear acceleration over the Coriolis force, expressed by the Rossby number, becomes small enough. The Coriolis force present in the non-inertial reference frame integral with the Earth has a two-dimensionalisation effect on the flow, reducing the direct energy cascade process and the energy dissipation [1, 2]. Moreover, the Coriolis force leads to the formation of columnar vortex structures and Ekman boundary layers close to the horizontal no-slip boundaries; these Ekman layers are responsible for an enhancement of the vertical mixing by pumping effects.

The Lagrangian viewpoint in turbulent diffusion is not only natural, but also practical. One- and two-particles dispersion studies have been introduced by Taylor and Richardson in the twenties, with important additional contributions by Batchelor few decades later on. In its recent review of the Lagrangian investigations of turbulence [3], Yeung points out the lack of experimental data; only in the last few years the development of particle tracking techniques allowed to access high resolution multi-particle statistics directly in the Lagrangian frame [4, 5].

The first aim of this work is to feed the fundamental investigation of turbulence with experimental data, giving further insight into the anisotropic effects of rotation on the basic characteristics of turbulence and on one- and two-particle turbulent dispersion in the inertial range, and quantifying this anisotropy through the comparison of the horizontal (normal to the rotation axis) components and the vertical ones.

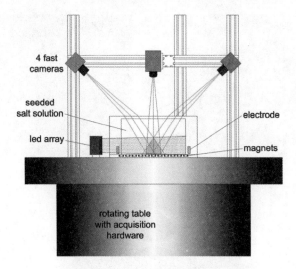

Fig. 1. Sketch of the setup on the rotating table: the electromagnetical forcing system and the PTV hardware are shown.

Series of experiments of electromagnetically forced turbulence are performed in a confined tank (500x500x300 mm in dimension), put on a rotating table (Fig. 1): the forcing acts in the bottom region of the container, inducing a highly 3D turbulent flow ($Re_\lambda \sim 200$) which decays along the upward vertical direction. The background rotation is varied between 0 and 5.0 rad/s, corresponding to a Rossby number in the forced bottom region between ∞ and 0.11.

The characterization of the turbulence with and without background rotation can be achieved through stereoscopic PIV (SPIV) measurements [6], which allow us to recover the three components of velocity on horizontal planes. Recent experiments reveal the turbulent decay with height in absence of background rotation, see Fig. 2. However SPIV does not allow for simultaneous velocity field measurements at distinct heights, thus preventing access to vertical correlations. Therefore, it is necessary to perform measurements inside a 3D domain: Particle Tracking allows us to recover the three velocity components for a measurement volume, so that gradients and correlations in all three directions can be accessed.

A 3D-PTV technique, based on the tracking code developed at ETH (Zurich) [7, 8], is used to extract trajectories and velocities in a cubic volume inside the container. The newly-designed PTV system consists of four high speed cameras and a LED array as a continuous light source, and is able to track hundreds of particles in a volume comparable with the integral scale of the flow, and with space- and time-resolutions adequate to resolve the Kolmogorov scales.

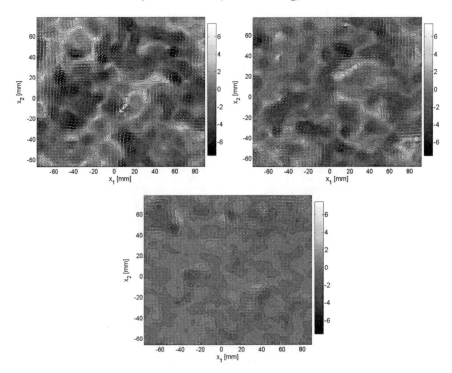

Fig. 2. Horizontal velocities (vectors) and vertical vorticity (colours, units in s^{-1}) at three vertical heights H, as obtained from SPIV measurements of a highly 3D stationary turbulent flow without background rotation — SPIV instead of PIV avoids strong perspective errors in the in-plane displacements. Upper left panel, H=20 mm; upper right panel, H=50 mm; lower panel, H=100 mm. The snapshots result from distinct measurements under identical conditions, but correspond to the same time after forcing is initiated.

References

1. L. Jacquin, O. Leuchter, C. Cambon, J. Mathieu: J. Fluid Mech. **220**, 1 (1990)
2. P. Orlandi: Phys. Fluids **9**, 1 (1997)
3. P.K. Yeung: Annu. Rev. Fluid Mech. **34**, 115 (2002)
4. S. Ott, J. Mann: J. Fluid Mech. **422**, 207 (2000)
5. M. Bourgoin, N. Ouellette, H. Xu, J. Berg, E. Bodenschatz: Science **311**, 835 (2006)
6. L.J.A. van Bokhoven, H.J.H. Clercx, G.J.F. van Heijst, R.R. Trieling: Proc. 11th European Turbulence Conference, Porto, Portugal (2007)
7. H.G. Maas, A. Gruen, D. Papantoniou: Exp. Fluids **15**, 133 (1993)
8. J. Willneff: A Spatio-Temporal Matching Algorithm for 3D Particle Tracking Velocimetry. PhD Thesis, Swiss Federal Institute of Technology, Zurich (2003)

Turbulent Thermal Convection in a Vertical Channel - Correlation Length and Turbulent momentum exchanges

M. Gibert, F. Chillà, B. Castaing

Laboratoire de Physique de l'École Normale Supérieure de Lyon, CNRS UMR5672, 46 Allée d'Italie, F-69007 Lyon, France
mgibert@ens-lyon.fr

1 Introduction

In a previous publication [1], we have introduced a new paradigm in convection: Turbulent Thermal Convection in a Vertical infinitely long Channel. Compared to the classical Rayleigh-Bénard (RB) case, this situation has the advantage of avoiding plates and thus, their neighborhood in which is usually concentrated most of the temperature gradient. In this respect, this new configuration should better correspond to many interesting situations such as convection in stars or planets, or more generally geophysical and astrophysical flows. It is also a good example of permanent mixing process induced by gravity.

2 Inertial Turbulent Thermal Convection

In a channel which is not infinite, but a very good approximation of this ideal case [1], we have measured the classical quantities that characterize buoyant flows: the Nusselt and the Reynolds numbers as a function of the Rayleigh and Prandtl numbers. When compared to the RB case, the ideal case of the infinite channel, with its assumed translational invariance along the vertical direction, also asks for slightly different definitions of those dimensionless numbers. The average vertical temperature gradient (β) will be very important to define them but we also need a length scale L. In these definitions, $Ra_n \propto \beta L^4$, $Nu \propto \beta^{-1}$ and $Re_n \propto L^2$ (see [1] for more details). The length L we choose is based on statistical measurements[1]. As shown in the figure 1 (a) where we have represented its evolution with respect to $\log(Re_n)$, this length is clearly flows' dependent. We have already found that all the influence of the viscosity

---

[1] $L = \theta/2\beta$ where θ is a temperature fluctuation measured inside the channel.

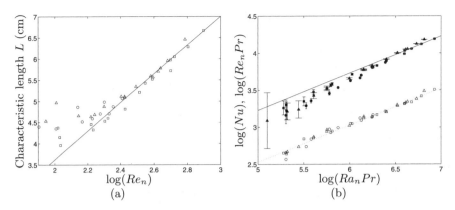

Fig. 1. Circles: $20°C(Pr = 7)$; triangles: $30°C(Pr = 5.3)$; squares: $40°C(Pr = 4)$. (a) Behavior of the characteristic length L versus $\log(Re_n)$, the line corresponds to $L/d = 0.14 \ln(Re_n) - 0.3$. (b) Full symbols: Nusselt number. Open symbols: Reynolds number. With Re_n and Ra_n **based on the characteristic length** L, the experiments at different Prandtl merge. The dotted line corresponds to $Re_n Pr = 1.07\sqrt{Ra_n Pr}$ and the full one to $Nu = 5.37\sqrt{Ra_n Pr}$.

is hidden in the behavior of this characteristic length L. Indeed, the figure 1 (b) enlightens clearly the inertial behavior of this kind of turbulent convection, which leads to the following equation :

$$Nu \propto Re_n.Pr$$

This relation between the turbulent heat flux and the momentum flux can be very interesting in a wide range of applications. Keeping in mind this idea, our work is now focused on the comparison of L to the temperature correlation length to enhance our understanding of this complex phenomenon.

3 Turbulent Momentum Exchanges

Using Particle Imaging Velocimetry, we are able to measure the velocity field inside our system : the channel. In figure 2 (a) we show a screen-shoot of the vertical component of the velocity field in the central (Oxz) plane of the channel. The velocity bin size of the colormap has been chosen in a way that one can see, despite the turbulent motion, that the flow is mainly going upwards on the left-hand side of the channel and downwards on the other hand. Very long time measurements have been done and we have been able to show that this velocity field topology reverses horizontally with a characteristic time longer than any other time in the system.

Based on these observations, we have made thorough investigations on these long time measurements to analyse this turbulent flow. To illustrate this topic,

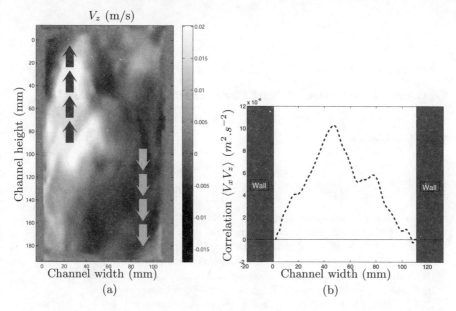

Fig. 2. (a) Velocity projected along the vertical direction in the central (Oxz) plane of the channel, (b) Correlation $\langle V_x V_z \rangle (x)$ versus the width of the channel.

Figure 2 (b) represents the correlation $\langle V_x V_z \rangle (x)$ averaged over the vertical direction in the central region of the channel where β (the mean temperature gradient) is experimentally known to be constant (measures reported in [1]) and averaged carefully over the time taking into account the reversal effects. This curve teaches us the very interesting result that momentum exchanges between the flow and the lateral walls of the channel are negligible compared to the exchange in between ascending and descending flows.

All those measures and observations allowed us to build a simple model, based on the mixing length theory, which describe properly the main components of this flow to go further in the analysis of this new paradigm in thermal turbulent convection.

References

1. M. Gibert, H. Pabiou, F. Chillá, B. Castaing: Phys. Rev. Lett. **96**, 084501 (2006)
2. E. Calzavarini, D. Lohse, F. Toschi, et Al.: Phys. Fluids **17**, 055107 (2005)
3. F. Perrier, P. Morat, and J. L. Le Mouel: Phys. Rev. Lett. **89**, 134501 (2002)
4. M. Debacq, J. P. Hulin, et Al.: Phys. Fluids **15**, 3846 (2003)
5. J. H. Arakeri, Conference & Euromech Colloquium #480 on High Rayleigh Number Convection, Trieste, Septembre 2006

Momentum and Heat Transfer in Turbulent Boundary Layers with External Grid Turbulence

Kouji Nagata[1], Yasuhiko Sakai[1] and Satoru Komori[2]

[1] Department of Mechanical Science and Engineering, Nagoya University,
Nagoya 464-8603, Japan. nagata@nagoya-u.jp
[2] Department of Mechanical Science and Engineering, Nagoya University,
Nagoya 464-8603, Japan. ysakai@mech.nagoya-u.ac.jp
[3] Department of Mechanical Engineering and Science, and Advanced Research
Institute of Fluid Science and Engineering, Kyoto University,
Kyoto 606-8501, Japan. komori@mech.kyoto-u.ac.jp

1 Introduction

Turbulent boundary layer (TBL) near a planer surface is encountered in many industrial and environmental flows. There are many situations where turbulent boundary layer evolves or lies beneath a turbulent free-stream. The purpose of this study is to investigate the momentum and heat transfer in isothermal and strong convective TBLs over a planer surface with external grid turbulence (i.e., free-stream turbulence) by means of laboratory experiments.

2 Experiments

The apparatus used was a wind tunnel with a glass test section of 5m long (x), 0.3m high (y) and 0.3m wide (z). The heating apparatus on the flat bottom wall consists of an aluminium plate (1mm), a silicon rubber heater and an insulator (3mm). The wall temperature was monitored using a resistance thermometer and controlled at the constant temperature using PID control unit. For the experiments with external turbulence, a turbulence-generating grid ($M = 50$mm) was installed upstream of TBLs to generate nearly isotropic external turbulence having $1.3 \sim 2.4\%$ turbulence levels. Two sets of experiments have been conducted: the first is isothermal boundary layer and the second is strong convective boundary layer with heating the bottom surface. In the latter case, temperature difference between the air flow and the surface was set to 60K, giving Obukhov length scale $-0.99 \sim -0.06$m (typical boundary layer thickness being approximately 0.1m). Instantaneous velocities and

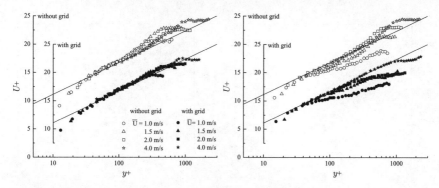

Fig. 1. Mean velocity profiles in (a) isothermal and (b) convective TBLs.

temperature are measured using a hot wire anemometry (DANTEC 55P61) and a constant-current resistance-thermometer (DANTEC 55C16).

3 Results

Figure 1 shows the mean velocity profiles. In both cases, the developments of the mean velocity field in the inner region of the TBL are insensitive to the external grid turbulence. On the other hand, the outer wake regions are affected by the grid turbulence. The results for the isothermal cases are consistent with previous studies (e.g., Evans 1973; Hancock & Bradshaw 1983).

Figure 2 shows the Reynolds stress profiles near the wall, normalised by the friction velocity. The Reynolds stresses are strongly suppressed by the grid turbulence in most part of the boundary layer except the near-wall region. The results for the isothermal cases are consistent with Hancock & Bradshaw (1983) for the case where the length scale of free-stream turbulence L is smaller than the boundary layer thickness δ. They suggested that the scale ratio L/δ as well as the intensity of free-stream turbulence is important to determine the turbulent structure in TBL. However, Evans (1973) found the increase of Reynolds stress for the case $L/\delta < 1$. The results suggest that the alternation of turbulent structure in TBL depends not only on turbulence level in the free-stream and length-scale ratio L/δ, but also on some other factors. In the strongly convective case of $\overline{U} = 1\text{m/s}$, the reduction of the Reynolds stress is not significant because of the strong convective motions in the boundary layer.

Figure 3 shows the profiles of vertical turbulent heat flux near the wall in the convective boundary layer, normalised by the friction velocity and friction temperature. The suppressions of vertical turbulent heat flux by the grid turbulence are also observed as well as the Reynolds stress. On the other hand, it is found that the skin friction and heat transfer at the surface increase under the effect of the grid turbulence.

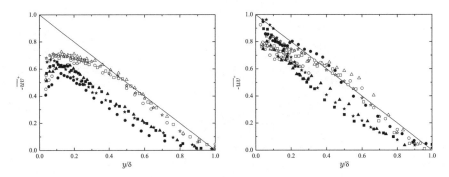

Fig. 2. Vertical distributions of the normalised Reynolds stress near the wall in (a) isothermal and (b) convective TBLs. Symbols as in Fig.1.

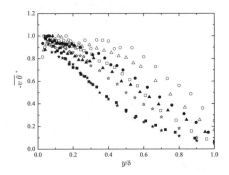

Fig. 3. Vertical distributions of the normalised vertical turbulent heat flux near the wall in the convective TBL. Symbols as in Fig.1.

Acknowledgements This study was partially supported by the JSPS Grants-in-Aid for Scientific Research (No.18686015) and the Center of Excellence for Research and Education on Complex Functional Mechanical Systems (COE program of the Ministry of Education, Culture, Sports, Science and Technology, Japan).

References

1. R. L. Evans: ARCCP, No.1282 (1973).
2. P. E. Hancock, P. Bradshaw: J. Fluid Mech., **205**, 45 (1989).

Study on Jet Mixing Rate Based on Controlled Jets

K. TSUJIMOTO[1], S. KARIYA, T. SHAKOUCHI and T. ANDO

Div. of Mech., Grad. Sch. of Eng., Mie Univ., 1577 Kurimamachiya-cho, Tsu, 514-8507 Japan tujimoto@mach.mie-u.ac.jp

1 Introduction

In various industrial applications, enhancement of heat transfer and mixing is required to be more improved. In particular, jet mixing which is fundamental technology in many industrial applications, have been investigated both experimentally and numerically. Recently, because of a rapid progress of computer technology, highly accurate calculation such as DNS (direct numerical simulation) are conducted, and the evolution of the detailed structures are tracked. In addition controlled jets are examined by DNS as well as experiment. However, so far it has not been enough to discuss about whether mixing itself is effective for the controlled jets. As most simple measure, it has been generally considered that jet width, turbulence energy, decay of centerline velocity of jet, *etc*, however, these measures do not always represent accurately the mixing rate. In the present paper, we prepare DNS data of the controlled jets and evaluate a new measure based on entropy .

2 Numerical procedure

The flow is assumed to be incompressible. Thus, the governing equations are the continuity, momentum and energy equations:

$$\nabla \cdot \mathbf{u} = 0 \tag{1}$$

$$\frac{\partial \mathbf{u}}{\partial t} + \omega \times \mathbf{u} = -\frac{1}{\rho}\nabla p + \nu \nabla^2 \mathbf{u} \tag{2}$$

$$\frac{\partial T}{\partial t} + \frac{\partial u_i T}{\partial x_i} = \frac{1}{RePr}\frac{\partial^2 T}{\partial x_i^2} \tag{3}$$

where $\mathbf{u}, \omega, \rho$, and ν denote velocity, vorticity vector, density, and kinematic viscosity of fluid, respectively. The nonlinear terms are written in the rotational form $\omega \times \mathbf{u}$ to conserve the total energy; thus, p represents the total pressure. The Cartesian coordinate system is employed[1] . Computational conditions such as the size of the computational domain, grid number, and Reynolds number are $(H_x, H_y, H_z) = (7D, 15D, 7D)$, where D is the nozzle diameter, $(L_x, L_y, L_z) = (256, 200, 256)$, Reynolds number $Re = 1500$ and Prandtl number $Pr = 0.707$, respectively. The spatial discretization involves a Fourier series expansion in the x and z directions and sixth-order compact scheme [2] in the streamwise direction. A top-hat profile of both velocity and temperature is imposed as an inflow boundary condition. In the present simulation, two cases of excitation pattern are examined: axial and helical disturbances are superimposed on the inlet velocity.

3 Mixing rate based on entropy

It is well-known that entropy, which is a measure expressing randomness, increases with homogenization. Thus entropy is expected to be correlate to the mixing rate. This idea has been proposed by Everson, et al. [3]. In the present paper, according to the literature [3], the temperature is correlates to the mixing state and finally the entropy, S is defied as follows:

$$S = T_t \cdot \ln(T_t) - \sum_i^M \{T_{(i)} \cdot \ln(T_{(i)})\} \tag{4}$$

where, $T_{(i)}$ means temperature at each grid point and T_t is summed on the plane perpendicular to the streamwise direction! %

4 Results

(a) No excitation   (b) Axial excitation   (c) Helical excitation
Fig. 1 Instantaneous vortical structure

In order to visualize the vortical structures, iso-contours of second invariance of velocity gradient tensor are shown in Fig. 1. In case of natural jet, vortex ring like structures upstream are regularly formed due to the Kelvin-Helmholtz instability and interact with each other, then the fine-scale vortices are generated after the break down. On the other hand, in case of axial excitation, strong vortex structure are formed compared with the no excitation case. In case of helical excitation, the helical structures are continuously formed for downstream and the break down rapidly occurs than the other cases. From these results, it is manifested that helical excitation induce th most mixing enhancement. As measure of mixing rate, jet width, turbulence energy and entropy are evaluated for jet mixing. Distributions of these quantities are shown in Fig.2. From Fig.2(a), irrespective control procedure, jet width abruptly expand downstream, and it is considerably different from the streamwise variation of visualized coherent vortices. As not shown here, the jet width of non-circular jets is unable to uniquely defined because of their anisotropic expansion. Fig. 2(b) shows the distribution of integrated turbulent kinetic energy on cross section. It is found that turbulence is attenuated downstream while the mixing state is mature. Also since the maximum value of helical excitation is lowest in all case, therefore the turbulence energy does not reflect the mixing state. On the other hand, since the distribution of entropy monotonically increase for downstream in Fig. 2(c), the entropy enables to correctly evaluate the mixing state.

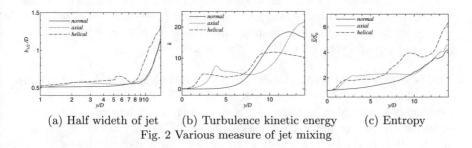

(a) Half wideth of jet   (b) Turbulence kinetic energy   (c) Entropy

Fig. 2 Various measure of jet mixing

5 Conclusion

The measures of mixing rate are investigated based on the DNS data of the controlled jets. As a result, it is found that entropy is good measure to describe the mixing state between different controlled jets and that the fluctuating entropy distribution enable to represent the highly mixed region and the production term is correlate to the vortical structure. These findings suggest that this measure is expected to contribute to the optimization of jet mixing.

References

1. Silva, C. B. and Metais, O., 2002, *Vortex control of bifurcating jets: A numerical study*, Phys. of Fluids **14**, pp.3798–3819.
2. Lele, S. K., 1992, Compact finite difference schemes with spectral-like resolution, J. of Comp. Phys. **103**, pp.16–42.
3. Everson, R., Manin, D. and Sirovich, L., 1998, Quantification of Mixing and Mixing Rate from Experimental Observations , *AIAA J.* **36**, pp.121–127.

Mixing study of a jet in crossflow using accurate thermal anemometry techniques

Jean-Paul MORO[1], Pierre FOUGAIROLLE[1], and Yves GAGNE[2]

[1] CEA Grenoble 17 rue des Martyrs - 38054 Grenoble Cedex 9, France
   Jean-Paul.Moro@cea.fr & Pierre.Fougairolles@cea.fr
[2] Université Joseph Fourier - BP 53 - 38041 Grenoble Cedex 9, France
   Yves.Gagne@hmg.inpg.fr

The objective of this experimental work is to focus on the turbulent mixing between a pipeflow exiting heated jet and a confined crossflow, with both kinematic and passive scalar measurements in order to characterize its statistical properties (energy spectra, pdf, skewness, anisotropy...).

The experimental set-up consists of a grid-turbulence flow in a rectangular wind tunnel (60x50cm^2 in section and 5m long) in which emerges a perpendicular rectangular pipeflow (8x5cm^2 section, equivalent to an hydraulic diameter D_h=6.2cm for a circular jet). Ranges of velocity are U=0-20m.s^{-1} and U_j=0-40m.s^{-1} for the crossflow and the jet respectively. Measurements are performed with classical hot-wire anemometry and cold-wire thermometry but with a peculiar care in the apparatus construction. A unique and reproductible technique allows us to make probes with Wollaston wire of 500nm diameter, which can be used to measure velocity fluctuations as well as temperature ones. To avoid bad tuning effects, we have chosen a constant voltage anemometer home-made in CEA (SSTH/LIEX) on the basis of the method developed by Sarma [1] and Comte-Bellot & al. Our thermo-anemometers lead to a signal to noise ratio up to 1000 for the velocity and a sensitivity of 0.015 °C for the temperature.

Temperature maps (in the scalar passive case, $\Delta T = T_j - T$=7.5 °C) with a momentum ratio $r = U_j/U$ of 10, measured 3D_h downstream of the jet inlet, are displayed on figure 1. It reveals the global signature of the counter-rotating vortex pair (CVP) analogously to the round jet case (cf. Smith and Mungal [2] and Fric [3]).

Figure 2 shows mean temperature profiles measured in the crossflow channel at several sections downstream of the jet inlet. We clearly observe that there is no self-similarity between these latters. That can be explained by the jet interaction with the opposite boundary wall since 8D_h. Even for distances under 5D_h, the shapes of the curves seem different.

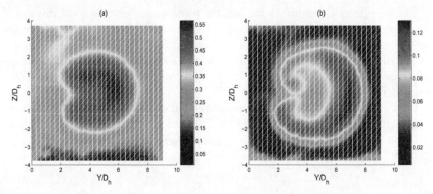

Fig. 1. Temperature field (normalized by ΔT) interpolated on 29x24 measurement points in the section located at $3D_h$ downstream of the jet inlet. a) Average value b) rms

Figure 3 shows the temperature pdf locally obtained on the center of the jet hot spot in the same sections. Despite the lack of similarity at large scales, the statistics exhibit some common features ; particularly, they are skewed whatever the distance from the jet inlet.

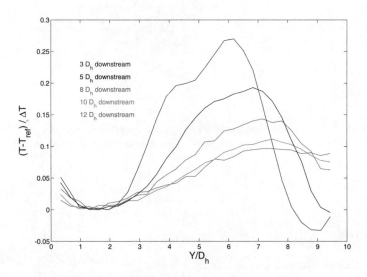

Fig. 2. Normalized mean temperature profiles measured on the center of the jet inlet $(Z/D_h = 0)$ at several sections downstream. Jet emerges from $Y/D_h = 0$ and the opposite wall is situated at $Y/D_h = 9.7$.

The merging of the curves for hot fluctuations suggests a good mixing of these ones. On the other side, the quasi-exponential negative wings reveals the existence of "cold" fluid structures coming from the crossflow, which are not yet uniformly mixed. These wing become longer as the cold fluctuations diffuse. Such statistics correspond to the typical "ramp", characteristic of a temperature mixing signal.

Further investigations will explore statistical properties of the external side of this transversal hot jet, especially its interaction with the crossflow channel wall, and of its internal side which is mainly affected by the wake of the jet inlet.

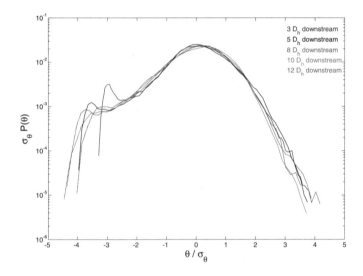

Fig. 3. Pdf of the temperature fluctuations θ normalized by its standard deviation, measured in the center of the jet hot spot at several sections downstream of the jet inlet.

References

1. G.R. Sarma : Review of scientific instruments Vol. **69** N ° **6** (1998)
2. S.H. Smith, M.G. Mungal : J. Fluid Mech. **357**, 83-122 (1998)
3. T.F. Fric : Structure in the near field of the transverse jet. PhD Thesis, California Institute of Technology (1990)

Experimental study of turbulent transport of particles in non-isothermal flows and formation of large-scale structures

A. Eidelman, T. Elperin, N. Kleeorin, I. Rogachevskii, I. Sapir-Katiraie

Department of Mechanical Engineering, Ben-Gurion University of Negev, Beer-Sheva, Israel `elperin@bgu.ac.il`

1 Introduction

The main goal of this communication is to review our recent experimental studies of turbulent transport of inertial particles and formation of large-scale structures in spatial distribution of particles. In particularly, we discuss a new phenomenon of turbulent thermal diffusion which has been predicted theoretically in [1]. The essence of this phenomenon is the appearance of a nondiffusive mean flux of particles in the direction of the mean heat flux. This phenomenon causes formation of large-scale inhomogeneities in the spatial distribution of particles that accumulate in the regions of minimum mean temperature of the surrounding fluid.

The mechanism of turbulent thermal diffusion for solid particles is as follows. The inertia causes particles inside the turbulent eddies to drift out to the boundary regions between eddies (i.e., regions with low vorticity or high strain rate and maximum fluid pressure) and accumulate there. Similarly, there is an outflow of particles from regions with minimum fluid pressure. In homogeneous and isotropic turbulence without large-scale external gradients of temperature, a drift from regions with increased (decreased) concentration of particles by a turbulent flow of fluid is equiprobable in all directions, and pressure (temperature) of the surrounding fluid is not correlated with the turbulent velocity field. Thus, there exists only turbulent diffusion of particles. The situation drastically changes in a turbulent fluid flow with a mean temperature gradient, whereby the turbulent heat flux $\langle \mathbf{u}\,\theta \rangle$ is not zero, i.e., fluctuations of fluid temperature θ and velocity $\mathbf{u}$ of the fluid are correlated. Fluctuations of temperature cause pressure fluctuations, which result in fluctuations of the number density of particles. Increase (decrease) of the pressure of the surrounding fluid is accompanied by accumulation (outflow) of the particles. Therefore, the direction of mean flux of particles coincides with that of the turbulent heat flux, i.e. $\langle \mathbf{u}\,n \rangle \propto \langle \mathbf{u}\,\theta \rangle \propto -\boldsymbol{\nabla} T$, where T is the mean fluid temperature and n are the fluctuations of particle number density. The mean flux of particles is directed to the minimum mean temperature, and the particles are accumulated in this region.

The effect of turbulent thermal diffusion has been detected experimentally in air flow in two experimental apparatuses: oscillating grids turbulence generator [2]-[4] and multi-fan turbulence generator [5]. These devices are capable of producing a confined homogeneous turbulent flow with a small mean velocity. We use Particle Image Velocimetry to determine the turbulent velocity field, a specially designed temperature probe with twelve sensitive thermocouples to measure the temperature field, and an Image Processing Technique based on an analysis of the intensity of Mie scattering to determine the spatial distribution of particles. We have performed experiments with two directions of the mean temperature gradient, for stably stratified fluid flow (the cooled bottom and heated top walls of the chamber) and for unstably stratified fluid flow (the heated bottom and cooled top walls of the chamber). In these experiments, it has been found that particles are accumulated in regions with minimum mean temperature.

2 Experimental set-up and results

In the experiments in the oscillating grids turbulence generator the test section is constructed as a rectangular chamber of dimensions $30 \times 60 \times 30$ cm^3. Pairs of vertically oriented grids with bars arranged in a square array (with a mesh size 5 cm) are attached to the right and left horizontal rods. This two grids system is capable of oscillating at a controllable frequency up to 20 Hz. A mean temperature gradient in the turbulent flow is formed with two aluminium heat exchangers attached to the bottom and top walls of the chamber which allow us to form a mean temperature gradient up to 200 K/m for the stably stratified flow and up to 110 K/m for the unstably stratified flow at a mean temperature of about 300 K. An incense smoke with sub-micron particles is used in our experiments which are produced by high temperature sublimation of solid incense particles.

We have determined the mean and the r.m.s. velocities, two-point correlation functions and an integral scale of turbulence from the measured velocity fields. The turbulent flow parameters in the oscillating grids turbulence generator are as follows: the r.m.s. velocity is $3.6 - 14$ cm/s depending on the frequency of grid oscillations, the integral scale of turbulence is $1.6 - 2.3$ cm, the Kolmogorov length scale is $380 - 660\,\mu$m. The measured r.m.s. velocity is several times higher than the characteristic mean velocity in the core of the flow. The effect of the gravitational settling of small particles $(0.5 - 1\,\mu$m$)$ is negligibly small since the terminal fall velocity of these particles is less than 0.01 cm/s. The accuracy of the measurements in these experiments $(\sim 0.5\%)$ is considerably higher than the magnitude of the observed effect $(\sim 10\%)$. We have demonstrated that even in strongly inhomogeneous three-dimensional mean temperature fields, the locations of regions with the minimum mean temperature are strongly correlated with the locations of the regions with the maximum particle number density due to the phenomenon of turbulent thermal diffusion.

Turbulent thermal diffusion is a fundamental phenomenon which should be studied for different types of turbulence and different experimental set-ups. The phenomenon of turbulent thermal diffusion has been also detected in multi-fan turbulence generator. A multi-fan turbulence generator includes eight fans (with rotation frequency of 2800 rpm) mounted in the corners of a cubic Perspex box of dimensions $40 \times 40 \times 40$ cm^3 and facing the center of the box. The turbulent flow parameters in the multi-fan turbulence generator are as follows: the r.m.s. velocity is $70 - 80$ cm/s, the integral scale of turbulence is $1.5 - 1.6$ cm, the Kolmogorov length scale is $100 - 110 \, \mu$m. In the experiments the maximum mean flow velocity is of the order of $0.1 - 0.2$ m/s. Measurements performed using different concentrations of particles in the flow have shown that particles are accumulated in regions with minimum mean temperature.

3 Conclusions

Our experiments have detected the effect of turbulent thermal diffusion. We have demonstrated that turbulent thermal diffusion occurs independently of the method of turbulence generation. The coefficient of turbulent thermal diffusion increases with increase of Reynolds number, e.g.., the value of the coefficient of turbulent thermal diffusion obtained in the experiments in multi-fan turbulence generator is larger than that obtained in the experiments in oscillating grids turbulence, where the Reynolds numbers are smaller than those achieved in the multi-fan turbulence generator. The developed theory [1, 6] is in agreement with the experimental results. The phenomenon of turbulent thermal diffusion can cause formation of the large-scale aerosol layers in the atmospheric turbulence with temperature inversions.

References

1. T. Elperin, N. Kleeorin, I. Rogachevskii: Phys. Rev. Lett. **76**, 224 (1996); Phys Rev E **55**, 2713 (1997).
2. A. Eidelman, T. Elperin, N. Kleeorin, A. Krein, I. Rogachevskii, J. Buchholz, G. Grünefeld: Nonlinear Processes in Geophysics **11**, 343 (2004).
3. J. Buchholz, A. Eidelman, T. Elperin, G. Grünefeld, N. Kleeorin, A. Krein, I. Rogachevskii: Experiments in Fluids **36**, 879 (2004).
4. A. Eidelman, T. Elperin, N. Kleeorin, A. Markovich, I. Rogachevskii: Nonlinear Processes in Geophysics **13**, 109 (2006).
5. A. Eidelman, T. Elperin, N. Kleeorin, I. Rogachevskii, I. Sapir-Katiraie: Experiments in Fluids **40**, 744 (2006).
6. T. Elperin, N. Kleeorin, I. Rogachevskii, D. Sokoloff: Phys Rev E **64**, 026304 (2001).

A theory of relative dispersion in homogeneous turbulence

Pasquale Franzese[1] and Massimo Cassiani[2]

[1] Dept. of Computational and Data Sciences, George Mason University, Fairfax, VA 22030, USA. pfranzes@gmu.edu
[2] Dept. of Civil and Environmental Engineering, Duke University, Durham, NC 27708, USA. mcassian@duke.edu

1 Introduction

We derive a differential equation for the mean square separation of passive particles from the centre of mass of a cluster in homogeneous isotropic turbulence. The derivation is based on a statistical diffusion theory of relative dispersion, and on the inertial subrange scaling form of the turbulent kinetic energy of separation.

The theory is consistent with Taylor's [11] absolute dispersion theory and with the Richardson–Obukhov [9, 7] relation $\langle r^2 \rangle = C_r \varepsilon t^3$, where $\langle r^2 \rangle$ is the mean square separation between particles, ε is the mean dissipation of kinetic energy and C_r is a constant. The present theory provides an estimate for C_r. Comparisons to DNS and laboratory observations are shown. Details of the derivation can be found in [3].

2 Governing equations

Position and velocity of a particle relative to the centre of mass of a cluster along an arbitrary axis are denoted by y_r and v_r, respectively.

The statistic $\langle y_r^2(t) \rangle$ satisfies the differential equation:

$$\frac{\mathrm{d}}{\mathrm{d}t} \langle y_r^2 \rangle = 2 \langle v_r^2(t) \rangle \int_0^t R_r(t, \tau) \mathrm{d}\tau \tag{1}$$

where angle brackets indicate averaging over an ensemble of realizations of the flow field, and we defined the relative velocity autocorrelation function $R_r(t, \tau) = \langle v_r(t) v_r(t - \tau) \rangle / \langle v_r^2(t) \rangle$. Ultimately, we obtain:

$$\frac{\mathrm{d}}{\mathrm{d}t} \langle y_r^2 \rangle = \frac{4 \langle v_r^2 \rangle^2 t}{\langle v_r^2 \rangle + C_o \varepsilon t} \tag{2}$$

where C_o is the Lagrangian velocity structure function constant. Equation (2) can be solved numerically, or analytically in the inertial subrange. It provides, among the others, the Richardson–Obukhov constant C_r which appears in the equation $\langle r^2 \rangle = C_r \varepsilon t^3$, namely

$$C_r = (18\sqrt{6} - 44)C_o \tag{3}$$

or $C_r \approx \frac{1}{11}C_o$. Assuming $C_o = 7$, we obtain $C_r = 0.64$.

The relation $C_k \propto C_o^{2/3}$ is also established, where C_k is the Eulerian velocity structure function constant. It follows that an explicit dependence of C_r and C_k on the Reynolds number Re_λ can be simply written as $\widetilde{C}_r = C_r f(Re_\lambda)$ and $\widetilde{C}_k = C_k f(Re_\lambda)^{\frac{2}{3}}$, where tilde indicates values at finite Reynolds number, and one can use Sawford's [10] estimate $f(Re_\lambda) = (1 + 7.5C_o^2 Re_\lambda^{-1.64})^{-1}$.

2.1 Comparisons with experiments and DNS

The present theory was tested with available data. $\widetilde{C}_r$ is plotted in figure 1 along with experimental [8] and DNS [5, 1, 2] data. The predicted value $C_r = 0.64$ is supported by the reported data. We plot in figure 2 the complete

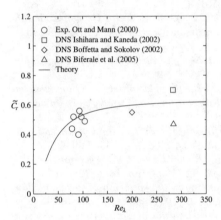

Fig. 1. $\widetilde{C}_r$ as a function of Re_λ.

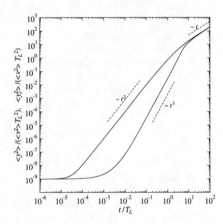

Fig. 2. Absolute and relative variances $\langle y^2 \rangle$ and $\langle y_r^2 \rangle$.

solution for $\langle y_r^2 \rangle$, along with Taylor's $\langle y^2 \rangle$. Note that $\langle y_r^2 \rangle$ has a unique representation as a function of t/T_L, likewise $\langle y^2 \rangle$. Also, the large time behaviour of $\langle y_r^2 \rangle$ is not a consequence of ad hoc assumptions, but is a natural result of the theory. Figure 3 shows the predicted $\widetilde{C}_k$ along with data from various DNS at several Reynolds numbers [13, 5, 4, 12]. Figure 4 shows the predicted $R_r(t, \tau)$ as a function of the single variable τ/t along with experimental data in two-dimensional turbulence [6] and DNS data [1]. All data clearly support the predicted dependence of $R_r(t, \tau)$ on the single variable τ/t.

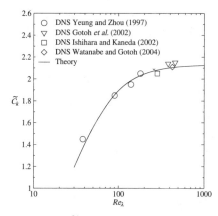

Fig. 3. $\tilde{C}_k$ as a function of Re_λ.

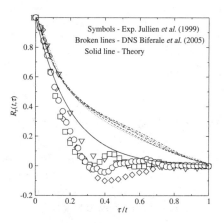

Fig. 4. $R_r(t, \tau)$ as a function of τ/t

References

1. L. Biferale, G. Boffetta, A. Celani, B. J. Devenish, A. Lanotte, and F. Toschi, *Lagrangian statistics of particle pairs in homogeneous isotropic turbulence*, Phys. Fluids **17** (2005), no. 11, 115101/1–115101/9.
2. G. Boffetta and I. M. Sokolov, *Relative Dispersion in Fully Developed Turbulence: The Richardson's Law and Intermittency Corrections*, Phys. Rev. Lett. **88** (2002), no. 9, 094501/1 – 094501/4.
3. P. Franzese and M. Cassiani, *A statistical theory of turbulent relative dispersion*, J. Fluid Mech. **571** (2007), 391–417.
4. T. Gotoh, D. Fukayama, and T. Nakano, *Velocity field statistics in homogeneous steady turbulence obtained using a high-resolution direct numerical simulation*, Phys. Fluids **14** (2002), no. 3, 1065–1081.
5. T. Ishihara and Y. Kaneda, *Relative diffusion of a pair of fluid particles in the inertial subrange of turbulence*, Phys. Fluids **14** (2002), no. 11, L69–L72.
6. M.-C. Jullien, J. Paret, and P. Tabeling, *Richardson Pair Dispersion in Two-Dimensional Turbulence*, Phys. Rev. Lett. **82** (1999), no. 14, 2872–2875.
7. A. M. Obukhov, *On the distribution of energy in the spectrum of turbulent flow*, Izv. Akad. Nauk USSR, Ser. Geogr. Geofiz. **5** (1941), no. 4–5, 453–466.
8. S. Ott and J. Mann, *An experimental investigation of the relative diffusion of particle pairs in three-dimensional turbulent flow*, J. Fluid Mech. **422** (2000), 207–223.
9. L. F. Richardson, *Atmospheric diffusion shown on a distance-neighbour graph*, Proc. Roy. Soc. Lond. Ser.-A **110** (1926), no. 756, 709–737.
10. B. L. Sawford, *Reynolds number effects in Lagrangian stochastic models of turbulent dispersion*, Phys. Fluids **3** (1991), no. 6, 1577–1586.
11. G. I. Taylor, *Diffusion by continuous movements*, Proc. Lond. Math. Soc. **20** (1921), 196–211.
12. T. Watanabe and T. Gotoh, *Statistics of a passive scalar in homogeneous turbulence*, New J. Phys. **6** (2004), no. 40, 1–36.
13. P. K. Yeung and Y. Zhou, *Universality of the Kolmogorov constant in numerical simulations of turbulence*, Phys. Rev. E **56** (1997), no. 2, 1746–1752.

Experimental Study of Hysteresis Phenomenon in Turbulent Convection

A. Eidelman, T. Elperin, N. Kleeorin, I. Rogachevskii, I. Sapir-Katiraie

Department of Mechanical Engineering, Ben-Gurion University of Negev,
Beer-Sheva, Israel gary@bgu.ac.il

1 Introduction

Large-scale coherent structures in turbulent convection are observed in the atmospheric convective boundary layers and numerous laboratory experiments in the Rayleigh-Bénard apparatus. The life-times of coherent structures are very long compared to the largest turbulent time-scales. Their properties differ from those of small-scale turbulence. As a result the turbulence and the coherent structures interact in practically the same way as the turbulence and the mean flow. These structures show more similarity in their behavior with regular flows than with turbulence. They can be identified as the motions, whose spatial and temporal scales are much larger than the characteristic turbulent scales.

The new mechanism of formation of the large-scale coherent structures in turbulent convection has been proposed recently in [1, 2]. It has been suggested that the redistribution of the turbulent heat flux plays a crucial role in the formation of the large-scale circulations in turbulent convection. In particular, two competitive effects, namely redistribution of the vertical turbulent heat flux due to convergence or divergence of the horizontal mean flows, and production of the horizontal component of the turbulent heat flux due to the interaction of the mean vorticity with the vertical component of the turbulent heat flux, cause the large-scale instability and formation of the large-scale coherent structures in turbulent convection (see [1, 2]). The modification of the turbulent heat flux results in strong reduction of the critical Rayleigh number (based on the eddy viscosity and turbulent temperature diffusivity) required for the excitation of the large-scale instability and formation of the large-scale coherent structures in turbulent convection [3].

The main goal of this communication is to describe the experimental study of large-scale circulations of turbulent thermal convection in air flow (the aspect ratios $A = 2 - 2.23$). In order to study large-scale circulations we use Particle Image Velocimetry to determine the turbulent and mean velocity fields, and a specially designed temperature probe with twelve sensitive thermocouples is employed to measure the temperature field. Coherent large-scale circulations of turbulent thermal convection in air have been studied experimentally in a rectangular box heated from below and cooled from above.

The hysteresis phenomenon in turbulent convection has been found by varying the temperature difference between the bottom and the top walls of the chamber (the Rayleigh number varies within the range of $10^7 - 10^8$). The hysteresis loop comprises the one-cell and two-cells flow patterns while the aspect ratio is kept constant. We have found that the change of the sign of the degree of the anisotropy of turbulence is accompanied by the change of the flow pattern [4]. The developed theory [1, 2] of coherent structures in turbulent convection is in agreement with the experimental observations. The observed coherent structures are superimposed on a small-scale turbulent convection.

2 Experimental set-up and results

The experiments are conducted in a rectangular chamber with dimensions $26 \times 26 \times 58$ cm in air flow. The side walls of the chamber are made of transparent Perspex with the thickness of 10 mm. A number of experiments have been also conducted with different additional thermal insulation of the side walls of the chamber in order to study whether a heat flux through the side walls affects the turbulent convective pattern. The temperature difference between the bottom and the top walls of the chamber is changed within a range of 5 to 80 K (the Rayleigh number varies within the range of Ra $\approx 10^7 - 10^8$). We have studied the detailed three-dimensional structure of the velocity field in the large-scale circulations.

Increasing the temperature difference from 5 to 15 K we observe first two-cell flow pattern with the downward motions in the central region of the chamber between two cells, then one-cell flow pattern within a range of the temperature difference of 20 to 30 K with the counterclockwise mean flow. Further increase of the temperature difference from 35 to 80 K results in two-cell flow pattern with the upward mean flow in the central region of the chamber between the two cells. Decreasing the temperature difference we observe two-cell flow pattern within a range from 80 to 45 K with the upward flow in the region between two cells, then one-cell flow pattern with clockwise mean flow within a range of the temperature difference of 40 to 20 K. Further decrease of the temperature difference from 15 to 5 K results in two-cell flow pattern with the downward mean flow in the central region of the chamber between the two cells.

In all experiments the transition from the two-cell to one-cell flow patterns has been found to be accompanied by the change of the sign of the degree of anisotropy χ of turbulent velocity field $\mathbf{u}$, which is defined as $\chi = (4/3)[\langle \mathbf{u}_y^2 \rangle / \langle \mathbf{u}_z^2 \rangle - 1]$. We have determined in the experiment the dependence of the degree of anisotropy χ of turbulent velocity field on the temperature difference between the bottom and the top walls of the chamber. The parameter χ is negative for the two-cell flow pattern and is positive for the one-cell flow pattern. The hysteresis phenomenon has been observed in dependence $\chi(\text{Ra})$.

The experimental observations can be explained by invoking the theory of formation of coherent structures developed in [1,2]. We take into account that the aspect ratio of the large-scale cell in the two-cell flow pattern observed in the experiment is approximately 1, while in the one-cell flow pattern it is approximately 2. We have found that the growth rate of the large-scale instability for the two-cell mode, γ_2, is larger than that for the one-cell mode, γ_1, for negative values of the degree of anisotropy χ of turbulent velocity field, and $\gamma_2 < \gamma_1$ for positive values of the parameter χ.

We have also measured distributions of the heat fluxes in the large-scale circulations. Thermal structure inside the large-scale circulation is inhomogeneous and anisotropic. The hot thermal plumes accumulate at one side of the large-scale circulation, and cold plumes concentrate at the opposite side of the large-scale circulation.

3 Conclusions

This study demonstrates that the anisotropy of turbulent velocity field plays a crucial role in the hysteresis phenomenon that has been found in our experiments in turbulent convection (the aspect ratios of the chamber is $A = 2 - 2.23$ and the Rayleigh number varies within the range of $\text{Ra} \approx 10^7 - 10^8$). The observed transition from the two-cell to one-cell flow patterns causes a drastic change of the degree of anisotropy χ of the turbulent velocity field from negative to positive values. This finding is in a good agreement with the theoretical predictions. We demonstrate that the redistribution of the turbulent heat flux plays a crucial role in the formation of coherent large-scale circulations in turbulent convection. The obtained results may be important in atmospheric turbulent convection and laboratory turbulent flows.

References

1. T. Elperin, N. Kleeorin, I. Rogachevskii, S. Zilitinkevich: Phys. Rev. E **66**, 066305 (2002).
2. T. Elperin, N. Kleeorin, I. Rogachevskii, S. Zilitinkevich: Boundary-Layer Meteorology **119**, 449 (2006).
3. T. Elperin, I. Golubev, N. Kleeorin, I. Rogachevskii: Phys. Fluids **18**, 126601 (2006).
4. A. Eidelman, T. Elperin, N. Kleeorin, A. Markovich, I. Rogachevskii: Experiments in Fluids **40**, 723 (2006).

Velocity and temperature derivatives in high Reynolds number turbulent flows in the atmospheric surface layer

G. Gulitski[1], M. Kholmyansky[1], W. Kinzelbach[2], B. Lüthi[2], A. Tsinober[1,3], and S. Yorish[1]

[1] Engineering Faculty, Tel Aviv University, 69978 Tel Aviv kholm@eng.tau.ac.il
[2] Institute of Environmental Engineering, ETH Zürich, CH-8093 Zürich
[3] Institute for Mathematical Sciences and Department of Aeronautics, Imperial College London

This is a report on a field experiment in an atmospheric surface layer which is a larger scale continuation of experiments (see [1, 2] and references therein) performed using a system providing explicitly the full set of velocity and temperature derivatives both spatial and temporal.

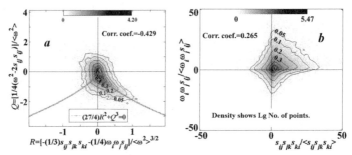

Fig. 1. Joint PDF of the second invariant, $Q = (\omega^2 - 2s_{ik}s_{ik})/4$, and the third invariant, $R = -s_{ik}s_{km}s_{mi}/3 + 3\omega_i\omega_k s_{ik}/4$, of the velocity gradient tensor (*a*) and joint PDF of $\omega_i\omega_j s_{ij}$ and $-4s_{ij}s_{jk}s_{ki}/3$ (*b*).

The experiment was performed in a valley to the south of St. Moritz at heights between 0.8 and 10 m with the Taylor micro-scale Reynolds number in the range $1.6 - 6.6 \cdot 10^3$. A brief description of facilities, methods and results reported before is given in [1, 2] and references therein. The main emphasis of this paper is given to the results not obtained before. Two typical examples include joint statistics of enstrophy and strain production and the tear-drop $R - Q$ plot (Fig. 1) and the statistical dependence between large and small scales (Fig. 2).

Fig. 3 and Fig. 4 exemplify a set of results concerning Lagrangian accelerations, $\mathbf{a} = D\mathbf{u}/Dt$, and its Eulerian components: the local acceleration,

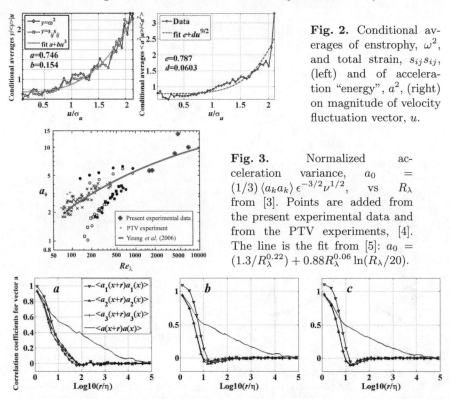

Fig. 2. Conditional averages of enstrophy, ω^2, and total strain, $s_{ij}s_{ij}$, (left) and of acceleration "energy", a^2, (right) on magnitude of velocity fluctuation vector, u.

Fig. 3. Normalized acceleration variance, $a_0 = (1/3)\langle a_k a_k\rangle\,\epsilon^{-3/2}\nu^{1/2}$, vs R_λ from [3]. Points are added from the present experimental data and from the PTV experiments, [4]. The line is the fit from [5]: $a_0 = (1.3/R_\lambda^{0.22}) + 0.88R_\lambda^{0.06}\ln(R_\lambda/20)$.

Fig. 4. The auto-correlations of vectors $\mathbf{a}$ (a), $\mathbf{a}_c$ (b), $\mathbf{a}_l$ (c) and their Cartesian components from field experiment. While all the vectors and their components are correlated over relatively short distances (typically 10 Kolmogorov lengths), their moduli are correlated over large distances of order 10^4 Kolmogorov lengths, which is comparable with the integral scale, just like for Lagrangian correlations both in DNS and laboratory experiments. This behavior is observed also in PTV experiments, [4].

$\mathbf{a}_l = \partial\mathbf{u}/\partial t$, and the convective one, $\mathbf{a}_c = u_k\partial\mathbf{u}/\partial x_k$ or $(\mathbf{u}\cdot\nabla)\mathbf{u}$, the issue known as random Taylor (sweeping decorrelation) hypothesis, and associated issues of geometrical statistics of accelerations, involving these and a variety of other Eulerian components of the total acceleration.

Finally Fig. 5 and Fig. 6 represent results concerning temperature with the emphasis on joint statistics of temperature and velocity derivatives. Apart of some conventional results these contain a variety of results concerning production of temperature gradients, $\mathbf{G} = \nabla\theta$: role of vorticity and strain, eigen-contributions, geometrical statistics, tilting of $\mathbf{G}$, comparison of the true production of $\mathbf{G}$ with its surrogate. Among the specific results of importance is that production of $\mathbf{G}$ is much more intensive in regions dominated by strain, whereas it is practically independent of the magnitude of vorticity.

The present experiments went far beyond the previous ones in two main respects. The first one is that all the data were obtained without invok-

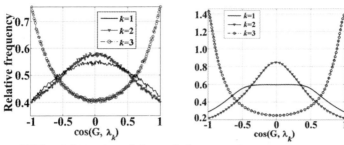

Fig. 5. PDFs of the cosine of the angle between the temperature gradient, **G**, and the eigenframe, λ_i, of the rate of strain tensor, s_{ij}. Field experiment (left); direct numerical simulation of Navier–Stokes equations (right).

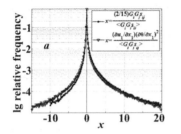 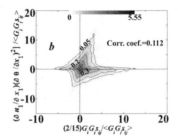

Fig. 6. PDFs of the true production rate of the temperature gradient, $-G_i G_k s_{ik}$, and its surrogate, $-(\partial u_1/\partial x_1)(\partial\theta/\partial x_1)^2$, (*a*) and their joint PDF (*b*). Though univariate PDFs of the production rate of the temperature gradient and its surrogate are very similar (which indicates the tendency to isotropy in small scales by this particular criterion), their joint PDF is not close to a bisector. This means that the true production rate of the temperature gradient is far from being fully represented by its surrogate.

ing the Taylor hypothesis and therefore a variety of results on fluid particle accelerations became possible. The second is that simultaneous measurements of temperature and its gradients with the emphasis on joint statistics of temperature and velocity derivatives are provided.

References

[1] B. Galanti, G. Gulitsky, M. Kholmyansky, A. Tsinober, S. Yorish, *Proceedings of the Tenth European Turbulence Conference*, H. Andersson, P.-Å. Krogstad, eds. (CIMNE, 2004), vol. X of *Advances in Turbulence*, pp. 267–270.
[2] M. Kholmyansky, A. Tsinober, S. Yorish, *Phys. Fluids* **13**, 311 (2001).
[3] A. Gylfason, S. Ayyalasomayajula, Z. Warhaft, *J. Fluid Mech.* **501**, 213 (2004).
[4] B. Lüthi, A. Tsinober, W. Kinzelbach, *J. Fluid Mech.* **528**, 87 (2005).
[5] P. K. Yeung, S. B. Pope, A. G. Lamorgese, D. A. Donzis, *Phys. Fluids* **18** (2006). Art. No. 065103.

Role of Turbulence for Droplet Condensation

Antonio Celani[1], Andrea Mazzino[23], Agnese Seminara[12], and Marco Tizzi[23]

[1] Centre National de la Recherche Scientifique, Institut Non Linéaire de Nice,
   1361 Route des Lucioles, 06560 Valbonne, France
   antonio.celani@inln.cnrs.fr, agnese.seminara@inln.cnrs.fr
[2] Università di Genova, Dipartimento di Fisica, Via Dodecaneso 33, I-16146
   Genova, Italia mazzino@fisica.unige.it, tizzi@fisica.unige.it
[3] Istituto Nazionale di Fisica Nucleare, Sezione di Genova, Via Dodecaneso 33,
   I-16146 Genova, Italia

Introduction

Despite the recent improvements in the comprehension of cloud development, many fundamental issues still must be fully understood. Measures in clouds reveal a broad size distribution of small droplets, while classical air-parcel models point to narrowing size spectra during the condensation stage [1]. Including the effect of entrainment of dry air inside the cloud [2] can only partially justify wide size spectra in the cloud boundaries, whereas the spreading of size distribution is measured in cloud cores as well. Other mechanisms have been proposed to explain such property: stochastic fluctuations in the vapour field [3], its local fluctuations due to droplet reaction on the vapour [4, 5] or the effects of developed turbulence on an ascending moist air-parcel [6, 7], with however very limited degree of spreading of the size distribution.

A recent simple model of condensation in clouds isolated a spreading mechanism [8], with the idea that droplets and vapour are correlated since they are advected by the same velocity field. Since every droplet experiences the same ambient conditions for a timescale comparable to the large-eddy turnover time τ_L, droplets belonging to very moist regions are able to grow faster than the ones correlated to less moist regions, thus spreading the size distribution and showing the importance of large-scale spatial fluctuations of the vapour field.

The present work aims at verifying the robustness of the identified mechanism in a more realistic framework, including thermal convection due to an imposed temperature gradient and the effect of droplet feedback on vapour.

Model

The model used in the two-dimensional numerical simulations consists of three equations for the Eulerian fields (vorticity ω, temperature fluctuation

ϑ around the mean profile $-Gz$ and supersaturation s) coupled with $3N$ equations for the Lagrangian evolution of the N droplets (trajectories $\boldsymbol{X}_i$ and radii R_i with $i = 1, \ldots, N$), assumed as tracer particles (according to [8]):

$$\begin{cases} \partial_t \omega + \boldsymbol{v} \cdot \boldsymbol{\partial}\omega = \beta \boldsymbol{g} \times \boldsymbol{\partial}\vartheta + \nu\partial^2\omega \\ \partial_t\vartheta + \boldsymbol{v} \cdot \boldsymbol{\partial}\vartheta = (G - \Gamma)w + \kappa\partial^2\vartheta \\ \partial_t s + \boldsymbol{v} \cdot \boldsymbol{\partial}s = A_1 w - \frac{1}{\tau_s}s + D\partial^2 s \end{cases} \qquad \begin{cases} \dot{\boldsymbol{X}}_i(t) = \boldsymbol{v}(\boldsymbol{X}_i(t), t) + \sqrt{2D}\boldsymbol{\eta}_i \\ \dot{R}_i^2(t) = 2A_3 s(\boldsymbol{X}_i(t), t) \end{cases}$$

where ν, κ and D are the kinematic viscosity, the thermal and the vapour diffusivities respectively, β is the thermal expansion coefficient, Γ is the adiabatic lapse rate, A_1 and A_3 are constant coefficients and $\boldsymbol{\eta}_i$ are independent white-noises.

In this work we introduce a convective turbulent Bolgiano regime due to the presence of the dry Boussinesq buoyancy term. Moreover the droplet evolution can affect the value of s via the s/τ_s coupling, where $\tau_s^{-1}(\boldsymbol{x}, t) \propto \sum_i R_i(t)$ (i running on the $N(\boldsymbol{x}, t)$ droplets being in a neighborhood of $\boldsymbol{x}$ at time t) is the *absorption frequency* field. As mentioned at the beginning, the previous results were based on the presence of a strong correlation between droplets trajectories and the vapour field. The local feedback introduced could in principle change this mechanism, thus varying the general picture drawn in [8]. Indeed, if correlation effects are still present, dry regions will be void of droplets, providing no vapour loss there; all droplets will be segregated in moist regions, where they will grow consuming the surrounding vapour, thus slowing down their growth. Therefore the air will become undersaturated on average and a mean-field-type argument will provide no mean growth of radii.

Results

We performed a series of high resolution (1024^2) Direct Numerical Simulations of model system, putting 10^6 droplets randomly in space once the stationary state is reached. Since we want to take into account the effect of large-scale fluctuations (droplets being in fact able to span a very large volume of the cloud), we choose to simulate the evolution of the whole cloud, not resolving the small-scale details and abandoning the air-parcel approach.

Results indicate that the correlation mechanism is still determining. Since therefore droplets populate only supersaturated regions (as shown in Fig. 1) the Lagrangian average of s is positive and $\langle R \rangle$ continue to grow for some τ_L, despite the constant negative trend of $\langle s \rangle$. Finally from Fig. 2 we can still observe a considerable spreading of droplet spectra. On the other hand the comparison with the case without local droplet feedback on vapour clarifies that all these effects are reduced: this feedback weakens but does not break down the spreading mechanism.

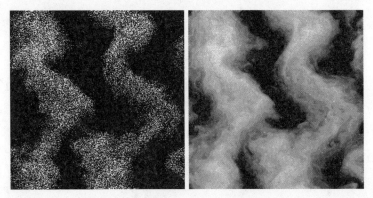

Fig. 1. Left panel: snapshot of the field $\tau_s^{-1}(\boldsymbol{x}, t)$; by definition τ_s^{-1} vanishes where no particles are present (black regions). Right panel: snapshot of the supersaturation field $s(\boldsymbol{x}, t)$ at the same time $t = 0.1\,\tau_L$; moist regions are represented in white. Droplet trajectories and vapour field are clearly strong correlated.

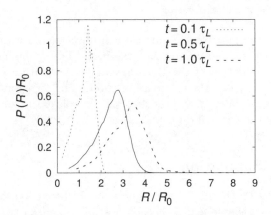

Fig. 2. Droplet size spectrum when droplet feedback on vapour is accounted ($R_0 = 4\,\mu$m): after one τ_L the droplet maximum size is of about $25\,\mu$m, more reasonable than in the case without feedback, where droplet radii reach even $100\,\mu$m.

References

1. S. Twomey: Pure Appl. Geophys. **43**, 1 (1959)
2. J. W. Telford: Atmos. Res. **40** (1996)
3. J. T. Bartlett, P. R. Jonas: Q. J. R. Meteorol. Soc. **98** (1972)
4. R. A. Shaw, W. C. Reade, L. R. Collins, J. Verlinde: J. Atmos. Sci. **55** (1998)
5. A. Jaczewski, S. P. Malinowski: Q. J. R. Meteorol. Soc. **131**, 609 (2005)
6. P. A. Vaillancourt, M. K. Yau, W. W. Grabowski: J. Atmos. Sci. **58** (2001)
7. P. A. Vaillancourt, M. K. Yau, P. Bartello, W. W. Grabowski: J. Atmos. Sci. **59** (2002)
8. A. Celani, G. Falkovich, A. Mazzino, A. Seminara: Europhys. Lett. **70**, 6 (2005)

Kinematic Simulation and Rapid Distortion Theory, analyses of one and two-particle diffusion in stably stratified and rotating turbulence

F. Nicolleau[1] and G. Yu[2]

[1] The Univ. of Sheffield, Dept of Mech. Eng., Mappin Street, S1 3JD, Sheffield, UK F.Nicolleau@Sheffield.ac.uk
[2] Engineering Department, Queen Mary, Univ. of London, Mile End Road, E1 4NS, London, UK g.yu@qmul.ac.uk

The properties of one-particle and particle-pair diffusion in rotating and stratified turbulence are studied by applying the Rapid Distortion Theory (RDT) to a Kinematic Simulation of the Boussinesq equation with a Coriolis term. We particularly emphasize the existence of two $t-$ and t^2-regimes of different nature in rotating turbulence, and the inconclusive role of the spatial structure in the capping of the vertical diffusion.

1 Kinematic Simulation

Kinematic Simulation (KS) is a Lagrangian model where a synthetic Eulerian velocity field $u(x,t)$ is assumed. It is then possible to track one, two or many particles and study their trajectories by integrating this velocity field. We use the KS developed in [1] for stratified and rotating turbulence.

2 Existence of two t-regimes generated by rotation.

When studying the one-particle rms diffusion in the horizontal plane $< \zeta_h^2 >$ as a function of time τ, one will eventually find linear relations. However, these τ-régimes are of very different natures in pure stratification and pure rotation:

i) In pure stratification, the τ-régime is the well-known random walk or Brownian motion that appears when the particle has been diffusing for longer than the turbulence characteristic time, it is independent of the strength of the stratification (N) and always appears at a time that scales with the turbulence time scale L/u'.

ii) In pure rotation the τ-régime is not non-linear or random walk by nature, it is independent of the turbulence characteristic time and appears when the particle has been diffusing for longer than $1/\Omega$ where Ω is the rotation rate.

Having that picture in mind, it is then easier to understand the main pattern of one-particle horizontal diffusion in a turbulent stratified and rotating flow as illustrated in figure 1.

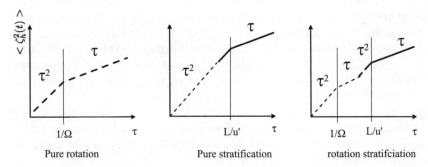

Fig. 1. Schematic scenario of superposition of rotation and stratification diffusion patterns. Dash lines show what can be predicted by the simplified Corrsin hypothesis.

This is confirmed in figure 2, when rotation is superimposed on stratification, but $\Omega < N$, the one-particle horizontal diffusion barely changes its pattern from the one in purely stratified turbulence. The effect of the superimposed rotation is therefore negligible, although the tendency to have a transition τ−régime between two τ^2−régimes can be already observed when $\Omega \simeq N$.

As Ω increases and $\Omega > N$, the diffusion in the early ballistic τ^2−régime is not affected but this régime is shortened, it ends with the rotation ballistic régime which length was shown to scale with $1/\Omega$. This ballistic régime is followed by a τ-régime the length of which increases with Ω. This τ-régime is the rotation-τ-régime, it is not a random walk régime and in this régime the particle has not forgotten its initial position. So that for longer times ($\sim 1/N$) when stratification waves develop, a pure stratification pattern - i. e. τ^2 up to L/u' and then a random walk - is superimposed onto the rotation pattern yielding that typical feature of a τ-régime in between two τ^2-régimes as sketched in figure 1.

3 Role of spatial structures

KS for stratified turbulence, as it is generated in this paper, does not contain any information about spatial structures such as layers or columns that are observed respectively for stratified turbulence and rotating turbulence. Recent and high-resolution results of such structures can be found in [2]. So how important are these structures in the prediction of particles' diffusion? First what matters in Lagrangian tracking is Lagrangian correlations not Eulerian's ones. So it is necessary to get accurate Lagrangian velocity correlation. Comparisons with DNS [3, 4] show that the KS without the Eulerian structures predicts accurately the diffusion for one and two particles for stratified, rotating, and stratified and rotating turbulence.

Conversely, it is worth noting that layered-structures cannot explain the main feature of diffusion in stratified turbulence that is that the vertical diffusion exhibits a plateau when $N\tau > 1$. Indeed the energy argument that $N^2 < \zeta^2(\tau) >$ must be bounded is valid with or without rotation as the Coriolis force does not work. Therefore whatever $N \neq 0$ there exists a plateau irrespective of the Eulerian spatial structures predominantly layer-like or column-like (or neither when $2\Omega = N$).

We can conclude that linear time-oscillations that are contained in KS are necessary and sufficient to predict accurately particles' diffusion in stratified and/or rotating turbulence whereas Eulerian structures such as layers or columns are neither sufficient nor necessary.

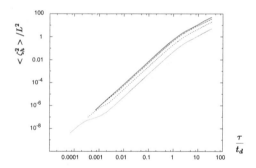

Fig. 2. Effects of rotation on stratification with $N = 500$. Non-dimensional one-particle mean square horizontal displacement $< \zeta_h^2 > /L^2$ is plotted as a function of τ/t_d from top to bottom $\Omega = 0$, 125, 500 and 2500.

References

1. F. Nicolleau, G. Yu, and J. C. Vassilicos, Fluid Dyn. Res. In Press (2006).
2. L. Liechtenstein, C. Cambon, and F. Godeferd, Journal of Turbulence **6**, 1 (2006).
3. Y. Kimura and J. R. Herring, Proc. of FEDSM99 **FEDSM99-7753** (1999).
4. C. Cambon, F. S. Godeferd, F. Nicolleau, and J. C. Vassilicos, J. Fluid Mech. **499**, 231 (2004).

Turbulent Flow Structure in the Similarity Region of a Swirling Jet

Abolfazl Shiri[1], Sara Toutiaei[2], and William K. George[3]

[1] Turbulence Research Laboratory, Department of Applied Mechanics, Chalmers University of Technology, SE-41296, Gothenburg, Sweden
   `abolfazl@chalmers.se`
[2] Turbulence Research Laboratory, Department of Applied Mechanics, Chalmers University of Technology, SE-41296, Gothenburg, Sweden
   `toutiaei@student.chalmers.se`
[3] Turbulence Research Laboratory, Department of Applied Mechanics, Chalmers University of Technology, SE-41296, Gothenburg, Sweden
   `wkgeorge@chalmers.se`

Several previous investigations of the near field of swirling jets have shown that these jets grow at a faster rate than non-swirling jets and experience significant changes in the turbulence quantities (c.f. Gilchrist, Naughton [1]). A recent study by Shiri et al.[2], however, showed that the growth rate enhancement does not persist in the far-field of a swirling jet flow with moderate swirl numbers (0.15 and 0.25). The results were shown to be consistent with the equilibrium similarity theory of Ewing [3] in which the mean swirl velocity was argued to decrease downstream as $1/(x - x_o)^2$, while the mean streamwise velocity decreased as $1/(x - x_o)$. In fact the only statistically significant effect of the swirl on the mean velocity for even the highest swirl number was a shift in the virtual origin (to $x/D_* = 0.75$ from -2.9) as shown in Figure 1.

The present investigation extends the previous study to include all three velocity components of the turbulence quantities at a swirl number of $S = 0.25$. Since only a two-component LDA was available, this was accomplished by making traverses in both the vertical and horizontal directions. All moments to third order were obtained, excepting those involving both the azimuthal and radial components simultaneously.

Some of the results are shown in Figures 2 and 3, which also show the earlier measurements of Hussein et al. [4] in the same jet (but without swirl). The second and third-order moments are quite close to the earlier non-swirling results. But unlike the mean velocity and spreading rate (which were nearly identical), the differences may be significant and a consequence of the swirl. As noted by George [5], if there were an effect of the source conditions on the similarity profiles, it is in the second and higher moment profiles where it would be expected to appear. On the other hand, the differences in second-

order moments near the centerline could also be accounted for by a slight
angular misalignment in either experiment ($< 0.5\,$deg).

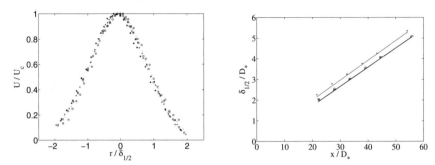

(a) Normalized mean stream-wise velocity profiles at $x/D = 20, 25, 30, 35, 40, 50$.

(b) Stream-wise variation of the velocity
profiles half-width. S: $0, 0.15, 0.25$; Slopes:
6.87, 6.86, 6.81; Virtual origins: 0.755,
0.754, -2.859

Fig. 1. Mean axial velocity

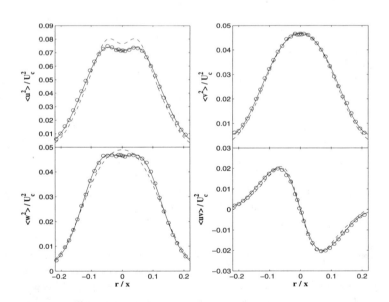

Fig. 2. Second-order moments (symbols) compared with non-swirling jet curve-fit
suggested in Hussein et al.[4] (dashed line).

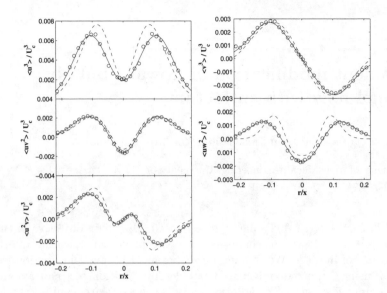

Fig. 3. Third-order moments (symbols) compared with non-swirling jet curve-fit suggested in Hussein et al.[4] (dashed line).

Acknowledgments The authors would like to acknowledge especially the assistance of Prof. Jonathan Naughton of the U. Wyoming for initiating and inspiring this experiment. This work was supported by Vetenskaprådet, the Swedish Research Foundation.

References

1. R.T. Gilchrist, J.W. Naughton: AIAA J., **43**, 741 (2005)
2. A. Shiri, W.K. George, J.W. Naughton: 36th AIAA Fluid Dynamics Conference and Exhibit, AIAA 2006-3367
3. D. Ewing: Proceedings of the 4th International Symposium on Engineering Turbulence Modeling and Experiments, Elsvier (1999)
4. H.J. Hussein, S.P. Capp, W.K. George: J. of Fluid Mech., **258**, 31 (1994)
5. W.K. George: The self-preservation of turbulent flows and its relation to initial conditions and coherent structures. In: *Advances in Turbulence*, (Hemisphere, NY, 1989) pp 39–73

Turbulent modification of upward bubbly channel flow with surfactant

Toshiyuki Ogasawara[1], Shu Takagi[1] and Yoichiro Matsumoto[1]

Department of Mechanical Engineering, University of Tokyo, Japan
ogasawara@fel.t.u-tokyo.ac.jp

Bubbly flow in a vertical channel is investigated. Our main objective of this research is to clarify the mechanism of the turbulence modification due to the presence of bubbles. We conduct our experiment at low void fraction less than 1% using 1mm mono-dispersed bubbles by addition of small amount of surfactant. The surfactant addition cause so-called Marangoni effect and change the boundary condition of the bubble surface, though do not change the turbulent structure directly in this case. Our previous study reveals that the slightly contaminated bubbles migrate toward the wall in upward turbulent flow, which produces the bubble clustering structure near the wall. On the other hand, highly contaminated bubbles do not have this tendency and uniformly distributed across the channel. The difference of migration behavior is well-explained with the discussion of lift force acting on bubbles and particles. This strong accumulation of bubbles near the wall changes the turbulent structure drastically. Once the bubble clusters are formed, they lift up the near-wall region and give a plug-like flow structure at the centre of the channel. Then, the turbulent fluctuation and the Reynolds stress in the liquid phase are very much reduced in this region.

1 Introduction

The early study on the measurement of bubbly flow was conducted by Serizawa et al.[1], in which local measurements of turbulent flow structure in pipe were made using a hot-film anemometer. It is also reported that the bubbles migrate toward the wall and the distribution of the local void fraction has saddle shape at low void fraction in upward pipe flow. Following these studies provided experimental information on: the local void fraction profile and its dependence on bubble size; the effect of initial bubble size and the bubble-induced liquid turbulence.

It is known that surfactants in water affect bubbles. The presence of surfactants in water avoid a bubble coalescence. The other effect explained by a Marangoni effect[2]. Though surfactant effect on single bubble motion has been fairly examined, less is known about surfactant effects on bubbly flow. Hear, we are interested in surfactant effect on a spatial distribution of bubbles in an upward channel flow and its dependence on a turbulent structure.

2 Experimental set-up

The experimental apparatus is shown in figure 1. The vertical channel has a thickness of $2H$=40mm and a width of 400mm. The channel height of 1600mm from the bubble generator to the test section is large enough for single-phase turbulent flow to be fully developed. The coordinate x denotes the stream-wise direction, y denotes the perpendicular direction from the wall and z dinotes the span-wise direction. Tap water is pumped vertically upwards through the channel. Air bubbles are generated from the bubble generator which is constructed with 474 stainless steel pipe of 0.07mm inner diameter. The bubble generator is installed above the inlet nozzle of the channel, which is located at x/H=80 downward from the test section.

The experimental conditions are shown in Table 1. 3-Pentanol (21ppm-168ppm) and Triton X-100 (2ppm) are used as surfactants and added into the liquid phase. The bulk Reynolds numbers $Re(=2HU_b/\nu)$, based on the characteristic length of the channel width $2H(=40\text{mm})$ and the bulk mean velocity of liquid phase U_b, are 1350 as a laminar flow, 4100 and 8200 as turbulent flows. The average void fraction f_g is set to be less than 1%.

The bubble motion and the spatial distribution of the bubbles are observed using high speed video camera (Motion Pro 10000, Redlake MASD, Inc.) and analysed by digital image processing. To measure the liquid phase flow

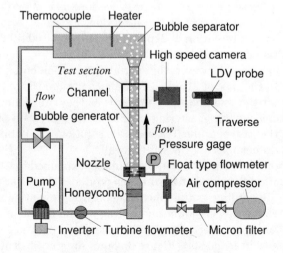

Fig. 1. Experimental apparatus

Table 1. Experimental conditions

Surfactants	3-Pentanol, Triton X-100				
Temperature []	25±0.5				
Reynolds number Re $[= 2U_b H/\nu]$	1350	4100		8200	
Reynolds number Re_τ $[= u_\tau H/\nu]$	–	147		260	
(Average void fraction [%])	(0.6)	(0.3)	(0.6)	(0.3)	(0.6)
Superficial velocity of gas injection [mm/s]	0.5	0.5	1.0	0.7	1.4

structure we apply a two-color fiber Laser Doppler Velocimetry (LDV) system previously described by So et al.[3]. In the present experiment, the Kolmogorov length and time scale in the test section are estimated to be the order of 100μm and 10ms.

3 Results

3.1 Surfactant effect on the lateral migration of bubbles

Figure 2 shows the snapshots of bubbly flow in different surfactant conditions: (a) Tap water, (b) 3-Pentanol solution of 21ppm and (c) Triton X-100 solution of 2pmm. Left ones are view from the front of the channel ($x - z$ plane) and right ones are view from the side of the channel ($x - y$ plane). The bulk Reynolds number is 8200 and superficial gas velocity is 1.4 mm/s in every cases. The average void fraction is about 0.6%, however, it increases with the stronger Marangoni effect due to the presence of surfactant. Figure 2 (a) shows the snapshot of bubbly flow without any addition of surfactant into the tap water. The range of the bubble size becomes from 0.6mm to 4mm because of the coalescences up to the test section. The large deformed bubbles move upwards with zigzag or spiral motion. Figure 2 (b) and (c) are in the cases of 3-Pentanol solution of 21ppm and Triton X-100 solution of 2ppm, respectively. Mono-dispersed 1mm bubbles are obtained by the prevention of the bubble coalescences. However, the spatial distribution of the bubbles is totally different. In the case of 21ppm 3-Pentanol solution, the bubbles strongly accumulated near the wall and move upwards as they are sliding on the wall. At the same time, bubble clustering phenomenon can be observed. On the other hand, in the case of 2ppm Triton X-100 solution, the bubbles distribute uniformly and do not migrate toward the wall.

The surfactant effect on a bubble can be explained by a Marangoni effect[2]. Nonuniform surfactant adsorption produces the shear stress on the bubble surface. Regarding as this phenomenon from a continuum scale view point, a surface velocity gets lower and a drag on the bubble increases. That is to say, the Marangoni effect changes the boundary condition on the bubble surface from free-slip to no-slip. Figure 3 shows drag coefficient on a almost spherical bubble in stationary fluid. Re_b is bubble Reynolds number and is

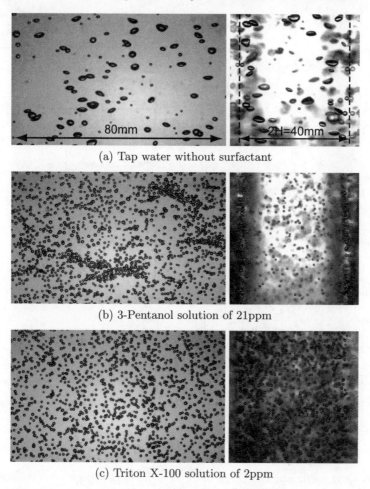

(a) Tap water without surfactant

(b) 3-Pentanol solution of 21ppm

(c) Triton X-100 solution of 2ppm

Fig. 2. Left side photographs show front ($x - z$ plane) view and right side photographs show side ($x - y$ plane) view. (Re=8200, f_g=0.6%)

defined using the bubble diameter and the relative velocity of a bubble. In 3-Pentanol solution, the bubble takes the intermediate value of C_D between the clean bubble and the rigid sphere. In Triton X-100 solution of 2ppm, C_D is almost as same as that of rigid sphere. It is clear from this that the bubble surface in 3-Pentanol solution can still slip, on the other hand, the bubble surface in Triton X-100 solution almost becomes no-slip condition.

Around Re_b=100, as is the case with our experiment, the lift coefficient of the rigid sphere is considerably less than that of the clean bubble[8][9][10]. In the case of 21ppm 3-pentanol solution, it is assumed that the lift coefficient has an intermediate value and it decreases as the surface condition approaches no-slip condition due to the increase of the concentration. In the

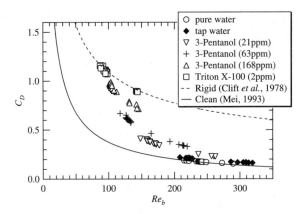

Fig. 3. Surfactant dependence on the drag coefficient C_D as a function of bubble Raynolds number Re_b.

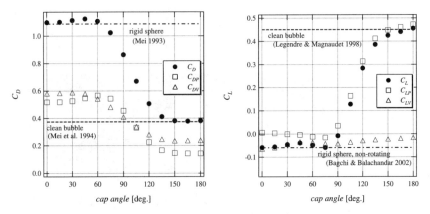

Fig. 4. Drag coefficient C_D (left) and lift coefficient C_L (right) for stagnant cap bubble in simple shear flow at $Re_b=100$. [11]

case of 2ppm Triton X-100 solution, the lift coefficient might become as same as that of rigid sphere. Fukuta et al.[11] reported numerical work on shear-induced lift on stagnant cap bubble. With the decrease of cap angle (free-slip region on a bubble surface), basically C_D increases and C_L decreases (Fig. 4). This decrease of the lift coefficient is a main reason why the tendency of the lateral migration of the bubbles toward the wall weakens with the effect of the stronger surfactant effect.

3.2 LDV measurement

In any surfactant solution in our expriment, the addition of surfactant do not affect the single-phase turbulece directly and fully developed turbulent flow can be obtained at the test section. Turbulence in the bubbly flow is modified due to the spatial distribution of bubbles in the channel. When the bubble strongly accumulated near the wall, the turbulent structure drastically changed compared to that of single-phase flow. The results of LDV measurement in the case of 21ppm 3-Pentanol solution are shown. Figure 5 shows the mean velocity profiles. Regardless of the flow state (laminar or turbulent) as single-phase flows, the velocity profiles of the corresponding bubbly flows are very similar each other. The mean velocity has a steep velocity gradient in the vicinity of the wall and a flat profile in wide center region. Figure 6 shows stream-wise and wall-normal dimensional turbulent fluctuation profiles, respectively. Dimensional fluctuations quantitatively coincide over the channel width except the near wall. It suggests that the fluctuations in this region are not related to the turbulence characteristics and are produced by the bubble motions in the flow which have small fluctuations.

Figure 7 shows the Reynolds stress profile of the liquid phase at $Re=4100$. In the case of bubbly flow, the fluctuations over most of the channel width are mainly supplied by the rising motions of bubbles. Although most of bubbles accumulate near the wall (3-Pentanol solution of 21ppm), small number of bubbles rise up in this region and create small fluctuations. Since these fluctuations are not a source of Reynolds stress, figure 7 clearly shows that the Reynolds stress in this region is much smaller than that of single-phase flows. From the turbulence theory of single-phase flow, the sum of viscous stress and Reynolds stress must be constant across the channel. Since this flow had a strong buoyancy effect near the wall where the many bubbles accumulate, this relation is not necessarily satisfied and actually the flow itself behaves more like a plug flow with a small Reynolds stress.

The present phenomena can be summarized as follows. In the case of 21ppm 3-Pentanol solution, a lift force acts on a bubble towards the wall. Due to this lateral force, bubbles form a high void fraction region near the wall and slide up it. At this stage, most of the bubbles are sliding up at the

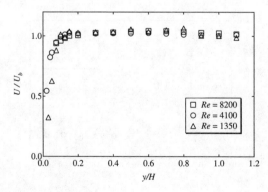

Fig. 5. Dependence of Reynolds number on mean velocity profiles. (21ppm 3-Pentanol solution, f_g=0.6%)

same distance from the wall. Once the bubble accumulation has occurred, the bubbles behave like a sheet of bubbles. This bubble sheet pulls up the surrounding liquid by its buoyancy. Then, the mean velocity profile of the liquid phase becomes steeper near the wall due to this driving force and the streamwise turbulent intensity in the vicinity of the wall is enhanced. Furthermore the mean velocity profiles of the liquid phase are flattened in the wide region of the channel center. This is because the large driving force from buoyant bubbles drastically changes the flow structure. Most of region (except in the vicinity of the wall) is suspended by the near-wall bubble sheet and a plug-like flow structure occurs. The turbulent fluctuations and Reynolds stress in the liquid are very much suppressed in this region. Thus, two regions separated by the bubble sheet have different origins for velocity fluctuations. In the narrow region between the wall and bubble sheet, the fluctuations are mainly

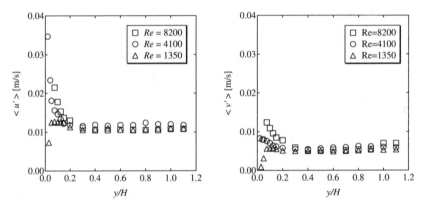

Fig. 6. Dimensional rms value of velocity fluctuation profiles. Left graph shows stream-wise fluctuation and right one shows wall-normal fluctuation. (21ppm 3-Pentanol solution, f_g=0.6%)

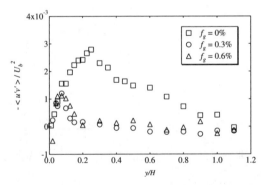

Fig. 7. Dependence of void fraction on Reynolds stress profiles. (21ppm 3-Pentanol solution, Re=4100)

produced by high shear rate turbulence. In the other regions, fluctuations are caused by the motion of bubbles rising in plug-like flow with much less fluctuations.

4 Conclusions

In low concentration 3-Pentanol solution, the bubble surface can still slip. Therefore, the lift force is generated on the bubble due to the presence of the mean-shear. Finally, the lateral migration occurs and the bubbles accumulate near the wall. On the other hand, in 2ppm Triton X-100 solution, the bubble surface almost becomes no-slip boundary. So, the shear-induced lift force acting on the bubble becomes very small. This is a dominant reason why the 1mm bubbles do not migrate toward the wall in upwardchannel flow.

Once the bubbles strongly accumulate near the wall, the turbulent structure drastically changes. The bubble sheet lifts up the surrounding liquid owing to buoyancy. Then, the mean velocity profile of the liquid phase becomes steeper near the wall due to this driving force and the stream-wise turbulent intensity in the vicinity of the wall is enhanced. Furthermore the mean velocity profiles of the liquid phase are flattened in the wide region of the channel center. This region is sustained by the bubble sheet near the wall and a plug-like flow structure developed. The turbulent fluctuation and Reynolds stress in the liquid phase are very much suppressed in this region. It is noted that the main contribution to the fluctuation observed in this region is due to the rising motion of bubbles, where the local void fraction in this region is much lower than that in the near-wall region.

References

1. A. Serizawa et al.: Int. J. Multiphase Flow **2**, 235-246 (1975)
2. V. G. Levich, *Physicochemical Hydrodynamics*, (Prentice Hall 1962)
3. S-H. So et al.: Exp. in Fluids **33**, 135-142 (2002)
4. T. R. Auton: J. Fluid Mech. **183**, 199-218 (1987)
5. R. Clift et al.: *Bubbles, drops, and particles*, (Academic Press 1978)
6. R. Mei: Int. J. Multiphase Flow **19**, 509-525 (1993)
7. P. G. Saffman: J. Fluid Mech. **22**, 385-400, (1965), and corrigendum, J. Fluid Mech. **31**, 624 (1968)
8. D. Legendre and J. Magnaudet: J. Fluid Mech. **368**, 81-126 (1998)
9. P. Bagchi and S. Balachander: Phys. Fluids **14**, 2719-2737 (2002)
10. R. Kurose and S. Komori: J. Fluid Mech. **384**, 183-206 (1999)
11. M. Fukuta et al.: Theoretical and Applied Mechanics Japan **54**, 227-234 (2005)

Turbulent clustering of inertial particles in the presence of gravity

E. Hascoët[1] and J.C. Vassilicos[2]

[1] Department of Aeronautics and Institute for Mathematical Sciences, Imperial College London e.hascoet@imperial.ac.uk
[2] Department of Aeronautics and Institute for Mathematical Sciences, Imperial College London j.c.vassilicos@imperial.ac.uk

Recent work by Chen et al (2006) [1], Goto and Vassilicos (2006) [2] and Goto and Yoshimoto (2006) [3] has confirmed the observation of Boffetta et al (2004) [4] that clustering of inertial particles in homogeneous isotropic turbulence has a multi-scale structure, and has also shown, in the case of two-dimensional inverse cascading turbulence, that this clustering is a direct reflection of the clustering of zero-acceleration points. The relation to the acceleration in 3D turbulence is not fully clear yet as shown by Goto and Yoshimoto (2006) [3] and Goto et al (2006, ETC11).

The works just mentioned neglect all effects of gravitational acceleration. Indeed, it is sometimes thought that gravity may not have an effect on clustering as it only adds a constant fall velocity to the particles. We integrate particle motions by solving

$$\frac{d}{dt}\mathbf{v}_p(t) = \frac{1}{\tau_p}\left(\mathbf{u}(\mathbf{x}_p(t), t) - \mathbf{v}_p(t)\right) + \mathbf{g}$$

for each particle, where $\mathbf{v}_p$ and $\mathbf{x}_p$ are the velocity and position vectors of particles. Here, we assume that (i) particles (rigid spheres) are very much heavier than the background fluid, (ii) the particle diameter is sufficiently small for the surrounding flow to be approximated by a Stokes flow and (iii) the number density of particles is dilute enough for collisions between particles and feedback to fluid motion to be negligible.

The velocity field $\mathbf{u}(\mathbf{x}, t)$ is obtained from a Direct Numerical Simulation of 2D inverse-cascading turbulence as in Goto & Vassilicos (2006) [2] (see also Goto et al 2006, ETC11). The turbulence has a well-defined -5/3 power law energy spectrum over a widely extending inertial range. We limit the present study to values of τ_p which are within or sufficiently close to the bounds of the inertial time-scale range.

Here, we start by showing that if the Froude number $Fr_p \equiv \tau_p g/u'$ (where τ_p is the relaxation or Stokes time of the particles, g the gravitational acceleration and u' the r.m.s. velocity of the turbulence) is significantly larger than 1, then the clustering of inertial particles is fundamentally different from when $Fr_p = 0$ or even when $Fr_p \ll 1$, which is the case that this paper mostly concentrates on. When $Fr_p \ll 1$, the inertial particle clusters are very similar to the zero-acceleration point clusters (see Figure 1). However, direct application of the random sweeping argument of Chen et al (2006) [1] and Goto and Vassilicos (2006) [2] implies that inertial particle clusters should rather resemble clusters of points where the acceleration vector $\mathbf{a} = \mathbf{g}$. Our numerical simulations show that the clusters of points where $\mathbf{a} = \mathbf{g}$ are in fact very similar to the clusters of zero-acceleration points (see Figure 2) as long that the gravitational acceleration is not too large compared to the acceleration variance. In fact, we find that the zero-acceleration clusters are similar to clusters of points where $\mathbf{a}$ equals some arbitrary constant vector of amplitude not too large compared to the acceleration variance. As a result we are led to visualise the acceleration field as composed of regions of high acceleration fluctuations where all moderate-constant-acceleration points may lie, interspersed between regions of relatively homogeneous values of $\mathbf{a}$. This property of the acceleration field's clusters is of a multiscale nature and is yet another manifestation of intermittency. It is an important property as we expect it to hold in 3D homogeneous isotropic turbulence as well as in other non-homogeneous and/or non-isotropic turbulent flows.

When $Fr_p \geq 1$, then the argument based on random sweeping of acceleration field clusters does not apply and a different clustering mechanism comes into play. In fact, in this regime, the clusters of inertial particles are qualitatively different and form columns when Fr_p is not too high.

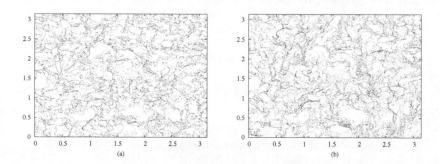

(a)   (b)

Fig. 1. (a) Spatial distribution of inertial particles for $St = \tau_p/\tau_\eta = 0.5$ and $Fr_p = 0.0$ in the $2\pi \times 2\pi$ domain of the 2D inverse-cascading turbulence. (b) Spatial distribution of inertial particles for the same Stokes number as in (a) but with $Fr_p = 0.5$. The distribution displays an almost similar clustering network as in (a).

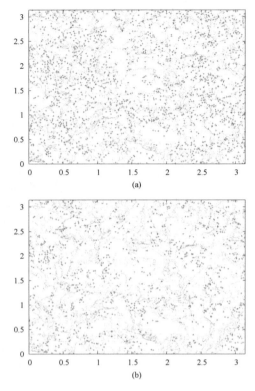

Fig. 2. (a) Spatial distribution of particles (red) for $St = \tau_p/\tau_\eta = 0.5$ and $Fr_p = 0.5$ on which is superimposed the distribution of zero-acceleration points (big blue). Zero-acceleration points cluster in the same way as inertial particles. (b) Same distribution of particles as in (a) but, here, acceleration points corresponding to $\mathbf{a} = (0, 100)$ are plotted (big blue). These acceleration points cluster in the same way as zero-acceleration points.

References

1. L. Chen, S. Goto & J. C. Vassilicos: J. Fluid Mech. **553**, 143-154 (2006)
2. S. Goto & J. C. Vassilicos: Phys. Fluids **18**, 115103 (2006)
3. H. Yoshimoto & S. Goto: submitted to J. Fluid Mech. (2006)
4. G. Boffetta, F. De Lillo & A. Gamba: Phys. Fluids **16**, L20-L23 (2004)

Acceleration measurements in turbulent-like flows

S. Ferrari, L. Rossi, J.C. Vassilicos

Imperial College London, Department of Aeronautics

Following the work of Davila and Vassilicos [1] and Goto and Vassilicos [2], who showed how the multi-scale distribution of stagnation points is related to the energy spectrum of the turbulence and how it determines pair-dispersion properties, Rossi et al. [3] developed a new laboratory experiment where a turbulent-like multiscale distribution of stagnation points could be imposed onto a quasi-two-dimensional (Q2D) laminar brine flow by multi-scale electromagnetic forces (see figure 1). They used the apparatus to show that such a turbulent-like laminar flow has a broad-band power-law energy spectrum as a result of the fractal structure of the distribution of stagnation points. Hascoet et al [4] showed that the exponent of this power law varies linearly with fractal dimension of this fractal structure, as in [1]. With the same apparatus, it was also shown [5] that such a turbulent-like distribution of stagnation points generates a Richardson-like pair-dispersion even though the flow is Q2D, laminar and steady. The link between the Eulerian multiscale stagnation point structure and the Lagrangian pair-dispersion properties exists in turbulence because of the high statistical coherence of stagnation points in the sense that, on average, they move very slowly compared to fluid elements and they have a life-time comparable to the integral time-scale (see [6, 7]) which is why it has been particularlly pertinent to start these experiments in steady flow.

Significant activity has been devoted recently to the study of the fluid acceleration field, e.g. [8, 9, 10, 11, 12, 13]. As explained in Pope's book (2000), many phenomenological and stochastic models of mixing rely on some understanding of or assumptions on the fluid acceleration field. We therefore report here a study of the acceleration properties of the multiscale Q2D laminar and steady flow of [3] which has various turbulent-like properties as mentioned above. Are the acceleration statistics in this flow also turbulent-like? And to what extent are these statistics mostly determined by the Eulerian intermittency of the turbulence and/or by the time-dependence and multiscale coherence of some parts of the turbulence field? The second question is particularly meaningful in the light of recent kinematic simulations [14] which have shown that the correlation signature of Lagrangian accelerations obtained by [13] in measurements of turbulent flows reflects the persistence of the underlying streamline structure, and that intermittency may influence them but is not their primary cause. In a general flow control context, acceleration measurements can also provide a tool to analyze and optimise various properties of a turbulent flow, for example mixing efficiency.

Given the multi-scale property of our flow, we have developed a new method to accurately measure acceleration at each scale using an adaptive scheme [15]. This method ensures very good quality of velocity measurements and good quality of acceleration measurements (typical relative errors smaller than 10% on acceleration) at all scales. In figure 2 we compare standard methods with the new Particle Tracking Velocimetry and Accelerometry (PTVA) method of [15]. Thus we are able to use such high quality measurements to compute the Eulerian fields as illustrated in figures 3 and 4 which are plots of the velocity and acceleration fields respectively. In fact the quality of these measurements is so good that we can compute the viscous term in the Navier-Stokes equation, see figure 5, thereby providing all the equation's terms, $\mathbf{a} = \frac{\partial \mathbf{u}}{\partial t} + (\mathbf{u} \cdot \nabla)\mathbf{u} = -\frac{1}{\rho}\nabla \mathbf{P} + \nu\nabla^2\mathbf{u} + \mathbf{f}$, where P is the pressure, ρ the fluid density, ν the fluid kinematic viscosity and $\mathbf{f}$ the multiscale electromagnetic forcing. The viscous term is found to be small compared to the measured acceleration in the vast majority of the flow, suggesting a direct balance between forcing and pressure gradients. The local power input/output is given by $\mathbf{u} \cdot \mathbf{a}$ and is also of direct access, see figure 6. One interesting result is that this power input/output is closely related to the pressure terms in these turbulent-like flows. The multi-scale forcing can be guessed from the plot of $\mathbf{u} \cdot \mathbf{a}$, but this plot is more subtle than a direct computation of $\mathbf{f} \cdot \mathbf{u}$ which gives a power distribution reflecting the distribution of the permanent magnets.

It has been proposed (see references on acceleration cited above) that the PDF of accelerations in turbulent flows is not gaussian but can instead be well fitted by $P(a) = C \exp(-[a^2]/[(1 + |a\beta/\sigma|^\gamma)\sigma^2]$. Measurements of acceleration PDFs in our turbulent-like yet laminar Q2D steady flow are given in figure 7. Both acceleration components appear approximately Gaussian around zero-acceleration values but exhibit fat tails at higher values of acceleration, as reported in [8]. The parameters turn out to be similar between the two experiments (with $\beta = 0.562$, $\sigma = 0.614$ and $\gamma = 1.27$ in the present case, and beta $\beta = 0.539$, $\sigma = 0.508$, $\gamma = 1.59$ in [8]). It may be tempting to conclude that this is yet another feature of real turbulence which can be captured by our multiscale, laminar and steady Q2D flow because of its turbulent-like properties, but we are careful not to draw such a conclusion too quickly. It is well-known that data can be fitted in many different ways, and in fact the tails of our acceleration PDFs may also be fitted by other functional forms, some of which we have derived by direct mathematical calculation for single and many multi-scale vortices or for models of eddies within eddies as in the multiscale cat's eye structure of our flows. More importantly, a time-dependence imposed by time-dependent multiscale electromagnetic forcing can potentially very significantly modify our PDF results and will also have an impact on the correlation signature of Lagrangian accelerations, which we also plan to measure and study on time for the conference. Our flow can be given different time and space properties at will and is therefore ideal for studies of what in the flow causes what in the output statistics.

References

[1] J Davila and J C Vassilicos. *Phys. Rev. Lett.*, 91:144501, 2003.
[2] S Goto and J C Vassilicos. *New J. Phys.*, 6:1–35, 2004.
[3] L Rossi, J C Vassilicos, and Y Hardalupas. *J. Fluid Mech.*, 558:207 – 242, 2006.

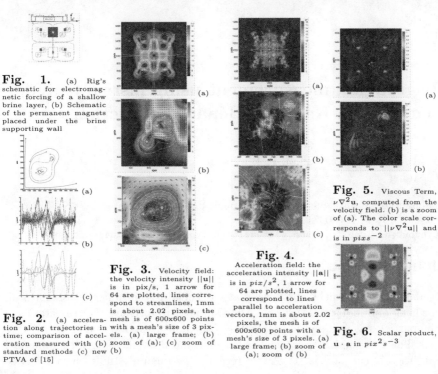

Fig. 1. (a) Rig's schematic for electromagnetic forcing of a shallow brine layer, (b) Schematic of the permanent magnets placed under the brine supporting wall

(a)

(b)

(c)

Fig. 2. (a) acceleration along trajectories in time; comparison of acceleration measured with (b) standard methods (c) new PTVA of [15]

(a)

(b)

(c)

Fig. 3. Velocity field: the velocity intensity $||u||$ is in pix/s, 1 arrow for 64 are plotted, lines correspond to streamlines, 1mm is about 2.02 pixels, the mesh is of 600x600 points with a mesh's size of 3 pixels. (a) large frame; (b) zoom of (a); (c) zoom of (b)

(a)

(b)

(c)

Fig. 4. Acceleration field: the acceleration intensity $||a||$ is in pix/s^2, 1 arrow for 64 are plotted, lines correspond to lines parallel to acceleration vectors, 1mm is about 2.02 pixels, the mesh is of 600x600 points with a mesh's size of 3 pixels. (a) large frame; (b) zoom of (a); zoom of (b)

(a)

(b)

Fig. 5. Viscous Term, $\nu\nabla^2 u$, computed from the velocity field. (b) is a zoom of (a). The color scale corresponds to $||\nu\nabla^2 u||$ and is in $pix s^{-2}$

Fig. 6. Scalar product, $u \cdot a$ in $pix^2 s^{-3}$

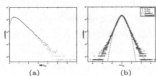

Fig. 7. PDF of the acceleration intensity (a) and of its two components (b)

(a)   (b)

[4] H Hascoet, L Rossi, and J C Vassilicos. "turbulent-like laminar flows sustained and controlled by multiscale body forces" submitted. *Submitted to J. Fluid Mech.*, 2007.

[5] L Rossi, J C Vassilicos, and Y Hardalupas. *Phys. Rev. Lett.*, 97(14), 2006.

[6] S Goto, D R Osborne, J C Vassilicos, and J D Haigh. *Phys. Rev. E*, 71:015301, 2005.

[7] D R Osborne, J C Vassilicos, K S Sung, and J D Haigh. *Phys. Rev E*, 74:036309, 2006.

[8] A La Porta, G A Voth, A M Crawford, J Alexander, and E Bodenschatz. *Nature*, 409:1017–1019, 2001.

[9] L Biferale and F Toschi. *J. of Turbulence*, 6(40), 2005.

[10] M Mordant, A N Crawford, and E Bodenschatz. *Physica D*, 193:245, 2003.

[11] Aringazin and Mazhitov. *Physical Letters A*, 313, 2003.

[12] Aringazin and Mazhitov. *Phys Rev. E*, 70:036301, 2004.

[13] N Mordant, P Metz, O Michel, and J-F Pinton. *Phys. Rev. Lett.*, 87:214501, 2001.

[14] D R Osborne, J C Vassilicos, and J D Haigh. *Phys. Fluids*, 17:035104, 2006.

[15] S Ferrari and L Rossi. "measurements of velocity and acceleration via ptva (particle tracking velocimetry and accelerometry) and its application to electromagnetically controlled quasi-two-dimensional multi-scale flows". *Submitted to Exp. Fluids, see also archive*, 2007.

Lagrangian measurement using instrumented particles in Rayleigh-Bénard convection

W. L. Shew, Y. Gasteuil, J.-F. Pinton, R. Volk, M. Gibert, F. Chillá, B. Castaing

Laboratoire de Physique de l'École Normale Supérieure de Lyon, CNRS UMR5672, 46 allée d'Italie F-69007 Lyon, France
pinton@ens-lyon.fr,yoann.gasteuil@ens-lyon.fr,sheww@mail.nih.gov

1 Introduction

Scalar mixing in turbulent flows plays a crucial role in uncountable natural, medical, and industrial systems. An example is to measure the trajectories of the particles as well as the properties of the flow along fluid particle trajectories. This Lagrangian approach has been advanced significantly with numerical and theoretical models [1]. A few recent experimental studies have successfully measured the trajectories of small tracer particles in turbulence [2]. In the spirit of the atmospheric balloons and ocean floaters, we have developed new instrumentation for making Lagrangian measurements of temperature in diverse fluid flows. A small neutrally buoyant capsule (dubbed "smart particle") is equipped with on-board electronics which measure temperature and transmit the data via a wireless radio frequency link to a desktop computer. To our knowledge we report the first measurement of Lagrangian temperature and heat transport in a laboratory scale convection experiment.

2 Experimental set-up

Figure 1 shows elements of the experiment. The smart particle consists of a $D = 21$ mm diameter capsule containing temperature instrumentation, an RF emitter, a battery. A resistance controlled oscillator is used to create a square wave whose frequency depends on the temperature of several thermistors. This square wave is used directly to modulate the amplitude of the radio wave generated by the RF emitter. The entire mobile circuit is powered with a coin cell battery and may be put in a low power standby mode using an externally applied magnetic field. The stationary parts of the system consist of an RF receiver, 2 RF amplifiers, a high speed data acquisition system, and a PC. The receiver is carefully tuned to demodulate the signal produced by the emitter. The receiver outputs a square wave identical to that generated

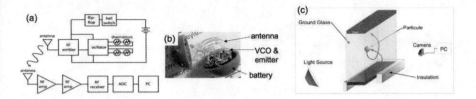

Fig. 1. (a) Block diagram of measurement system. (b) Photo of assembled capsule. Three (out of four) of the thermistance are visible about the equator of the particle. (c) sketch of the Rayleigh-Bnard convection cell and particle position detection using a digital camera.

by the resistance controlled oscillator. The frequency of this square wave, and hence temperature, is recovered on-the-fly using a Labview algorithm.

Simultaneously the particle trajectory is recorded by a digital camera, so as to monitor the horizontal and vertical coordinates of the particle. As a result, we perform synchronous measurements of the particle position and local temperature as the particle is carried by the fluid motion.

The flow is a Rayleigh-Bénard setup, discribed in details in [3], and sketched in Fig.1(c). The cell is $H = 40$ cm tall, $L = 40$ cm wide, $E = 12$ cm thick and filled with water. A temperature difference $\Delta T = 18°C$ is imposed by heating the bottom and regulating the temperature of the top wall. The large scale flows has a period of about 100 s.

3 Results

Fig. 2(a)shows time traces of the temperature fluctuations in the flow (the sampling time is 50 ms). The upper trace (red) corresponds to a stationary temperature probe located near the upper wall, and the bottom one (blue) to a stationary probe near the cold wall. The black trace in-between is the signal send by the smart particle entrained by the flow motions (the particle is made neutrally buoyant at the mean fluid density in the bulk of the cell). One clearly identifies the convective motion of the particle, which comes quasi-periodically in contact with the hot and cold plates and then mixes its acquired temperature with the fluid. The power spectrum of the entire time trace is shown in Fig.2(a): the low frequency peak is consistent with the large scale convection roll. The dynamics of the measurement is over 80dB, and compares well with usual thermometry (the red trace is the spectrum of the signal recorded by a stationary thermistor).

We now take advantage of our simultaneous recording of the particle position and temperature to define a local Nusselt number:

$$Nu^L(t) = 1 + \frac{L}{\kappa \Delta T} \theta(t) \cdot v_z(t) ,$$

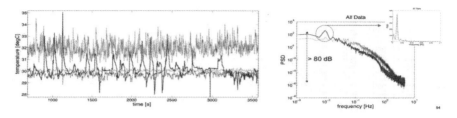

Fig. 2. (a) Temperature measured by the moving particle (black), compared to data from stationary probes near the hot plate (red) or cold plate (blue) (b) correponding power spectra. The periodicity of the large scale convection roll is evidenced in the inset.

where κ is the fluid's thermal diffusivity, $\theta(t) = \langle T \rangle$ is the local temperature fluctuation of the particle's temperature and $v_z(t)$ its vertical velocity. The trajectories shown in figure 3(a) show that the particle explores most of the flow volume, save for the center of the flow. Heat exchanges are largest as the particle moves away from the end plates. The statistic of the local exchanges is very non Gaussian, as revealed by the probability density function (PDF) plot in figure 3(b).

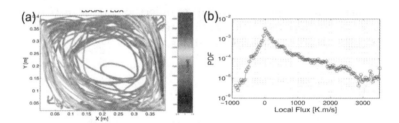

Fig. 3. (a) Motion of the tracer particle with a color coding corresponding to the heat flux; (b) PDF of $Nu(t)$.

References

1. P.K. Yeung, *An. Rev. Fluid Mech.* **34**, 115 (2002); G. Falkovich, K. Gawedzki, M. Vergassola, *Rev. Mod. Phys.* **73**, 913 (2001); N. Mordant, E. Lévêque, J.-F. Pinton, *New J. Phys.* **6**, 116 (2004); L. Biferale at al. *Phys. Fluids* **17**, 021701 (2005); L. Chevillard et al. *Phys. Rev. Lett.* **95** 064501 (2005)
2. S. Ott, J. Mann, *J. Fluid Mech.*, **422** 207 (2000); La Porta A.L., *et al.*, *Nature*, 409, 1017, (2001); N. Mordant et al. *Phys. Rev. Lett.*, **87**(21), 214501, (2001); H. Xu et al. *Phys. Rev. Lett.* **96**, 024503 (2006)
3. M. Gibert et al. *Phys. Rev. Lett.* **96**, 084501 (2006). F Chillá et al., *Nouvo Cimento* **15**, 1229 (1993).

Simultaneous Lagrangian and Eulerian velocity measurements in a round jet

P. Gervais[1], M. Bourgoin[2], C. Baudet[2], and Y. Gagne[2]

[1] LEPTAB, Av. Michel Crépeau, 17042 La Rochelle, France pgervais@univ-lr.fr
[2] LEGI, 1025, rue de la piscine, 38041 Grenoble, France bourgoin@hmg.inpg.fr
[3] LEGI, 1025, rue de la piscine, 38041 Grenoble, France baudet@hmg.inpg.fr
[4] LEGI, 1025, rue de la piscine, 38041 Grenoble, France gagne@hmg.inpg.fr

Lagrangian description of turbulent flows is well suited for mixing problems. Very few experiments have been conducted in turbulent flows until the past few years [2, 1, 3], most of data coming from DNS of Lagrangian motion [4]. Recent experiments in high-Reynolds number flows [1, 3] have been performed in the Von-Kármán swirling flow, which have no mean velocity and so, make the measurement of Eulerian correlations difficult. Thus, no clear comparison of Eulerian and Lagrangien velocity statistics could be made. We report measurements obtained in a turbulent round air jet ($R_\lambda = 320$). It has the advantages of having a mean velocity and, being non-confined, allows easier comparisons of Eulerian and Lagrangian statistics. We used Helium-filled soap bubbles (2 mm diameter) as fluid particle tracers, with the same density as the surrounding air.

The bubble velocity was then measured, one bubble at a time, by the setup presented on figure 1. Two acoustical sinusoidal waves (resp. 110 kHz and 120 kHz) are emitted by two transducers (diameter 24 cm). The waves are scattered by the bubble, the scattered wave being recorded by two receivers (diameter 24 cm). Upon scattering, the wave frequency changes, in proportion to the velocity (Doppler effect). This setup enables to get simultaneously the three velocity components along time, provided the bubble is in the intersection of the emitter and receivers beams (which defines a measurement volume of the order of the integral scale). The data obtained take the form of a set of finite-duration signals (around 100 ms each), each being a independent realization of a Lagrangian signal. The temporal dynamic is large enough to get inertial time scales down to the Taylor time scale. Thanks to the mean velocity, simultaneous Eulerian measurement of the longitudinal velocity component was made with a classical hot-wire anemometer.

We observe that the lagrangian velocity probability density functions are Gaussian for all components, despite the mean-flow inhomogeneity. Eulerian and Lagrangian measurements show very close one-point probability density functions for the longitudinal component. Longitudinal and transverse Eule-

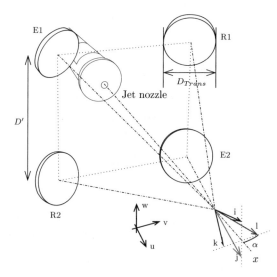

Fig. 1. Jet nozzle surrounded by four acoustical transducers. Four bubble velocity projections are measurable (i,j,k,l).

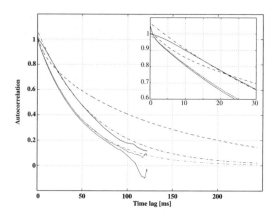

Fig. 2. Temporal autocorrelation. Solid lines from bottom to top: transverse Lagrangian, longitudinal Lagrangian. Eulerian is the topmost dashed curve, the other dashed curves are exponential fits.

rian velocity components in a turbulent jet are correlated. This is not true in the lagrangian case, for which velocity component are each decorrelated one from another, up to the measurement uncertainties.

Whereas flow inhomogeneity does not seem to influence one-point statistics, several biases exist for Lagrangian velocity autocorrelation. To be able to compute correlation for large time lags, one needs long trajectories, which spans a few integral length scales. Due to mean-flow inhomogeneity, velocity signals are not stationary any more. Stationarity is recovered by removing a local mean velocity and dividing by the standard deviation of the local velocity [4]. Results are shown on figure 2 (solid lines). Curves exhibit shapes close to exponentials. This is consistent with results obtained in the Von-Kármán flow [1].

Eulerian autocorrelation has been computed as well, so that Lagrangian and Eulerian autocorrelation functions could be properly compared for the first time (to our knowledge). Streamwise Lagrangian integral time scale is found to be smaller than the corresponding Eulerian one, by a factor ranging from 1.4 to 1.7, slightly dependent on the distance from the nozzle (figure 2)

We have also performed simultaneous Eulerian and Lagrangian measurements in order to compute the cross-correlation between Eulerian and Lagrangian longitudinal velocity. Correlation levels are significant (figure 3), even for distances between the measurement volume and the hot-wire much larger than the integral length scales, and times larger than both Eulerian and Lagrangian time scales. Furthermore the Eulerian velocity is found to be decorrelated with both transverse Lagrangian velocities.

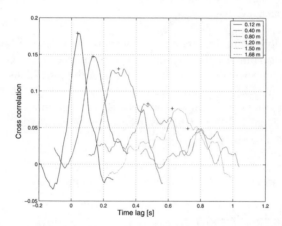

Fig. 3. Cross correlation between Lagrangian and Eulerian longitudinal velocity, for different distances between measurement volume and hot-wire.

References

1. Nicolas Mordant, Pascal Metz, Olivier Michel, and Jean-Franois Pinton. Measurement of lagrangian velocity in fully developed turbulence. *Physical Review Letters*, 87(21):214501, 2001.
2. Søren Ott and Jakob Mann. An experimental investigation of the relative diffusion of particle pairs in three-dimensional turbulent flow. *Journal of Fluid Mechanics*, 422:207–223, 2000.
3. Greg A. Voth, Arthur la Porta, Alice M. Crawford, Jim Alexander, and Eberhard Bodenschatz. Measurement of particle accelerations in fully developed turbulence. *Journal of Fluid Mechanics*, 469:121–160, 2002.
4. P. K. Yeung. Lagrangian investigations of turbulence. *Annual Review of Fluid Mechanics*, 34:115–142, 2002.

Stochastic Analysis and New Insights into Turbulence

J. Peinke[1], A. Nawroth[1], St. Lück[2], M. Siefert[3] and R. Friedrich[4]

[1] ForWind, Institute of Physics, Carl-von-Ossietzky University of Oldenburg, D-26111 Oldenburg, Germany peinke@uni-oldenburg.de
[2] FLSH Gymnasium Gaibach, D-97332 Volkach stephan.lueck@web.de
[3] Institute of Physics, University of Postdam, D-14415 Potsdam siefert@stat.physik.uni-potsdam.de
[4] Institute for Theoretical Physics, University of Münster, D-48149 Münster, fiddir@uni-muenster.de

We present a more complete analysis of measurement data of fully developed, local isotropic turbulence by means of the estimations of Kramers- Moyal coefficients, which provide access to the joint probability density function of increments for n- scales including intermittency effects [1]. In this contribution we report on new findings based on this technique and based on the investigation of many different flow data over a big range of Re numbers.

In particular we show:

(1) an improved method to reconstruct from given data the underlying stochastic process in form of a Fokker-Planck equation,

$$
-r\frac{\partial}{\partial r}p(\mathbf{u}, r|\mathbf{u_0}, r_0) = \tag{1}
$$

$$
\left(-\sum_{i=1}^{n} \frac{\partial}{\partial u_i} D_i^{(1)}(\mathbf{u}, r) + \sum_{i,j=1}^{n} \frac{\partial^2}{\partial u_i \partial u_j} D_{ij}^{(2)}(\mathbf{u}, r) \right) p(\mathbf{u}, r|\mathbf{u_0}, r_0).
$$

(Velocity increments are denoted by vectors $\mathbf{u}$, $\mathbf{u_0}$ is the velocity increment on scale r_0, i labels the components, we fix $i = 1$ for the longitudinal and $i = 2$ for the transverse increments.) In particular the limited-memory Broyden-Fletcher-Goldfarb-Shanno algorithm for constraint problems (L-BFGS-B) and alternatively the Kullback Leibler distance are used to minimize the distance between the numerical solutions of the Fokker-Planck equation and the empirical probability density functions. Thus the drift and diffusion term of the Fokker-Planck equation are estimated properly. [2]

(2) It is shown that a new length scale, the Einstein- Markov coherence length l_{mar}, for turbulence can be defined, which corresponds to a memory effect

in the cascade process. This coherence length can be seen as an analogue to the mean free path length of a Brownian motion. For length scales larger than this coherence length the complexity of turbulence can be treated as a Markov process. This Einstein- Markov coherence length scales with $Re^{1/2}$ and it is closely related to the Taylor micro-scale, λ, as shown in Fig. 1, see also [3].

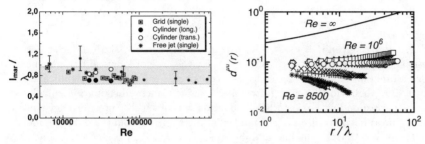

Fig. 1. left: Ratio l_{mar}/λ versus Reynolds number Re for different flows. Squares: grid turbulence ($M = 5$ cm, $x = 1.6$ m), circles: cylinder wake ($D = 2$ cm, $x = 2$ m) and asterix: free jet (air and helium experiment). For some data points the error bars are shown exemplarily (after [3]). – right: Intermittency coefficients $d_2^{uu}(r, Re)$ exhibit a strong dependence on the Re- number with a tendency towards the limiting value $d_2^{uu}(r, Re \to \infty)$ (after [4]).

(3) It is shown that the stochastic process of a cascade will change with the Re-number and that it has some non-universal contributions. Especially we study the Re number-dependence of the drift and diffusion coefficients:

$$D^{(1)}(u, r, Re) = d_1^u(r, Re)u \qquad (2)$$

$$D^{(2)}(u, r, Re) = d_2(r, Re) + d_2^u(r, Re)u + d_2^{uu}(r, Re)u^2. \qquad (3)$$

A clear Re number dependency of the d_2^{uu} term, as shown in Fig. 1 is found, further details see [4] . From this experimental finding we conclude that the cascade process performs clear changes with the Re- number. A limiting case of large Re- numbers (defined by the fulfilling of the third order structure function law) is not reached even for such high Re- numbers as 10^6.

(4) For longitudinal and transversal velocity increments we present the reconstruction of the two dimensional stochastic process equations, which shows that the cascade evolves differently for the longitudinal and transversal increments. A different "speed" of the cascade for these two components can explain the reported difference for these components. The rescaling symmetry is compatible with the Kolmogorov constants and the Kármán equation and give new insight into the use of extended self similarity (ESS) for transverse increments, see Fig. 2 and [5]

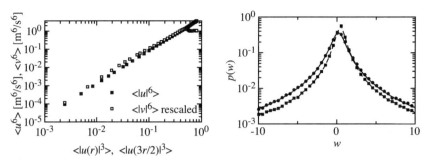

Fig. 2. left: Longitudinal, $< u^6 >$, and transversal, $< v^6 >$, sixth order structure functions plotted as a function of $< u(r)^3 >$ for the case of longitudinal increments and as a function of $< u(3r/2)^3 >$ for the transversal case. Note no significant different scaling behavior is found (after [5]). – right: Cauchy distribution calculated from the corresponding Fokker-Planck equation without using intermittency contributions (lines) in comparison to the multiplier distribution from the data (symbols). $r_1 = L$, $r_2 = L/2$ (straight line and circles), $r_n = 2\lambda$, $r_{n+1} = \lambda$ (dotted line and squares). .

(5) Knowing the stochastic cascade process as $\partial_r u_r = \ldots$ it is easy to calculate also the stochastics of multipliers defines as $w := u_{r_1}/u_{r_2}$. As u_r is given by a stochastic process it is evident that the quotient to two stochastic variables leads to a Cauchy statistics. It is interesting to note that for this often discussed statistics of the multiplier not intermittency correction has be be taken into account. In Fig. 2 the multiplier distributions were obtained from Gaussian approximations of $p(u,r|u',r')$, using only the linear contribution of the drift term $D^{(1)}$ and the constant term in the diffusion term $D^{(2)}$, for further details see [7]

References

1. Ch. Renner, J. Peinke & R. Friedrich: J. Fluid Mech. **433**, 383 (2001).
2. D. Kleinhans, R. Friedrich, A.Nawroth, and J. Peinke: Phys. Lett. A **346**, 42 (2005); A. P. Nawroth, J. Peinke, D. Kleinhans, and R. Friedrich, *Estimation of Fokker-Planck equations through optimisation* (in preparartion).
3. St. Lück, Ch. Renner, J. Peinke, and R. Friedrich: Phys. Lett. A **359**, 335 (2006).
4. Ch. Renner, J. Peinke, R. Friedrich, O. Chanal, and B. Chabaud: Phys. Rev. Lett. **89**, 124502 (2002).
5. M. Siefert and J. Peinke: J. of Turbulence **7**, (No 50) 1-35 (2006)
6. A.P. Nawroth and J. Peinke: Phys. Lett. A **360**, 234 (2006)
7. M. Siefert and J. Peinke, *Complete Multiplier Statistics Explained by Stochastic Cascade Processe*, Phys. Lett. (submitted).

On the Deficiency of Structure Functions as Inertial Range Diagnostics

P.A. Davidson[1] and P.-Å. Krogstad[2]

[1] University of Cambridge, Cambridge, CB2 1PZ, UK
[2] Norwegian University of Science and Technology, N-7491 Trondheim, Norway

In the limit of infinite Reynolds number, Re, Kolmogorovs two-thirds and five-thirds laws are formally equivalent. However, for the sorts of Reynolds numbers encountered in terrestrial experiments, or numerical simulations, it is invariably easier to observe the five-thirds law. We explain that this is because the second-order structure function is a poor diagnostic, mixing information about energy and enstrophy, and about small and large scales. It is shown that, as a result, the form of the structure function in the inertial range is not a simple power law, but rather a combination of two powers, of the form $\langle (\Delta v)^2 \rangle (r) = a + br^2 + cr^{2/3}$, where the coefficients a, b and c are functions of Re. It is only as $Re \to \infty$ that a pure two-thirds law is obtained. Similar problems arise for higher-order structure functions, which also display combined power laws.

1 Introduction

It is known that the second-order, longitudinal structure function, $\langle (\Delta v)^2 \rangle (r)$, is a poor diagnostic tool, mixing information from large and small scales and information about energy and enstrophy [1],[2]. This can be seen by looking at the relationship between $\langle (\Delta v)^2 \rangle (r)$ and the energy spectrum, $E(k)$, in isotropic turbulence, where it may shown [2] that,

$$\frac{3}{4} \langle (\Delta v)^2 \rangle (r) \approx \int_{\pi/r}^{\infty} E(k) dk + (r/\pi)^2 \int_{0}^{\pi/r} k^2 E(k) dk. \qquad (1)$$

This has a simple physical interpretation in terms of the properties of the velocity increment, $\Delta v = u_x (\mathbf{x} + r\hat{\mathbf{e}}_x) - u_x (\mathbf{x}) = (u_x)_B - (u_x)_A$. Any eddy in the vicinity of A or B whose size s is much smaller than r will contribute to either $(u_x)_A$ or $(u_x)_B$, but not to both. Hence eddies for which $s << r$ will make a contribution to $\langle (\Delta v)^2 \rangle$ of the order of their kinetic energy. On the other hand, eddies whose size is much greater than r will make a contribution to $\langle(\Delta v)^2\rangle(r)$ of the order of $r^2\left(\partial u_x/\partial x\right)^2$, which is consistent with Eq. (1).

Evidently, $\langle(\Delta v)^2\rangle(r)$ is a very leaky filter, mixing information across the scales. This suggests that, in the inertial range, $\langle(\Delta v)^2\rangle(r)$ may take the form of a combined power law of the form $\langle(\Delta v)^2\rangle(r) = a + br^2 + cr^{2/3}$, where the quadratic term tracks the enstrophy of the eddies while the $r^{2/3}$ term tracks the variation of energy with scale. This is readily confirmed using a simple model problem, as we now show.

2 A Simple Model Problem

Consider an artificial field of turbulence composed of a random sea of Gaussian eddies whose size s lies in the range $\ell < s < L$, and whose energy distribution, $\hat{E}(s)$, is specified (statistically) in accordance with K41, $s\hat{E}(s) \sim s^{2/3}$. Then it is readily confirmed [2] that,

$$12\pi^{1/2}E(k) = \int s\hat{E}(s)(ks)^4 \exp\left[-(ks)^2/4\right] ds. \tag{2}$$

Perhaps it is not surprising that a reasonable approximation to $E(k)$ given by Eq. (2) is $E(k) = \alpha k^{-5/3}$, $\ell/\pi < k^{-1} < L/\pi$, and $E(k) \approx 0$ for all other k. Note that, from the empirical dissipation law $\epsilon \sim u^3/L$, $u = <u_x^2>^{1/2}$, we expect the ratio of L/ℓ to be related to the Taylor micro-scale, λ, by $\gamma = L/\ell = \beta R_\lambda^{3/2}/15^{3/4}$, where β is of order unity.

Let us now turn to the structure function. From Eq. (1) the form of $\langle(\Delta v)^2\rangle(r)$ corresponding to $E(k) = \alpha k^{-5/3}$, $\ell/\pi < k^{-1} < L/\pi$, is

$$\frac{\langle(\Delta v)^2\rangle(r)}{2u^2} \approx \frac{1 - \gamma^{-4/3}}{2(\gamma^{2/3} - 1)}\left(\frac{r}{\ell}\right)^2 \quad , \quad r < \ell \tag{3}$$

$$\frac{\langle(\Delta v)^2\rangle(r)}{2u^2} \approx \frac{3(r/\ell)^{2/3} - 2 - \gamma^{-4/3}(r/\ell)^2}{2(\gamma^{2/3} - 1)} \quad , \quad \ell < r < L \tag{4}$$

and $\langle(\Delta v)^2\rangle(r) = 2u^2$ for $r > L$. As anticipated above, we find a mixed power law in the inertial range.

We may compare Eq. (3) and (4) with measurements made in grid turbulence. In is necessary only to specify β. Just such a comparison is shown in Fig. 1, in which we have chosen $\beta = 0.34$. The comparison is surprisingly good considering our model is so simple.

It is readily confirmed that higher-order structure functions suffer from similar problem, mixing information from different scales. For example, the analog of Eq. (1) for $\left\langle(\Delta v)^4\right\rangle(r)$ is

$$\left\langle(\Delta v)^4\right\rangle(r) \approx (16/3)\,[\text{contribution to energy squared from below scale } r]$$

$$+ r^4 \times \left[\text{contribution to }(\partial u_x/\partial x)^4\text{ from above scale } r\right]$$

The fact that structure functions exhibit mixed power laws in the inertial range has clear implications for the interpretation of measured anomalous scaling exponents.

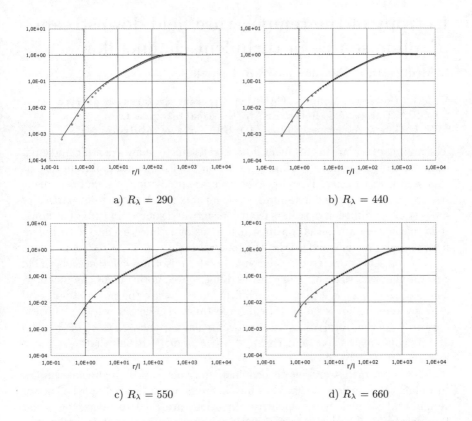

a) $R_\lambda = 290$ b) $R_\lambda = 440$

c) $R_\lambda = 550$ d) $R_\lambda = 660$

Fig. 1. Comparisons between measured structure functions with equations (3)-(4). Symbols are measurements and lines the theoretical predictions.

References

1. P.A. Davidson and B.R. Pearson: Identifying turbulent energy distribution in real, rather than Fourier, space. *Phys. Rev. Let.*, **95**, 21 (2005)
2. P.A. Davidson *Turbulence, An introduction for Scientists and Engineers* (Oxford University Press, 2004)

Isotropy of the temperature field downstream of a line source in turbulent channel flow

L. Mydlarski[1], L. Danaila[2], and R.A. Lavertu[3]

[1] McGill University, Montréal, CANADA – laurent.mydlarski@mcgill.ca
[2] CORIA, Université de Rouen, FRANCE – danaila@coria.fr
[3] W. L. Gore & Associates, Elkton, MD, USA – rlavertu@wlgore.com

The mixing of scalars in turbulent flows is relevant to many scientific and environmental phenomena, including combustion, environmental pollutant dispersion, and heat transfer. Therefore, the accurate prediction of such phenomena requires a thorough understanding of scalar mixing. A problem of particular interest is the diffusion of scalars from concentrated sources in turbulent flows. It is relevant to any situation in which a scalar is injected into a turbulent flow at scales much smaller than the integral scale.

The present work studies the small-scale anisotropy of the scalar field generated downstream of a concentrated line source in turbulent channel flow. The fact that passive scalars in turbulent flows violate the postulate of local isotropy remains an issue of critical importance [1]. Herein, we focus on the anisotropy of the (total) dissipation rate of scalar variance ($\varepsilon_\theta \equiv \kappa \langle (\partial\theta/\partial x_i)^2 \rangle$) and its components (particularly $\varepsilon_{\theta_x} \equiv \kappa \langle (\partial\theta/\partial x)^2 \rangle$). Recall that $\varepsilon_\theta / \varepsilon_{\theta_x} = 3$ in a locally isotropic flow.

The experiments are performed in a fully-developed, high-aspect-ratio, turbulent channel flow facility [2]. Fine Nichrome wires (heated by a DC power supply and aligned in the spanwise direction) are used to inject the scalar (temperature). The latter is measured by means of cold-wire thermometry. Simultaneous (longitudinal and transverse) velocity measurements were made by means of hot-wire anemometry. A schematic of the experiment is given in figure 1 and the experimental conditions are given in Table 1.

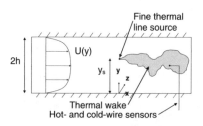

$U_c(= \langle U \rangle_{y/h=1})$	5.2 m/s
$u_{rms_{y/h=1}}$	0.21 m/s
$u_\star$	0.26 m/s
$Re = U_c h/\nu$	10,400
$Re_\tau = u_\star h/\nu$	520
$Re_{\lambda_{y/h=1}} = u_{rms}\lambda/\nu$	58

$y_s/h = 0.067$	$y_s^+ = 35$
$y_s/h = 0.17$	$y_s^+ = 87$
$y_s/h = 1.0$	$y_s^+ = 520$

Fig. 1. A schematic of the experiment. The channel width $(2h)$ is 6 cm.

Table 1. The experimental conditions.

Given that u, v and θ are all simultaneously measured, the various terms in the scalar variance budget:

$$\frac{\partial}{\partial t}\left\langle \frac{1}{2}\theta^2 \right\rangle + U_i \frac{\partial}{\partial x_i}\left\langle \frac{1}{2}\theta^2 \right\rangle = -\frac{\partial T}{\partial x_i}\langle u_i\theta\rangle - \frac{\partial}{\partial x_i}\left\langle \frac{1}{2}u_i\theta^2 \right\rangle + \kappa \frac{\partial^2}{\partial x_i^2}\left\langle \frac{1}{2}\theta^2 \right\rangle - \varepsilon_\theta$$

can be estimated and used to infer the (total) scalar dissipation rate (ε_θ), with some assumptions (steady state, homogeneity in z, high Reynolds number, etc.). A typical example of the data used to infer ε_θ is shown in figure 2, where a contour plot of $\langle v\theta^2\rangle$ ($= \langle u_2\theta^2\rangle$) – used to estimate the transverse turbulent diffusion of $\langle \theta^2\rangle$ – is plotted. A typical result is shown in figure 3, which plots the different (non-negligible) components of the scalar variance budget for a centreline source $(y_s/h = 1.0)$. In this particular instance, which corresponds to measurements downstream of a centreline source $(y/h = y_s/h = 1.0)$, it is clear that the budget is well approximated by a balance between the downstream decay of $\langle \frac{1}{2}\theta^2\rangle$ and its dissipation for $x/h > 7$.

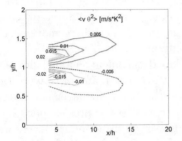

Fig. 2. Contour plot of $\langle v\theta^2\rangle$ $(= \langle u_2\theta^2\rangle)$ in the thermal plume emanating from a line source located at the channel centreline $(y_s/h = 1.0)$.

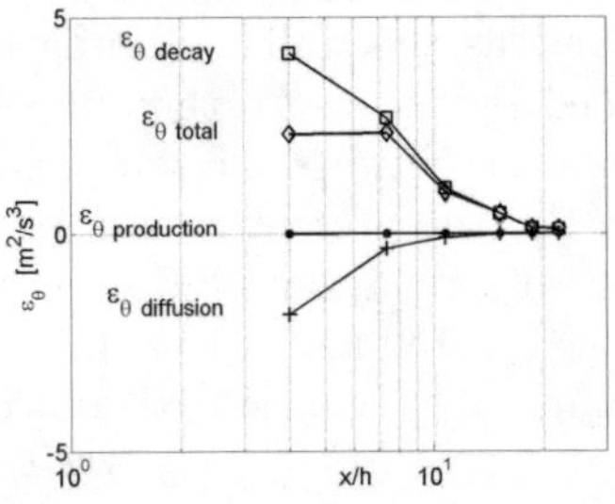

Fig. 3. Contributions to ε_θ: $-U_i\frac{\partial}{\partial x_i}\langle \frac{1}{2}\theta^2\rangle(\square)$; $-\frac{\partial}{\partial x_i}\langle \frac{1}{2}u_i\theta^2\rangle(+)$; $-\frac{\partial T}{\partial x_i}\langle u_i\theta\rangle(\bullet)$; and the total $\varepsilon_\theta(\Diamond)$. Measurements are made at $y/h = 1.0$.

The longitudinal component of ε_θ (ε_{θ_x}) is estimated from its definition (see above) in conjunction with Taylor's hypothesis. This permits the calculation of $\varepsilon_\theta/\varepsilon_{\theta_x}$ (the ratio of the total scalar dissipation rate to its longitudinal component). Contour plots of $\varepsilon_\theta/\varepsilon_{\theta_x}$ are shown in figures 4-6 for thermal fields emitted from line sources located at $y_s/h = 0.067$, 0.17, and 1.0, respectively.

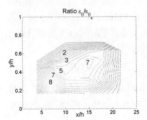 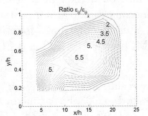 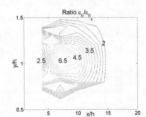

Fig. 4. Contour plot of $\varepsilon_\theta/\varepsilon_{\theta_x}$ for $y_s/h = 0.067$.

Fig. 5. Contour plot of $\varepsilon_\theta/\varepsilon_{\theta_x}$ for $y_s/h = 0.17$.

Fig. 6. Contour plot of $\varepsilon_\theta/\varepsilon_{\theta_x}$ for $y_s/h = 1.0$.

Small-scale anisotropy of the scalar field is clearly observed in figures 4-6 as $\varepsilon_\theta/\varepsilon_{\theta_x}$ takes on values larger than 3. Note that the loci of maximum anisotropy correspond to those of maximum scalar variance. For near-wall sources (figures 4 and 5), the peak in the anisotropy moves to larger y/h as the downstream distance (x/h) increases and the plume widens – similar to the profiles of θ_{rms}. In [2], maximum scalar variance was found to occur where "flapping" of the plume was significant. Such a correlation between the flapping of the plume and the anisotropy of the scalar dissipation rate validates the notion of Rosset et al. [3] that small-scale anisotropy (especially $\varepsilon_{\theta_y}/\varepsilon_{\theta_x}$) is amplified at interfaces between the plume and the ambient fluid.

Furthermore, it can be observed that the maximum anisotropy varies non-monotonically with y_s/h. For sources at $y_s/h = 0.067$, 0.17 and 1.0, $(\varepsilon_\theta/\varepsilon_{\theta_x})_{max} \approx 8$, 5.5 and 6.5, respectively. A possible explanation for the non-monotonic variation of $(\varepsilon_\theta/\varepsilon_{\theta_x})_{max}$ is as follows. When moving from the centreline of the channel towards the wall, the turbulence intensity increases. Therefore, so will the intensity of the mixing, thus decreasing the anisotropy. However, as the wall is approached, the plume growth is inhibited (being bounded by the wall), thus impeding the mixing. Hence there may exist a competition between mechanisms that amplify and destroy anisotropy.

Lastly, the present results in fully-developed turbulent channel flow are compared with those of Rosset et al. [3] in turbulent boundary layers (for which $R_\lambda = 130$ and $y_s/\delta = 0.3$, where δ was the boundary layer thickness). At comparable (normalized) downstream distances to the present ones (i.e., similar $x^* \equiv (x/U)/(h/u_{rms}) \approx t/t_L$), the anisotropy in boundary layers was similar to or smaller than that observed herein. For $x^* \approx 0.15$, Rosset et al. [3] observe $\varepsilon_\theta/\varepsilon_{\theta_x} \approx 4$, which corresponds to their farthest downstream distance. (In the present work, $x^* \approx 0.15$ corresponds to roughly $x/h = 4$, our furthest upstream location.) Assuming that the scalar field anisotropy should decay in the downstream direction (and that the low values of $\varepsilon_\theta/\varepsilon_{\theta_x}$ at $x/h = 4$ are artifacts of being at the edge of our domain in which we calculate finite differences of turbulence statistics), it would appear that the anisotropy in turbulent channel flow is generally larger than that in turbulent boundary layers. This is consistent the lower turbulence intensity (u_{rms}/U) in the present work than in Rosset et al.'s turbulent boundary layer (4% vs. 8% at the centrelines of the respective flows). Consequently, the magnitude of the turbulent mixing will also be weaker here, leading to larger anisotropies.

L.D. acknowledges generous financial support from the ANR (Grant 05-BLAN-0242-01) and PRI 'Echangeurs multi-fonctionnels.' Support has been graciously provided to L.M. by the NSERC and FQRNT (Canada).

References

1.Warhaft, Z, 2000. Passive scalars in turbulent flow. *Ann. Rev. Fl. Mech.* **32**, 203.

2.Lavertu, R.A. & Mydlarski, L., 2005. Scalar mixing from a concentrated source in turbulent channel flow. *J. Fluid Mech.* **528**, 135.

3.Rosset, L. et al., 2001. Anisotropy of a thermal field at dissipative scales in the case of small-scale injection. *Phys. Fluids* **13**, 3729.

One-particle dispersion in turbulent convection

A. Bistagnino[1], G. Boffetta[1], and A. Mazzino[2]

[1] Dipartimento di Fisica Generale and INFN, Via P.Giuria 1, 10125 Torino (Italy)[†]
[2] Dipartimento di Fisica, Via Dodecaneso 33, 16146 Genova (Italy)

The statistical properties of particle tracers advected by turbulent flows are one of the main topics in turbulent research. A largely looked-at object is the statistics of Lagrangian velocity differences $\delta \boldsymbol{v}(t) = \boldsymbol{v}(t) - \boldsymbol{v}(0)$. In hydrodynamic turbulence its behaviour, coincidentally close to diffusion, is at the basis of stochastic models of turbulent dispersion. We report here the results of a work focused on Lagrangian structure functions in two-dimensional free convection [1]. Owing to its two-dimensional character, this system displays an inverse cascade of energy with an exponent coherent with the Bolgiano–Obukhov theory of convection in Boussinesq approximation [2]. The Lagrangian behaviour of the system can be derived by the Bolgiano scaling and is largely different from that predicted by the classical K41 theory, but a careful statistical analysis is needed to observe it.

Our study is based on direct numerical simulations (DNS) of statistically stationary free convection, with a resolution $N = 1024^2$. The system is described by the following set of partial differential equations [3]:

$$\partial_t \omega + \mathbf{v} \cdot \nabla \omega = +\nu \Delta \omega - \beta \nabla T \times \mathbf{g}$$
$$\partial_t T + \mathbf{v} \cdot \nabla T = \kappa \Delta T \tag{1}$$

where $\omega = \nabla \times \boldsymbol{v}$ is the vorticity, T is the temperature field, $\boldsymbol{g}$ is the gravitational acceleration, β is the thermal expansion coefficient and ν and κ are respectively viscosity and molecular diffusivity. To obtain a stationary state a mean temperature profile $\langle T(\boldsymbol{r}, t) \rangle = \boldsymbol{G} \cdot \boldsymbol{r}$ is kept constant ($\boldsymbol{G}$ points in the direction of gravity). This acts as a forcing on the velocity field at all scales, and formally sets the Bolgiano scale to zero.

The Bolgiano–Obukhov theory of convection postulates a constant transfer rate of temperature fluctuations ϵ_θ from larger to smaller scales. Assuming balance between buoyancy and the inertial term in (1) one obtains the prediction for the scalings of δT and δv with r [4]. From this we find the veocity structure functions to obey:

[†] E-mail contact: `bistagni@to.infn.it`

$$S_p(r) = \langle (\delta_r v)^p \rangle \sim (\epsilon_r r)^{p/3} \sim r^{\zeta_p} \tag{2}$$

with $\zeta_p = hp = 3p/5$. It is worth remarking that, while the temperature field is highly intermittent, this is not the case for the velocity field [3]. The Bolgiano scaling of velocity has been recently observed in laboratory experiments [5].

The prediction for the statistics of the Lagrangian velocity increments is the following: let us consider the velocity v at each point as the superposition of contributions coming from eddies of all sizes. If we assume a range of scales with a scaling exponent h for the velocity field, the characteristic turnover time of an eddy of scale r is $\tau_r \simeq \tau_L (r/L)^{1-h}$, if τ_L is the turnover time of large-scale eddies. This implies that the scale of the eddies decorrelating in a time t is $r \simeq L(t/\tau_L)^{1/1-h}$.

Eddies with characteristic turnover time much lower than t are decorrelated and will bring no contribution to velocity variations over a time t. We can estimate the contribution to $\delta v(t)$ of eddies with $\tau_r \simeq t$ as:

$$\delta v(t) \simeq v_L (r/L)^h \simeq v_L (t/\tau_L)^q . \tag{3}$$

Eddies with very large characteristic times still have to be considered. When $t \ll \tau_L$ their contribution is differentiable, so that $\delta v \simeq (\partial_t v_L)t$. The total fluctuation is then given by these two effects:

$$\delta v(t) \simeq \tau_L (\partial_t v_L)(t/\tau_L) + v_L (t/\tau_L)^q . \tag{4}$$

When $t \ll \tau_L$, the leading term is the one with the minimum exponent. In the classical K41 theory we have $h = 1/3$, $q = 1/2$ and thus the dominant contribution comes from the local term. This leads to the well-known diffusive-like behaviour $\delta v \sim t^{1/2}$. In the case of free convection $h = 3/5$ ($q = 3/2$) and so it is the infrared term in (4) to dominate. The analysis of velocity fluctuations through Lagrangian structure functions is unable to disentangle the two effects and only shows the large-scale sweeping.

Equation (4) is an example of a signal with superimposed *more than smooth* fluctuations. It has been recently shown that this kind of signal can be analyzed on the basis of exit-time statistics [6]. This analysis is based on the time it takes for a tracer to observe a fixed change δv in its velocity [7].

Let us now consider equation (4) again. The differentiable part is leading everywhere, except when the prefactor $\partial_t v_L$ vanishes. $v_L(t)$ is a *more than smooth* signal with $1 \leq q \leq 2$ and thus its derivative is a one-dimensional self-similar signal with scaling exponent $\xi = q - 1$. The derivative will vanish on a fractal set of dimension $D = 1 - \xi = 2 - q$. Therefore, the probability to observe the component $O(t^q)$ is equal to the probability of picking a point on a fractal set of dimension D:

$$P(T \sim \delta v^{1/q}) \sim T^{1-D} \sim (\delta v)^{1-1/q} . \tag{5}$$

The moments of exit-time statistics must then be computed by use of this probability. The result is the following bifractal prediction:

$$\langle T^p(\delta v)\rangle \sim \delta v^{\chi(p)} \ , \quad \text{with} \quad \chi(p) = \min(p, \frac{p}{q} + 1 - \frac{1}{q}) \ . \tag{6}$$

Low-order moments are dominated by the large scales, but for $p > 1$ the Bolgiano contribution gets stronger and stronger. Figure 1 shows a few exit-time moments, which indeed have a power law scaling. The bifractal spectrum is shown in figure 2, in excellent agreement with (6).

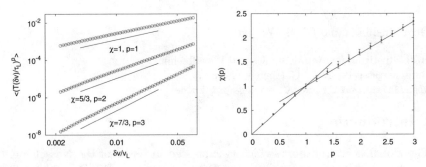

Fig. 1. Moments of order $p = 1,2$ and 3 and the slope χ predicted by (6).

Fig. 2. Scaling exponents $\chi(p)$ for the Lagrangian velocity differences δv in the inertial range. The squares represent the actual exponents and have been computed by fitting the average velocity increments in the inertial range. The error bars have been estimated by changing the fitting interval and observing the variations in χ. The continuous lines are the bifractal prediction (6).

In conclusion, we have analyzed the Lagrangian velocity structure functions in two-dimensional turbulent convection. The prediction coming from the Bolgiano–Obukhov scaling is shown to be quite different from that of the standard K41 theory. The unusual behaviour of the velocity increments can be understood in terms of *more than smooth* signals when analyzed via exit-time statistics. Only this approach is able to disentagle the effects of local and nonlocal contributions to the Lagrangian velocity fluctuations.

References

1. A. Bistagnino, G. Boffetta, A. Mazzino: Physics of Fluids **19**, 011703 (2007)
2. E.D. Siggia: Annu. Rev. Fluid Mech. **26**, 137 (1994)
3. A. Celani, A.Mazzino, M.Vergassola: Phys. Fluids **13**, 2133–2135
4. A. Monin, A. Yaglom: *Statistical fluid mechanics* (MIT Press, Cambridge MA 1975)
5. J. Zhang, X. L. Wu, K .Q.Xia: Phys. Rev. Lett. **94**, 174503 (2005)
6. L. Biferale, M. Cencini, A. Lanotte et al: Phys. Rev. Lett. **87**, 124501 (2001)
7. V. Artale, G. Boffetta, A. Celani et al: Phys. Fluids **9**, 3162 (1997)

Spatial distribution of the heat transport in turbulent Rayleigh–Bénard convection

Olga Shishkina and Claus Wagner

DLR - Institute for Aerodynamics and Flow Technology,
Bunsenstrasse 10, 37073 Göttingen, Germany
Olga.Shishkina@dlr.de, Claus.Wagner@dlr.de

1 Introduction

The global heat transport, which is expressed in terms of the Nusselt number Nu, and its dependency on the Rayleigh number Ra, Prandtl number Pr and aspect ratio Γ has been the subject of numerous studies of turbulent Rayleigh-Bénard convection (RBC) [1]. To improve the understanding of the $Nu(Ra, Pr, \Gamma)$ dependency, we investigate local heat fluxes, the thermal dissipation rates, relations between them and geometrical characteristics of the thermal plumes.

2 Governing equations and numerical method

The governing dimensionless equations for the Rayleigh–Bénard problem are

$$\mathbf{u}_t + \mathbf{u} \cdot \nabla \mathbf{u} + \nabla p = \Gamma^{-3/2} Ra^{-1/2} Pr^{1/2} \Delta \mathbf{u} + T \mathbf{e}_z, \quad \nabla \cdot \mathbf{u} = 0, \quad (1)$$

$$T_t + \mathbf{u} \cdot \nabla T = \Gamma^{-3/2} Ra^{-1/2} Pr^{-1/2} \Delta T \quad (2)$$

with $\mathbf{u}$ the velocity vector, T the temperature, $\mathbf{u}_t$ and T_t their time derivatives and p the pressure. The main parameters are $Ra = \alpha g H^3 \Delta/(\kappa\nu)$, $Pr = \nu/\kappa$ and the aspect ratio $\Gamma = D/H$ with H the height and D the diameter of the container, α the thermal expansion coefficient, g the gravitational acceleration, Δ the temperature difference between the bottom and the top plates, ν the kinematic viscosity and κ the thermal diffusivity, $\mathbf{e}_z$ the unit vector in the vertical z-direction. The temperature varies between $+0.5$ at the bottom and -0.5 at the top plate. On the adiabatic lateral wall $\partial T/\partial r = 0$ and on all solid walls the velocity field vanishes due to impermeability and no-slip conditions.

To investigate turbulent RBC we conducted DNS for $Ra = 10^5$, 10^6 and 10^7 and LES utilizing the tensor-diffusivity model [2] with the top-hat filtering for $Ra = 10^8$. In all simulations cylindrical containers of $\Gamma = 5$ filled with air ($Pr = 0.7$) are considered. The simulations were performed with a fourth order

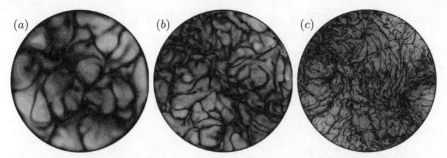

Fig. 1. Horizontal distribution of the instantaneous temperature field for $z = H/(2Nu)$, $Ra = 10^6$ (a), 10^7 (b), 10^8 (c), $Pr = 0.7$ and the aspect ratio $\Gamma = 5$. The colour scale ranges from white (negative values) to black (positive values).

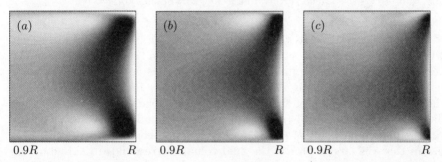

0.9R   R   0.9R   R   0.9R   R

Fig. 2. Vertical distribution of the mean heat flux $< \Omega >_{t,\varphi}$ near the vertical wall for $Ra = 10^6$ (a), $Ra = 10^7$ (b) and $Ra = 10^8$ (c). The colour scale as in Fig. 1.

accurate finite volume method [3] developed for solving (1–2) in cylindrical coordinates (z, φ, r) on staggered non-equidistant grids with (110, 512, 192) nodes clustered in the vicinity of the rigid walls.

3 Analysis of the simulations data

Analysing DNS- and LES-data of turbulent RBC it is shown that the regions of clustered thermal plumes increase and move apart with growing Ra. In presence of the vertical walls they settle close to these walls. The roots of the plumes which form in the vicinity of the horizontal walls become thinner and their number increase with growing Ra. In Fig. 1 instantaneous temperature fields in (φ, r)-planes located close to the heated bottom plate are shown for different Ra, where the roots of the thermal plumes are reflected in black.

Although the time- and area-averaged heat flux equals Nu in any horizontal cross-section of the domain, the local heat flux $\Omega = \Gamma^{1/2} Ra^{1/2} Pr^{1/2} u_z T - \Gamma^{-1} \frac{\partial T}{\partial z}$ is not constant (Fig. 3 a). In regions of large Ω-values in the vicinity of the bottom or the top plates the fluid moves predominantly from the centers

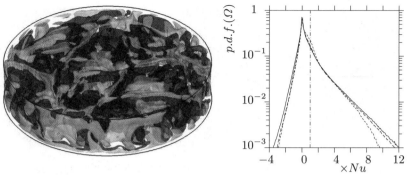

Fig. 3. Left: regions of the local heat flux $\Omega < 0$ (dark) and $\Omega \geq 2Nu$ for $Ra = 10^6$, $\Gamma = 5$. Right: p.d.f. of the local heat flux Ω, for $Ra = 10^8$ (——), $Ra = 10^7$ (– – –) and $Ra = 10^6$ (- - - -). The vertical line (– · –) corresponds to $\Omega = Nu$.

of the plume caps towards their borders in all possible horizontal directions, while in the center horizontal cross-section these regions correspond to the vertical movement of the fluid through the plume stems, around which the fluid can move rotationally. Analysing the Ω-density function (Fig. 3 b), it is shown that the portion of the fluid which corresponds to negative Ω-values increases with Ra and fills up to $1/3$ of the volume for $Ra = 10^8$. For any Ra the maximum point of the Ω-density function coincides with zero. This means that the largest part of the fluid volume corresponds to Ω fluctuating not around Nu, but around zero. The spread of the tails of the Ω-distributions depends strongly on the distance from the plates and widens in the bulk. In up to $1/3$ of the center horizontal cross-section the Ω-values are negative, while at the bottom plate they lie mainly in the interval $[0, 2Nu]$. With growing Ra zones of high values of the time averaged Ω move closer to the cell corners (see Fig. 2). The highest values of the mean heat flux are reached at a distance $z = Nu^{-1}H$ from the horizontal plates.

Further, analysing the thermal dissipation rates ε_θ as in [4] we conclude that both the portion of the whole domain, which corresponds to relatively small ε_θ-values, and the contribution to the volume averaged thermal dissipation rate ε_θ from these parts of the domain increase with Ra. This supports the conjecture by Grossmann & Lohse [5] that the turbulent background part of the volume averaged thermal dissipation rate ε_θ dominates for high Ra.

References

1. E. Bodenschatz, W. Pesch & G. Ahlers: Annu. Rev. Fluid Mech., **32**, 709 (2000)
2. A. Leonard: Adv. Geophys., **18**, 237 (1974)
3. O. Shishkina, C. Wagner: Computers and Fluids, **36**, 484 (2007)
4. O. Shishkina, C. Wagner: J. Fluid Mech. **546**, 51 (2006)
5. S. Grossmann, D. Lohse: J. Fluid Mech. **407**, 27 (2000)

Non-Oberbeck-Boussinesq effects in turbulent Rayleigh-Bénard convection

Francisco Fontenele Araujo[1], Siegfried Grossmann[2], and Detlef Lohse[1]

[1] Department of Applied Physics and J. M. Burgers Centre for Fluid Dynamics, University of Twente, 7500 AE Enschede, The Netherlands
`f.fontenelearaujo@utwente.nl`
`d.lohse@utwente.nl`

[2] Department of Physics, Phillips-University of Marburg, Renthof 6, D-3502 Marburg, Germany
`grossmann@physik.uni-marburg.de`

Thermally driven turbulence usually involves significant variations in the properties of the fluid, such as viscosity, thermal conductivity, thermal expansivity, and specific heat. In particular, when the fluid is strongly heated from below and cooled from above, such variations may even break the top-down symmetries of the velocity, temperature, and density profiles. To characterize non-Oberbeck-Boussinesq (NOB) effects of this nature, we have measured the temperature T_c at the center of the convection container. In this way, the top-down asymmetry between thermal boundary-layers can be conveniently described by the deviation of T_c from the mean temperature $T_m = (T_t + T_b)/2$ between the temperatures of the bottom (T_b) and top (T_t) plates.

We have developed a theory [1, 2] for calculating the difference $T_c - T_m$ as function of $\Delta = T_b - T_t$ that is based on boundary-layer equations with variable transport properties. Two different fluids have been considered: (i) water [1] and (ii) gaseous ethane under high pressure [2]. In the water case, we have found $T_c > T_m$, indicating that the top thermal-layer becomes thicker than its counterpart at the bottom plate. In contrast, $T_c < T_m$ has been observed in gaseous ethane, since the top thermal-layer becomes thinner. In both cases, our theoretical results are in reasonable agreement with experimental measurements (cf. figure 1). Nevertheless, NOB effects on the Nusselt number are not predicted by our theory.

References

1. G. Ahlers, E. Brown, F. Fontenele Araujo, D. Funfschilling, S. Grossmann, D. Lohse: J. Fluid Mech. **569**, 409–445 (2006).
2. G. Ahlers, F. Fontenele Araujo, D. Funfschilling, S. Grossmann, D. Lohse: Phys. Rev. Lett. **98**, 054501 (2007).

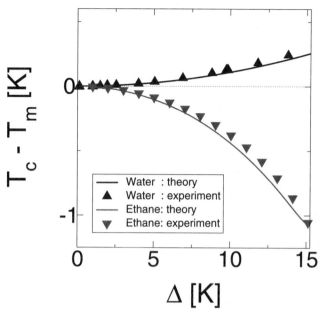

Fig. 1. $T_c - T_m$ as function of the temperature difference Δ between the bottom and top plates. Solid lines: theory. Triangles: experiments [1, 2]. In both cases, the mean temperature was held constant at $T_m = 40°C$. The ethane measurements [2] were made at constant pressure $P_m = 48.269$ bars. The Rayleigh number spans the range from 10^8 to 10^{11}.

Wall Shear stress measurements in the atmosperhic surface layer

Ivan Marusic, Jason Monty, Nicholas Hutchins, and Min Chong

Walter Bassett Aerodynamics Laboratory, Mechanical and Manufacturing
Engineering, University of Melbourne, Victoria 3010 Australia.
montyjp@unimelb.edu.au

The difficulty of measuring mean wall shear stress, $\overline{\tau_w}$, in a turbulent boundary layer is a constant hinderance to the experimental researcher. Hence it is
not surprising that a considerable amount of research into shear stress measurement techniques has been conducted [6]. Measurements of the fluctuating
component of the shear stress, τ'_w, are often more difficult but can provide
interesting information. In fact, from this study of atmospheric τ'_w data, valuable contributions have been made to the physical understanding of turbulence
over an unprecedented range of Reynolds numbers.

The shear stress measurements were made at the unique SLTEST (Surface Layer Turbulence and Environmental Science Test facility) site on the
great salt lakes of Utah, pictured in figure 1. Winds over the site are known
to remain strong and consistent for extended periods. Upstream of the measurement site, the surface is extremely flat and smooth over many kilome-
tres [5]. The geophysically driven air flow is therefore thought to share important characteristics with common wind-tunnel boundary layers, albeit at
three orders of magnitude higher Reynolds number. Thus, another goal was
to further understand similarities that may exist between the SLTEST surface layer and the wind tunnel boundary layer. For such comparisons, neutrally buoyant conditions are required; figure 2 confirms that these conditions
were present throughout the night. From extensive sonic anemometer measurements under these conditions, turbulence statistics were calculated which
exhibit laboratory-boundary-layer-like behaviour.

Beyond sonic anemometery, the present work followed on from that of
Heuer & Marusic [3] who developed a floating-element-type shear stress sensor, specifically designed to measure τ'_w in the atmospheric surface layer. By
determining peaks in the two-point correlations of shear stress and streamwise
velocity (with neutral buoyancy), [3] showed that a characteristic inclination
angle of around 15° existed. This result is more conclusively shown from the
current analysis and an illustration is provided in figure 3. This figure displays
contours of the shear stress-velocity correlation over a range of wall-normal

Fig. 1. Photograph of the SLTEST measurement site in Utah, USA.

distance and streamwise separation. It is interesting to note that Brown &
Thomas[2] performed a very similar experiment in a low Reynolds number
laboratory boundary layer. Their conclusions were strikingly similar, i.e., that
structures of approximately 12–13° inclination characterise the flow. The literature contains a number of other low Reynolds number studies also finding
similar angles [1, 7, 4]. It is therefore confirmed that the characteristic struc-

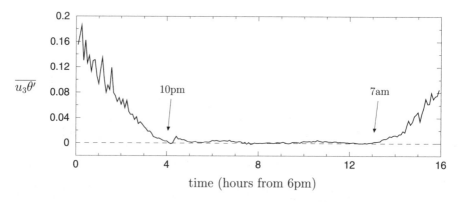

Fig. 2. Heat flux distribution throughout the evening, night and morning. u_3 is the
wall-normal velocity fluctuation and θ' is the temperature fluctuation. Zero heat
flux indicates neutrally buoyant conditions.

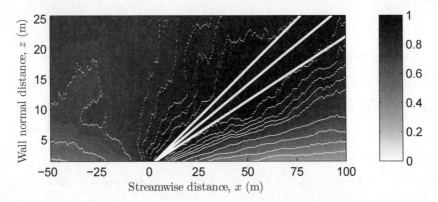

Fig. 3. Normalised τ'_w-u' correlation contours for neutrally buoyant conditions. Selected contour lines are marked with white dotted lines for clarity. Solid white lines indicate angles of 12°, 15° and 18° (note the figure axes are of different scale).

ture angle maintains a constant value over orders of magnitude Reynolds number range.

There are many other interesting results which will be presented but could not be included here for brevity. These include wall shear stress statistics, spanwise shear-velocity correlation, non-neutral buoyancy effects and comparisons with low Reynolds number numerical simulations. The compilation of all the results gives us an insight into turbulent flow structure from low to extremely high Reynolds numbers.

References

1. R. J. Adrian, C. D. Meinhart, and C. D. Tomkins. Vortex organization in the outer region of the turbulent boundary layer. *J. Fluid Mech.*, 422:1–54, 2000.
2. G. L. Brown and A. S. W. Thomas. Large structure in a turbulent boundary layer. *Phys. Fluids*, 20(10):S243–252, 1977.
3. W. D. C. Heuer and I. Marusic. Turbulence wall-shear stress sensor for the atmospheric surface layer. *Meas. Sci. Tech.*, 16:1644–1649, 2005.
4. I. Marusic and G. J. Kunkel. Streamwise turbulence intensity formulation for flat-plate boundary layers. *Phys. Fluids*, 15(8), 2003.
5. M. M. Metzger and J. C. Klewicki. A comparative study of near-wall turbulence in high and low Reynolds number boundary layers. *Phys. Fluids*, 13:692–701, 2001.
6. J. W. Naughton and M. Sheplak. Modern developments in shear-stress measurement. *Prog. in Aero. Sci.*, 38:515–570, 2002.
7. C. E. Wark and H. M. Nagib. Experimental investigation of coherent structures in turbulent boundary layers. *J. Fluid Mech.*, 230:183–208, 1994.

Point (Sonic Anemometer) Measurements in a Gusty Wind Over Complex Terrain

L.M.F. Ribeiro[1,3] and J.M.L.M. Palma[1,2]

[1] CEsA — Centro de Estudos de Energia Eólica e Escoamentos Atmosféricos
[2] Faculdade de Engenharia da Universidade do Porto
[3] Instituto Politécnico de Bragança
    frolen@ipb.pt

Introduction

Sonic and cup anemometer measurements were made at 42 m (a.g.l.) in a region close to a cliff, 100 m above the sea level, in a peninsula exposed to incoming wind from the ocean. Sonic transducer shadow effect originated a mean velocity correction of 3.23% and, due to large turbulence intensity and gust factors (0.34 and 2.12 respectively), cup anemometer deviated from sonic measurements by 2.3%. However, the major set-back refers to the statistical unsteadiness of the data, impairing traditional statistical and spectral analysis based on FFT algorithms for the detection of isolated events.

Methodology

The unstationarity of the flow lead to the adoption of two different methodologies for identification of a smaller data set that could be analysed by conventional methods. Data sub-sets were selected by visual inspection and by acceleration criteria. Visual inspection proved to be unfruitful. By conditioning the analysis to the data that presented acceleration values below ±2 ms^{-2} we obtained a 45-min data subset, though being non-stationary, it was the steadier obtained from the whole data series. Wavelet [1] and spectral analysis revealed the existence of periodic events with an oscillation period from 2 to 3-min. From the acceleration criteria a 3 hr data subset, that presented strong transient characteristics, was also analysed with the same tools. The same events were detected from wavelet analysis although superimposed by larger scale events.

An alternative path on data processing was adopted with fruitful results. Based on the ogive function [2], (integral) spectral analysis of the momentum flux, (equation 1)

$$Og_{u'w'} = \int_{\infty}^{f_0} Co_{u'w'}(f)df \qquad (1)$$

for the identification of stationary data sets, we encountered a time interval of 40-min, i.e. to 2.3 times the integral time scale of the flow. To increase statistical significance of the results the defined time block of 40-min was applied for the analysis of the whole series divided into 32×40-min time intervals.

Results

Combined techniques such PDF intermittency, flow visualization, spectral, wavelet and quadrant analysis were then explored [3]. From PDF intermittence methods we detected 3 types of flow: Type A, a steadier one (9.4% of the measurements) with no reverse flow; Type B, a moderate to strong flow (84.3% of the measurements) where up to 15% of the flow is reversed; and Type C, the most complex flow, where the reverse flow surpassed 15%, figure 1.

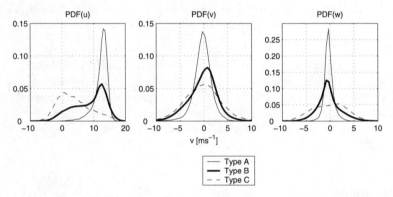

Fig. 1. Longitudinal u, transversal v and vertical w velocity component PDF of flow types A to C.

For each flow type, flow visualization, spectral, wavelet and quadrant analyses were applied unveiling different features of the turbulent flow field. Two major features were observed: a periodic behavior of the flow superimposed by coherent structures. The periodic behavior was confirmed by spectral and wavelet analysis, where eddies presented consistently energy peaks circa 2.9 min oscillation period, figure 2.

The coherent structures, although being detected by wavelet analysis, were identified and analyzed with quadrant analysis [4] [5]. It was found that flow unsteadiness was from gusts, sweeps and ejections but for different flow conditions. For the most frequent type of flow, Type B, coherent structures were mainly sweeps and gusts, with 75% of the turbulent flux transport, while for

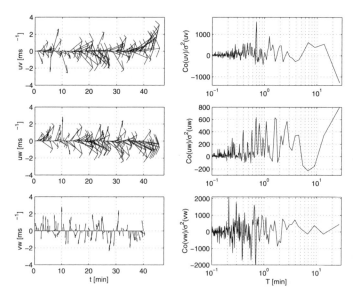

Fig. 2. Type B flow visualization and periodicity. Each arrow correspond to a 30 s mean on the uv, uw and vw plane accompanied by the cospectra of each series.

Type C ejections were dominant, 90% . The events changed in scale when mean advection conditions were altered and were the cause of large vertical velocity components (-18.75 and +22.29 ms^{-1}). Consequences for measurements were relevant and justify the large mean sonic corrections for sonic velocity measurements and cup anemometer overestimation due to large inclination and unsteadiness of the flow [6].

References

1. C. Torrence and G. Compo. A practical guide to wavelet analysis. *Bulletin of the American Meteorological Society*, 79(1):61–78, January 1998.
2. S. P. Oncley. *Flux Parametrization Techniques in the Atmospheric Surface Layer*. PhD thesis, University of California, 1989.
3. L. M. F. Ribeiro. *Sonic Anemometer and Atmospheric Flows over Complex Terrain*. PhD thesis, University of Porto, 2005.
4. B. Shiau and Y. Chen. Observation on wind turbulence characteristics and velocity spectra near ground at the coastal region. *Journal of Wind Engineering and Industrial Aerodynamics*, (90):1671–1681, 2002.
5. P-A. Krøgstad and J. H. Kaspersen. Structure Inclination Angle in a Turbulent Adverse Pressure Gradient Boundary Layer. *Journal of Fluids Engineering*, 124, 2002.
6. K. H. Papadopoulos, N. C. Stefanatos, U. S. Paulsen, and E. Morfidiakis. Effects of turbulence and flow inclination on the performance of cup anemometers in the field. *Boundary-Layer Meteorology*, (101):77–107, 2001.

Atmospheric surface layer turbulence over water surfaces and sub-grid scale physics

Elie Bou-Zeid[1], Hendrik Huwald[2], Ulrich Lemmin[3], John S. Selker[4], Charles Meneveau[5], and Marc B. Parlange[6]

[1] Ecole Polytechnique Fédérale de Lausanne `eliebz@jhu.edu`
[2] Ecole Polytechnique Fédérale de Lausanne `hendrik.huwald@epfl.ch`
[3] Ecole Polytechnique Fédérale de Lausanne `ulrich.lemmin@epfl.ch`
[4] Oregon State University at Corvallis `selkerj@engr.orst.edu`
[5] Johns Hopkins University `meneveau@jhu.edu`
[6] Ecole Polytechnique Fédérale de Lausanne `marc.parlange@epfl.ch`

1 Introduction

Numerous experimental and numerical investigations have focused on the study of surface layer turbulence over land; this has resulted in improved understanding of coherent structures, similarity relations, and various turbulence features controlling land-atmosphere interaction. However, comparable measurements of turbulence in the surface layer over water surfaces have been far less common. Developing our understanding of air-water interaction and turbulence over water surfaces is crucial for improving simulations of environmental turbulence in the lower atmosphere, the upper ocean, and lakes. This in turn will help in enhancing evaporation models and understanding the hydrologic cycle and its interaction with global atmospheric circulation [1].

The Lake-Atmosphere Turbulent EXchanges (LATEX) field measurement campaign was designed to address these issues. The experiment took place on a platform situated in Lake Geneva, Switzerland (exposed to a 30 km long wind fetch) from August through October, 2006. The primary instrumentation consisted of (1) a vertical array of four sonic anemometers and four open-path H_2O/CO_2 analyzers both measuring at 20 Hz, (2) a Raman scattering fiber-optic temperature profiler (1 meter above the water surface and 2 meters below), and (3) a lake current profiler. Other supporting measurements included: surface temperature, net radiation, relative humidity, and wave height and speed (Fig. 1). The next section of this paper analyzes the diurnal cycle of surface fluxes from LATEX. Then, we investigate the dynamics and models of small scale turbulence and the implications for large-eddy simulations (LES) of turbulent atmospheric flows over water surfaces.

2 Air-lake exchanges

The measured latent heat flux (LE) was always positive at our site, i.e. lake water is continuously evaporating without an apparent diurnal cycle. On the other hand, the sensible heat (H) tends to be positive during the night and early morning when the air is cooler than the water. At mid day and in the afternoon, as the air heats up over land and flows over the lake, H tends to become negative. The Bowen ratio (H/LE) shows a well defined diurnal cycle (Fig. 2) with a positive peak around 11:00 a.m. The scatter in the plot is not surprising since we plot the data from the entire experimental period versus the time of day without considering relevant factors such as net radiation or wind speed which vary substantially over the course of the three month experiment. Note the small absolute values of the Bowen ratio indicating that evaporation is generally a more important source of heat exchange at the surface than sensible heat. The consequence is that the ABL is almost always unstably stratified even when the water is colder than the air; this is due to evaporation that decreases the density of near-surface air.

Fig. 1. vertical array setup (left) and upwind fetch (right) at LATEX

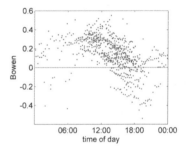

Fig. 2. Diurnal variation of the Bowen Ratio

3 Sub-grid scale models for water vapor fluxes

With the high frequency water vapor measurements from the gas analyzers, we study SGS fluxes of latent heat, apparently for the first time. We also assess the applicability, over water surfaces, of SGS models developed over solid surfaces. Fig. 3 depicts the comparison of the SGS dissipation of TKE $(-\tau_{ij}S_{ij})$, where τ_{ij} is the SGS stress tensor and S_{ij} is the resolved rate of strain tensor [2] and the dissipation based on second and third order longitudinal structure functions, $(D_{u,u}$ and $D_{u,u,u})$[3, 4]. The two dissipation estimates match well with values computed from the structure functions slightly exceeding SGS values. The same comparisons were made for the dissipations of temperature and water vapor variance and good agreement was also found for the scalars.

The eddy-viscosity needed in SGS model of the Smagorinsky type was computed and good agreement was found with values computed dynamically

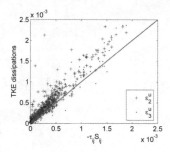

Fig. 3. TKE dissipations comparison

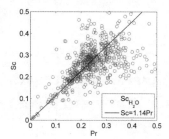

Fig. 4. Pr. and Sc. for SGS turbulence

in LES [5] or a priori from field expriments [6]. We also compute the SGS turbulent Prandtl number (Pr) and the SGS turbulent Schmidt number for water vapor (Sc). We found fluctuating values of Pr and Sc that seem to depend on turbulence intensity and thermal stratification of the flow. An interesting result was the good correlation between values of Pr and Sc suggesting that turbulent transport of heat and moisture are well correlated. However, as depicted in (Fig. 4) we found the ratio Sc/Pr to be about 1.14 suggesting that, under the current experimental conditions, turbulent transport of heat is more efficient than turbulent transport of water vapor.

4 Summary and Conclusion

The LATEX field measurement campaign was designed to further our understanding of air-water exchanges and atmospheric turbulence dynamics over water surfaces. Analysis of surface fluxes underlines the importance of latent heat flux which was found to be higher than sensible heat flux. Computations for sub-grid scale eddy viscosity models show similarity to results obtained over land. This is the first experimental setup that allows the computation of SGS fluxes and Schmidt numbers for water vapor. We found a good correlation between Prandtl number and Schmidt number for water vapor suggesting perhaps that only one of the two numbers needs to be computed dynamically in LES; the other can then be obtained from $Sc = 1.14Pr$.

References

1. W. Brutsaert: *Evaporation into the atmosphere*, (Reidel, Dordrecht, 1982)
2. C. Meneveau and J. Katz : Annu. Rev. Fluid Mech. **32**, 1, (2000)
3. S. Pope: *Turbulent Flows*, (Cambdrige University Press, Cambdrige, 2000)
4. J.D. Albertson, M.B. Parlange, G. Kiely, W.E. Eichinger: J. Geophys. Resear. **102**, 13423-13432, (1997)
5. E. Bou-Zeid, C. Meneveau, M.B. Parlange : Phys. Fluids **07**, 025105, (2005)
6. J. Kleissl, M.B. Parlange, C. Meneveau : J. Atmos. Sci. **61**, 2296-2307, (2004)

SNOHATS: Stratified atmospheric turbulence over snow surfaces

Marc B. Parlange[1], Elie Bou-Zeid[2], Hendrik Huwald[3], Marcelo Chamecki[4], and Charles Meneveau[5]

[1] Ecole Polytechnique Fédérale de Lausanne `marc.parlange@epfl.ch`
[2] Ecole Polytechnique Fédérale de Lausanne `eliebz@jhu.edu`
[3] Ecole Polytechnique Fédérale de Lausanne `hendrik.huwald@epfl.ch`
[4] Johns Hopkins University `chamecki@jhu.edu`
[5] Johns Hopkins University `meneveau@jhu.edu`

1 Introduction

Stably stratified turbulence presents particular challenges both from an experimental and a modeling perspective. The damping of the turbulence due to flow stratification and the presence of features such as gravity waves, and Kelvin-Helmholtz instabilities complicate the the application of turbulence similarity theories and the formulation of turbulence models. From an LES perspective, the main problem is that the classic parameterizations of the subgrid scales are often found to be inadequate for stable conditions. To guide the improvement of SGS modeling under stable conditions and, more generally, to understand turbulence dynamics under stable stratification and its interaction with other flow featuers, the Snow-Horizontal Array Turbulence Study (SnoHATS) field study was held at the extensive "Plaine-Morte" glacier in the Swiss Alps (3000 m) from February to April 2006. The snow cover provided stable stratification of the flow over long periods. Two horizontal arrays of 3D sonic anemometers were deployed to allow two dimensional filtering and computation of the three-dimensional strain rate tensors (Fig. 1).

Fig. 1. Side view of the 12 sonics array (left) and upwind fetch (right)

2 SubGrid Scale Turbulence

The SnoHATS setup was designed to measure a wide range of turbulence scales and allow the evaluation of the sub-grid scale (SGS) stress tensor $\tau_{ij} = \widetilde{u_i u_j} - \widetilde{u}_j \widetilde{u}_j$ [1]. This term needs to be parameterized in LES and its direct field measurement for a-priori studies is helpful in assessing the accuracy of various parameterizations especially under conditions known to be challenging to LES such as stably stratified boundary layers. We first examine eddy-viscosity closure models, specifically the Smagorinsky model [2], by computing the values of the model constant c_s that match the measured and modeled SGS dissipations [4]. The general results from SnoHATS agree with previous findings in [4]: as the stability (Δ/L) increases the model coefficient decreases (Fig. 2). Similarly, the SGS Prandtl number (Pr) was plotted versus Δ/L in (Fig. 3); the figure shows a clear trend of increasing Pr despite some scatter of the data. Note the increase in Pr around $\Delta/L = 0.5$ which coincides with a marked decrease in c_s. In practice, the coefficient determined from the experiment is c_s^2/Pr; this is the coefficient appearing in the expression of eddy diffusivity of heat in Smagorinsky type models; this coefficient actually decreases at $\Delta/L = 0.5$ but not as fast as c_s^2. This suggests that all turbulent transport efficiencies decrease with increasing stability; however, the efficiency of turbulent momentum transfer decreases faster than that of heat. Other studies [5] under unstable atmospheric conditions clearly show that Pr decreases as Δ/L becomes negative (unstable condition). The data from the two experiments appear to indicate that atmospheric stability has a significant influence on the relative efficiency of momentum and heat transport by turbulence: increasingly stable conditions reduce the efficiency of heat transport (relative to momentum) while increasingly unstable conditions increase the relative efficiency of heat transport.

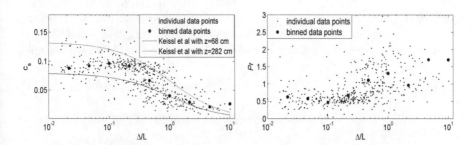

Fig. 2. c_s as a function of Δ/L **Fig. 3.** Pr as a function of Δ/L

Despite the general agreement with previous literature, several periods were detected where the turbulence and SGS dynamics were not consistent with our classic understanding of ABL flows. For example, increasing stability sometimes caused an increase in the turbulent kinetic energy; during those periods the Smagorinsky coefficients reflected the atmospheric stability i.e. the coefficients and the SGS viscosity were decreasing despite an increase in TKE. Other interesting results also showed that the TKE tends to decrease as the temperature variance increases in contrast with the trends under unstable conditions. Data from this experiment and from a lake experiment under unstable conditions [5] are shown in Fig. 3.

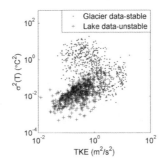

Fig. 4. TKE versus temperature variance for stable and unstable atmospheres

3 Summary and Conclusion

Experimental results under stable atmospheric stratification from the Sno-HATS experiment were presented. Investigation of the subgrid scale fluxes indicate that increasing stability reduces turbulent transport efficiencies. The efficiency of momentum transport is reduced faster that that of heat transport. SGS models should be able to take that into account. TKE was also found to be negatively correlated with the variance of temperature under stable conditions; in contrast to the positive correlation of the two parameters under unstable conditions. The above results point to the need to reconsider some of the basic questions about turbulence and SGS physics in stable flows: How is turbulence affected by other features of stably stratified flows? What is the main function of the subgrid scales under a strong stable stratification?

References

1. P. Sagaut: *Large Eddy Simulation for Incompressible Flows* - 3rd ed., (Springer-Verlag, 2006)
2. S. Pope: *Turbulent Flows*, (Cambdrige University Press, Cambdrige, 2000)
3. C. Meneveau and J. Katz : Annu. Rev. Fluid Mech. **32**, 1, (2000)
4. J. Kleissl, M.B. Parlange, C. Meneveau : J. Atmos. Sci. **61**, 2296-2307, (2004)
5. E Bou-Zeid, H. Huwald, U. Lemmin, J.S. Selker, C. Meneveau, and M.B. Parlange: *Atmospheric surface layer turbulence over water surfaces and sub-grid scale physics*, presentation at 11th European Turbulence Conference, Porto, Portugal, 25-28 June 2007.

The effect of persistent separation in turbulent relative dispersion: self-similar telegraph equation

Kentaro Kanatani, Takeshi Ogasawara and Sadayoshi Toh

Division of Physics and Astronomy, Graduate School of Science, Kyoto University, Kyoto 606-8502, Japan
kanatani@kyoryu.scphys.kyoto-u.ac.jp

1 Introduction

In turbulent relative dispersion, an anomalous dispersion so-called Richardson's law $\langle r^2 \rangle \propto t^3$ is believed to be well established. However, the asymptotic form of the probability density function (PDF) $P(r,t)$ of the separation r of a particle pair as well as its governing equation has not been settled yet. This is because the inertial ranges achieved by the experiments and the direct numerical simulations (DNSs) so far are not wide enough to confirm them.

In order to study the relative dispersion in the inertial range, we limit our attention to the early stage of the evolution of particle pairs whose initial separations are also in the inertial range, using DNS of 2D thermal convection (2DFC) turbulence. Then, we will show that Eq.(1) can well describe at least the evolution of the front edge of the PDF.

2 Results

We have derived the following self-similar telegraph equation as a governing equation of the PDF[1]:

$$\frac{T_c(r)}{\lambda} \frac{\partial^2 P}{\partial t^2} + \frac{\partial P}{\partial t} = \frac{\partial}{\partial r} \left[D(r) r^{d-1} \frac{\partial}{\partial r} \left(\frac{P}{r^{d-1}} \right) \right] + \sigma \frac{\partial}{\partial r} [v(r) P], \qquad (1)$$

where $T_c(r)$ represents a characteristic time scale, $D(r)$ the turbulent diffusion coefficient, $v(r)$ the relative velocity, and λ and σ are parameters characterizing the turbulent field. The coefficients $T_c(r)$ and $D(r)$ are assumed to obey the following scaling laws: $T_c(r) = \breve{A}^{-1} r^s$, $D(r) = \breve{A} \lambda^{-1} r^{2-s}$. Here $\breve{A}$ is a dimensional constant and s a scaling exponent; $s = 2/3$ for Kolmogorov scaling and $s = 2/5$ for Bolgiano-Obukhov scaling.

The self-similar telegraph equation includes the effects of **the persistency and the self-similarity** of the separation as well as the random one which is described by Richardson's diffusion equation with self-similar drift term. We assume that the relative velocities of the particle pairs are governed by $v(r) \equiv \frac{dr}{dt} = \check{A}r^{1-s}$[1, 2]. This means that the front of the PDF moves at a finite speed and the maximum separation r_{max} exists and scales as $r_{max} \propto t^{1/s}$. It should be noted that the scaling of r_{max} is the same as that of standard deviation $\langle r^2 \rangle^{1/2}$. In fact, Eq.(1) has the similarity solution as shown in Fig.1. The slowly-separating particle pairs are well approximated by the Richardson's diffusion equation. On the other hand, the distribution of fast-separating ones is drastically affected by the persistency.

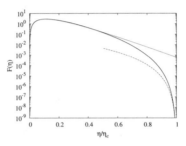

Fig. 1. The similarity solutions of Eq. (1) (solid line) and Richardson's diffusion equation with self-similar drift term (dashed line). The dot-dashed line is the asymptotic form of the former around the front. $s = 2/5$, $d = 2$, $\lambda = 5.2$ and $\sigma = 0.083$.

In this study, we deal with the turbulent relative dispersion in 2DFC turbulence. The basic equations of the 2DFC system are

$$\partial_t \boldsymbol{u} + (\boldsymbol{u} \cdot \nabla)\boldsymbol{u} = -\nabla p/\rho_0 + \nu \triangle \boldsymbol{u} - \alpha g T \boldsymbol{e}_g, \tag{2}$$

$$\partial_t T + (\boldsymbol{u} \cdot \nabla)T = \kappa \triangle T, \tag{3}$$

where $\boldsymbol{u}$, T and p represent solenoidal velocity, temperature and pressure fields, respectively. $\boldsymbol{e}_g$ is the unit vector in the gravity-direction. The entropy $S \equiv \frac{1}{2} \int T^2 d\boldsymbol{x}$ cascades from larger to smaller scales similar to the energy cascade in 3D-NS turbulence [3]. This leads the Bolgiano-Obukhov (BO) scaling: $E(k) \propto k^{-11/5}$ and $S(k) \propto k^{-7/5}$. We carried out DNS of 2DFC system by using the 4th order Runge-Kutta and pseudo-spectral method with resolution of 2048^2.

Fig.2 shows the time evolution of the PDF of the separation starting from the initial condition $P(r,0) = \delta(r - r_0)$ where r_0 is $100\Delta x = 100 \times (2\pi/2048)$. The lower and upper ends of the inertial range are estimated as $r = 80\Delta x$ and $r = 320\Delta x$, respectively by exit time statistics[4]. Fig.3 shows the PDF at $t = 0.2 = 0.823\tau_\theta$. [1]

---

[1] τ_θ is the dissipation time scale of Bolgiano-Obukhov scaling.

3 Concluding remarks

Even in this transient stage, the PDF normalized by $\langle r^2 \rangle^{1/2}$ seems to approach the similarity solution at least in the tail. The dashed line in Fig.3 is the asymptotic form of the similarity solution displayed in Fig.1. These results confirm that the front of separation PDF moves at a finite speed and the tail of the PDF is approximated by the similarity solution. The dependence of the PDF on the initial separations, the detailed comparison of the results obtained by the DNS to the time-dependent solutions of the self-similar telegraph equation and the experimental results recently reported by Ouelette et al.[5] will be presented at the conference.

This work was supported by the Grant-in-Aid for the 21st Century COE "Center for Diversity and Universality in Physics" from the Ministry of Education, Culture, Sports, Science and Technology (MEXT) of Japan. The numerical calculations were carried out on SX8 at YITP in Kyoto University.

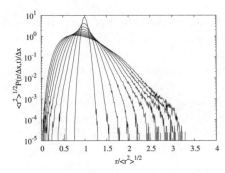

Fig. 2. The evolution of the separation PDF from $t = 0.04 = 0.165\tau_\theta$ to $t = 0.4 = 1.65\tau_\theta$, normalized by $\langle r^2 \rangle^{1/2}$.

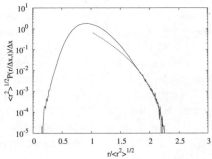

Fig. 3. The separation PDF at $t = 0.2 = 0.823\tau_\theta$ (solid line), fitted with the asymptotic form of the similarity solution around the front (dashed line).

References

1. T. Ogasawara and S. Toh, J. Phys. Soc. Jpn, **75**, 083401 (2006).
2. I. M. Sokolov: Phys. Rev. E **60** 5528 (1999).
3. S. Toh and E. Suzuki, Phys. Rev. Lett, **73**, 1501 (1994).
4. T. Ogasawara and S. Toh , J. Phys. Soc. Jpn. **75** 104402 (2006).
5. M. Bourgoin et al., Science, **311**, 835 (2006); N. Ouellette et al., New J. Phys., **8**, 109 (2006).

Statistics of Acceleration Field Motions in 2D Inverse-Cascading Turbulence

Frédéric Schwander[1], Erwan Hascoët[1,2], and J. Christos Vassilicos[1,2]

[1] Department of Aeronautics
Imperial College London
SW7 2AZ London, United Kingdom
frederic.schwander@imperial.ac.uk
[2] Institute for Mathematical Sciences
SW7 2PG London, United Kingdom
e.hascoet@imperial.ac.uk, j.c.vassilicos@imperial.ac.uk

Turbulence is characterized by its variability in time. Nevertheless, it posseses a degree a persistence that is made clear by the fact that, in isotropic turbulence, the Lagrangian autocorrelation time is systematically larger than its Eulerian counterpart [1]. The underlying reason is that the time fluctuations of the flow in a Eulerian frame has two components, one coming from the sweeping of spatial fluctuations by flows on larger scales, the other from the internal distortion of the fluctuation itself, and the time-scale associated with the sweeping is the fastest. Hence studying the dynamics of the distortion of the field needs removing sweeping effects by considering Lagrangian quantities.

One such quantity is the acceleration. It is a signature of the time-dependence of the local velocity field in a frame moving with that velocity, and it was proposed that it defines the motion of this frame, so that the statistics of this motion can be used to define and quantitatively demonstrate the multiscale persistence of the turbulence [2]. Following this idea, the velocity of points with zero acceleration was studied in [3], and it was found that this velocity becomes increasingly coincident with the local flow velocity as the ratio of outer to inner length scales L/η of the turbulence is increased, effectively showing the growing importance of sweeping over distortion in accounting for Eulerian statistics as the turbulence becomes stronger.

We expand the latter study by investigating the motion of the acceleration field in a Direct Numerical Simulation of forced, two-dimensional, isotropic, homogeneous, turbulence in an inverse-cascading regime, with a $k^{-5/3}$ spectrum spanning over one decade. Three quantities are considered here: the local fluid velocity $\mathbf{u}$, the velocity of the acceleration field $\mathbf{V}_a$, defined as:

$$\mathbf{V}_a = \frac{\mathrm{d}\mathbf{s}}{\mathrm{d}t} \tag{1}$$

where $s(t)$ defines the trajectory of a point with constant acceleration, and their difference $\boldsymbol{\xi} = \mathbf{u} - \mathbf{V}_a$.

The PDF of $\mathbf{V}_a$ for zero-acceleration points is of particular interest: it has extended tails for values larger than several times u_{rms}, with non-negligible probability density, which should correspond to rare, intense events, where the acceleration undergoes fast displacements and its time-history can no longer be described as a sheer advection process. Nevertheless, the core of the PDF, where most of the energy of the distribution is contained, can be superimposed on that of the local flow velocity. In particular, as should be expected, the velocity has a Gaussian distribution, and this feature is also shared by the core of the PDF of $\mathbf{V_a}$. The similarity between the acceleration field velocity

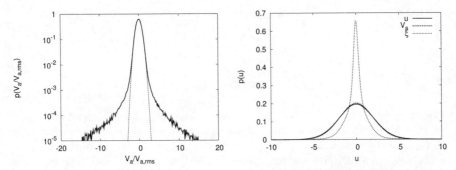

Fig. 1. Probability density of $\mathbf{V}_a$ (left), and probability densities of $\mathbf{u}$, $\mathbf{V}_a$ and $\boldsymbol{\xi}$ for $\mathbf{a} = 0$. The dotted curve on the left corresponds to a Gaussian distribution. The $\mathbf{V}_a$ distribution matches it for $|\mathbf{V}_a| < 3V_{a,rms}$. The right-hand graph shows the almost perfect collapse of the PDF of $\mathbf{V}_a$ and that of $\mathbf{u}$, as well as the marked peaking of the $\boldsymbol{\xi}$ distribution.

and the flow velocity can be appropriately quantified by observing the PDF of $\boldsymbol{\xi}$. The latter is strongly peaked at the origin (Figure 1 right), showing the dominant correspondence between the flow velocity and the acceleration velocity. It had already been observed that its standard deviation decreases with regard to the flow r.m.s. velocity as L/η is increased, in agreement with Kolmogorov scaling, and we recover the result obtained in [3]:

$$< \boldsymbol{\xi}^2 |\mathbf{a} = 0 >= C u_{rms}^2 (L/\eta)^{-2/3} \qquad (2)$$

These results effectively confirm the adequacy of the picture of the acceleration being swept, on average, along fluid trajectories for points with zero-acceleration.

We investigate here the relevance of this idea for non-zero values of the acceleration, by sampling the field at points where $|\mathbf{a}| = 2a_{rms}$. Our results indicate that the advection velocity of the acceleration vector is independent of the acceleration. Its distribution remains isotropic (Figure 2 left), has the same

Gaussian core and extended tails as previously, and is identical as for zero-acceleration points. In opposition to that, the PDFs of **u** and $\boldsymbol{\xi}$ differ strongly in the directions along the acceleration vector and across it: the statistics in the longitudinal direction are most reminiscent of their equivalent for zero-acceleration points (although the standard deviation depends on the modulus of the acceleration), but the transverse PDFs acquire a bi-modal shape, the peaks being more marked for $\boldsymbol{\xi}$ (Figure 2 right).

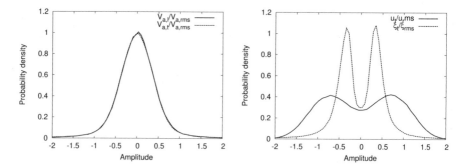

Fig. 2. Probability distribution function of the longitudinal and transverse components of $\mathbf{V}_a$ (left), and probability distributions for the transverse components of **u** and $\boldsymbol{\xi}$ (right) for $|a| = 2a_{rms}$. It is clear that as far as the acceleration velocity is concerned, the transverse and longitudinal directions are equivalent. The bi-modal structure of the distributions of **u** and $\boldsymbol{\xi}$ in the trasnverse direction is evident.

This study shows the relevance of the work by Tennekes [1] for the motion of the acceleration: except for events with small probability, the acceleration is swept at a velocity independent of the acceleration itself, with a Gaussian probability. The discrepancy between the transverse and longitudinal directions for the other two velocities shows the existence of a correlation between them and the local acceleration.

References

1. H. Tennekes: J. Fluid Mech. **67** 561 (1975)
2. S. Goto *et al.*: Physical Review E **71** 015301 (2006)
3. L. Chen, S. Goto and J.C. Vassilicos: J. Fluid Mech. **553**, 143 (2006)
4. P.K. Yeung and S.B. Pope: J. Fluid Mech. **207** 531 (1989

Probability density function (PDF) and filtered density function (FDF) methods for turbulent scalar dispersion in incompressible flows

M. Cassiani[1], J. D. Albertson[1] and P. Franzese[2]

[1] Duke University, Department of Civil and Environmental Engineering, Durham, NC 27708, USA, massimo.cassiani@duke.edu, john.albertson@duke.edu
[2] George Mason University, Department of Computational and Data Sciences, Fairfax, VA, 22030, USA, pfranzes@gmu.edu

1 Introduction

Concentration fluctuations are of importance in a number of fields from industrial to environmental. Applications include flammability of substances, toxic gases effects on human, malodorous materials dispersion and interaction between turbulence and chemistry. Several methods have been proposed to deal with the scalar fluctuations of passive scalars [9, 5, 15, 2, 8]. The picture is further complicated when chemical reactions are involved. An approach that has proven to be particularly useful in this case is based on the PDF transport equation of scalar quantities [11]. Pope [12] introduced the concept of filtered density function (FDF) connecting the PDF and large eddy simulation (LES). More recently the FDF approach has been further developed [6, 14] and applied to a few actual problems [10]. There exists many cases, especially related to atmospheric applications, where the scalar source length scale (σ_o) is smaller than the length scales of the dispersion domain by a factor ranging from 10^{-3} to 10^{-4} and more. In these cases the PDF and FDF approaches can be applied using some specialized techniques [3, 4].

2 Methods and results

The LES code used here is a modified version of the one developed by [1] to simulate the atmospheric boundary layers. It is a pseudo-spectral code using a Poisson equation for the pressure in order to obtain a divergence-free velocity field. The Adam–Bashfort scheme is used for the time advancement. Here the sub-grid viscosity has been treated trough the mixed-scale model (MSM) of Sagaut [13]. The Lagrangian PDF Monte Carlo method is based on a system of ordinary and stochastic differential equations describing the fluid particle characteristics. The numerical scheme of Cassiani et al. [4] was used with a seven blocks structured grid nested around the scalar source and a procedure of particles splitting/erasing to reduce the statistical and deterministic errors. The most resolved grid had a resolution about 64 times finer than the resolution used in the LES. The evolution of the concentration carried by each particle was modeled trough an Interaction by Exchange with the Mean (IEM) micro-mixing model. The necessary micro-mixing time scale was related to the source time scale in the finer level of the nested grid while the relation proposed by [6] was used in the coarse cells. A locally averaged parameterized value was used in the other grid levels. Comparisons with wind

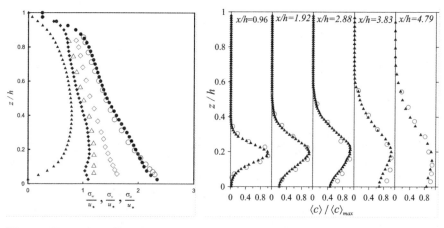

Fig. 1. Vertical profile of the standard deviation of fluctuating velocity components. z is the elevation above ground and h is the boundary layer height. Resolved LES velocity (solid symbols) and experiments (open symbol).

Fig. 2. Vertical profile of the mean concentration normalized by the local maximum for five different downwind position. FDF-IEM model (solid symbols) and experiments (open symbol)

tunnel measurements [7] for several scalar statistics is proposed. The scalar tracer was emitted by a 8.5 mm tube in a 1.2 m high boundary layer. The horizontal extension of the LES domain was 7.53 m in both directions. The resolution used in the LES velocity field simulations was 32x32x40. The standard deviations of the three components of the resolved velocity are reported in fig. 1 with the corresponding measurements of [7]. Fig. 2 shows the model results for the prediction of the mean concentration at several downwind distances and elevations. Fig. 3 shows the computed and measured concentration PDF. The results obtained using a Stochastic Interaction by Exchange with the Conditional Mean (SIECM) for the micro-mixing and parameterized velocity statistics are also shown (PDF-SIECM). These comparisons show the

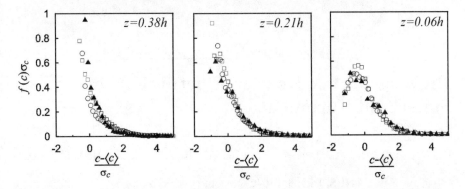

Fig. 3. Normalized concentration PDF, $f(c)\sigma_c$, as a function of the normalized concentration for three elevations above ground at the downwind distance $x = 4.79h$. σ_c is the concentration standard deviation and $\langle c \rangle$ is the mean concentration. Measurements (open circles), FDF-IEM model (solid triangles), PDF-SIECM model (open squares).

overall good model results therefore qualifying the FDF simulation method as a useful tool in the prediction of scalar statistics for applications involving under-resolved scalar sources in LES.

References

1. J.D. Albertson, M.B. Parlange. Water Resour. Res. **35**, 2121–2132 (1999).
2. M. Cassiani, U. Giostra: Atmospheric Environment **36**, 4717-4724, (2002).
3. M. Cassiani, P. Franzese P, U. Giostra: Atmospheric Environment **39**, 8, 1457-1469 (2005)
4. M. Cassiani, A. Radicchi, J.D. Albertson, U. Giostra: J. Comput. Phys, *in press*
5. P.C. Chatwin, P.J. Sullivan: J. Fluid. Mech. **91**, 337–355 (1979).
6. P.J. Colucci, F.A. Jaberi, P. Givi, S.B. Pope: Phys. Fluids **10**, 2, 499–515 (1998).
7. J.E. Fackrell, A.G. Robins: J. Fluid. Mech. **117**, 1-26 (1982).
8. P. Franzese: Atmospheric Environment **37**, 1691-1701 (2003).
9. F.A. Gifford: Advances in Geophysics **6**, 117-137 (1959).
10. H. Pitsch: Annual Review of Fluid Mechanics **38**, 453–482 (2006).
11. S.B. Pope: Progress in Energy Combustion Science **11**, 119-192 (1985).
12. S.B. Pope: *In proceedings of the 23rd Symposium (international) on Combustion* (The Combustion Institute, Pittsburgh, 1990) pp 591–612.
13. P. Sagaut: *Large Eddy Simulation for Incompressible Flows.* Springer Verlag, 426 pp.
14. M.R.H. Sheikhi, T.G. Drozda, P. Givi, S.B. Pope: Phys. Fluids **15**, 8 2321-2337 (2003)
15. D. J. Thomson: J. Fluid. Mech. **210**, 113-153 (1990).

Probing Vortex Density Fluctuations in Superfluid Turbulence

P.-E. Roche[1], B. Chabaud[1], O. Français[2], L. Rousseau[2], and H. Willaime[3]

[1] Institut Néel - CNRS, BP166, 38042 Grenoble Cedex 9, France
[2] Groupe ESIEE, BP99 - 93162 Noisy le Grand Cedex, France
[3] ESPCI, 10 rue Vauquelin, 75231 Paris Cedex 05, France

1 Introduction

The flow dynamics of a superfluid can be described by a viscousless tangle of quantized vortex lines [1, 2, 3, 4]. In spite of these unusual mechanical properties, strong similarities have been reported between superfluid (or "quantum") turbulence and its Navier-Stockes counterpart. Most of them were detected by macroscopic sensors probing the largest scales (integral length) of the flow. At smaller scales, the zero-viscosity and quantization is expected to ruin this similarity between the two turbulences but probing the inertial range is already an experimental challenge [5]. In a keystone experiment, Maurer and Tabeling measured local fluctuations in a bulk flow of helium with and without superfluid[6]. In both case, they found a Kolmogorov-like spectrum for velocity in the inertial range, which has been confirmed since by numerical simulations [7, 8, 9].

Recently, we micro-machined and operated a local probe of the vortex line density fluctuations (projected on a plane). It provides a second direct experimental characterization of the inertial range of superfluid turbulence. The flow, the operation of this probe and the first physical results are presented in another paper[10]. On figure 1 (left), we illustrate these results with power spectra of the vortex line density. In the present paper, we present further details on micro-machining of the probe. To our knowledge, this is the first fully micro-machined sensor used in a cryogenic turbulence experiment.

2 Flow and Probe

The flow

The flow is confined in a cryostat and continuously powered by a centrifugal pump (see figure 1). Turbulence is probed in a 23 mm-diameter pipe, located upstream from the pump. Velocity can span the range $0.25 - 1.3\,m/s$. The

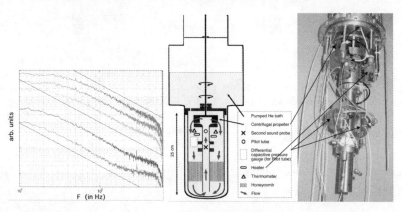

Fig. 1. (Left) Power spectrum of the projected vortex line density for a mean velocities from 0.25 (bottom curve) up to 0.78 m/s (top curve). Saturations at high frequencies result from the noise floor and filtering. Solid lines are -5/3 power laws. (Center and Right) The closed loop of He.

measurements are performed in ^{4}He near 1.6 K. In such conditions, 84% of helium is superfluid, the rest being a normal (viscous) fluid superimposed to the superfluid. More details are presented in [10].

The probe's micro-machining

The principle of operation of the sensor is presented in [10]. In short, it consists in an open second-sound resonator inserted across the flow. Second sound (thermal waves) is known to be attenuated by the vortex lines which are polarized in the plane of propagation[1]. Thus, a time-resolved measurement of the standing wave amplitude provides a direct measurement of the total length of the (projected) vortices within the fluid element advected through the resonator. This signal is somehow analog to an enstrophy defined using a projected vorticity.

Both mirrors of the cavity are $15\,\mu m$ thick silicon beams separated by a $250\,\mu m$ gap. The length and width of both beams is $1.5\,mm$ x $1\,mm$ and they are facing each other with a lateral positioning within a few tenths of mm typically. A granular Al film is deposited over a $0.8\,mm$ square area at the tip of one beam. This film is used as transition edge superconducting thermometer. Facing this thermometer, a chromium heating film is deposited on the tip of the other beam. Between both beams, a third silicon spacer sets the gap of the cavity. On both sides of it, contact pads provide electrical connections of both beams to the electronic circuit.

The micro-machining process is given in figure 2. SOI wafers (4 inches diameter) have been used with the following typical thicknesses of the 3 layers: $15\mu m$, $0.5\mu m$ and $500\ \mu m$. Figure 3 (left and center) shows the wafer with

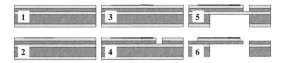

Fig. 2. Steps of the micro-machining process of one beam (the other beam is machined following a similar procedure). 1-Electrical isolation (thermal oxidation, 400 nm), 2-contact pads (Cr 150 nm, Au 500 nm), 3-Heater (Cr, 500 nm), 4-DRIE front side (15 μm), 5-DRIE back side (500 μm), 6-oxide, cleaning...

beams of various sizes. The picture on figure 3 shows the resonator after assembly.

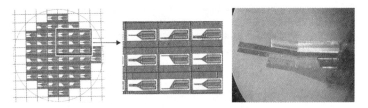

Fig. 3. (Left) Schematics of a region of the wafer, at the end of the micro-machining process. (Right) Picture of the tip of the probe after assembly.

acknowledgments

We thanks P. Tabeling, C. Lemonias, E. André, T. Fournier, B. Hébral and B. Castaing for their inputs. We acknowledge financial support from the IPMC, the Région Rhônes-Alpes and of the ANR (TSF project).

References

1. R. J. Donnelly: *Quantized Vortices in Helium-II* (Cambridge Studies in Low Temperature Physics, Cambridge University Press, Cambridge 1991)
2. S. K. Nemirovskii and W. Fiszdon:Rev. Modern Phys. **67**, 37 (1995)
3. C.F. Barenghi et al. (Ed.): *Quantized Vortex Dynamics and Superfluid Turbulence* (Springer, Lecture Notes in Physics 571, 2001)
4. W. F Vinen and J. J Niemela:J. Low Temp. Phys. **128**, 167 (2002)
5. W. F Vinen:J. Low Temp. Phys. **124**, 101 (2001)
6. J. Maurer and P. Tabeling:Europhys. Lett. **1**, 29 (1998)
7. C. Nore et al.:Phys. Fluids **9**, 2644 (1997)
8. D. Kivotides et al.:Europhys. Lett. **57**,845 (2002)
9. M. Kobayashi and M. Tsubota:Phys. Rev. Lett. **94**, 065302 (2005)
10. to appear in Europhys. Lett (2007)

Contribution of Coherent and Incoherent Vorticity Fields to High Reynolds Number Homogeneous Isotropic Turbulence : a Wavelet Viewpoint

Katsunori Yoshimatsu[1], Naoya Okamoto[1], Kai Schneider[2], Marie Farge[3] and Yukio Kaneda[1]

[1] Department of Computational Science and Engineering, Graduate School of Engineering, Nagoya University, Chikusa-ku, Nagoya 464-8603, Japan
`yosimatu@fluid.cse.nagoya-u.ac.jp`, `okamoto@fluid.cse.nagoya-u.ac.jp`, `kaneda@cse.nagoya-u.ac.jp`
[2] MSNM–CNRS & CMI, Université de Provence, 39 rue Frédéric Joliot-Curie, 13453 Marseille Cedex 13, France
`kschneid@cmi.univ-mrs.fr`
[3] LMD–IPSL–CNRS, Ecole Normale Supérieure, 24 rue Lhomond, 75231 Paris Cedex 05, France `farge@lmd.ens.fr`

Summary. A wavelet-based method to extract coherent vortices is applied to data obtained by direct numerical simulations of three-dimensional incompressible homogeneous isotropic turbulence for different Taylor microscale Reynolds numbers, ranging from $R_\lambda = 167$ to 732. We find that the coherent vortices well preserve statistics of the total flow. The incoherent flow is structureless and noise like. The percentage of wavelet coefficients representing the coherent vortices decreases as the Reynolds number increases.

A wavelet-based method to extract coherent vortices, proposed in [1, 2], is applied to data obtained by direct numerical simulations (DNSs) of three-dimensional incompressible homogeneous isotropic turbulence performed on the Earth Simulator [3, 4]. We use four datasets for $k_{\max}\eta \simeq 1$ at different resolutions, from 256^3 up to 2048^3, which correspond to different Taylor microscale Reynolds numbers, from $R_\lambda = 167$ to 732. Here $k_{\max}$ is the maximum wavenumber retained in each of the DNSs, and η is the Kolmogorov length scale.

The wavelet-based extraction method assumes that coherent vortices are what remain after denoising, without requiring any template of their shape. Hypotheses are only made on the noise which, as the simplest guess, is considered to be additive, Gaussian and white. An orthogonal wavelet decomposition is applied to the vorticity field. A threshold depending on the enstrophy and the resolution of the field, which are both known *a priori* splits the wavelet coefficients into two sets. The coherent (incoherent) vorticity is reconstructed from few (most) wavelet coefficients whose moduli are larger (smaller) than the threshold.

The aim of this work is to examine the coherent and incoherent contribution to statistics of the turbulent flow, and also the dependence of compression rate (the percentage of wavelet coefficients representing the coherent vorticity) on the Reynolds number.

In Figs. 1, 2 and 3, we present the case $R_\lambda = 471$. We observe that the coherent vorticity represented by 2.9% of the wavelet coefficients well retains the vortex tubes observed in the total vorticity field. In [5], we observed a strong scale-by-scale correlation between the coherent and total velocity fields over the scales retained by the data with $R_\lambda = 471$. In contrast, the incoherent vorticity is structureless without any vortex tubes left.

The probability density functions (PDFs) of velocity and vorticity of the total, coherent and incoherent flows are depicted in Fig. 2. All velocity PDFs exhibit parabola-like shapes. The total and coherent velocity PDFs coincide well, while the incoherent one has a strongly reduced variance. The PDFs of the total and coherent vorticity, both almost superimpose, show a stretched exponential behavior which illustrates the intermittency due to the presence of coherent vortices. The PDF of the incoherent vorticity has an exponential shape with a reduced variance compared to the PDFs of the total and coherent vorticity.

Figure 3 shows the compensated energy spectra of the total, coherent and incoherent flows. The spectrum of the coherent flow is identical to that of the total flow all along the inertial range. The shoulder of the energy spectrum of the total flow with a maximum around $k\eta \sim 0.13$ is well retained in the coherent flow. For the incoherent flow, we observe that $E(k)$ scales as k^2, which corresponds to an energy equipartition.

In [6], we found that as the Reynolds number increases, the compression rate improves from 3.6% for $R_\lambda = 167$ to 2.6% for $R_\lambda = 732$ and that the coherent velocity fields preserve the nonlinear dynamics of the flow in the inertial range. It is conjectured that the number of degrees of freedom N to compute fully-developed turbulent flows could be reduced in comparison to the estimation based on the Kolmogorov theory, i.e. from $N \propto R_\lambda^{9/2}$, to $R_\lambda^{3.9}$.

References

[1] M. Farge, K. Schneider, N. Kevlahan: Phys. Fluids **11** 2187 (1999)
[2] M. Farge, K. Schneider, G. Pellegrino et al: Phys. Fluids **15** 2886 (2003)
[3] M. Yokokawa, K. Itakura, A. Uno et al: Proc IEEE/ACM SC2002 Conf, Baltimore (2002) http://www.sc-2002.org/paperpdfs/pap.pap273.pdf
[4] Y. Kaneda, T. Ishihara, M. Yokokawa et al: Phys. Fluids **15** L21 (2003)
[5] K. Yoshimatsu, N. Okamoto, K. Schneider et al: Wavelet-based extraction of coherent vortices from high Reynolds number homogeneous isotropic turbulence.

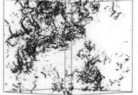

Fig. 1. Isosurfaces of the vorticity modulus for the total (left), coherent (middle) and incoherent (right) flow fields (from Ref. [5]). The values of the isosurfaces are $|\boldsymbol{\omega}| = \omega_m + 3\sigma_\omega$ for the total and coherent vorticity and $2(\omega_m + 3\sigma_\omega)/5$ for the incoherent one. ω_m and σ_ω are the mean value of $|\boldsymbol{\omega}|$ and the standard deviation of $|\boldsymbol{\omega}|$, respectively. Subcubes of size 256^3 are visualized.

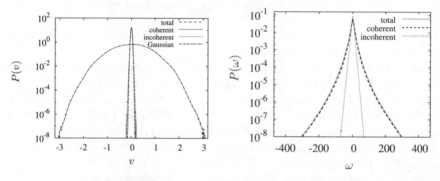

Fig. 2. PDFs of velocity (left) and vorticity (right).

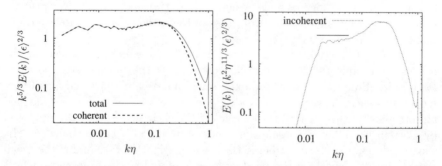

Fig. 3. Compensated energy spectra of the total and coherent flows (left), and of the incoherent flow (right). Here $\langle \epsilon \rangle$ is the mean energy dissipation rate per unit mass of the total flow.

In: IUTAM Symposium on Computational Physics and New Perspectives in Turbulence, ed by Y. Kaneda, Springer-Verlag (to appear).

[6] N. Okamoto, K. Yoshimatsu, K. Schneider et al: (submitted).

The effect of shear on anisotropic fluctuations in a homogeneous shear flow

P.Gualtieri[1], B. Jacob[2], C.M. Casciola[1] and R. Piva[1]

[1] Dipartimento di Meccanica e Aeronautica, Università di Roma *La Sapienza*, Via Eudossiana 18, 00184 Roma, Italy p.gualtieri@caspur.it

[2] INSEAN, Via di Vallerano 128, Roma Italy b.jacob@insean.it

Introduction The largest scales in turbulent flows are directly affected by the geometry of the forcing mechanism and are strongly anisotropic. Usually, for large enough Reynolds number, it is assumed that anisotropic effects are progressively lost during the process of energy cascade towards the smaller scales where isotropy is eventually recovered. Based on these ideas Kolmogorov theory provides the framework of many closure models for RANS and LES see e.g. [1]. This scenario is well established when the energetics of the flow are dominated by the inertial energy transfer i.e. at scale separations small enough with respect to the typical length scales where production of turbulence kinetic energy occurs. In fact, the presence of a strong shear alter this picture as it is observed in the high shear region of wall bounded flows. Actually close to solid walls, the range of scales where energy transfer prevails over production is limited by the distance from the wall and the isotropy recovery range eventually becomes very narrow.

Purpose of this contribution is to provide a systematic analysis of the decay rate of anisotropic velocity fluctuations via the $SO(3)$ decomposition. Both Large Eddy Simulation and experimental data are addressed for the homogeneous shear flow, which sets ideal conditions to deal with shear induced anisotropies. In this context the concept of the shear scale originally proposed by Corrsin has allowed to gain new understanding of shear turbulence. The shear scale $L_S = \sqrt{\bar{\epsilon}/S^3}$ – $\bar{\epsilon}$ being the mean dissipation rate and S the mean shear – ideally separates the production-dominated scales from those where inertial transfer takes over. Strong anisotropic effects on the statistics of turbulent fluctuations are expected in the range $r > L_S$ while for separations $r < L_S$ isotropy recovery might take place. The position of the shear scale within the inertial range is controlled by the shear parameter $S^* = S\langle u^2 \rangle/\bar{\epsilon} = (L_0/L_S)^{2/3}$ and the Corrsin parameter $S_c^* = \nu(S/\bar{\epsilon})^{1/2} = (\eta/L_S)^{2/3}$.

Statistics of anisotropic fluctuations Isotropic and anisotropic contribution to the statistical behavior of relevant turbulent observables can be systematically analyzed by using the $SO(3)$ technique, see e.g. [3]. For instance, we consider the longitudinal structure functions $S^{(n)}(\mathbf{r}) = \langle(\delta\mathbf{u}\cdot\hat{\mathbf{r}})^n\rangle$ where $\delta\mathbf{u}$ is the velocity difference between two points separated by a distance r in the direction $\hat{\mathbf{r}}$. Their projection on the different sectors of the $SO(3)$ decomposition reads

$$S^{(n)}_{jm}(|\mathbf{r}|) = \int_\Omega S^{(n)}(\mathbf{r})\, Y^*_{jm}(\hat{\mathbf{r}})d\Omega \tag{1}$$

where $Y_{jm}(\hat{\mathbf{r}})$ are the spherical harmonics and Ω denote the solid angle. Following this procedure isotropic and anisotropic contributions are disentangled resulting in *pure* scaling laws for the projections $S^{(n)}_{jm} \propto |\mathbf{r}|^{\zeta^{(n)}_{jm}}$ where $\zeta^{(n)}_{jm}$ are the scaling exponents of the structure functions in the jth sector of the $SO(3)$ decomposition. Once a proper validation through comparison with experiments is achieved, see [2], numerical data can be used to evaluate scaling laws in the different sectors of the $SO(3)$ decomposition via eq. (1).

In fact, we find that a strong shear is able to modify the scaling laws of the isotropic sector which, under weak shear keeps its classical isotropic scaling, see figure 1 (left). The same effect is also found on the scaling laws in the anisotropic sectors. For instance, in figure 1 (right), are shown the $j = 2$ projections both for the high ad low shear simulations. Under weak shear the anisotropic projections vanish at a relatively fast rate. Under strong shear, the decay rate of the anisotropic components is considerably reduced and they might keep a significant amplitude up to viscous scales. Actually the process of isotropy recovery at small scales is governed by the difference of the scaling exponents $\zeta^{(n)}_{jm} - \zeta^{(n)}_{00}$ which provides the decay rate of the anisotropic

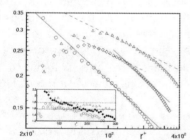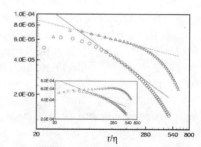

Fig. 1. Left: isotropic projection for the sixth order longitudinal structure function $S^{(6)}_{00}/r^2$. High shear $S^* = 7$ (circles), Low shear $S^* = 3$ (triangles) and intermediate shear $S^* = 5$ (diamonds). The slope of the solid line is -0.42 corresponding to $\zeta(6) = 1.58 \pm 0.08$. The slope of the dashed line is -0.22, i.e. $\zeta(6) = 1.78 \pm 0.08$. In the inset local slopes. Right: projection for the sixth order longitudinal structure in the $j = 2$ sector. Data are compensated by their dimensional scaling $S^{(6)}_{2-2}/r^{8/3}$. High shear (circles), Low shear (triangles). The slope of the solid line is -0.80 corresponding to $\zeta^{(6)}_{2-2} = 1.87$. The slope of the dashed line is -0.27, i.e. $\zeta^{(6)}_{2-2} = 2.3$. In the inset same plot for the $n = 2$ moment.

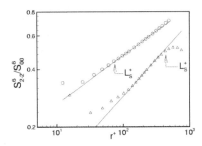 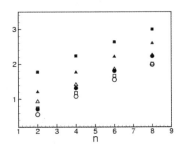

Fig. 2. Left: $\mathcal{S}^{(6)}_{2-2}/\mathcal{S}^{(6)}_{00}$. High shear data (circles), low shear data (triangles), the slope of the solid lines is 0.32 and 0.62 for the two datasets respectively. Right: exponents for the leading components of the structure functions of order $n = 2, 4, 6, 8$, in the sector $j = 0$ (circles), $j = 2$ (triangles) and $j = 4$ (squares). Open symbols give the exponents in the high shear case, filled symbols in the low shear case.

projections as compared to the isotropic one according to the expression

$$\mathcal{S}^{(n)}_{jm}/\mathcal{S}^{(n)}_{00} \propto (|\mathbf{r}|)^{\zeta^{(n)}_{jm} - \zeta^{(n)}_{00}}.$$

The values of the scaling exponents in the different sectors appear to be crucial. This ratio is plotted in figure 2 (left) where the anisotropic component of the sixth order structure function in the $j = 2$ sector is compared against its isotropic projections. From the data it emerges that anisotropies are recovered at a relatively slower rate in high shear conditions. In fact, we find that the intensity of the shear affects the hierarchy of exponents leading to a slower recovery of isotropy in high shear conditions. The fitted scaling exponents are given in figure 2 (right) where both high shear data (open symbols) and low shear data (filled symbols) are reported (see caption). As the shear intensity is increased the exponents in the anisotropic sectors are much closer to scaling exponents in the isotropic sector. Moreover the hierarchy in the anisotropic sectors i.e. $\zeta^{(n)}_2 < \zeta^{(n)}_4$ found in low shear flows – see [3] – is found to be altered by the shear intensity. These results are consistent with recent DNS findings in similar conditions [4].

Conclusions We find that a strong shear is able to alter the hierarchy of the scaling exponents both in the isotropic and anisotropic sectors. This implies a slower isotropy recovery rate at smaller scales in these conditions.

References

1. S. Pope, Cambridge University Press, 2000
2. C.M. Casciola, P. Gualtieri, B. Jacob, R. Piva: Phys. Rev. Lett. A **95**, 2005
3. L. Biferale, I. Procaccia: Phys. Reports, **414**, 2005
4. C.M. Casciola, P. Gualtieri, B. Jacob, R. Piva: Submitted to Phys. Rev. Lett.

Azimuthal Velocity Correlations in an Axisymmetric Far Wake

Murat Tutkun[1], Peter B. V. Johansson[2] and William K. George[3]

[1] Turbulence Research Laboratory, Department of Applied Mechanics, Chalmers University of Technology, SE-41296 Gothenburg, Sweden
`murat.tutkun@chalmers.se`
[2] Turbulence Research Laboratory, Department of Applied Mechanics, Chalmers University of Technology, SE-41296 Gothenburg, Sweden
`peter.johansson@chalmers.se`
[3] Turbulence Research Laboratory, Department of Applied Mechanics, Chalmers University of Technology, SE-41296 Gothenburg, Sweden `wkgeorge@chalmers.se`

1 Introduction

The azimuthal velocity correlations in an axisymmetric far wake are experimentally studied in this paper via the analysis of power spectral densities of the turbulent fluctuations together with the two-point cross-correlations and cross-spectra. The experiment extends the previous experimental results of Johansson and George (2006) [3] to include all three components of velocity. The main aim of this paper is to investigate the basis of the dominance of azimuthal mode-2 for axisymmetric turbulent far wake.

A wind tunnel experiment has been performed to measure the axisymmetric wake generated with a circular disk. A hot-wire rake of 12 cross-wires was used to perform multi-point simultaneous measurements. The rake was in two parts (one fixed and one rotating) with 6 probes on each. Two sets of experiments were performed in order to obtain first the streamwise and azimuthal components of the velocity, and then the streamwise and radial components of the velocity. Non-measured components of the Reynolds stress tensor were computed using the spectral continuity equations. Further details on the experiment setup and details of implementation of spectral continuity equation can be found in Tutkun et al. (2006)[5].

Previously the proper orthogonal decomposition (POD) analysis of axisymmetric turbulent wake based only on the measurement of streamwise velocity fluctuations showed that azimuthal mode-1 was dominant in the near wake and azimuthal mode-2 dominant was dominant in the far wake [3]. Similar results were documented for the axisymmetric turbulent jet [4, 1]. On the other hand, azimuthal mode-1 was found to be dominant one for both near and far wake of axisymmetric turbulent jet when all components of the

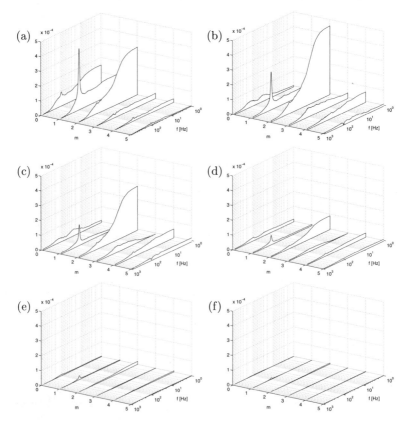

Fig. 1. Two-point cross-spectrum, $\sum_1^3 S_{k,k}(r,r';m;f)$: (a) r= $2\Delta r$, r$'$ = $2\Delta r$; (b) r= $3\Delta r$, r$'$ = $3\Delta r$; (c) r= $4\Delta r$, r$'$ = $4\Delta r$; (d) r= $5\Delta r$, r$'$ = $5\Delta r$; (e) r= $6\Delta r$, r$'$ = $6\Delta r$; (e) r= $7\Delta r$, r$'$ = $7\Delta r$.

Reynolds stress tensor included in the analysis [2, 6]. The same approached was implemented for the axisymmetric turbulent wake by considering the full Reynolds stress tensor, and azimuthal mode-2 was found to be the dominant one in the far wake.

2 Results and discussions

Figure 1 shows the total power spectral distribution over the span of the far wake as a function of azimuthal mode number and frequency. The largest amount of turbulent kinetic energy is always carried by the azimuthal mode-2 across the wake.

Figure 2 shows the modal decomposed two-point cross-correlations for the full Reynolds stress tensor. Most of the energy is concentrated at the azimuthal

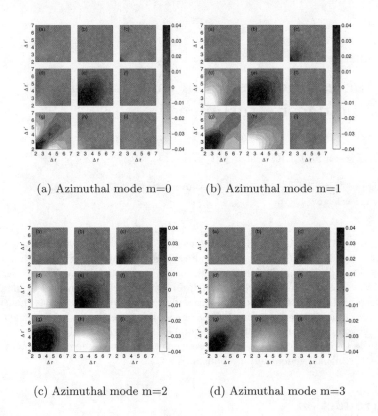

(a) Azimuthal mode m=0   (b) Azimuthal mode m=1

(c) Azimuthal mode m=2   (d) Azimuthal mode m=3

Fig. 2. Modal decomposed two-point cross-correlations: (a) $B_{w,u}(r,r';m)$, (b) $B_{w,v}(r,r';m)$, (c) $B_{w,w}(r,r';m)$, (d) $B_{v,u}(r,r';m)$, (e) $B_{v,v}(r,r';m)$, (f) $B_{v,w}(r,r';m)$, (g) $B_{u,u}(r,r';m)$, (h) $B_{u,v}(r,r';m)$, (i) $B_{u,w}(r,r';m)$

mode-2 because of streamwise velocity fluctuations. Therefore, unlike the jet, the three component decomposition results are in agreement with the earlier results based only on the streamwise velocity fluctuations.

References

[1] S. Gamard, D. Jung, and W. K. George. *J. Fluid Mech.*, 514:205, 2004.
[2] M. O. Iqbal and F. O. Thomas. *J. Fluid Mech.*, 571:281, 2007.
[3] P. B.V. Johansson and W. K. George. *J. Fluid Mech.*, 555:387, 2006.
[4] D. Jung, S. Gamard, and W. K. George. *J. Fluid Mech.*, 514:173, 2004.
[5] M. Tutkun, P. B. V. Johansson, and W. K. George. *AIAA-2006-3300*, 2006.
[6] M. Wänström, W. K. George, and K. E. Meyer. *AIAA-2006-3368*, 2006.

Mixing characteristics in buoyancy-driven, variable density turbulence

Daniel Livescu and J.R. Ristorcelli

Los Alamos National Laboratory, Los Alamos, NM 87545, USA
livescu@lanl.gov

Summary. We examine the mixing induced by buoyancy generated motions in an unstably stratified field composed of two incompressible miscible fluids with different densities, as occurs in variable density Rayleigh-Taylor instability. The statistically homogeneous case is considered as a unit problem for variable density turbulence. It involves both the transition to turbulence and the decay of turbulence as the friction forces overcome buoyancy generation. Diverse mixing metrics and their dependence on the Atwood, Reynolds, and Schmidt numbers are used to study the homogenization of the two fluids leading to the subsequent reduction of the net buoyancy force and the turbulence generation mechanism.

1 Introduction

Mixing to molecular scale in the presence of turbulence induced stirring is an important process in many practical applications. In general, the fluids participating in the mixing have different densities and we refer to such flows as variable density (VD) flows. In these flows, the specific volume changes in both time and space depending on the amount of each fluid in the mixture and the resulting velocity field is not divergence free even for constant density fluids. VD mixing is encountered in atmospheric and ocean flows, combustion and many flows of chemical engineering interest, astrophysical flows, etc.

Here we consider a simple form of multi-material mixing which involves two miscible fluids with different microscopic densities [1], in the presence of a constant acceleration, as occurs in the Rayleigh-Taylor (RT) instability. The current investigation focuses on the nonlinear dynamics and statistics of buoyantly driven turbulence in the statistically homogeneous configuration. The problem is an extension of the buoyantly generated turbulence in a Boussinesq fluid studied in [2] and in a VD fluid examined in [3].

2 Problem formulation

The equations describing the mixing between two miscible fluids with different microscopic densities, ρ_1 and ρ_2, are the Navier-Stokes and continuity equations. For binary mixing, the mass fraction of the two fluids can be recovered uniquely from the density. The non-dimensional instantaneous equations, with usual notations, can be written as (see [1]):

$$\rho_{,t} + (\rho u_j)_{,j} = 0 \tag{1}$$

$$(\rho u_i)_{,t} + (\rho u_i u_j)_{,j} = -p_{,i} + \tau_{ij,j} + \frac{1}{Fr^2}\rho g_i \tag{2}$$

$$u_{j,j} = -\frac{1}{Re_0 Sc}\ln\rho_{,jj} \tag{3}$$

with $\tau_{ij} = \frac{1}{Re_0}(u_{i,j} + u_{j,i} - \frac{2}{3}u_{k,k}\delta_{ij})$. In VD turbulence with arbitrary boundary conditions, the two first order moments, the mean pressure gradient, $P_{,i}$, and the mean specific volume, V, are dynamical variables evolving as the mixing proceeds. For periodic boundary conditions though, the mean pressure gradient can be determined up to a constant gradient which is a free parameter. This is chosen such that the energy conversion of potential to kinetic energy is maximized [1] and leads to the maximally unstable flow in this configuration.

Equations (1)-(2) are solved in a triply periodic domain, on up to 1024^3 meshes, using a pseudo-spectral algorithm. The equations are time advanced using the pressure projection method. The numerical algorithm introduces several improvements over the existing approaches, as detailed in [1]. Thus, the pressure step is treated exactly, without additional temporal integration errors.

3 Variable density mixing: mixing measures and progress

The flow starts with zero solenoidal velocity in a non-premixed state. Small and large scale fluctuations in the velocity and density fields are generated by nonlinear interactions, as the fluids move under the action of the body force, increasing the turbulent kinetic energy. Simultaneously, the fluids in contact molecularly mix. The amplification of the fluctuating strain field by baroclinic generation increases the interfacial area between the two fluids and the molecular mixing is accelerated (figure 1a). As the fluids become molecularly mixed, the viscous forces overcome the buoyancy forces reduced by mixing and the turbulence begins to decay. At some late time only the relatively large scale regions survive; the small scale density fluctuations have been smoothed out by molecular diffusion (figure 1b)

The rate of conversion of potential energy into kinetic energy, as well as between Favre mean and turbulent kinetic energies, is mediated by the mass flux so that the mass flux is likely the most important quantity to predict in lower dimensional models [1, 4]. The transport equation for the mass flux:

$$\bar\rho\langle\rho' u_i'\rangle_{,t} = -\langle\rho' v'\rangle P_{,i} + \bar\rho\langle v' p_{,i}'\rangle - \bar\rho\langle u_i' u_{j,j}'\rangle + \langle v'\tau_{ij,j}'\rangle, \tag{4}$$

shows that the production of the mass flux directly depends on the fluid configuration as it is proportional to the density specific volume correlation. For a Boussinesq fluid, the energy conversion is performed through a different mechanism and the density variance mediates the production of the mass flux. Thus, the specific volume density covariance explicitly appears in the dynamical equations and is a better measure of mixing as it relates to the flow evolution than the density variance for VD flows. The

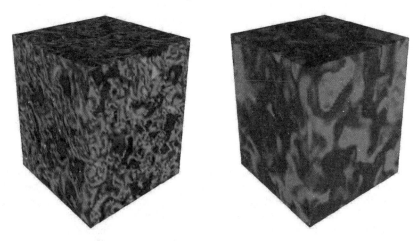

Fig. 1. Typical density field at a) maximum density variance and b) late time.

latter was introduced as a mixing metric in [5] and is usually used in RT turbulence. Nevertheless the density variance has some physical significance for VD flows: the normalized density variance can be related to the variance of the excess reactant in an infinitely fast hypothetical chemical reaction. In this framework, one can also define a metric for the fully mixed fluid [6]. It can be shown that the two measures yield similar results only under special circumstances. In particular, they can not be the same at the center and edges of a RT mixing layer. Neither measure though can express the amount of pure fluid in the flow, which is related to the tails of the density PDF, although all usual mixing metrics can be computed from the PDF.

Mixing rate is related to the density gradient appropriately normalized for each mixing metric. The density derivative PDF shows that mixing progresses nonuniformly in space. In addition, at early times the buoyancy production acts at all scales leading to anisotropy at all scales. This is reflected in the anisotropy of the density derivatives and further increases the spatial nonuniformity of the mixing rate.

References

1. D. Livescu, J.R. Ristorcelli: submitted to J. Fluid Mech., LA-UR-06-7190 (2006).
2. G.C. Batchelor, V.M. Canuto, J.R. Chasnov: J. Fluid Mech. **235**, 349 (1992).
3. S. Sandoval: The dynamics of variable density turbulence. Ph.D. Thesis, University of Washington, Seattle (1995).
4. D. Livescu, J.R. Ristorcelli: Characteristics of buoyancy-driven variable density turbulence. In: *Proceedings of the 10th International Workshop on the Physics of Compressible Turbulent Mixing*, ed by M. Legrand, M. Vandenboomgaerde (Paris, 2006).
5. D.L. Youngs: Phys. Fluids **A 3**, 1312 (1991).
6. A.W. Cook, P.E. Dimotakis: J. Fluid Mech. **343**, 69 (2001).

Turbulent Von Kármán Swirling Flows

S. Poncet[1], R. Schiestel[2], and R. Monchaux[3]

[1] MSNM-GP, UMR 6181, Technopôle Château-Gombert, 38 rue F. Joliot-Curie, 13541 Marseille - France `poncet@l3m.univ-mrs.fr`
[2] IRPHE, UMR 6594, Technopôle Château-Gombert, 49 rue F. Joliot-Curie, 13384 Marseille - France `schiestel@irphe.univ-mrs.fr`
[3] Service de Physique de l'Etat Condensé / GIT, CEA Saclay, 91191 Gif sur Yvette - France `romain.monchaux@cea.fr`

We investigate the turbulent Von Kármán flow generated by two counter-rotating flat or bladed disks. Numerical predictions based on a Reynolds Stress Model (RSM) are compared to velocity measurements performed at CEA [2]. This flow is of practical importance in many industrial devices such as in gas-turbine aeroengines. From an academic point of view, this configuration is often used for studying fundamental aspects of developed turbulence and especially of magneto-hydrodynamic turbulence.

1 Geometrical model and numerical approach

The Von Kármán geometry is composed of two counter-rotating disks ($R = 92.5$ mm) enclosed by a stationary cylinder ($R_c = 100$ mm) (Fig.1). The interdisk spacing H can vary in the range $10 - 180$ mm. We use bladed disks (n blades of height h) to ensure inertial stirring or flat disks for viscous stirring. The rotation rates Ω_1, Ω_2 can be increased up to 900 rpm. The main flow is controlled by three parameters: the ratio between the two rotation rates $\Gamma = -\Omega_2/\Omega_1$, the aspect ratio of the cavity $G = H/R_c$ and the Reynolds number $Re = \Omega_1 R_c^2/\nu$. In the following, $G = 1.8$ and $\Gamma = -1$.

Our numerical approach is based on one-point statistical modeling using a low Reynolds number second-order full stress transport closure (Reynolds Stress Model, RSM) sensitized to rotation effects and already validated for $\Gamma = 0$ and a wide range of G and Re [1]. To model straight blade effects, we add a volumic drag force in the equation of V_θ the tangential velocity component: $f = n\rho C_D(\Omega_{1,2}r - V_\theta)|\Omega_{1,2}r - V_\theta|/(4\pi r)$, where ρ is the fluid density, $C_D = 0.5$ the drag coefficient and r the local radius. The procedure is based on a finite volume method using staggered grids for mean velocity components with axisymmetry hypothesis. A 120^2 (resp. 160^2) mesh in the (r, z) frame is used in the smooth (resp. bladed) disk case. About 20000 iterations (several hours on the bi-Opteron 18 nodes cluster of IRPHE) are necessary to obtain the numerical convergence of the calculation.

2 Results

The mean flow is decomposed into two poloidal cells (fig.1) in the (r,z) plane. In the viscous stirring case, the flow structure is of Batchelor type close to the periphery of the cavity at $r/R_c = 0.81$ (fig.2a) and so exhibits five distinct zones: two thin boundary layers on each disk, a shear layer at mid-plane and two cores on either side of this layer. For an inertially driven flow, the mean flow is divided into three main regions (fig.2b): a very intense shear layer at mid-plane and two fluid regions close to each bladed disks. At the top of the blades, there is a strong decrease of $|V_\theta|$ interpreted as the wake of the blades. The turbulence is mainly confined in the equatorial plane (fig.3a,1) and is found to be almost isotropic in that region for viscous stirring. When one imposes inertial stirring, the normal components are increased by a factor 10 (fig.2b) and a much higher shear stress is obtained around $z^* = 0$. The weak discrepancies between the two approaches are attributed to the appearance of strong coherent structures observed in the experiments [2].

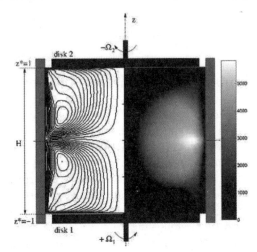

Fig. 1. Sketch of the cavity in the smooth disk case. Computed (left) streamlines and (right) iso-turbulence Reynolds number $Re_t = k^2/(\nu\epsilon)$ for $Re = 6.28 \times 10^5$.

3 Conclusion

We proposed an easy and efficient way to model the effects of impellers on the turbulent Von Kármán flow. A parametric study according to the flow

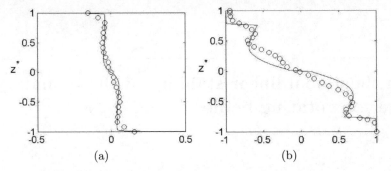

Fig. 2. Axial profiles of $V_\theta/(\Omega_1 r)$: (a) viscous stirring ($Re = 6.28 \times 10^5$, $r/R_c = 0.81$), (b) inertial stirring ($Re = 2 \times 10^5$, $n = 8$, $h/R_c = 0.2$, $r/R_c = 0.4$). (lines) RSM, (○) LDV data of [2].

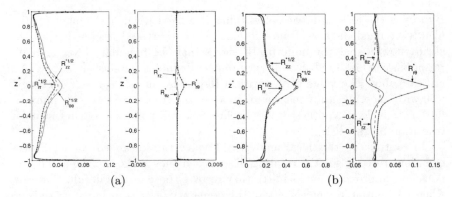

Fig. 3. Axial profiles of the Reynolds stress tensor at $r/R_c = 0.81$: (a) viscous stirring ($Re = 6.28 \times 10^5$), (b) inertial stirring ($Re = 2 \times 10^5$, $n = 8$, $h/R_c = 0.2$) (RSM). (○) LDV data of Ravelet [2] for $R_{\theta\theta}^{*1/2}$.

control parameters has been performed for viscous and inertial stirrings. In particular, we highlighted three main transitions: Batchelor / Stewartson, Batchelor / torsional Couette, one-cell / two-cell regimes. Some calculations for curved blades are still in progress.

References

1. S. Poncet, M.P. Chauve, R. Schiestel: Phys. Fluids **17(7)**, 075110 (2005)
2. F. Ravelet: Bifurcations globales hydrodynamiques et magntohydrodynamiques dans un écoulement de Von Kármán turbulent. PhD Thesis, École Polytechnique, Palaiseau (2005)

On the global linear stability of the boundary layer on rotating bodies

S. J. Garrett[1] and N. Peake[2]

[1] Department of Mathematics, University of Leicester, UK
  `Stephen.Garrett@mcs.le.ac.uk`
[2] Department of Applied Mathematics & Theoretical Physics, University of Cambridge, UK

By taking the local approach of working at a fixed Reynolds number (equivalently at fixed distance from the axis of rotation) and assuming that the steady flow is spatially uniform, [1] shows that the boundary layer on a rotating disk is locally absolutely unstable at Reynolds numbers in excess of a critical value. The value of the critical Reynolds number agrees exceedingly well with experimentally measured values of the transition Reynolds number, leading to a clear hypothesis that absolute instability plays a role in turbulent transition on the disk.

In contrast to this local analysis, [2] solve the linearised Navier–Stokes equations directly for the rotating disk. When they make the same homogenous flow approximation as in [1], they recover those results in full. However, when the spatial inhomogeneity of the boundary layer is included there is no evidence of an unstable global oscillator in the long-term response.

In order to address this discrepancy between the local results and the numerical simulations of the full inhomogeneous flow, we consider the linear global modes of the rotating disk/cone boundary layer.

1 Formulation

For a cone with general half angle, ψ_0, we define a non-dimensional coordinate system consisting of S measured along the surface of the cone from the apex, z in the direction normal to the cone surface and ϕ the azimuthal angle.

The solution of the resulting steady boundary-layer flow equations (see [3]) has the similarity form $U = Rf(\tilde{z})$, $V = Rg(\tilde{z})$, $W = h(\tilde{z})\sqrt{\sin\psi_0}$, $P = 0$, with $\tilde{z} = z\sqrt{\sin\psi_0}$ and

$$f'' - f^2 - hf' + g^2 = 0, \ g'' - 2fg - hg' = 0, \ 2f + h' = 0, \tag{1}$$

subject to non-slip conditions applied at the surface of the cone, $z = 0$. These equations are exactly the similarity equations for the flow over a rotating disk (see [4]). The cone angle has therefore been scaled out of the steady-flow equations.

To derive the unsteady-perturbation equations we consider the velocity and pressure components in the form of the axisymmetric steady flow plus an unsteady perturbation of the form

$$\bar{u}(S, z, \phi, t) = \tilde{u}(S, z) \exp\left(in\phi - i\omega t + \frac{i}{\epsilon} \int^S k(S')dS'\right). \tag{2}$$

The form of this perturbation is consistent with the procedure for analysing weakly non-parallel flows which evolve slowly in the streamwise direction, due to [5]. Here n must be an integer in order to enforce periodicity in the azimuthal direction around the axis of symmetry. We require n large, and choose the preferred scaling $n = \bar{n}/\epsilon$, with $\bar{n} = O(1)$ and ϵ the ratio of the characteristic boundary-layer thickness to the characteristic size of the body.

It is possible to rescale the resulting perturbation equations by writing $\omega_c = \omega/\bar{n}$, $S_c = S(\sin \psi_0)^{3/2}/\bar{n}$ and $k_c = k/(\sin \psi_0)^{1/2}$, leading to

$$(w_c - k_c f S_c - g)\left\{\frac{\partial^2 \bar{w}}{\partial z^2} - \left(k_c^2 + \frac{1}{S_c^2}\right)\bar{w}\right\} + (k_c S_c f'' + g'')\,\bar{w} = 0, \tag{3}$$

so as to eliminate $\bar{n}$. Note that $\bar{w}$ is the perturbing velocity in the normal direction.

We see that the cone half angle has been completely scaled out of both the steady (1) and unsteady (3) equations, and so we need to consider only the stability of the rotating disk from which the behaviour of the cone can be directly inferred.

2 Results & Conclusion

[5] show that the long-time behaviour of weakly non-parallel flow is governed by the behaviour of the global mode of complex frequency ω_G. If $\mathrm{Im}(\omega_G) > 0$, the global mode is unstable and hence the flow will be globally unstable; whereas if $\mathrm{Im}(\omega_G) < 0$, the global mode is damped and the flow will be globally stable. The global-mode frequency is determined as follows:

1. For each real S_c we look for a pinch in the complex k_c plane, i.e. for points of zero group velocity, formed by the coalescence of modes from opposite halves of the complex k_c plane. This provides the complex local absolute frequency, $w_c = w_c^\circ(S_c)$, along the real S_c axis.
2. We search for a pinch point in $w_c^\circ(S_c)$, which in general will occur at complex S_c and will therefore necessitate analytical continuation off the real S_c axis. In other words, we find a saddle point $\partial w_c^\circ/\partial S_c$, and then verify that the S_c contour can be deformed off the real axis so as to lie along the steepest descent contour through this saddle. Once these conditions

have been satisfied, the global-mode frequency simply corresponds to the frequency of this double k_c–S_c pinch.

Such a saddle point has been located at $S_c = S_c^s$ with $k_c, \omega_c = k_c^s, \omega_c^s$ by solving (3) numerically in the complex S_c plane. Note that the similarity form of the steady flow allows trivial analytical continuation into the complex S_c plane. For the rotating disk we have found a saddle point with $\mathrm{Im}(\omega_c^s) < 0$, demonstrating that the global mode on the rotating disk is damped in agreement with the numerical results of [2]. No other such pinch points have been found in the S_c plane. We can infer immediately that the global mode on a cone of arbitrary half angle is also damped.

In our linearised analysis unsteady viscous, streamline curvature and Coriolis terms are absent, and the assertion that the region of local absolute instability does not lead to global instability can therefore be made on the basis of the simplest representation of the unsteady flow and does not require solution of the full linearised Navier–Stokes equations.

We suggest that our results do not imply that absolute instability has no role in the transition to turbulence on the rotating disk, and so do not contradict the original hypothesis of [1]. In order to see this one must include nonlinearity: [6] and [7] have shown that a self-excited, nonlinear global mode will always exist in the presence of a region of local absolute instability. This is in contrast to linear theory, since we have seen here that the region of local absolute instability on the disk is not sufficient to support an unstable linear global mode. [8] has further shown that this nonlinear global mode can undergo secondary instability very close to the convective-absolute boundary, providing a possible route to turbulence.

References

1. R. J. Lingwood, Absolute instability of the boundary layer on a rotating disk: J. Fluid Mech. **299**, 17–33. (1995)
2. C. Davies & P. W. Carpenter, Global behaviour corresponding to the absolute instability of the rotating-disk boundary layer: J. Fluid Mech. **486**, 287–329. (2003)
3. L. Rosenhead, Laminar Boundary Layers, Oxford. (1963)
4. Th. Von Kármán, Über laminare und turbulente Reibung: Z. Angew. Math. Mech. **1**, 233–252. (1921)
5. P. A. Monkewitz, P. Huerre & J–M Chomaz, Global linear stability analysis of weakly non-parallel shear flows: J. Fluid Mech. **251**, 1–20. (1993)
6. B. Pier, P. Huerre & J–M Chomaz, Bifurcation to fully nonlinear synchronized structures in slowly varying media: Physica D. **148**, 49–93. (2001)
7. B. Pier & P. Huerre, Nonlinear self-sustained structures and fronts in spatially developing wake flows: J. Fluid Mech. **435**, 359–381. (2001)
8. B. Pier, Fully nonlinear waves and transition in the boundary layer over a rotating disk: Advances in Turbulence IX, Proceedings of the 9th European Turbulence Conderence, ed by I. P. Castro & P. E. Hancock (2002)

The modulated dissipation rate in periodically forced turbulence

Robert Rubinstein[1] and Wouter Bos[2]

[1] NASA Langley Research Center, USA r.rubinstein@larc.nasa.gov
[2] Ecole Centrale de Lyon, FRANCE wouter.bos@ec-lyon.fr

Steady state homogeneous isotropic turbulence perturbed by small periodic modulations of the force amplitude [1] was originally investigated with a view to finding resonance-like response in the modulated energy. More generally, this problem provides an ideal test case to study time-dependent statistics in turbulence. We focus on the dynamics of the small scales as revealed by the amplitude and phase of the modulated dissipation rate.

To formulate the problem analytically, a small periodic perturbation $\varepsilon \bar{P}(\kappa) \cos(\omega t)$ is added to a time-independent production spectrum $\bar{P}(\kappa)$. Under this periodic forcing, the turbulence statistics eventually become periodic: the turbulent kinetic energy is $k(t) = \bar{k} + \tilde{k} \cos(\omega t + \phi_k)$ and the dissipation rate is $\epsilon(t) = \bar{\epsilon} + \tilde{\epsilon} \cos(\omega t + \phi_\epsilon)$. The oscillatory statistics are characterized by the phase averages $\bar{k}$ and $\tilde{\epsilon}$, and by the phase shifts with respect to the production, ϕ_k and ϕ_ϵ. In the low frequency limit $\omega \approx 0$, the modulations follow the forcing quasi-statically, so that $\phi_k \approx \phi_\epsilon \approx 0$. In the high frequency limit, the turbulence cannot respond to the rapid oscillations; in this 'frozen' turbulence limit, elementary arguments give $\tilde{k} \sim \tilde{\epsilon} \sim \omega^{-1}$ and $\phi_k \sim \phi_\epsilon \sim \pi/2$. Our concern is the intermediate frequency range, where complex behavior not suggested by elementary arguments occurs.

Periodically forced turbulence was simulated using the EDQNM closure, which reproduces existing numerical and experimental results with good accuracy. The phase average $\tilde{\epsilon}$ and phase shift ϕ_ϵ are shown in the figures as functions of modulation frequency ω for a range of Reynolds numbers. At all Re, the quasistatic and frozen turbulence limits are recovered, but it is evident that recovery of the frozen turbulence limit $\tilde{\epsilon} \sim \omega^{-1}$ is delayed by increasing the Reynolds number: instead, as Re increases, $\tilde{\epsilon}$ converges to a ω^{-3} scaling range. It will be shown that this scaling range is the result of distant interactions propagating the oscillatory disturbance to the small scales; it contradicts elementary arguments, which instead suggest confinement of the oscillatory disturbance to the production range. We can also note that the phase shift ϕ_ϵ shows no trend to a high Re asymptote.

The behavior of ϕ_ϵ is at variance with standard arguments based on Kolmogorov theory, according to which high Reynolds number turbulence should become asymptotically independent of Reynolds number. Since many subgrid models assume such an 'equilibrium' of small scales, this observation could have implications for subgrid modeling in periodically forced flows.

We outline an analysis of this problem using simplified closures. Write the spectrum perturbion due to the periodic forcing as $\tilde{E}(\kappa)\cos(\omega t + \psi(\kappa))$ and define $\tilde{F}(\kappa) = \tilde{E}(\kappa)\cos\psi(\kappa)$ $\tilde{G}(\kappa) = \tilde{E}(\kappa)\sin\psi(\kappa)$. They satisfy

$$-\omega\tilde{G}(\kappa) = \tilde{P}(\kappa) - \mathcal{L}[\tilde{F}(\kappa)] - 2\nu\kappa^2\tilde{F}(\kappa)$$
$$-\omega\tilde{F}(\kappa) = \mathcal{L}[\tilde{G}(\kappa)] + 2\nu\kappa^2\tilde{G}(\kappa).$$

where $\mathcal{L}$ denotes the energy transfer linearized about the steady state solution.

A lowest order approximate solution is obtained by balancing the leading order terms in ω, so that $\tilde{F}(\kappa) \approx 0$, $\tilde{G}(\kappa) \approx \tilde{P}(\kappa)/\omega$. This result suggests the confinement of the periodic perturbations to the production region. It does not depend on $\mathcal{L}$. To compute corrections, we use the model [3]

$$T(k) = C\frac{\partial}{\partial\kappa}\left\{\int_0^\kappa d\mu\mu^2 E(\mu)\int_\kappa^\infty dp E(p)\theta(p) - \int_0^\kappa d\mu\mu^4\int_\kappa^\infty dp\frac{E(p)^2\theta(p)}{p^2}\right\}$$

with the algebraic closure for the time-scale $\theta(\kappa) = [\kappa^3 E(\kappa)]^{-1/2}$. This model is an improved Heisenberg model consistent with existence of equipartition ensembles and both forward and backward transfer of energy.

The corrections of lowest order give $\tilde{F}(\kappa) = \omega^{-2}\mathcal{L}[\tilde{P}(\kappa)]$ and $\tilde{G}(\kappa) = -\omega^{-1}\tilde{P}(\kappa) + \omega^{-3}\mathcal{L}^2[\tilde{P}(\kappa)]$. All terms in the linearized transfer $\mathcal{L}$ vanish whenever the production vanishes, with one exception, the term

$\sqrt{\tilde{E}(\kappa)/\kappa^3}\int_0^\kappa d\mu\,\mu^2\tilde{P}(\mu)$. It permits the oscillatory disturbance to propagate to all scales of motion, contrary to the conclusions of the elementary analysis leading to the lowest order solution.

We find $\tilde{\epsilon}\cos(\phi_\epsilon) \sim \omega^{-2}\kappa_P^2\tilde{P}\bar{\epsilon}^{1/2}\nu^{1/2}$ and $\tilde{\epsilon}\sin(\phi_\epsilon) \sim \omega^{-1}2\nu\kappa_P^2\tilde{P} + \omega^{-3}\kappa_P^2\tilde{P}\bar{\epsilon}$. Evidently, there is a competition between the limits $\omega \to \infty$ and $\nu \sim Re^{-1} \to 0$. The limit $\omega \to \infty$ at fixed Re will indeed recover the elementary result $\tilde{\epsilon} \sim \omega^{-1}$, but at fixed large ω, the limit $Re \to \infty$ gives instead $\tilde{\epsilon} \sim \omega^{-3}\kappa_P^2\tilde{P}\bar{\epsilon}$. The ω^{-3} range for $\tilde{\epsilon}$, which is very evident in the figure, is a nontrivial consequence of *distant interactions* in wavenumber space; it is absent in calculations using local energy transfer models like the Leith model and is not suggested by elementary arguments. The phase is $\tan\phi_\epsilon \approx \omega\nu^{1/2}\bar{\epsilon}^{-1/2} + \omega^{-1}\nu^{-1/2}\bar{\epsilon}^{1/2}$ indicating a complex joint dependence on ω and Re in general.

Markovianization is a questionable assumption in the presence of rapidly oscillating forcing [4]. A preliminary non-Markovian analysis suggests that the lowest order corrections become $\tilde{F}(\kappa) = 0$ and $\tilde{G}(\kappa) = -\omega^{-1}\tilde{P}(\kappa) + \omega^{-3}\mathcal{L}^2[\tilde{P}(\kappa)]$: the correction of order ω^{-2} is suppressed and the phase relations are altered. Further investigation is in progress.

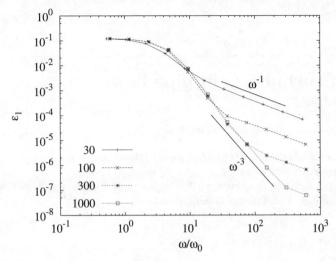

Fig. 1. Closure results for modulated dissipation amplitude as a function of forcing frequency for various Reynolds numbers. $\omega_0 \propto \bar{\epsilon}/\bar{k}$ is the 'critical frequency.'

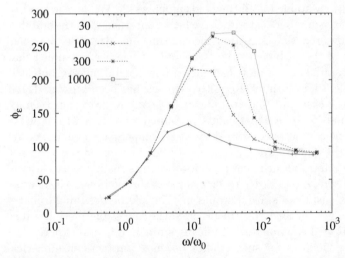

Fig. 2. Phase shift between forcing and modulated dissipation for various Reynolds numbers

References

1. A. von der Heydt, S. Grossmann, and D. Lohse, Phys. Rev. E **67**, 046308 (2003).
2. R. Rubinstein, T. Clark, D. Livescu, and L.-S. Luo, J. Turbul. **5**, 049 (2004).
3. R. Rubinstein and T. Clark, Theor. Comput. Fluid Dyn. **27**, 249 (2004).
4. Y. Kaneda, private communication (2006).

Lifetime of turbulence in pipe flow

Björn Hof[1], Wilco Tax[2], and Jerry Westerweel[2]

[1] School of Physics and Astronomy, The University of Manchester,Manchester M13 9PL, UK. bjorn.hof@mancester.ac.uk
[2] Laboratory of Aero & Hydrodynamics, Delft University of Technology, 2628 CA Delft, The Netherlands j.westerweel@tudelft.nl

1 Introduction

In practice pipe flows typically turn turbulent once the Reynolds number exceeds 2000 [5], stability considerations on the other hand suggest that the laminar flow is linearly stable for all Reynolds numbers [2]. While over the last 120 years or so a large number of studies have been concerned with the stability of the laminar flow and the transition mechanism, our study is concerned with the stability of the turbulent state. Generally it has been assumed that turbulence is a sustained state and that once a flow has become turbulent it will remain turbulent for all times. Recently evidence has been presented that this may not be the case and that turbulence in pipe flow, very surprisingly, has a finite lifetime [1]. The experiments presented were carried out in an extremely long pipe and were confirmed by direct numerical calculations.

Other numerical [6]and experimental [4]studies in shorter pipes on the other hand suggest that the turbulent pipe flow becomes sustained at a critical Reynolds number, Re_{crit}. In order to determine if turbulence in shear flows is a transient or a sustained state it would in principle be desirable to carry out studies for longer lifetimes but in practice the accessible lifetime range in experimental as well as numerical studies is limited. While in experiments the length of pipe facilities for such studies becomes impractical, numerical studies are generally limited to relatively short pipes with periodic boundary conditions [6, 3]. We here suggest an alternative way to determine if turbulence in pipe flow is a transient or sustained even in pipe set ups of moderate length.

2 Results

Numerical as well as experimental data for pipe flow suggests that at least for $Re < 1800$ turbulence has a finite lifetime [3] [4] [1] [6]. The lifetime of these disordered transients follows an exponential distribution: $P(t, RE) = exp((t-t_0)\tau^{-1}(Re))$, where τ is the characteristic lifetime and t_0 is the intial

time period required for turbulence to form after the disturbance has been
applied to the laminar flow at t=0.

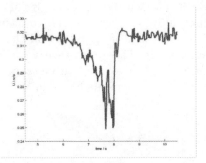

Fig. 1. Laser Doppler measurement of a turbulent event.

A question of much recent debate is if the lifetime of the turbulent state
diverges to infinity or remains finite. The fomer would imply that the chaotic
saddle becomes an attractor at a critical Reynolds number, Re_{crit}. This view
was supported in earlier studies where it was reported that τ^{-1} decreases
linearly with Re, and becomes zero at a finite value of Re. However in these
studies the initial formation time of turbulence t_0 had not been taken into
account and as pointed out by [1] a re analysis of this data indeed reults in
an exponential decay of τ^{-1} where zero is only approached assymptotically.
Hence this behaviour has been confirmed experimetnally in an extremely long
pipe (L/d=7500) and in further numerical simulations in [1]. A more recent
numerical study [6] and a previous experimental study [4] however suggest a
linear decay of τ^{-1} and hence a diverging lifetime of turbulence at a critical
point.

We here present experiments carried out in an extremely accurately controlled pipe of relatively short length. Particular care has been taken to control
the flow rate within 0.1% and with the alignment of the high precision bore
pipe. While in our earlier study the uncertainty in the pipe diameter with
$4mm \pm 1.5\%$ was relatively large, the larger diameter pipe ($D = 10mm$) used
in the present study has an accuracy better than $\pm0.1\%$). Similar to [1] the
experimental procedure was to create a turbulent puff upstream at a perturbation point and then to determine at a fixed distance downstream if the
puff survived its downstream journey (in this Reynolds number regime turbulence is advected at the bulk velocity in the streamwise direction). While
earlier studies relied on flow visualisation [4] and changes in the velocity profile [1] to distinguish laminar from turbulent flow we here measured velocities
with a LDA system (figure 1). Differing perturbation mechanisms were used
to create turbulence and we found that the lifetime statistics of the turbulent puffs are independent of the initial perturabtion. As shown in figure 2
the probability distribution found in our measurement follows an S-shaped

curve. This suggests an exponential scaling of τ^{-1} with Re and therefore that turbulence at these Reynolds numbers has a finite lifetime.

In comparison to our earlier study [1] the data set is shifted by 2.5% in Reynolds number, which is within the error margins of determining the absolute value of Re. Importantly however the relative scaling of the probability with Re is S-curved (see caption of figure 2) and the shape is identical to that found in our earlier study. This result clearly confirms that inverse lifetimes decay exponentially and hence that turbulence at least in the regime accessible in experimental measurements has a finite lifetime.

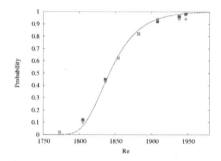

Fig. 2. Probability of turbulence to survive a time of L/D=380 as a function of Re. The curve is the probability distribution found in [1]. The presrent data clearly follows the same S-shaped curve (shifted by 2.5%), confirming that $\tau - 1$ decreases exponentially with Re and not linearly as suggested in other studies [4, 6].

References

1. B.Hof, J. Westerweel, T.M. Schneider, and B. Eckhardt. *Nature*, 443:55–62, 2006.
2. P.G. Drazin and W.H. Reid. *Hydrodynamic Stability.* Cambridge University Press, 1981.
3. H. Faisst and B. Eckhardt. *J. Fluid Mech.*, 504:343, 2004.
4. J. Peixinho and T. Mullin. *Phys. Rev. Lett.*, 93:094501, 2006.
5. O. Reynolds. *Philos. Trans. R. Soc. London*, 174:935, 1883.
6. A.P. Willis and R.R. Kerswell. *Phys. Rev. Lett.*, 98:014501, 2007.

Dynamics at the Edge of Chaos in Pipe Flow

Tobias M. Schneider and Bruno Eckhardt

Fachbereich Physik, Philipps-Universität Marburg, D-35032 Marburg, Germany
`tobias.schneider@physik.uni-marburg.de`

The stability of the laminar profile in pipe flow implies that the laminar profile and the turbulent dynamics coexist, raising the question about the border between the two. This border can be studied by tracking the time evolution of trajectories which neither decay nor grow towards the turbulent state. Independent of the choice of initial perturbation states in the border converge to a single state, which is dominated by a pair of downstream vortices.

1 Introduction

The transition to turbulence in pipe flow is puzzling because of the absence of linear instabilities that can trigger the transition. Many experiments have hence focussed on the determination of the 'double threshold' in Reynolds number and perturbation amplitude that has to be crossed in order to obtain turbulence [1]. Applications of dynamical systems theory and systematically tailored laboratory experiments and numerical simulations have led to new insight into the transition [2]. There is growing evidence that transient turbulent motion is generated by a strange chaotic saddle in the state space of pipe flow [3]. Nonlinear travelling wave solutions have been detected, both in numerical simulations and in experiment [4, 5, 6]. They correspond to unstable periodic orbits that are embedded in the chaotic saddle around which turbulent dynamics can form. In the system's state space there is therefore a domain that generates the turbulent dynamics. It coexists with the stable laminar profile and its basin of attraction in another domain of state space. This coexistence naturally raises the question of the 'boundary' between both domains. Since the motion on one side of this boundary is chaotic, it is visible as the *edge of chaos* in lifetime plots [7].

In this work we apply tools developed in [7] to analyze the *edge of chaos* that is located at the boundary between laminar and turbulent motion and we extract and characterize the dynamical objects that dominate and 'define' the boundary.

2 Probing the Edge of Chaos

Since a perturbation has to exceed a threshold in amplitude in order to trigger turbulence, an initial perturbation starting out on one side of the border will swing up to the turbulent region, whereas one on the other side will decay to the laminar profile. This can be used to detect points on the edge of chaos and to trace out the dynamics within the border.

Operationally, we detect the boundary by adding a perturbation of fixed structure but varying amplitude to the flow and numerically following its time evolution. If the amplitude is small, the perturbations smoothly decays. If the amplitude is large enough the flow will become turbulent. Inbetween both initial conditions lies one that will not decay nor reach the turbulent state. It lives in the edge of chaos.

We have numerical evidence that the edge of chaos is located on a smooth but folded surface in state space [8]. Crossing this surface from one side to the other then marks the transition to turbulence. The folds in the surface can be seen in high resolution studies of the critical amplitude required to trigger perturbations: the crossing of a fold when increasing the Reynolds number can lead to jumps in the amplitude. The jumps are small and hence do not affect the overall $1/Re$ scaling seen experimentally [9].

3 The Edge States

Numerical experiments and theoretical arguments suggest that the edge trajectories collapse onto invariant structures that are relative attractors within the edge of chaos. In the simplest case, the relative attractors or *edge states* can be saddle points (then the stable manifold coincides with the edge of chaos). For pipe flow the invariant edge state that attracts all trajectories in the boundary has a simple topological structure, shown in Fig. 1. The flow field is dominated by two high-speed streaks which are located off-center. It shows no discrete rotational symmetry, does not settle down to a simple travelling wave and shows intrinsic chaotic dynamics of vortical structures in its center region (see Fig 2). Therefore, the edge state is chaotic. It is universal in the sense that all trajectories in the boundary are attracted to the same structure. An example of the evolution of a complex initial field towards the edge state is presented in [10]. The global structure of the state also seems to vary only little with the Reynolds number. No bifurcations or drastic changes in the edge state have been observed for Reynolds numbers between 2160 and 4000.

The significance of the edge state lies in its guarding role for the transition to turbulence. The edge state 'creates' the surface a trajectory has to cross to turn turbulent. If indeed there is only one state (up to the obvious rotational and translational degrees of freedom), control strategies could probably be

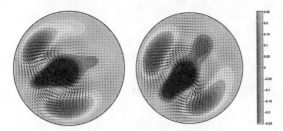

Fig. 1. Time-averaged cross sections of the edge state at Re = 2160 (left) and Re = 2875 (right). The out-of-plane components are shown in color, the in-plane components as vectors.

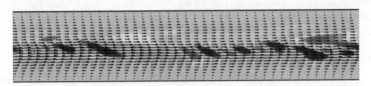

Fig. 2. Instantaneous snapshot of the edge state at Re = 2875. Cross section along the pipe axis and the symmetry axis between the vortices. The chaotic nature of the edge state shows up in the non-periodic structures along the axis.

applied to either encourage or prevent this state, in order to achieve or prevent relaminarization, respectively.

While the exact travelling wave solutions [4, 5, 6] are believed to form the backbone of the chaotic saddle supporting turbulence, the transition is mediated by the edge state presented here. The relations between the different states follow from the intersections of their stable and unstable manifolds, as in other dynamical systems.

We thank the *Deutsche Forschungsgemeinschaft* for support.

References

1. L. Boberg and U. Brosa, Z. Naturforsch. **43a**, 697 (1988).
2. B. Eckhardt, T. M. Schneider, B. Hof and J. Westerweel, Annu. Rev. Fluid Mech., **39**, 447–468
3. H. Faisst and B. Eckhardt, J. Fluid Mech. **504**, 343–352
4. H. Faisst and B. Eckhardt, Phys. Rev. Lett. **91**, 224502 (2003).
5. H. Wedin and R. R. Kerswell, J. Fluid Mech. **508**, 333 (2004).
6. B. Hof, C. W. H. van Doorne, J. Westerweel, F. T. M. Nieuwstadt, H. Faisst, B. Eckhardt, H. Wedin, R. R. Kerswell, and F. Waleffe, Science **305**, 1594 (2004).
7. J. Skufca, J.A. Yorke, and B. Eckhardt, Phys. Rev. Lett. **96**, 174101 (2006).
8. T. M. Schneider, B. Eckhardt and J. A. Yorke, (submitted).
9. B. Hof, A. Juel and T. Mullin, Phys. Rev. Lett. **91**, 244502 (2003).
10. T. M. Schneider and B. Eckhardt, Chaos, **16**, 041103 (2006).

Transition and Transition Control in a Square Cavity

Peter J. Schmid

Laboratoire d'Hydrodynamique (LadHyX), CNRS-École Polytechnique
F-91128 Palaiseau, France
peter@ladhyx.polytechnique.fr

We study instabilities and the transition to turbulence in a square cavity. This is a generically complex yet sufficiently simple geometry that helps in the development and testing of numerical and algorithmic methods for the analysis and control of flow instabilities.

1 Introduction and Background

The accurate description of disturbance behavior in complex geometries poses a great challenge. At the same time, industrial applications, such as, for example, the flow in a combustion chamber, could greatly benefit from a more thorough understanding of the underlying transition and instability mechanisms, and the effective manipulation of flows in these types of geometries relies heavily on an efficient description of the most dominant processes. The departure from the generic and well-studied configurations (among them, channel flow, pipe flow, boundary layers and their common variations) quickly leads into new territory, both in the analysis of stability and transition characteristics and in the design of control schemes. Standard techniques become difficult to generalize and their validity — often extended by asymptotic methods — are of limited use or locally constrained. In many cases, the full dynamics has to be described within a global point of view.

Global stability analyses for flows in complex geometries are becoming more commonplace, but their resulting large matrix sizes have put considerable strain on computational resources. Direct methods, the method of choice for simple problems, become prohibitively expensive, and iterative schemes have thus to be employed to extract the global stability modes. The Arnoldi scheme [1] has been particularly successful in this respect. It is based on an approximation of the high-dimensional system matrix by projecting it onto a lower-dimensional Krylov subspace. In this way, the dominant eigenvalues (and corresponding eigenvectors) of the full system can be computed rather efficiently. Convergence can be improved upon in various ways by shifting, mapping, restarting, locking and purging techniques, but in most cases this is accomplished at the expense of computational efficiency due to the necessity of additional matrix inversions.

Based on recent advances in hydrodynamic stability theory, it has been recognized that the dynamics of many wall-bounded shear flows is better described by a superposition of normal modes rather than by a single (the least stable) mode. Even though the spectrum resides entirely in the stable half-plane, transient effects can cause energy amplification that subsequently may trigger nonlinear saturation followed by secondary instabilities [2]. The same view point should hold for global modes, that is, the superposition of mutually non-orthogonal global modes may result in substantially different perturbation dynamics than is predicted by the global spectrum. Previous studies along this line have reproduced the convective instability of wave-packets in a variety of flows [3, 4, 5].

Once the inherent dynamics of a fluid dynamical system has been probed, it is a small conceptual step to formulate the associated control problem. In effect, one needs to ask the question for appropriate conditions at the wall (referred to as blowing/suction velocities) such that a prescribed cost functional is optimized. Information about the flow is gathered at a small set of sensor locations. This information is then used to compute a control strategy that is optimal and, in the presence of noise, robust [6]. Control problems are greatly gaining in popularity since their technological potential (for example, in drag reduction, instability suppression or mixing enhancement) has readily been recognized.

In general, two different, but related, approached are followed: the *continuous* technique based on a variational principle which results in the formulation and solution of an adjoint system of governing equations, or the *discrete* method based on the solution of a Riccati equation for the control and estimation gains. For large problems in complex geometries, both approaches quickly reach the limit of computational resources or lose favorable convergence properties. It thus becomes necessary to resort to approximate methods adopted from iterative techniques for the solution of large-scale linear systems. Two approaches are commonly used: (i) the formulation of the full flow control problem, followed by its iterative solution, or (ii) the reduction of the full flow control problem to a smaller (but equivalent) one, followed by a direct solution of the resulting equations. In the second case, the important question of how to reduce the system and of what constitutes equivalence between the full and the reduced problem needs to be addressed. In particular, fluid structures that play a dominant role in describing the inherent (uncontrolled) disturbance behavior, may be irrelevant when it comes to representing the dynamics between actuator input and sensor output.

In this article we will touch upon some of the above-mentioned questions by studying the incompressible flow in a cavity and by attempting its control. Even though this flow configuration cannot do justice to the geometric complexity encountered in many industrial applications, it contains some relevant features that are often observed in more complex geometries, and it thus provides a valuable testbed to develop techniques and intuition.

2 Example: Flow in a Square Cavity

We consider the incompressible flow over a square cavity. A sketch of the flow configuration is given in Figure 1(a). The Reynolds number is based on the uniform inflow velocity U_∞ and the depth/width of the cavity. For sufficiently large Reynolds numbers the flow exhibits oscillatory behavior which is caused by the pressure feedback from the downstream edge of the shear layer to its upstream edge.

2.1 Numerical Method

A two-dimensional Navier-Stokes solver based on finite differences forms the foundation for the computations in this study. A fractional-step method is employed, where the advective terms are evaluated using a higher-order Godunov-type method. The diffusive term is treated implicitly, and the resulting linear system is solved using a geometric multigrid method. The pressure Poisson equation is solved using a geometric multigrid method as well, after which the preliminary velocity field is rendered divergence-free.

2.2 Steady State

Before we consider the linear evolution of disturbances, we need to establish a steady based flow, i.e. a solution of the nonlinear Navier-Stokes equations. To this end we implement the *selective frequency damping* method [7]. This method adds a damping term to the Navier-Stokes equations which drives the solution toward a specific velocity field. This second velocity field, in turn, is governed by a simple damping process, where the remaining difference between the two velocity fields is driving the second velocity field. As time progresses, the nonlinear Navier-Stokes equations will be driven toward their steady state, even in the presence of an unstable mode. Two governing parameters which determine the amount of external driving in both the Navier-Stokes and the auxiliary equation assure the rapid convergence toward a steady solution. The steady flow field is then used as a base state, and only perturbations about this base state will be considered. The computed steady state for a Reynolds number of $Re = 5000$ is shown in Figure 1(b). A weakly diffusing shear layer is forming on top of the cavity which contains a clockwise vortex.

A snapshot of the full nonlinear simulation is depicted in Figure 1(c) and shows the instability of the shear layer as well as vortical structures near the

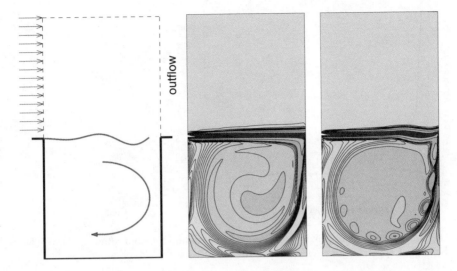

Fig. 1. (a) Sketch of geometry for flow in a square cavity. (b) Steady mean field, displayed by vorticity contours, for $Re = 5000$ obtained by the selective frequency damping (SFD) technique. (c) Instantaneous vorticity field for $Re = 5000$.

downstream edge of the shear layer and near the start of the recirculating cavity vortex. These dominant structures should be part of the disturbance dynamics and should be reflected in the global modes that correspond to their growth rate.

2.3 Linear Solver and Global Modes

The linearized equations are solved using the same techniques mentioned above. The solutions are then further processed by an iterative subspace technique. The global modes are extracted using the Arnoldi method. This method constructs an orthonormalized sequence of flow fields onto which the linear disturbance dynamics is projected. This results in a lower-dimensional representation of the full system. The eigenvalues of this reduced system (also referred to as the Ritz values) approximate the eigenvalues of the full system, and the accuracy can be improved by increasing the dimension of the Krylov space, i.e., by adding to the orthonormalized basis fields.

For the case of the cavity flow we increased the dimension of the Krylov space up to seventy. The convergence of the spectrum can be monitored by tracking the residual norm of the Arnoldi method. The convergence history is displayed in Figure 2(a). We observed a rapid drop of the residual norm to a level of nearly 10^{-7} which illustrates the efficiency of the method. The approximate spectrum, i.e. the set of Ritz values, for the largest Krylov space (seventy fields) is shown in Figure 2(b). For the chosen parameter combination

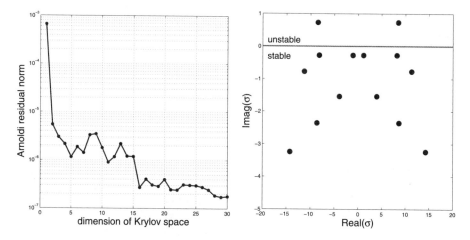

Fig. 2. (a) Convergence history of the Arnoldi residual for cavity flow at $Re = 5000$. (b) Global spectrum of cavity flow at $Re = 5000$.

($Re = 5000$) two Ritz values are found in the unstable half-plane. The growth rate is established as $\omega_i \approx 0.728$, the frequencies are identified as $\omega_r \approx \pm 8.47$, which is easily verified numerically.

The associated global modes are shown in Figure 3(a,b,c) for three different global eigenvalues.

The least stable global mode (see Figure 3(a)) shows a characteristic vorticity pattern along the main shear layer, but also captures (to a lesser extent) the unstable behavior of the vortex sheet that forms in the cavity. The other two modes (see Figures 3(b,c)) also capture the instability of the main shear layer, but put less emphasis on the disturbance dynamics within the cavity.

2.4 Multiple Global Modes

Owing to the recognition that the interaction of multiple modes captures the short-term disturbance dynamics, it follows that a superposition of the extracted global modes may describe the transient and non-modal behavior of perturbations [8]. Denoting the global modes by $\Phi_i(x, y)$ we have

$$\mathbf{u}(x, y, t) = \sum_{i=1}^{N} a_i(t)\Phi_i(x, y).$$ (1)

The dynamical system (i.e. the linearized Navier-Stokes equations) then becomes a system of N decoupled ordinary differential equations for the expansion coefficients $a_i(t)$.

$$\dot{\mathbf{u}} = \mathcal{P}\mathbf{A}\mathbf{u} \qquad \longrightarrow \qquad \dot{\mathbf{a}} = \hat{\mathbf{A}}\mathbf{a}$$ (2)

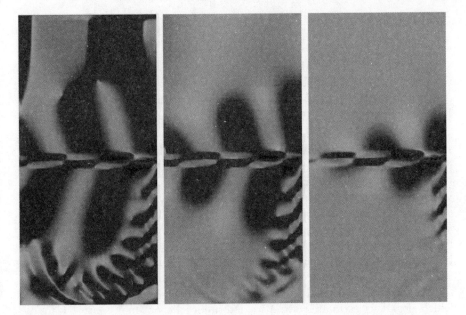

Fig. 3. Vorticity fields of three global modes: (a) least stable (unstable) global mode for $Re = 5000$, (b,c) damped global modes for $Re = 5000$.

where **a** denotes the N-dimensional vector of expansion coefficients, $\mathcal{P}$ represents the projection step onto solenoidal velocity fields (via a pressure Poisson equation), and the new system matrix $\hat{\mathbf{A}}$ is given as $\mathbf{V}^{-1}\mathbf{A}\mathbf{V}$ with $\mathbf{V}$ as the matrix whose columns contain the global modes $\Phi_i, i = 1, ..., N$.

This new, reduced system can now be analyzed as to its potential for transient energy amplification. To this end the matrix exponential of a diagonal matrix containing the eigenvalues is evaluated using an energy norm based on the extracted global modes. The norm of this matrix exponential represents the maximum energy amplification any initial condition can reach over a specified period of time. A particular optimal initial condition can easily be recovered by a singular value decomposition (SVD). For rapidly decaying global modes only a small number of them has to be included in the projection of the dynamics onto a reduced basis. In this case, the dynamics of the small system should approximate the full dynamics well.

2.5 Flow Control

For the flow control problem, we have to augment our problem by an external forcing **f** as well as a vector of measurements **y**. We get

$$\dot{\mathbf{u}} = \mathcal{P}\mathbf{A}\mathbf{u} + \mathbf{B}\mathbf{f} \tag{3}$$

$$\mathbf{y} = \mathbf{C}\mathbf{u} \tag{4}$$

with $\mathbf{B}$ and $\mathbf{C}$ governing the location and type of control and measurement, respectively. The task is then to linearly express the control input $\mathbf{f}$ in terms of the state vector $\mathbf{u}$, i.e. $\mathbf{f} = \mathbf{K}\mathbf{u}$ with $\mathbf{K}$ as the control gain. The control gain $\mathbf{K}$ is determined via a Riccati equation.

Under realistic conditions, the dependence of the control $\mathbf{f}$ on the *entire* state vector $\mathbf{u}$ is impractical or unfeasible. Instead, the control gain $\mathbf{K}$ is taken to depend on an estimate of the state vector, denoted by $\hat{\mathbf{u}}$, which satisfies the same differential equation and whose values at the sensor locations, i.e. $\hat{\mathbf{y}} = \mathbf{C}\hat{\mathbf{u}}$, match the measured values $\mathbf{y}$ as closely as possible. This auxiliary problem is known as the estimation problem. The influence of the measurement error $\hat{\mathbf{y}} - \mathbf{y}$ is incorporated in the estimation problem in the form of a linear driving term whose coefficient matrix is the estimation gain $\mathbf{L}$. Again, the estimation gain $\mathbf{L}$ can be determined by solving a Riccati equation. The full control problem then consists of an estimation problem which reconstructs an estimated state $\hat{\mathbf{u}}$ from the actual measurements $\mathbf{y}$ and a control problem which uses this estimated state to determine the optimal control $\mathbf{f}$ in the original problem.

2.6 Model Reduction

For anything but simple test problems numerical and algorithmic difficulties are encountered when the two Riccati equations (for the control and estimation gain) have to be solved. The availability of global modes, which have already been used in the description and analysis of the inherent system dynamics, may suggest to reduce the full problem by a global eigenfunction expansion. In this case the reduced system matrices are

$$\hat{\mathbf{A}} \longrightarrow \mathbf{V}^{-1}\mathbf{A}\mathbf{V} \qquad \hat{\mathbf{B}} \longrightarrow \mathbf{V}^{-1}\mathbf{B} \qquad \hat{\mathbf{C}} \longrightarrow \mathbf{C}\mathbf{V}. \tag{5}$$

For this system, the calculation of the resulting Riccati equations can be performed using standard (direct) numerical methods. Despite this advantage, caution has to be exercised regarding the suitability of the global mode basis for the control problem.

Global modes typically capture the fastest growing flow structures as they appear in the uncontrolled and unforced dynamical system $\dot{\mathbf{u}} = \mathbf{A}\mathbf{u}$. It is conceivable, however, that these flow structures react only weakly to external control efforts or leave an exceedingly weak signal at the sensor locations. In other words, what needs to be addressed is the controllability and the observability of global modes before they can successfully be adopted as an expansion basis for the controlled dynamical system. Even if some global modes are sufficiently controllable, and others are sufficiently observable, the two sets have to overlap to allow efficient communication between actuators and sensors.

An ideal set of basis functions should maximize both controllability and observability and thus describe the information flow between input and output in an optimal manner. The singular vectors of the product of input-to-state and state-to-output mapping satisfy this constraint. Their calculation, however, involves the solution of large-scale linear algebra problems which is currently only possible using iterative techniques [9].

3 Summary and Conclusions

Direct numerical simulations have recently advanced to a level that allows the accurate description of complex flow phenomena in complex geometries. However, the analysis of dominant fluid structures as well as the objective extraction of relevant instability mechanisms from available data is somewhat less developed. The embedding of direct numerical simulation codes into an iterative subspace algorithm represents a promising direction in gaining insight into the underlying flow physics of a great many fluid dynamical processes. Standard techniques, learned from simpler geometries, can provide guidelines for their generalized use for complex geometries.

The successful control of flow instabilities in complex geometries will continue to pose great challenges to numerical methods and computational resources. A first attempt at tackling these difficulties is the reduction of the full problem to a more manageable size. Model reduction methods will thus develop into a key technology in the design of feedback control strategies, where the preservation of the actuator-sensor map will play a pivotal role. As in the case of global stability theory, the merging of direct numerical simulations and iterative linear algebra methods may yield efficient algorithms and pave the way to the successful control of fluid flow in complex geometries.

References

1. W. Edwards, L. Tuckerman, R. Friesner, D. Sorensen: J. Comp. Phys. **110**, 82–102 (1994)
2. P.J. Schmid: Ann. Rev. Fluid Mech. **39**, 129–162 (2007)
3. C. Cossu, J.-M. Chomaz: Phys. Rev. Lett. **78**, 4387–4390 (1997)
4. P.J. Schmid, D.S. Henningson: J. Fluid Mech. **463**, 163–171 (2002)
5. J.-M. Chomaz: Ann. Rev. Fluid Mech. **37**, 357–392 (2005)
6. J. Kim, T.R. Bewley: Ann. Rev. Fluid Mech. **39**, 383–417 (2007)
7. E. Åkervik, L. Brandt, D.S. Henningson, J. Hœpffner: Phys. Fluids **18** (2006)
8. E. Åkervik, J. Hœpffner, U. Ehrenstein, D.S. Henningson: J. Fluid Mech. (in press, 2007)
9. C.W. Rowley: Int. J. Bifurc. Chaos **15**, 997–1013 (2005)

DNS on drag reduction by the injection of dilute polymer solution into a buffer region in turbulent water channel flow

Shogo Tatsumi and Yoshimichi Hagiwara

Department of Mechanical and System Technology, Graduate School of Science and Technology, Kyoto Institute of Technology, Matsugasaki, Sakyo-ku, Kyoto, 606-8585 Japan yoshi@kit.ac.jp

1 Introduction

Heterogeneous mixing of a dilute polymer solution with turbulent water flow is of interest for local and temporal reduction of turbulent friction drag. Tiederman carried out measurements on the duct flow with an injection of polymer solution from a slot on one wall[1, 2]. He obtained not only turbulence modifications but also (1) a streamwise increase in the drag reduction level to its maximum at a distance from the slot and (2) a decrease in the drag reduction level further downstream. These changes in the drag reduction have not yet been predicted.

We considered that these are caused by the aggregation process of the polymers and the disintegration process of the aggregated polymers. Thus, we developed a model representing these processes[3]. In the present study, we carry out DNS for the flow, taking account of the injection effect and dealing with the higher Reynolds number case.

2 Computational procedure

In our model, spotty regions of concentrated polymers were represented by beads. When one spotty region was brought closer to another spotty region by surrounding flow, these regions were then connected by many linking polymers from the background, low-concentration region. These linking polymers were represented by a connecting spring and a connecting dashpot in parallel. When the distance between two spotty regions, which had already been linked with each other, became long, the spring and dashpot were removed.

The computational domains were a box of $6\pi h \times 2h \times \pi h$ (hereafter called case 1) and a box of $2\pi h \times 2h \times \pi h$ (case 2). The Reynolds numbers Re_τ were 150 in case 1 and 300 in case 2. The numbers of grid points were $192 \times 96 \times 64$

in case1 and 128^3 in case 2. Furthermore, the beads with a normal velocity were introduced into the flow from a narrow region on one wall near the inlet in case 1. The two-way coupling algorithm was adopted for the interaction between the models and surrounding flow. The schemes used in the present study were the same as those adopted in our previous study.

3 Results and discussion

Figure 1 indicates the shear stress profiles in cases 1 and 2. The Reynolds shear stress in section A ($911 < x^+ < 941$) of case 1 and case 2 is decreased particularly in the buffer region by the models.

Figure 2 shows the streamwise change in the local wall shear stress in case 1. The circles show the experimental result obtained by Tiederman. In his experiment, the wall shear stress was increased by the injection of polymer solution. The drag reduction level increased (i.e. the wall shear stress decreased) downstream to its maximum, and decreased further downstream. The solid line obtained from our computation shows near agreement with the experimental results. Thus, the aggregation and disintegration processes are found to play an important role in the streamwise change in the drag reduction. One of the reasons for underestimation of drag reduction is that the diffusion of the model in the buffer region is more noticeable than experiment.

Figure 3 exhibits the snapshots for the lower part of cross sections in cases 1 and 2. The gray closed curves show the crosscut of hairpin vortices, the white regions show high-speed streaks, the black regions show low-speed streaks, and white dots show the beads in the models. It is found that some models are located close to the vortices. The vortices were not developed, and the models were deformed. Occasionally, the models were disintegrated by the strong vortices. The energy of the vortices was used for deforming the models. Some other models are located inside the streaks in the figure. The ejection and sweep were attenuated by the models. These are the main causes of decreases in the Reynolds shear stress.

4 Conclusions

The direct numerical simulation was carried out for turbulent flow with the injection of polymer solution from a slot on a wall. The main conclusions obtained are as follows.

1. The increase and successive decrease in the wall shear stress due to the injection of polymer solution from a slot are predicted by the model for polymer aggregation and disintegration.
2. The model is also effective for the prediction of turbulence modification in the case of the higher Reynolds number.
3. The coherent structure is attenuated by the models.

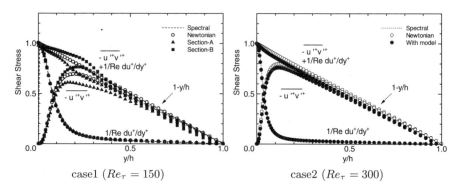

case1 $(Re_\tau = 150)$ case2 $(Re_\tau = 300)$

Fig. 1. Shear stresses

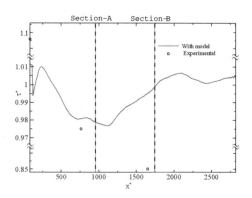

Fig. 2. Streamwise change in wall shear stress

case1 $(Re_\tau = 150)$ case2 $(Re_\tau = 300)$

Fig. 3. Hairpin vortices, low-speed streaks, high-speed streaks and models

References

1. W. G. Tiederman, T. S. Luchik, D. G. Bogard: J. Fluid Mech. **156**, 419 (1985)
2. T. S. Luchik, W. G. Tiederman: J. Fluid Mech. **190**, 241 (1988)
3. G. Oba, Y. Hagiwara: Turbulent drag reduction due to polymer aggregation and its breakup by near-wall coherent structure. In: *Advances in Turbulence X*, ed. by H. I. Andersson and P. –Å. Krogstad (CIMNE, Barcelona 2004) pp 699-702

Velocity-Gradient Modification in Particle-Laden Turbulent Channel Flows

Maarten J. Bijlard and Luís M. Portela[†]

Multi-Scale Physics Department, Delft University of Technology
Prins Bernhardlaan 6, 2628 BW Delft, The Netherlands
[†]Corresponding author: L.Portela@tudelft.nl

In particle-laden turbulent flows the interaction between the particles and the turbulence can promote a large modification in the turbulence characteristics of the flow. These effects, already significant in homogeneous isotropic turbulence, can become much larger in wall-bounded flows. There exists a limited amount of work on the Reynolds-stresses and turbulence kinetic energy modification (e.g., [2]), which is mostly dominated by the large scales, however, very little is known about the small-scale structure modification. In this work, we study the modification of the two non-zero invariants of the velocity-gradient tensor, which is mostly dominated by the small scales, in a channel flow laden with small heavy spherical particles. We use standard point-particle Eulerian-Lagrangian DNS simulations with two-way coupling and elastic bouncing at the walls. We consider Stokes drag without gravity, and, in order to isolate the effects of the particle-turbulence interactions, we do not consider inter-particle interactions (i.e., the particles ignore each other). Details of the numerical method can be found in Portela and Oliemans [3]. Several simulations were performed, in order to analyze the effect of the Reynolds number, particle mass-loading and particle relaxation-time..

In figure 1 is shown the modification of the invariant-map (normalized joint PDF of the second and third invariants of $\mathbf{A} \equiv \nabla \mathbf{V}$, respectively, Q_A and P_A), for one of the cases considered: Reynolds number based on the channel width and wall-shear velocity equal to $Re_\tau = 360$, particle relaxation-time, in wall units, equal to $\tau_p^+ = 58$ (particle diameter, in wall units, equal to $D_p^+ = 0.36$, and particle-fluid density ratio equal to $\rho_p/\rho = 8000$), and particle mass-loading (total mass of the particles divided by the total mass of the fluid) equal to $\phi_m = 0.65$. Similarly to [1], the values were normalized using the mean centerline velocity and the channel half-width, and the invariant-maps were determined for different regions of the channel, according to the distance to the wall, in wall units: viscous sublayer ($0 < z^+ \leq 5$), buffer layer ($5 < z^+ \leq 35$), log-law region ($35 < z^+ \leq 150$), and outer-region ($150 < z^+ < \frac{1}{2}Re_\tau$). We computed the Eulerian invariant-maps (sampled in space), and two Lagrangian invariant-maps, sampled at the particles moving towards the wall ($W_p < 0$), and at the particles moving away from the wall ($W_p > 0$).

We can see that in the viscous sublayer the particles promote a large modification in the invariant-map, with a stretching in the direction of Q_A. This indicates that the small-scale structure becomes more two-dimensional, switching between high-vorticity and high-strain regions. This behavior is closely associated with the well-known accumulation of the particles into elongated streaks very near the wall (e.g, [4]), making the turbulence more streamwise-aligned and with sharper contrasts between the streamwise-elongated low-speed and high-speed regions. In the buffer and log-law regions (and also in the outer-region, not shown here), the modification promoted by the particles is quite different, narrowing the invariant-map, but with a much smaller modification in its shape than in the viscous sublayer. Except very close to the wall, where there exists an extremely-high accumulation of particles into streamwise streaks, the modifications in the invariant-map do not appear to be strongly related with the preferential concentration of the particles. They appear to be mostly "indirect effects" of the overall particle-turbulence interaction. For example, this can be seen when we compare the Eulerian and Lagrangian invariant-maps: we can note that both Lagrangian invariant-maps are not very different from the Eulerian invariant-map.

The picture that emerges from the invariant-maps, and from the analysis of snapshots of the two invariants together with the distribution of the particles (not shown here), is that there exist two competing effects: (i) the modification promoted by the particle-streaks, and (ii) "indirect effects" of the overall particle-turbulence interaction. The first effect is dominant close to the wall and leads to a highly intermittent small-scale structure, particularly in the location of the particle-streaks. The second effect is dominant further away from the wall and leads to a smoother and weaker small-scale structure, but without a significant change in the shape of the invariant-map; i.e., the invariant-maps keep its characteristic tear-drop shape. It is interesting to note that the first effect can be felt quite far from the wall. For example, this can be seen when comparing both Lagrangian invariant-maps: we can note that the invariant-maps sampled at the particles moving away from the wall are broader, indicating a stronger, more intermittent, small-scale structure.

References

1. H.M. Blackburn, N.N. Mansour, B.J. Cantwell: J. Fluid Mech. **310**, 269 (1996)
2. Y.M. Li, J.B. McLaughlin, K. Kontomaris, L. Portela: Phys. Fluids **13**(10), 2957 (2001)
3. L.M. Portela, R.V.A. Oliemans: Int. J. Num. Meth. Fluids **43**, 1045 (2003)
4. D.W.I. Rouson, J.K.E. Eaton: J. Fluid Mech. **428**, 149 (2001)

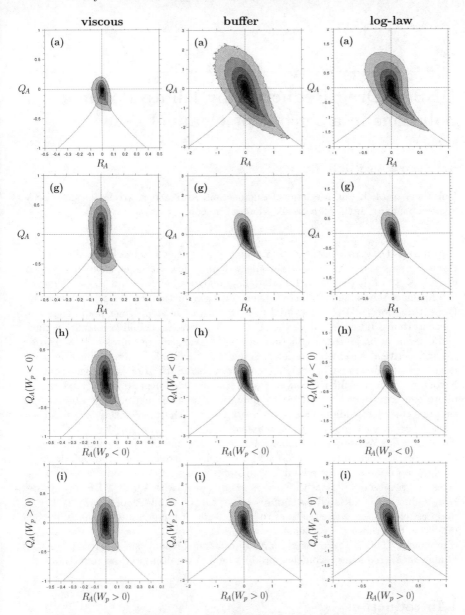

Fig. 1. Invariant-maps in three regions of the channel, for $Re_\tau = 360$, $\tau_p^+ = 58$, and $\phi_m = 0.65$. Left column: viscous sublayer ($0 < z^+ \le 5$). Center column: buffer layer ($5 < z^+ \le 35$). Right column: log-law region ($35 < z^+ \le 150$). First row from the top (a): unladen flow invariant-map. Second row from the top (g): laden flow Eulerian invariant-map. Third row from the top (h): Lagrangian invariant-map sampled at the particles moving towards the wall. Fourth row from the top (i): Lagrangian invariant-map sampled at the particles moving away from the wall. The gray-scales denote a logarithmic-decade, ranging over six decades.

DNS of Drag Reduction by Dilute Polymer Solutions in a Turbulent Channel Flow

R. Akhavan* and D. H. Lee

University of Michigan, Mechanical Engineering, Ann Arbor, MI 48109-2125, USA
*corresponding author, e-mail: raa@umich.edu

Summary. Drag reduction by dilute polymer solutions is investigated by DNS. Computations were performed in turbulent channel flows at a base Reynolds number of $Re_{\tau_o} \sim 230$ using a mixed Eulerian/Lagrangian scheme with a FENE-P dumbbell model of the polymer. The full range of drag reduction from onset to Maximum Drag Reduction (MDR) is captured in DNS with realistic polymer parameters and concentrations. Investigation of the effect of polymer extensibility, relaxation time, and concentration on drag reduction shows the Weissenberg number (We_{τ_o}) to be the most critical parameter in determining the magnitude of drag reduction, with the other parameters having only a secondary effect. Saturation with concentration, beyond which increasing the concentration has no added benefit, is observed at very dilute concentrations ($\beta \approx 0.95$). A polymer of relaxation time λ is observed to suppress turbulent eddies of size $\ell < O(\lambda u_\tau)$. Virk's MDR asymptote is approached when ℓ becomes comparable to the channel half width, or when $We_{\tau_o} \sim O(Re_{\tau_o})$. At even higher Weissenberg numbers, and in the absence of a background noise, the flow begins to relaminarize on viscous time scales. Introduction of an infinitesimal background noise on these relaminarized states re-triggers the production of turbulence and results in a flow with features in agreement with Virk's MDR asymptote. Drag reduction is initiated by suppression of turbulent eddies of size $\ell < O(\lambda u_\tau)$. Suppression of these eddies reduces the magnitude of the pressure-strain term, thus inhibiting the transfer of turbulence kinetic energy from the streamwise to the other components. This, in turn, leads to a drop in the Reynolds shear-stress in regions of high shear, reducing turbulence production and resulting in drag reduction.

1 Introduction

The phenomenon of drag reduction by dilute polymer solutions was discovered over fifty years ago by Toms[1]. Nevertheless, a clear understanding of the mechanism of polymer drag reduction is not yet at hand. Two principal theories have been proposed to explain these phenomena. One is the so-called "time criteria" suggested by Lumley[2], in which drag reduction is attributed to the enhanced extensional viscosity of the polymer. This enhanced viscosity is assumed to suppress turbulent eddies whose time scale is shorter than both the relaxation time (λ) of the polymer and $(\nu_{eff}/\epsilon)^{1/2}$, where ϵ is the rate of turbulence production and ν_{eff} is the effective viscosity of the polymer solution and is proportional to the polymer concentration. The maximum drag reduction which can be achieved by a given polymer is thus limited, not by the polymer concentration, but by the relaxation time of the polymer. The second theory, suggested by DeGennes[3], assumes that it is the elastic energy (per unit volume) stored in the polymer which is responsible for drag reduction. Once this elastic energy becomes comparable to the energy density at a given turbulent scale, the kinetic energy which would normally cascade into that scale is redirected into the polymer, thus interrupting the normal dynamics of turbulence and leading to drag reduction. In this picture, the maximum drag reduction which can be achieved by a given polymer is determined by the concentration at which unstretched polymer coils begin to overlap.

It has proved difficult to verify either of these theories by direct experiments. In the recent years, DNS has become an alternative tool of research. However, DNS results to date have also remained controversial, with some advocating extensional viscosity [4], and others advocating elastic theory [5] as the mechanism of drag reduction. The objective of the present study is to use results from DNS to clarify the mechanism of polymer drag reduction.

2 Results

The studies were performed in turbulent channel flows, using a mixed Eulerian/Lagrangian scheme employing standard pseudo-spectral methods for the hydrodynamics and a backward-tracking Lagrangian particle method[6] for the polymer dynamics. Computations were performed in channels of size $10h \times 5h \times 2h$ and $40h \times 10h \times 2h$ with resolutions of $128 \times 128 \times 129$ and $512 \times 256 \times 129$, respectively. The simulations were performed at $Re_{\tau_o} \sim 230$ for $10 \leq We_{\tau_o} \leq 200$ and employed a FENE-P dumbbell model of the polymer with realistic polymer parameters ($b = 45,000$) and concentrations ($nk_BT/\rho u_{\tau_o}^2 = 1.14 \times 10^{-3}$), corresponding to PEO of $M_w = 5 \times 10^6$ and $c = 30$wppm. In addition, parametric studies were performed for $4500 \leq b \leq 450,000$ and $1.14 \times 10^{-5} \leq nk_BT/\rho u_{\tau_o}^2 \leq 1.14 \times 10^{-2}$ to investigate the effect of polymer extensibility and concentration.

The results are shown in Figure 1. Onset of drag reduction is observed at $We_{\tau_o} \approx 10$. With increasing We_{τ_o}, drag reduction increases until Virk's MDR asymptote is approached at $We_{\tau_o} \approx 100$. The flow statistics show good agreement with the available experimental data at corresponding levels of drag reduction. At $We_{\tau_o} = 100$, the flow is still turbulent with a non-zero Reynolds shear stress, in agreement with experiments of [8]. For $We_{\tau_o} > 100$, and in the absence of a background noise, the flow begins to relaminarize on slow (viscous) time scales. The turbulence statistics during the intermediate states leading to this relaminarized state show near-zero Reynolds shear stress in agreement with experiments of [7]. Introduction of an infinitesimal background

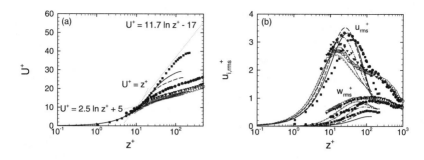

Fig. 1. Mean velocity profiles and turbulence intensities obtained in DNS at low to moderate We_{τ_o} compared to experiments. ---, Newtonian; $\cdots$, $We_{\tau_o} \approx 10$; $-\cdots-$, $We_{\tau_o} \approx 20$; $-\cdot-$, $We_{\tau_o} \approx 35$; — —, $We_{\tau_o} \approx 70$; —, $We_{\tau_o} \approx 100$; open circle, Warholic et al.[7] Newtonian; filled inverted triangle, Warholic et al.[7] $14\%DR$; filled circle, Warholic et al.[7] $33\%DR$; filled square, Ptasinski et al.[8] $63\%DR$.

noise on these relaminarized states re-triggers the production of turbulence and results in a flow with features in agreement with Virk's MDR asymptote.

Investigation of the effect of the polymer extensibility and concentration shows these to have only a secondary effect. The saturation concentration, beyond which increasing the concentration has no added benefit, is reached at $nk_BT/\rho u_{\tau_o}^2 \sim 10^{-3}$ ($\beta = 0.96$) for $We_{\tau_o} = 35$, and $nk_BT/\rho u_{\tau_o}^2 \sim 10^{-4}$ ($\beta = 0.98$) for $We_{\tau_o} = 150$. Analysis of the energetics of the flow shows that the polymer induces drag reduction by suppressing turbulent eddies of size smaller than $\sim O(\lambda u_\tau)$. Suppression of these eddies reduces the magnitude of the pressure-strain term, thus inhibiting the transfer of turbulence kinetic energy from the streamwise to the other components. This, in turn, reduces the magnitude of the Reynolds shear-stress in regions of high shear, leading to a drop in the net turbulence production and resulting in drag reduction. Overall, these results confirm the "time-criteria" theory suggested by Lumley[2].

References

1. B. Q. Toms: In Proc. 1st Int. Congress on Rheology **2** (1949) 135–141
2. J. L. Lunley: Annu. Rev. Fluid Mech. **1** (1969) 367–384
3. P. G. De Gennes: Physica **140A** (1986) 9–25
4. P. K. Ptasinski, B. J. Boersma, F. T. M. Nieuwstadt, M. A. Hulsen, B. H. A. A. Van Den Brule, J. C. R. Hunt: J. Fluid Mech. **490** (2003) 251–291
5. T. Min, J. U. Yoo, H. Choi, D. D. Joseph: J. Fluid Mech. **486** (2003) 213–238
6. P. Wapperom, R. Keunings, V. Legat: J. Non-Newtonian Fluid Mech **91** (2000) 273–295
7. M. D. Warholic, H. Massah, T. J. Hanratty: Exps Fluids **27** (1999) 461–472
8. P. K. Ptasinski, F. T. M. Nieuwstadt, B. H. A. A. Van Den Brule, M. A. Hulsen: Flow, Turbulence and Combustion **66** (2001) 159–182

Turbulent mixing in the atmospheric boundary layer: from flat terrain to narrow valley

Charles Chemel[1] and Jean-Pierre Chollet[2]

[1] Centre for Atmospheric and Instrumentation Research, University of Hertfordshire, College Lane Campus, Hatfield, Herts AL10 9AB, United Kingdom; e-mail: `c.chemel@herts.ac.uk`
[2] Laboratoire des Ecoulements Géphysiques et Industriels, CNRS / UJF / INPG, BP 53, 38 041 Grenoble Cedex 9, France; e-mail: `chollet@ujf-grenoble.fr`

Mixing processes involved in the atmospheric boundary layer (ABL) are of primary interest for air quality modelling. Ground surface heating by solar radiation induces convection throughout a mixed layer, which is usually topped by a stably-stratified layer. Large-eddy simulation (LES) makes it possible to detail various flow regimes and to assess the impact on mixing processes over a broad range of length and time scales. The present paper first deals with the ABL over flat homogeneous terrain and then over complex mountainous terrain. In both cases LESs are performed with the ARPS model [1], a non hydrostatic code with a detailed soil model and the capability of accommodating a complex topography. Careful comparisons with on site measurements have already been undertaken in the literature (see for instance [2]). The main objective of this study is to discuss further the interactions which lead to either a hindered or enhanced turbulent mixing.

1 Thermally-driven mixed layer over flat terrain

Over flat terrain the evolution of the ABL is mainly driven by convection. The turbulent motions which are responsible for mixing interact with the large scale motions in a way that basically tends to transfer kinetic energy to the smaller scales. A high-resolution LES run initialized with a commonly-used 0900 EST (Local Time) sounding of Day 33 of the Wangara experiment [3] is used to investigate the mixed-layer dynamics. The horizontal resolution is 20 m. The vertical resolution is also chosen to be 20 m within the mixed-layer, whilst being refined to 5 m within the capping inversion layer.

The subgrid-scale model is based on subgrid kinetic energy with a self adjusting length scale according to local stability. Within the mixed layer, computed spectra exhibit a $k^{-5/3}$ behaviour typical of quasi-homogeneous isotropic turbulence at least for a significant range of scales which are explicitly resolved by the LES. Potential temperature spectra exhibit the same behaviour although the compensated spectrum $E_\theta(k)\,k^{5/3}$ suggests more intricate scale interactions for scalar turbulent mixing. As in any flow driven by thermal convection, convective cells develop as coherent structures. Because of high Reynolds numbers, shear associated with these cells triggers a variety of structures. The Q criterion is used in Fig. 1a, to highlight such coherent eddies. Strong eddies develop in updrafts which vanish within the stably-stratified atmosphere above while intermittently disturbing it. This is the basic mechanism for the erosion of the capping inversion layer and for some mixing within the interfacial layer (see [4] for a comprehensive discussion on mixing at the interface).

2 On the mixed layer over mountainous terrain

Complex topography makes mixed and stable capping layers major features of mountain meteorology. These layers are determined by thermal convection as well as its interactions with both slope- and valley-winds. These interactions are found to significantly affect the boundary-layer structure and its evolution in narrow valleys such as the Chamonix valley, which has been considered in the POVA project [2]. Numerical simulations have been performed with the same code ARPS but using a slightly coarser resolution (namely a horizontal resolution of 300 m and a vertical resolution ranging from about 30 m near the ground surface up to 500 m at the top of the domain) in order to accommodate the whole valley with surrounding mountain ridges.

The orography prevents the convective structures from fully developing horizontally, while enhancing vertical motions. Turbulence is also affected by midscale topographic features such as tributary valleys (e.g. the *Mer de Glace* glacier in the Chamonix valley). Slope winds as well as topographic narrowing induce air mass export up to the free atmosphere as from early in the morning and the opposite as from late in the afternoon, with small-scale turbulence produced near the ground surface because of wind shear. The erosion of the capping stable layer from convective mixing is indirectly enhanced by shear from valley winds. Up-slope winds in the morning contribute to the erosion of the capping layer along the slopes. Model results are compared with measurements from a DIAL Lidar system, which probes the vertical distribution of particles (see Fig. 1b). As expected for the behaviour of quasi passive scalar quantities, the thickness of the (thermo-) dynamic layer appears to decay faster in the late afternoon than the thickness associated to particles.

3 Conclusions

Under typical summertime conditions during daylight hours, thermal convection drives the mixing of the air mass. The convectively-driven mixed layer is

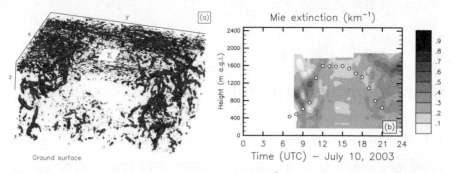

Fig. 1. (a) 3D view of isosurfaces of $Q = 0.0015$ s^{-2} for a convective cell. (b) Time-height evolution of Mie extinction from Lidar DIAL system measurements. Mixing depths computed from the model results are indicated as white dots ($\circ$).

characterized by large-scale turbulent structures, which may be evidenced by appropriate eduction methods, as well as smaller structures that have a broad range of scales with a quasi homogeneous isotropic turbulence behaviour. Taylor's hypothesis holds as it can be inferred from wave number and frequency spectra. The rather weak coupling between large and inertial scales suggests that LES is an appropriate technique to compute such flow fields even at coarse resolution and that usual subgrid-scale models might be sufficient. Scalar spectra do not exhibit such inertial features. Hence, more sophisticated subgrid-scale models, which are able to mimic interactions between different range of scales, would be considered at least when tracer dispersion needs to be accurately simulated.

In a narrow valley, the mixing processes are nearly the same as over flat terrain. The scales which contain most of the energy depend intrinsically on thermal convection. Nonetheless the valley geometry constrains the horizontal extension of the convective structures. In the morning and late afternoon, slope winds contribute to vertical exchange through direct bypassing effects. These topographic-driven mechanisms can even exceed the classical turbulent contribution a model using a coarse resolution would captured. This study shows that LES makes it possible to realistically drive turbulent scales.

References

[1] M. Xue, K. K. Droegemeier, V. Wong: Met. Atm. Phys. **75**, 161 (2000)
[2] G. Brulfert, C. Chemel, E. Chaxel, J.-P. Chollet: Atm. Chem. Phys. **5**, 2341 (2005)
[3] C. Chemel, C. Staquet, J.-P. Chollet: J. Atm. Sci. Submitted (2006)
[4] C. Chemel, C. Staquet: J. Fluid Mech. In revision (2007)

GCM representation of turbulence on Jupiter

L. C. Zuchowski[1], Y. H. Yamazaki[1], and P. L. Read[1]

University of Oxford, Clarendon Laboratory, AOPP, Parks Road, OX1 3PU
Oxford, UK zuchowski@atm.ox.ac.uk

Introduction

Yamazaki et al. (2004) have investigated the hydrodynamic stability of observed subtropical jets in Jupiter's Northern and Southern hemisphere as an initial value problem by use of the Oxford Planetary Unified model System (OPUS). Using an extended form of the Hydrodynamic Primitive Equations, OPUS is capable of including 30 vertical levels from 0.1 bar to 10 bar. Initial fields in their study had been derived by extending the zonal Voyager profile [6] with a vertical profile constituting an eigenmode to the vertical structure equation [2]. They found that in both hemispheres the prominent jets were unstable at various scales. Vortices of various sizes were generated.

Here we present results from a follow-up study to the work by Yamazaki et al. (2004). We have initiated OPUS with temperature and thermally balanced wind profiles recently obtained by Cassini's Composite InfraRed Spectrometer (CIRS) during the Jupiter fly-by in 2000 [3]. The initial profiles extend down to an atmospheric depth of ≈ 0.5 bar and thus significantly reduce the domain over which arbitrary assumptions have to be made. While no momentum

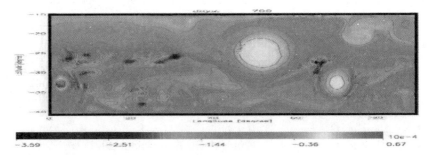

Fig. 1. Instantaneous flow fields as simulated by OPUS in the upper troposphere for run FF2. The image show maps of Ertel's Potential Vorticity at θ=243 K.

Table 1. Properties of singular vortices in the Great Red Spot band. In the BF forcing case only the bottom model level was forced while the FF cases featured forcing throughout the entire vertical domain.

Run name	BF	FF1	FF2
Forc. Constant [day]	0.01	200	400
Cent. Lat. [degree]	27	24	22
Lat. Size [degree]	8	10	10
Long. Size [degree]	10	15	15
Zon. Velocity [m/s]	42	8	23
Lifetime [day]	140	500	180

forcing was employed by Yamazaki et al. (2004), we have conducted a sensitivity study examining the effects of different approaches to maintaining the zonally-symmetric pattern of zonal velocity on the atmospheric configurations obtained. Furthermore a cloud scheme has been developed for OPUS and was used to illustrate the atmospheric configurations obtained by indicating the corresponding motion of ammonia ice and ammonia vapor clouds.

Results and Discussion

By improving the quality of the initialization and momentum forcing in the model we obtained a more realistic representation of turbulence in Jupiter's Southern hemisphere, including the formation of isolated, slowly drifting vortices in the Great Red Spot band (figure 1). Table 1 lists their average properties for different time constants and forcing algorithms. In accordance with previous works on vortex development longer lifetimes and slower drift velocities were found for well preserved atmospheric zonal flow-structures as induced by more effective forcing [1].

Additionally multiple short-lived smaller eddies developed at the latitudes of Jupiter's White Ovals (figure 1). During the S2 run, these smaller anticyclones occasionally rose into the Great Red Spot band and were swallowed by the singular vortex. In accordance with observed swallowing events on Jupiter the merging did not increase the vorticity or spatial extension of the vortex. The significantly longer lifetime of this vortex in comparison to those created in other forcing cases might also be related to enhanced sustainment by these indigestions.

Vortex generation took place at about 37° S regardless of the later location of the eddy. In the initial zonal wind profile this latitude corresponds to the shear zone between a westward and eastward jet.

The singular anticyclones were comparable to the Great Red Spot in size and vertical structure as well, featuring a cold core area down to ≈ 0.04 bar followed by a warm anomaly down to ≈ 0.5 bar [4].

Using OPUS' cloud scheme we were able to investigate the ammonia cloud

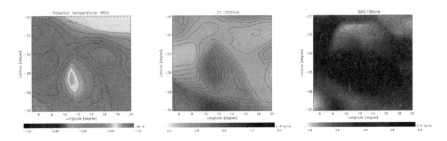

Fig. 2. Ammonia clouds of the Southern hemisphere anticyclones. a) Map of Ertel's potential vorticity at $\theta = 243$ K b) NH_3-vapor mixing ratio at 0.07 bar c) NH_3-ice mixing ratio at 0.5 bar

structure of the Southern hemisphere eddies. The vortices showed clear signatures in both ammonia ice and ammonia vapor mixing ratios. They were characterized by a dense stratospheric ammonia vapor layer (figure 2(b)), a tropospheric layer of high ammonia ice content (figure 2(c)) and a region of anomalous thin vapor below 700 bar. The occurrence of stratospheric haze, dense tropospheric cloud layer and thin deeper layer is in correspondence with the three layer model deduced from observational results [5].

Although instabilities occurred in the Northern hemisphere jets our model did not generated stable eddies at these latitudes. This is in accordance with the observed state on Jupiter, where the Northern hemisphere possesses no equivalent to the Great Red Spot. However the Northern hemisphere zonal profiles were only restrictively realistic as the 23° N high-speed westward jet was found to be extremely unstable in all runs.

It is envisioned that further improvements of OPUS (in particular the inclusion of moist circulations) will lead to a more realistic representation of Jovian turbulence.

References

1. R. K. Achterberg and A. P. Ingersoll: J. Atm. Sci. **51**, 541-562 (1994)
2. R. K. Achterberg and A. P. Ingersoll: J. Atm. Sci. **46**, 2448-2462 (1989)
3. F. M. Flasar, V. G. Kunde, R. K. Achterberg et al.: Nature **427**, 132-135 (2004)
4. F. M. Flasar, B. J. Conrath, J. A. Pirraglia et al.: J. Geophys. Res. **86**, 8759-8767 (1981)
5. A. A. Simon-Miller, P. J. Gierasch, R. F. Beebe et al.:Icarus. **158**, 249-266 (2002)
6. A. A. Simon: Icarus **144**, 29-39 (1999)
7. Y. H. Yamazaki, D. R. Skeet and P. L. Read: Planetary and Space Science **52**, 423-445 (2004)

Vorticity and divergence spectra in the upper troposphere and lower stratosphere

Erik Lindborg

Department of Mechanics, KTH. SE-100 44 Stockholm `erikl@mech.kth.se`

We show that the horizontal two-point correlations of vertical vorticity and horizontal velocity divergence can be constructed from previously measured velocity structure functions in the upper troposphere and lower stratosphere. For the two-point vertical vorticity correlation, Q_{zz}, we derive the relation

$$Q_{zz}(\rho, z) = -\frac{1}{2}\left[\frac{1}{\rho}\frac{\partial D_{\rho\rho}}{\partial \rho} - \frac{1}{\rho^2}\frac{\partial}{\partial \rho}\left(\rho^2\frac{\partial D_{\phi\phi}}{\partial \rho}\right)\right].\tag{1}$$

where $D_{\rho\rho}$ and $D_{\phi\phi}$ are the structure functions including the radial and azimuthal velocity components respectively, with the vertical defining the z-axis. For the two point correlation of horizontal divergence of horizontal velocity, P, we derive the corresponding relation

$$P(\rho, z) = -\frac{1}{2}\left[\frac{1}{\rho}\frac{\partial D_{\phi\phi}}{\partial \rho} - \frac{1}{\rho^2}\frac{\partial}{\partial \rho}\left(\rho^2\frac{\partial D_{\rho\rho}}{\partial \rho}\right)\right].\tag{2}$$

In figure 1, we see a plot of the Q_{zz} and P calculated from the structure functions calculated by Lindborg [1] from aircraft data. In the mesoscale range, up to separations of about 100 km P and Q_{zz} have the form $\rho^{-4/3}$. At larger, synoptic scales, P is close to zero while Q displays a logarithmic dependence. This synoptic functional dependence is consistent with the theory of quasi-geostrophich turbulence as formulated by Charney [2].

We also demonstrate how the spectra of Q_{zz} and P can be calculated. In figure 2 we see a plot of the spectra of The vorticity spectrum has a minimum around $k = 10^{-2}$ cpkm (cycle per kilometers) corresponding to wave lengths of 100 km. For smaller wave numbers it displays a k^{-1}-range and for higher wave numbers, corresponding to mesoscale motions, it grows as $k^{1/3}$. The horizontal divergence spectrum is of the same order of magnitude as the vorticity spectrum in the mesoscale range and show similar inertial range scaling. We argue that these results show that the mesoscale motions are not dominated by internal gravity waves. Instead, we suggest that the dynamic origin of the $k^{1/3}$-range is stratified turbulence. However, in contrast to Lilly [3], we argue

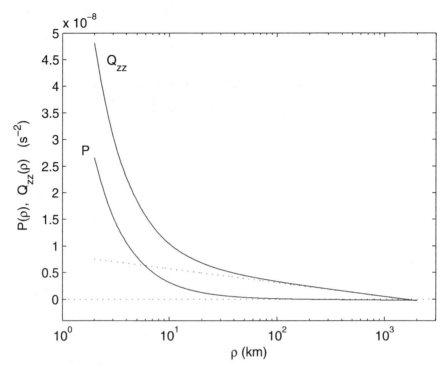

Fig. 1. Two-point correlation functions of vertical vorticity and horizontal divergence for separation between 2 and 2000 km, in the upper troposphere and lower stratosphere. The dotted lines are the zero-line and the logarithmic part of Q_{zz}: $-q_2 \log \rho + q_3$.

that stratified turbulence is not a phenomenon associated with an upscale energy cascade, but with a downscale energy cascade.

The results will be compared with the results from an extensive numerical study of stratified turbulence which recently has been performed by Lindborg & Brethouwer [4].

References

1. E. Lindborg: J. Fluid. Mech. **388**, 259-288 (1999)
2. J.G. Charney: J. Atmos. Sci **28**, 1087-1095 (1971)
3. D.K. Lilly, D.K.: J. Atmos. Sci. **40**, 749-761 (1983)
4. E.Lindborg & G. Brethouwer: J. Fluid. Mech. Submitted

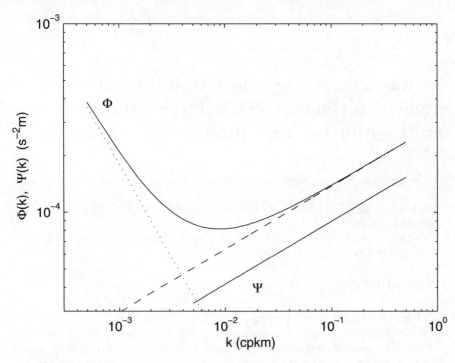

Fig. 2. The spectra of vertical vorticity, Φ, and horizontal divergence, Ψ, in the upper troposphere and lower stratosphere. The dashed line represents $\sim k^{1/3}$ and the dotted line represents $\sim k^{-1}$. k is here measured in cpkm, which means that $1/k$ is the corresponding wave length measured in km.

On the accuracy of velocity and velocity gradient turbulence statistics measured with multi-sensor hot-wire probes

P.V. Vukoslavčević[1], N. Beratlis[2], E. Balaras[2], and J.M. Wallace[2]

[1] Masinski Fakultet, University of Montenegro, 81000 Podgorica, Montenegro.
petav@cg.ac.yu
[2] University of Maryland, Dept. Mech. Eng., College Park, MD, 20742, USA.
balaras@umd.edu

1 Introduction

This paper examines the accuracy of important velocity and velocity gradient based statistical properties of turbulence measured with multi-sensor hot-wire probes. For example, to determine the vorticity vector and the strain rate tensor it is necessary to simultaneously measure the instantaneous velocity gradient tensor. To do this with hot-wire anemometry requires at least three arrays of three sensors each that are separated in the y and z directions. To provide some redundancy and more accuracy, sometimes three arrays of four sensors are used for a total of 12 sensors. Determination of the three streamwise gradients is possible using Taylor's hypothesis [1]. Several investigators made such measurements in turbulent boundary layers with nine-sensor probes [2, 3, 4]. Others used twelve-sensor probes in turbulent boundary layers, grid flows and wakes [5, 6, 7].

To date however, little is known about how the accuracy of the statistics measured with these probes depends on their spatial resolution. To address this question we have used a highly resolved turbulent channel flow Direct Numerical Simulation (DNS) to simulate the responses of such probes for various sensor separations.

2 Twelve-sensor probe operation and simulation

A sketch and photo of the twelve sensor probe of Vukoslavčević and Wallace [7] is shown in fig. 1. The characteristic probe dimensions are the array separations, d_y and d_z, and the prongs separation, h.

The velocity components cooling the sensors can be expressed in terms of the components at the geometrical center of the probe and of the gradients

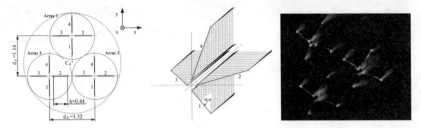

Fig. 1. Sketch of the front view of twelve-sensor probe, schematic view of one of its four-sensor arrays and a photo of probe. Typical dimensions in mm.

in the cross-stream directions. Using these velocity components in Jogensen's expression for the effective velocity, an expression for the response of the sensors to the flow is determined. This represents 12 nonlinear algebraic equations with three velocity components and six cross-stream gradient unknowns. The coefficients of this equation are found from a calibration procedure. In a physical experiment the effective velocity can be found from the voltage drop across each sensor using King's Law or from a polynomial fit of the data. An iterative algorithm is then used to solve the system of equations [7].

To investigate spatial resolution effects on such measurements, we simulated probe sensors as points on the computational grid of a turbulent channel flow DNS at $R_\tau = 200$. To have the highest possible resolution we used a minimal domain size that permitted the existence of one low and one high speed streak (see e.g. [8]). A uniform grid of approximately 32 million points was used in all directions with $\Delta x^+ = \Delta y^+ = \Delta z^+ = 1$. At the channel wall and centerline, one viscous length is about 0.5 and 2.3 Kolmogorov lengths, respectively. Using calibration coefficients from a physical experiment and the effective cooling velocities from the simulation, the sensor response equations were solved for the 'virtual' probe. The resulting statistics for several array separations were compared to those from the DNS.

3 Results

In Fig. 2a the distributions of the rms values of the three components of the velocity fluctuations are shown. The results from virtual probes with sensor array separations of $d^+ = 2, 4, 8$ and 12 are compared to the DNS data. The laboratory measurements of Ong & Wallace [4] using a nine-sensor probe in a turbulent boundary layer with $R_\tau \approx 540$ are also compared. In these laboratory measurements, the array separations of the probe were $d^+ = 6.9$, corresponding to about 7 Kolmogorov lengths at $y^+ = 15$. The peak of the streamwise rms value is attenuated approximately 15% for the virtual probe with a sensor separation of $d^+ = 12$. There is some attenuation of the spanwise rms values with increasing array separations but very little of the wall-normal values. In Fig. 2b the distributions of the rms values of the three components

590   Vukoslavčević et al.

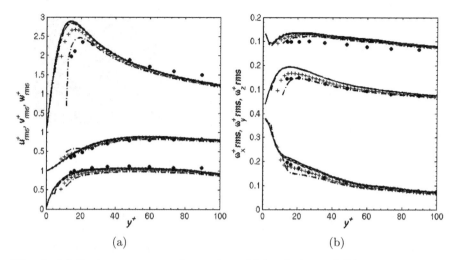

(a)   (b)

Fig. 2. (a) Root mean square fluctuation of (a) velocity and (b) vorticity. ——— DNS; • boundary layer experiment of [4]; ········ $d^+ = 2$; --- $d^+ = 4$; + $d^+ = 8$; —·— $d^+ = 12$.

of the vorticity fluctuations are shown. The attenuation here is greatest for the wall-normal and spanwise components.

4 Conclusions

Attenuation of the velocity and vorticity component rms measurements with twelve-sensor probes that have array separations of $d^+ \leq 4$ are negligible. There is a small attenuation for $d^+ = 8$ which becomes significant for array separations for $d^+ = 12$.

References

1. G.I. Taylor: Proc. Roy. Soc. London **164**, 15 (1938).
2. P. Vukoslavčević, J. M. Wallace & J.-L. Balint: J. Fluid Mech. **228**, 25 (1991).
3. A. Honkan, & Y. Andreopoulos: J. Fluid Mech. **350**, 29 (1997).
4. L. Ong & J.M. Wallace: J. Fluid Mech. **367**, 291 (1998).
5. A. Tsinober, E. Kit, & T. Dracos: J. Fluid Mech. **242**, 169 (1992).
6. B. Marasli, P. Nguyen, & J. M. Wallace: Exp. Fluids **15**, 209 (1993).
7. P. Vukoslavčević, & J. M. Wallace: Meas. Sci. Technol. **7**, 1451 (1996).
8. Jimenez, J. & Moin, P. J. Fluid Mech. **225**, 213-240 (1998).

A stochastic SGS model with application to turbulent channel flow with a passive scalar

Linus Marstorp, Geert Brethouwer and Arne V. Johansson

Linné Flow Centre, KTH Mechanics, Sweden

Summary. A new computationally inexpensive stochastic Smagorinsky model which allows for backscatter is proposed. The inclusion of stochastic backscatter is shown to improve some statistics of the passive scalar field in LES of fully developed channel flow.

Introduction

The interaction between the resolved scales and subgrid scales give rise to random fluctuations in the SGS stress (Leslie and Quarini [3] Kraichnan [2]). Consequently, the subgrid-scale stress and the subgrid-scale scalar flux extracted from DNS contain stochastic noise that cannot be modelled by a deterministic subgrid-scale model. The purpose of this study is to continue the work by Marstorp et al. [4] and develop a computationally cheap stochastic model for the subgrid scales of a passive scalar field and validate this new model in turbulent channel flow. The idea behind the present stochastic subgrid model is to model the random influence of the subgrid stress and flux by a stochastic process with the aim to improve the description of the smallest resolved scales. We make use of the solution to the Langevin equation

$$dX(x,t) = aX(x,t)dt + b\sqrt{2a}dW(x) \qquad (1)$$

where a and b are constants and $dW(x)$ are spatially and temporary independent random numbers with the normal distribution $N(0, \sqrt{dt})$. The solution $X(x,t)$ to (1) is a stationary process with zero mean, $E[X(x_0,t)] = 0$, and constant variance, $V[X(x_0,t)] = b^2$. The time scale of the process can be characterised by the decay rate of the correlation $E[X(x_0,t)X(x_0,t_0+t)]/V_X = exp(-at)$. It follows that the time scale of the process, $\tau_X = 1/a$, decreases with increasing values of a. Our stochastic model is based on the Smagorinsky model in which the eddy viscosity is constructed from the filter scale Δ and a velocity scale $\Delta|\tilde{S}_{ij}|$. We modify the velocity scale to include a stochastic part induced by the random SGS motions $\Delta(1+X)|\tilde{S}_{ij}|$ resulting in a stochastic eddy viscosity

$$\nu_T = C_s^2(1 + X(x,t))\Delta^2|\tilde{S}_{ij}| \tag{2}$$

where C_s is the Smagorinsky constant. The eddy diffusivity, κ_T, for the subgrid-scale scalar flux in the present LES is modelled using a constant $Pr_T = 0.35$ and thus, κ_T also contains stochastic noise. A more extensive description of the model can be found in Marstorp et al. [4].

Simulations and results

LES of turbulent channel flow at the wall friction Reynolds number $Re_\tau = 265$ represented on $64 \times 97 \times 48$ grid-points was performed. The passive scalar is constant at the walls with a higher temperature at the upper wall and the Prandtl number is $Pr = 0.71$. The model parameters are chosen as $a = \langle|\tilde{S}_{ij}|^3\rangle^{1/3}$ and $b = 2$, which implies a significant amount of backscatter. The results of the LES with the stochastic model, using $C_s = 0.14$ and van Driest damping near the wall, are compared to LES using the standard Smagorinsky model with the same near wall damping.

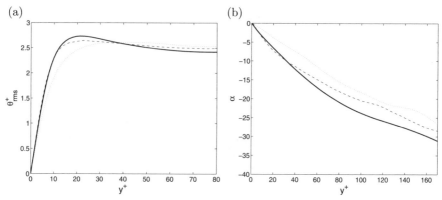

Fig. 1. (a): $\sqrt{\langle\theta'^2\rangle}^+$ (b): $\alpha = \tan^{-1}(\langle v'\theta'\rangle/\langle u'\theta'\rangle)$ *Stochastic model, dashed line; Smagorinsky model, dotted line; DNS, thick solid line*

The scalar field is affected by the inclusion of stochastic backscatter. Figure 1a shows that the scalar variance is larger near the wall and that the near wall peak is more pronounced than with the standard eddy diffusivity model. Moreover, the direction of the scalar flux, $\alpha = \tan^{-1}(\langle v'\theta'\rangle/\langle u'\theta'\rangle)$, is improved by the inclusion of the stochastic term, see figure 1b. This improvement is likely due to the improved description of the anisotropy of the velocity fluctuations, see Marstorp et al. [4]. Figure 2 shows an instantaneous slice of the fluctuating scalar. We see that the length scale of the near wall structures is reduced and that the near wall scalar fluctuations are more intense. The subgrid-scale statistics are also affected by the stochastic backscatter. The variance of the scalar variance SGS dissipation is increased and its length scale is reduced by the stochastic backscatter.

(a)

(b)

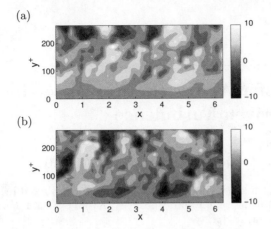

Fig. 2. *Fluctuating scalar normalised in wall units. (a): Smagorinsky model. (b): Stochastic model*

Conclusions and outlook

Because of the controllable timescale and variance of X the new model provides for significant backscatter without leading to numerical instabilities. The stochastic backscatter improves both grid-scale and SGS statistics. Further investigations are needed to include a more detailed analysis of the implications for the scalar field. The effect of system rotation on the scalar mixing will also be investigated, because it has been observed that rotation has a large influence on scalar transport in shear flows [1]. Accurate descriptions of scalar variance and the fluctuations of the dissipation are important for turbulent combustion [5]. The stochastic model may here offer an interesting approach to this challenging area for LES-computations.

References

1. Brethouwer G. The effect of rotation on rapidly sheared homogeneous turbulence and passive scalar transport. Linear theory and direct numerical simulations, *J. Fluid Mech.* 542:305- 342, 2005
2. Kraichnan,R.H. Eddy viscosity in two and three dimensions. *J. Atmos. Sci* 33:1521-1536, 1976
3. Leslie D.C., Quarini G.L. The application of turbulence theory to the formulation of subgrid modelling procedures. *J. Fluid Mech.* 91:65-91, 1979
4. Marstorp L, Brethouwer G, Johansson AV. A stochastic SGS model with application to turbulent flows and scalar mixing. To appear in *Phys. of Fluids*, 2007
5. Pitsch H. Large-Eddy Simulation of turbulent combustion. *Annual rev. of Fluid Mech.* 38:453-482, 2006

Heat Transfer Across the Air-Water Interface in Wind-Driven Turbulence

Shuhei OHTSUBO, Kenji TANNO and Satoru KOMORI

Department of Mechanical Engineering and Science and Advanced Institute of Fluid Science and Engineering, Kyoto University, Kyoto 606-8501 Japan
komori@mech.kyoto-u.ac.jp

1 Introduction

Global and local climate changes have been predicted using coupled ocean-atmosphere models in which heat and mass transfer between atmosphere and oceans plays an important role. The heat and mass transfer across the air-sea interface has not been well understood, while the heat and mass fluxes have been assumed to be proportional to wind-speed over the air-sea interface in the climate models[1]. However, it is found that mass transfer across the sheared air-water interface is controlled by organized surface-renewal events in turbulent water flow below the interface and the mass transfer coefficient is not correlated only by wind-speed[2]. The heat flux has also been determined using a bulk method based on the proportionality with wind-speed, but the proportional relationship between heat transfer and turbulent motion has not been fully investigated in wind-driven turbulence. Therefore, the purpose of this study is to clarify both the heat transfer mechanism across the air-water interface in wind-driven turbulence by measuring the heat transfer coefficient on the water side and turbulent motions near the air-water interface.

2 Experiments

Figure 1 shows the schematic of a wind-wave tank. To keep the water temperature constant, water circulation system with an electric heater was used. The air-water bulk mean temperature difference was fixed to around 15K during the measurements. Instantaneous vertical velocity and temperature in the water flow were simultaneously measured using a laser-Doppler velocimetry(LDV) and a cold film I-probe operated by a constant current temperature bridge. The measurements were carried out in the free-stream wind-speed range of U_∞=3.1~17.3m/s.

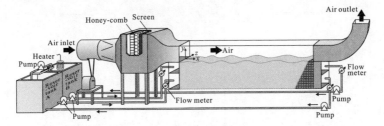

Fig. 1. Schematic of a wind-wave tank.

3 Results and discussion

Figure 2 shows the vertical distribution of the vertical turbulent heat flux $\overline{v\theta}$ at the free-stream wind-speed of $U_\infty=9.6\text{m/s}$. The $\overline{v\theta}$ increases as the distance approaches to the air-water interface and it reaches the maximum value just below the interface. This peak value of $\overline{v\theta}$ equals to the total heat flux across the air-water interface[3]. Therefore, the heat transfer coefficient on the water side h_L was estimated by

$$h_L = \frac{\overline{v\theta}|_{max}}{T_L - T_i}. \tag{1}$$

Here T_L is the water bulk mean temperature and T_i is the surface mean temperature. Figure 3 shows the heat transfer coefficient on the water side h_L against the wind-speed U_∞. The h_L increases with U_∞ but it shows a kink in the middle wind-speed region of $8\text{m/s}< U_\infty <10.5\text{m/s}$. This suggests that the conventional assumption that the heat flux across the air-sea interface is proportional to wind-speed can not be acceptable. To clarify the relationship between heat transfer and turbulence structure near the air-water interface, the frequency of the appearance of surface-renewal events was estimated by applying a variable-interval time averaging(VITA) technique to time records of the vertical turbulent heat flux $v\theta$. Figure 4 shows the relation between h_L and the frequency of the appearance of surface-renewal events f_S. The h_L increases in proportion to f_S in the whole wind-speed region. The result suggests that heat transfer across the sheared air-water interface is dominated by surface-renewal events near the air-water interface in the water flow.

References

1. T. V. Blanc: Variation of bulk-derived surface flux, stability, and roughness results due to the use of different transfer coefficient schemes. In: *J. Phys. Oceanogr.*, **15** (1985), pp. 650-669.

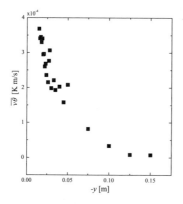

Fig. 2. Vertical distribution of the vertical turbulent heat flux $\overline{v\theta}$ at the free-stream windspeed of $U_\infty = 9.6\text{m/s}$.

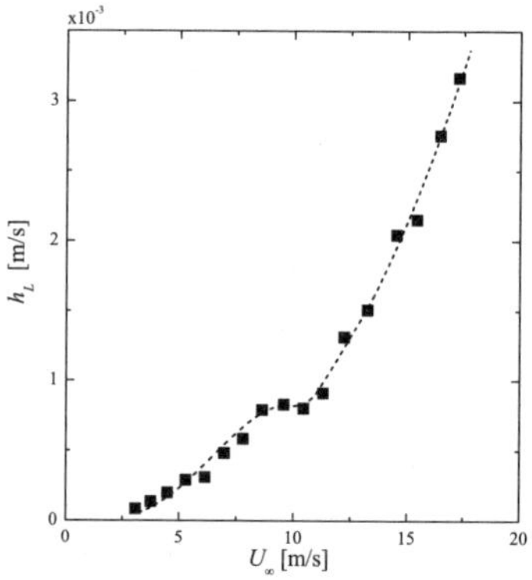

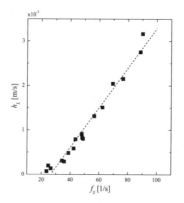

Fig. 3. Heat transfer coefficient on the water side h_L versus U_∞.

Fig. 4. Heat transfer coefficient on the water side h_L versus the frequency of the appearance of surface-renewal events f_S.

2. S. Komori, R. Nagaosa & Y. Murakami: Turbulence structure and mass transfer across a sheared air-water interface in wind-driven turbulence. In: *J. Fluid Mech.*, **249** (1993), pp. 161-183.
3. B. M. Howe, A. J. Chambers, S. P. Klotz, T. K. Cheung, & R. L. Street: Comparison of profiles and fluxes of heat and momentum above and below an air-water interface. In: *Trans. ASME.*, **104** (1982), pp. 34-39.

Attenuation of turbulent flow separation on a wavy wall by a compliant surface

Hui Zhang, Naoki Yoshitake and Yoshimichi Hagiwara

Department of Mechanical and System Technology, Graduate School of Science and Technology, Kyoto Institute of Technology, Matsugasaki, Sakyo-ku, Kyoto, 606-8585 Japan yoshi@kit.ac.jp

1 Introduction

The reduction of friction drag by swimming dolphins has been focused on recently not only in marine biology and bio-mimicry but also in fluids engineering and naval engineering. This is because of saving fuel consumption by boats, ocean liners and submarines. The present authors have paid attention to the skin folds among several possible drag-reduction mechanisms [1], because we took photos of skin folds of bottlenose dolphins, which swam swifty.

In the present study, we conduct an experiment for turbulent water flow over an angled wavy wall. We measure velocity profiles near the wall and estimate the wall shear stress from the profiles. Two types of surfaces are tested; bare aluminum and silicon rubber sheet. The silicon-rubber flat wall is also used as a reference.

2 Experimental procedures

Figure 1 shows the apparatus. Water in the constant-head upper tank (1) was introduced into chamber (3) through a valve. The water in the chamber flowed into the open channel of 2000 mm in length and 270 mm in width. The height of the perforated weir (6) was adjusted in order to control the water depth and velocity distribution. The Reynolds number based on the bulk mean velocity and the water depth (=35 mm) was 7200 and 8800.

The wavy wall consists of a thin silicon-rubber sheet, a thin adhesive film and an aluminum wavy base. The wave amplitude and wavelength are 0.75 mm and 20 mm, respectively. The angle between the channel axis and the ridgeline of the wavy wall was set at 80 degrees.

The flow was visualized with fluorescent particles whose diameter was smaller than 0.10 mm. Figure 2 indicates the optic arrangement and image capturing system. The laser light was expanded with cylindrical lenses to obtain a light sheet. The light sheet illuminated the flow from above.

Fluorescence induced from the tracer particles was captured by a color CMOS camera through an optical filter located in front of the camera. The optical axis of the camera was set parallel to the ridgelines of the wavy wall so that the valley of the wall illuminated by the laser light sheet was completely observed. In order to reduce the effect of light refraction at the channel side wall in front of the camera, a prism was attached to the surface of the side wall. The PTV technique based on the velocity gradient tensor method [2] was applied for obtaining velocity vectors from the images.

The total drag was estimated from the measured strains of thin metal beams, which supported the walls, with strain gauges.

3 Results and discussion

Figure 3 shows the mean velocity distribution in the uphill and downhill regions near the valley. The velocity in the case of the aluminum wall is very low near the surface. Thus, the wall shear stress is low in the valley region. This is due to the flow separation. On the other hand, the velocity in the case of the compliant wall is high near the surface. This shows the flow separation is attenuated by the compliant wall. The wall shear stress in the case of the aluminum wall in the other regions is higher than that in the case of the compliant wall.

Figure 4 depicts the turbulence intensities. The intensities in the case of the compliant wall are slightly lower than those in the case of the aluminum wall. This indicates that the compliant wall attenuates turbulence.

It was found that the total drag for the silicon-rubber wavy wall was higher than that for the silicon-rubber flat wall. However, the increasing rate was much lower than that estimated from the DNS results [3] for non-angled rigid wavy wall.

4 Conclusions

The measurements were conducted on the velocity and drag for the turbulent flow over the angled wavy wall in the open channel. The main conclusions obtained are as follows:

1. The silicon rubber attenuates slightly flow separation in the valley of the wavy wall. The silicon rubber also attennuates the near-wall turbulence.
2. The increasing rate of the total drag is lower than that estimated for non-angled rigid wavy wall.

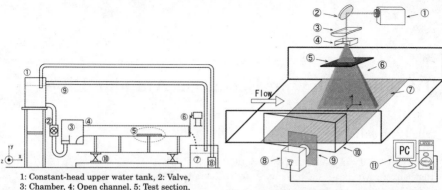

1: Constant-head upper water tank, 2: Valve,
3: Chamber, 4: Open channel, 5: Test section,
6: Movable weir, 7: Lower water tank,
8: Immersed pump, 9: Overflow outlet, 10: Jack

1: Laser light source, 2: Mirror, 3: Plano-convex lens,
4: Plano-concave lens, 5: Slit, 6: Laser light sheet,
7: Angled wavy wall, 8: CMOS camera, 9: Optical
filter 10: Prism, 11: PC

Fig. 1. Apparatus

Fig. 2. Optics arrangement and image-capturing system

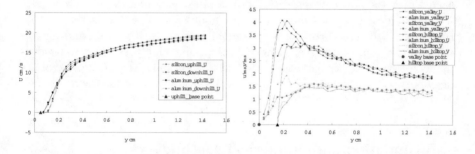

Fig. 3. Mean velocity distribution

Fig. 4. Turbulence intensities

References

1. H. Zhang et al.: Reduction of friction drag by angled wavy silicon-rubber wall as a model of dolphin skin. In: *Proc. 3rd Int. Symp. on Aero Aqua Biomechanisms,* (2006) Paper No. S43, 8p.
2. M. Ishikawa et al.: Experiments in Fluids, **29**, 519 (2000)
3. H. A. Tuan et al.: Immersed boundary method for simulating turbulent flow over a wavy channel. In: *Book of Extended Abstracts of Whither Turbulence Prediction and Control,* (2006) 116

Turbulent Flow in Eccentric Annular Pipe

Nikolay Nikitin

SES, University of Southampton, Southampton, UK;
Institute of Mechanics, Moscow State University, Moscow, Russia.
e-mail: nvnikitin@mail.ru

The results of direct numerical simulations of turbulent flow through a pipe with an eccentric annular cross-section are presented. The simulations were performed at Re = 4000 (where the Reynolds number Re is based on the bulk velocity and hydraulic diameter) using an energy-conserving finite-difference method. It was found that two types of flow develop depending on the geometric parameters. In the flow of first type, turbulent fluctuations are observed over the entire cross-section of the pipe, including the narrowest gap, where the local Reynolds number is only about 500. The flow in the another type is divided into turbulent and laminar regions (in the wide and narrow parts of the gap, respectively). Eccentric annular pipe presents a convenient model for study inhomogeneous turbulent flow, peculiarities of the flow in the neighborhood of the laminar/turbulent interface and the related entrainment phenomenon.

1 Formulation and Numerical Method

Consider a pressure-gradient driven fluid flow along the gap between two parallel but eccentric cylinders of different radii R_i and $R_o > R_i$ and eccentricity $e < R_o - R_i$ (see the sketch in Fig. 1(a)).

The Navier-Stokes equations for the incompressible fluid with no-slip boundary conditions on the rigid walls and periodic condition in the streamwise direction z are solved using curvilinear bipolar coordinates (ξ, η) in the cross-sectional plane of the pipe:

$$x = -H(\xi, \eta) \sinh \xi, \quad y = H(\xi, \eta) \sin \eta, \quad H(\xi, \eta) = \frac{c}{\cosh \xi - \cos \eta}, \quad (1)$$

where $c = [(R_o^2 - R_i^2 - e^2)^2/4e^2 - R_i^2]^{1/2}$. The new coordinates vary in the range $\eta \in [0, 2\pi]$ and $\xi \in [\xi_i, \xi_o]$, where $\xi_i = \ln[(1 + c^2/R_i^2)^{1/2} - c/R_i]$ and $\xi_o = \ln[(1 + c^2/R_o^2)^{1/2} - c/R_o]$. The surface $\xi = \xi_i$ coincides with the surface of the inner cylinder, and $\xi = \xi_o$, with the surface of the outer cylinder. The plane $\eta = 0$ intersects the pipe along the wide part of the gap, and the plane $\eta = \pi$, along its narrow part.

Numerical solutions are performed using finite-difference energy-conserving method for arbitrary orthogonal curvilinear coordinates [1] with overall 3-rd order-accurate semi-implicit Runge-Kutta scheme for time advancement [2]. Implementation of the method to bipolar coordinates is given in [3]. The discrete Poisson equation for the pressure after FFT in z-direction is solved iteratively using a combination of the cyclic reduction and the conjugate-gradient methods (for details see [3]).

2 Results

In what follows, all the linear sizes are scaled by the mean gap $\delta = R_o - R_i$ and the velocities are scaled by the bulk velocity U_b which was held constant during the simulations. Here are presented the results obtained in two runs with the same $\mathrm{Re} = 2\delta U_b/\nu = 4000$ (here, 2δ is hydraulic diameter, ν is the fluid viscosity) but with different geometric parameters of the pipe: $(R_i, R_o, e) = (1, 2, 0.5)$ (case A) and $(R_i, R_o, e) = (2, 3, 0.5)$ (case B). In each case, the streamwise period was 2π, and computational mesh consisted of $64 \times 256 \times 64$ nodes in the radial (ξ), angle (η) and axial (z) directions, respectively. The mesh was stretched both in the radial and angle directions to provide the grid clustering in the dynamically most active regions, namely in the near-wall regions and in the wide-gap part of the pipe (see Fig. 1(a)).

Qualitatively different turbulent flow regimes were obtained in two considered cases. In the first configuration turbulence occupies the entire cross-section, while in pipe B the region of turbulent flow coexists with the region of laminar flow (see Fig. 1(b,c)).

The flows in a statistically steady turbulent regime are characterized by a considerable increase in the mean shear stress τ_w as compared with the corresponding laminar flows. The resistance coefficient $C_d = 2\tau_w/(\rho U_b^2)$ is increased by a factor of 2.4 (from 4.41×10^{-3} to 1.05×10^{-2} in case A and by a factor of 2.0 (from 4.38×10^{-3} to 8.91×10^{-3} in case B, respectively. The velocity profiles in turbulent regime become more homogeneous in the core of the flow both in the radial and angle directions with much steeper near-wall gradients. The maximal mean velocity reduces from 2.37 to 1.42 in pipe A and from 2.41 to 1.43 in pipe B.

The largest velocity fluctuations in turbulent flow are observed in the near-wall region of the wide cross-section, where they reach 0.23 in pipe A and 0.25 in pipe B (see Fig. 1(b,c)). In pipe A significant velocity fluctuations (about 0.12) are observed even in the narrowest part of cross-section, while in pipe B they are virtually missing (less than 0.02). Thus, the region of turbulent flow coexists with the region of laminar flow in this case. It should be noted, that the local Reynolds number, based on the local gap width and velocity averaged over this local gap varies from about 3600 in the wide part of the pipe to about 500 in the narrow part in both cases.

The existence of considerable turbulent fluctuations in the narrow gap of pipe A, where the local Reynolds number is less than 500 suggests that the turbulent structures transferring turbulence from wide to narrow part of cross-section are global in character. This interpretation is supported by the shape of the secondary motions (secondary motions of Prandtl's 2-nd kind) in a cross-sectional plane. The secondary motion in pipe A exhibits a pair of pronounced counter-rotating vortices in each half of the pipe cross-section. These vortices transfer the high-momentum fluid from the wide to the narrow part of the pipe along the midline of the gap and return the slow-momentum fluid along the pipe walls. The maximum velocity in secondary motion is about 0.011 and is observed near the inner wall. The secondary motion in pipe B is weakly pronounced (with a maximum velocity of 0.005) and have no clear structure.

Acknowledgments

This work was supported by Russian Foundation for Basic Research, project no. 05-01-00607.

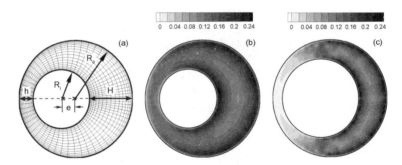

Fig. 1. (a), sketch of the eccentric pipe's cross-section and computational mesh (every 4-th line in each direction is shown). Contours of rms velocity fluctuations at Re = 4000 and $e = 0.5$: (b), fully turbulent flow $((R_i, R_o) = (1, 2))$; (c), partly turbulent flow $((R_i, R_o) = (2, 3))$.

References

1. N. Nikitin: J. Comput. Phys. **217**, 759 (2006)
2. N. Nikitin: Int. J. Numer. Meth. Fluids **51**, 221 (2006)
3. N. Nikitin: Comput. Math. Math. Phys. **46**, 489 (2006)

Quasi-normal hypothesis revised

Alberto Maurizi

CNR-ISAC, via Gobetti 101, I-40129 Bologna, Italy `a.maurizi@isac.cnr.it`

Introduction

Two-points statistics of turbulence velocity are known to be markedly non-Gaussian at least for separations in the inertial subrange (see, e.g., Vincent and Meneguzzi, 1991), where third-order moments of velocity differences are responsible for energy transfer subrange Monin and Yaglom (1975). On the other hand, single point statistics in non-homogeneous, non-isotropic flows, have long been recognised to have non-normal distributions (Champagne et al., 1976, among others).

The so called "quasi-normal" (QN) hypothesis formulated by Millionshchikov Millionshchikov (1941) for homogeneous hisotropic turbulence, assumes that fourth-order cumulants are zero, regardless of any other quantity. This assumption is frequently used even in different conditions than that for inertial subrange separations. If separation is large compared to the Eulerian scale of turbulence, then the hypothesis is trivially verified. Nonetheless, if separation is zero, QN turns to be an hypothesis on one-point statistics that, even very recently, have been taken as a reference (see Losch, 2004, for instance).

Although this hypothesis is seems the least biased when no information on high order moments is available, it will be shown that some results from purely statistical ground push to a riconsider it in a different perspective. Furtermore, new experimental measurements will be shown to give further insight.

Results from theory of statistics

Consider a single centred continuous variate. If variance is used as a normal measure for moments, the third- and fourth-order cumulants turn out to be the skewness S and the kurtosis K, respectively.

One result of probability theory leads to state that the quantities K and S must satisfy the relationship (Kendall and Stuart, 1977)

$$K \geq S^2 + 1.$$
(1)

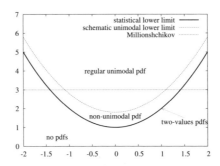

Fig. 1. The S-K space and PDF properties.

This makes the hypothesis of $K = 3$, which is consequence of the Million-shchikov hypothesis for one-point univariate, clearly non-realisable.

Another, usually disregarded, statistical fact, is that the relationship $K = S^2 + 1$ strictly holds for any two-value process. In fact, from the Gauss-Winckler inequality, it can be shown that for a unimodal symmetric probability density function (PDF) it must be $K \geq 1.8$ and that for non-symmetric PDFs ($S \neq 0$)

$$K \geq 1.8 + f(S, m),\tag{2}$$

where m is the normalised mode and the function f can be argued to be $\mathcal{O}(S^2)$, at least for small S. As two value velocity implies infinite accelerations, by extension, a highly bi-modal PDF cannot be considered as a reliable representation of turbulence statistics. This further limits the range of applicability of QN.

QN hypothesis revised

Given the parabolic structure of the S-K space, the "zero-order" approximation of K must be a function of S^2. This can be expressed defining the new set of variables $(S, \tilde{K})$ with $\tilde{K} = K(S^2 + 1)^{-1}$. This new variable incorporates the zero-order effect due to the parabolic structure of the relationship and constitutes a basis for parameterisation development.

The results shown so far can be used to formulate a more general parameterisation $\tilde{K} = \sum_{p=0}^{N} \alpha_p S^p$ where $N = 2n$ for realisability. The zero-order effect is then given by $\tilde{K} = \alpha_0$ with a loosely-defined Millionshchikov-equivalent hypothesis given by $\alpha_0 = 3$.

Figure shows data of longitudinal and vertical components of velocity a turbulent boundary layer in the S-$\tilde{K}$ space. Curves represent the QN hypothesis, the revised QN and different order of approximation for a data measured in a wind tunnel experiment specifically designed to compute high-order statistics. It is worth noting that, according to Alberghi et al. (2002), even symmetric pdf present a marked non-zero fourth-order cumulant.

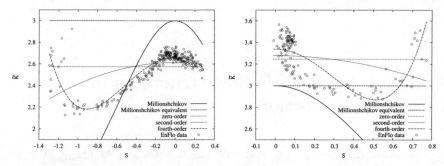

Fig. 2. $S - \tilde{K}$ representation of wind tunnel data data for longitudinal component (left panel) and component normal to the surface (right panel).

Conclusions

The idea underlying the Millionschikov hypothesis was presumably that to make the less biased choice with respect to the small amount of available information. In fact, it has been proven here that, this hypothesis is not the least biased due to the topology of the S-K space.

An alternative hypothesis based on the intimate structure of the third- and fourth-order normalised cumulants space can be formulated and prove to capture better the data behaviour. Being based on pure statistical ground, it fails to capture features that are clearly connected to dynamical features. Nevertheless dynamical theories could be expressed in terms of expansion with respect to the revised hypothesis to asses its validity.

References

Alberghi, S., A. Maurizi, and F. Tampieri, 2002: Relationship between the vertical velocity skewness and kurtosis observed during sea-breeze convection. *J. Appl. Meteorol.*, **41**, 885–889.

Champagne, F., Y. Pao, and I. Wygnanski, 1976: On the two-dimensional mixing region. *Journal of Fluid Mechanics*, **74**, 209–250.

Kendall, S. M. and A. Stuart, 1977: *The Advanced Theory of Statistics*, vol. 1, 4th ed., C. Griffin & Co., London.

Losch, M., 2004: On the validity of the millionshchikov quasi-normality hypothesis for open-ocean deep convection. *Geophysical Research Letters*, **31**.

Millionshchikov, M., 1941: Theory of homogeneous isotropic turbulence. *Dokl. Akad. Nauk SSSR*, **32**, 611–614.

Monin, A. and A. Yaglom, 1975: *Statistical fluid mechanics*, vol. II, MIT Press, Cambridge, 874 pp.

Vincent, A. and M. Meneguzzi, 1991: The spatial structure and statistical properties of homogeneous turbulence. *J. Fluid Mech.*, **225**, 1–20.

Coherent large–scale flow structures in turbulent convection

C. Resagk, R. du Puits, E. Lobutova, A. Maystrenko, and A. Thess

Dept. of Mech. Eng., Ilmenau University of Technology, 98684 Ilmenau, Germany
christian.resagk@tu-ilmenau.de

1 Introduction

Thermal convection appears in a large variety of fields like geophysics, metallurgy, heat exchange, and room air conditioning. A widely used model experiment to investigate natural thermal convection is the Rayleigh-Bénard (RB) cell, a cavity filled with air, heated from the bottom and cooled from above [1]. In the last years there have been a lot of contributions published

Fig. 1. Cylindrical experimental facility "Barrel of Ilmenau" (left) and the rectangular cell "ILKA" (right)

about experimental and theoretical investigations of temperature fields and heat transfer in RB convection, but only a few papers about the global velocity field [2][3]. Up to now there is no experimental method to analyze the properties of the full three–dimensional flow field in large–scale RB cells. Despite of the strongly turbulent behavior of the velocity and the temperature field we can find coherent structures in the boundary layers and in the bulk region of the cells.

2 Experiment

We report about experimental investigations of coherent large–scale flow structures in turbulent RB convection in a large cylindrical cell called "Barrel of Ilmenau" (BOI)(7m in diameter and 6.3m high)[4][5] and in a rectangular cell called "ILKA"(2.5m long, 0.5m wide, and 0.5m high) at Rayleigh numbers between $Ra \approx 10^7$ and $Ra \approx 10^{11}$ and at aspect ratios 2 and 5 (see figure 1). Large scale spatial flow pattern are measured by means of optical flow measurements like stereoscopic particle image velocimetry (SPIV) and 3D particle tracking velocimetry (3DPTV), whereas the temporal oscillations of the velocity and the temperature in the boundary layer are analyzed with laser Doppler velocimetry (LDV) and micro thermistor probes.

3 Results

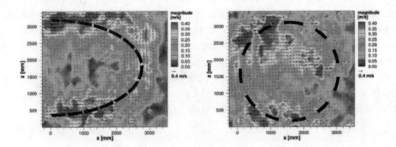

Fig. 2. SPIV velocity fields (half cross section) in the BOI at aspect ratio 2 show different coherent flow structures: (a) single roll, and (b) two roll or ring structure

In Fig. 2 a typical example of a large–scale flow structure in the cylindrical cell at aspect ratio 2 (diameter/height) is shown. At this geometry the large–scale flow pattern (wind) tends to pass from a large single roll into a two roll structure [5]. This process is very unstable and therefore the SPIV velocity vector plots demonstrate in the half cross section of the cell at one time still a single roll, but at another time a two roll pattern. Similar properties of the mean flow can be found in the smaller rectangular cell "ILKA" (see figure 3). Preliminary flow visualization measurements with laser light sheet and helium filled soap bubbles as tracer particles show also a two roll large–scale flow structure. One can see this effect also in the temperature profiles in the boundary layer under the cooling plate. If we compare the rms data from a central position with a peripheral position large fluctuations indicate a strong up–flow between two rolls and small fluctuations represent a homogenous horizontal flow under the wall, respectively [6]. For a complete 3D analysis in the cylin-

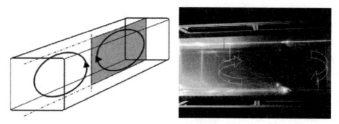

Fig. 3. Large–scale structures in the rectangular cell "ILKA" with aspect ratio 5, schematic sketch of two roll structure (left) and visualization of one roll in the half cross section (right)

drical cell the 3DPTV technique based on image processing of four cameras and tomographic reconstruction was developed and tested in a rectangular cavity of 3.4m x 3.6m x 4.0m. After detailed calibration measurements this technique has been applied to the large–scale cylindrical cell at aspect ratio 2 in order to verify the transition from the single roll to the dual roll structure measured with SPIV. As tracer particle helium filled soap bubbles with about 5mm diameter are used. The coherent large-scale flow structures which one can find in both RB cells also influences the properties of the velocity and temperature fields close to the cooling plate. The autocorrelation function of the velocity time series shows characteristic time scales depending on the aspect ratio and Rayleigh number. The periodical fluctuation of the angular velocity component can be simulated by an enhanced model of viscose fluid flow in an elliptic cavity based on a dynamical model [7].

4 Acknowledgments

This work is supported by the German Research Foundation (TH 497 and RE 1066), and by the Helmholtz Association of German Research Centers.

References

1. Castaing, B., Gunaratne, G., Heslot, F., Kadanoff, L., Libchaber, A., Thomae, S., Wu, X.-Z., Zaleski, S., Zanetti, G.: J. Fluid Mech. **204**, 1–30 (1989).
2. Sun, C., Xia, K.-Q., Tong, P.: Phys. Rev. E, **72**, 026302 (2005).
3. Eidelman, A., Elperin, T., Kleeorin, N., Markovich, A., Rogachevskii, I.: Exp. in Fluids **40**, 723–732 (2006).
4. du Puits, R., Busse, F.-H., Resagk, C., Tilgner, A., Thess, A.: J. Fluid Mech. **572**, 231–254 (2007).
5. du Puits, R., Resagk, C., Thess, A.: Phys. Rev. E, **75**, 016302 (2007).
6. Maystrenko, A., Resagk, C., Thess, A.: Phys. Rev. E, (2007), submitted.
7. Resagk, C. du Puits, R., Thess, A., Dolzhansky, F. V., Grossmann, S., Araujo, F. F., Lohse, D.: Physics of Fluids **18**, 095105 (2006).

Large-scale Behaviour of Turbulent Convection Governed by Low-dimensional Fixed-points

M. K. Verma[1,4], J. J. Niemela[2], K. Kumar[3], S. Paul[3], and D. Carati[4]

[1] Department of Physics, I I T Kanpur, Kanpur-208016, India
mkv@iitk.ac.in
[2] I. C. T. P., Strada Cosetiera 11, I-34100 Trieste, Italy
[3] Department of Physics, I I T Kharagpur, Kharagpur-721302, India
[4] Physique Statistique et Plasmas, Université Libre de Bruxelles, B-1050
Bruxelles, Belgium

A large scale flow, known as the "mean wind" [1,2] is a robust feature of turbulent convection. Recently, extensive studies of its properties have been performed in experiments using low temperature helium gas [1], where the Prandtl number, Pr, of the gas was approximately 0.7 and the Rayleigh number was in the range $10^8 - 10^{13}$. Niemela et al. [1] observed that the large-scale flow circulates in a particular direction for a period of time, and then it abruptly switches direction. The large-scale velocity (V) follows a scaling relationship with Rayleigh number (R) as $V \propto \sqrt{R}$. Similar features have been observed in many other simulations and experiments including dynamo experiments.

There are several theoretical attempts to explain the experimental findings of Niemela et al. Sreenivasan et al. [3] modelled the mean wind using thermal plumes that rise and descend. Araujo et al. [4] generalised the Lorenz equations to model the motion of plumes in Sreenivasan et al.'s theory [3].

In this paper we will show that the observed properties of the wind are similar to those observed in the Lorenz model. We argue that the large-scale modes of convective flow are captured quite well by the Lorenz model. Lorenz truncated the Fourier series expansion of velocity and temperature fields of Rayleigh Bénard (RB) convection under free-slip, and considered only three small wavenumber modes, w_{101}, θ_{101}, and θ_{002}, where the three indices denote the wavenumbers in x, y, and z directions respectively. The direction of gravity is along z direction.

The solution of the Lorenz equations are expressed in terms of $r = R/R_c$, where $R_c = 27\pi^4/4 \sim 657$ is the critical Rayleigh number at which the convection starts. For $1 < r \leq 470/19$, the property of the convective rolls can be characterized by the Lorenz fixed points ($x_c \propto \sqrt{r-1}$); in this case the rolls move either clockwise or counter-clockwise with a constant speed. After $r = 470/19$, the dynamics becomes chaotic. In the chaotic regime the velocity field fluctuates around the Lorenz fixed point, and it switches directions randomly. The time evolution of mode w_{101} is quite similar to the timeseries

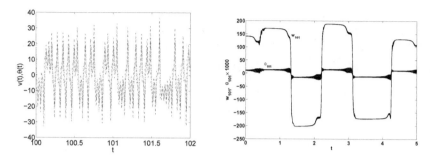

Fig. 1. Timeseries of velocity (solid) and temperature fluctuations (dashed) in (a) Lorenz model for $r = 50$, in (b) direct numerical simulation for $r = 830$.

observed in convection experiments. For $r \gg 1$ the scaling laws of the Lorenz model is quite similar to the experimental scaling law.

To explore the validity of the Lorenz equations in convective turbulence, we performed direct numerical simulation (DNS) of RB flows for *free-slip* boundary condition. Our simulation was done on 64^2 and 128^2 grids with $Pr = 6.8$ and $r = 1 - 850$ [5]. We observe that the large scale modes $w_{101}, \theta_{101}, \theta_{002}$ are constants for r up to 70, after which they become oscillatory until $r = 650$. These modes then show period doubling and quasiperiodicity for $660 < r < 750$. The simulation shows chaotic behaviour for $r = 790$ and above. In chaotic regime, the values of the large-scale modes flip signs as observed in the Lorenz model. These modes fluctuate around the mean value between two flips. Fig. 1b illustrates the chaotic flipping of the modes in 2D simulations up to $r = 850$. The chaotic behaviour of the modes has similarity with the chaos observed in the Lorenz model.

We computed the mean of the absolute values of w_{101}, θ_{101}, and θ_{002} and found them to be in the same range as the values of the Lorenz fixed point. A curve $w_{101} \sim R^{0.7}$ fits reasonably well with the numerical $w_{101} - R$ curve for $r > 500$. The value of θ_{002} saturates at -0.158, which is same as asymptotic θ_{002} of the Lorenz model. The mode θ_{101} appears to saturate in simulations, and has a different behaviour compared to the Lorenz model. Overall, the Lorenz fixed point appears to describe the large-scale modes of numerical simulations reasonably well.

We investigated the energy budget of these modes using energy transfer diagnostics [6]. Mode θ_{101} receives energy from the heat source through $\Re[u_z(101)\theta(101)]$, where $\Re$ stands for the real part. This energy is transferred to high wavenumber modes through energy cascade. In DNS this quantity is computed as $T_{tot}(101) = \Re[\mathrm{nlin}_{101}\theta_{101}]$, where nlin_{101} is the nonlinear term of θ_{101}. In Lorenz model the above transfer reduces to $T_{Lorenz}(101) =$

$\Re[2\pi u_{z101}\theta_{101}\theta_{002}]$ due to the mode truncation. When we compare these two quantities numerically, we find that $T_{tot}(101) \sim T_{Lorenz}(101)$ within 5-10%. Hence most of the energy of θ_{101} mode is going to θ_{002} mode. The energy budget of θ_{101} appears to lend credence to the low-dimensional behaviour of the experimental and numerical results.

The dynamics and flip of convective rolls can be summarized in terms of large-scale flows as follows. Mode θ_{100} oscillates around the value of the Lorenz fixed point with increasing amplitude with time. The velocity mode w_{101} is forced by θ_{101}, hence it is in approximate phase with the oscillating θ_{101}. The amplitude of these modes drops rather suddenly after sometime. After this reduction, the modes continue from their original values or their corresponding flipped values (Lorenz fixed points). The flip takes place only when θ_{101} and w_{101} cross the origin, which corresponds to stoppage of the rolls. The above transitions are abrupt, and they bear resemblances to the breakdown phenomena like earthquakes, discharges etc.

In this paper we attempt to explain some of the observed large-scale behaviour in experiments and simulations using a low-dimensional model referred to as Lorenz model. Diagnostics using direct numerical simulations appear to show that the energy transfer in large-scale modes are captured reasonably well by the Lorenz equations. We need to perform more exhaustive DNS (both 2D and 3D) to come to a definite conclusion. However the present results encourage us to investigate low-dimensional behaviour of convective flows. We note that the results here may also have bearing on geo- and solar magnetohydrodynamic flows.

Acknowledgments: We thank K. R. Sreenivasan for useful comments and suggestions.

References

1. J. J. Niemela, L. Skrbek, K. R. Sreenivasan, R. Donnaly: Turbulent Convection at Very High Rayleigh Number, Nature **404**, 837 (2000).
2. R. Krishnamurti and L. N. Howard: Large-scale Flow Generation in Turbulent Convection, Proc. Nat. Acad. Sci., **78**, 1981 (1981).
3. K. R. Sreenivasan, A. Bershadskii, and J. J. Niemela: Mean Wind and its Reversal in Thermal Convection, Phys. Rev. E **65**, 56306 (2002).
4. F. F. Araujo, S. Grossmann, and D. Lohse: Wind Reversals in Turbulent Rayleigh-Bénard Convection, Phys. Rev. Lett. **95**, 084502 (2005).
5. O. Thual: Zero-Prandtl-number convectivon, J. Fluid Mech. **240**, 229 (1990).
6. M. K. Verma, K. Kumar, and B. Kamble: Mode-to-mode energy transfers in convective patterns, Pramana, **67**, 1129, 2006.

Structure Formation in Homogeneous Rotating Turbulence

P. J. Staplehurst[1], P. A. Davidson[1] and S. B. Dalziel[2]

[1] Department of Engineering, University of Cambridge, Trumpington Street, Cambridge, CB2 1PZ, UK
[2] Department of Applied Mathematics and Theoretical Physics, University of Cambridge, Wilberforce Rd, Cambridge, CB3 0WA, UK

Since the discovery of columnar structures within a field of rotating turbulence [3], there have been relatively few laboratory experiments which have helped to explain this phenomena. We introduce a new laboratory experiment and use two-point correlations to show that these structures are formed by linear inertial wave propagation.

1 Introduction

We are interested in freely decaying rotating turbulence where inertial forces are able to create a chaotic field of vorticity before the Coriolis force becomes significant. As the turbulence decays, the presence of the background rotation produces columnar structures whose formation has traditionally be attributed to non-linear mechanisms. Both two-point closure models and computer simulations have suggested that a non-linear transfer of energy is required for the turbulence to evolve towards a two-dimensional state [1].

Nonetheless, recent laboratory experiments have shown that from an inhomogeneous cloud of turbulence, columnar structures emerge, which are created by linear inertial wave propagation [2]. In this paper, we turn our attention to the evolution of a homogeneous cloud of rotating turbulence and see that linear dynamics again play an important role.

2 Experimental Methods and Results

To investigate how the Coriolis force creates these columnar structures, approximately homogeneous, isotropic turbulence in solid body rotation was generated by lowering a grid through a rotating tank of water. The tank was 45 cm square by 60 cm deep and filled to a depth of 50 cm. To reduce the effects of surface waves, a perspex sheet was placed 5 cm below the surface of the water, giving a working depth of 45 cm. The turbulent flow was initiated

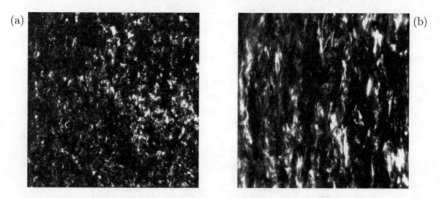

Fig. 1. Images of the flow taken after initiation of the turbulence with $\Omega = 2\text{rad/s}$. (a) $2\Omega t = 8.9$, (b) $2\Omega t = 25.6$. The images are 34cm square.

by lowering the grid with mesh size, $M = 32$ mm and 64% porosity, through the working depth at a constant speed, $U = 190\,\text{mms}^{-1}$, giving a Reynolds' number, $Re = UM/\nu$, of 6000. Fifty experimental runs were conducted for rotation rates of $1\,\text{rads}^{-1}$ and $2\,\text{rads}^{-1}$. A schematic of the apparatus is given in [4], and for brevity shall not be included here.

The turbulence was visualised by adding approximately 2 ml of Pearlescence to the water. Pearlescence consists of neutrally buoyant reflective flakes that tend to become aligned with regions of intense strain, and provide an effective way of visualising the turbulent structures. Images recorded after the grid had reached the bottom of the tank are shown in Fig. 1. At early times the turbulence is approximately isotropic (Fig. 1a), but as the flow evolves and energy is dissipated, the Coriolis force creates anisotropic columnar structures that dominate the large-scale motions (Fig. 1b).

To provide a quantitative measure of the how these structures develop along the direction of bulk rotation, two-point correlation statistics, based on the light intensity, were computed according to,

$$\Pi\left(r\right) = \langle I\left(\mathbf{x}\right) I\left(\mathbf{x} + r\mathbf{e}_z\right)\rangle / \langle I\left(\mathbf{x}\right) I\left(\mathbf{x}\right)\rangle \tag{1}$$

where I is the recorded light intensity, r is the correlation length and $\mathbf{e}_z$ is the unit vector along the rotation axis. The correlation curves for $\Omega = 2\,\text{rads}^{-1}$ are displayed in Fig. 2a. As the flow evolves, the large-scale structures become elongated and the area underneath these curves increases. If the flow is evolving according to linear dynamics and these structures are created by linear wave propagation, then the area under these curves should increase linearly with time. To test this hypothesis the solid curves in Fig. 2a, corresponding to later times when rotation is significant, have been normalised by time and plotted in Fig. 2b. Indeed, it is observed that the curves do collapse onto each other and hence the correlations are growing linearly with time. Furthermore, if the linear dynamics are creating these columnar structures, one would ex-

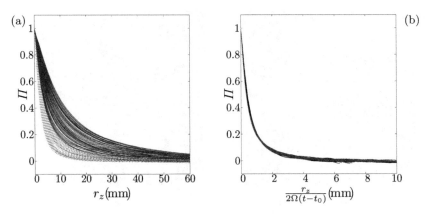

Fig. 2. Correlation curves plotted in intervals of $2\Omega t = 0.44$, t_0 is a virtual origin in time. (a) $\cdots$ $\Omega = 2\text{rads}^{-1}$, $1.5 < 2\Omega t < 8.9$. $\cdots$ $\Omega = 2\text{rads}^{-1}$, $8.9 < 2\Omega t < 25.6$. (b) — $\Omega = 1\text{rads}^{-1}$, $6.9 < 2\Omega t < 15.2$. $\cdots$ $\Omega = 2\text{rads}^{-1}$, $8.9 < 2\Omega t < 25.6$.

pect that the rate of growth is proportional to the bulk rotation rate. This is also displayed by Fig. 2b for the correlation axis has been normalised by Ω, and the results from the two experiments, with $\Omega = 1$ and $2\,\text{rads}^{-1}$, collapse on to each other.

The development of the flow during this proposed linear regime is shown in Fig. 1. Figure 1a shows the turbulent structure when linear dynamics begin to become significant, and the collapse in the two-point correlation curves is observed up to the time corresponding to Fig. 1b.

3 Conclusions

Laboratory results have shown that once the Coriolis force is dynamically significant, it is linear inertial wave propagation that is responsible for the formation of columnar structures.

References

1. C. Cambon, N. N. Mansour, and F. S. Godeferd. Energy transfer in rotating turbulence. *J. Fluid. Mech.*, 337, 1997.
2. P. A. Davidson, P. J. Staplehurst, and S. B. Dalziel. On the evolution of eddies in a rapidly rotating system. *J. Fluid. Mech.*, 557, 2006.
3. E. J. Hopfinger, F. K. Browand, and Y. Gagne. Turbulence and waves in a rotating tank. *J. Fluid. Mech.*, 125, 1982.
4. P. J. Staplehurst, P. A. Davidson, and S. B. Dalziel. On the large-scale evolution of rotating turbulence. *IUTAM*, 2006.

Differential diffusion in double-diffusive stratified turbulence

Hideshi Hanazaki[1] and Kazuhiro Konishi[1]

Department of Mechanical Engineering and Science, Kyoto University, Sakyo-Ku, Kyoto 606-8501, Japan

1 Introduction

Differential diffusion of salt and heat in doubly stratified turbulence is investigated, where "doubly stratified" means that the fluid is stratified by salt and heat, both of which affect the fluid motion through the buoyancy force. Since these two scalars have different molecular diffusivities, "differential diffusion" would lead to different vertical scalar fluxes[1]. This problem has recently attracted significant attention in oceanography since it has strong link to the turbulent diffusivity coefficients of heat and salt in the ocean.

We have considered here a decaying homogeneous turbulence, and investigated the effects of the Prandtl number Pr of heat and the Schmidt number Sc of salt by direct numerical simulations (DNS). (These parameters are assumed to be $Pr = 1$ and $Sc = 6$ in this study to save computer resources.). A significant result is that the small-scale structure of salt, which has small molecular diffusivity, does "not" appear in this double-diffusive system.

2 Results and discussions

The results could be most clearly identified in the visualisation of pan-cake structures of $|\partial S/\partial z|$ (S: salinity fluctuation, z: vertical coordinate) which constitute most of the horizontal "vorticity" or the molecular diffusion of scalar in stratified turbulence [2]. In the doubly-diffusive fluid (Fig. 1), the length scale of salt ($Sc = 6$) is comparable to velocity and temperature ($Pr = 1$), while in the ordinary stratified fluid (Fig. 2), passive scalar ($Sc = 6$) has much smaller scales. This would occur since the salt exhanges its potential energy with the kinetic energy, and the length scale of its variance is strongly "regulated" by the length scale of velocity (i. e., kinetic energy) and temperature (i. e., potential energy of heat).

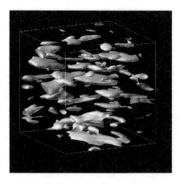

Fig. 1. Isosurfaces of $|\partial S/\partial z|$ in double-diffusive stratified turbulence, where S is the salt ($Sc = 6$).

Fig. 2. Isosurfaces of $|\partial S/\partial z|$ in ordinary stratified turbulence, where S is the passive scalar ($Sc = 6$).

Time development of the spectra of kinetic energy u, temperature T and salinity S also support these ideas (Fig. 3). At low wave numbers, there is a periodic exchange of energy between the vertical kinetic energy (VKE) and potential energy (PE). On the other hand, at high wave numbers, i.e. at small scales, the energy is transferred only from PE to VKE, where PE is accumulated initially at small scales due to the nonlinear cascade processes. As far as $PE > VKE$ holds, this one-way energy transfer would continue, approaching the agreement of spectra between salt and kinetic energy. This process does not occur in the high-Sc passive scalar field (Fig. 4), showing that the scaling due to Batchelor[3] is applicable to the passive scalar, but it may not be applied to active scalar with potential energy.

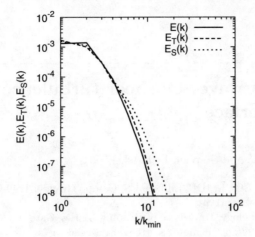

Fig. 3. Energy spectra when S ($Sc = 6$) is a active scalar. $E(k)$: kinetic energy, $E_T(k)$: temperature ($Pr = 1$), $E_S(k)$: salt ($Sc = 6$).

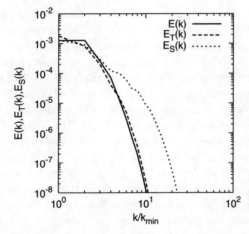

Fig. 4. Energy spectra when S ($Sc = 6$) is a passive scalar. $E(k)$: kinetic energy, $E_T(k)$: temperature ($Pr = 1$), $E_S(k)$: passive scalar ($Sc = 6$).

References

1. H. Hanazaki: Effects of initial conditions on the passive and active scalar fluxes in unsteady stably stratified turbulence. Phys. Fluids **15**, 841–848 (2003).
2. G. K. Batchelor: Small-scale variation of convected quantities like temperature in turbulent fluid. Part 1. General discussion and the csase of small conductivity. J. Fluid Mech. **5**, 113–133 (1959).

Generation of waves by shear turbulence at an air–water interface

Miguel A. C. Teixeira[1] and Stephen E. Belcher[2]

[1] University of Lisbon, CGUL, IDL, Edifício C8, Campo Grande, 1749-016 Lisbon, Portugal, mateixeira@fc.ul.pt
[2] Department of Meteorology, University of Reading, Earley Gate, P.O. Box 243, Reading RG6 6BB, United Kingdom, s.e.belcher@reading.ac.uk

1 Introduction

Despite having been studied for a long time, ocean wave generation processes cannot be considered fully understood at present. This is because an experimental evaluation of these processes is extremely difficult, hence the existing theories are hard to test and validate. Phillips [5] and Miles [4] proposed the first mathematically consistent wave generation theories, the former explaining the initiation of the waves by a resonance with the turbulence in the airflow and the latter their subsequent amplification by an inviscid critical-level instability mechanism. More recently, Belcher and Hunt [1] addressed the effects of turbulence, calculating the wave amplification due to the non-separated sheltering mechanism initially proposed by Jeffreys [3]. Unlike wave amplification, wave initiation has received relatively little attention since Phillips' study, apart from a few experimental investigations that seem to confirm the resonance mechanism qualitatively, but are not totally conclusive. One of the difficulties associated with the theory of [5] is that wave growth is specified in terms of the pressure spectrum of the turbulence, a quantity that is hard to measure. In the present study, Phillips' theory is extended by allowing surface waves to be initiated by turbulent pressure fluctuations associated with a constant-shear flow, calculated using rapid distortion theory (RDT).

2 Theoretical approach

Consider an initially flat and quiescent air-water interface at $z = 0$. Two model situations will be treated. In one of them the constant-shear flow and the associated turbulence will be assumed to exist in the water. In this case the shear not only produces the pressure fluctuations that drive the waves (by interacting with the turbulence) but it also distorts the waves themselves. In the second case to be considered, both the shear flow and the turbulence exist in the air. This is the situation originally envisaged by [5]. In both cases, RDT (e.g. [2]) can be used to relate the pressure fluctuations that occur in [5]'s theory or in its modification to a flow in the water, and the corresponding velocity fluctuations. Apart from the wave linearization already present in Phillips' theory, RDT implies a linearization of the turbulence itself, which in acceptable in regions where the turbulence is strongly distorted by external forcings (i.e. a strong shear), such as in boundary layers.

Assuming that the turbulence is homogeneous far from the air-water interface and initially isotropic, and characterized by a given energy spectrum, with velocity and length scales, u and l, various statistics of the surface waves may be calculated as a function of Γ (the shear rate), u and l (see [6]).

3 Results and discussion

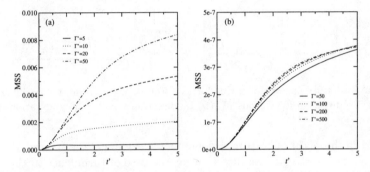

Fig. 1. MSS of surface waves against dimensionless time $t' = \Gamma t$ for various dimensionless shear rates $\Gamma' = \Gamma l/u$ for (a) turbulence in the water with $u = 5\,\mathrm{cm\,s^{-1}}$ and $l = 5\,\mathrm{cm}$, (b) turbulence in the air with $u = 0.5\,\mathrm{m\,s^{-1}}$ and $l = 0.5\,\mathrm{m}$.

Fig. 1 shows the time evolution of the mean-square-slope (MSS) of surface waves for turbulence in the water (Fig. 1a) and turbulence in the air (Fig. 1b), for turbulent pressure fluctuations of the same magnitude. This constraint, which can be implemented by appropriately defining Γ, u and l in each case, is suggested by the fact that the shear stress is constant across an air-water interface driven by the wind. It can be seen that the MSS is much larger for turbulence in the water, and much more dependent on the shear rate, than for turbulence in the air. This behaviour is explained by the fact that the decorrelation time of the turbulent pressure is inversely proportional to the shear rate Γ. The shear rate is larger in an air flow than in a water flow possessing pressure fluctuations of similar magnitude. Since, as [5] demonstrated, the growth rate of surface waves is proportional to the integral time scale, a water flow is considerably more efficient in generating waves than an air flow.

The dependence of the MSS growth on the shear rate for the case of turbulence in the water (that does not occur for turbulence in the air) arises because the waves with the lowest MSS are not resonant. This is also borne out by the fact that these curves tend to become horizontal, indicating a finite period of wave growth. The slowing down of the MSS growth as time progresses (that can be observed both in Fig. 1a and Fig. 1b) can be explained by the stretching of the turbulent vorticity in the streamwise direction, which is a well-known feature of shear turbulence. This makes the pressure forcing move to progressively lower wavenumbers.

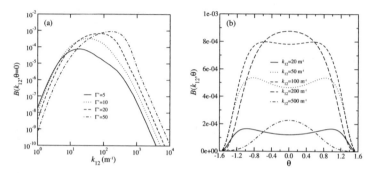

Fig. 2. Curvature spectra of waves for $u = 5\,\mathrm{cm\,s}^{-1}$, $l = 5\,\mathrm{cm}$ at $t' = 5$ (a) along streamwise direction for various Γ', (b) by direction for $\Gamma' = 50$ and various k_{12}.

Fig. 2a shows the curvature wavenumber spectrum of the generated surface waves along the streamwise direction for the case of turbulence in the water. It can be seen that the peak of the spectrum moves to higher wavenumbers as the shear rate increases, which is due to the air-water interface being excited at higher frequencies. In Fig. 2b, the angular distribution of the wave energy for different wavenumbers can be seen. Whereas at high wavenumbers the spectrum peaks at the direction $\theta = 0$, at lower wavenumbers it broadens and becomes bimodal, peaking at two symmetric values of θ. Lower wavenumbers are excited by larger eddies existing at a larger distance from the air-water interface, and that are advected at a faster speed, resonating with surface waves that propagate at an angle to the mean flow.

References

1. S. E. Belcher, J. C. R. Hunt: *J. Fluid Mech.*, **251**, 109–148 (1993).
2. P. A. Durbin: PhD thesis, University of Cambridge (1978).
3. H. Jeffreys: *Proc. R. Soc. Lond.* **A 107**, 189–206 (1925).
4. J. W. Miles: *J. Fluid Mech.*, **3**, 185–204 (1957).
5. O. M. Phillips: *J. Fluid Mech.*, **2**, 417–445 (1957).
6. M. A. C. Teixeira, S. E. Belcher: *Dyn. Atmos. Oceans*, **41**, 1–27 (2006).

Modeling of Multipoint Correlations in Turbulent Flows

H. Chang[1] and A. Bhattacharya[2] and S. C. Kassinos[3] and R. D. Moser[1]

[1] University of Texas at Austin, Austin TX 78735, USA
[2] University of Illinois at Urbana-Champaign, Urbana IL 61801, USA
[3] University of Cyprus, Nicosia 1678, Cyprus

Multipoint velocity correlations have long been used to characterize turbulence, and more recently, have been found to be important to modeling in large eddy simulation (LES) [1]. Two promising developments in the modeling of two-point and three-point correlations ($R_{ij}(\mathbf{x}, \mathbf{r}) = \langle v_i(\mathbf{x})v_j(\mathbf{x} + \mathbf{r}) \rangle$ and $T_{ijk}(\mathbf{r}, \mathbf{s}) = \langle v_i(\mathbf{x})v_j(\mathbf{x} + \mathbf{r})v_k(\mathbf{x} + \mathbf{s}) \rangle$, respectively) are described in this paper. The two approaches, and some preliminary results, are briefly described below.

Modeling Two point Correlations: A finite dimensional model for R_{ij} anisotropy is proposed based on the structure tensors developed by Kassinos et al [2] in the context of RANS modeling. These second rank tensors are single point moments of derivatives of fluctuating streamfunctions and are related to integrals of the two point correlation over separations $\mathbf{r}$, and therefore contain information about the anisotropy of R_{ij}. For homogeneous turbulence, the independent structure tensors are given by the componentality (or Reynolds Stress) $B_{ij} = \epsilon_{ipq}\epsilon_{jts}\langle \psi'_{q,p}\psi'_{s,t} \rangle$, dimensionality $Y_{ij} = \langle \psi'_{n,i}\psi'_{n,j} \rangle$ and strophylosis Q_{ijk}[**]. While R_{ij} is not uniquely determined by the structure tensors, our modeling approach is to use the theory of invariants [4] to formulate the most general linear representation of R_{ij} in terms of the structure tensors, i.e. $\Delta R_{ij}(\mathbf{x}, \mathbf{r}) = \frac{R_{ij}(\mathbf{x},\mathbf{r})-B_{ij}}{q^2} = F_{ij}(\mathbf{r}, \mathbf{b}, \mathbf{Q}, \mathbf{y})$, where $q^2 = B_{kk}$, $b_{ij} = B_{ij}/q^2 - \delta_{ij}/3$, $y_{ij} = Y_{ij}/q^2 - \delta_{ij}/3$. Under these assumptions, and with the additional assumption that the dependence on r is a power law, we obtain (dependence on $\mathbf{x}$ will be implicit in the rest of the section):

$$\Delta R_{ij}(\mathbf{r}) = r^{\alpha_I}[a_1\delta_{ij} + a_2\hat{r}_i\hat{r}_j] \tag{1}$$
$$+r^{\alpha_b}[a_3 b_{ij} + a_4\hat{\mathbf{r}}\cdot\mathbf{b}\cdot\hat{\mathbf{r}}\delta_{ij} + a_5\hat{\mathbf{r}}\cdot\mathbf{b}\cdot\hat{\mathbf{r}}\hat{r}_i\hat{r}_j + a_6(\hat{r}_i(\hat{\mathbf{r}}\cdot\mathbf{b})_j + \hat{r}_j(\hat{\mathbf{r}}\cdot\mathbf{b})_i)]$$
$$+r^{\alpha_y}[a_7 y_{ij} + a_8\hat{\mathbf{r}}\cdot\mathbf{y}\cdot\hat{\mathbf{r}}\delta_{ij} + a_9\hat{\mathbf{r}}\cdot\mathbf{y}\cdot\hat{\mathbf{r}}\hat{r}_i\hat{r}_j + a_{10}(\hat{r}_i(\hat{\mathbf{r}}\cdot\mathbf{y})_j + \hat{r}_j(\hat{\mathbf{r}}\cdot\mathbf{y})_i)]$$
$$+r^{\alpha_Q}[a_{11}(\epsilon_{imk}Q_{klj} + \epsilon_{jmk}Q_{kli})\hat{r}_l\hat{r}_m + a_{12}(\hat{r}_j\epsilon_{ink} + \hat{r}_i\epsilon_{jnk})Q_{klm}\hat{r}_l\hat{r}_m\hat{r}_n]$$

[**] $Q_{ijk} = \left(Q^*_{ijk} + Q^*_{jki} + Q^*_{kij} + Q^*_{kji} + Q^*_{jik} + Q^*_{ikj}\right)/q^2$, where $Q^*_{ijk} = -\langle u'_j\psi'_{i,k} \rangle$

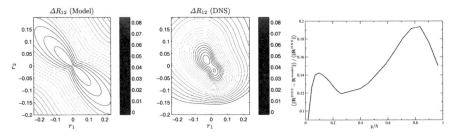

Fig. 1. *Left and center:* Comparison of $\Delta R_{ij}(\mathbf{x}, \mathbf{r})$, at $r_3 = 0$, with correlations calculated from DNS of turbulent channel flow at $Re_\tau = 940$, with the reference point $\mathbf{x}$ at a distance of $y^+ = 220$ from the wall. *Right:* Fractional error between the model and DNS at different y/h locations, h being the channel half-width. The error is calculated using the norm given (for any $T_{ij}(\mathbf{r})$) by $\|T\| = \sqrt{\int_V T_{ij} T_{ij} d\mathbf{r}}$.

Here $\hat{\mathbf{r}} = \mathbf{r}/r$ and α_s, $s \in \{I, b, y, Q\}$ are power-law indices. The number of free constants $(a_1 - a_{12})$ are reduced to 4 by enforcing the continuity constraint (i.e. $\frac{\partial \Delta R_{ij}(\mathbf{r})}{\partial r_j} = 0$) and a self-consistency constraint, which requires that when $\mathbf{Q}$, $\mathbf{b}$ or $\mathbf{y}$ are zero, the values of the respective tensors calculated directly from the representation F_{ij} should be zero. We fit the representation to DNS data over a space spanned by the 4 free constants, 4 power law indices and 10 free components of $\mathbf{y}$ and $\mathbf{Q}$. The representation captures some features of the exact correlation, like the inclination of the principle axis of the iso-contours and the shape of the isocontours (fig 1, left and center), but is not able to capture viscous effects due to the power law assumption or the effect of inhomogeneity at large separations. Thus, the error between the exact and model representation is larger for reference points near the wall and the center of the channel (fig 1, right), due to strong inhomogeneity and viscous effects respectively. Further improvements to account for both inhomogeneity and viscosity are being pursued.

Modeling the isotropic 3-point 3rd-order velocity correlation: Proudman & Reid [3] determined the general form for the Fourier transform Φ_{ijk} of $T_{ijk}(\mathbf{r}, \mathbf{s}) = \langle v_i(\mathbf{x}) v_j(\mathbf{x} + \mathbf{r}) v_k(\mathbf{x} + \mathbf{s}) \rangle$, for stationary incompressible isotropic turbulence. Inverse Fourier transforming their expression, we get the analogous equation in real space:

$$T_{ijk}(\mathbf{r}, \mathbf{s}) = P_{im}^t P_{jn}^s P_{kp}^r [\delta_{np} \partial_m^s \psi(r, s, t) + \delta_{mp} \partial_n^r \psi(t, r, s) + \delta_{mn} \partial_p^s \psi(t, s, r)] \quad (2)$$

where the separation vectors are interrelated $\mathbf{t} = \mathbf{r} - \mathbf{s}$, and spatial derivatives are denoted $\partial_i^r = \left.\frac{\partial}{\partial s_i}\right|_{\mathbf{r}}$, $\partial_i^s = \left.\frac{\partial}{\partial r_i}\right|_{\mathbf{s}}$, $\partial_i^t = -\left.\frac{\partial}{\partial r_i}\right|_{\mathbf{s}} - \left.\frac{\partial}{\partial s_i}\right|_{\mathbf{r}}$, the operators $P_{ij}^\alpha = \delta_{ij} \partial_k^\alpha \partial_k^\alpha - \partial_i^\alpha \partial_j^\alpha$ and symmetries require $\psi(r, s, t) = -\psi(s, r, t)$. This form for T_{ijk} is a linear combination of seventh derivatives of ψ, where ψ is a scalar function of scalar separations r, s and t. Further, for inertial range separations, the Kolmogorov 4/5 law implies a linear dependence on separation for the two-point third order correlation. Assuming that ψ is a polynomial in r, s

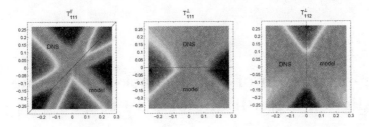

Fig. 2. Comparison of 3-point correlation model with DNS data at $Re_\lambda = 164$.

and t, consistency with the $4/5$ law then requires that the polynomial have overall order eight. Thus, we may consider ψ of the form $(r^a s^b - r^b s^a)t^c$, where $a+b+c = 8$. If the exponents are non-negative, there is a 20-dimensional space of possible T_{ijk} expressions, in which there is only a 5-dimensional subspace of T_{ijk} that are non-zero, non-singular, and continuous.

To determine a representation for T_{ijk} within this space that is consistent with isotropic turbulence, we perform a least-squares fit to DNS data for forced isotropic turbulence at $Re_\lambda = 164$. Because of the large amount of data (T_{ijk} is a 27-dimensional function of two vectors), we only fit to data for **r** and **s** parallel and orthogonal, but weighted to approximate an integral over all **r** and **s**. The least-squares fitting results in a coefficient of determination $R^2 = 1 - \frac{\|T^{DNS} - T^{model}\|^2}{\|T^{DNS}\|^2} = 0.96$. This indicates that our analytical model, composed of just 5 functions, describes the DNS data quite well. Figure 2 shows plots of several non-zero components of T_{ijk}. Colors indicate the magnitude of T_{ijk} as a function of r (horizontal axis) and s (vertical axis). The superscript on T_{ijk} indicates whether **r** and **s** are parallel ($//$) or orthogonal ($\perp$). Half of the r–s plane is DNS data; the other half is model data, with a reflection symmetry line dividing them. The model and DNS are very similar in magnitude and shape. There is a minor discrepancy for r and s near zero, which is due to viscous effects not represented in the model.

References

1. Langford, J., Moser, R.: Optimal LES formulations for isotropic turbulence. Journal of Fluid Mechanics **398** (1999) 321–346
2. Kassinos, S., Reynolds, W., Rogers, M.: One-point turbulence structure tensors. Journal of Fluid Mechanics **428** (2001) 213–248
3. Proudman, I., Reid, W.: On the Decay of a Normally Distributed and Homogeneous Turbulent Velocity Field. Philosophical Transactions of the Royal Society of London. Series A, Mathematical and Physical Sciences **247**(926) (1954) 163–189
4. Spencer, A.: Theory of invariants. Volume 1. (1971) 239–353

Over-prediction of energy back-scatter due to misaligned eigen-frame of SGS tensor

Beat Lüthi[1,2], Søren Ott[2], Jacob Berg[2], and Jakob Mann[2]

[1] Risø Nat. Lab. DK-4000 Roskilde
[2] Institute of Environmental Engineering, ETH Zürich, CH-8093 Zürich
   beat.luethi@ifu.baug.ethz.ch

An important aspect of turbulence is the interaction of large scales with small scales. This interaction needs to be modelled in large eddy simulations (LES). Of particular importance is the so called local energy flux, $\Pi = \tau_{ij} S_{ij}$, with the subgrid-scale stress (SGS) tensor $\tau_{ij} = \overline{u_i u_j} - \bar{u}_i \bar{u}_j$, and the coarse grained rate of strain tensor $S_{ij} = \frac{1}{2} \left(\frac{\partial \bar{u}_i}{\partial x_j} + \frac{\partial \bar{u}_j}{\partial x_j} \right)$. We present an analysis of the measured energy flux, Π_m, of data that is obtained from a particle tracking velocimetry (PTV) experiment and from direct numerical simulation (DNS). The PTV and DNS experiment are described in [2] and [3] and the method to obtain coarse grained derivatives from particle trajectories follows [6]. We stress that it is the first time SGS data is obtained from a PTV experiment.

We compare Π_m with the flux $\Pi_{nl} = \tau_{ij}^{nl} S_{ij}$, where τ_{ij}^{nl} is obtained from the so called 'nonlinear model', $\tau_{ij}^{nl} = c_{nl} \Delta^2 \frac{\partial \bar{u}_i}{\partial x_k} \frac{\partial \bar{u}_j}{\partial x_k}$. The nonlinear model is essentially a first order Taylor expansion of the subgrid flow field and is capable to predict the behaviour of the SGS tensor quite well [4]. However, it is well known that the nonlinear model leads to an inevitable blow-up in simulations, because it overestimates the phenomenon of back-scattering, where energy is transferred from small towards large scales.

In our analysis of the SGS tensor we diagonalize the matrix τ_{ij} as $D = V^{-1} T V$ where D and V are the corresponding eigenvalues and eigenvectors of τ. It allows to study in detail how differences between measured and modelled SGS come about. It turns out that especially for the largest eigenvalue there is no one to one correspondence but only a close to linear relationship. This leads to over-prediction of extreme events. In figure1(a) we show eigenvalue-PDFs as obtained from measurements, from the nonlinear model, and from a linear correction. Recently[5] introduced two parameters to characterize the state of the tensor τ_{ij}. In figure1(b) we show how the linear correction affects the so called dilatational parameter. Investigating the orientation of the eigenvectors we find that in addition to a too weak alignment of the most extensive eigenvector τ_1 of τ_{ij}^{nl} with the compressing strain axis λ_3 [7] there are yet two additional systematic misalignments. They turn out

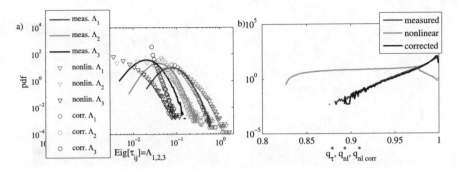

Fig. 1. (a) PDFs of eigenvalues of τ_{ij} as obtained from measurements, nonlinear model and corrected model. (b) PDFs of SGS state parameter q^*.

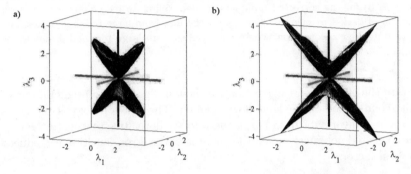

Fig. 2. Relative orientation of the eigenvector τ_1 with respect to the most compressing, intermediate and most extensive eigenvectors, λ_3=blue, λ_2=green, λ_1=red, of the rate of strain tensor. The blue surfaces reflect the probabilities for given orientations of τ_1, a) measured b) nonlinear model.

to be just as relevant for over estimating back-scattering. As an example we show here in figure 2(a) and (b) the different orientations of τ_1^m and τ_1^{nl} with respect to the eigen-frame of strain, $\lambda_{1,2,3}$. The spherical PDF in (a) has a smaller opening angle and is more axis-symmetric than in the nonlinear case (b). Generally we find that all misalignments of the nonlinear model favor additional back-scattering. We will present results that show how the nature of these misalignments is fairly robust and almost completely in-sensitive to both, the filtering scale and, even more important, to the type of flow. In the case of the PTV data the flow possesses a mean strain where the DNS flow is isotropic.

These findings offer interesting possibilities from a SGS modelling point of view. We show that the systematic prediction errors, $\Delta\Pi = \Pi_{nl} - \Pi_m$, of the nonlinear model significantly decrease if the SGS eigen-frame orientation is corrected for the above mentioned misalignments. The procedure is equivalent to a more general form of the so called 'mixed model' proposed by [1]. To

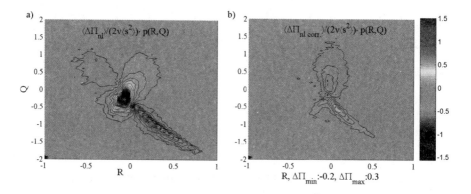

Fig. 3. Normalized prediction error density $\Delta\Pi/\langle 2\nu s^2\rangle \cdot p\,(R,Q)$ for a) the nonlinear model and b) for the corrected nonlinear model. Yellow to red colours denote over prediction of back-scattering or to weak energy flux from large to small scales, and light to dark blue colors show where energy flux from large to small scales is too strong.

illustrate the effect of this correction we show in figure 3 the prediction error density conditioned on the local flow topology. The flow topology is expressed with the two invariants R and Q of the velocity derivative tensor. In the case of the corrected nonlinear model the maxima of the error densities are reduced by several factors.

References

1. J. Bardina, J.H. Ferziger, and W.C. Reynolds. Improved subgrid scale models for large eddy simulation. *AIAA Paper, 80-1357,* 1980.
2. Jacob Berg, Beat Lüthi, Jakob Mann, and Søren Ott. An experimental investigation: backwards and forwards relative dispersion in turbulent flow. *Phys. Rev. E,* 74(1):016304, 2006.
3. L. Biferale, G. Boffetta, A. Celani, B. J. Devenish, A. Lanotte, and F. Toschi. Multiparticle dispersion in fully developed turbulence. *Physics of Fluids,* 17(11):111701, 1–4, 2005.
4. Vadim Borue and Steven A. Orzag. Local energy flux and subgrid-scale statistics in three-dimensional turbulence. *J. Fluid Mech.,* 366:1–31, 1998.
5. Sergei G. Chumakov. Statistics of subgrid-scale stress states in homogeneous isotropic turbulence. *J. Fluid Mech.,* 562:405–414, 2006.
6. Beat Lüthi, Jacob Berg, Søren Ott, and Jakob Mann. Lagrangian multi-particle statistics. *Proc. Euromech Coll.-477, Twente,* 2006.
7. Bo Tao, Joseph Katz, and Charles Meneveau. Statistical geometry of subgrid-scale stresses determined from holographic particle image velocimetry measurements. *J. Fluid Mech.,* 457:35–78, 2002.

Large eddy simulation of turbulent separated flow over a three-dimensional hill

M. García-Villalba[1], T. Stoesser[2], D. von Terzi[1], J. G. Wissink[1], J. Fröhlich[3], and W. Rodi[1]

[1] Institute for Hydromechanics, University of Karlsruhe villalba@ifh.uka.de
[2] School of Civil and Environmental Engineering, Georgia Institute of Technology
[3] Institute for Technical Chemistry and Polymer Chemistry, University of Karlsruhe

1 Introduction

The turbulent flow separation from a three-dimensional curved body is a complex problem which plays an important role in practical applications. In recent laboratory studies Simpson and co-workers [1, 2, 3] investigated extensively the separated flow over and around a 3D hill at high Reynolds number using LDV measurement techniques, oil flow visualisation and hot-wire anemometry. The examinations reveal the complex flow physics associated with the geometry of the hill. Complex separation occurs on the leeside and the evolving vortical structures merge into two large counter-rotating streamwise vortices downstream. There is also some evidence of low frequency spanwise meandering of the vortices in the wake.

The laboratory configuration mentioned above was simulated using LES. It consists of flow over and around an axisymmetric hill of height $H = 78$ mm and base-to-height ratio of 4; the approach-flow turbulent boundary-layer has a thickness of $\delta = 0.5H$. The Reynolds number of the flow based on the free-stream velocity $U_{ref} = 27.5$ m/s and the hill height H is $Re = 1.3 \cdot 10^5$.

2 Numerical model

The LES was performed with the Finite Volume Code LESOCC2 [4]. This solves the incompressible 3D time-dependent Navier-Stokes equations on body-fitted curvilinear block-structured grids using second order central differences for the discretisation of the convective and viscous fluxes. Time advancement is accomplished by an explicit, low-storage Runge-Kutta method.

The geometry of the computational domain is shown in Fig. 1. The size of the domain is $20H \times 3.2H \times 11.7H$ in streamwise, wall-normal and spanwise directions, respectively. The grid consists of $770 \times 240 \times 728$ cells in these

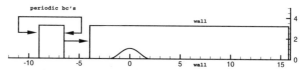

Fig. 1. Sketch of the computational domain and inflow generator.

directions. The inflow conditions are obtained by performing simultaneously a separate periodic LES of channel flow in which the mean velocity is forced to assume the experimental vertical distribution using a body-force technique [5]. The length of the channel is $1.8H = 3.6\delta$ and the number of cells in streamwise direction is 110. The cost of the precursor simulation is, therefore, $1/8$ of the total cost. A no-slip condition is employed at the bottom wall while the Werner-Wengle wall function is used at the top wall, so that the boundary layer there is not well resolved. Free-slip conditions are used at the lateral boundaries and convective conditions at the exit boundary. The quality of the grid resolution is judged by determining the cell size in wall units. The centre of the wall-adjacent cell is located at $y_1^+ \sim 2$. The streamwise and spanwise cell sizes in wall units are roughly 70 and 30, respectively. These values are just within the limits of the recommendations given by Piomelli & Chasnov [6] for wall resolving LES. As for the quality of the inflow conditions, the mean streamwise velocity, turbulent kinetic energy and shear-stress profiles (not shown here) are in good agreement with experimental data.

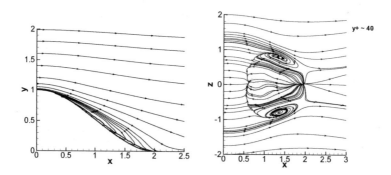

Fig. 2. Left, streamlines in midplane. Right, streamlines at $y^+ = 40$.

3 Results

An illustration of the mean flow obtained in the simulation is displayed in Fig. 2 by showing streamlines of the flow in the symmetry plane (left) and in a plane close to the hill wall at $y^+ \sim 40$ (right) With respect to the experiments, the thickness of the recirculation zone is well predicted although the separation occurs somewhat earlier in the simulation. The flow topology is also rather well predicted with the two counter-rotating vortices appearing roughly at the same location as in the experiment.

A visualization of the instantaneous coherent structures of the flow is displayed in Fig. 3. It shows an iso-surface of pressure fluctuations; the color represents the y-coordinate. In animations of the flow, coherent structures are observed to form in the lee of the hill and are convected downstream. Many of them have the shape of a hairpin vortex although due to the high level of turbulence they are usually rapidly deformed. It happens frequently that structures appear only in one side of the wake. In the instant observed in the picture all the structures are in the left part of the wake, while no activity is observed in the right part. There are also instants in which structures appear everywhere in the wake. Time signals of velocity recorded in the near wake show pronounced peaks when one of these big structures crosses the recording point.

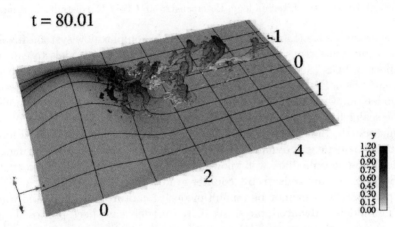

Fig. 3. Iso-surface of pressure fluctuations.

References

1. R.L. Simpson, C.H. Long, and G. Byun. Study of vortical separation from an axisymmetric hill. *Int. J. Heat Fluid Flow*, 23:582–591, 2002.
2. R. Ma and R.L. Simpson. Characterization of turbulent flow downstream of a three-dimensional axisymmetric bump. In *Proc. 4th Int. Symposium on Turbulence and Shear Flow Phenomena. Williamsburg. USA*, 2005.
3. G. Byun and R. L. Simpson. Structure of three-dimensional separated flow on an axisymmetric bump. *AIAA J.*, 44(5):999–1008, 2006.
4. C. Hinterberger. *Dreidimensionale und tiefengemittelte Large-Eddy-Simulation von Flachwasserströmungen*. PhD thesis, University of Karlsruhe, 2004.
5. C.D. Pierce. *Progress-variable approach for large-eddy simulation of turbulent combustion*. PhD thesis, Stanford University, 2001.
6. U. Piomelli and J. R. Chasnov. Large eddy simulation: theory and applications. In *Turbulence and transition modelling*, pages 269–331. Kluwer, 1996.

Large Eddy Simulations of Passive-scalar Mixing using a Tensorial Eddy Diffusivity-based SGS-Modeling

Y. Huai[1], B. Kniesner[2], A. Sadiki[1] and S. Jakirlić[2]

[1] Chair of Energy and Power Plant Technology, sadiki@ekt.tu-darmstadt.de
[2] Chair of Fluid Mechanics and Aerodynamics, s.jakirlic@sla.tu-darmstadt.de
Darmstadt University of Technology, Petersenstr. 30, 64287 Darmstadt, Germany

The accuracy of LES (Large Eddy Simulation) in complex flow systems involving heat and mass transfer relies on the capability of SGS models to capture the effect of SGS (Sub-Grid Scale) turbulent fluxes on the resolved velocity and scalar fields. A common approach uses the so-called linear eddy viscosity models for the SGS stress tensor and the linear eddy diffusivity models for the scalar flux vector. While these models require the specification of two parameters (the Smagorinsky coefficient and the SGS Schmidt/Prandtl number) by utilizing, e.g. a dynamic procedure, their weaknesses in capturing the physics to be described are well known. With regard to the scalar flux, effects like SGS scalar flux anisotropy, counter-gradient or Sc-dependency of fluid transport processes cannot be (simultaneously) captured by these and other existing models. Advanced, physically-based models have been proposed only rarely, see e.g., the review of Jaberi and Colucci (2003).

In the present study, the capability of a new SGS model (so-called anisotropy model) developed in [2] in describing turbulent mixing processes in complex configurations is demonstrated. In particular, this model includes a nonlinear formulation for the eddy diffusivity tensor in terms of the strain-rate along with an SGS time scale that can be linked to the local Schmidt/Prandtl number. First "a priori" and "a posteriori" tests, both based on highly resolved experimental data in jet configurations characterized by high Re-numbers for which no DNS data are available, have been conducted in [2]. In this work, more complex flows are considered. First, the passive-scalar mixing with $Sc > 1$ occurring in a number of important applications such as heat/mass transport in liquids, pollutant and nutrient dispersion or associated with various biological processes, is investigated. Here, a spatially developing turbulent mixing layer as investigated experimentally in [3] is considered. The measurements were performed in a mixing channel with a cross section of $60x60mm^2$, Fig. 1. The first part of the channel is divided into two smaller feed channels with a cross section of $20x60mm^2$ by an obstruction of size $20x60x330mm^3$.

The fluid in the channel is water at room temperature. The experimental data are available in the x-z-planes, yielding the U and W velocity components in x and z direction, respectively. Fig. 2 shows the normalized streamwise velocity (w) at different axial positions compared with experimental data. The results obtained exhibit good agreement at selected locations along the axial direction. In order to obtain a clear idea of the SGS scalar flux model influence, different SGS scalar flux models (Table 1) are evaluated. They were coupled with the same SGS stress model with respect to the velocity field. The results presented in Fig. 3 relate to the mean mixture fraction F. Because the considered linear eddy diffusivity models are based on the same assumption, their use resulted in the almost same outcome with respect to the mean quantities. Therefore, only the results obtained with the dynamic model are presented. The new anisotropy model is implemented in conjunction with the dynamic procedure for determination the model coefficient (Case 4). The comparison at the position 3 shows that there is no important difference between model results because just at this position the mixture behaves uniformly. The mixing process here is characterized by a weak intensity (scalar variances become smaller). At the positions 1 and 2 differences are much more pronounced because of the important contribution of the subgrid-scales in the near field of the mixing layer being characterized by strong anisotropy. In general, the anisotropy model led to the improved predictions with respect to some important details. The computational time used for the computations with the new model is about 1.8 times longer than in the case of the eddy diffusivity model.

A second type of mixing at low Sc occurs often in combustion systems in which the performance of the new model has also to be evaluated. Here a tubo-annular combustion chamber, as shown in Fig. 4, is considered without chemical reactions. Computations have been performed by using LES method and a hybrid LES/RANS - HLR (Reynolds-averaged Navier-Stokes) model (Jakirlic et al., 2006) focussing on the mixing between central non-swirling stream ($Re_m = 23500$) and a swirling co-axial jet discharging from an annular inlet section ($Re_c = 49530$) in the near-field of the flue in a range of swirl intensities: $S = 0.0, 0.6$ and 1.0. Reference LDA (inlet section including central and annular pipes, Fig. 4) and PIV (combustor) measurements are performed by Palm et al. (2005, 2006). Prior to considering the flow within the flue, an intensive study of the flow structure entering the combustor was conducted accounting for the simulations of the entire inlet section including swirl generator system. (Fig. 5a). LES simulations were performed by applying both the Smagorinsky model and its dynamic variant. The solution domain consisting of the flue, swirler with annular pipe and central pipe is discretized by a Cartesian grid ($N_x x N_y x N_z$) comprising about 3.3 Mio. cells in total. Figs. 5 show some selected results obtained by different computational methods. All simulations were performed with an in-house code FASTEST3D based on the block-structured finite-volume method (a second-order central differencing scheme for spatial discretization; the second-order Crank-Nicolson method for time discretization).

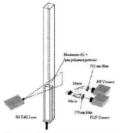

Figure 1: Schematic drawing of the experimental setup and the PIV/PLIF system of the mixing layer configuration

SGS model	Case 1	Case 2	Case 3	Case 4
SGS stress	Dynamic	Dynamic	Dynamic	Dynamic
SGS scalar flux	No model	Eddy Diffusivity	Dynamic	Anisotropy

Table 1: Different SGS scalar flux models coupling with dynamic SGS stress model

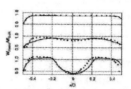

Figure 2: Normalized streamwise velocity at different positions versus experimental data (□LES ... EXP)

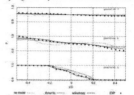

Figure 3: Mean mixture fraction with different SGS scalar flux models versus experimental data

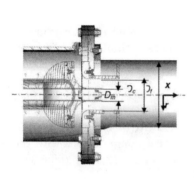

Figure 4: Combustion chamber model

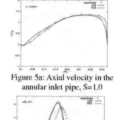

Figure 5a: Axial velocity in the annular inlet pipe, S=1.0

Figure 5b: Axial velocity in the near field of the combustor flue, S=1.0

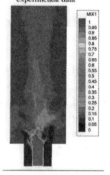

Figure 5c: Snap shot of the mixture fraction field, S=1.0

References

1. F. A. Jaberi, P.J. Colucci: Large eddy simulation of heat and mass transport in turbulent flows. Int. J. Heat and Mass Transfer, **46**, 19, pp. 1827–1840 (2003)
2. Y. Huai: Large Eddy Simulation in the scalar field. Doctoral thesis, Darmstadt University of Technology (2005)
3. Y. Huai, A. Sadiki, S. Pfadler, M. Löffler, F. Beyrau, A. Leipertz , F. Dinkelacker: Experimental Assessment of Scalar Flux Models for Large Eddy Simulations of Non-Reacting Flows. *5th Int. Symp. on THMT*, Dubrovnik, Croatia, September 25-29 (2006)
4. L.K. Hjertager, B.H. Hjertager, N.G. Deen, T. Solberg: Measurement of turbulent mixing in a confined wake flow using combined PIV and PLIF. Can. J. Chem. Eng., **81**, 6, pp. 1149–1158 (2002)
5. R. Palm, S. Grundmann, S. Jakirlic, C. Tropea: Experimental investigation and modelling of flow and turbulence in a swirl combustor. *4th Int. Symp. on TSFP*, Williamsburg, USA, June 27-29 (2005)
6. R. Palm, S. Grundmann, M. Weismüller, S. Saric, S. Jakirlic, C. Tropea: Experimental characterization and modelling of inflow conditions for a gas turbine swirl combustor. Int. J. Heat and Fluid Flow, **27**, 5, pp. 924–936 (2006)
7. S. Jakirlic, B. Kniesner, S. Saric, K. Hanjalic: Merging near-wall RANS models with LES for separating and reattaching flows. FEDSM2006-98039 (2006)

Near-Wake Decaying Turbulence Behind a Cross-bar

Lyazid Djenidi[1] and Philippe Lavoie[2]

[1] Discipline of Mechanical Engineering, University of Newcastle, Newcastle, 2308 NSW, Australia lyazid.djenidi@newcastle.edu.au
[2] Department of Aeronautics, Imperial College London, London SW7 2AZ, UK p.lavoie@imperial.ac.uk

1 Introduction

The present paper reports on a direct numerical simulation of a turbulent wake behind a cross-bar made of two square cylinders in a bi-plane configuration. The motivation to study such flow in relation to grid-generated tubulence is that the cross-bar arrangement can be seen as the "unit element" of a grid made of perpendicular bars. This arrangement allows allows to isolate some of the geometrical factors of the grid that are responsible for the inequality in the lateral velocity fluctuations observed behind a grid [1]. It thus makes possible to study the cross-bar wake alone (its generation and decay), which would be difficult in the actual grid turbulence where the wakes interact quickly behind the grid. The emphasis of the work is on the near-field region of the cross-bar wake, where the individual wakes of each bar interact strongly.

The present study is an exploratory one aimed, firstly, at documenting the wake of a cross-bar and, secondly, studying the turbulence decay in a such wake. In relation to the first point of the aim, it is interesting to document the manner by which the wake is generated and how anisotropy is created in the flow. The study of the turbulence decay is carried out with the perspective to compare it with that of grid turbulence.

2 Numerical Procedure

The direct numerical simulation (DNS) is carried out using the lattice Boltzmann method (LBM). The basic idea of the LBM is to construct a simplified kinetics model that incorporated the essential physics of microscopic average properties, which obey the desired (macroscopic) Navier-Stokes equations [2].

The computational domain, which includes the cross-bar, is Cartesian and has $800 \times 120 \times 120$ (or $80D \times 12D \times 12D$, D is the *diameter* of the square cylinders) mesh points. The mesh increments in the three directions are equal

$(\Delta x = \Delta y = \Delta z = 0.1D)$. The cross-bar is placed at a distance of $6D$ after the inlet where a constant velocity, U_1, is imposed. Periodic conditions are applied on the lateral sides of the computational domain and a convective boundary condition is imposed at the outlet. The Reynolds number $R_D(= U_1 D/\nu)$ is about 1600. To prevent numerical instabilities in the region of the cross-bar a LES scheme, with a Smagorinsky model (the filter size is equal to Δx) is used to help dissipate those instabilities.

3 Results

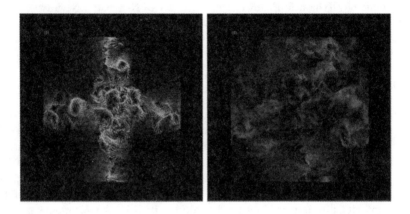

Fig. 1. Velocity field in the (z, y) plane; Left: $x/D = 1$, Right: $x/D = 14$

Figure 1 shows the velocity field in the (z, y) plane at two positions downstream of the cross-bar ($x/D = 3$ and 14; $x = 0$ is at the cross-bar). Close to the cross-bar ($x/D = 3$) vortical motions develop along the span of the two bars, which reflects concentrations of steamwise vorticity (w_x) marking the presence of streamwise vortical stuctures. These latter inevitably interact leading to turbulence production. For $x/D = 14$, the velocity field is similar to that found in DNS results of a grid-generated turbulence [3].

Figure 2 presents the variations along the centreline of the normal Reynolds stresses, $\langle u^2 \rangle$, $\langle v^2 \rangle$ and $\langle w^2 \rangle$, normalized by the upstream velocity, U_0 (u, v and w are the x, y and z-components of the velocity, respectively). Also shown are the experimental data of grid turbulence [1] ; x is normalised by the bar diameter D and not M, the mesh size of the grid). The experimental data are taken at two vertical locations behind the grid along the x-direction: $(z/M = 0.5, y/M = 0)$ and $(z/M = 0.5, y/M = 0.5)$, where $z/M = y/M = 0$ is taken at the centre of a grid node, i.e. crossing of two bars. The data for the two lateral positions collapse into one profile for $x/D \geq 20$; no distinction is made for the data between these two locations for $x/D \geq 100$.

Just behind the cross-bar, $\langle u \rangle^2$ is larger than $\langle v^2 \rangle$ which in turn is larger than $\langle w^2 \rangle$ until about $x/D = 0.6$. Then $\langle w^2 \rangle$ becomes the largest and $\langle u^2 \rangle$ the smallest. Notice the relatively sharp increase in $\langle v^2 \rangle$ and $\langle w^2 \rangle$, with a peak occuring at about $x/D = 3$. The x/D region of high values of $\langle v^2 \rangle$ and $\langle w^2 \rangle$ is comprised between 1 and 6. Conversely to the numerical data, the experimental data show that $\langle u^2 \rangle$ is always bigger than $\langle v^2 \rangle$. However, measurements are needed along the axis of a grid node (i.e. $z/M = y/M = 0$) to provide a better comparison with the cross-bar in the region $0 \leq x/D \leq 10$. Note that it is quite remarkable that the cross-bar data come into alignment (although not quite perfectly) with the grid-turbulence data for $x/D \geq 40$.

$\langle w^2 \rangle$ is larger than $\langle v^2 \rangle$ because of the present arrangement of the bars: the second bar promotes more $\langle w^2 \rangle$ than $\langle v^2 \rangle$. Since the velocity fluctuations in the z-direction are related to the vorticity in the direction perpendicular to that axis, here ω_y, one can then expect to observe a difference in the lateral velocity fluctuations close to the cross-bar.

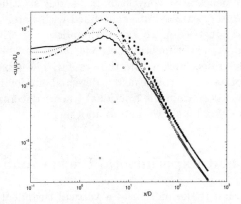

Fig. 2. Variations of the Reynolds stresses along the centreline. Solid line: $\langle u^2 \rangle$, doted line: $\langle v^2 \rangle$, dot-dashed line: $\langle w^2 \rangle$. Symbols: experiments[1]; circles: $z/M = 0.5, y/M = 0.0$; squares: $z/M = 0.5, y/M = 0.5$; solid symbols: $\langle u^2 \rangle$, open symbols: $\langle v^2 \rangle$

Finally, the variation of the turbulent kinetic energy and its decays revealed that along the centreline (not shown) the turbulence becomes almost homogenueous and isotropic.

References

1. Lavoie, 2006, Effect of initial conditions on decaying grid turbulence, *PhD thesis*, University of Newcastle, Australia.
2. U. Frisch, B. Hasslacher & Y. Pomeau: Phys. Rev. Lett. **56**, 1505 (1986).
3. L. Djenidi: J. Fluid Mech. **552**, 13 (2006).

Penetrative Convection in Stratified Fluids: Velocity Measurements by Image Analysis Techniques

Antonio Cenedese[1], Valentina Dore[2], and Monica Moroni[3]

DITS - "Sapienza" University of Rome, Via Eudossiana 18, 00184,Rome (Italy)
[1]antonio.cenedese@uniroma1.it, [2]valentina.dore@uniroma1.it,
[3]monica.moroni@uniroma1.it

The phenomenon of penetrative convection in a stably stratified fluid has been reproduced in laboratory employing a tank filled with water and subjected to heating from below. To be able to assess the adequacy of existing theories of flow and transport, experimental methods must be able to obtain 3D-Lagrangian particle trajectories within the system. The mixing layer growth and the transport of a passive tracer field is detected employing the transilient matrix, whose elements contain the fractional tracer concentration moving in a given time interval from a subvolume to another.

1 Theoretical and Experimental Background

Penetrative convection is the motion of a vertical turbulent plume or dome into a fluid layer of stable density and temperature stratification when the plume has enough momentum to extend into that fluid layer for a significant distance from the original interface. In its initial stages, convection is organized in coherent structures persisting over time. Subsequently the flow becomes turbulent and the structures break up. Penetrative convection is of importance in several areas of geophysical fluid dynamics, most notably in the lower atmosphere, the upper ocean, and lakes. 3D Particle Tracking Velocimetry (3DPTV) has been used for the first time to reconstruct tracer trajectories rather than more traditional 2D techniques to examine the effect of convective-driven perturbation at the mixed layer when no vertical shear occurs. The model used for laboratory experiments is a tank with glass sidewalls of dimension $40 \times 40 \times 41 \, cm^3$ in the two horizontal and vertical directions respectively. The working fluid is stably stratified distilled water. Pollen is used as passive tracer. The fluid is heated from below, to simulate the solar radiation effects and to cause penetrative convection. Thermocouples are employed to measure temperature inside the tank. When studying turbulent convective phenomenon such penetrative convection, dispersion is mostly due to transport by large organized structures while diffusion can be neglected. A way of dealing with the dispersion topic is through the formulation of the transilient turbulence theory, which employs a Lagrangian description of transport, based on tracking of a species passively advected by the main flow. The main output of the theory is the transilient matrix that accounts for all the mixing processes resolved by the grid spacing, from smallest eddy traced by the pollen particles to the medium and large coherent structures over the entire mixed layer depth.

2 Quantifying Mixing

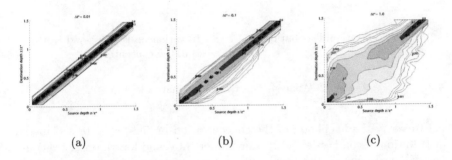

(a)   (b)   (c)

Fig. 1. Contour plots of the evolving transilient matrix. The contour levels are 0.001, 0.003, 0.03, 0.1, 0.3, 0.6

The temporal evolution of the transilient matrix is exhibited in Fig. 1. Although the spatial extent of each bin increases with Δt^* due to the increase of the mixing layer height, the matrices are plotted with respect to the origin and destination depths in order to simplify their physical interpretation. The destination index increases from the bottom to the top. Height will then increase upwards. For small Δt^* ($\Delta t^* < 0.05$), particles will have moved only a short distance or they still belong to their starting position. The matrix, in its graphical representation, presents darker zones along the diagonal (probability close or equal to 1), corresponding to absence of mixing. For intermediate time lags Δt^* ($\Delta t^* = 0.05$ to 0.2) the darker area starts spreading out mainly below the matrix diagonal, evidencing a slow mixing process occurring downwards, corresponding to the negative value of the velocity probability density function mode [4]. For large time lags Δt^* ($\Delta t^* > 0.6$), particles spread out almost uniformly within the mixing layer and the transilient matrix assumes uniform values in that area. Fig. 2(a) presents a cross section through the transilient matrix. The concentration begins as a function at the source depth, marked with a continuous line, and progressively more disperse curves correspond to later times. The dotted line marks the boundary of convection zone. If the mixing occurred via classical dispersion in an infinite domain, cross sections through the matrix would yield Gaussian which would progressively decrease in amplitude and increase in dispersion with time [3].

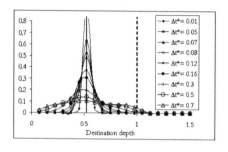

(a) concentration distributions

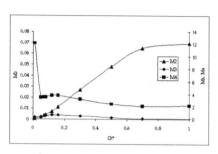

(b) variance, skewness and Kurtosis of the concentration

Fig. 2. Analysis of a transilient matrix section

To verify the likelihood of the concentration profiles, $c(z)$, to a Gaussian distribution, second order (M2), third order (M3) and fourth order (M4) moments are computed. The third and fourth order moments are normalized by $M2^{3/2}$ and $M2^2$ and will be referred to as M3* and M4*. Fig. 2(b) presents the behavior of three moments as a function of the non-dimensional time. The increasing trend of the second order moment implies an increasing volume of fluid interested by the presence of the scalar. The third order moment shows how the concentration distribution moves away from symmetry while the dispersion, related with the mixing layer growth, is going on. Finally the concentration profiles tend toward a symmetrical distribution for long time intervals. The distribution curves are never Gaussian, as the kurtosis value, which always differs from 3, demonstrates.

References

1. J.H. Cushman, M. Park, N.Kleinfelter, M. Moroni: Geophys. Res. Letter. **32**, 19 (2005)
2. J.W. Deardorff, G.E. Willis, D.K. Lilly: J. Fluid. Mech. **35**, 1 (1969)
3. M.S. Miesch, A. Brandenburg, E.G. Zweibel: Physical Review E **61**, 1 (2000)
4. M. Moroni, A. Cenedese: Nonlinear Process in Geophysics **13**, 3 (2006)

Enstrophy, Strain and Scalar Gradient Dynamics across the Turbulent-Nonturbulent Interface in Jets

Carlos B. da Silva and José C. F. Pereira

Instituto Superior Técnico, Pav. Mecânica I, 1º andar/LASEF, Av. Rovisco Pais, 1049-001, Lisboa, Portugal Carlos.Silva@ist.utl.pt

1 Introduction

The mechanism of turbulent entrainment takes place in a sharp interface region separating turbulent and non-turbulent flow which is present in many turbulent flows such as wakes, shear layers and jets. The goal of the present work is to analyse detailed dynamics of the enstrophy, strain and scalar gradient across this turbulent/non-turbulent (T/NT) interface. For this purpose direct numerical simulations (DNS) of turbulent plane jets are used. The simulations are made with a Navier-Stokes solver using pseudo-spectral methods and 3rd order Runge-Kutta time scheme described in da Silva and Pereira[1]. The present simulations include the computation of an advected or passive scalar field with Schmidt number $Sc = 0.7$ and Lagrangian particle tracking. The resolution is about 30 million grid points and the Taylor scale Reynolds number reaches $Re_\lambda \approx 140$ at the centre of the shear layer. The results presented here consist in statistics made using 10 instantaneous fields taken from the turbulent regime. The T/NT interface can be defined using either the vorticity norm as in Bisset *et al.*[2] or using a scalar or concentration field as in Westerweel *et al.*[3]. The present work uses the scalar field with a threshold of $\Theta = 0.1$ to define the interface.

2 Results

Figure 1 (a) shows mean profiles of the terms from the enstrophy transport equation $\frac{1}{2}\Omega_i\Omega_i$, using "classical" averages as in reference [1]. As can be seen, the mean advective and mean viscous diffusion terms are negligible for all values of y/H, even at the borders of the turbulent/non-turbulent region ($|y|/H \approx 1.5$), and the mean dynamics of the enstrophy is dominated by a global balance between enstrophy production and viscous dissipation. A similar picture is observed for the square of the scalar gradient transport equation $\frac{1}{2}G_iG_i$.

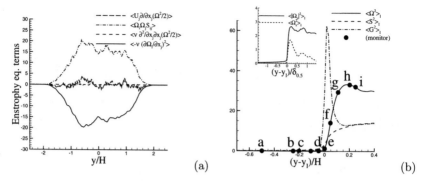

Fig. 1. (a) Mean profiles of terms from the enstrophy equation $\frac{1}{2}\Omega_i\Omega_i$, in the turbulent plane jet; (b) Enstrophy, strain and square of the scalar gradient as a function of the distance from the T/NT interface. The letters "a" to "j" represent particular locations in relation to the T/NT interface position y_I ("e" corresponds to $y - y_I = 0$). The spanwise vorticity component Ω_z and its norm $|\Omega_z|$ are also shown.

Figure 1 (b) shows several mean profiles as a function of the distance from the T/NT interface. We denote these means as $<>_I$ to distinguish them from the classical means $<>$. The procedure to obtain these means is very similar to the one used in reference [3]. The enstrophy Ω^2 and strain S^2 increase quickly as the turbulent region is approached and the Ω_z vorticity and its modulus $|\Omega_z|$ display a small peak just at the start of the turbulent region as in references [2] and [3]. The square of the scalar gradient G^2 on the other hand has a very intense peak very close to the T/NT interface.

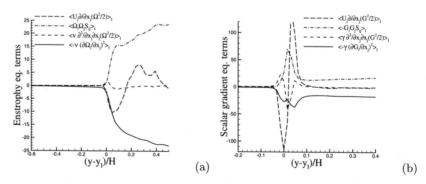

Fig. 2. Mean profiles as a function of the distance from the T/NT interface $y - y_I = 0$: (a) Terms from the enstrophy equation $\frac{1}{2}\Omega_i\Omega_i$; (b) Terms from the square of the scalar gradient equation $\frac{1}{2}G_iG_i$.

Figures 2 (a,b) show mean profiles of the enstrophy and square of the scalar gradient across the T/NT interface, respectively. In the fully developed turbulent shear layer region $((y - y_I)/H > 0.5)$ the advective and viscous diffusion terms are negligible, and there is basically a strong balance between production and molecular dissipation of both enstrophy and square of the scalar gradient. The differences occur very close to the T/NT interface $y - y_I = 0$. Indeed the advection is very important for both equations near $y - y_I = 0$, particularly for the scalar. Moreover, for the scalar, in strong contrast to the "classical" averages all terms attain extremes very close to the interface (notice that even the molecular diffusion term has some importance). This suggests that there are indeed big differences between the process of entrainment of velocity and of a given passive scalar. In isotropic turbulence the different dynamics of the enstrophy and square of scalar gradient have been studied extensively since the paper by Ruetsch and Maxey [4]).

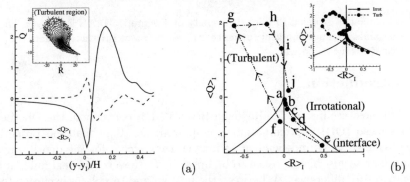

Fig. 3. (a) Mean profiles of the 2nd and 3rd invariants of the velocity gradient tensor Q and R as a function of the distance from the T/NT interface (Joint PDF in the turbulent region also shown); (b) Trajectories of the mean invariants $< Q >_I$ and $< R >_I$ in the (R,Q) map across the T/NT interface.

Finally, figures 3 (a,b) show mean profiles of the 2nd and 3rd invariants of the velocity gradient tensor in relation to the T/NT interface, and their position in the (R,Q) map. It is interesting to see that right at the interface R has a maximum and Q has a minimum, and that this situation changes very quickly afterwards. The (R,Q) map shows that as the irrotational flow approaches the T/NT interface the flow tends to be more and more strain dominated with a predominance of compression over vortex stretching. After the T/NT interface is crossed the flow evolves quickly into an opposite situation *i.e.* vorticity dominated with a predominance of vortex vortex stretching over vortex compression. The interface region itself is characterised mainly by vortex compression and bi-axial strain.

References

1. C.B. da Silva and J.C.F. Pereira: Phys. Fluids **16** (12), 4506 (2004)
2. D.K. Bisset, J.C.R. Hunt and M.M. Rogers: J. Fluid Mech. **451**, 383 (2002)
3. J. Westerweel, C. Fukushima, J.M. Pedersen and J.C.R. Hunt. Phys. Review Lett. **95**, 174501 (2005)
4. G. R. Ruetsch and M. R. Maxey. Phys. Fluids A **3** (6), 1587 (1991)

Numerical study of Non-Oberbeck-Boussinesq effects on the heat transport in turbulent Rayleigh-Bénard convection in liquids

Kazuyasu Sugiyama[1], Enrico Calzavarini[1] and Detlef Lohse[1]

[1]Department of Applied Physics, University of Twente, 7500 AE Enschede, The Netherlands. k.sugiyama@tnw.utwente.nl, e.calzavarini@tnw.utwente.nl, d.lohse@tnw.utwente.nl

1 Introduction

In most studies on the Rayleigh-Bénard (RB) convection, the Oberbeck-Boussinesq (OB) approximation is employed, i.e., fluid material properties are assumed to be independent of temperature except for the density in the buoyancy term which is taken to be linear in T. However, in real fluids if the temperature difference, Δ, between the bottom and top plates is chosen to be large, deviations from the OB approximation may be relevant.

Here, we investigate broken symmetry features and heat-flux modifications due to the Non-Oberbeck-Boussinesq (NOB) effects in the two-dimensional RB turbulence for water and for glycerol. We perform direct numerical simulations (DNS) by solving the equation set for liquid, which consists of the incompressible ($\partial_i u_i = 0$) Navier-Stokes equation

$$\rho_m(\partial_t u_i + u_j \partial_j u_i) = -\partial_i p + \partial_j(\eta(\partial_j u_i + \partial_i u_j)) + g(\rho_m - \rho)\delta_{i3}, \quad (1)$$

and the heat-transfer equation

$$\rho_m c_{p,m}(\partial_t T + u_j \partial_j T) = \partial_j(\Lambda \partial_j T). \quad (2)$$

The dynamic viscosity, $\eta(T)$, the heat conductivity, $\Lambda(T)$, and the temperature dependent density in the buoyancy term, $\rho(T)$, are all temperature dependent, with given empirical relations [1]. As justified in [1], we assume the density and the isobaric specific heat capacity c_p in the material time derivative terms to be constant, their values (ρ_m and $c_{p,m}$) are set at the mean temperature T_m among the bottom and top plates . We vary the Rayleigh number Ra up to 10^8 and the level of the *non-Boussinesqness* Δ up to 60K. Comparison with a recent NOB Boundary Layer theory (NOB-BL) [1] is on the focus of the present study.

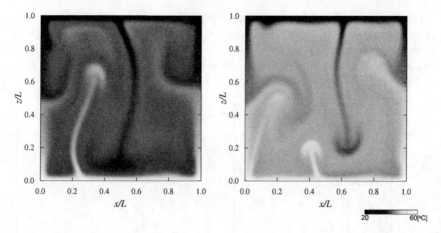

Fig. 1. Snapshots of the temperature field for glycerol at $Ra = 10^7$ and $T_m = 40°C$. The left panel corresponds to the OB case, the right one to the NOB case with $\Delta = 40K$. Temperature color scale is the same in the two panels.

2 Results and discussion

Typical temperature snapshots from glycerol simulations are shown in figure 1. As already observed in experiments [1, 2], NOB convection is characterized by a sensible enhancement of the bulk temperature and top-bottom asymmetric variations of the thermal BLs.

In figure 2 we show the behavior of the center temperature T_c as a function of the thermal gap Δ both for water and glycerol. For water, interestingly enough, T_c beyond $Ra = 10^5$ is rather independent of Ra and shows good agreement with NOB-BL theory. In glycerol instead, only a qualitative trend is attained by the NOB-BL prediction.

The mean heat-flux behavior in NOB case as compared to OB is addressed by looking at the Nusselt ratio behavior Nu_{NOB}/Nu_{OB} at changing the level of non-Boussinesqness (Δ). In particular, we decompose Nu_{NOB}/Nu_{OB} into the product of two terms, corresponding to different effects. (i) the relative change of the thermal BL thicknesses λ^{sl} based on the temperature slope at the plate

$$F_\lambda = 2\lambda^{sl}_{OB}/(\lambda^{sl}_t + \lambda^{sl}_b), \tag{3}$$

and (ii) the T_c shift

$$F_\Delta = (\kappa_t(T_c - T_t) + \kappa_b(T_b - T_c))/(\kappa_m \Delta), \tag{4}$$

where the subscripts t and b represent the top and bottom plates respectively. The relations between $Nu_{NOB}/Nu_{OB} = F_\lambda \cdot F_\Delta$, F_λ, F_Δ and Δ at $Ra = 10^8$ are reported in figure 3. Particularly for water, the DNS results are in quite good agreement with experiments [1], which were indeed carried out at higher Rayleigh numbers ($10^8 \leq Ra \leq 10^{10}$). In the water case $F_\lambda \approx 1$, which is a basic phenomenological assumption of NOB-BL theory, appears to be here a good approximation. $F_\lambda \approx 1$ indicates that the NOB Nusselt number variation in water is mainly due to the T_c shift. On the other hand, for glycerol, the Nusselt number modification is governed by the change of boundary layers, and F_λ sensibly depends on Δ.

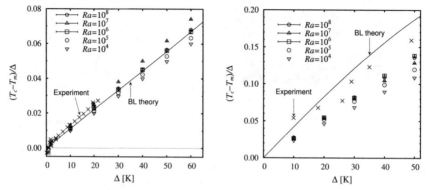

Fig. 2. Deviation of the center (T_c) from the mean (T_m) temperature normalized by the thermal gap (Δ), $(T_c - T_m)/\Delta$ versus Δ for water (left) and for glycerol (right), both at fixed $T_m = 40°C$ for various values of Ra. The symbols × show the available experimental data for water [1] and for glycerol [2]. The line is NOB-BL prediction.

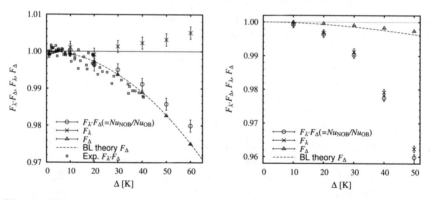

Fig. 3. Nusselt number ratio $Nu_{NOB}/Nu_{OB} = F_\lambda \cdot F_\Delta$, F_λ, and F_Δ versus Δ for water (left) and for glycerol (right), both at fixed $Ra = 10^8$ and $T_m = 40°C$. Symbols ◯, × and △ show the DNS results. Lines show the NOB-BL prediction. On right panel also the experimental data for water (□) are shown [1].

References

1. Ahlers, G., Brown, E., Fontenele Araujo, F., Funfschilling, D., Grossmann, S. and Lohse, D.: J. Fluid Mech. **569**, 409–445 (2006).
2. Zhang, J., Childress, S. and Libchaber, A.: Phys. Fluids **9**, 1034–1042 (1997).

Ultimate regime of convection: search for a hidden triggering parameter

F. Gauthier, B. Hébral, J. Muzellier and P.-E. Roche

Institut Néel, CNRS et Université J.Fourier
BP166, 38042 Grenoble Cedex 9, France

Introduction

In 1962, R. Kraichnan [1] predicted a transition to a new regime of convection for high Rayleigh number, the so called *ultimate regime of convection*. This regime is characterized by the most efficient thermal transfer of all those predicted in convection. Recently, Rayleigh-Bénard experiments in cryogenics conditions [2, 3, 4] report apparently contradictory results on the existence of a transition compatible with this regime for Rayleigh numbers above 10^{12}. Here, we report the first steps of a systematic investigation on the conditions in which the transition can be obtained.

Experimental setup

The reference cell chosen for comparison with others is the one used by Chavanne et al. [3]: cylindrical cell of aspect ratio 0.5, cryogenic helium as the working fluid, copper top and bottom plates and a 0.5 mm thick stainless-steel side wall. The Prandtl number is in the region $0.7 < Pr < 20$.

The first tested parameter is the conductivity of the plates in order to test the phenomenological model proposed by Chillà and colleagues

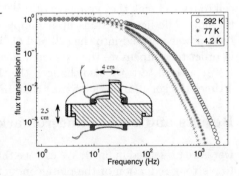

Fig. 1. Magnetic field transfer function through the brass plate.

[5]. The test consists in replacing the bottom copper plate of the Rayleigh Benard cell with a brass plate which has a thermal conductivity about 250 times lower than a copper one. Further details are given in [6]. For reference we present here a method used in the Grenoble's group since 1999 to

measure the thermal conductivity of plates [this method was suggested to us by L. Puech and P.E Wolf]. We measure the ratio between the electrical conductivity of the plate at room and liquid helium temperature. The thermal conductivity is well-known at room temperature and not very dependent on the impurities and annealing; using Wiedemann-Franz equation ($\kappa_{T_1}/\sigma_{T_1} T_1 = \kappa_{T_2}/\sigma_{T_2} T_2$) we then calculate the thermal conductivity at liquid helium temperature (κ is the thermal conductivity, σ the electrical conductivity and T the temperature). The electrical conductivity measurement is based on a Foucault current approach. Two coils are placed on both sides of the plates (see insert in fig. 1). Varying the input voltage frequency of one coil, we measure the currents induced in the other one for different temperatures. It can be shown with Maxwell equations that we can deduce the value of the RRR[1] rescaling the frequency on the curves on figure 1. For the brass plate the RRR=$\sigma_{4.2K}/\sigma_{292K}$ = 2.38 (RRR=180 for the copper plate). At 4.2 K we find a thermal conductivity of 4.0 W/m.K (brass plate) and 1090 W/m.K (copper plate).

A second tested parameter is the influence of the flanges at the interface between plates and the side wall. Indeed the flanges for the Grenoble cells (see fig. 2.) differ from the Oregon one [4] .

Starting from Grenoble flanges design we added a heater ring on the outside of the side wall within 1 cm above the bottom plate and a temperature regulated copper ring to extract a controled fraction of heat from the side wall within 1 cm from the top plate. We also put a local heater on the side wall of the cell to try to break the symmetry of

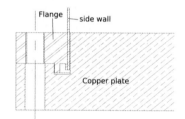

Fig. 2. Detail of the flange assembly for the Grenoble cell.

the flow. The figure 3 sums up a sample of the working conditions (P is the total power injected into the cell and Q_0 is the power injected in the top ring in order to compensate the extracted power from the side).

As a third test, we also changed the thickness of the side wall, which was 4 times thicker (2 mm vs 0.5mm) than in the reference cell [7].

Results and concluding remarks

Data are presented in fig. 4. We can see that the transition occurence is robust to a strong reduction of the plate conductivity, an axisymetric parasitic heat flux on the side wall near the plate, an axisymetric break by a lateral heat flux on the side wall and a significant thickness increase of the side wall. These parameters in the range of values explored here can't explain the Grenoble-Oregon controversy.

---

[1] Residual Resistivity Ratio ($\sigma_{T_1}/\sigma_{T_2}$).

ed copper ring to ex-
roled fraction of heat
wall within 1 cm from
e. We also put a local
side wall of the cell   Fig. 2. Detail of the flange assembly for
ak the symmetry of the   the Grenoble cell.

ble ?? sums up a sample of the working conditions (P is the total
ed into the cell and Q_0 is the power injected in the top ring in
npensate extracted power from the side). We also changed the
the side wall, which was 4 times thicker (2 mm vs 0.5 mm) than
nce cell [7].

Copper plate

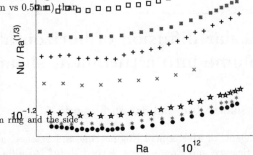

serie	1	2	3	4	5
top ring	Q_0	Q_0	Q_0	$0.97Q_0$	$0.97Q_0$
side heater	0	0	0.3P	0	0
bottom ring	0	0.3P	0	0	0.3P
bottom plate	P	0.7P	0.7P	P	0.7P

Fig. 3. Different working conditions for the top and bottom ring
and the side heater.

Resistivity Ratio ($\sigma_{T_1}/\sigma_{T_2}$).

Fig. 4. Compensated Nusselt number $Nu/Ra^{1/3}$ (on a log scale) vs Ra for the brass plate (+), the thick side (x) and for serie 1(full circles), 2(full squares), 3(open squares), 4(full stars), 5(open stars), see figure 3.

Acknowledgments

We thank the Région Rhône-Alpes for support.

References

1. R. Kraichnan: Phys. Fluids **5**, 1374 (1962)
2. X. Z. Wu: Along a road to developed turbulence: free thermal convection in low temperature helium gas. Thesis, University of Chicago, Chicago (1991)
3. X. Chavanne et al.: Phys. Rev. Lett.**79**, 3648 (1997)
4. J.J. Niemela et al.: Nature. **404**, 837 (2000)
5. F. Chillà et al.: Phys. Fluids **16**, 2452 (2004)
6. P.-E. Roche et al.: Phys. Fluids **17**, 115107 (2005)
7. P.-E. Roche et al.: European Physical Journal B **24**, 405 (2001)

Scalar diffusion of horizontally released heated plume into a turbulent shear flow

Yasumasa Ito[1] and Satoru Komori[1]

[1] Dept. of Mechanical Engineering and Science, and Advanced Institute of Fluid Science and Engineering, Kyoto Univ., Yoshida-honmachi, Sakyo-ku, Kyoto 606-8501, Japan, ito@mech.kyoto-u.ac.jp

[2] Dept. of Mechanical Engineering and Science, and Advanced Institute of Fluid Science and Engineering, Kyoto Univ., Yoshida-honmachi, Sakyo-ku, Kyoto 606-8501, Japan, komori@mech.kyoto-u.ac.jp

1 Introduction

Turbulent diffusion of a plume with different density from ambient flow often appears in various kinds of environmental and industrial flows. Most of the studies have referred to flows with the situations where the heated jet is vertically released into a static flow and the jet is developed mainly by thermal convection. However, the flow is often changed by disturbances of ambient flow. In particular, mean fluid shear based on the velocity gradient of the ambient flow can strongly influence on the scalar diffusion of plume. The purpose of this study is, therefore, to investigate the turbulent diffusion of horizontally released heated plume in a sheared flow.

2 Experiments

Figure 1 shows the schematic of the experimental apparatus. The test apparatus used was a rectangular water tunnel of 1.5 m in length and 0.1 × 0.1 m in cross-section. A turbulence-generating grid was installed at the entrance to the test section. The initial velocities of the upper and lower layers of ambient flow were set to 0.165 m/s and 0.085 m/s, respectively. A plume is released from a nozzle installed at the center to the test channel. The diameter of the nozzle is 0.008 m and plume was injected from 0.1 m downstream the entrance to the test channel. Experiments were carried out for the isothermal case and weakly and strongly heated cases. The water temperature of plume was set to 289K for the isothermal case and they were set to 10K and 18K higher than that of ambient flow for the heated cases. The instantaneous streamwise and vertical velocities and concentration of plume were simultaneously measured by a LDV and LIF technique, respectively.

3 Results

Figure 2 shows the vertical distribution of the normalized turbulent vertical mass flux, $\overline{vc}$, at $x/d =12.5$ and 37.5. Since the density of the plume is smaller than that of the ambient flow, the upper layer is unstably stratified and the lower layer is stably stratified. The value of $\overline{vc}$ in the upper layer in the heated case is larger than that in the isothermal case and $\overline{vc}$ in the lower layer in the heated case is smaller than that in the isothermal case at $x/d=12.5$. These results depend on the vertical temperature gradient. However, the values of $\overline{vc}$ in the heated case is smaller than that in the isothermal case in the upper layer and larger than that in the isothermal case in the lower layer at $x/d=37.5$. These trends are completely opposite to those in the upstream region.

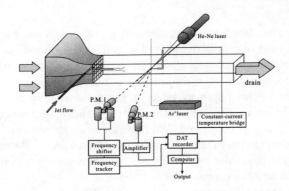

Fig. 1. Schematic of experimental apparatus.

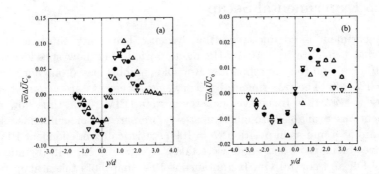

Fig. 2. Vertical distributions of the vertical mass flux at (a) $x/d =12.5$ and (b) $x/d =37.5$. $\triangledown$, isothermal case; $\bullet$, weakly heated case; $\triangle$, strongly heated case.

Estimation techniques in a turbulent flow field

Julie Ausseur, Jeremy Pinier and Mark Glauser

Syracuse University 149 Link Hall Syracuse Ny 13244, USA `jausseur@syr.edu`

Part of our continuous efforts to implement an effective closed-loop feedback control of the flow over a NACA 4412 airfoil is the obtention of an accurate estimate of the actual flow state. An elaborate controller combines both prediction and measurement techniques to obtain a precise estimation of the control variable. In this paper we focus on the measurement process. We first present the different candidates for the control variable, and describe the low-dimensional techniques employed. We then focus on the estimation methods that will be incorporated in the controller in order to access the state variable from the pressure real-time measurements. The investigation of the dynamics in the flow field and the correlations between the variables at stake reveal the benefits that a spectral estimation approach will bring.

1 The Experimental Setup

The experiment is conducted in the Syracuse University subsonic closed-loop wind tunnel on a NACA-4412 model airfoil. The flow speed is set at U_∞=10 m/s, and the corresponding Reynolds number based on chord length is Re=135,000. The experimental velocity measurements are acquired using stereoscopic Particle Image Velocimetry system (PIV). Real-time flow measurements are available through 11 unsteady pressure sensors embedded along the chord at a mid-span position. $T = 1000$ statistically independent PIV velocity vector maps or snapshots were taken at $\alpha = 10$, 12, 14, 16, and 18°, control Off and control On. Instantaneous PIV snapshots taken at $\alpha = 14^0$ and $\alpha = 16^0$ without control show that at $\alpha = 14^0$, the flow is still attached to most part of the airfoil surface, resulting in significant lift performances and small drag forces. As the angle of attack is increased, the boundary layer separation that initiated at the trailing-edge, progresses upstream of the airfoil. Passed $\alpha = 15^0$ the turbulent boundary layer is no longer able to counteract the adverse pressure gradient and a massive separation occurs from the leading edge.

2 The Control Approach

Our goal is to estimate the flow state over the NACA-4412 airfoil at each time instant in order to be able to control the flow using piezoelectric actuators and satisfy our control objective. It has been shown several times that the energy-cost is considerably higher in a process that attempts to reattach a flow that has already separated from a surface. Therefore our low-cost control objective for this closed-loop approach is to always keep the flow attached to the airfoil.

2.1 Control Design

Actuation system and control diagram

The actuating system consists of 14 small oscillatory slot jets near the leading edge of the airfoil, produced by vibrating piezoceramic discs located in individual cavities under the surface of the airfoil. The major elements of the feedback control are displayed on Fig. 1. Using a National Instruments SCXI/PXI signal conditioning/data acquisition platform we are able to operate our real-time control at $10kHz$.

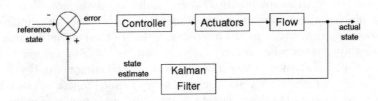

Fig. 1. Control diagram

The Kalman filter [3] has been used for a wide range of estimation problems. We propose to use it here as a linear estimator for the flow state. This combines a measurement model that assumes a stochastic linear relationship between the flow measurements and the control variable as well as a process equation that models the dynamics of the state variable. The controller can be of different types. A simple proportional controller, as implemented by Pinier *et al.*[1] sends a signal to the actuators that is proportional to the feedback error.

3 Estimation Techniques

The velocity data obtained from the PIV system is not time-resolved. However the PIV trigger was purposely recorded in order to synchronize pressure measurements with the PIV velocity measurements. The high pressure sampling rate allows for the computation of the normalized pressure/velocity cross-correlations. The idea is to observe the spatial regions of highest correlation at the time of the PIV trigger - i.e. when the pressure and the velocity are time-aligned ($\tau = 0$) - and to investigate the benefits to be gained from including the surrounding time events in the computations. Figure 2 presents the spatial correlations for the single time $\tau = 0$ and compares it to the local maximum correlation:

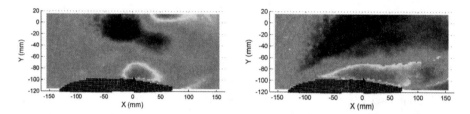

Fig. 2. Single-time (left) and maximum (right) normalized cross-correlations of pressure sensor location $\frac{x}{c} = 0.64$ and U velocity at $\alpha = 16°$ control off

Strong correlations of up to 60% are found on both plots. The single-time plot displays maximum correlation values in a zone surrounding the corresponding pressure sensor. The spatial extent of the correlations becomes considerably greater when a range of τ containing the PIV trigger is taken into account. This indicates the importance of investigating not only the single time $\tau = 0$ at which velocity and pressure were sampled together, but a larger time period surrounding this τ, as significant additional amount of information on the velocity field will be retrieved from the pressure time series. Therefore, results from the implementation of the spectral linear stochastic measurement technique [2] as compared to the single time linear stochastic measurement will be shown.

References

1. Pinier, J., J. Ausseur, M. Glauser, and H. Higuchi: 2007/01/, 'Proportional Closed-Loop Feedback Control of Flow Separation'. *AIAA Journal* **45**(1), 181 – 190.
2. Tinney, C., F. Coiffet, J. Delville, A. Hall, P. Jordan, and M. Glauser: 2006, 'On spectral linear stochastic estimation'. *Experiments in Fluids* **41**(5), 763 – 775.
3. Welch, G. and G. Bishop: 1995, 'An Introduction to the Kalman Filter'. Technical report, Chapel Hill, NC, USA.

Turbulence generated by fractal grids in the wind tunnel

Rich E. E. Seoud[1] and J C Vassilicos[2]

[1] Imperial College London- Department of Aeronautics.
richard.seoud@imperial.ac.uk
[2] Imperial College London - Department of Aeronautics and Institute for
Mathematical Sciences. j.c.vassilicos@imperial.ac.uk

Multiscale flow control is a new concept whereby turbulent eddies are passively or actively forced into a flow by a distribution of objects or actuators of various sizes (as in the fractal objects and grids of [6],[1] and [3] or where eddies of various sizes are fully controlled so as to maintain a flow that is laminar but with turbulent-like properties (as in the multiscale electromagnetic forcing of [5]) or where, in general, multiscale distributions of eddies are forced up and down in intensity and/or position and/or time whether actively or passively in an overall laminar, transitional or turbulent state. The simultaneous manipulation of coexisting eddies of a broad range of sizes can dramatically alter interscale energy transfers as well as the dynamics, stretching and alignments of vorticity and strain rates [2] but also pressure and acceleration fields and thereby momentum transfers in particular. Here we focus on planar fractal grids used in a wind tunnel to generate turbulence. These grids generate broad range turbulence energy spectra and unusual but controllable turbulence build-up and decay rates. They also present potential for many applications particularly because they allow independent control of pressure drop and turbulence intensity.

1 Grid technology

Detailed laboratory measurements and scaling studies carried out by [3] and ourselves with a total of twenty one planar fractal grids pertaining to three different fractal families (fractal cross, fractal I and fractal square grids) in two different wind tunnels have shown that turbulence decay can be controlled with specific parameters defining multiscale grids such as the fractal dimension D_f and the number of fractal iterations (which together determine an effective mesh size M_{eff}) and the ratio t_r of largest to smallest bar thicknesses on the grid. Specifically, in the case of fractal cross grids, the turbulence intensity

u'/U scales and decays as $(u'/U)^2 = t_r^2 C_P f(x/M_{eff})$ where x is the streamwise distance from the grid and C_P is a normalised static pressure drop across the grid. In the case of fractal I grids, $(u'/U)^2 = t_r(T/L_{max})^2 C_P f(x/M_{eff})$ (T is the tunnel cross-sectional width and L_{max} is the maximal length on the grid). In the case of fractal square grids, the turbulence intensity first increases till a distance $x_{peak} = 75 t_{min} T / L_{min}$ (where t_{min} and L_{min} are the minimal bar thicknesses and lengths on the grid) is reached downstream from the grid beyond which the turbulence decays fully within a distance l_{turb} from x_{peak} which is also controllable, for example by the mean speed U. Note that if you double the smallest thickness of the bars (millimetres) on the fractal grid you double the distance downstream (metres) where the turbulence builds up. This is unprecedented behaviour in fluid flows.

2 Measurements and Results

The bulk of the present contribution is concerned with turbulence generated by fractal square grids (see Fig. 1(a)) in a wind tunnel of test-section 0.46m 0.46m 4.0m. Measurements were made via hot wire anemometry. Free stream velocities range from 7 to 22 m/s, single wire and x-wire measurements have been taken on the centre line as well as off centre line in the decay region $x > x_{peak}$. Various velocity profiles have been obtained for the purpose of documenting the flow as completely as possible, in particular its small scale and large scale isotropy properties and absence/presence of turbulence production. Reynolds numbers Re_λ based on the Taylor microscale are exceptionally high for such a small wind tunnel and range between 0(100) and 1200 in the decay region. A particularly striking observation is that fractal square grids can modify turbulence decay to the point that the integral and the Taylor time scales remain approximately constant decay (Fig. 1(b)). This implies an exponential turbulence decay in homogeneous isotropic turbulence, and our decay results fit such an exponential very well. Nevertheless, energy spectra have well defined broad $-5/3$ power law shapes over a broad range of times during decay. The only way in which these results can be made to be consistent with each other theoretically, assuming homogeneous isotropic turbulence decay, is for the spectra obtained at different stages of decay to collapse with the use of only one length scale, [4]. This proves possible here because the integral and Taylor scales remain proportional to each other during decay and is observed in our measurements (Fig 1(c)). Finally, these observations and the hypothesis of homogeneous isotropic turbulence imply that the non-dimensionalised kinetic energy dissipation rate per unit mass is inversely proportional to Re_λ, something which also agrees very convincingly with our measurements (Fig. 1(d)). Hence, fractal square grids generate a homogeneous and isotropic turbulence beyond x_{peak} which becomes increasingly non-dissipative as the Reynolds number increases. This is clearly unlike classical grid-generated homogeneous isotropic turbulence.

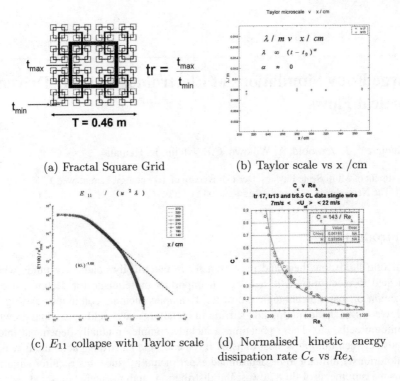

(a) Fractal Square Grid   (b) Taylor scale vs x /cm

(c) E_{11} collapse with Taylor scale   (d) Normalised kinetic energy dissipation rate C_ϵ vs Re_λ

Fig. 1.

References

1. Staicu A., Mazzi B., Vassilicos J. C., and van de Water W. Turbulent wakes of fractal objects. *Phys. Rev. E*, 67(6), 2003.
2. Mazzi B. and Vassilicos J. C. Fractal generated turbulence. *J. Fluid Mechanics*, 502:65–87, 2004.
3. Hurst D. and Vassilicos J. C. Scalings and decay of fractal generated turbulence. *Phys. Fluids*, 2007.
4. George W. K. The decay of homogeneous isotropic turbulence. *Physics of Fluids*, 4(7), 1992.
5. Rossi L., Vassilicos J. C., and Hardalupas Y. Electromagnetically controlled multi-scale flow. *J. Fluid Mechanics*, 558:207–242, 2006.
6. D. Queiros-Conde and J.C. Vassilicos. Turbulent wakes of 3d fractal grids. In J.C. Vassilicos, editor, *Intermittency in Turbulent Flows and other dynamical systems*, pages 136–167. Cambridge Press, 2001.

Large Eddy Simulations of Electromagnetically Driven Vortical Flows

S. Kenjereš[1], J. Verdoold, A. Wibowo, C.R. Kleijn, K. Hanjalić

Department of Multi Scale Physics, Delft University of Technology, Lorentzweg 1, 2628 CJ Delft, The Netherlands. S.Kenjeres@tudelft.nl

1 Introduction

Predicting fluid flow, heat and mass transfer in electrically conductive fluids when subjected to electromagnetic fields is an important prerequisite for design and optimization of many technological process. Examples include: continuous casting of steel, electromagnetic mixing and stirring in metallurgy, arc-welding, crystal growth, aluminium cells, etc. These phenomena include complex mutually dependent interactions between fluid flow, heat transfer, turbulence and electromagnetic fields which are notoriously difficult to be determined experimentally, since this requires simultaneous measurements of fluid flow, scalar distributions and magnetic fields and electric currents. In order to provide insights into these interactions, we performed combined numerical simulations and experimental studies for simplified configurations where many of these physical phenomena can be isolated and studied in detail, see Fig. 1-above. In this generic setup, the electromagnetic forcing is imposed by interactions between electric fields generated by two electrodes in the upper part of the side walls and different combinations of permanent magnets. The permanent magnets are located under the horizontal wall. The resulting Lorentz force for this situation can be simply expressed as $\mathbf{F_L} = \mathbf{E} \times \mathbf{B}$ and is constant in time. The working fluid is water with $7\% Na_2 SO_4$ electrolyte solution in order to enhance its electric conductivity. The upper and lower plate can be kept on different temperatures making it possible to study effects of the electromagnetic forcing on the local heat transfer. In this paper we will address isothermal situations and we focus on comparisons between performed PIV measurements and LES results.

2 Results

In this study we performed well-resolved large eddy simulations (LES) of the electromagnetically driven flows (with different orientations, distributions and strengths of the permanent magnets) without heat transfer. The intensity of flow forcing can be easily controlled by changing the intensity of the electrical current supplied over the electrodes. For the considered range of currents, $0.5 \le I \le 10$ A, the resulting flow is in the transitional regime ($1.5 \times 10^3 \le Re \le 10^4$), i.e. turbulent fluctuations for higher intensities of electric currents have been observed at specific locations in the proximity of the magnets. As such, this investigation represents a continuation of the previous research of Verdoold et al. [1] where initial PLIF flow visualisations were performed but without velocity measurements. In contrast to the work of Rossi et al. [2] where typical two-dimensional flow features were studied, here we focus on genuinely three-dimensional flow patterns. Two different types of subgrid (SGS) models have been applied: the magnetically extended Smagorinsky model (in order to account for additional magnetic reduction of turbulent viscosity), ($\nu_t = \nu_s exp \left[- (\sigma/\rho) (C_m \Delta)^2 |B_0|^2 / \nu_s \right], \nu_s = (C_s \Delta)^2 (S_{ij} S_{ij})^{1/2}$) by Shimomura [3], and the dynamically adjustable Smagorinsky model by Lilly [4]. The LES results presented here were performed with the Shimomura [3] subgrid model. Numerical results have been compared with originally performed PIV and LDA measurements along characteristic vertical and horizontal planes. The time evolutions of vortical structures for a configuration with two magnets, obtained from LES, are shown in Fig. 1. This configuration is introduced in order to create centrally located swirling flow (draft tornados) and consequently to enhance mixing in the horizontal planes. It can be seen that after an initial period where well defined and coherent flow structures are clearly visible, in latter time instants fluctuations take over - making the flow significantly turbulent in specific flow regions. The configuration with three magnets differs from the previous 2-magnets setup not only in the number of magnets but also in the orientation of the magnets with respect to the electrodes (now rotated over 90^0). This is done in order to produce an enhanced vertical motion that can disturb the horizontal boundary layers (pump-in/pump-out mechanism). Note that a jet-like motion is created (flow from right to left) in contrast to the swirling flow pattern when 2-magnets are considered. By comparing PIV and LES profiles of the horizontal velocity at different vertical locations, we concluded that a good agreement between simulations and experimental profiles is obtained. This confirms that the newly developed integrated Navier-Stokes/Maxwell solver can be used with confidence for other parametric studies. Future studies will involve heat transfer and configurations with more magnets and with time-dependent alternate electric currents.

References

1. J. Verdoold, L. Rossi, M.J. Tummers, K. Hanjalić: "Towards electromagnetic control of thermal convection", Proceedings of the ISFCMV conference, Sorrento, Italy, Eds. G. M. Carlomagna and I. Grant, 214-1 (2003)
2. L. Rossi, J. C. Vasisilicos, Y. Hardalupas: "Electromagnetically controled multi-scale flows", J. Fluid Mech. **558**, 207 (2006)
3. Y. Shimomura: "LES of magnetohydrodynamic turbulent channel flows under a uniform magnetic field", Phys. Fluids A, **3**, 3098 (1991)
4. D. K. Lilly: "A proposed modification of the Germano subgrid-scale closure method", Phys. Fluids A, **4**, 633 (1992)

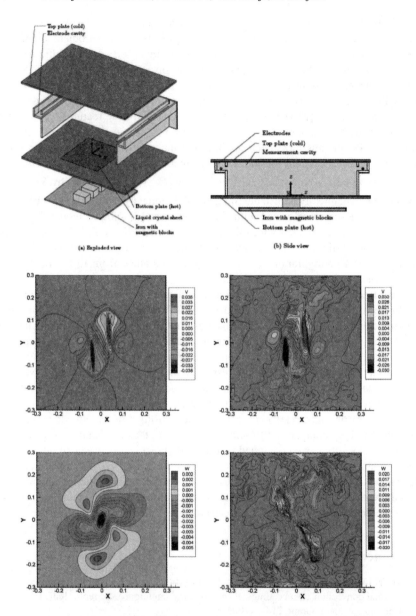

Fig. 1. Schematic of the simulated setup (top row) and time evolution of the vortical structures (middle and bottom row). Middle row: contours of horizontal velocities in the horizontal plane close to the bottom wall ($z/H = 0.03$). Bottom row: contours of the vertical velocities in the horizontal mid plane ($z/H = 0.5$). In the middle and bottom row, the left picture shows the initial flow configuration. The right picture shows an instantaneous realisation of the fully developed flow.

Skin-friction Drag Reduction via Steady Streamwise Oscillations of Spanwise Velocity

Maurizio Quadrio[1], Claudio Viotti[1], and Paolo Luchini[2]

[1] Politecnico di Milano, Dipartimento di Ingegneria Aerospaziale, Via La Masa 34, 20156 Milano - Italy maurizio.quadrio@polimi.it, viotti@aero.polimi.it
[2] Universitá di Salerno, Dipartimento di Meccanica, Fisciano (SA) - Italy luchini@unisa.it

Reducing the skin-friction drag in turbulent wall flows has seen a growing interest in recent years, owing to potential energetic and environmental advantages. Passive techniques (like riblets) are not yet in widespread use, notwithstanding their applicative appeal; most of the strategies currently under investigation are active techniques. One of the simplest and most interesting amongst active approaches is the oscillating-wall technique [1], where the wall moves according to:

$$w(t) = A \sin\left(\frac{2\pi}{T}t\right).$$ (1)

Here w is the spanwise (z) component of the velocity vector, t is time, A is the oscillation amplitude and T its period. The forcing (1) is purely temporal and is known [2] to yield a maximum drag reduction P_{sav} of about 40% when T has the optimal value of $T_{opt}^+ = 125$; the total energetic budget P_{net}, that accounts for both the saving and the cost of moving the wall against the viscous resistence of the fluid, can be positive and up to 7%.

The obvious drawback of the oscillating wall is the presence of moving parts. Since near the wall a turbulent flow is known [3] to possess a well-defined convection velocity, namely $U_c^+ \approx 10$, the possibility exists that the fluid senses the same spanwise lagrangian acceleration either when the wall oscillates with period T, or a steady sinusoidal distribution of wall blowing is applied with wave length $\lambda_x = U_c T$.

Hence in this paper the following, purely spatial forcing is investigated numerically via DNS:

$$w(x) = A \sin\left(\frac{2\pi}{\lambda_x}x\right),$$ (2)

where x is the streamwise coordinate. A related forcing has been addressed in [4] in the context of Lorentz force control. As observed in [5], however, these two forcings may yield different outcomes, owing to the impossibility for the body force to alter the velocity at the wall.

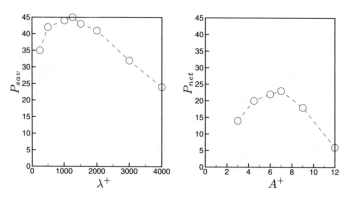

Fig. 1. Left: variation of P_{sav} vs. λ^+ at $A^+ = 12$ for the forcing (2). Right: variation of P_{net} vs A^+ for $\lambda^+ = \lambda^+_{opt} = 1250$.

We employ for the analysis the parallel DNS code described in [6]. A number of simulations is carried out at $Re_\tau = 200$ (based on the friction velocity u_τ) to assess the effects of (2). 320 Fourier modes in both streamwise and spanwise directions, as well as 160 points in the wall-normal direction are used. Each simulation runs for more than 8000 viscous time units.

Paralleling the existence of an optimal T for (1), an optimal λ_x for (2) is revealed by our simulations. This is shown in fig. 1a, where at $A^+ = 12$ P_{sav} reaches a maximum at $\lambda^+_x = \lambda^+_{x,opt} = 1250 \approx 10T^+_{opt}$. This confirms the direct link between (1) and (2), given by the convective nature of the flow.

When the wall oscillates, turbulence interacts with the transverse Stokes layer [7]. Its oscillating shear breaks the quasi-coherent pattern of turbulence-sustaining wall structures. Similarly, with spanwise blowing (2) the flow develops a locally steady w profile that shows x-periodic variations strongly reminding the temporal oscillations of the Stokes layer.

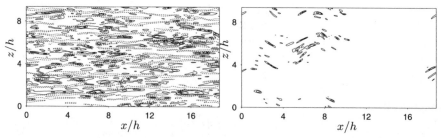

Fig. 2. Contours of streamwise velocity fluctuations u'^+ at $y^+ = 5$. The flow is from left to right. Contours are from $u'^+ = 2$ by 2 (continuous) and from $u'^+ = -2$ by -2 (dashed). Wall units with actual u_τ.

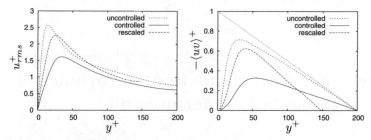

Fig. 3. Comparison of r.m.s. values of streamwise velocity (left) and Reynolds stresses (right) between the uncontrolled and controlled flow. Wall units are computed both with u_τ of the uncontrolled flow and with the actual u_τ.

The forcing (2) achieves a slightly better drag reduction in comparison to (1). More important, however, is the much lower power input required (see fig.1b), that gives a maximum net power saving of 24% at $A^+ = 12$, compared to a net power loss of 39% for the oscillating wall.

Fig.2 compares a snapshot of a flow field with a corresponding one in the uncontrolled case, and highlights the disruption imparted by the forcing (2) on the turbulence structure: low-speed streaks almost disappear. This can be observed also by looking at statistical quantities (see fig.3). Structural changes are not an obvious effect of decreasing the flow Reynolds number. As expected, drag reduction reduces turbulence intensities; the r.m.s of velocity fluctuations, as well as the Reynolds stresses present a reduced peak moved towards the centerline, indicating a thickening of the viscous sublayer.

The success of the steady forcing (2) in reducing turbulent drag suggests that a suitably designed rough surface may lead to similar effects.

References

1. W.J. Jung, N. Mangiavacchi, and R. Akhavan. Suppression of turbulence in wall-bounded flows by high-frequency spanwise oscillations. *Phys. Fluids A*, 4 (8):1605–1607, 1992.
2. M. Quadrio and P. Ricco. Critical assessment of turbulent drag reduction through spanwise wall oscillation. *J. Fluid Mech.*, 521:251–271, 2004.
3. M. Quadrio and P. Luchini. Integral time-space scales in turbulent wall flows. *Phys. Fluids*, 15(8):2219–2227, 2003.
4. T. W. Berger, J. Kim, C. Lee, and J. Lim. Turbulent boundary layer control utilizing the Lorentz force. *Phys. Fluids*, 12(3):631–649, 2000.
5. H. Zhao, J.-Z. Wu, and J.-S. Luo. Turbulent drag reduction by traveling wave of flexible wall. *Fluid Dyn. Res.*, 34:175–198, 2004.
6. P. Luchini and M. Quadrio. A low-cost parallel implementation of direct numerical simulation of wall turbulence. *J. Comp. Phys.*, 211(2):551–571, 2006.
7. G.E. Karniadakis and K.-S. Choi. Mechanisms on Transverse Motions in Turbulent Wall Flows. *Ann. Rev. Fluid Mech.*, 35:45–62, 2003.

Direct numerical simulation on turbulent flow around a regularly deforming film

Kousuke Takashima[1], Shuhei Koyama[2] and Yoshimichi Hagiwara[1]

[1] Department of Mechanical and System Engineering, Graduate School of Science and Technology, Kyoto Institute of Technology, Matsugasaki, Sakyo-ku, Kyoto, 606-8585 Japan yoshi@kit.ac.jp

[2] Mitsubishi Electric Corp, Wakayama, 640-8319, Japan

1 Introduction

It is thought that the membranous structure of life in water, such as algal blades and fish gills, has obtained a mechanism for reducing pressure drag and friction drag. However, the drags of filmy structure in turbulent flow are poorly understood. One of the present authors carried out a direct numerical simulation on a film which deformed sinusoidally, and found that the Reynolds shear stress and the total shear stress in the case of deformable film were lower than those respectively in the case of flat film[1]. In the present study, we carry out computations for a deforming film with higher amplitude than that in our previous study in order to investigate drags acting on the film.

2 Computational procedure

It was assumed that several films were parallel in the direction normal to the films. The two-dimensional deformation of films was assumed to be identical so that the distance between two adjacent films, $2h$, was unchanged. The deformation was assumed to be regular and periodical. A piece of the film surrounded by the others was dealt with. The thin region shown with the dotted curves in Fig. 1 was considered. The film was located in the center of the upper and lower boundaries of the region. Then, the upper (lower) surface of the film was converted to the inner surface of lower (upper) boundary for the computational domain. The maximum amplitude A_{max}^+ was set equal to 3 (hereafter called Case 1) and 4 (hereafter called Case 2). The ratio of A_{max}^+ to the wavelength was 0.0032 and 0.0042. This is much lower than the critical value of 0.02, over which the separation regime is generated in the valley of a wavy wall. The period of deformation was 60 wall units.

An unsteady generalized curvilinear coordinate was introduced in order to transform non-uniform grid points in the physical space to uniform grid points for the Cartesian coordinate in the computational domain. The equation of continuity and the momentum equations in the computational domain are expressed as follows:

$$\frac{1}{J} \cdot \frac{\partial JU^j}{\partial \xi^j} = 0, \frac{\partial u_i}{\partial t} + U^k \frac{\partial u_i}{\partial \xi^k} = -\frac{1}{\rho} \cdot \frac{\partial \xi^k}{\partial x_i} \cdot \frac{\partial p}{\partial \xi^k} + \frac{\nu}{J} \cdot \frac{\partial}{\partial \xi^k} \cdot \left(J \frac{\partial \xi^k}{\partial x_m} \cdot \frac{\partial \xi^l}{\partial x_m} \cdot \frac{\partial u_i}{\partial \xi^l} \right)$$

where J is the 3D transformation Jacobian. A collocated grid system was used for obtaining the discretized form of the spatial derivatives of the equations. The fourth-order central difference schemes were adopted for the time integration of convection and viscous terms[2]. The 3rd-order Runge-Kutta method was used for the time advancement. The pressure Poisson equation was solved with FFT and the Gaussian elimination method. The Reynolds number, $Re_{\tau 0}$, were 150.

3 Results and discussion

Figure 2 indicates the Reynolds-shear stress profiles. According to the numerical study on the flow over the fixed wavy wall carried out by De Angelis et al.[3], the Reynolds shear stress increases with the ratio of amplitude to wavelength. The result in Case 2 shows agreement with their result. However, the result in Case 1 shows a decrease in the stress. This is because the turbulence modification in the valley was moderate in Case 1.

Figure 3 exhibits the budget of turbulent kinetic energy. The absolute values of the terms for the governing equation of turbulent kinetic energy increase with the amplitude of wall deformation. This is consistent with the result obtained by De Angelis et al.[3]. It is striking that the viscous diffusion becomes predominant and balances with the dissipation in the near-wall region. This is caused by the time change in the linear sublayer thickness around the valley and hilltop of the deforming wall.

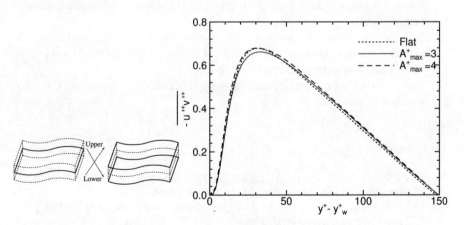

Fig. 1. Physical domain   **Fig. 2.** Reynolds shear stress

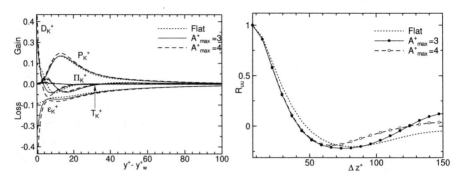

Fig. 3. Budget of turbulent kinetic energy   **Fig. 4.** Cross-correlation coefficient

Kim et al.[4] determined the average spacing of low-speed streaks in the transverse direction as twice of the transverse length, Δz, between the positions where the cross-correlation coefficient for the streamwise fluctuating velocity in the transverse direction, R_{uu}, takes its maximum and its minimum. Figure 4 depicts the cross-correlation coefficient. As the wave amplitude increases, the values of Δz are found to decrease, thus the spacing of the steaks become narrower. This shows that the coherent structure becomes finer as the wave amplitude increases.

4 Conclusions

The direct numerical simulation was carried out for turbulent flow along the regularly deforming wall. The main conclusions obtained are as follows:

1. The computational procedure with the unsteady generalized curvilinear coordinate, the collocated grid system and the schemes for solving the equations are effective for simulating the turbulent flow around the deforming walls.
2. Due to the change in the gradient of streamwise velocity, turbulence is modified in the lower part of the buffer region. The modification is more noticeable when the amplitude is high.
3. The coherent structure becomes finer by the wall deformation as the wave amplitude increases.

References

1. Y. Hagiwara, T. Imamura, A. Taki: J. of Turbulence, **3**, article no. 010 (2002)
2. Y. Morinishi: *Trans. Japan Soc. Mech. Eng.* **62** 4098 (1996)
3. V. De Angelis, P. Lombardi, S. Banerjee, *Physics of Fluids*, **9**, 2429 (1997)
4. J. Kim, P. Moin, R. Moser, J. Fluid Mech. **177** 133 (1987)

Progress in Large Eddy Simulation modeling of temporally and spatially complex land-atmosphere interactions

C. Meneveau[1], V. Kumar[2], S. Chester[3], and M. B. Parlange[4]

[1] Johns Hopkins University, Baltimore, USA meneveau@jhu.edu
[2] Johns Hopkins University, Baltimore, USA vijayant@jhu.edu
[3] Johns Hopkins University, Baltimore, USA chester@jhu.edu
[4] Ecole Polytechnique Federal de Lausanne, Switzerland marc.parlange@epfl.ch

1 Introduction

Large Eddy Simulation (LES) of the atmospheric boundary layer (ABL) is used to study several temporally and spatially complex phenomena. First, temporal complexity is considered by simulating an entire daily cycle of the turbulent atmospheric boundary layer. Secondly, spatial complexity is considered by simulating flow over multi-scale, fractal-like, objects using a new technique, called Renormalized Numerical Simulations (RNS). In both applications, the goal is to utilize notions of scale-invariance in a practical, predictive fashion.

2 Simulating a temporally complex diurnal cycle

While there is an abundance of LES studies of isolated, quasi-steady daytime (convective) and nighttime (stable) ABLs, there is a dearth of the same when it comes to LES studies of the diurnal evolution of ABL. The difficulties associated with simulating an entire diurnal cycle of the ABL stem from the extreme disparities between the nature of the two main parts of the diurnal ABL: The daytime convective boundary layer (CBL) and the nocturnal stable boundary layer (SBL). Specifically, the varied nature of these disparities makes it impossible for an *a priori* estimate based SGS model to smoothly adapt to the change in flow physics and this has led to a significant interest in using SGS models that do not require ad-hoc tuning of parameters (known as "dynamic models"[4]). Thus, a critical test of the performance of LES can be the simulation of the diurnal cycle of the ABL. Below we present the results obtained from LES of a diurnal ABL using a Lagrangian scale-dependent dynamic SGS model (for more details, see Refs. [10, 11]) and comparisons with an extensive field experimental study (the HATS 2000 experiment - henceforth referred to as H2K [6, 9]).

2.1 Simulation setup: LES of diurnal ABL cycle

The LES of the diurnal cycle of ABL is performed over a domain of size 4 km × 4 km × 2 km discretized using 160 nodes in each direction totalling to 4.1 million points. The horizontal resolution is $\Delta_x = \Delta_y = 25$ m while the vertical resolution is $\Delta_z = 12.5$ m. The code uses pseudo-spectral discretization in horizontal planes and centered second-order differentiation in the vertical direction. The data collected during H2K have been used to generate boundary conditions. Specifically, the heat flux boundary condition is a smoothed version of the 24 hour heat flux data observed on 6/9/2000, H2K (Fig 1a). The geostrophic wind is set to $(U_g, V_g) = (8,0)$ ms^{-1} during the convective part of the day and is gradually decreased to $(6,0)$ ms^{-1} during the nocturnal regime and is based on the observed temporal behavior of near surface wind velocity. Figure 1b shows the comparative evolution of friction velocity obtained from LES and H2K. The comparison between LES and H2K for the friction velocity is qualitatively good in terms of the temporal trend although the discrepancy seems to increase in the unstable part of the day.

2.2 Structure of ABL turbulence: Impact of stability

The onset of nocturnal boundary layer dynamics in the diurnal ABL is triggered by the rapid collapse of the convective boundary layer after sunset. The nearly-vanishing turbulent stresses lead to the onset of a dampened inertial oscillation which leads to the subsequent appearance of an inversion-layer wind maxima also known as low-level jet (LLJ) [1]. The evolution of the LLJ can be clearly seen in Figure 1c where the LLJ has a broad maximum at 2300 hours which evolves into a well-defined narrow peak at 0300 hours. Figure 2a shows the diurnal evolution of the vertical profile of Smagorinsky coefficient C_s obtained from the LES results. C_s in addition to being a function of space and time, is strongly affected by the atmospheric stability[9]. The tuning-free scale-dependent model reacts rapidly to the change in the atmospheric stability. The C_s attains higher values during the reign of the mixed layer while its value is greatly reduced during the stable, nocturnal regime. The coefficient has smaller values both in the near-surface region and in the overlying inversion layer albeit for altogether different reasons (i.e. enhanced shear for the former and dampened turbulence for the latter).

In order to further assess the behavior of computed variables as a function of atmospheric stability, we investigate the behavior of plane-averaged C_s as a function of stability parameter Δ/L (where L is the Obukhov length) for the entire diurnal cycle at different vertical levels (Δ/z). The empirical relation developed in *Kleissl et al.* (2004) relating C_s to height and stability given by

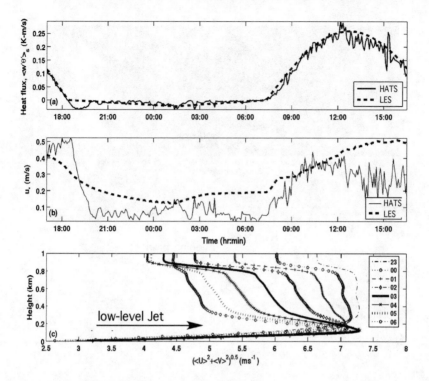

Fig. 1. (a): Observed diurnal cycle of surface heat flux in HATS 2000 and the smoothed surface heat flux boundary condition in LES. (b): Diurnal evolution of friction velocity, u_* from LES and H2K observations. The thin solid line represents the H2K observations while the dashed line represents the LES results. (c): Evolution of low-level jet in the stable boundary layer. The legend displays the hour corresponding to the plotted velocity profile.

$$C_s^{\Delta} = C_0 \left[1 + R(\frac{\Delta}{L}) \right]^{-1} \left[1 + \left(\frac{C_0}{\kappa} \frac{\Delta}{z} \right)^n \right]^{-1/n} \tag{1}$$

is used as a diagnostic reference for the observed behavior ($R(x) = 0$, $x < 0$; $R(x) = x$, $x \geq 0$ is the ramp function). Note that since C_s is plotted as a function of Δ/L, the points belonging to similar stability regime are lumped together irrespective of their chronological occurrence during the cycle. C_s as a function of Δ/L is in reasonable agreement with the empirical relation given by equation 1 as can be seen in Figure 2b for $\Delta/z = 0.142$ during the evening transition. For the morning transition, the agreement is not good due to the fact that the Obukhov length L is a physically meaningful parameter only in the immediate vicinity of the ground. At $\Delta/z = 0.142$ it takes some time for convective plumes to reach that height and so there is a delay in the increase of the coefficient ("hysteresis") on the morning branch of the curve.

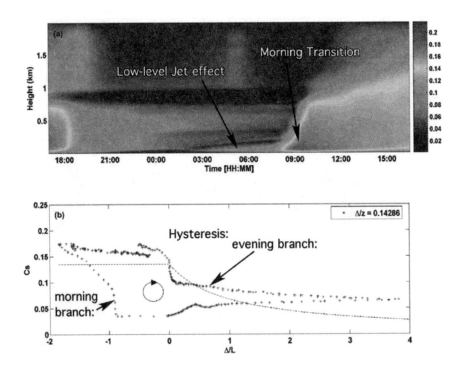

Fig. 2. (a) Diurnal evolution of Smagorinsky coefficient C_s (averaged over horizontal planes) using the Lagrangian scale-dependent SGS model. (b) Smagorinsky coefficient C_s as a function of Δ/L for $\Delta/z = 0.142$.

These observations underscore the attractiveness of the dynamic approach in which C_s does not need to be prescribed a-priori. Also the Lagrangian averaging technique allows to simulate temporally evolving flows in which regular time-averaging would not be feasible.

3 Simulating spatially complex flow using RNS

Spatial complexity is considered by simulating flow over multi-scale, fractal-like, objects. Fractals display large scale-disparity and complexity while being amenable to simple and standardized description. Hence, they offer an elegant idealization of the actual interactions between turbulence and boundaries in practical applications where boundaries are characterized by multiple length-scales. Prior experimental work on turbulent wakes induced by fractal grids has been performed[14, 15]. In terms of numerical simulations that explicitly resolve the fractal shapes, using LES of flow over prefractal shapes with increasing numbers of branch generations, the dependence of the tree drag on the inner cutoff scale of the fractal has been quantified[2]. The convergence of the drag coefficient towards a value that is independent of inner cutoff-scale was shown to be very slow. In order to address this fundamental difficulty and avoid the need to resolve all the small-scale branches of the fractal, a new numerical modeling technique called Renormalized Numerical Simulation (RNS), was introduced. In analogy with the dynamic model in LES[4], RNS models the drag of the unresolved geometry elements using drag coefficients as measured from the drag forces generated by the resolved branches and unresolved elements as modeled in previous iterations of the procedure. The RNS technique and its convergence properties have been tested by means of a series of simulations using different levels of resolution. The Renormalized Numerical Simulation (RNS) technique was introduced [2] as a way to compute the forces that develop between a high-Reynolds number flow and a fractal boundary, in a computationally affordable way. The RNS technique was shown to be stable and capable of predicting the total drag of certain idealized fractal trees. The increase of the drag of these idealized fractal trees with increases in their fractal dimension D was also quantified.

The method presented in [2] was restricted to trees with all of their branches perpendicular to the bulk flow and with branch orientations restricted to 90 or 45 degrees with the horizontal direction. In this presentation, we present a generalization of the above work by quantifying the effects of considering angles other than 90 and 45 degrees. Specifically, we wish to determine if the method is able to capture the expected changes in net drag. Such changes are expected to arise from the fact that depending on the angle the branches are exposed to different incoming velocity magnitudes due to the mean shear next to the ground.

3.1 Geometry of Fractal Trees

The fractal trees are placed in an infinite single span-wise row on the ground, resulting in a developing flow in the streamwise direction. The bulk flow is along the x-direction, and the trees are spaced a distance L in the spanwise (y) direction. The plane $z = 0$ is the ground. As in [2] for simplicity we consider only trees where the fractal construction is confined to a plane perpendicular to the main flow direction (for more details and a generalization to 3-D trees, see [3]). The trunk (generation $g = 0$) and branches ($g > 0$) are square cylinders to reduce Reynolds-number effects by fixing separation points, and all branches at a given generation have the same size. The trees are self-similar with a constant scale ratio $r < 1$ between successive branch generations. The diameter d_g and length l_g of a branch at generation $g \geq 0$ are related to those at the next generation $g + 1$ by $l_{g+1} = rl_g$ and $d_{g+1} = rd_g$. The number of sub-branches per branch is $N_B = 3$, and the scale ratio between branch generations is $r = 0.48$. It follows that the similarity fractal dimension [12] is $D = \log N_B / \log r^{-1} \approx 1.50$. Two orientations of the branches (80^0 and 100^0 instead of 90^0 of [2]) will be considered, as shown in Figure 3.

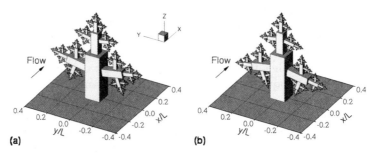

Fig. 3. (a) Sketch of fractal trees to be simulated in LES using RNS. (a) has an angle of 80^0 between branch-generations (upward branches), while (b) has 100^0 orientation, leading to down-ward pointing branches.

3.2 RNS and computational technique

The basic code is the same LES code described in section 2. To prescribe the inlet condition, a precursor simulation is used and a sponge region is used at the outlet to allow using periodic boundary conditions. In the bulk of the flow, the Smagorinsky SGS model is used for simplicity in these initial applications of RNS. In RNS one distinguishes between resolved and unresolved branches. In our LES code (see section 2), we use the immersed boundary method (IBM)[13]. The trees in which up to generation g is resolved are represented by a force field (per unit volume)

$$\mathbf{f}(\mathbf{x}) = \mathbf{r}(\mathbf{x}) + \sum_{\beta} \mathbf{f}_{\beta}(\mathbf{x}), \qquad (2)$$

where $\mathbf{r}(\mathbf{x})$ is the force from the IBM, the sum is over all unresolved branches β at generation $g + 1$, and $\mathbf{f}_{\beta}(\mathbf{x})$ is a distributed force field (per unit volume) representing effects of these branches and all higher-generation descendant branches of the fractal attached to them. RNS is used to determine the force fields $\mathbf{f}_{\beta}(\mathbf{x})$ at positions corresponding to the unresolved branches at generations $g + 1$ and above. RNS relates the total force total on the fluid due to the generation-$(g+1)$ branch β, and its descendants, to the total force due to the (resolved) generation g parent branch b and *its* descendants. The relationship involves a drag coefficient C_d determined by measuring the force at the larger scales and minimizing the error associated with predicting that force using the drag coefficient determined during a prior iteration (time-step) of the procedure. The basic equation[2] is of the form

$$C_d^m(g) = -\frac{2\sum_{b(g)} \left[\mathbf{F}_b^m (C_d^{m-1}) \cdot \mathbf{V}_b^m \right] |\mathbf{V}_b^m| A_b}{\rho \sum_{b(g)} |\mathbf{V}_b^m|^4 A_b^2}, \qquad (3)$$

where $\sum_{b(g)}$ signifies that the sum includes every branch b at the last resolved generation g, $\mathbf{V}_b^m$ is the spatially averaged velocity vector in a region in the neighborhood of branch b and its descendants, and $A_b = ld/(1 - N_B r^2)$ is the projected area of branch b and its descendants as seen by the bulk streamwise flow. The iterations in Eq. (3) at time 0 are initialized using $C_d^0(g+1) = 0$. The force field is replicated using the same geometric construction rules of the fractal. It also includes a spatial smoothing step as explained in detail in [2].

3.3 Results

Simulations are performed and the turbulence and forces generated by the fractal trees display stationary statistical behavior. Contours of the streamwise mean velocity field obtained from RNS for the two geometries tested are shown in Fig. 4. Darker regions of decreased velocity on the cross-stream vertical planes are visible at positions of unresolved branches. The slowed flow regions resemble the tree geometries (see Fig. 3). A global measure of the tree drag

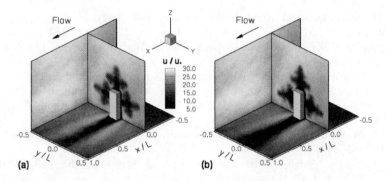

Fig. 4. Mean velocity as predicted by RNS and LES using $g = 0$ (i.e. only the trunk is resolved). Dark regions show where the mean velocity is low and visualizes the distributed force acting in the region of the unresolved fractal branches.

is given by the time-averaged effective drag coefficient $C_T = \bar{F}_T/(\frac{1}{2}\rho U^2 L^2)$. Here L^2 is the frontal flow domain area and $\bar{F}_T$ is the time-averaged total x-direction drag force the fluid applies on the tree, and U is the domain-mean streamwise velocity. From the two simulations, the main results are:

$$C_T(\theta = 80^0) = 0.364, \quad \text{and} \quad C_T(\theta = 100^0) = 0.334 \tag{4}$$

The 8.2% decrease in C_T in tilting the branches from $\theta = 80^0$ to 100^0 can be explained as follows: $\theta = 100^0$ has some of the tree's branches closer to the ground, in slower moving fluid, resulting in less force on the tree.

This illustrates that RNS can be used to predict the effects of changing unresolved geometry parameters, even though the resolved geometry is the same in each of the simulations. As a measure of how active the RNS force is in a simulation, we have checked that the unresolved force accounts for more than 75% of the total force in each case.

4 Conclusions

Modeling complex flows using Large Eddy Simulation requires advanced subgrid modeling strategies. It has been shown that the use of dynamical information at the resolved scales can be used in simulations of temporally complex flows, such as a diurnal cycle of the atmospheric boundary layer. A generalization of the dynamic model to determine drag coefficients associated with unresolved branches of a scale-invariant object (fractal tree in a boundary layer flow) has also been tested and shown to lead to meaningful results.

Acknowledgements: Financial support from the National Science Foundation (WCR-0233646, ATM-0222238) is gratefully acknowledged.

References

1. A. K. Blackadar: Bull. Amer. Meteo. Soc. **38**, 283 (1957)
2. S. Chester, C. Meneveau and M. B. Parlange. Modeling turbulent flow over fractal trees with Renormalized Numerical Simulation. *J. Comput. Phys.*, in press (2007).
3. S. Chester and C. Meneveau. Renormalized Numerical Simulation of Flow over Planar and Non-Planar Fractal Trees. *J. Env. Fluid Mech.*, submitted (2007).
4. M. Germano, U. Piomelli, P. Moin, and W. H. Cabot: Phys. Fluids A **3**, 1760 (1991).
5. A. A. M. Holtslag and F. T. M. Nieuwstadt: Boundary-Layer Meteorology, **36**, 201 (1986).
6. T. W. Horst, J. Kleissl, D. H. Lenschow et al: J. Atmos. Sci. **61**, 1566 (2004).
7. J. C. Kaimal, J. C. Wyngaard, D. A. Haugen et al: J. Atmos. Sci. **33**, 2152, (1976).
8. J. Kleissl, C. Meneveau, and M. B. Parlange: J. Atmos. Sci. **60**, 2372 (2003).
9. J. Kleissl, C. Meneveau, and M. B. Parlange: J. Atmos. Sci. **61**, 2296 (2004).
10. J. Kleissl ,V. Kumar, V., C. Meneveau and M. B. Parlange: Water Resour. Res. **42**, W06D09, (2006).
11. V. Kumar, J. Kleissl, C. Meneveau and M. B. Parlange: Water Resour. Res. **42**, W06D09, (2006).
12. B. B. Mandelbrot. *The Fractal Geometry of Nature* (W. H. Freeman, New York, 1982).
13. R. Mittal and G. Iaccarino. Immersed boundary methods. *Annu. Rev. Fluid Mech.* **37** 239–261 (2005).
14. D. Queiros-Conde and J. C. Vassilicos. Turbulent wakes of 3D fractal grids. In J. C. Vassilicos, ed., *Intermittency in turbulent flows*, 136–167 (2001).
15. A. Staicu, B. Mazzi, J. C. Vassilicos and W. van de Water. Turbulent wakes of fractal objects. *Phys. Rev. E* **67**. 066306 (2003).

Two Dimensional Polar Beta Plane Turbulence

G. Carnevale[1], A. Cenedese[2], S. Espa[2], M. Mariani[2]

[1] Scripps Institution of Oceanography , California University- La Jolla, California (U.S.) gcarnevale@ucsd.edu
[2] D.I.T.S. , Universita di Roma 'La Sapienza', Rome (IT) stefania.espa@uniroma1.it

The evolution of a two-dimensional turbulent decaying flow is experimentally and numerically analysed in a rotating system considering the effect of the change of the Coriolis force with latitude.

1 Introduction

Large scale flows in the ocean and in the atmosphere of the Earth and other planets are significantly influenced by rotation and stratification. Both of these effects cause the inhibition of the vertical velocity; as a consequence, motion on these scales can be represented using a quasi-two-dimensional approximation. Also, turbulent flows subjected to rotation develop spectral anisotropy [3]. The dynamics of 2D turbulence in the presence of strong rotation differs from 2D classical turbulence in at least two main respects: the inverse energy cascade is arrested at a characteristic wavelength [8], known as Rhines scale, above which the flow is predicted to show high anisotropy; opposite sign vortices organize along the meridional axis forming zonal jet like structures [9] observed in several geophysical systems. The argument to explain the energy transfer towards zonal modes in beta-plane turbulence [8] is based on a competition between nonlinear and beta terms in the quasi-geostrophic vorticity equation [6]. The associated characteristic scale separating the regions of wave-vector space where either beta or nonlinear effects dominate respectively, *i.e.* the Rhines scale , can be expressed in terms of rms velocity and beta coefficient: $k_{Rh} = \sqrt{\beta/U_{rms}}$. Thus k_{Rh} represents the arrest scale of the inverse cascade, a soft barrier to energy transfer towards smaller wave numbers. The presence of this barrier leads to a steep power law spectrum for $k > k_{Rh}$: $E(k) \propto k^{-5}$. This evidence has been confirmed by numerical simulations in different domain geometry [4, 10] and laboratory experiments [1, 7, 2].

2 The Model

The experimental set-up consists of a plexiglass tank L=33 cm × W=33 cm, filled with a 10 mm layer of an electrolyte solution of water and NaCl and placed flat on a rotating table (counter-clockwise sense of rotation in order to simulate flows in the northern hemisphere). The flow is forced by superposing an electric field and a magnetic field generated by permanents magnets covering the area of a circle on a metallic plate below the bottom surface of the tank. The forcing induces the continuous formation of opposite signed vortical structures whose characteristic length scale ($\sim$ 2 cm here) is related to the magnet size and to the distance between them. In order to reproduce in a laboratory rotating frame a similar effect of the quadratic variation of f with y, we used the parabolic profile assumed by the free surface of the fluid under rotation [1]. After the forcing is switched off, flow images are acquired using a standard video camera (acquisition frequency=25 Hz, resolution 720 × 570 pixels) placed orthogonally to the tank surface and co-rotating with the system. Images are then digitized and post-processed using a Feature Tracking technique to reconstruct the velocity field evolution in a Lagrangian frame of reference. After a resampling procedure, the vector fields are interpolated to a regular grid of 128 × 128 mesh points. A series of experiments characterized by increasing rotation rates ($0\,\mathrm{s}^{-1} \leq \Omega \leq 3.1\,\mathrm{s}^{-1}$), and, therefore, by different values of β and of the Ekman friction, λ_E, has been performed. Results corresponding to highest Ω will be shown. The experimental results are supported by direct numerical simulation of the quasi-geostophic vorticity equation [5] solved on a square 128 × 128 domain with no-slip boundary conditions. As initial condition, the vorticity field measured immediately after the forcing has arrested (t=0) is considered. Rotation and Ekman dissipation effects are modeled using values of Ω and λ_E which slightly differ from the experimental parameters.

3 Results

In figures 1, we show snapshots of velocity and potential vorticity fields obtained experimentally (a) and numerically (b). The pictures show the flow pattern: a weak anticyclonic circulation forms at the center of the domain while a cyclonic jet-like zonal current forms around the edge of the central anticyclonic region. This structure can be highlighted by the azimuthally averaged zonal velocity *vs.* the distance r (not shown) which is peaked in correspondence of the jet. Moreover, the jet region is subjected to a wave-like perturbation, a Rossby waves, which induces the formation of meanders associated with vortices. Experimental (a) and numerical (b) energy spectra computed averaging velocity vectors in time and over direction in wave-number space, are plotted in figures 2. Both spectra show a peak near $k \sim 10^0\,\mathrm{cm}^{-1}$, close to the theoretical estimate of k_{Rh}, and the slope approximates the k^{-5} scaling.

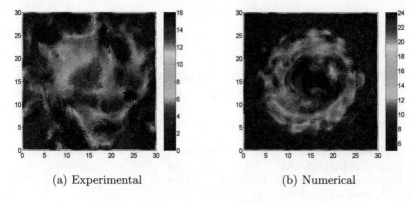

(a) Experimental   (b) Numerical

Fig. 1. Velocity and Potential Vorticity Fields

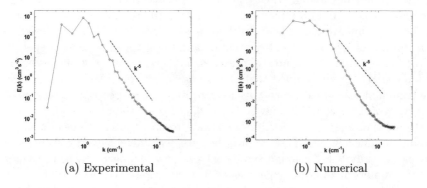

(a) Experimental   (b) Numerical

Fig. 2. Energy Spectra E(k)

References

1. Y. D. Afanasyev and J. Wells: Geophys. Astrophys. Fluid Dyn. **99-1**, 1 (2005)
2. J. Aubret, S. Jung, H.L. Swinney: Geophys. Research Letters **29**, 1876 (2002)
3. A. Cheklov, S.A. Orszag, S. Sukoriansky, B. Galperin, OI. Staroselsky: Physica D **98**, 321 (1995)
4. J. Y.-K. Cho and L. M. Polvani: Phys. Fluids **8**, 1531 (1996).
5. P. Orlandi: *Fluid Flow Phenomena : A Numerical Toolkit* (Kluwer Academic, Dordrecht 1999)
6. J. Pedlosky: *GeophysicalFluidDynamics* (Springer, Berlin 1979)
7. J. Juul Rasmussen, O. E. Garcia, V. Naulin, A. H. Nielsen, B. Stenum, L. J. A. van Bokhoven and S Delaux: Phys. Scr. **T122**, 4451 (2006)
8. P.B. Rhines: J. Fluid Mech. **69**, 417 (1975)
9. P.B. Rhines: Chaos **4**, 313 (1994)
10. G.K. Vallis and M.E. Maltrud: J. Phys. Oceanogr. **23**, 1346 (1993).

Laboratory study of gravity wave turbulence

Petr Denissenko[1], Sergei Lukaschuk[1] and Sergey Nazarenko[2]

[1] Fluid Dynamics Laboratory, the University of Hull, HU6 7RX, UK
S.Lukaschuk@hull.ac.uk
[2] Mathematics Institute, The University of Warwick, Coventry, CV4 7AL, UK
snazar1@yahoo.co.uk

Energy spectra of random gravity surface waves and their probability density functions (PDF) contain important information about nonlinear mechanisms of the wave interaction. There are two most celebrated theories related to different mechanisms of energy dissipation and the shape of the energy spectra in the inertial (universal) interval of scales. The first theory was suggested by Phillips (PH) [1] who argued that sharp wave crests of breaking waves play the dominant role in the short-wave asymptotics of the spectrum. The second theory, by Zakharov and Filonenko (ZF) [2], considers the wave energy scaling in the inertial range as the result of four-wave resonant interactions of random weakly nonlinear waves. Both approaches predict different energy scalings in the inertial interval. Although significant progress has been reached recently both in numerical and field experiments [3, 4], an unresolved issue still remains which of these theories, and under what conditions, could be applicable to the random waves generated in laboratory wavetanks. For example, Toba's experiment [5] which is cited as supporting ZF theory was done using wind wave forcing which is spread over the entire frequency range. Thus, there was no well-defined inertial range and no statistical isotropy due to high tank aspect ratio (typical for wind-wave tunnels), - both conditions are used in deriving ZF spectrum. In order to achieve the statistical isotropy and to obtain a wide inertial range of scales one should use the flumes with the horizontal aspect ratios closer to one and with a forcing localized at low frequencies. As far as we know, no attempts were made before to study nonlinear evolution of random waves in such flumes, and here we report the results of the first experiment of this kind.

We conducted the experiments with the surface gravity waves in a rectangular tank with dimensions 12 x 6 x 1.5 meters filled with water up to the depth of 0.9 meters. The wavemaker consists of 8 piston-type vertical paddles of width 0.75 m covering the full span of one short side of the tank. The amplitude, frequency and phase can be set for each paddle independently allowing to control directional distribution of the generated waves. In this experiment, we used sinusoidal excitation with the frequency 1.1 Hz and the wave vector

k_m ($|k_m| = 4.9m^{-1}$) at angle $15°$ to the flume axis such that the resulting wavefield is a composition of multiply reflected waves. The main control parameter in the experiment is the amplitude of wavemaker oscillations. We measured the surface elevation using two handmade wire capacitance probes positioning in the middle part of the flume. The average wave amplitude A can be expressed as the RMS of the wave height η: $A = \sqrt{\langle (\eta - \eta_0)^2 \rangle}$. The degree of nonlinearity can be characterized by the mean slope at the energy containing scale, i.e. $\gamma = k_m A$. The experiments covered the range of wave amplitudes from very weak waves with mostly smooth surface and occasional seldom wavebreaking, $A \approx 1.3$ cm and $\gamma \approx 0.052$, to very strong wave amplitudes characterized by a choppy surface with the numerous wavebraking events, $A \approx 5.3$ cm and $\gamma \approx 0.21$.

The typical wave spectra are shown in Figures 1a,b. The spectrum in Figure 1a corresponds to small excitation amplitudes and is much steeper than the spectrum in 1b corresponding to a strong excitation. For the wavefield of minimal amplitude the slope $\nu \approx 6.5 \pm 0.5$, which is in qualitative agreement with the prediction $\nu = 6$ made in [7] for the critical spectrum where the nonlinear resonance broadening is of the same magnitude as the mean spacing between the discrete k-modes. Thus, we indirectly confirm that the k-space discreteness (caused by the finite flume size) plays a defining role in shaping the frequency spectrum at low wave excitations. It was shown in [7] that in any large but finite box weak turbulence and ZF spectrum are realized if the wave intensity is strong enough so that the nonlinear resonance broadening is much greater than the spacing of the k-grid in the finite box. It implies a condition on the minimal angle of the surface elevation $\gamma > 1/(kL)^{1/4}$. We see that for small wave intensity this condition is marginally satisfied (up to the factor of 5), $\gamma \approx 0.074 \sim 1/(k_m L)^{1/4} \approx 0.4$.

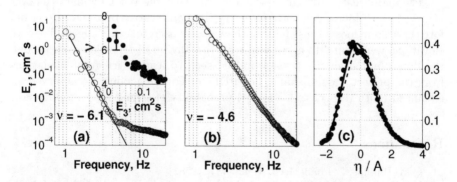

Fig. 1. Power spectra of wave elevation at wave intensities (a) A=1.85 cm and (b) A=3.6 cm. Inset shows the spectral slopes versus the wave intensity. (c) PDF of the wave elevation with the Gaussian (dashed line) and Tayfun (continuous line) distribution fits at the rms wave amplitude 3.2 cm

At large wave field intensities, one can see much better scaling behavior with significantly smaller scatter or uncertainty in the slope values, 1a. The width of the scaling range in this case reaches one decade in ω which is significantly greater than the scaling ranges observed in numerics and in field observations. There is a range of intensities where PH slope $\nu = 5$ is observed, and we note that wave breaking events were common for such intensities. At higher intensities, the slopes approaching $\nu = 4$ are observed as it was predicted by both ZF and Kuznetsov theories [2, 6]. However, the water surface was visibly very choppy with numerous frequent wavebreaking and high values of the surface slope, $\gamma > 0.15$ that rules out the weak nonlinearity assumption which is the basis of ZF theory [2]. Kuznetsov theory [6] is more likely to be relevant to these conditions, because it derives the slope ($\nu = 4$) from considering strongly nonlinear wavecrests with sharp 1D ridges and the speed of which is nearly constant while they pass the gauge.

The inset in Fig.1a shows the slopes as a function of wave intensity measured at frequency 3 Hz that can be considered as a low frequency boundary of the inertial interval. One can see that the slopes trend toward $\nu = 4$ as the wave intensity increases. However, no plateau were registered in our experiments either for $\nu = 5$ or $\nu = 4$. Such a change of slope could be explained by changing the fractal dimension of the wavecrest ridges from a set of $1D$ lines giving $\nu = 4$ at larger amplitudes (Kuznetsov) to more complex fractal curves at the lower intensities with $1 < D < 2$ giving $4 < \nu < 5$.

PDFs of the surface elevation measured for a weak and strong wave intensities are shown in Figure 1c. There is a good qualitative agreement of those PDFs with Tayfun distribution. However, we notice rather irregular deviations, especially near the PDF maximum corresponding to probability of waves with less-than-mean intensities.

The major feature of our experiments was that the WTT regime was never achieved: with increasing wave intensity the nonlinearity becomes strong before the system looses sensitivity to the k-space discreteness. PDFs of the wave elevations were found to agree well with the Tayfun distribution even though the flume finite-size effects were indeed seen to lead in irregular deviations from the perfect Tayfun shape. For PDFs of Fourier modes, we observed an enhanced (with respect to Gaussian) probability of strong wave amplitudes.

References

1. O.M. Phillips, J. Fluid Mech. **4**, 426-434, 1958.
2. V.E.Zakharov, N.N. Filonenko, J. Appl. Mech. Tech. Phys. **4** , 506-515 (1967).
3. N. Yokoyama, J. Fluid Mech. **501**, 169-178 (2004).
4. P.A. Hwang and D.W. Wang, J. Phys. Ocean. **30**, 2753-2787 (2000).
5. Y. Toba, J. Oceanogr. Soc. Jpn. **29**, 209 (1973).
6. E.A. Kuznetsov, JETP Letters, **80**, 83-89 (2004).
7. S.V. Nazarenko, J. Stat. Mech. L02002 (2006).

Quantification of the Discretization Effects in the Representation of Key Inertial-wave Interactions in Rotating Turbulence

Lydia Bourouiba

McGill University, Montréal, Québec, Canada, lydia.bourouiba@mail.mcgill.ca

1 Inertial waves: continuous vs. discrete wavenumbers

Rotation influences geophysical and engineering flows amongst many others. Its strength is only appreciable if it is large compared to the nonlinear term. The Rossby number, $Ro = U/fL$, is a dimensionless measure of the relative size of these terms. Here, f is twice the background rotation rate and U and L are characteristic length and velocity scales, respectively.

Nonlinear rotating flows are characterized by complex interactions involving 2D structures and waves. When rotation is strong, Ro becomes a small parameter and the dynamics of the 2D structures are slow compared to that of the waves. Thus, a multiple timescale expansion can be used. The leading order of the expansion admits linear inertial-wave solutions with frequencies $\omega_{s_{\mathbf{k}}} = s_{\mathbf{k}} k_z/k$ (axis of rotation is arbitrarily chosen to be $\hat{\mathbf{z}}$), where $s_{\mathbf{k}} = \pm 1$, $k = |\mathbf{k}|$ and $\mathbf{k} = (k_x, k_y, k_z)$ is the wavenumber in Fourier-space. The zero-frequency modes correspond to 2D structures invariant along $\hat{\mathbf{z}}$ ($k_z = 0$). Modes with non-zero-frequencies ($k_z \neq 0$) are referred to as 3D wave modes. Weakly nonlinear wave interactions were shown to contribute on the slow timescale only for a particular type of modes satisfying an additional condition of resonance [7]:

$$\omega_{s_{\mathbf{k}_1}} + \omega_{s_{\mathbf{k}_2}} + \omega_{s_{\mathbf{k}_3}} = 0, \quad \text{and} \quad \mathbf{k}_1 = \mathbf{k}_2 + \mathbf{k}_3. \tag{1}$$

When only modes satisfying (1) are active, a decoupling between 2D and 3D modes can be derived ([7] in periodic domains). This is a consequence of the zero interaction coefficient for resonant interactions involving one 2D mode and two 3D modes, noted $2 \rightarrow 33$ following the notation in [1].

Statistical theories and models assume infinitely large domains and timescale separation. [2] suggested that two-dimensionalization cannot rigorously be reached in unbounded domains. Experimental investigations of wave phenomena are done in bounded domains and numerical simulations usually assume periodic boundaries. In infinite domains, the components of the wavevector solutions to (1) are real numbers, whereas they are restricted to be integers in bounded and periodic domains. The importance of the discreteness of the wavenumbers in finite domains was first realized by [4]. Numerous theoretical and numerical studies followed (e.g. [5]). In the case of capillarity waves, [3] showed that a very low level of nonlinearity induces an insufficient capture of near-resonances for discrete wavenumbers. This comes from the non-existence of a resonant triad for capillary waves when wavenumbers are discrete [4]. Thus, frozen turbulence is observed numerically, unless sufficient nonlinear frequency shift is present. This nonlinear frequency shift induces a broadening of the resonant condition, giving

$$|\omega_{s_{\mathbf{k}_1}}(\mathbf{k}_1) + \omega_{s_{\mathbf{k}_1}}(\mathbf{k}_2) + \omega_{s_{\mathbf{k}_3}}(\mathbf{k}_3)| < O(Ro), \qquad (2)$$

instead of (1). The case of the inertial waves is different from that of capillarity waves. Similarly to gravity waves examined in [6], a subset of integer-value wavenumber solutions to (1) still exist for inertial waves. We investigate discretization effects for rotating flows with a focus on the $2 \to 33$ near and exact resonances due to the key role that these interactions play in differentiating between results obtained in continuous and bounded domains.

2 Discretization issues and finite-size effects

We count the near-resonant interactions (with non-zero interacting coefficient) satisfying (2) in a discrete domain. We use a right circular cylindrical truncation with height $2k_t$ and radius k_t (used in [1] due to the anisotropy of $\omega_{s_{\mathbf{k}}}$). The number of interactions as a function of Ro is shown in figure 1 for $k_t = 40$. The effect of the discretization is to introduce a minimum non-zero value of nonlinear broadening Ro, denoted by Ro_{min}.

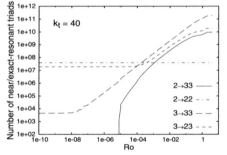

Fig. 1. Number of near and exact resonances of type $2 \to 33$ (and other types not discussed here) for $k_t = 40$.

Ro_{min} is directly linked to the cubic mesh $\Delta_{\mathbf{k}}$ used in the computational domain. For nonlinear frequency shift below Ro_{min}, no near-resonant interactions $2 \to 33$ are resolved. In the numerical domain examined here, $Ro_{min,2\to33} = 8 \times 10^{-6}$ for $k_t = 40$ (figure 1). In numerical simulations, the effects due to these near-resonances could only be captured if $Ro > Ro_{min}$.

There may be an infinite number of integer-value wavenumber triad solutions to (1) (can be shown), but the truncation k_t reduces the possible contribution of some interactions. That is, both the shape and limit imposed by the domain's truncation, k_t, affect the number of resolved near and exact resonances. The number of near-resonant $2 \to 33$ is displayed in figure 2 for truncations $k_t = 20, 30$ and 40. One way to investigate the effect of k_t, given a fixed $\Delta_\mathbf{k}$, is to renormalize the obtained numbers (figure 2 left) by the total number of all possible solutions to (1) (i.e. k_t^5 for $2 \rightarrow 33$ interactions, when $\Delta_\mathbf{k} = 1$). This leads to figure 2 right. For a nonlinear broadening sig-

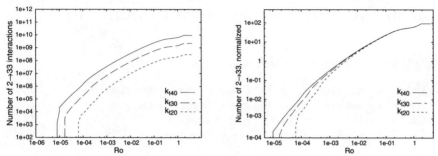

Fig. 2. Number of near-resonant $2 \rightarrow 33$ interactions as function of Ro. The right figure has been renormalized by the total number of modes in the discrete domain.

nificantly larger than the estimated Ro_{min}, the discrepancies in the collapse of the curves gives an estimation $Ro_{k_t} > R_{min}$ of the effect the truncation. When increasing k_t from 20 to 30, a doubling of the number of $2 \rightarrow 33$ near-resonant interactions is observed at $Ro_{k_t} \approx 1.6 \times 10^{-3}$. When increasing k_t from 30 to 40, the doubling occurs at $Ro_{k_t} \approx 3.1 \times 10^{-4}$ (figure 2 right).

We first investigated the effects of the discrete domain's mesh, $\Delta_\mathbf{k}$, on the number of near resonances, followed by the intricate effect of truncation, k_t. Due to the building assumptions of weak-turbulence, statistical results obtained/derived in infinite domains also have limited validity. Thus, investigations of rotating turbulence in finite and infinite domains could be complementary. The combination of Ro_{min}, Ro_{k_t} and the actual nonlinear broadening of the flow, Ro, is key when assessing the regimes, the interpretations and making accurate comparisons between rotating flow properties obtained from statistical theories and numerical simulations. The investigation of the nonlinear frequency shift for rotating simulations in periodic domains is the focus of our current work.

References

1. L. Bourouiba and P. Bartello: J.Fluid Mech., *in review*
2. C. Cambon, B. Rubinstein & F. Godeferd: New J. Phys. **6**, 73 (2004)
3. C. Connaughton, S. Nazarenko, A. Pushkarev: Phys. Rev. E **63** 046306 (2001)
4. E.A. Kartashova: AMS Transl, **182** (2), 95 (1998)
5. A.N. Pushkarev and V.E. Zakharov: Physica D **135** 98 (2000)
6. M. Tanaka, N. Yokoyama: Fluid Dynam. Res. **34** 199 (2004)
7. F. Waleffe: Phys.Fluids A, **4** 350 (1992)

Anisotropy and universality in Solar Wind turbulence. Ulysses spacecraft data.

A. Bigazzi[1], L. Biferale[2], S.M.A. Gama[3], and M. Velli[4]

[1] INAF and Dip. Fisica, Univ. Roma Tor Vergata, Rome, Italy.
[2] Dip. Fisica and INFN, Univ. Roma Tor Vergata, Rome, Italy
[3] CMUP and DMA, Univ. Porto, Porto, Portugal
[4] JPL, California Institute of Technology, Pasadena (CA), USA.

Summary. A novel (SO(3)) analysis of anisotropies in Solar Wind turbulence (Ulysses) shows the absence of a complete recovery of isotropy in magnetic field fluctuations at the smallest scales and persistence of intermittent, anisotropic contributions. Quantities mixing isotropic and anisotropic fluctuations, such as longitudinal structure functions, may thus be affected by uncontrolled errors.

1 Anisotropy of field fluctuations

Suppression of turbulence has long been observed in 2nd order longitudinal and transverse structure functions (SF) in the in the direction aligned with the large–scale mean field. Ulysses has been sampling Solar Wind (SW) plasma measuring velocity and magnetic field (MF) while orbiting the Sun for the first time on a polar orbit, from 1 A.U., where mean MF is almost parallel to the stream, to 6 A.U., where MF is almost perpendicular to the stream, an ideal dataset to study anisotropies.

Several MHD models incorporate at various levels the asymmetry of spectral indices in the field-aligned and transverse directions. A critical bibliography on SW turbulence and anisotropy, may be found in our longer paper [1] and in [4] . Statistics of MF fluctuations is reconstructed via the nth order correlation, $S^{(n)}_{\alpha_1,\dots,\alpha_n}(\boldsymbol{r})$, which depends on separation $(\boldsymbol{r})$,

$$S^{(n)}_{\alpha_1,\dots,\alpha_n}(\boldsymbol{r}) = \langle \delta_{\boldsymbol{r}} B_{\alpha_1} \delta_{\boldsymbol{r}} B_{\alpha_2} \cdots \delta_{\boldsymbol{r}} B_{\alpha_n} \rangle, \tag{1}$$

where $\delta_{\boldsymbol{r}} B_\alpha \equiv B_\alpha(\boldsymbol{x}+\boldsymbol{r}) - B_\alpha(\boldsymbol{x})$, and is the main quantity directly available from spacecraft data. Brackets $\langle \cdot \rangle$ in (1) indicate average over the locations $\boldsymbol{x}$. In (1) homogeneity is assumed, but not isotropy. This correlation function includes both *isotropic* and *anisotropic* contributions:

$$S^{(n)}_{\alpha_1,\dots,\alpha_n}(\boldsymbol{r}) = S^{(n),iso}_{\alpha_1,\dots,\alpha_n}(\boldsymbol{r}) + S^{(n),aniso}_{\alpha_1,\dots,\alpha_n}(\boldsymbol{r}). \tag{2}$$

For $n = 2$ and $\alpha_1 = \alpha_2$, we get the 2nd order SF, connected to the energy spectrum $E_{\alpha,\alpha}(\boldsymbol{k}) = \langle|\hat{B}_\alpha(\boldsymbol{k})|^2\rangle$ via a Fourier transform. Another widely used form of (1) is the longitudinal SF, obtained by projecting all field increments along the separation versor, $\hat{\boldsymbol{r}}$: $S_L^n(r) = \langle(\delta_{\boldsymbol{r}}\boldsymbol{B} \cdot \hat{\boldsymbol{r}})^n\rangle$. SO(3) decomposition of correlations (1), makes it possible to analize the anisotropic structure of MF fluctuation. However, the whole field in a 3D volume is required. Spacecraft data are instead inherently one-dimensional. However, it is still possible to extract those correlation functions, such as $S_{xy}^{(2)}(r_x)$ and $S_{xz}^{(2)}(r_x)$, whose isotropic contributions are, by symmetry, zero [2]. Their measure shall therefore quantify the degree of anisotropy of MF fluctuations. For a recent review on SF decomposition in hydrodynamics, for experimental and numerical data analysis see [4].

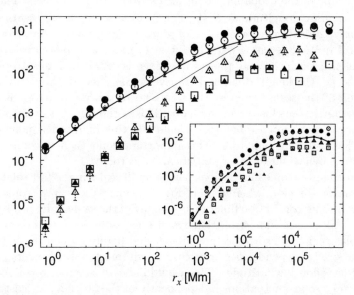

Fig. 1. Second order longitudinal, transverse and *purely anisotropic* SF. Low latitude dataset. Upper three curves: longitudinal and tranverse SF: solid line — $S_{xx}^{(2)}$; empty cirles $\circ$ $S_{yy}^{(2)}$; filled circles $\bullet$ $S_{zz}^{(2)}$. Errorbars are superimposed on — $S_{xx}^{(2)}$. Reference slope has angular coefficient of 0.7. Lower curves: purely anisotropic SF: $S_{xy}^{(2)}$, $\blacktriangle$ filled triangles; $S_{xz}^{(2)}$, $\triangle$ empty triangles; $S_{yz}^{(2)}$, $\square$ empty squares. Errorbars are superimposed on $\triangle$ $S_{xz}^{(2)}$. Inset: fourth order SF, longitudinal, transverse and *purely anisotropic*. Solid line, — $S_{xxxx}^{(4)}$; empty circles $\circ$ $S_{yyyy}^{(4)}$; filled circles $\bullet$ $S_{zzzz}^{(4)}$. *Purely anisotropic* SF are: $S_{xyyy}^{(4)}$, $\blacktriangle$ filled triangles; $S_{xzzz}^{(4)}$, $\triangle$ empty triangles; $S_{yzzz}^{(4)}$, $\square$ empty squares.

Let us therefore compare the undecomposed 2nd order SF with its anisotropic content. In Fig.1 we plot the longitudinal SF of 2nd order, $S_{x,x}^{(2)}(r_x)$ and the two transverse SF in the directions perpendicular to the $\hat{\boldsymbol{x}}$ axis, $S_{yy}^{(2)}(r_x)$ and $S_{zz}^{(2)}(r_x)$. All these functions have both isotropic and anisotropic contribution:

$$S_{\alpha,\alpha}^{(2)}(r_x) = S_{\alpha,\alpha}^{(2),iso}(r_x) + S_{\alpha,\alpha}^{(2),aniso}(r_x). \tag{3}$$

The two *purely anisotropic* 2nd order SF $S_{xy}^{(2)}(r_x)$ and $S_{xz}^{(2)}(r_x)$, are plotted in the same figure. We notice that *anisotropic* correlations have a smaller amplitude than the full correlation functions. This suggests that the isotropic contribution in the decomposition (2) is dominant. Moreover, anisotropic curves decay is slightly faster than full correlation, at small scales: isotropic fluctuations become leading proceeding to small scales, although very slowly. This is consistent with the recovery-of-isotropy assumption in some MHD models However, it is important to also control higher order statistical objects, i.e. the whole shape of the probability density distribution, at all scales. The inset of Fig. 1 compares longitudinal, $S_{xxxx}^{(4)}(r_x)$, transverse, $S_{\alpha\alpha\alpha\alpha}^{(4)}(r_x)$ (with $\alpha = y, z$) and *purely anisotropic* correlations of *fourth order* (see caption). Now the situation is quite different. First, the intensity of some *purely anisotropic* components are much closer to those with mixed isotropic and anisotropic contributions, i.e. the longitudinal and transverse SF. Second, the decay rate as a function of the scale is almost the same: no recovery of isotropy is detected for fluctuations of this order any more. This is the signature that anisotropy is mainly due to intense but rare events affecting high order moments more than 2nd order moments [3]. This important conclusion is confirmed by a consistent behaviour of other statistical indicators [1].

Strong anisotropic fluctuations persist at all scales in the fast solar wind. In the equatorial region, where data of Fig. 1 belong to, the anisotropic contents of fourth order correlation function is roughly of the same order as its isotropic part, at all scales, indicating that small scale isotropy is never achieved. In the polar region, anisotropies are smaller and highly fluctuating in time, but with a spatial dependencies compatible, within statistical errors, with the one observed at low latitudes. This would indicate some universal features of anisotropic solar fluctuations independently of the latitude, at least for what concerns their scaling properties. Our results therefore point toward a crucial role played by anisotropic fluctuations in the small scales statistics.

References

1. Bigazzi A.,Biferale L.,Gama S.M.A,Velli M.: ApJ, **638**, 499 (2006)
2. S. Kurien and K. R. Sreenivasan, Phys. Rev. E (2000) **62**, 2206.
3. L. Biferale and F. Toschi, Phys. Rev. Lett. (2001) **86**, 4831.
4. Biferale, L. & Procaccia, I., 2005, Phys. Rep. 414, Issues 2-3, 43.

Turbulence budgets in the wind flow over homogeneous forests

J.C. Lopes da Costa[1,3], F.A. Castro[1,3], A. Silva Lopes[1,2], and
J.M.L.M. Palma[1,2]

[1] CEsA — Centro de Estudos de Energia Eólica e Escoamentos Atmosféricos
[2] Faculdade de Engenharia da Universidade do Porto
[3] Instituto Superior de Engenharia do Porto
  loc@isep.ipp.pt

Introduction

The understanding of the flow inside and above forest canopies is an important
issue in the study of atmospheric flows, e.g. [1, 2]. Many canopy models have
been proposed within the framework of the k-ε RANS model. These models
differ basically in the way they define the source/sink terms of the transport
equations of the turbulence kinetic energy and its rate of dissipation,

$$S_k = \rho C_z \left(\beta_p |\mathbf{u}|^3 - \beta_d |\mathbf{u}|k \right); \; S_\varepsilon = \rho C_z \left(C_{\varepsilon 4} \beta_p \frac{\varepsilon}{k} |\mathbf{u}|^3 - C_{\varepsilon 5} \beta_d |\mathbf{u}|\varepsilon \right)$$

using different constants β_p, β_d, $C_{\varepsilon 4}$ and $C_{\varepsilon 5}$ (cf. Katul et al. [3]), aiming to
reproduce the canopy turbulence production and the turbulence dissipation
"short-cut" described by Finnigan [1].

Canopy Model	β_p	β_d	$C_{\varepsilon 4}$	$C_{\varepsilon 5}$
A - Katul et al.[3], Sanz [4]	1.0	5.1	0.9	0.9
B - Green [5]	1.0	4.0	1.5	1.5
C - Svensson et al. [6]	1.0	0.0	1.95	0.0
D - Liu et al. [7]	1.0	4.0	1.5	0.6

Appraisal of RANS based turbulence models are usually performed by
comparing experimental or numerical data of a variable of interest. Good
numerical data includes DNS or even LES results. Here, we present a set of
results from a range of canopy models available in the published literature
(A, B, C and D), and compare their ability to predict the turbulence kinetic
energy budgets, against the LES results based on a sub-grid dynamic model,
with a Lagrangian approach.

Results and Conclusions

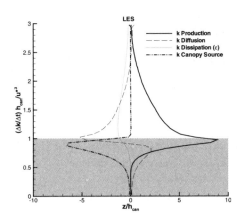

Fig. 1. LES turbulence kinetic energy budget terms. Results normalised by the canopy height, h_{can}, and the friction velocity at $z = h_{can}$, u^*.

The effects of the canopy are restricted to a mixing layer of about one canopy height above the canopy top (fig. 1). The canopy is a source of both additional shear, which increases the production of turbulence kinetic energy, and dissipation that are balanced by the production and diffusion terms inside the canopy.

The interaction among these three physical phenomena – production, dissipation and diffusion – is far from being captured by all models under study (fig. 2): A and B are qualitatively right, whereas C and D are unable to mimic the main physical phenomena and their relative magnitude. In the case of model C, because $\beta_d = 0$, the canopy source term in the k equation (S_k) is always positive. However, a β_d different from zero is not the only condition for a good prediction of the turbulence budget, as can be seen from the results of Model D, also with a positive contribution associated with the canopy source term.

A comparison, not included here, of the four canopy model based on the mean velocity, turbulence kinetic energy and eddy viscosity profiles, revealed in different degrees the aforementioned differences in the budget behaviours. The profiles as predicted by model C did not differ much from the LES results; at least to the extent that the analysis above would suggest.

The methodology presented here, because it was based on the analysis of each individual term in the k transport equation, enabled a much clearer comparison of all models and the conclusion that model A is the most adequate.

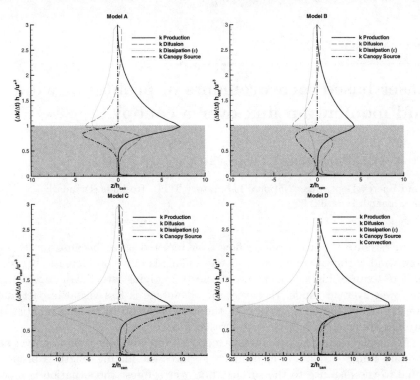

Fig. 2. RANS calculations turbulence kinetic energy budget terms for canopy models A, B, C and D.

References

1. J. Finnigan. Turbulence in plant canopies. *Annual Review of Fluid Mechanics*, 32:519–571, 2000.
2. J. C. Lopes da Costa, F. A. Castro, J. M. L. M. Palma, and P. Stuart. Computer simulation of athmosferic flows over real forests for wind energy resource evaluation. *Journal of Wind Engineering and Industrial Aerodynamics*, 94:603–620, 2006.
3. G. G. Katul, L. Mahrt, D. Poggi, and Christophe Sanz. One- and two-equation models for canopy turbulence. *Boundary-Layer Meteorology*, 113:81–109, 2004.
4. C. Sanz. A note on $k - \varepsilon$ modelling on a vegatation canopy. *Boundary-Layer Meteorology*, 108:191–197, 2003.
5. R. S. Green. Modelling turbulent air flow in a stand of widely-spaced trees. *Phoenics J.*, 5:294–312, 1992.
6. U. Svensson and K. Häggkvist. A two-equation turbulence model for canopy flows. *Journal of Wind Engineering and Industrial Aerodynamics*, 35:201–211, 1990.
7. J. Liu, J. M. Chen, T. A. Black, and M. D. Novak. $e - \varepsilon$ modelling of turbulent air flow downwind of a model forest edge. *Boundary-Layer Meteorology*, 72:21–44, 1998.

Laser based measurements of profiles of wind and momentum flux over a canopy

Jakob Mann, Ferhat Bingöl, Ebba Dellwik and Ole Rathmann

Wind Energy Dept., Risø National Laboratory/DTU, Roskilde, Denmark
jakob.mann@risoe.dk

We investigate the atmospheric flow over forested areas because more and more wind turbines are situated in these. This is quite problematic because the forest causes higher turbulence intensity implying larger structural loads and lower average winds. However, the advantage is less visual impact and lower prices of the land, so despite the drawbacks an increasing number of turbines are erected in or close to forests.

Atmospheric flow over a porous canopy is quite different from flow over an ordinary rough surface. Over ordinary surfaces the mixing length is proportional to the distance to the surface, but over a forest the situation is more complicated. Also the correlation coefficient of the horizontal and vertical velocity fluctuations is larger right over the forest compared to an ordinary surface [2]. The purpose of this work is to quantify these differences by an atmospheric field experiment using sophisticated laser anemometry.

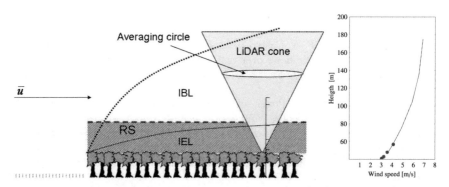

Fig. 1. Right: Sketch of the field experiment with a met mast and laser anemometer situated 1 – 2 km from the forest edge. Left: Wind profile from in situ anemometers (dots) and the laser (line).

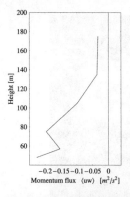
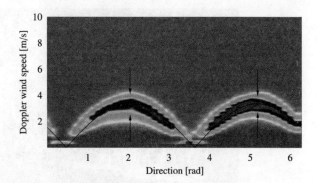

Fig. 2. Left: Profile of momentum flux. Right: Average Doppler spectra as a function of the beam direction ($z = 105$ m).

Wind turbines today reach heights of 150 m, but making long term measurements at these heights using a conventional meteorological tower is impractical and expensive. A homodyne laser Doppler anemometer based on fiber optics components has therefore been developed to measure remotely the wind speeds at these heights [3]. The instrument only needs to receive 1 backscattered photon out of 10^{12} emitted to assess the radial velocity at the point of focus. Assuming the flow field to be roughly homogeneous over the averaging circle (see figure 1) with a mean (u, v, w) the radial velocity in the direction of the laser beam v_r is (because the half opening angle of the cone is $\approx 30°$)

$$v_r = \left| \frac{1}{2} u \cos\theta + \frac{1}{2} v \sin\theta + \frac{\sqrt{3}}{2} w \right|,$$

where θ is the horizontal angle from the downwind direction. From this equation all three velocity components can be obtained through fitting the equation to the Doppler spectra, as shown as the thin, black curve in figure 2. The fluctuations in the upwind ($\theta = \pi$) and downwind ($\theta = 0$) directions are

$$\sigma^2(v_{r,up}) = \frac{1}{4}\sigma_u^2 + \frac{3}{4}\sigma_w^2 - \frac{\sqrt{3}}{2}\langle u'w' \rangle$$

$$\sigma^2(v_{r,down}) = \frac{1}{4}\sigma_u^2 + \frac{3}{4}\sigma_w^2 + \frac{\sqrt{3}}{2}\langle u'w' \rangle$$

so subtracting these equations the momentum flux $\langle u'w' \rangle$ can be obtained. $\pm\sigma(v_{r,up})$ and $\pm\sigma(v_{r,down})$ are indicated by arrows in figure 2, and it can be seen that $\sigma(v_{r,up})$ at direction ≈ 2 is the larger implying a negative $\langle u'w' \rangle$. The whole flux profile is also given in figure 2.

We want to evaluate recent models for flow above canopies, both k-ε models [4] and analytic [1], in light of the application to wind energy.

Currently, we have mounted the lidar horizontally right over the canopy in order to assess the degree of inhomogeneity over the crowns (see figure 3).

Fig. 3. Experimental site seen from above with the platform for the lidar to the right.

We do that to investigate the representativeness of profile measurements close to the top of the forest.

References

[1] S. E. Belcher, N. Jerram, and J. C. R. Hunt. Adjustment of a turbulent boundary layer to a canopy of roughness elements. *J. Fluid Mech.*, 488:369–398, 2003.
[2] M. R. Raupach, J. J. Finnigan, and Y. Brunet. Coherent eddies and turbulence in vegetation canopies: The mixing layer analogy. *Boundary-Layer Meteorol.*, 56:163–195, 1996.
[3] David A. Smith, Michael Harris, Adrian S. Coffey, Torben Mikkelsen, Hans E. Jørgensen, Jakob Mann, and Régis Danielian. Wind lidar evaluation at the Danish wind test site Høvsøre. *Wind Energy*, 9(1–2):87–93, 2006.
[4] Andrey Sogachev and Oleg Panferov. Modification of two-equation models to accound for plant drag. *Boundary-Layer Meteorol.*, 2006. available online.

Anomalous one- and two-particle dispersion in anisotropic turbulence

F.S. Godeferd, L. Liechtenstein and C. Cambon

LMFA UMR 5509, École Centrale de Lyon, France
fabien.godeferd@ec-lyon.fr

We study the dispersion of fluid particles in homogeneous turbulence rendered anisotropic by external forces, as in rotating stably stratified flows, which are found in the atmosphere or the ocean. Two-particle dispersion plays a key role in environmental studies, so that parameterizing the anisotropy magnitude and providing accurate models for vertical and horizontal dispersion is a timely issue. Stably stratified and/or rotating flows have a well-known Eulerian structure: flat vertically sheared horizontal structures in strongly stratified cases; elongated Taylor-vortex like structures along the rotation axis when rotation is dominant. Previous studies (*e.g.* [2]) have shown that non linearity is a key ingredient for the generation and long-time persistence of this anisotropy. Inertio-gravity waves. correspond to exact linear solutions but are only marginally active in the Eulerian anisotropy creation. However, the one-particle dispersion of linearly interacting inertio-gravity waves—wave turbulence—presents the same anomalous features as observed in full direct numerical simulations (DNS) ([3]). We here study mechanisms coupling Eulerian and Lagrangian anisotropies, through Lagrangian particle trajectories.

Two methods are used: Direct Numerical Simulations, for obtaining the full nonlinear Lagrangian response of turbulence upon injection of particle pairs; we also derive an analytical estimate for anisotropic two-particle dispersion by combining the approximation for particle pair dispersion proposed by [1] and Rapid Distortion Theory (RDT), based on linearized Navier-Stokes equations.

Methods and results

The problem is multi-parametric: Reynolds, Rossby, Froude non dimensional numbers for turbulence, rotation and stratification. The initial separation of particle pairs, relative to the Kolmogorov or to the vertical stratification length scale, can also modify the long time behavior of two-particle dispersion.

We perform high resolution DNS of stably stratified turbulence (512^3), following fluid particles trajectories, and compute one- and two-particle dispersion. These full nonlinear results are compared to a linear model, based on the implicit integral formula of [1], in which we introduce the analytical prediction for two-time velocity correlations coming from the linear Rapid Distortion Theory.

The two-particle relative displacement tensor is $\Delta_{ij}(t) = \langle \delta x_i(t)\delta x_j(t)\rangle - \langle \delta x_i(0)\delta x_j(0)\rangle$; δx_i is the two-particle separation in the i-th direction, with initial separation $\delta x_i(0)$. Ishihara and Kaneda's approximation yields the implicit equation $\Delta_{ij}(t) = \mathcal{F}(\Delta_{\alpha\beta}(s'), \hat{Q}_{ij}(\boldsymbol{k}, s'; s))$ where $\hat{}$ is the Fourier transform, and $\mathcal{F}$ an integral functional operator, with three integrals: over three-dimensional $\boldsymbol{k}$–space, twice over time to account for Lagrangian history.

Solving the implicit equation requires estimates for the Lagrangian two-point two-time velocity correlation $Q_{ij}(\boldsymbol{r}, s'; s) = \langle v_i(\boldsymbol{x} + \boldsymbol{r}, s'; s)v_j(\boldsymbol{x}, s'; s')\rangle$, with $\boldsymbol{v}(\boldsymbol{x}, s'; s)$ the velocity at time s of a fluid particle that was at $\boldsymbol{x}$ at time s'. In their work devoted to isotropic turbulence only, $\mathbf{Q}$ was evaluated from Lagrangian Renormalized Approximations or Taylor expansions. Here, we retain the vertical/horizontal anisotropy of dispersion, and we assimilate $\mathbf{Q}$ to the Eulerian two-point two-time velocity correlation tensor, given analytically by RDT (see refs. in [3]), as a function of the prescribed spectrum $E(k)$.

The model provides the full tensor $\Delta_{ij}(t)$. We plot the horizontal and vertical displacement statistics $\Delta_{hor}(t) = (\Delta 11(t) + \Delta_{22}(t))/2$ and $\Delta_{vert}(t) = \Delta_{33}(t)$ on figure 2 for different parameters. Figure 1 shows them from DNS for various initial separations. It clearly shows that the plateau observed on vertical dispersion can appear in two distinct stages when the separation is small enough. This is retrieved in the linear model (fig. 2b), in an increasingly marked manner with larger initial separations. On both figures, and on fig. 2(a), the horizontal dispersion initially follows the ballistic t^2 law which seems to accelerate to t^3 or more, before retrieving the t^2 behavior at long times (as expected in isotropic turbulence).

Fig. 2(a) shows the dispersion for rotating/stratified cases at small initial particle pair separation. Three stages are clearly marked on the vertical dispersion curves, which ultimately reach a second plateau, at a faster pace when rotation is smaller. This confirms the competing roles of rotation and stratification for vertical dispersion. In rotation-dominant cases (not shown here), rotation prevails and the trend is reversed.

Two cases of fig. 2(a), with the same parameters, correspond to a different choice of the spectral distribution $E(k)$ chosen in the RDT model. One spectrum is narrow banded, the other extends towards the smaller dissipative scales, while keeping the same peak location, and hence almost the same global kinetic energy (i.e. larger inertial range and higher Reynolds number). It is clear from fig. 2 that Δ_{vert}'s second plateau is reached later for the case with larger spectral extent, reflecting the importance of turbulent characteristic length scales in parameterizing vertical stratified dispersion. The same is observed in DNS.

From all the DNS results we obtained for two-particle dispersion, no obvious unique non dimensionalization was found to scale the later stage plateau of vertical dispersion. Although the Eulerian mechanisms for anisotropic structuration of stratified rotating turbulence are now well known, the highly non-linear character of the dynamics still eludes single scalings, especially for Lagrangian laws and time-scales for particle pair dispersion.

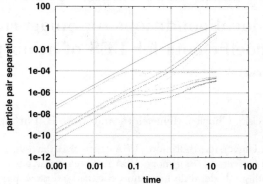

Fig. 1. Ensemble average of the separation distance of particle pairs computed by DNS for stably stratified turbulence, at different initial separations. Vertical separation curves exhibit a plateau, horizontal separation is faster. The ballistic t^2 short-time regime (Taylor 1921) is observed on all curves.

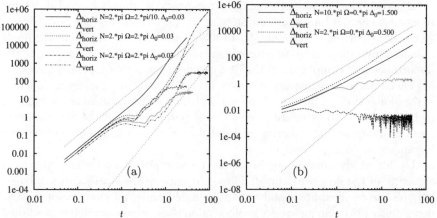

Fig. 2. Analytical results of two-particle displacement, normalized by initial separation, for various stratification and rotation parameters N and Ω. Δ_{horiz}: horizontal separation; Δ_{vert}: vertical direction. (a) Small initial separation, moderate stratification with rotation. Bottom and middle labels have the same parameters, respectively with a narrow spectrum and a large one. (b) Larger initial separations and strongly stratified case. (Dotted lines for t^2 and t^3 laws.)

References

1. ISHIHARA, TAKASHI & KANEDA, YUKIO 2002. *Phys. Fluids*, 14(11).
2. LIECHTENSTEIN, L. , GODEFERD, F. , & CAMBON, C. 2005. *J. Turb.*, (6), 1–18.
3. LIECHTENSTEIN, L. , GODEFERD, F. , & CAMBON, C. 2006. *Int. J. of Heat and Fluid Flow*, 27, 644–652.

Effects of local conditions on Smagorinsky and dynamic coefficients for LES of atmospheric turbulence

Marcelo Chamecki[1], Charles Meneveau[2], and Marc B. Parlange[3]

[1] Johns Hopkins University, Baltimore, USA `chamecki@jhu.edu`
[2] Johns Hopkins University, Baltimore, USA `meneveau@jhu.edu`
[3] Ecole Polytechnique Federal de Lausanne, Switzerland `marc.parlange@epfl.ch`

1 Introduction

Parameterization of subgrid-scale (SGS) phenomena is a critical component in large eddy simulation (LES) of the atmospheric boundary layer. Effects due to mean shear and buoyancy compromise the performance of the SGS parameterizations currently available [5]. In addition, the extremely high Reynolds number limits the resolution affordable near the ground surface, increasing the importance and complexity of the SGS phenomena in this region. The resulting effects upon SGS parameters have been characterized mostly as function of the global state of the flow. However, in LES mean quantities are not known and are difficult to estimate in complex situations. Ideally, SGS parameters must be expressed as function of local flow variables that characterize the instantaneous flow phenomena.

In this study several dimensionless parameters characterizing the local structure of the flow are defined. These parameters are locally defined, based only on the resolved fields and remain bounded under all circumstances. The dependence of the Smagorinsky coefficient on these local parameters is studied *a priori* from field data measured in the atmospheric surface layer (HATS data set [3]). Dependences on locally defined parameters are expected to improve the SGS model by sensitizing it to local flow conditions. Capitalizing on the findings of the *a priori* test, a new formulation for the dynamic model [2] is proposed, where the Smagorinsky coefficient is regarded as a function of the local structure of the turbulence.

2 Results

Results from the *a priori* analysis show various important and inter-related trends, such as significant increases of the coefficient in regions of large strain-rate self-amplification and vortex stretching. The well known changes in the

value of Smagorinsky coefficient with atmospheric stability [4, 5] are captured
by a locally defined version of the gradient Richardson number (Ri_*). As in [4]
the distance from the surface is characterized by Δ/z. The joint dependences
on these two parameters are shown in Fig. 1(a). dependences on the strain
state parameter (S^*) are shown in Fig. 1(b). Definitions and results for the
entire set of parameters are described in [1].

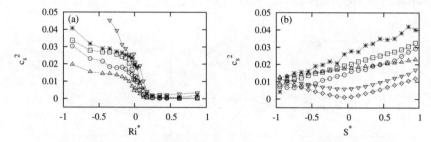

Fig. 1. Values of the Smagorinsky coefficient from the HATS data as a function of
(a) Ri^* for different values of Δ/z (with ∇: $\Delta/z = 0.5$; $*$: $\Delta/z = 1$; $\square$: $\Delta/z = 2$; $\circ$:
$\Delta/z = 2$; $\triangle$: $\Delta/z = 4$ and (b) S^* for different atmospheric stability and $\Delta/z = 1$
(with $*$: strongly unstable; $\square$: unstable; $\circ$: weakly unstable; $\triangle$: weakly stable; ∇:
stable; $\diamond$: strongly stable).

In order to avoid fitting a complicated curve to the data and determining
an *ad-hoc* form of the optimal dependence of the Smagorinsky coefficient on
the set of parameters, the Germano identity [2] can be used. If the coefficient
is regarded as a function of the local structure one can write $c_s^2 = c_s^2(\mathcal{S})$, where
$\mathcal{S}$ is a set of parameters describing the local structure of the turbulence (e.g.
in Fig. 1(a), $\mathcal{S} = \{Ri^*, S^*\}$). Following the minimization approach proposed
by Lilly [6] the error in the Germano identity from closing the SGS stress with
the Smagorinsky model at both scales becomes:

$$\varepsilon_{ij} = \mathcal{L}_{ij} + 2[c_s(\widetilde{\overline{\mathcal{S}}})\gamma\Delta]^2|\widetilde{\overline{\mathbf{S}}}|\widetilde{\overline{S}}_{ij} - 2\overline{[c_s(\widetilde{\mathcal{S}})\Delta]^2|\widetilde{\mathbf{S}}|\widetilde{S}_{ij}}. \tag{1}$$

In Eq. (1) S_{ij} is the strain-rate tensor, $\mathcal{L}_{ij}$ is the Leonard stress tensor, and
tilde and bar represent grid and test-filtering. Then the conditional mean
squared error $\langle\varepsilon_{ij}\varepsilon_{ij} \mid \widetilde{\overline{\mathcal{S}}}\rangle$ can be minimized iteratively:

$$[c_s^{n+1}(\widetilde{\overline{\mathcal{S}}})]^2 = \frac{2\Delta^2\langle[c_s^n(\widetilde{\mathcal{S}})]^2|\widetilde{\mathbf{S}}|\widetilde{S}_{ij}|\widetilde{\overline{\mathbf{S}}}|\widetilde{\overline{S}}_{ij} \mid \widetilde{\overline{\mathcal{S}}}\rangle - \langle\mathcal{L}_{ij}|\widetilde{\overline{\mathbf{S}}}|\widetilde{\overline{S}}_{ij} \mid \widetilde{\overline{\mathcal{S}}}\rangle}{(\gamma\Delta)^2\langle|\widetilde{\overline{\mathbf{S}}}|^4 \mid \widetilde{\overline{\mathcal{S}}}\rangle} \tag{2}$$

The proposed dynamic formulation given by Eq. (2) is tested *a priori*
using DNS of neutrally buoyant, isotropic turbulence. Results show that the
iterative approach converges and reproduces the correct trends. The example
for S^* is shown in Fig. 2(a), where the measured coefficient (i.e. obtained using

the small scale information not available during the simulation) is compared to the iterative solution of Eq. (2). The coefficient obtained by the traditional dynamic approach is shown for comparison. Note the similarity between the measured coefficient (solid line) and the same results from HATS for near neutral atmospheric stability (circles and triangles in Fig. 1(b)). The iterative procedure is further illustrated in Fig. 2(b), where the initial condition ($c_s^2 = (0.17)^2$) and the first five iterations from Eq. (2) are shown. Note that after 3 iterations the trend has already converged.

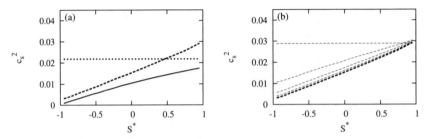

Fig. 2. Values of the Smagorinsky coefficient as a function of S^* for DNS of isotropic turbulence (a) *a priori* optimal curve (solid line), traditional dynamic model (dotted line) and iterative approach using Eq. (2) (dashed line) and (b) first 5 iterations.

3 Conclusions

A priori analysis of atmospheric turbulence data shows dependences of the optimal value of the Smagorinsky coefficient upon several dimensionless parameters characterizing the local structure of the turbulent flow. The enforcement of the Germano identity in a conditionally average sense together with an iterative solution of the resulting equation provides a methodology to evaluate these dependences during the simulation. The proposed model is shown to converge to the desired results within a few iterations in *a priori* analysis of DNS of isotropic turbulence.

References

1. M. Chamecki, C. Meneveau, and M. B. Parlange: J. Atmos. Sci. (in press)
2. M. Germano, U. Piomelli, P. Moin, and W. H. Cabot: Phys. Fluids A **3**, 1760 (1991)
3. T. W. Horst, J. Kleissl, D. H. Lenschow et al: J. Atmos. Sci. **61**, 1566 (2004)
4. J. Kleissl, C. Meneveau, and M. B. Parlange: J. Atmos. Sci. **60**, 2372 (2003)
5. P. J. Mason: Quart. J. Roy. Meteor. Soc. **120**, 1 (1994)
6. D. K. Lilly: Phys. Fluids A **4**, 633 (1992)

Asymptotic behaviour of the shearless turbulent kinetic energy mixing

D.Tordella, M.Iovieno and P.R.Bailey

Dipartimento di Ingegneria Aeronautica e Spaziale, Politecnico di Torino, Corso Duca degli Abruzzi 24, 10129 Torino, Italy. daniela.tordella@polito.it

1 Introduction and Experimental Rationale

A turbulent shearless mixing layer is generated by the interaction of two homogeneous isotropic turbulent fields, see sketch in fig. 1(left). The mixing is characterized by the absence of a mean velocity gradient, so that there is no production of turbulent kinetic energy, but has inhomogeneous statistics due to the presence of gradients of turbulent kinetic energy and integral scale. It was first experimentally investigated in [1] and [2] by means of passive grid generated turbulence and later numerically in [3] and [4] and more recently in [5] and [6]. All these studies considered a decaying turbulent mixing.

In the present study, we consider mixings in which the integral scale is homogeneous. The aim is to show the strong intermittent behaviour of such a configuration that in the past was considered almost Gaussian. This interpretation was based on laboratory observations carried out without a sufficiently high kinetic energy gradient, and on the other hand was motivated by the absence of either kinetic energy production or an integral scale variation, two typical sources of intermittency.

In passive grid experiments the gradients of integral scale and kinetic energy are intrinsically linked. In numerical, see [5], or active grid experiments these two parameters can be independently varied. In the present study, numerical experiments have been performed to determine the intermittency as a function of the energy ratio only, which is possible by keeping the integral scale uniform across the mixing. Another aim of this numerical experiment is also to reach the asymptotic condition where the kinetic energy ratio E_1/E_2 goes to infinity (see fig. 1(right) where E_1/E_2 is up to 10^4 and fig. 2, where E_1/E_2 reaches the value 10^6). This last condition is relevant in applications concerning the diffusion of a turbulent field in a region of still fluid.

Navier-Stokes equations are numerically solved with a fully dealiased Fourier-Galerkin pseudospectral method [7]. The computational domain is a parallelepiped with periodic boundary conditions in all directions of dimensions $2\pi \times 2\pi \times 2n\pi$, where $n = 2$ and 4 to check for any interference due to the finite

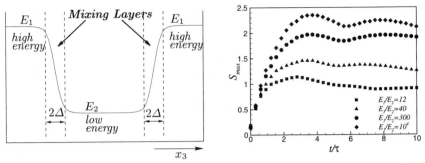

Fig. 1. Left: sketch of the shearless turbulent mixing in a periodic domain. Reference frame: x_1, x_2 normal to the mixing direction x_3. The high energy (E_1) and low energy (E_2) regions are separated by mixing layers of conventional thickness $\Delta(t)$ defined as in [2] and [5]; $\Delta(0) \approx 1/40$ of the domain width. Right: time evolution of the maximum of the skewness $S = \overline{u_3^3}/\left(\overline{u^3}\right)^{3/2}$ for various energy ratios.

nature of the domain. Initial conditions are generated by matching two homogeneous turbulence fields of differing energies, this ensures that the two flows have the same spectrum and thus no initial gradient of scales is introduced. Statistics are obtained by averaging over planes normal to the mixing.

2 Results and discussion

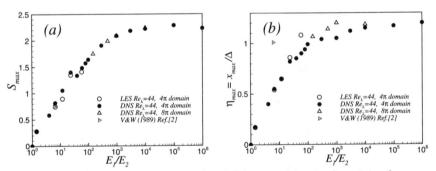

Fig. 2. (a) Maximum of the skewness and (b) its position in the mixing layer as a function of the initial energy ratio. Note that data from [2] have non-unity integral scale ratio ($\ell_1/\ell_2 \simeq 1.5$) so are not perfectly comparable with the present results.

Results from numerical simulations show that the mixing layer is highly intermittent and has a self-similar stage of decay. The skewness distribution of the velocity fluctuations normal to the mixing layer, that is the component in the direction of the flow of turbulent kinetic energy, is a principal indicator of intermittent behaviour. Skewness vanishes in homogeneous isotropic turbulent flows and thus it remains zero in the field external to the mixing. Fig. 1(right) shows the time evolution of the maximum of the skewness for three

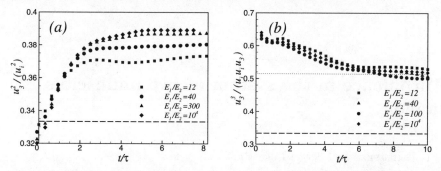

Fig. 3. Anisotropy of the turbulent kinetic energy and of the third order moments at the centre of the mixing layer. Thick horizontal dashed lines indicate isotropic reference values.

simulations with energy ratios between 12 and 10^4. During the initial eddy turnover times $\tau = \ell/E_1^{1/2}$, ℓ being the integral scale (homogeneous through the mixing), the skewness increases steadily and reaches an asymptote after about $2.5 - 3$ eddy turnover times. At this point the mixing layer enters a self-similar stage of evolution. The value of the maximum is a function of the energy ratio as depicted in fig. 2(a). For values of E_1/E_2 lower than 10^2 it scales almost linearly with the logarithm of the energy ratio, which is in fair agreement with the scaling exponent of 0.29 found in [5].

The same behaviour can be seen in the mixing penetration, defined as the normalised position of the maximum of skewness, see fig. 2(a).

Fig. 3 shows the anisotropy of the second and third order moments. For the second order moments deviations of up to 17% from the isotropic value of $1/3$ are visible. For third order moments, that is turbulent kinetic energy flow in the mixing direction, a very different behaviour is observed. About half of the total kinetic energy flow $\overline{u_i u_i u_3}$ ($i = 1, 2, 3$) was contributed by the velocity fluctuations u_3 in the direction of the mixing. This is seen to be a common feature for varying the kinetic energy ratio. Anisotropy is in agreement with the similarity analysis carried out in [6].

We wish to acknowledge the support of CINECA, HLRS and BSC computing centres in providing access to their resources and technical assistance.

References

1. B.Gilbert. *J. Fluid Mech.* **100**, 349–365, (1980).
2. S.Veeravalli, Z.Warhaft. *J. Fluid Mech.* **207**, 191–229, (1989).
3. D.A.Briggs, J.H.Ferziger, J.R.Koseff, S.G.Monismith. *J. Fluid Mech.* **310**, 215–241, (1996).
4. B.Knaepen, O.Debliquy, D.Carati. *J. Fluid Mech.* **414**, 153–172, (2004).
5. D.Tordella, M.Iovieno. *J. Fluid Mech.*, **549**, 441-454, (2006).
6. M.Iovieno, D.Tordella. Submitted to *Phys. Fluids*, (2007).
7. M.Iovieno, C.Cavazzoni, D.Tordella *Comp. Phys. Comm.*, **141**, 365–374, (2001).

Turbulence in the system of two immiscible liquids

Petr Denissenko and Sergei Lukaschuk

Laboratory of Fluid Dynamics, University of Hull, P.Denissenko@gmail.com

Energy dissipation in two-phase turbulent flows involves an additional mechanism associated with the deformation of interfaces [1]. This raises a question about applicability of the conventional turbulent theory to the case of multiphase flows. A two-phase liquid-liquid system provides a simple example where knowledge of the turbulent statistics is important for describing mixing and emulsification. While static properties and stability of emulsions has been extensively studied by post-process analysis, not much is known on the dynamics and statistics of turbulence in the liquid-liquid systems [2]. One of the major obstacles in measuring turbulence in two-phase systems is that multiple light scattering by interfaces produces turbidity within the medium and prevents the application of optical techniques. We overcame this problem by precise matching of the liquids refractive indices. Here we present the first results on velocity statistics and phase distribution in a system of two immiscible liquids.

Our experimental cell is a closed rectangular vessel $8 \times 8 \times 12$ cm where liquid is stirred by a pair of coaxial counter-rotating two-bladed impellers (Fig.1a). Configuration of the impellers was chosen to maximize their impact on the fluid motion while minimizing breaking of the droplets. We use a pair of immiscible liquids, the silicone oil and the water-glycerol mixture with the viscosity $\nu \approx 0.03$ cm^2/s. Refractive indices of the liquids were coarsely matched by choosing the ratio of water to glycerol (approximately 60:40) and then finely tuned by adjusting the temperature. For visualization, the water-glycerol mixture was stained with a fluorescent dye Rhodamine 6G. The cell was illuminated by a lasersheet generated by a pulsed Nd YAG laser with the wavelength of 532 nm. Images were captured by a CCD camera equipped with a fluorescence filter. One component of the fluid velocity was simultaneously measured at two points by Laser Doppler Anemometry (LDA). Seeding particles were added to both oil and water-glycerol phases. Velocity spectra were calculated using the non-uniform time sampling intervals because the seeding particles tend to be trapped by the interfaces [2].

A sample of the acquired image is shown in Fig.1b. The images were processed to find the boundaries of droplet sections. The surface energy contained in the fluid was calculated as the total surface area of the droplets times the constant of surface tension $\sigma \approx 0.04$ N/m. We use the assumption that diameter of the visible cross-section is the same as diameter of the droplet. To estimate the energy of the droplets' deformations, shapes of the droplet boundaries were expanded to angular Fourier harmonics. The energy of the nth harmonic was calculated as the difference of the surface area of a distorted droplet and the area of a sphere of the same volume:

$$E_n = (k_n a_n)^2 \cdot \sigma \cdot r_{droplet}^2 \tag{1}$$

Here k and a are the harmonic wavenumber and amplitude. The equation (1) provides an estimate of E_n since we only assess the distortion of a particular droplet cross-section. The energy of droplets' deformation was calculated as the sum of energies of all harmonics with the wavelength greater than 10 image pixels (130 μm) at all droplets detected in all acquired images. Few to several hundred images with several thousand droplets were analyzed at each flow regime and the obtained value of energy was normed using the total volume of detected droplets. The surface energy and the energy of droplet deformation are plotted against the energy of turbulent pulsations defined from LDA measurements (Fig.2a). Observe that the surface energy is several times greater than the energy of turbulent pulsations.

In addition to the energy scale and the viscous scale as in continuous fluid, a capillary scale is expected to appear in a two-fluid system [1]. We define it as the mean wavelength of the droplet surface deformations calculated as

$$l_c = \left(\sum E_n\right)^{-1} \sum \lambda_n E_n \tag{2}$$

where the wavelength assigned to the nth harmonic is the circumference of the droplet section divided by the angular wavenumber of the harmonic. In Fig.2b we plot l_c versus the energy of turbulent pulsations.

To assess the influence of the interfaces on the flow structure we analyze the energy spectrum and the two-point correlations of fluid velocity. LDA measurements showed significant, by an order of magnitude, decrease of the energy of pulsations at the scales corresponding to l_c and shorter in the two-phase flow compared to that in the flow of pure oil. The two-point measurements have not revealed significant changes in the velocity spatial correlation at the scale of l_c.

References

1. M. Chertkov, I. Kolokolov, and V. Lebedev: Phys. Rev. E **71**, 055301(R) (2005)
2. Bernard P. Binks: Emulsions - Recent Advances in Understanding. In: *Modern Aspects of Emulsion Science*, ed by Bernard P. Binks (The Royal Society of Chemistry, Cambridge 1998) pp 1–55

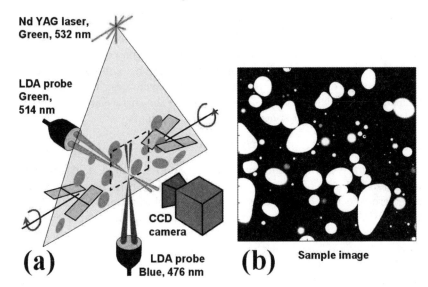

Fig. 1. (a) The flow of silicone oil (80%) and the water-glycerol mixture (20%) is driven by a pair of counter-rotating impellers. Impellers' outer diameter is 55 mm, rotation speed $3 \div 8$ rps. Matching of the refractive indices of liquid phases enables the use of LDA for velocity measurement and the use of lasersheet for analysis of droplet shapes. (b) A section of fluorescent-dyed water-glycerol droplets suspended in silicone oil.

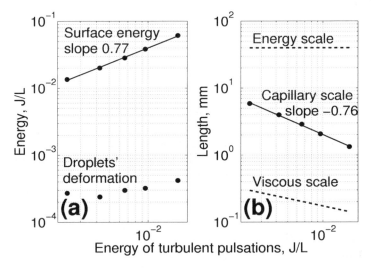

Fig. 2. (a) Dependence of the energy stored in droplets' surface and the energy stored in deviation of droplets' surface from spherical on the energy of turbulent pulsations. (b) Dependence of the capillary scale (the mean wavelength of the droplet surface deformation) on the energy of pulsations. The energy scale (propellers' diameter) and the viscous scale are indicated by dotted lines.

Diffusion in time-dependent laminar flow with multi-scale Eulerian flow topology

P. Kewcharoenwong, L. Rossi and J.C. Vassilicos

Department of Aeronautics and Institute for Mathematical Sciences,
Imperial College London, SW7 2AZ, United Kingdom.
pk1@ic.ac.uk, l.rossi@ic.ac.uk and j.c.vassilicos@ic.ac.uk

In this work, a new class of unsteady laminar flows is introduced and studied. These flows are generated in the laboratory using a shallow layer of brine and controlled by multi-scale electromagnetic forces resulting from the combination of an electric current and a fractal magnetic field created by a fractal permanent magnet distribution (see figure1). These flows are laminar yet turbulent-like in that they have multi-scale streamline topology in the shape of "cat's eyes" within "cat's eyes" or 8 within 8 (figure1) similar to the known schematic streamline structure of two-dimensional turbulence [1, 3]. This multi-scale topology is invariant over a broad range of Reynolds numbers, Re_{2D} from 600 to 9900 [7] and the flows have a power-law energy spectrum $E(k) \sim k^{-p}$ over a broad range of electromagnetically forced scales $2\pi/L < k < 2\pi/\eta$ where p is controlled by the imposed fractal dimension, D_s of the laminar multi-scale streamline structure's stagnation points. Specifically, p is a decreasing linear function of D_s and D_s can be set by the multi-scale electromagnetic control scheme [2, 4]. Re_{2D} and L/η are independent parameters in these multi-scale forced flows.

Diffusion in turbulent-like multi-scale laminar flows with steady forcing has previously been studied [7, 8] and found to possess a Richardson-like behaviour comparable to $\langle \Delta^2(t) \rangle \sim t^\gamma$ with $\gamma \approx 3$ [8] as a result of the cumulative effect of all multi-scale stagnation points. We follow [7] and generate time-dependent quasi-two-dimensional (Q2D) laminar flows with turbulent-like multi-scale Eulerian flow topology in the laboratory and investigate their diffusion properties which are important as measures of stirring and mixing. Although it is known that turbulence is unsteady, recent works by [3, 5] have suggested that there are regions characterised by the so-called "stagnation points" within turbulent flows which are statistically persistent in the ways that they are long-lived compared to turbulent time-scales and slowly moving in a particular frame where there is no mean flow. Whilst [7] and [8] merely focus on steady flows which are relevant in the sense that their multi-scale distribution of stagnation point is persistent, we produce controlled unsteady flows where stagnation points move (or not, according to the chosen forcing) more or less slowly compared to the rest of the flow. This is an important novelty of this work compared to [7]. For this, the unsteadiness is introduced to the flows by means of time-dependent forcing, specifically time-dependent electrical current (e.g. in figure2). Even more chaotic properties and a higher rate of diffusion are expected in the turbulent-like multi-scale laminar flows with time-dependent forcing.

We thus generate various time dependencies of the forcing, e.g. frequency, mean intensity and magnitude, so as to excite different flow scales. Each flow scale has its own length-scale and time-scale which vary with different flow/forcing intensities. We estimate time-scales via the characteristic turnover times (for each length-scale) at different flow/forcing intensities by various approaches using the magnet sizes, the sizes of the streamline attaching to the stagnation points, strain rate and energy spectrum. In figure 3, these time-scales (coloured areas) are used to represent the typical timescales associated to the large scale, medium scale and small scale. We are able to target and force flow scales "one by one" and/or all together by adjusting the frequency of the forcing and its amplitude as illustrated by the vertical bar in figure 2 corresponding to the forcing of the dye visualisation in figure 4 which clearly shows that the medium and the large scales are excited by this forcing. Various flow measurements including dye visualisation and Particle Tracking Velocimetry (PTV) are performed and post-processed. The techniques developed provide highly resolved Eulerian velocity fields in space and time. An example of preliminary result of time-dependent Eulerian velocity fields is illustrated in figure 5.

We are then able to investigate diffusion in our turbulent-like unsteady multi-scale laminar flows by tracking fluid elements numerically in the highly resolved velocity fields. In general, diffusion in turbulence follows Richardson's law of particle pair dispersion $\langle \Delta^2(t) \rangle \sim t^3$. On the contrary, chaotic advection type of diffusion with exponential pair separation $\Delta(t) \approx \Delta_0 e^{Lt}$ is what is normally found in unsteady laminar flows [6, 9]. Among the questions we are currently addressing is the type of diffusion we will obtain according to the selected time dependent forcing. We explore the impact of our time-dependent properties on various Eulerian and Lagrangian statistics and seek to identify the spatio-temporal structure of our flows and its consequences for efficient mixing. Other flow properties such as wavenumber and frequency energy spectra will also be extracted from the flow and will be used to explore/confirm possible relationships between: γ, p and D_s following the approach suggested in [1].

The outcome of the work provides the analysis of multi-scale flows which are a tool for the controlled study, in the laboratory, of complex flow properties related to turbulence and mixing with potential applications as efficient mixers as well as in geophysical, environmental and industrial fields.

References

[1]   J.C.H. Fung and J.C. Vassilicos: Phys.Rev. E **61**, 1677-1690, (1998).
[2]   J. Davila and J.C. Vassilicos: Phys. Rev. Lett. **91** (14), 144501, (2003).
[3]   S. Goto and J.C. Vassilicos: New Journal of Phys. **6** (65), 1-35, (2004).
[4]   Hascoet et al.: Submitted to J. Fluid Mech, (2006).
[5]   Osborne et al.: Phys. Rev. E **74**, 036309, (2006).
[6]   J.M. Ottino: The kinematics of mixing: stretching, chaos, and transport (Cambridge University Press, Cambridge, United Kingdom, 1989).
[7]   Rossi et al.: J. Fluid Mech. **558**, 207-242, (2006).
[8]   Rossi et al.: Phys. Rev. Lett. **97** 144501, (2006).
[9]   J.C. Vassilicos: Phil. Trans. R. Soc. Lond. A **360**, 2819-2837, (2002).

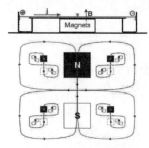

Fig. 1. Rig's schematic for electromagnetic forcing of a shallow brine layer (side view) and Schematic of the permanent magnets placed under the brine supporting wall (top view).

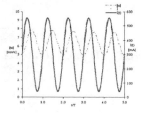

Fig. 2. Preliminary result of mean velocity from PTV of time-dependent flow, $\|u\|$ in mm/s and I in mA. • are measured electrical current which is well appointed by forcing: $I(t)=0.298+0.258\sin(2\pi t/T)$ with T =12.5s.

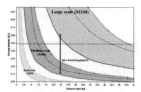

Fig. 3. Approximated time scales characterising 3 length scales (according to the size of the magnets M10, M40, M160) at various forcing intensity (current).

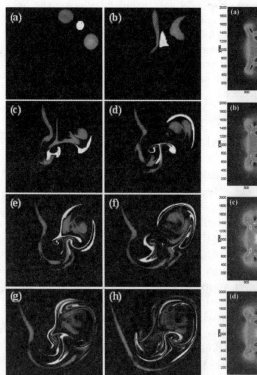

Fig. 4. Dye visualisation of an unsteady multi-scale laminar flow (quarter of flow) with time-dependent forcing: $I(t)=0.4+0.25\sin(2\pi t/T)$ where T =15s. Figure (a) - (h) show flow at the at t/T= 0, 1, 2, 3, 4, 5, 6 and 9 respectively.

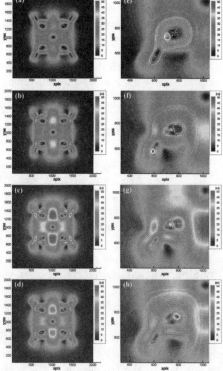

Fig. 5. Preliminary result of Eulerian velocity fields extracted from adaptive PTV measurements with time-dependent forcing of period T. Figure (a,b,c,d) and (e,f,g,h) show velocity fields with time-increment of T/4 of the entire flow and quarter of flow respectively.

Dynamics of Scalar Injection in Freestream Turbulence

E. Sanz, C. Nicot, R. Point, and F. Plaza

L.M.F.A., UMR C.N.R.S. 5509, Ecole Centrale de Lyon- Université Lyon I - INSA Lyon, 36, avenue Guy de Collongues, F-69314 ECULLY Cedex, FRANCE
fplaza@ec-lyon.fr

Summary. We report an experiment about the process of temperature injection from a heated patch, situated in a flat plate boundary layer, submitted to freestream turbulence. We measure simultaneously the incoming turbulent velocity, the heat flux released in the flow by the controlled constant-temperature heated patch and the outgoing temperature. The heat flux signal and the outgoing temperature shows a $f^{-7/3}$ temporal spectral scaling, in agreement with the prediction for scalar flux in homogeneous turbulence. Cross-correlations shows that the outgoing scalar field is anti-correlated with the incoming turbulence and the heat flux, showing the importance of slow events in scalar injection.

Introduction and experimental setup

The role played by injection in passive scalar statistics, through the formation of the "large scales" has already been stressed [1, 2]. We attempt here to evaluate directly this factor. We designed an experiment where the passive scalar is temperature, injected by a inhouse-built hot film maintained at constant temperature. It acts as a heat flux sensor that measures the total heat transfer released in the flow over a large surface but with a fast resolution. It consists in a metal film, whose electrical resistance is kept constant by electronic conditioning, so that the electric power supplied to the sensor equals the total heat flux lost in the flow. The experimental details of the sensor design and fabrication have already been given elsewhere [3]. The sensor is positioned in a profiled support which is placed in a freestream turbulent air jet (fig.1). The shape of this support enables the transition to turbulence from its downstream edge, even when placed in a laminar outer flow. Thus, the effects observed here are not due to the triggering of the transition by the outer turbulence, but to the outer fluctuations upon the already developed turbulent boundary layer. The nozzle of the jet is squared (12 cm side), and the support is placed at a distance of 9 nozzle sides. The sensor itself is at 10 cm from the support downstream edge. The size of the patch (2.42 cm x 2.42 cm) is comparable to the

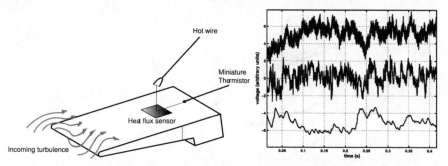

Fig. 1. left: sketch of experimental set-up; right: typical signals obtained by the heat flux sensor (up), the hot wire probe (middle) and the probe temperature (down)

transverse integral length of the incoming turbulence. To complete the analysis of the injection phenomenon, a hot wire is placed directly above the hot film, far enough not to be contaminated by the temperature wake of the film, but at a distance smaller than the transverse integral scale of the incoming turbulence. A miniature thermistor (130-micrometer diameter) is placed in the thermal wake of the film, in order to collect the temperature fluctuations released by the hot film into the flow.

Results and discussion

Figure 1 presents typical signals from the three sensors and figure 2 typical results concerning temporal power spectra and cross-correlation between the ingoing longitudinal velocity, measured by the hot wire, the whole heat flux and the outcoming temperature measured by the thermistor. The incoming velocity shows a spectral scaling of $f^{-5/3}$ (f is the temporal frequency), in agreement with the Taylor hypothesis and Kolmogorov classical scaling. The heat flux signal does not exhibit the same scaling but a power law closer to $f^{-7/3}$. First, it indicates that all dynamical structures in the incoming turbulence participate to the heat transfer. Moreover, this scaling is in agreement with the prediction for scalar flux in turbulence [4]. Concerning the temperature released in the wake of the heated patch, a even clearer $f^{-7/3}$ scaling is observed. This indicates that these temperature fluctuations are totally monitored by the heat transfer process from the patch, no turbulent mixing mechanism has time to develop at this stage of scalar dynamics. This scaling is quite different from the f^{-1} dependence sometimes observed in the close wake of a scalar source in line emission experiments [5] or from the $f^{-5/3}$ scaling observed in many passive scalar experiments [1]. As shown in figure 2, there is a clear correlation between velocity and heat flux, with a delay corresponding to a characteristic time formed with the boundary layer thickness δ and the friction velocity U_τ (see [3]). The value and delay of this correlation maximum

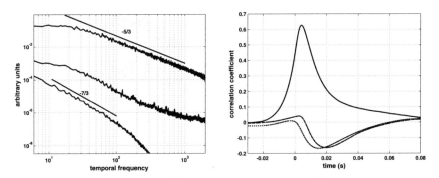

Fig. 2. Left: temporal power spectrum of the three signals; up: velocity, with the $f^{-5/3}$ scaling above; middle: heat flux; down: temperature. Right: cross-correlation coefficients; plain line : heat flux vs velocity; dashed line: temperature vs velocity; dotted line: temperature vs heat flux.

is dependent on the vertical distance between the hot wire and the heat flux sensor. Note that this distance is quite superior to the turbulent boundary layer thickness [3]. That is consistent with the already observed fact that the outer structure penetrate inside the turbulent boundary layer [6] and drive the transfer process . The temperature signal is anticorrelated with the others signals, showing that the slower structures catch the higher temperature after the convection process.

As a conclusion, we observe several evidence that enlighten the process of scalar injection in a boundary layer submitted to freestream turbulence : all the incoming turbulent scales participate to the process; the time scaling of the heat flux and outgoing temperature is coherent with classical prediction on scalar flux spectrum; the slower events in the incoming velocity provide higher scalar values to the flow. A more complete work with full experimental details and systematic study will be provided shortly. We acknowledge A. Effernelli, M. Teissieux, P. Dutheil for technical help, J.-L. Leclercq and P. Cremillieu from INL for the lending of clean room facilities, B. Houx and ISYMECA platform, and W. Bos for fruitful discussions.

References

1. Z. Warhaft: Annu. rev. Fluid Mech. **32**,203 (2000).
2. A. Celani et al: Phys. Fluids **13**(6),1768 (2001).
3. E. Sanz, C. Nicot, R. Point and F. Plaza in: *Proceedings 13th International Heat Transfer Conference* (Begel house, inc. publisher, 2006) pp 248.
4. J.L. Lumley: Phys. Fluids **10**(4), 855 (1967).
5. E. Villermaux, C. Innocenti and J. Duplat: Phys. Fluid **13**(1), 284 (2001).
6. G. Charnay, J. Mathieu and G. Comte-Bellot: Phys. Fluids **19**(9),1261 (1976).

Direct Numerical Simulation of Pulsating Turbulent Channel Flow for Drag Reduction

Kaoru Iwamoto[1,2], Naoaki Sasou[1], and Hiroshi Kawamura[1]

[1] Dept. of Mech. Eng., Tokyo Univ. of Science,
   2641 Yamazaki, Noda, Chiba 278-8510, Japan
[2] Present address: Dept. of Mech. Sys. Eng., Tokyo Univ. of A & T,
   2-24-16 Nakacho, Koganei, Tokyo 184-8588, Japan
   iwamotok@cc.tuat.ac.jp

1 Introduction

Development of efficient turbulence control techniques for drag reduction and heat transfer augmentation is of great importance from the viewpoint of energy saving and environment impact mitigation. Pulsating turbulent flows occur in many engineering applications. However, pulsating turbulent flow control such as for drag reduction has rarely been studied. The only investigations concerning the drag reduction are showed in Refs. [1, 2, 3]. In addition, the flow control for reduction of pumping power needed to drive a flow has never been reported.

In the present study, direct numerical simulation of pulsating turbulent channel flow is performed to examine the possibility of drag reduction and pumping power reduction. The mean pressure gradient is changed pulsatingly, not same as those in the most previous papers: the flows were driven by a sinusoidal time-oscillatory mean pressure gradient. The final goal of the present work is to optimize time-varying mean pressure gradient to minimize skin friction drag or pumping power.

2 Numerical Method

The numerical method used in the present study is almost the same as that of Ref. [4]; a pseudo-spectral method with Fourier series is employed in the streamwise (x) and spanwise (z) directions, while a Chebyshev polynomial expansion is used in the wall-normal (y) direction. The Reynolds number $Re_{\tau 0}$ is 110, which is based upon the wall friction velocity $u_{\tau 0}$ calculated by the mean pressure gradient for the steady case, the channel half-width δ and the kinematic viscosity ν. Hereafter, u, v, and w denote the velocity components in the x-, y-, and z-directions, respectively. Superscripts $(+)$ and $(*)$ represents quantities non-dimensionalized with $u_{\tau 0}$ and ν, and with $u_{\tau 0}$ and δ, respectively.

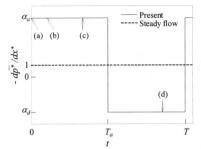

Fig. 1. Change of mean pressure gradient. (a)-(d) indicate the phases described in Fig. 4.

Table 1. Parameters in the change of mean pressure gradient.

Index	α_a	α_d	T^+	T_a/T	Symbol
C-1	5	-3	352	0.5	○
C-2	5	-3	528	0.5	△
C-3	5	-3	704	0.5	□
C-4	5	-3	880	0.5	◇
C-5	5	-3	1056	0.5	+

The mean pressure gradient is changed pulsatingly in time as shown in Fig. 1. Here, t is the time, T the time period for one cycle, T_a the acceleration time period, α_a the constant amplitude for acceleration and $\alpha_d(< \alpha_a)$ the one for deceleration. These parameters are summarized in Table 1. Note that α_a, α_d and T_a/T are fixed in this paper, which should be optimized in future research. The statistics described below are calculated through ensemble averaging after the fully-developed periodic flow field is obtained.

3 Results and discussion

Fig. 2 shows the cycle-averaged skin friction coefficients $[C_f]_T$. Here, $[Re_m]_T$ is the cycle-averaged bulk Reynolds number based upon the bulk mean velocity and the channel width. In the case of $T^+ = 352$, $[C_f]_T$ is almost consistent with that of the steady flow. In the cases of the longer time period, on the other hand, $[C_f]_T$ is decreased with increasing T. The same trend can be observed for the pumping power (not shown here).

We introduce the analytical relation between the Reynolds shear stress distribution and the skin friction coefficient almost same as Ref. [5] to evaluate the reduction mechanism of the skin friction coefficient as:

$$[C_f]_T = \frac{12}{Re_b} + 24 \left(\frac{Re_\tau}{Re_b}\right)^2 \frac{1}{T} \int_0^T \int_0^1 (1 - y^*) \overline{-u'^+v'^+} dy^* dt, \qquad (1)$$

where the first term on the RHS is the laminar contribution, which is identical to the well-known laminar solution, while the second term is the turbulent contribution, which is a weighted integral of the Reynolds shear stress distribution. As is noticed from Eq. (1), the decrease of $[C_f]_T$ is mainly due to the change of the weighted Reynolds shear stress distribution as shown in Fig. 3. The profiles in the acceleration period are almost smaller than that of the steady flow especially near the wall ($y^+ \sim 25$), indicating that the decrease of $[C_f]_T$ is principally owing to the decrease of the weighted Reynolds shear stress near the wall in the acceleration period.

Fig. 4 shows instantaneous flow fields for $T^+ = 528$. The phases are shown in Fig. 1. In the acceleration period, both the streaky structures and the

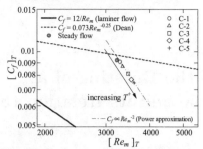
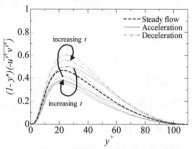

Fig. 2. Cycle-averaged skin friction coefficient $[C_f]_T$.

Fig. 3. Weighted Reynolds shear stress distribution for $T^+ = 528$.

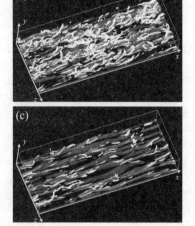
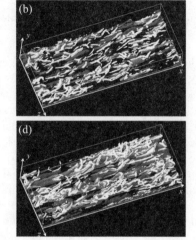

Fig. 4. Instantaneous flow fields for $T^+ = 528$. Contours of the streamwise velocity fluctuation u' and the second invariant of the deformation tensor Q are visualized: gray, $u'^+ = \pm3$; white, $Q^+ = -0.02$. (a)$t/T = 0$; (b)$t/T = 0.13$; (c)$t/T = 0.33$; (d)$t/T = 0.86$.

vortices first become weakened (Figs. 4(a)-(b)). Subsequently, the shapes of the streaky structures are changed to long and smooth, while the vortices are continuously weakened (Figs. 4(b)-(c)). In the middle phase of deceleration period, these structures become complicated (Figs. 4(c)-(d)) and the turbulent kinetic energy is increased (not shown).

Acknowledgement. This work was conducted in Research Center for the Holistic Computational Science (Holcs) supported by MEXT.

References

1. Z. X. Mao & T. J. Hanratty: AIChE J. **40**, 1601 (1994)
2. C. R. Lodahl, B. M. Sumer & J. Fredsøe: J. Fluid Mech. **373**, 313 (1998)
3. H. Kawamura, K. Honma & Y. Matsuo: Proc. Turbulence and Shear Flow and Phenomena **1**, 175 (1999)
4. J. Kim, P. Moin & R. Moser: J. Fluid Mech. **177**, 133 (1987)
5. K. Fukagata, K. Iwamoto & N. Kasagi: Phys. Fluids **14**, L73 (2002)

Numerical Simulations of the Bursting of a Laminar Separation Bubble and its Relation to Airfoil Stall

O. Marxen, D. You, and P. Moin

Center for Turbulence Research, Stanford University, Stanford, CA 94305-3035, USA olaf.marxen@stanford.edu

A laminar separation bubble (LSB) can originate if an initially laminar boundary layer is subject to a sufficiently strong adverse pressure gradient, and transition to turbulence takes place in the detached shear-layer. LSBs can occur on the surface of slender bodies, e.g. airfoils at high angles of attack, and may strongly affect their performance, such as lift and drag.

Short and long laminar separation bubbles can be distinguished based on the bubble length in comparison to the chord of an airfoil [1]. Under certain conditions, for slight changes in the flow conditions short bubbles can break-up into long ones, so-called bubble bursting [2].

Stall of an airfoil at high angles of attack corresponds to a flow condition where the boundary layer on the suction side separates at some distance from the leading edge (LE) and does not reattach in the mean before the trailing edge. Two different types of stall are commonly observed: leading-edge stall, associated with laminar separation in the vicinity of the leading edge, and trailing-edge stall (typically) associated with turbulent separation further downstream along the chord. Thus, in case of leading-edge stall we have a long LSB while in case of trailing-edge stall a short LSB might or might not be present in the LE region in addition to the turbulent separation bubble.

A model problem for such a situation is given by the laminar boundary layer on a flat plate subject to a strong adverse pressure gradient. Direct numerical simulation (DNS) results for such a model problem shall here qualitatively be compared to large-eddy simulations (LES) of flow around a NACA 0015 airfoil at 17 degree angle of attack. The objective is a better understanding of the role of laminar separation for airfoil stall.

Two different situations are considered for both, the airfoil and the flat-plate: unforced and controlled with a zero net mass flux (ZNMF) actuation with a low amplitude. The actuator is placed around the changeover of the pressure gradient from favorable to adverse, which is very close to the LE for the airfoil. The forcing frequency is selected guided by linear stability theory, respectively.

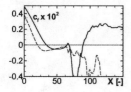 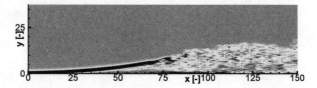

Fig. 1. Left: Skin-friction coefficient c_f from DNS with (solid) and without (dashed) forcing. Right: Instantaneous spanwise vorticity without forcing during bursting.

DNS of Laminar Separation on a Flat Plate

DNS are based on a high-order discretization scheme. Some details of the simulation method as well as more results can be found in [3]. At the inflow, a Blasius boundary layer is prescribed. The global Reynolds number is based on the displacement thickness around $x=0$ and amounts to $Re_{\delta*}=600$. Transition is triggered via 3-d small-amplitude ($\hat{v}_{wall}/U_\infty=10^{-4}$) wall blowing/suction upstream of separation at a frequency $F^+=f\,L_{ref}/U_\infty=47.75$, chosen guided by a local linear stability analysis.

Forcing an oblique pair of waves causes the detached shear layer to quickly reattach after transition in the mean (Fig. 1, left), so that a short separation bubble develops. However, a switching-off disturbance input results in a long LSB after some time. While transition still occurs in the shear layer quickly after separation as for the short LSB, it is unable to reattach the flow immediately (Fig. 1, right).

LES of an Airfoil Configuration

LES computations were carried out using CDP, a code based on a second-order finite volume scheme with a dynamic Smagorinsky turbulence model. Details of the numerical method and configuration can be found in [4], even though a different chord Reynolds number of $Re_c=100,000$ and a different ZNMF actuation is applied here. Around the LE the flow is forced via 2-d wall blowing/suction ($A/U_\infty=10^{-2}$) at a frequency $F^+=f\,L_{ref}/U_\infty=40.0$, chosen again guided by a local linear stability analysis. Forcing is prescribed in the region $x/c \in [0,0.01]$ with a velocity perpendicular to the surface distributed as $A/U_\infty \cdot (\tilde{x}^7 - 3\tilde{x}^5 + 3\tilde{x}^3 - \tilde{x})/0.238$ for $\tilde{x}=200(x/c - 0.005)$.

With forcing only a short laminar separation bubble develops in the vicinity of the LE on the surface of the airfoil (Fig. 2, left), while without forcing the laminar flow separates without immediate reattachment (Fig. 2, right), causing LE stall (Fig. 3). Unlike in the DNS, we see immediate re-separation causing a turbulent separation bubble downstream of the LSB.

Conclusion

A low-amplitude unsteady forcing at the wall is applied in both, DNS of a flat plate and LES of an airfoil. In both cases, with forcing a short LSB develops,

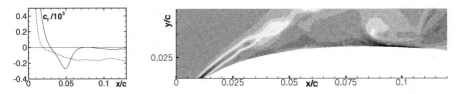

Fig. 2. Left: Skin-friction coefficient c_f from LES with (solid) and without (dashed) forcing. Right: Instantaneous spanwise vorticity without forcing.

Fig. 3. Isosurfaces of the instantaneous spanwise vorticity in the leading-edge region of the airfoil: Results from LES with (left) and without (right) forcing.

while without forcing a long LSB occurs. In case of the airfoil, this latter situation corresponds to LE stall. Bursting from a short to a long LSB is observed in DNS if disturbance input is switched off.

Our results indicate that bubble bursting and thus LE stall can be prevented by high-frequency ZNMF actuation. It remains to be determined whether this can prevent stall or whether it leads to trailing-edge stall.

Acknowledgments. O. Marxen acknowledges financial support by the Deutsche Forschungsgemeinschaft under grant MA 3916/1-1 and computational time by the HLRS Stuttgart, project *long_lsb*. Furthermore, he thanks U. Rist and M. Kloker, IAG, Uni Stuttgart for providing the DNS code *n3d*. D. You and P. Moin gratefully acknowledge support from Boeing company.

References

1. P. R. Owen and L. Klanfer: *On the laminar boundary layer separation from the leading edge of a thin airfoil*, No. Aero 2508, Royal Aircraft Establishment, UK (In: A.R.C. Technical Report C.P. No. 220, 1955)
2. M. Gaster: *The structure and behaviour of laminar separation bubbles*, AGARD CP-4 (1966) pp 813–854
3. O. Marxen and D.S. Henningson: *Numerical Simulation of the Bursting of a Laminar Separation Bubble*, **AIAA-2007-0538** (2007)
4. D. You and P. Moin, *Large-eddy simulation of flow separation over an airfoil with synthetic jet control* (In Annual Research Briefs 2006, Center for Turbulence Research, Stanford University, 2006) pp. 337–346

Fluctuations in the bluff body wake – modelling an ultrafast aircraft thermometer

K. Bajer[1], B. Rosa[2], K.E. Haman[1], and T.S. Szoplik[1]

[1] University of Warsaw, Inst. of Geophysics, ul. Pasteura 7, 02-093 Warszawa, Poland, kbajer@fuw.edu.pl
[2] University of Delaware, Dept. Mech. Eng., 221 Spencer Lab., Newark, DE 19716-3140, bogdanr@udel.edu

The ultrafast aircraft thermometer, built for measuring temperature in clouds at flight speeds up to 100 m/s, employs a 2.5 μm thick platinum-coated tungsten wire as a sensing element. Temperature measurements made in a wind tunnel and during flights show noise that may be related to the von Kármán vortex street generated behind the shield that protects the sensing element against the impact of cloud droplets. The temperature fluctuations are caused mainly by adiabatic compressions and decompressions in vortices in the shield's wake. To reduce both the level of turbulence and the amount of water collected on the shield a suction is applied through the slits in its sides. To design an optimal shape of the shield that minimizes the noise and the probability of droplet collision with the sensor we perform 2D and 3D DNS of the air flow and of the trajectories of droplets of various sizes and initial positions. We also analyze the influence on the temperature distribution of the irreversible dissipation of energy due to air viscosity. This is found to have small but measurable effect. We discuss the effects associated with sampling and processing of the analogue signal obtained from the sensing wire. Our results quantitatively explain the nature of the measured aerodynamic noise.

The designers of the fast thermometers aim at measurements with millimeter spatial resolution. To design a shield that will allow low noise measurements at frequencies higher than 10 kHz, we have to understand the nature of the observed aerodynamic disturbances. In this study we analyze turbulence in the neighborhood of the sensing element. We focus on the dominant effect of suction on the von Kármán vortices or their 'seeds', that is, the weak instabilities in the formation region [3]. Our computations show that those are suppressesed by suction. To compare our numerical simulations with the temperature measurements we convert our computed pressure distributions into temperature. As a first approximation we assume the adiabatic relation. However, in the regions of strong shear, that is, in a boundary layer on the shield's surface, viscous heating becomes significant. Its rate is equal to $\Phi = 2\mu e_{ij}e_{ij}$ where e_{ij} is the of strain tensor and $\mu = 1.47 \times 10^{-5}\,\mathrm{kg\,m^{-1}\,s^{-1}}$ is the dynamic viscosity of air. The total amount of heat released in a fluid element

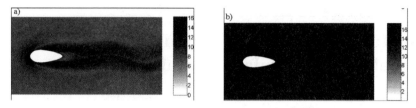

Fig. 1. The rate of viscous heating with suction *off* (left) and suction *on* (right).

reaching the sensor is given by the Lagrangian integral over the path line of the element, $\int_0^\theta \Phi \, dt$, where θ is the time that the element spends in the region of strong shear. We have computed this integral for selected particles of fluid that hit the sensor.

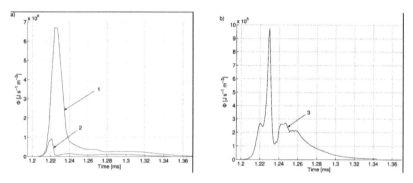

Fig. 2. The rate of viscous heating Φ along selected pathlines with suction *off* (left) and with suction *on* (right). Viscous heating is 123, 17 and 17 $\mathrm{J\,m^{-3}}$ respectively.

Droplets suspended in a cloud through which the thermometer is moving from time to time hit the sensing element. A signature of every such collision is a short-lasting drop of about 1.5 K in the recorded temperature caused by the wet-bulb effect. We computed the trajectories of droplets starting at the inlet plane. When the suction is *off*, larger droplets are deflected by the shield and do not hit the sensor, but the small ones can be entrained in the von Kármán vortices and follow a rather erratic path going through the point where the 2.5 μm thick wire sensor is located. When the suction is *on* large droplets again are unlikely to hit the sensor, but the smaller ones come very close as the suction deflects the boundary layer in the wake toward the shield's symmetry axis.

In conclusion, the effect of suction on the flow field is twofold. First, at the Reynolds numbers that the thermometer is operated, suction practically eliminates the von Kármán vortices in the region where the sensing wire is located. The wake of the shield is stable and there are virtually no adiabatic temperature fluctuations associated with the pressure drops in the cores of the vortices. Such improvement will be possible with new designs of both the electronics and the shield of the thermometer. shed from the shield and passing the sensor.and data acquisition system Second, suction diverts the inner part of the boundary layer into the slit. This inner part is a region of strong shear and, therefore, a region where intensive viscous heating takes place. When the suction is *on*, much of the air that is heated in the boundary layer in the front part of the shield is removed through the slits and never reaches the wake. Our computations show that without suction this thin layer of warmer air appears in the undulating wake, thus, increasing the amplitude of the recorded fluctuations.

The suction also affects the probability of droplets colliding with the sensor in two ways. The first is the removal of water from the shield's surface. This dramatically improves the operation of the thermometer as it reduces, or even eliminates, the water flowing along the surface to the rear stagnation point, and then detaching and forming secondary droplets well positioned to hit the sensor.

The second, weaker effect is the stabilization of the wake. On the one hand, this eliminates the wandering of small droplets in the von Kármán vortices. On the other hand, the suction deflects the droplets toward the axis of the wake, thus, reducing the width of the droplet-free 'channel' behind the shield. We have also established the general dependence of the average pressure drop and pressure fluctuations on the aircraft's speed. This helps to assess the systematic temperature shift and the level of noise. The simulations help to understand the spectrum of the recorded signal. If the system recorded a continuous signal, as we can do in the simulations, then the spectrum obtained in the flow without suction would be dominated by two frequencies - the basic frequency of the vortex shedding and its double. These frequencies need to be computed so that the signal processing can be optimized to ensure the least possible attenuation of them. In fact, the system records a sampled signal and, therefore, the spectrum contains additional peaks, which are the effect of aliasing. The simulations clarify the pattern of this aliasing, and this is essential for the analysis of the experimental data. To understand the small scales of cloud turbulence we need to 'clean' the signal by removing all frequencies that are associated with the instruments we use, both the 'true' ones and the 'spoof' frequencies due to aliasing.

1. K. Haman, S.P. Malinowski, B.D. Struś, R. Busen & A. Stefko 2001 Two new types of ultrafast aircraft thermometer. *J. Atmos. Oceanic Technol.* **18**, 117-134
2. B. Rosa, K. Bajer, K. Haman, & T. Szoplik 2005 Theoretical and experimental characterization of the ultrafast aircraft thermometer: reduction of aerodynamic disturbances and signal processing. *J. Atmos. Oceanic Technol.* **22**, 988-1003.
3. C.H.K. Williamson (1996) Vortex dynamics in the cylinder wake. *Annu. Rev. Fluid Mech.* **28**, 477-539.

Velocity and Wall Pressure Correlations Over a Forward Facing Step

G. Aloisio[1], R. Camussi[2], A. Ciarravano[1], F. Di Felice[1], A. Di Marco[2], M. Felli[1], E. Fiorentini[1], and F. Pereira[1]

[1] INSEAN - Italian Ship Model Basin - Rome, Italy, `f.pereira@insean.it`
[2] DIMI, University 'Roma 3' Rome, Italy, `camussi@uniroma3.it`

This work describes a study of the flow field and wall pressure fluctuations induced by quasi-two-dimensional incompressible turbulent boundary layers (TBL) over a forward-facing step (FFS). Geometrical singularities, e.g. on the exterior surfaces of aircrafts, are largely responsible for flow detachment, resulting in increased aerodynamic sound radiation and interior noise generation. Knowledge of the physical mechanisms is therefore important to understand the source of the pressure fluctuations and to develop noise-control strategies. To correlate the flow structures of the TBL to the induced pressure field on the wall, time-resolved PIV measurements of the flow field in the TBL are coupled with wall measurements of the sound pressure levels (SPL). Autospectra on a pressure signal recorded next to the FFS are shown in Fig. 1a, where the -1 and -7/3 scalings are seen to coexist. The SPL along the wall is reported in Fig. 1b. The most efficient pressure source is the unsteady reattachment point of the recirculation bubble downstream of the FFS, located beween 1.5h and 2.5h from the vertical wall, depending on Re_h. This point is accurately identified from the whole field velocity measurements, see Fig. 1c.

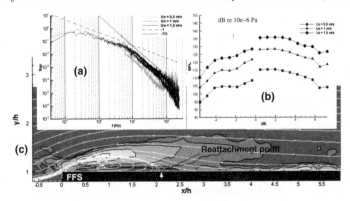

Fig. 1. (a) Wall pressure spectra at x/h=-0.4h versus U_0; (b) SPL versus x/h and U_0; (c) Downstream mean vorticity field, $Re_h = 26300$

Time evolution of Thorpe profiles corresponding to atmospheric soundings

J. L. Cano and P. López González-Nieto

Departamento de Física de la Tierra, Astronomía y Astrosfísica II
Facultad de Ciencias Físicas, Universidad Complutense, 28040 Madrid, Spain
jlcano@fis.ucm.es azufre2@hotmail.com

The authors investigate the time evolution of atmospheric potential temperature profiles obtained by means of balloon soundings during field experiments. Our focus is on characterizing the small scale turbulence–generated by overturns–and its evolution. This small scale turbulence is identified by Thorpe displacements d_T. Vertical mixing can sometimes be quantified by measurements of the Thorpe scale L_T. Thorpe profiles are found by comparing the observed potential temperature profile $\theta(z)$ and the monotonic potential temperature profile $\theta_m(z)$which is constructed by reordering $\theta(z)$ to make it gravitationally stable.

The time evolution of overturns is studied from the up and down atmospheric profiles. This up and down potential temperature profiles–and their corresponding up and down Thorpe displacements–give us an approximate evolution of overturns which it has been measured directly. The down Thorpe profile at a fixed height compared with the up Thorpe profile at the same height provide a real idea of the turbulent stage of overturns. The authors also investigate the time evolution of Thorpe displacements during a day cycle.

References

1. T. Dillon: J. Geophys. **87**, 9601 (1982)
2. P. S. Galbraith, D. E. Kelley: J. Atmos. Ocean. Tech. **13**, 688 (1996)
3. P. L. González-Nieto: *La mezcla turbulenta por convección gravitatoria: modelización experimental y aplicación a situaciones atmosféricas* Ph. D. Thesis, Universidad Complutense, Madrid (2004)
4. J. Piera, E. Roget, J. Catalán: J. Atmos. Ocean. Tech. **19**, 1390 (2002)
5. W. D. Smyth, J. N. Moum, D. R. Caldwell: J. Phys. Oceanogr. **31**, 1969 (2001)
6. H. Tennekes, J. Lumley: *A First Course in Turbulence* (MIT press, 2000)
7. S. Thorpe: Philos. Trans. Roy. Soc. London A **286**, 125 (1977)

Divergent and rotational modes in stratified flows

G. Brethouwer and E. Lindborg

Department of Mechanics, KTH, SE-100 44 Stockholm, Sweden
geert@mech.kth.se

To identify vortices and internal waves in stratified flows, Helmholtz decomposition of the velocity can be employed

$$\boldsymbol{u} = \nabla_h \times \Psi \boldsymbol{e}_z + \nabla_h \Phi + u_z \boldsymbol{e}_z, \tag{1}$$

where Ψ is a stream function, Φ a velocity potential, ∇_h the horizontal divergence operator and u_z and $\boldsymbol{e}_z$ are the vertical velocity component and unit vector respectively. The first term on the right-hand-side is the rotational part and the second term the divergent part. In flows with mainly vertically oriented vortices the rotational part is dominating whereas in flows with mainly internal waves the divergent part is dominating. The timescale ratio of the 'fast' waves and the 'slow' horizontal vortical motions can be estimated as $T_{fast}/T_{slow} \sim u/(Nl_v) = F_v$ where u is a horizontal velocity scale, l_v a vertical length scale, N the Brunt-Väisälä frequency and F_v is the vertical Froude number. It is often assumed that $F_v \to 0$ in strongly stratified flows implying that vortices and waves have separate time scales, suggesting weak interactions.

Scaling analysis suggests $l_v \sim u/N$ and hence $F_v \sim 1$ when the dynamics of the stratified flow determines l_v. This led to the hypothesis of three-dimensional but strongly anisotropic stratified turbulence with a forward energy cascade and an inertial range with a horizontal $k_h^{-5/3}$-spectrum. The scaling analysis and hypothesis were supported by recent numerical simulations (Lindborg 2006, Brethouwer *et al.* 2007). The scaling ($F_v \sim 1$) suggests that in stratified turbulence the divergent and rotational modes have a similar timescale. This means that a decomposition into divergent and rotational modes is not very fruitful. Instead strong nonlinear interactions between rotational and divergent modes are expected and they are expected to have similar energy. The presence of strong interactions also implies that stratified turbulence produced by forcing in either divergent or rotational modes will be dynamically similar.

We have performed numerical simulations of strongly stratified and homogeneous flows with hyperviscosity to address the questions: (*i*) Are the divergent and rotational part of similar magnitude? (*ii*) Does large-scale forcing in either divergent or rotational modes lead to the same dynamics and under what conditions?

Results are presented showing that the inertial range spectra are very similar in simulations with forcing of rotational modes and divergent modes, suggesting similar dynamics, if in the latter simulations the large-scale dynamics obeys $F_v \sim 1$, but deviations are found when this condition is not fulfilled. The imposed vertical length scale by the forcing plays thus a crucial role.

On the secondary Kelvin-Helmholtz instability in a $3D$ stably stratified mixing layer

Denise M. V. Martinez[1], Edith B. C. Schettini[2], and Jorge H. Silvestrini[3]

[1] Fundação Universidade Federal do Rio Grande - FURG, Av. Itália, Km 8 -
962001-900, Rio Grande, RS, Brasil, denisevmartinez@yahoo.com.br
[2] Universidade Federal do Rio Grande do Sul - UFRGS, Av. Bento Gonçalves,
9500 - 91501-970 - Porto Alegre, RS, Brasil, bcamano@iph.ufrgs.br
[3] Pontifícia Universidade Católica do Rio Grande do Sul, Av. Ipiranga, 6681 -
90619-900 - Porto Alegre, RS, Brasil,jorgehs@pucrs.br.

The present work investigates the occurrence of secondary instabilities in the baroclinic layer in a $3D$ stably stratified temporal mixing layer in the context of Direct Numerical Simulation (DNS). Two different secondary instabilities are found in the baroclinic layer: one originated near the core region of the KH vortex, called near-core instability, which propagates towards the baroclinic layer, and another of KH type in the baroclinic layer itself. The development of the secondary instabilities in the baroclinic layer depends on the stratification degree of the flow characterized by the Richardson number (Ri), on the Reynolds number (Re) and on the imposed initial conditions. Also are focused the development of streamwise vortices and its interactions with the secondary KH structures. Typical Richardson numbers ranging from 0.07 to 0.167 are considered while the Reynolds number is kept constant and equal to 500. The Navier-Stokes equations, in the Boussinesq approximation, are solved numerically using a sixth-order compact finite difference scheme to compute the spatial derivatives, while the integration in time is performed with a third-order low-storage Runge-Kutta method. It is observed that the near-core instability appears due to the formation of a negative vorticity layer generated between two co-rotating positive vortices. The production of negative vorticity inside the vortex core is rapidly followed by the growth of the secondary KH instability. This instability only developed in the baroclinic layer if the negative vorticity is of magnitude comparable with the positive vorticity. The intensity of the negative vorticity layer depends on the Richardson and Reynolds number and defines the occurrence or not of the secondary KH structures in the flow. The occurrence of secondary KH instability in the baroclinic layer changes the dynamics of the flow and apparently speed up the transition to turbulence of a strongly stratified mixing layer.

Aerodynamic Flow Control of a Free Airfoil

Daniel P. Brzozowski, John R. Culp, and Ari Glezer

Woodruff School of Mechanical Engineering, Georgia Institute of
  Technology, Atlanta, GA, 3033

Transitory separation control on a free-pitching airfoil is effected in wind tunnel experiments using a pulsed fluidic actuation. Momentary aerodynamic forces are engendered by exploiting the receptivity of the separated flow *to actuation inputs on time scales that are significantly shorter than the characteristic advection time over the separated flow domain* (Amitay and Glezer, [1] and [2]). The actuation results in brief flow attachment that is associated with large-scale vorticity generation, accumulation, and transport, and consequently leads to changes in the circulation and aerodynamic forces and moments on the airfoil (e.g., Brzozowski and Glezer [3]). Actuation is applied by a momentary jet produced by a combustion-based actuator array where the characteristic duration of actuation pulse is $O[0.05U_0/c]$.

In the present work, the wing model is mounted on a 1-DOF traverse which allows for pitching motions while the model is trimmed and compensated for moment of inertia via a feedback-controlled torque motor. Transitory aerodynamic forces are demonstrated with emphasis on implementation of closed-loop flow control.

References

1. Amitay, M. and A. Glezer: *Int. Journal of Heat and Fluid Flow*, 23, 690-699, 2002.
2. Amitay, M. and A. Glezer: *Experiments in Fluids*, 40, 329, 2006.
3. Brzozowski, D. and A. Glezer: AIAA Paper 2006-3024, 2006.

Performance of Reynolds-Averaged Turbulence Models for Unsteady Separated Flows with Periodic Blowing and Suction

Masashi Yoshio and Ken-ichi Abe

Dep. of Aero. and Astro., Kyushu University, Nishi-ku, Fukuoka 819-0395, Japan
m_yoshio@aero.kyushu-u.ac.jp, abe@aero.kyushu-u.ac.jp

1 Introduction

Flow blowing/suction is one of effective strategies to control separation. In this study, to assess the predictive performance of current turbulence models, several Reynolds-Averaged Navier-Stokes (RANS) models are applied to flow fields over a hump with and without flow blowing/suction[1].

2 Results and Discussion

The predicted results are decomposed according to a triple decomposition methodology to investigate the model performance in detail.

$$f(x,t) = F(x) + \tilde{f}(x,t) + f'(x,t) \tag{1}$$

$F(x)$, $\tilde{f}(x,t)$ and $f'(x,t)$ are the time-mean value, the coherent periodic component and the random turbulence, respectively. Figure 1 shows some representative results, where the time-mean and rms surface pressure coefficients (C_p and $C_{\tilde{p}}^*$) are compared. The computational results generally correspond with the experimental data, though there remain several areas to be improved.

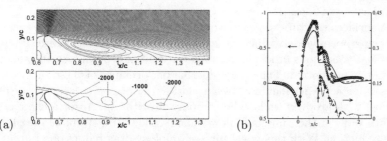

Fig. 1. Some results: (a) streamlines and vorticity map for oscillatory control case; (b) comparison of pressure coefficients, Exp. ($\diamond$ Baseline C_p $\times$ Control C_p $\triangle$ Control $C_{\tilde{p}}^*$), AJL[2] (- Baseline C_p - -Control C_p ---Control $C_{\tilde{p}}^*$) where $C_{\tilde{p}}^* = \sqrt{\overline{\tilde{p}^2}}/(1/2\rho U_\infty^2)$.

References

1. T.B. Gatski and C. Rumsey, Langley Research Center Workshop: CFD Validation of Synthetic Jets and Turbulent Separation Control, NASA (2004)
2. K.Abe, Y.-J.Jang, M.A. Leschziner, IJHFF 24 (2003) 181-198.

Turbulent Dissipation in Drag Reduced Flows

Bettina Frohnapfel, Jovan Jovanović and Antonio Delgado

Institute of Fluid Mechanics, Cauerstr. 4, 91058 Erlangen, Germany
bettina@lstm.uni-erlangen.de

All energy that is used to drive a flow is eventually dissipated into heat. In a turbulent flow field the averaged total dissipation rate is composed of the direct dissipation and the turbulent dissipation. At large Reynolds numbers that are of interest in engineering applications the turbulent dissipation outweighs the direct dissipation. For internal flows the dissipation, which can be expressed in terms of the friction coefficient c_f, needs to be reduced to obtain drag reduction. At high Reynolds numbers the largest contribution to c_f originates from the turbulent dissipation which is given by the gradient of the velocity fluctuations $\epsilon = \nu \partial u_i / \partial x_k \times \partial u_i / \partial x_k$. Analytical considerations of the velocity fluctuations at the wall, where ϵ reaches its maximum and where practical flow control is acting, show that ϵ at the wall will be minimized if the turbulent fluctuations are forced to fulfill conditions of local axisymmetry. This finding is confirmed through the analysis of the turbulent dissipation for existing DNS data of turbulent wall bounded flows such as channel, pipe and boundary layer flows [1, 2].

A surface structure with grooves aligned in flow direction is presented. Inside the grooves velocity fluctuations in the spanwise direction are suppressed such that conditions of local axisymmetry are fulfilled and the turbulent dissipation is minimized. The drag reducing performance of these surface structures is tested experimentally by measuring the pressure drop $\Delta p / \Delta l$ over a channel with grooved walls. The comparison with $\Delta p / \Delta l$ for a channel with smooth walls at the same bulk flow velocity allows to determine the obtained drag reduction. With this novel surface topology drag reduction of up to 25% is realized.

References

1. J. Jovanović, R. Hillerbrand: Thermal Science **9(1)**, 3 (2006)
2. B. Frohnapfel, P. Lammers, J. Jovanović, F.Durst: J. Fluid Mech., accepted (2007)

Instabilities of a barotropic rotating shear layer

Ana Aguiar and Peter Read

AOPP - Clarendon Lab., Parks Road, University of Oxford,
OX1 3PU Oxford, U.K. aguiar@atm.ox.ac.uk

The detached shear layers (Stewartson layers) occur tangential to one or two differentially rotating sections of adjustable height. Above a critical value of radial stress the flow is driven to an unstable limit, beyond which it develops a chain of vortices organized in an azimuthal wave. The present work follows from previous studies, whose results were not entirely conclusive, given the difference in observations from independent studies (Hide & Titman, 1967; Früh & Read, 1999; Hollerbach, 2003).

We have been studying such a shear layer with flat, stepped or conical end walls in a cylindrical domain, using both laboratory and numerical experiments. Hence, the effect of a topographically-induced planetary vorticity gradient is explored, together with a range of configurations which gathers all the previous studies in an attempt to interpret the asymmetry observed in some cases but not in others. The flow is forced mechanically, by one or two axial disks which rotate with an additional speed ω (positive or negative) with respect to the rotation rate of the tank Ω. Experimental velocity fields (in the $r\theta - plane$) are obtained, whereas the detailed structure of an axisymmetric form of the shear layer is resolved in the $rz - plane$, using a 2D Navier-Stokes computational model. In most cases, the onset of instability and the patterns observed are found to be in qualitative agreement with theoretical predictions for the instability criterion, for both flat and conical end walls. Surprisingly, for opposite values of forcing and with flat end walls, we observe only a *weak* asymmetry in the azimuthal wavenumber even when using a single disk immersed (nearly at mid-height) in the fluid bounded by whole rigid end walls. This suggests that the strong asymmetry reported in Hide & Titman (1967) could be linked with the rather thick disk used by those authors. Further experimental measurements will allow us to look into the detailed dynamics of the instability, and enable us to understand more clearly the wave mode selection processes and criteria that take place. We will extend the study to assess the effect of using a disk of different thickness immersed in the fluid. Also, we intend to seek evidence for the influence of topography on the stability of the patterns obtained when using conical end walls of either sign of slope.

Nonlinear development of Klebanoff modes in a laminar boundary layer

Pierre Ricco[1] and Xuesong Wu[2]

[1] Institute for Mathematical Sciences, Imperial College pierre.ricco@ic.ac.uk
[2] Department of Mathematics, Imperial College x.wu@ic.ac.uk

Experiments show that free-stream turbulence may form boundary layer streaks, i.e. "Klebanoff modes", which may breakdown causing bypass transition. Small-amplitude streaks have been studied by the unsteady *linearized* boundary *region* (BR) equations. We consider here free-stream turbulence of moderate level, so that the streaks evolve nonlinearly. The parameters are ϵ, the free-stream disturbance amplitude and $R_\Lambda = U_\infty \Lambda / \nu$, where U_∞ is the mean free-stream velocity and Λ is the spanwise length scale. The free-stream fluctuations are modelled by a pair of convective gusts and we take $k_1 = \mathcal{O}(R_\Lambda^{-1})$ and $\epsilon R_\Lambda = \mathcal{O}(1)$, i.e. the streaks are governed by the *nonlinear* unsteady BR equations. Nonlinearity has a stabilizing effect and the streamwise velocity profiles (see figure 1) take both positive and negative values w.r.t. the Blasius flow near the wall, while only smaller velocities than the unperturbed laminar flow occur near the free-stream. Inflection points in the $\eta - z$ plane are detected which could be presursors of inviscid instability.

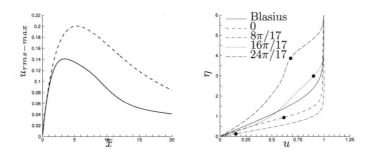

Fig. 1. Left: effect of nonlinearity on maximum r.m.s. of streamwise velocity ($k_1 = 0.05$, $R_\Lambda = 400$, $\epsilon = 0.01$). Solid lines: nonlinear case, dashed lines: linearized case. Right: Streamwise velocity profiles at different phases at $z = 0$ and $\bar{x} = k_1 x = 2$. $k_1 = 0.01$, $R_\Lambda = 400$, $\epsilon = 0.01$. Black dots = inflections points.

Primary instability of a rotating spherical Couette flow with a radial stratification and a radial buoyancy

Mathieu Jenny[1] and Blaise Nsom[1]

LIME / IUT de Brest - UBO - BP 93169 - 29231 BREST CEDEX 3
mathieu.jenny@univ-brest.fr, blaise.nsom@univ-brest.fr

For geophysical problems, the Taylor-Couette model allows to study the mixing and the flow in the ocean (see [1]). Indeed, the equatorial area can be approximated to a rotating cylinder. Although the Taylor-Couette flow has been extensively investigated since the Taylor's work, the buoyancy in the radial direction has never been taken into account because the laboratory gravity imposes an axial stratification. In a previous study, the authors have developed an effective numerical method to address a parametric study of the primary instability of a radially stratified Taylor-Couette. Depending on the Froude number and the Schmidt number, *i.e.* the gravity strength and the massic diffusivity, three unstable modes were identified [3]. In the present work, we consider a spherical Couette flow which improves the geometric representation into the geophysical model. For this configuration, the stability has been studied in the homogeneous case [2], but not again, in a radially stratified sketch. More, the effect of a radial buoyancy is not fully described in the bibliography. First results shows that the spherical flow is more stable than the cylindrical flow as the critical Taylor number is higher. More, the instability mode is modified at a given Froude and Schmidt number. For instance, the primary instability of the cylindrical Couette flow is non-axisymmetric while the instability mode of spherical flow is axisymmetric at Fr=100 and Sc=100.

References

1. E.V. Ermanyuk and J.B. Flór. Taylor-couette flow in a two-layer stratified fluid: instabilities and mixing. *Dyn. Atmos. Oceans*, 40:57–69, 2005.
2. Rainer Hollerbach, Markus Junk, and Christoph Egbers. Non-axisymmetric instabilities in basic state spherical couette flow. *Fluid Dynamics Research*, 38:257–273, 2006.
3. M. Jenny and B. Nsom. Primary instability of a taylor-couette flow with a radial stratification and radial buoyancy. *Phys. Fluids*, under revision, 2007.

Onset of Turbulence in T-jet Mixers

Ertuğrul Erkoç[1], Ricardo J. Santos[2], and José Carlos B. Lopes[3]

[1] Laboratory of Separation and Reaction Engineering, Faculdade de Engenharia da Universidade do Porto erkoc@fe.up.pt

[2] Laboratory of Separation and Reaction Engineering, Faculdade de Engenharia da Universidade do Porto rsantos@fe.up.pt

[3] Laboratory of Separation and Reaction Engineering, Faculdade de Engenharia da Universidade do Porto and Fluidinova lopes@fe.up.pt

T jet mixers are widely used for mixing purposes mainly associated to fast chemical reactions, which demand intensive mixing with extreme short contact times. For this type of mixer there are two distinct flow regimes: for lower Reynolds numbers, Re<100, the two jets impinge on the mixing chamber axis and flow outwards in parallel paths from their original injection sides; for Re≥120 the flow is chaotic with strong oscillations of the jets impingement point and formation of vortices. In the present work the oscillations in the flow field, downstream the jets impingement point, are observed to continuously grow with Re up to Re=120, which is clearly seen in Fig. 1 from the maps of the x component of dimensionless turbulence intensity. The maps were obtained from Particle Image Velocimetry, PIV. From the maps it is clearly seen that the oscillations of the jets, after the impingement point, continuously grow mainly at the regions farther from the two well balanced forces, the opposite jets, that can constrain the flow instabilities. After Re>115 the main centre of flow oscillation becomes the jets impingement point. In this work it is proven that in T jet mixers the original source of instability for flow regime transition grows upwards the mixing chamber, in the opposite direction to the jets pathway, until it reaches the jets impingement point. Once the jets engage on strong oscillatory behaviour the source of flow instability is the jets impingement.

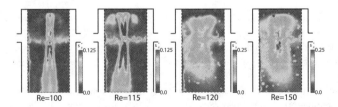

Fig. 1. Maps of turbulence intensity from PIV measurments

Transitional Flow in Annular Rotating Cavity

Ewa Tuliszka-Sznitko and Artur Zielinski

Poznan University of Technology Institute of Thermal Engineering, ul. Piotrowo 3, 60-965 Poznan, Poland sznitko@sol.put.poznan.pl

Summary. In the present paper the 3D DNS and LES computations are performed to study the isothermal and not-isothermal transitional flow between two co- and counter-rotating disks enclosed by two rotating cylinders. The investigation of the laminar-turbulent transition process inside the inter-disk 3D rotating flow is of great interest for the internal aerodynamics of engines. Computations have been performed for the wide range of the Reynolds numbers and for different end-wall conditions including thoughflow. The flow is described by 3D Navier-Stokes, energy and continuity equations. To take into account the buoyancy effects induced by the involved body forces the Boussinesq approximation is invoked. Numerical computations are based on a pseudo-spectral Chebyshev-Fourier method [1]. The time scheme is semi-implicit and second-order accurate. For the large eddy simulation we used a version of the dynamic Smagorinski eddy viscosity model proposed by Meneveau [2]; in this version Germano identity was averaged for some time along fluid pathlines. For the not-isothermal flow we adopted Lilly [3] approach to Lagrangean model. For the higher Reynolds number results obtained using DNS and LES methods were compared (we compared the instability structures, the time history of wall shear stress, radialwise velocity fluctuation profiles..). We considered cavities of the aspect ration ranging from 9 to 11 (curvature parameter Rm=1.5-3.0). Attention was focused on instability structures which appeared in the different stages of laminar-turbulent transition. For counter-rotating cases (for sufficiently large negative value of rotational rate of the disks) we observed the instability pattern which corresponds to the negative spirals described by Gauthier [4]. Our not-isothermal computations have shown the effectiveness of radial cooling. Distributions of the Nusselt number along the radius of disks are given.

References

1. Serre E., Tuliszka-Sznitko E., Bontoux P., Coupled numerical and theoretical study of the flow transition between a rotating and stationary disk, *Phys. of Fluids*, **16**, 3, (2004)
2. Meneveu Ch., Lund T., Cabot,W., A Lagrangean dynamic subgrid-scale model of turbulence, *J. Fluid Mech.*, **319**, 353–385, (1996)
3. Lilly D., A proposed modification of the Germano subgrid-scale closure method, *Phys. of Fluids A 4*, 633–653, (1992)
4. Gauthier G., Gondret P., Moisy F., Rabaud M., Instabilities in the flow between co- and counter-rotating disks, *J. Fluid Mech.*, **473**, 1–21, (2002)

Nonlinear evolution of the zigzag instability in a stratified fluid

Axel Deloncle, Paul Billant, and Jean-Marc Chomaz

LadHyX, Département de Mécanique, CNRS Ecole Polytechnique, 91128 Palaiseau, France `axel.deloncle@ladhyx.polytechnique.fr`

In a strongly stratified fluid, a columnar counter-rotating vortex pair is subject to the zigzag instability which bends the vortices and ultimately produces layers. We have investigated the nonlinear evolution of this linear instability by means of DNS. We show that the instability grows exponentially without nonlinear saturation and therefore produces rapidly intense vertical shear. The instability growth is only stopped when vertical viscous effects become dominant. This occurs when $F_h^2 Re = O(1)$ with F_h the horizontal Froude number and Re the Reynolds number. No secondary zigzag instabilities or shear instabilities have been observed. This means that the zigzag instability is a mechanism capable of directly transferring the energy from large scales to small vertical scales where it is dissipated without any cascade.

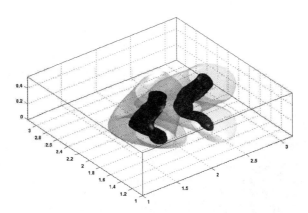

Fig. 1. Nonlinear simulation of counter-rotating vortices subject to the zigzag instability.

Compressibility effects in the Rayleigh-Taylor instability for miscible fluids

Marc-Antoine Lafay, Benjamin Le Creurer and Serge Gauthier

CEA/Bruyères-le-Châtel, B.P. 12, F-91680 Bruyères-le-Châtel
marc-antoine.lafay@cea.fr, serge.gauthier@cea.fr

We report on linear stability analysis and 2D numerical simulations carried out with the complete Navier-Stokes equations. The linear stability analysis is performed within the normal mode framework. The numerical method uses a self-adaptive multidomain method [2]. We obtain dispersion curves for various values of the model parameters (Re, Sc, Pr, γ compressibility, Sr stratification and confinement). It has been shown that this physical model exhibits a cut-off wave number beyond which the RT flow is neutraly stable. We have also shown the opposite effects of Sr and γ on the linear growth rate.
Numerical simulations are initialized thanks to the linear stability analysis [1].

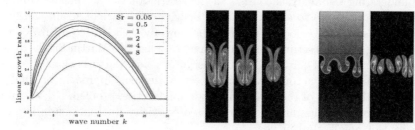

Fig. 1. *Left*: dispersion curves for several values of the Sr parameter. RT flow is stabilized when stratification of the density profile is increasing. *Center*: vorticity field of 2D nonlinear simulations initialized with a single mode perturbation for 3 values of the stratification (Sr $= 0.05, 0.5, 1$). *Right*: exemple of density and vorticity fields of a 2D nonlinear simulation initialized with a multimode perturbation.

References

1. M.-A. Lafay *et al.* Compressibility effects in RTI. 10th *International Workshop on the Physics of Compressible Turbulent Mixing*, to be published, 2007.
2. S. Gauthier *et al.* A self-adaptive domain decomposition method with Chebyshev method. *Int. J. Pure and Appl. Math.*, 24(4):553–577, 2005.

Instanton Theory of Turbulent Vorticity Fluctuations

L. Moriconi

Instituto de Física, Universidade Federal do Rio de Janeiro
C.P. 68528, Rio de Janeiro, RJ — 21945-970, Brasil
moriconi@if.ufrj.br

We investigate statistical properties of vorticity fluctuations in fully developed turbulence, which are known, both from numerical and real experiments, to exhibit a strong intermittent behavior. Taking as the starting point the usual stochastic Navier-Stokes equations, where a random force term correlated at large length scales is introduced, we obtain in the high Reynolds number regime a closed analytical expression for the probability distribution function (pdf) of an arbitrary component of the vorticity field. More specifically, we have found that the tails (and leading order corrections) of the vorticity pdf are reasonably well described by a Student t-distribution [1]. The central idea underlying the analysis consists in the restriction of the velocity configurational phase-space to a particular sector where the rate of strain and the rotation tensors can be locally regarded as slow and fast degrees of freedom, respectively. This prescription is implemented along the Martin-Siggia-Rose functional framework [2, 3], whereby instantons and perturbations around them are taken into account within a steepest-descent approach, originally devised for the study of the turbulent random advection problem [4]. We find good support from the numerical analysis of vorticity fluctuations in homogeneous isotropic turbulent flows [5].

References

1. L. Moriconi: Phys. Rev. E **70**, R25302 (2004).
2. P.C. Martin, E.D. Siggia, and A.H. Rose: Phys. Rev. A **8**, 423 (1973).
3. R. Phythian: J. Phys. A: Math. Gen. **10** (1977).
4. G. Falkovich, I. Kolokolov, V. Lebedev, and A. Migdal: Phys. Rev. E **54**, 4896 (1996).
5. M. Farge, G. Pellegrino, and K. Schneider: Phys. Rev. Lett. **87**, 054501 (2001).

Closure for anisotropic homogeneous turbulence as the problem of analytical and scaling properties of spectral tensors

S. R. Bogdanov

Karelian State Pedagogical University, Petrozavodsk, Russia

Spectral tensors of 2-point velocity correlations possess some general properties which give the heuristic impulse not only for deep physical insight into the structure and nature of turbulence, but for working out the alternative (relative to semi - empirical) closure strategies.

First of these properties is scaling: turbulence is treated as the critical system with universal dependence of spectral functions on wavenumber k and position $\mathbf{x}$ through the product kl. "Correlation radius" may depend on position $\mathbf{x}$ as well as orientation $\boldsymbol{\theta} \equiv \mathbf{k}/k$ of the wave vector $\mathbf{k}$.

Another basic property, stressed yet in [1], is the behavior of correlations in the low-wavenumber limit $kl \to 0$. Following the "tensorial volume" concept [2] it's assumed here that $F_{ij} \equiv \int < u_i(\mathbf{x}) u_j(\mathbf{x} + \mathbf{r}) > exp(-i\mathbf{k}\mathbf{r}) d\mathbf{r}$ are finite in the mentioned limit.

Both assumptions may be written as: $F_{ij}(\mathbf{x}, \mathbf{k}) = f_{ij}(\mathbf{x}, \boldsymbol{\theta})\varphi(kl)$ and similar ones for higher - order correlations. "Orientation amplitudes" $f_{ij}(\mathbf{x}, \boldsymbol{\theta})$ and scale $l(\mathbf{x}, \boldsymbol{\theta})$ form the set of "governing" parameters of turbulence, they play the role of "slow" variables. The closed set of equations for these variables is derived directly from exact spectral equations Reynolds stress tensor and other one-point characteristics are easily obtained by calculating the corresponding integrals over spherical shell of unit radius in $\mathbf{k}$- space. In particular, the decay laws $< u^2 > \sim x^{-6/5}, l \sim x^{2/5}$, compatible with experimental data, were derived for isotropic grid turbulence. For general homogeneous distortion simple analytical solution for fully developed turbulence is also available.

References

1. Saffman P.G. Coherent structures in turbulent flow. - Lecture Notes in Physics, 1981, vol. 136
2. S.C.Kassinos, W.C.Reynolds. Tensorial volume of turbulence revisited. - Phys. Fluids A 2 (9), 1990, p.p. 1669-1677.

On the modelling of subgrid-scale enstrophy transfer in turbulent channel flows

G. Hauët, C. B. da Silva, and J. C. F. Pereira

Instituto Superior Técnico, Pav. Mecânica I, 1º andar/LASEF, Av. Rovisco Pais, 1049-001, Lisboa, Portugal

An important issue in LES concerns the role played by the subgrid-scale (SGS) models on the quasi streamwise vortices near the wall. In order to be able to analyse this we extend the analysis of da Silva and Pereira [1] into a turbulent channel flow with $Re_\tau = 180$. We start by analysing the *SGS enstrophy dissipation*: $\frac{\partial^2 \tau_{kp}}{\partial x_j \partial x_p}$, that describes the effect of the SGS stresses τ_{ij} on the evolution of the resolved enstrophy field. From about $y^+ \approx 30$ until the centre of the channel the SGS enstrophy dissipation exhibits a small negative value (enstrophy forward scatter), as occurs in isotropic turbulence or in turbulent plane jets[1], but as the distance from the wall decreases the SGS enstrophy dissipation increases (in modulus) attaining a minimum at about $y^+ = 14$. In the viscous sublayer the SGS enstrophy dissipation provides a mean positive value, implying a mean backward enstrophy transfer. In this region the Smagorinsky model seems to perform well, unlike the dynamic Smagorinsky model. It was found that at the centre of the channel the JPDFs are quite similar to the ones obtained by da Silva and Pereira [1] *e.g.* all eddy-viscosity models considered are able to provide some (though small) enstrophy backscatter, and the correlation coefficient between the "real" (filtered DNS) and modelled SGS enstrophy dissipation is about 50% - very close to the values obtained in reference [1]. Surprisingly, at the buffer layer (about $y^+ = 14$) in the present channel flow, the "real" and modelled SGS enstrophy transfer displays very good results both for the forward and backward "modes" for the Smagorinsky model. It was found that this has to do with a link between the SGS enstrophy dissipation and the regions of high speed streaks. Ongoing work is aimed at analysing the impact of these mechanisms for SGS modelling, as well as in analysing the performance of several SGS models in reproducing the quasi-streamwise vortices from the buffer layer. More results can be found in Hauet *et al.* [2].

References

1. Carlos B. da Silva and José C. F. Pereira. The effect of subgrid-scale models on the vortices computed from large-eddy simulations. *Phys. Fluids*, 2004.
2. Gwenaël Hauët, Carlos B. da Silva, and José C. F. Pereira. The effect of subgrid-scale models on the near wall vortices: a-priori tests. *(submitted to Phys. Fluids)*, 2007.

Group-Theoretical Model of Developed Turbulence and Renormalization of the Navier-Stokes Equation

V. L. Saveliev[1] and M. A. Gorokhovski[2]

[1] Institute of Ionosphere, Almaty 050020, Kazakhstan `saveliev@topmail.kz`
[2] LMFA UMR 5509 CNRS Ecole Centrale de Lyon, 36 avenue Guy de Collongue, 69131 Ecully Cedex, France `mikhael.gorokhovski@ec-lyon.fr`

In relation to turbulence, we can *roughly* express Kadanoff's idea of "block picture" for the spin field in Ising's model as follows: If, instead of turbulent field $\mathbf{v}(\mathbf{r})$, we consider a field $\langle \mathbf{v} \rangle_{\sigma}(\mathbf{r})$ that is averaged on the scale σ, then the last one will "resemble" the original turbulent field $\mathbf{v}(\mathbf{r})$. The exact sense of "resemble" must be defined by the group of transformations for both fields and equations for these fields. In our paper [1], we proposed to associate the phenomena of stationary developed turbulence with the special self-similar solutions of the Euler equation. The self-similar solution implies its dependence on time through the parameter of the space symmetry transformation only. From this model it follows that the change of the scale of averaging from σ_0 to σ is equivalent to the composition of scaling, rotation, and translation transformations. We call this property a renormalization-group invariance of averaged turbulent fields. Assuming that on the small length scale σ_0 the turbulent velocity field can be approximated as the sum of a smooth velocity field and a random isotropic field, we averaged the Navier-Stokes equation over this small scale using our averaging formula. Then, the renormalization-group invariance provides an opportunity to transform the averaged Navier-Stokes equation over a small scale σ_0 to any scale σ. It is important to stress that we have shown that the turbulent viscosity in LES equation appeared not as a result of averaging of the nonlinear term in the Navier-Stokes equation, but from the molecular viscosity term with the help of renormalization-group transformation. Moreover, the Group-Theoretical Model of developed turbulence explains why the reduced description of turbulence is possible at all.

References

1. V.L. Saveliev and M.A. Gorokhovski, Phys. Rev. E **71**, 016302 (2005)

Scaling Properties of the Subgrid-scale Energy Dissipation in Large Eddy Simulation

Sergei G Chumakov

Center for Nonlinear Studies, Los Alamos National Laboratory
Los Alamos NM 87544, U.S.A.
chumakov@lanl.gov

We performed direct numerical simulation of forced homogeneous isotropic turbulence to obtain information about scaling of the SGS energy $k_s = (\overline{u_i u_i} - \bar{u}_i \bar{u}_i)/2$, its dissipation rate $\epsilon_s = 2\nu(\overline{\partial_j u_i \partial_j u_i} - \partial_j \bar{u}_i \partial_j \bar{u}_i)$ and their scaling. We found k_s and ϵ_s to be log-normally distributed and highly correlated. The SGS dissipation ϵ_s plays an important role in the subgrid-scale energy budget and thus the quality of modeling of ϵ_s is crucial for one-equation LES models.

We found that the assumption $\epsilon_s \sim k_s^\gamma$ holds well for the values of SGS Reynolds number $R_\Delta \equiv \sqrt{k_s}\Delta/\nu$ up to 5000 (maximum for our simulations). However, the value of γ was not found to be constant, nor was it found to depend on R_Δ but rather to depend on the proximity of the LES filter size Δ to the forcing length scale.

None of the observed scalings were close to $\epsilon_s \sim k^{3/2}$ which is widely used for modeling ϵ_s in the current literature. We found $\gamma \approx 1/2$ for Δ close to forcing scale, which corresponds to results by Chumakov and Rutland [2]; for smaller values of Δ the value of γ is close to 1, in accordance with Meneveau and O'Neal [3]; in the inertial range for both sets of data, γ varies between 0.6 and 0.9, which is partly corroborated by results of Borue and Orszag [1] obtained using DNS with hyperviscosity. It should be noted that we do not see a visible plateau at $\gamma \approx 2/3$, as would be expected based on Refined Similarity Hypothesis.

References

1. V. Borue and S. A. Orszag. Kolmogorov's refined similarity hypothesis for hyperviscous turbulence. *Phys. Rev. E*, 53(1):R21–R24, 1996.
2. S. G. Chumakov and C. J. Rutland. Dynamic structure subgrid-scale models for large eddy simulation. *Int. J. Numer. Meth. Fluids*, 47:911–923, 2005.
3. C. Meneveau and J. O'Neil. Scaling laws of the dissipation rate of turbulent subgrid-scale kinetic energy. *Phys. Rev. E*, 49(4):2866–2874, 1994.

On restraining the convective subgrid-scale production in Burgers' equation

Joop Helder and Roel Verstappen

Institute of Mathematics and Computing Science, University of Groningen,
P.O.Box 800, 9700AV Groningen, The Netherlands;
J.A.Helder@math.rug.nl, R.W.C.P.Verstappen@rug.nl

Most turbulent flows cannot be computed directly from the Navier-Stokes equations, because the convective term $C(u,v) = (u \cdot \nabla)v$ produces too many scales of motion. In quest of a dynamically less complex mathematical formulation, we consider regularizations $C_n(u,v)$ of the nonlinearity. Hereby, the regularizations conserve the energy, enstrophy (2D) and helicity by construction. This yields $C_n = C + \mathcal{O}(\epsilon^n)$, where ϵ is the filter length, and $n = 2, 4, 6$. The regularized system is more amenable to approximate numerically, while its solution approximates the large-scale dynamical behavior of the Navier-Stokes solution. The evolution of the vorticity ω is given by the usual equation, where the vortex stretching term becomes $C_n(\omega, u)$. By analyzing the regularized interactions, the filter length is determined such that the production of smaller sales by means of vortex stretching stops at the grid-size δ.

As a first step, the method is applied to Burgers' equation with Re = 50. Here, the analysis is relatively easy, while important aspects of 3D Navier-Stokes remain. To illustrate the results, energy spectra are shown below for $n = 4$. Clearly, without the model the physics are not captured correctly, as the energy is not dissipated enough at the high wavenumbers. With the model, a power law of k^{-20} for small scales is found, for different values of δ.

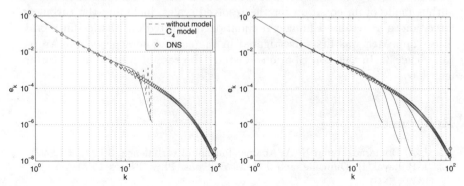

Fig. 1. *Left: Steady state energy spectra, with and without the model. Right: Steady state energy spectra for different values of the grid-size δ.*

A LES-Langevin model for turbulence

R. Dolganov[1], B. Dubrulle[2], and J.-P. Laval[1]

[1] Laboratoire de Mécanique de Lille, CNRS UMR 8107, Bld Paul Langevin,
   F-59655 Villeneuve d'Ascq, France rostislav.dolganov@gmail.com
[2] Groupe Instabilité et Turbulence, SPEC/DRECAM/DSM, CNRS URA 2464,
   CEA Saclay, F-91191 Gif sur Yvette, France bdubrulle@cea.fr

The rationale for Large Eddy Simulation is often rooted in our inability to handle all the degrees of freedom of a large Reynolds number turbulent flow. The price to pay is of course a need for parametrisation (the so-called SubGrid Scale models, or SGS)[1]. The backscatter modeling leads to the development of "stochastic" strategies, where the discarded small-scale motions are reproduced by a set of random numbers, mimicking either a random force or synthetic velocity fields[2].

We propose a new model of turbulence for use in large-eddy simulation (LES)[3]. The turbulent force, represented here by the turbulent Lamb vector, is divided in two parts. The contribution including only subfilter fields is deterministically modeled through a classical eddy-viscosity. The other contribution including both filtered and subfilter scales is dynamically computed as solution of a generalized (stochastic) Langevin equation. This equation is derived using Rapid Distorsion Theory (RDT) applied to the subfilter scale. The general friction operator therefore includes both advection and stretching by the resolved scale. The stochastic noise is derived as the sum of contribution from the energy cascade and a contribution from the pressure. The LES model is thus made of an equation for the resolved scales, including the turbulent force, and a generalized Langevin equation integrated on a twice-finer grid. These approach was first investigated in the case of homogeneous isotropic turbulence[3]. In this contribution we study the case of wall bounded turbulent flow.

References

1. J. A. Domaradzki and N. A. Adams: Journal of turbulence, **3**(024), 1 (2002)
2. J.-P. Laval, B. Dubrulle, and J.C. McWilliams: Phys. Fluids **15**(5), pp 1327–1339 (2003)
3. J.-P. Laval, B. Dubrulle: Eur. Phys. J. **B** 49, pp 471–481 (2006)

Quality assessment of inlet boundary conditions and domain size for fully compressible LES of wall-jet turbulent mixing

Guido Lodato, Pascale Domingo, and Luc Vervisch

INSA - Rouen, UMR-CNRS-6614-CORIA
lodato@coria.fr, domingo@coria.fr, vervisch@coria.fr

Summary. The present work aims at studying impinging jets using Large Eddy Simulation. LES is performed with the fully compressible form of the Navier Stokes equations and statistical results are compared to measurements [1] to assess the quality of inlet forcing and the effect of numerical acoustic perturbations generated at the boundaries.The complexity of the flow field's structures and the additional difficulties arising from abandoning the hypothesis of incompressibility, make this problem a particularly tough test bench for boundary conditions. A homemade parallel solver based on an explicit FV 4th order centered skew-symmetric-like scheme [2] is used. SGS terms can be closed by either the FSF [3] or WALE [4] models.

Two major problems relevant to the boundary conditions are observed: (a) the strong impact of the chosen inlet profile; (b) the effect of numerical perturbations generated at the outflow boundaries. The inlet profile shape highly affects the jet's spreading before the stagnation point and the amount of kinetic energy the jet core can retain before reaching the wall. Furthermore, the shear layer thickness has a strong impact on the level of turbulence which is generated before the stagnation point and convected downstream in the wall-jet region.

The outlet boundaries have shown too much reflection: the LODI assumption [5] has demonstrated unsuccessful and the highly tridimensional structures are not able to cross the boundaries with acceptably low level of reflection. The level of reflection can be drastically reduced by taking into account transverse convective terms [6] in the computation of incoming wave amplitude variations; in addition, suitable compatibility conditions [7] on the edges and the corners of the computational domain have to be implemented in order to properly handle transverse convection on these regions, thus preventing the development of diagonal numerical perturbations.

1. D. Cooper et al: Int. J. Heat & Mass Trans. **36**, 10 (1993)
2. F. Ducros et al: J. Comput. Phys. **161**, (2000)
3. F. Ducros et al: J. Fluid Mech. **326**, (1996)
4. F. Nicoud, F. Ducros: Flow, Turb. Comb. **62**, 3 (1999)
5. T. Poinsot, S. Lele: J. Comput. Phys. **101**, (1992)
6. C. Yoo et al: Comb. Theory & Modeling **9**, 4 (2005)
7. M. Valorani, B. Favini: Num. Meth. PDE's **14**, 6 (1998)

Parametric study of LES subgrid terms in turbulent phase separation flows

J. Larocque[1], S. Vincent[2], D. Lacanette[2], P. Lubin[2], J.P. Caltagirone[2], and P. Sagaut[3]

[1] CEA-CESTA BP2 33114 Le Barp France `jerome.larocque@cea.fr`
[2] TREFLE 16 avenue Pey Berland 33607 Pessac France `vincent@enscpb.fr`, `lacanette@enscpb.fr`, `lubin@enscpb.fr`, `caltagirone@enscpb.fr`
[3] LMM 4 place Jussieu 75252 Paris cedex 5 France `sagaut@lmm.jussieu.fr`

The numerical simulation of turbulent multiphase flows is of particular interest for academic and industrial problems such as spray formation, energy production, material processes or environment predictions. The single phase turbulence models can be used for simulating multiphase flows when the characteristic length scale of the interfacial structures are smaller than the Kolmogorov scale or when the space scale of interfaces or free surfaces are larger than the smaller turbulent structures. As soon as time and space scales of turbulence and interface are comparable, the Direct Numerical Simulation (DNS) requires too expensive computer resources in most configurations and the Large Eddy Simulation (LES) of multiphase flows becomes an interesting approach. It allows leading accurate unsteady simulations while modeling the smaller flow scales. However, the coupling between turbulence and phase filtering has to be explicitly taken into account. Indeed, when the filtered two-phase flow Navier-Stokes equations are considered, specific two-phase subgrid terms appear. Thus, LES models must be proposed to deal with these fluid mechanic problems.

The objectives of the present work are to provide an a priori estimate of the specific subgrid LES terms occurring in a turbulent two-phase flow. 3D simulations of a water/oil phase separation in a cubic cavity are carried out without turbulence modeling. These "DNS" results are filtered in order to quantify and classify the subgrid terms. Two filters F2 (small) and F4 (large) are used on a compact support. A parametric study is investigated for varying surface tension coefficient, density or dynamic viscosity ratios.

CVS of turbulent compressible mixing layers using adaptive multiresolution methods

Olivier Roussel[1] and Kai Schneider[2]

[1] TCP, Kaiserstr. 12, Karlsruhe, Germany. roussel@ict.uni-karlsruhe.de
[2] MSNM-GP, 38 rue Joliot-Curie, Marseille, France. kschneid@cmi.univ-mrs.fr

The Coherent Vortex Simulation (CVS) method aims at computing the organized part of turbulent flows, i.e. the coherent vortices, while modelling the incoherent background flow [1]. A non-linear wavelet filtering is used to separate coherent from incoherent contributions. Here we extend the CVS method, originally developed for incompressible flows, to compressible flows. The evolution of the coherent part is computed using a finite volume scheme on a locally refined grid, while the incoherent part is discarded. As example, we apply the CVS method to compute a time-developing three-dimensional turbulent mixing layer in the weakly compressible regime. We find that only 17.9 % wavelet coefficients contain around 98.3 % of the energy and 93.4 % of the enstrophy. Concerning the CPU time, it only represents 29.0% of the one required by the DNS [2].

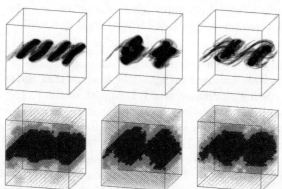

Fig. 1. CVS of a 3D compressible mixing layer, $Ma = 0.3$, $Re = 200$, maximal resolution $N = 128^3$. Top: isosurfaces of vorticity Bottom: adaptive grids.

References

1. M. Farge, G. Pellegrino, and K. Schneider. *Phys. Rev. Lett.*, 87(5), 2001.
2. O. Roussel and K. Schneider. Coherent vortex simulation of weakly compressible turbulent flows, 2007. preprint.

URANS and Seamless Hybrid URANS/LES : the forced turbulent temporal mixing layer

S. Carpy[1] and R. Manceau[1,2]

[1] Laboratoire d'Etudes Aérodynamiques, UMR-6609 CNRS, University of Poitiers/CNRS/ENSMA, Site SP2MI, Boulevard Marie et Pierre Curie, Téléport 2 - BP 30179 - 86962 Futuroscope Chasseneuil Cédex, France
sabrina.carpy@lea.univ-poitiers.fr

[2] remi.manceau@lea.univ-poitiers.fr

A seamless hybrid RANS/LES model is compared with URANS and LES, in a forced temporal mixing layer at Re=800 based on the momentum thickness. 2D-URANS-computations show that the Rotta+IP RSM is globally able to reproduce the Kelvin-Helmoltz-like vortices, whereas the standard k-ε is not. This study, together with a previous study [1], suggests that linear eddy-viscosity models are not suitable to reproduce the complex interaction between resolved, large scale, unsteady structures and the underlying turbulence. However, 3D-computations with RSM show that there is no fine-scale motion, but only the most amplified wave length: that augurs ill for the performance of URANS in flows where there are no strong, large-scale forcing. Therefore, a seamless hybrid RANS/LES approach based on the so-called PITM [2] is adopted. The main difference with the traditionnal Rotta+IP RSM lies in the modified coefficient in the dissipation equation: $C_{\epsilon_2}^* = C_{\epsilon_1} + \frac{C_{\epsilon_2} - C_{\epsilon_1}}{1 + \beta_0 \eta_C^{2/3}}$ with $\eta_c = \kappa_c L$, where L is the turbulence length scale and κ_c the cutoff wavenumber. On a fixed mesh, imposed variations of κ_c show that the subfilter energy k_{sgs} decreases when κ_c increases, which is consistent the PITM theory. Then, we compare two different meshes with κ_c linked to the cell size. Refinement permits to solve a larger range of structures and to approach LES solution (the balance resolved/modelled energy is controlled by the model). Our results show that on a relatively coarse mesh, PITM is better than LES, which is very interesting to reduce the cost of the computation. Moreover, while URANS is able to reproduce the 2D dynamics, the PITM is also able to reproduce the 3D aspects.

References

1. S. Carpy and R. Manceau. Turbulence modelling of statistically periodic flows: Synthetic jet into quiescent air. *Int. J. Heat Fluid Flow*, 27(5):756–767, 2006.
2. B. Chaouat and R. Schiestel. A new partially integrated transport model for subgrid-scale stresses and dissipation rate for turbulent developing flows. *Phys. Fluids*, 17:1–19, 2005.

Thermal Boundary Layers Simulations Under Adverse Pressure Gradients

Guillermo Araya, Kenneth Jansen and Luciano Castillo

Rensselaer Polytechnic Institute, Troy, NY, 12180, US arayaj@rpi.edu

A new approach for generating realistic turbulent hydrodynamic/thermal information at the entrance of a computational domain is presented. Large Eddy Simulations (LES) are performed over a straight-walled diffuser that induces an adverse pressure gradient flow.

1 Some Previous Results

We extend the rescaling-recycling method, proposed by Lund et al. [1], to thermal turbulent boundary layers in pressure gradient flows. The scaling laws for the thermal field are based on the investigation performed by Wang and Castillo [2]. In our simulations, the non-dimensional pressure parameter, β, remained approximately constant (0.085-0.11) along the streamwise direction and the range for Re_θ is 382-506. An isothermal wall is considered and the molecular Pr number is 0.71. Figure 1 shows a comparison of present simulations with experimental and numerical data in zero-pressure gradient flows. The effects of pressure gradient are more evident in the outer region of thermal fluctuations.

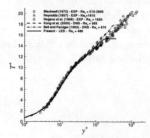 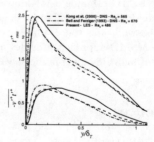

Fig. 1. Mean temperature (left) and rms temperature fluctuations together with normal heat fluxes (right) in wall units

References

1. T. Lund and X. Wu and K. Squires. Generation of turbulent inflow data for spatially-developing boundary layer simulations. J. Comp. Phys. **140**, 233–258 (1998)
2. X. Wang and L. Castillo. Asymptotic solutions in forced convection turbulent boundary layers. J. of Turbulence **4**, 1–18 (2003)

A Multi-scale, Multi-domain Approach to Wall-Modelling for LES of High Reynolds Number Wall-Bounded Turbulence

R. Akhavan* and M. U. Haliloglu

University of Michigan, Department of Mechanical Engineering,
Ann Arbor, MI 48109-2125, USA
*corresponding author, e-mail: raa@umich.edu

Summary. A new approach to wall-modelling for large-eddy simulation (LES) of high Reynolds number wall-bounded turbulent flows is presented. The proposed method utilizes the quasi-periodicity of the turbulence structures in the near-wall region to compute the near-wall region at high resolution in a minimal flow unit large enough to accommodate only one packet of turbulent vortical structures. This minimal flow unit, which is of size $L_x^+ \sim \pi Re_\tau$ wall units in the streamwise direction, $L_y^+ \sim 1500 - 2000$ wall units in the spanwise direction, and $L_z^+ \sim 200 - 250$ wall units in the wall-normal direction, is then repeated periodically (in the near-wall region) or quasi-periodically (as it is passed to the outer layer) and matched to a full-domain but coarse-resolution LES in the outer layer. The proposed multi-scale, multi-domain (MSMD) method has been implemented in LES of turbulent channel flow using a patching collocation spectral domain-decomposition method. Large-eddy simulations were performed at $Re_\tau \sim 1000, 2000, 5000$ and $10,000$ with a resolution of $32 \times 64 \times 17$ in the near-wall region and $32 \times 64 \times 33$ in the outer layer, independent of the Reynolds number. The Nonlinear Interactions Approximation Model (NIAM) of Haliloglu & Akhavan 2004 [1] was used as the subgrid-scale model. The predictions were found to be in good agreement with Dean's correlation [2], the law of the wall, and the available experimental data [3, 4]. These results suggest that the MSMD approach provides a simple and computationally efficient method for LES of high Reynolds number wall-bounded turbulent flows.

References

1. M.U. HALILOGLU & R. AKHAVAN, *DLES-V Proc.* (2004) 39–48.
2. R.B. DEAN, *J. Fluids Eng.* **100** (1978) 215–223
3. M.A. NIEDERSCHULTE, R.J. ADRIAN & T.J. HANRATTY, *Exp. in Fluids* **9** (1990) 222–230
4. G. COMTE-BELLOT, *Ph.D. Thesis, University of Grenoble* (1963)

RANS Modelling of Turbulent Flows Driven by a Travelling Magnetic Field

P.A. Nikrityuk[1], K. Eckert and R. Grundmann

Institute for Aerospace Engineering, Dresden University of Technology, D-01062
Dresden, Germany
nikrityuk@tfd.mw.tu-dresden.de

This work is devoted to the numerical study of the turbulent regime of a liquid
metal flow driven by Traveling magnetic field (TMF), the magnetic forcing
parameter of which, F, is in range $(1\text{-}100)F_{cr}^{3D}$ with F_{cr}^{3D} given by Grants
& Gerbeth [1]. The low-Re formulation of a standard k-ω turbulence model
was used to calculate the turbulent flow parameters. The computations were
performed for enclosed cylindrical containers of aspect ratio (height / radius)
$A = H_0/R_0$ equals to 2. The results of simualtions are shown in Fig. 1.
We found that for $F > F_{cr}^{3D}$ the maximum of u_z, scaled with ν/R_0, depends
on the forcing parameter as $0.7F^{1/2}$, which is in good agreement with our
analytical studies.

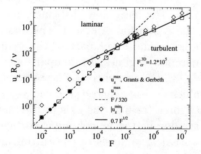

Fig. 1. The maximum and minumum of axial velocity vs. forcing number F for an
enclosed cylinder with $H_0/R_0 = 2$.

References

1. I. Grants, G. Gerbeth: J. Crystal Growth **269**, 630 (2004)

Direct Numerical Simulation of Turbulent Flow in Travelling Magnetic Fields

Kristina Koal[1], Jörg Stiller, and Roger Grundmann

Technische Universität Dresden, Institute for Aerospace Engineering (ILR)
01062 Dresden, Germany
[1]kristina.koal@tu-dresden.de

Our work, presented here, is focused on the numerical investigation of turbulent flows caused by travelling magnetic fields (TMF). Previous work about TMF is constrained to the laminar or transient regime, e.g. [2]. At this conference, we present the results of direct numerical simulation (DNS) of TMF driven turbulent flows for forcing parameters up to 132-times supercritical.

The flow is governed by the incompressible Navier-Stokes equations including an upwards driving Lorentz force $f = \frac{1}{2}Fr^2 e_z$, where r and z are the radial and axial coordinates and F the forcing parameter [2]. Low-frequency / low-induction and low interaction conditions are assumed. The simulations are done with a spectral element - Fourier spectral method [1]. DNS was carried out for a closed cylinder with an aspect ratio equal to 1 and $F \leq 1.6 \cdot 10^7 \approx 132F_c$. The used grid consists of approximately 40 million degrees of freedom (DOF). The basic flow is strongly meridional; only fluctuating currents exist in the azimuthal direction. The main roll of an averaged, turbulent flow is centered in the lower part of the cylinder and not in the midplane like in laminar flow regimes. In comparison with flows driven by rotating magnetic fields [3], the presence of fluctuations is much more dominating. The largest values appear where the mean velocity is reverse to the direction of the Lorentz force (nearby the axis in the main roll). The regions of highest turbulence are dominated by columnar vortices.

A comprehensive statistical analysis of the turbulence behaviour will be presented at the poster.

References

1. BLACKBURN, H.M. and SHERWIN, S.J.: *J. Comp. Phys.*, 197 (2004) 759–778.
2. GRANTS, I. and GERBETH, G.: *J. Cryst. Growth*, 269 (2004) 630–638.
3. STILLER, J., FRAŇA, K., and CRAMER, A.: *Phys. Fluids*, 18 (2006) 074105.

Turbulent pipe flow in a transverse magnetic field: A comparison between PIV measurement and DNS

Junichi Takeuchi[1], Shin-ichi Satake[2], Tomoaki Kunugi[3], Takehiko Yokomine[4], Neil B. Morley[1], and Mohamed A. Abdou[1]

[1] University of California, Los Angeles takeuchi@fusion.ucla.edu morley@fusion.ucla.edu abdou@fusion.ucla.edu
[2] Tokyo University of Science satake@te.noda.tus.ac.jp
[3] Kyoto University kunugi@nucleng.kyoto-u.ac.jp
[4] Kyushu University yokomine@ence.kyushu-u.ac.jp

Magneto-hydrodynamic turbulent pipe flow has been investigated by experimental measurements and numerical simulations as a part of US-Japan JUPITER-II collaboration. The objective of the present study is to elucidate the mechanism of MHD turbulence reduction by comparing the experimental results obtained by particle image velocimetry (PIV) with Direct Numerical Simulations (DNS) under the same flow conditions.

The MHD turbulent pipe flow experiments has been conducted using an aqueous KOH solusion as a working fluid. The test section is an 8 m long acrylic pipe with 89 mm inner diameter, and the visualization section are attached at 6.8 m from the inlet where the flow is fully developed. The maximum 2.1 Tesla magnetic field is applied for 1.4 m in streamwise length. PIV technique has been applied using specially designed optical componets for magnetic field condition. The measurements are performed at Re = 11000 with variable Hartmann number ($Ha = B_0 R(\sigma/\rho\nu)^{1/2}$) up to 20.

The DNS have also performed at the same flow and magnetic parameters. The detail of the procedures are described by Satake et al (2002).

PIV measurements yield velocity data in the plane parallel to the flow and fielddirection, and the turbulence statistics are compared with the DNS database. Mean velocity profiles normalized by centerline velocity are in good agreement. The profiles show the Hartmann effect, which flattens the velocity profile by accelerating the flow in the near-wall region and decelerating in the core region. The streamwise and the radial component of the velocity fluctuations decrease with increasing Hartmann number in the core region, whereas the near-wall peak of the streamwise velocity fluctuations remain in the same height. The Reynolds shear stress also decreases with increasing Hartmann number. The same tendencies appear in both cases.

Direct numerical simulations of the turbulent Hartmann flow in cylindrical ducts

I. E. Sarris[1], Y. Detandt[2], C. Toniolo[1], A. Viré[1], M. Kinet[1], D. Carati[1], G. Degrez[2], and B. Knaepen[1]

[1] Statistical and Plasma Physics Unit, Université Libre de Bruxelles, Belgium
[2] Aero-Thermo-Mechanics Department, Université Libre de Bruxelles, Belgium

Magnetohydrodynamic turbulent flows are often encountered in engineering fields. For laboratory and industrial scale flows of conducting fluids (except plasmas) the magnetic Reynolds number, R_m, is much less than unity, which means that the induced magnetic field is very small, and the Quasi-Static approximation is valid. This approximation is here incorporated into an hydrodynamic Spectral/Finite Element code named SFELES [1].

For the MHD channel flow, usually referred to as the Hartmann flow, experiments [2] show a similarity: several statistical properties only depend on the modified Reynolds number (R), the ratio of the hydrodynamic Reynolds number to the Hartmann number. Large eddy simulations of the Hartmann flow [3], show that the Stuart number (N) may also be a similarity parameter. Although these studies have used different ducts, the critical values of R for transition to turbulence were consistently found in the range $250 < R < 500$.

Direct numerical simulations (DNS) of the Hartmann flow are presented, in a closed cylindrical duct similar to the one used by [2] for several R values and $Ha < 20$. Results show that for weak magnetic fields a fully developed turbulent flow is observed. For low but not laminar R numbers, turbulent fluctuations are mostly confined close to the Hartmann and to the side walls, while an almost stratified flow exists at the center of the duct. Because of the low Reynolds numbers used here, turbulent structures from the sides and the Hartmann layers are strongly correlated. Assembling the simulations for several Re numbers and by increasing Ha we individuated the critical R numbers. It appears that for the $Ha < 20$ regime, laminarization occurs at different R numbers depending on both Re and Ha.

References

1. D.O. Snyder, G. Degrez: Intl. J. Num. Methods Fluids **41**, 1119 (2003)
2. P. Moresco, T. Alboussière: J. Fluid Mech. **504**, 167 (2004)
3. I.E. Sarris, S.C. Kassinos, D. Carati: submitted, (2007)

Numerical simulation of a longitudinal Lorentz force flowmeter for turbulent flows in a circular pipe

Bernard Knaepen[1], André Thess[2], Evgeny Votyakov[2], and Oleg Zikanov[3]

[1] Université Libre de Bruxelles, Belgium, bknaepen@ulb.ac.be
[2] Ilmenau University of Technology, Germany, thess@tu-ilmenau.de
[3] University of Michigan - Dearborn, USA, zikanov@umich.edu

Abstract

A Lorentz force flowmeter is a device for the contactless measurement of flow rates in electrically conducting fluids and has a wide variety of potential applications in metallurgy, semiconductor crystal growth and glassmaking. In order to successfully develop Lorentz force flowmeters it is important to answer the following question: Given a magnet system and a velocity field, what is the force acting on the magnet system? The goal of the present work is to analyse the sensitivity of a simplified longitudinal Lorentz force flowmeter in which the magnetic field is produced by a single coil located outside of a circular pipe. We concentrate here our attention on the influence of turbulence on the measurements and we show that velocity fluctuations do not strongly affect the performance of the flowmeter. An illustration of our numerical results is

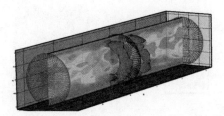

Fig. 1. Contour of the Lorentz force intensity. The sliced plane shows contours of the streamwise velocity intensity.

contained in figure 1 where an isocontour of the Lorentz force is presented, highlighting the fact that the longitudinal flowmeter considered samples preferentially the flow in the vicinity of the wall.

Experimental investigation of time-dependent flow driven by a travelling magnetic field

A. Cramer, J. Pal, C. Zhang, S. Eckert, and G. Gerbeth

Institute of Safety Research, Forschungszentrum Dresden–Rossendorf,
D-01314 Dresden, Germany a.cramer@fzd.de

The present work is devoted to both mean flow and turbulence in an electrically conducting melt subjected to a travelling magnetic field (TMF). Grants&Gerbeth (J. Crystal Growth, **269**(2004), 630-638) derived the analogue F (forcing parameter) of the magnetic Taylor number Ta, the latter being the dimensionless measure for the strength of the driving force in a rotating magnetic field, for the TMF. Velocity measurements were done in an eutectic GaInSn melt confined in a perspex cylinder ($2R = H = 60$mm) with the meanwhile established ultrasonic Doppler velocimetry (UDV). The results in Figure 1 indicate that F reasonably measures the strength of the Lorentz force in magnetic coil systems with different metrics. Besides evaluating turbulence spectra from series of UDV profiles, statistical data were also recorded with electric potential difference probes (described comprehensively in Flow Meas. Instrum. **17**(2006), 1-11), which are shown in Figure 2.

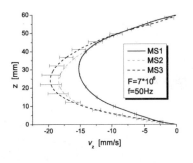
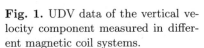
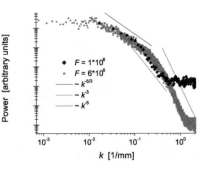

Fig. 1. UDV data of the vertical velocity component measured in different magnetic coil systems.

Fig. 2. Turbulence spectra measured with potential probes.

A non local shell model for MHD turbulence.

Rodion Stepanov[1] and Franck Plunian[2]

[1] Institute of Continuous Media Mechanics, Korolyov 1, 614061, Perm, Russia
   rodion@icmm.ru
[2] Laboratoire de Geophysique Interne et de Tectonophysique BP 53, 38041
   Grenoble Cedex 9, France Franck.Plunian@ujf-grenoble.fr

We derive a new shell model of magnetohydrodynamic (MHD) turbulence in which the energy transfers are not necessary local. The model is defined by the following set of equations

$$\dot{U}_n = ik_n \sum_{m=1}^{N} [Q_n^m(U,U,a) + Q_n^m(B,B,-a)] - \nu k_n^2 U_n + F_n,$$

$$\dot{B}_n = ik_n \sum_{m=1}^{N} [Q_n^m(U,B,b) + Q_n^m(B,U,-b)] - \eta k_n^2 B_n,$$

$$Q_n^m(X,Y,c) = T_m[c_m^1 X_{n+m}^* Y_{n+m+1} + c_m^2 X_{n-m}^* Y_{n+1} + c_m^3 X_{n-m-1} Y_{n-1}].$$

The parameters ν and η are respectively the kinematic viscosity and the magnetic diffusivity, F_n is the forcing of turbulence, and $k_n = \lambda^n$. For $N = 1$ and $B_n = 0$, we recognize the local Sabra model. The additional non-local interactions for $N \geq 2$ correspond to all other possible triad interactions except the ones involving two identical scales. The coefficients a_m^i and b_m^i are defined form conservation laws of the total energy $E = \frac{1}{2} \sum_n |U_n|^2 + |B_n|^2$, magnetic helicity $H_B = \frac{1}{2} \sum_n (-1)^n k_n^{-1} |B_n|^2$ and cross helicity $H_C = \frac{1}{2} \sum_n (U_n B_n^* + B_n U_n^*)$ in the limit $\nu = \eta = F_n = 0$.

In MHD turbulence, we investigate the fully developed turbulent dynamo for a wide range of magnetic Prandtl numbers in both kinematic and dynamic cases. At small P_m we find large scale magnetic energy obtained from smaller scale kinetic energy. We understand such a non-local energy transfer as a mean-field "alpha"-effect. At large P_m we find strong energy transfer from kinetic scales belonging to the inertial range and close to the viscous scale towards smaller magnetic scales. This non-local transfer corresponds to the usual kinematic picture in which the vortices of maximum shear (close to the viscous scale) generate magnetic field of much smaller scales. In addition we find that this magnetic energy is then transferred back locally to kinetic scales in the viscous range where it is lost by viscous dissipation. Both local and non local energy transfers are clearly identified.

Compressibility effects on the return to isotropy of homogeneous anisotropic turbulence

Matthieu Crespo, Stephane Jamme, and Patrick Chassaing

ENSICA, 1 Place Emile Blouin 31056 Toulouse Cedex 5, France
mcrespo@ensica.fr, jamme@ensica.fr, chassain@ensica.fr

This study aims at shedding some light on the influence of the compressibility on the anisotropic turbulence decay with a numerical approach. We solve the full three-dimensional Navier-Stokes equations using a finite difference approach. The inviscid part is resolved using a fifth-order Weighted Essentially Non-Oscillatory scheme (WENO). Viscous terms contributions are computed with a sixth-order accurate compact scheme. A third-order Runge Kutta algorithm is used to advance in time.

The well known "return to isotropy" phenomenon consists in energy transfers among normal components of the Reynolds stress tensor. In the budget equations, the fluctuating pressure-deformation correlation term is responsible for the energy redistribution. Previous studies have analyzed compressibility effects regarding isotropic flows. They have shown that the root mean square value of pressure fluctuations is correlated with the turbulent mach number M_t and the parameter χ that controls the energy ratio between the compressible and solenoidal part of the velocity field. Moreover, they have shown that for $M_t > 0.4$, eddy shocklets may occur and imply a faster decrease of the turbulent kinetic energy.

A parametric study depending on the initial values of M_t, χ and the initial anisotropic state has been achieved. We have observed similar evolutions than in the compressible isotropic situation. For $M_t = 0.6$, we have found that eddy shocklets occur in our simulations as the turbulent kinetic energy decrease faster than for low M_t values. The probability density function of the local Mach number confirms that a non-negligible fraction of the flow field is locally supersonic. Concerning the return to isotropy trajectories, we have been confronted to a transient evolution in the early stage of our simulations. We were not able to determine a specific impact of compressibility for two-component initial anisotropic states on the return to isotropy trajectories. Nevertheless, the return to isotropy rate could be exploited. We could see that increasing M_t implies a faster return to isotropy while the opposite is true when increasing χ. Moreover, the return to isotropy rate is more affected by the variation of M_t than the changes in χ.

Large eddy simulation of a tunnel fire using two step combustion chemistry

R J A Howard[1], N Peres[1], D Toporov[2] and A C M Sousa[1]

[1] Mechanical Engineering Department, University of Aveiro, Aveiro 3810-193, Portugal rhoward@mec.ua.pt
[2] RWTH Aachen University, Germany

Large eddy simulation (LES) methods have been developed to understand accidental fires in road tunnels better. The basis is a rapid, single step reaction model adapted for LES [1]. Validation based on the Offenegg tunnel fire experiment [2] showed that the inclusion of gas radiation losses [3] are essential to produce realistic results. The mole fractions of each of the product gases in the radiation model were extracted from a scalar representing all the products [2]. Instead of this, a two step mechanism for laminar flames [4] that was applied to (RANS) turbulent flames [5] is further adapted for LES here. The ultimate objective is the explicit modelling of the gases and radicals that are present in sufficient quantities to be resolved by LES. Other gases and radicals present in smaller quantities can then be lumped together and modelled using approaches such as those based on CHEMKIN.

References

1. R. J. A. Howard and D. Toporov, The eddy dissipation combustion model developed for large eddy simulation, submitted to *Combustion and Flame*, Elsevier, ref **C4207**
2. R. J. A. Howard, N. Peres and A. C. M. Sousa, Large eddy simulations of a tunnel fire, submitted to *Combustion and Flame*, Elsevier, ref **C4247**
3. R. S. Barlow, A. N. Karpetis and J. H. Frank, Scalar profiles and NO formation in laminar opposed flow partially premixed mathane/air flames, Combustion and Flame, No. 127 (2001) pp. 2102-2118, Elsevier.
4. Westbrook, C. K. and F. L. Dyer. Simplified reaction mechanisms for the Oxidation of Hydrocarbon fuels in flames. Combustion Science and Technology. Taylor and Francis, **27**, 31–43, 1981.
5. Ye T. H., J. Azevedo, M. Costa and V. Semiao. Co-combustion of pulverized coal, pine shells and textile wastes in a propane fired furnce: measurements and predictions. Combustion Science and Technology. Taylor and Francis, **176**, 12:2071–2104, 2004.

On fast chemical reactions and singular vortices advecting multi-scale concentration fields

Denis Martinand

Laboratoire de Physique, École Normale Supérieure de Lyon, F-69364 Lyon Cedex 07, France denis.martinand@ens-lyon.fr

The depletion of two reactants involved in a fast second order reaction is addressed analytically and numerically. The concentration fields of the reactants satisfy advection–diffusion–reaction equations. In the case of fast reaction, the reactants tend to become segregated by the depletion of the reactant locally in default. The reaction speed integrated over the physical domain, V, can then be approximated by integrating the diffusive flux of reactants along $\mathcal{B}$, the boundary separating the so formed patches of reactants.

The configuration addressed consists of a steady two-dimensional vortex, with angular velocity $\dot{\theta} = \mathrm{Da}^{-1} r^{-\beta-1}$, advecting initially self-similar concentration fields. This initial self-similarity takes the form of fields presenting a prescribed Hölder exponent h and $\mathcal{B}$ a prescribed box-counting dimension $D_{\mathcal{B}}$. Algebraic scaling laws $V \propto t^{\mu}$ are sought, with μ depending on β, h and $D_{\mathcal{B}}$.

As the velocity field stretches and folds $\mathcal{B}$ and the patches of reactants, a spiral winds up within an inner radius, where the fluctuations of concentration are smoothed-out by diffusion, and an outer radius, where $\mathcal{B}$ is noticeably stretched by the vortex. Two situations are to be considered, depending on the relative singularities of the vortex and initial conditions. For $D_{\mathcal{B}} - h > 2\beta/(\beta+1)$, V is governed by the increase of the length of $\mathcal{B}$ as the outer radius grows and $\mu = (h - D_{\mathcal{B}})/2 + 2/(\beta+1)$. For $D_{\mathcal{B}} - h < 2\beta/(\beta+1)$, V is governed by the accumulation of the plies of $\mathcal{B}$ close to the inner radius and $\mu = 3(4 + h - D_{\mathcal{B}})/2(\beta+2) - 1$.

Numerically, the advection–diffusion–reaction equations are integrated using a fourth order Runge–Kutta compact scheme and initial concentration fields obtained by fractional Brownian motion, leading to a good agreement with the predicted scaling law for the temporal evolution of V. Algebraic scaling laws are also predicted and observed for the dependencies of V with the Peclet and Damköhler numbers.

This work has been supported by the EU Research Training Network "Fluid Mechanical Stirring and Mixing: the Lagrangian Approach" HPRN-CT-2002-00300.

Front surfaces in turbulent premixed flames

G. Troiani[1,2], M. Marrocco[2] and C.M. Casciola[1]

[1] Dipartimento di Meccanica e Aeronautica, Università di Roma *La Sapienza*
   guido.troiani@uniroma1.it
[2] ENEA C. R. Casaccia, Roma Italy

The characteristics of flame fronts in turbulent premixed combustion are experimentally analyzed. The flame front is detected by laser-induced fluorescence (LIF) of the radical OH and a PIV system is used to measure the velocity field. Different regimes are achieved by varing the equivalence ratio, while keeping the other parameters fixed. Conceptually the combustion in the wrinkled flame regime, is confined to a thin laminar sheet (flamelet) wrinkled by the turbulent flow. Following Damköhler, the turbulent burning velocity S_T is proportional to the laminar burning velocity S_L, multiplied by the ratio of the instantaneus to averaged flamelet surface. The multiscale nature of the surface suggests the application of geometrical concepts from fractal theory to describe the amount of wrinkling in a range of scales confined between an inner (ϵ_i) and an outer (ϵ_o) cut-off.

Here a number of experiments have been carried out. We varied u'/S_L by modifying the equivalence ratio Φ, thus S_L, while keeping the fluctuation intensity constant. For each case, the fractal dimension D of the front has been estimated by a box-counting technique. D is substantially independent from u'/S_L confirming previous results (Gülder et al., *Comb. and Flame*, 120 2000). The value we find $D \simeq 2.24$ is well below the limit corresponding to passive scalar iso-surfaces ($D = 2.37$) (Sreenivasanet al., *J.F.M.*, 173 1986). The inner cut-off ranges in a verry narrow band. However when made dimensionless with the laminar flame thickness, which strongly depends on the composition of the flame, it exhibits the scaling law $\epsilon_i/\delta_L \propto Ka^{-1/2}$, where the Karlovitz number $Ka = (\delta_L/\eta)^2$, with δ_L the flame thickness and η the Kolmogorov scale. These results, which at the present stage are still to be considered preliminary, are encouraging for the assesment of the relationship between turbulent burning velocity and turbulence intensity u'/S_L. Despite the fact that the model does not take into account the effect of the strain rates which may sensibly alter S_L, we are fairly confident that most physical aspects are correctly captured by the expression we have evaluated experimentally in view of possible applications to closures for turbulent premixed combustion.

Dispersion of heavy spheroidal particles in 3D turbulent-like flows

A. Domínguez[1], P. Chhabra[2] and H.J.H. Clercx[1]

[1] J.M. Burgerscentre, Fluid Dynamics Laboratory., Department of Physics, Eindhoven University of Technology, P.O. Box 513, 5600 MB Eindhoven, The Netherlands A.Dominguez@tue.nl,H.J.H.Clercx@tue.nl
[2] Dept. Mechanical Eng., Indian Institute of Technology Guwahati, North Guwahati, Guwahati-781039, India chhabra@iitg.ernet.in

The motion of small non-spherical heavy particles is important in environmental and industrial processes. The dynamics of atmospheric aerosols is an example in the environmental area where this kind of particles are involved.

Using the general formulation of hydrodynamic force and torque for a spheroid, we derive and solve the 2-D equations for the translating-rotating particles in shear flows. Further, we extend the idea to turbulent flows by using a velocity field obtained through 2D Kinematic Simulations (2D-KS). Care was taken to consider particles smaller than η, to have only linear velocity profiles and small Re_p; to neglect the unsteady forces the density ratios used were $\gamma = \rho_p/\rho_f \gg 1$. Beside the torques acting in these spheroidal particles we also account for two forces: Stokes drag and gravity.

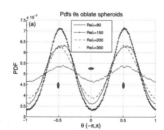

Fig. 1. PDFs of θ for oblate spheroids ($W_s = 0.15 ms^{-1}$, $l = 14\mu m$) at different Reynolds Numbers.

In order to verify these results we will couple the the equations of motion for heavy spheroidal particles with a 3D turbulent-like velocity field and with a fully resolved DNS for validation purposes. We expect to find similar results to those found in the 2D kinematic case.

The Route towards Isotropy in a Turbulent Jet

M. Falchi and G.P. Romano
DMA, University of Roma, Italy, romano@dma.ing.uniroma1.it

The investigation of turbulent jets continues to attract interests despite the simple geometry. This mainly depends on the "elusivity" of the problem to rather crude or even sophisticated simplifications. It is unclear which approximation is realistic and efficient especially for small-scales and where and in which conditions (Reynolds number, IC) it is valid.

Here, recovering of isotropy in a jet is considered. Experiments are performed using PIV-LIF in a water jet at Reynolds numbers from 5000 to 40000. To consider in detail isotropy recover, mean square derivative ratios are computed and shown in the figure on the left

$$K_1 = \frac{\overline{\left(\partial v / \partial x\right)^2}}{\overline{\left(\partial u / \partial x\right)^2}}, \quad K_2 = \frac{\overline{\left(\partial u / \partial y\right)^2}}{\overline{\left(\partial u / \partial x\right)^2}}, \quad K_3 = \frac{\overline{\left(\partial v / \partial y\right)^2}}{\overline{\left(\partial u / \partial x\right)^2}}, \quad K_4 = \frac{\overline{\left(\partial u / \partial y\right)\left(\partial v / \partial x\right)}}{\overline{\left(\partial u / \partial x\right)^2}} \tag{1}$$

Isotropic values are $K_1=2$, $K_2=2$, $K_3=1$ and $|K_4|=0.5$: data at $x/D \approx 10$ are quite close but not perfectly equal to these. Isotropic (ε_1), homogeneous (ε_2), axial-symmetric (ε_3, ε_4) forms of turbulent kinetic energy dissipation are also presented on the right, (non dimensional by D and U^3). Values increases up to $x/D \approx 9$ and then decrease in agreement with a $(x/D)^{-4}$ law.

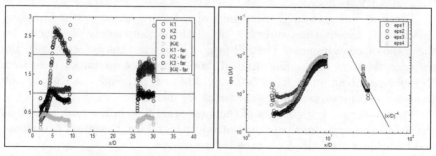

Mean square derivatives ratios (on the left) (the isotropic values are reported as horizontal lines). Turbulent kinetic energy dissipations (on the right).

Experimental study of turbulence in a counter-rotating flow

Robert Morris[1] and Timothy B Nickels[2]

[1] Cambridge University Engineering Department, Trumpington Street,
   Cambridge, UK. rm445@cam.ac.uk
[2] Cambridge University Engineering Department, Trumpington Street,
   Cambridge, UK. tbn22@cam.ac.uk

In order to understand the nature of turbulence at high Reynolds number, many researchers have used an enclosed flow with counter-rotating impellers. This is a convenient and compact laboratory apparatus, in which a region of intense turbulence is created between the impellers.

The aim is often to produce isotropic, homogeneous turbulence, in order to verify theories of turbulence. However, the isotropy in such apparatus has been measured in few experiments, and there have been few studies of the full flow field.

We present an experimental study of the flow field in this type of apparatus, with emphasis on the large-scale flow behaviour. Measurements of isotropy and the integral length scale of turbulence were made, and the effect of changes in the geometry of the apparatus on flow behaviour studied.

A mixing tank with typical configuration was built: a water-filled cylinder of height 0.5m and diameter 0.3m, fitted with bladed discs as impellers. The impellers are driven by stepper motors, allowing accurate control. A range of configurations, including different impellers and the placement of baffles in the flow to increase mixing, have been tested. Neutrally buoyant silver-coated glass microspheres were used to mark the flow, illuminated with a light sheet from an Nd:YAG laser.

The flow was measured using high speed stereoscopic particle imaging velocimetry, allowing measurement of all three components of velocity on the plane of measurement. The full field of flow was thus characterized, the frequency content of the flow studied, and turbulence quantities measured. Reynolds numbers up to $R_\lambda = 500$ were achieved, although the Kolmogorov timescale becomes too small to resolve as R_λ increases.

This poster presents images of the large-scale flow in the tank as well as the results of measurements of the isotropy, integral length scale of turbulence and turbulence statistics for a range of geometries.

Stability of upward and downward dusty-gas flows in a vertical channel

Boronin S.A. and Osiptsov A.N.

Institute of Mechanics of Lomonosov Moscow State University, Michurinskii pr. 1, 117192 Moscow, Russia. boroninsa@imec.msu.ru

The problem of hydrodynamic stability of plane-parallel particle-laden flows was considered previously using mostly the Saffman formulation [1, 2], in which the phase velocity slip in the basic flow was neglected and only the Stokes drag on the particles was taken into account. We modify the Saffman formulation by taking into account the phase velocity slip in the basic flow due to the action of the gravity force and evaluate the effect of non-Stokes components of the interphase momentum exchange, i.g. Saffman, Archemedes, virtual mass forces, on the flow stability with the reference to a dusty-gas flow in a vertical channel.

In the first part, we consider the linear stability of upward and downward dusty channel flows in the presence of the gravity force. The two-fluid model [3] with the incompressible carrier phase and the Stokes drag on the particles is used. In the basic flow, the gas velocity is given by the Poiseuille profile, the particle concentration is uniform, and the constant phase velocity slip is equal to the particle gravitational settling velocity. After the linearization, the problem is reduced to a modified Orr-Sommerfeld equation with several extra terms. The eigenvalues are calculated numerically using an orthonormalization method. A parametric study of the neutral curves is performed for a wide range of variation of the particle inertia parameter, the Froude number, and the particle mass concentration. It is shown that, for both upward and downward flows, the instability region in the "Reynolds number–wave number" plane is bounded for all finite values of the Froude number. A range of governing parameters exists, over which the presence of settling particles makes the flow stable with respect to small disturbances.

In the second part, we evaluate the effect of different non-Stokes components in the interphase momentum exchange on the channel flow stability in the absence of gravity. The work is supported by RFBR grant N 05-01-00502.

References

1. P.G. Saffman: J. Fluid Mech. **13**, 120 (1962)
2. E.S.Asmolov, S.V. Manuilovich: J. Fluid Mech. **365**, 135 (1998)
3. F.E. Marble: Ann. Rev. Fluid Mech. **2**, 397 (1970)

Effect of cationic surfactant, linear polymer chain, and their complexes on turbulent wall shear stress

Anuvat Sirivat and Siriluck Suksamranchit

Petroleum and Petrochemical College, Chulalongkorn University
anuvat.s@chula.ac.th, anuvat.s@chula.ac.th

Turbulent drag reduction in Couette flow was investigated in terms of wall shear stress in aqueous solutions of a nonionic polymer, polyethylene oxide (PEO), a cationic surfactant, hexadecyltrimethylammonium choride (HTAC), and their complexes. Consistent with literature data, drag reduction occurs for PEO solutions above a critical molecular weight, $0.91 \times 10^5 < M_c < 3.04 \times 10^5$. Maximum drag reduction occurs at an optimum concentration, c^*_{PEO}, which scales inversely with molecular weight, and the maximum drag reduction increases with molecular weight. For aqueous HTAC solutions, wall shear stress decreases with increasing HTAC concentration and levels off at an optimum concentration, c^*_{PEO}, comparable to the critical micelle concentration. For HTAC/PEO mixtures, the critical PEO molecular weight for drag reduction decreases, interpreted as due to an increase in hydrodynamic volume because of binding of HTAC micelles to PEO. Conductivity and surface tension data for PEO-HTAC in aqueous solution indicate that salt stabilizes binding of HTAC micelles to the polymer. Dynamic light scattering analysis indicates an increase in hydrodynamic radius for HTAC micelles in aqueous salt solution. In contrast, salt reduces the hydrodynamic radius of PEO-HTAC complexes. Minimum wall shear stress in aqueous HTAC solutions occurs at an optimum HTAC concentration, close to CMC, and this optimum concentration value decreases with increasing ionic strength, suggesting a lowering of the CMC in turbulent flow. For aqueous PEO-HTAC mixtures, the minimum wall shear stress occurs at an optimum PEO concentration smaller than that of pure PEO solutions, and this optimum concentration value increases with ionic strength. Our findings provide evidences that the turbulent wall shear stress does not always scale inversely with the hydrodynamic volume of the polymer-surfactant complex.

Direct Computation of Liquid Sheets in a Compressible Gas Medium

Xi Jiang, George A. Siamas and Luiz Wrobel

Mechanical Engineering, School of Engineering & Design, Brunel University, Uxbridge UB8 3PH, UK

An Eulerian approach with mixed-fluid treatment has been used to study the flow fields of liquid sheets in a compressible gas environment. A mathematical formulation was developed which is capable of representing the two-phase flow system with the gas phase treated as compressible and liquid as incompressible, where the volume of fluid (VOF) method has been adapted to take into account the gas compressibility. The gas-liquid two-phase flow system has been examined by direct solution of the governing equations using highly accurate numerical methods. A sixth-order non-dissipative finite difference scheme (*Padé*) was employed for the spatial differentiation, while a third-order Runge-Kutta scheme was used for the time advancement.

A thin liquid sheet present in the shear layer of a compressible gas jet was firstly investigated. The simulations showed that the dispersion of the liquid sheet is dominated by the vortical structures formed at the jet shear layer due to the Kelvin-Helmholtz instability. An annular liquid jet in a compressible gas medium was also investigated, in order to achieve a better understanding on the flow physics of such flow by providing detailed information on the flow field. The numerical simulation showed that the dispersion of the annular liquid jet is characterised by a recirculation zone adjacent to the nozzle exit. Without applying perturbation at the domain inlet, vortical structures develop at the downstream locations of the flow field due to the Kelvin-Helmholtz instability. The flow becomes more energetic at progressive downstream locations with the dominating frequencies becoming smaller.

Effect of bubbles on the turbulence modification in a downward gas-liquid pipe flow

Oleg Kashinsky[1], Pavel Lobanov[1], Maxim Pakhomov[1], Vyacheslav Randin[1], and Viktor Terekhov[1]

[1]Kutateladze Institute of Thermophysics SB RAS, Acad. Lavrent'ev Avenue 1, 630090 Novosibirsk, Russia, kashinsky@itp.nsc.ru

An experimental and numerical study of the local structure of downward gas-liquid flow in a vertical pipe with 20 mm i.d. is reported. Experiments were performed using an electrodiffusion technique in combination with electrical conductivity measurements. Two different gas-liquid mixers were used to produce gas-liquid flow with various bubble sizes. A model based the Eulerian representation for both phases were developed to numerically predict the structure of downward gas-liquid flow. It was observed, that near the pipe wall, the axial fluctuations of the liquid velocity in the two-phase flow were significantly suppressed as compared to a single-phase. Significant reduction of wall shear stress fluctuations depending on the mean bubble size was observed. The presence of small bubbles resulted in the laminarization of the near-wall region of the flow. The model proposed gives a good prediction of a local structure of a downward bubbly flow.

References

Nakoryakov VE, Kashinsky ON, Burdukov AP, Odnoral VP (1981) Local characteristics of upward gas-liquid flows. Int. J. Multiphase Flow 7:63-81

Kashinsky ON, Lobanov PD, Pakhomov MA, Randin VV, Terekhov VI (2006) Experimental and numerical study of downward bubbly flow in a pipe. Int. J. Heat Mass Transfer 49:3717-3727

One equation model for turbulence pipe flow with second order viscoelastic corrections

B. Sadanandan[1],R. Sureshkumar[1] and F. T. Pinho[2,3]

[1] Department of Energy, Environmental and Chemical Engineering, Washington University in Saint Louis, MO-63130, USA
balraj.sadanandan@intel.com,suresh@che.wustl.edu
[2] CEFT, Faculdade de Engenharia da Universidade do Porto Rua Dr. Roberto Frias s/n, 4200-465 Porto, Portugal fpinho@fe.up.pt
[3] Universidade do Minho, Largo do Paço, 4704-553 Braga, Portugal

This work investigates qualitatively the role of a second order correction to the Newtonian constitutive equation upon pipe flow turbulence. The corrective term is proportional to the first normal stress difference coefficient and the turbulence model used here is a modified Prandtl one-equation $k - l$ model accounting for the new elastic term in the constitutive equation.

The stress of the fluid is split into a Newtonian solvent component and a polymer stress component $\tau_{ij,p}$, which is expressed as $\tau_{ij,p} = \eta_p \dot{\gamma}_{ij} - \eta_p \lambda \dot{\gamma}_{ij_{(1)}}$. It contains a viscous contribution $(\eta_p \dot{\gamma}_{ij})$ and an elastic contribution dependent on the polymer relaxation time (λ) and $\dot{\gamma}_{ij_{(1)}}$ denotes the upper convected derivative of the deformation rate tensor $\dot{\gamma}_{ij}$.

The corresponding time-average momentum equation for fully-developed pipe flow has a new elastic stress that incorporates a second order one point correlation that required adequate treatment.

In excess of the terms found for Newtonian fluids, here due to the Newtonian solvent and polymer viscous contribution, the transport equation for the turbulent kinetic energy (k) contains two new contributions associated with the fluctuating elastic stress: the so-called elastic turbulent diffusion $(\partial \overline{u_i \tau_{i2}^e}/\partial y)$ and the correlation between the fluctuating elastic stresses and the fluctuating rate of strain $(\overline{\tau_{ij}^e \partial u_i/\partial x_j})$, here denoted polymer work (P_C).

The paper discusses the developed closures and results of predictions are presented to show the effects of Reynolds and Deborah numbers. The amount of drag reduction (DR) is commensurate with the upward shift in the velocity profile relative to the Newtonian law of the wall and towards Virk's asymptote. As DR increases, k^+ decreases and its peak moves away from the wall, both features seen in DR with polymer solutions. The viscous dissipation and the production of turbulence P_k both decrease with De, whereas the polymer work (P_C) increases with De.

Formulation of the settling velocity of small particles initially situated inside a vortex

U. Sánchez[1] and M. J. Moreno-López[2]

[1] Fluid Mechanics Division, University of Huelva, Campus La Rábida, 21819-Palos de la Frontera, Huelva (Spain) urbano.sanchez@dcaf.uhu.es
[2] Chemical Engineering Department, University of Huelva, Campus El Carmen, 21071-Huelva (Spain) moreno@uhu.es

The Average Settling Velocity of particles $< V_z^* >$ is an interesting feature in many practical situations. For example, it is a suitable condition to design mixture devices or particle separation devices. In a previous work Sánchez & Dávila (Ercoftac Conference on Small Particles in Turbulence, Seville 2002) settled down (from the equation of motion) a theoretical formulation of the time that small heavy particles initially situated inside a 3D vortex need to escape, *Escape Time*: $T_e^* = K \Gamma^{2*}/\tau_p^* V_t^{*4} \sin^4 \theta$. Considering the viscous response time of the particles $\tau_p^* = \left(\rho_p^*/18\nu^*\rho^*\right) d_p^{*2}$ and the terminal velocity of the particles in still flow $V_t^* = \tau_p^* g^* = \left(\rho_p^* g^*/18\nu^*\rho^*\right) d_p^{*2}$ the Escape Time can be written as $T_e^* = K_e \left(18^5 \mu^{*5} \Gamma^{*2}/\rho_p^{*5} g^{*4} \sin^4 \theta\right) d_p^{*-10}$ with Γ^* circulation of the vortex, θ angle between the axis of the vortex and the Z axis, ρ_p^* particle density, ρ^* fluid density, ν^* fluid kinematic viscosity, d_p^* particle diameter, g^* gravity, μ^* fluid viscosity and K_e an empirical constant.

The development of a theoretical formulation for the Settling Velocity similar to that for the Escape Time is not possible. For that reason, we carry out a numerical simulation of the Average Settling Velocity of a mesh of 500 uniformly distributed particles situated inside the trapping region of a 3D vortex. From the plot of the Dimensionless Average Settling Velocity $< V_z >= \frac{1}{N} \sum_{i=1}^{i=N} (z_{it} - z_{i0})/t$ versus the Average Escape Time $< T_e^* >$ it can be seen that for sufficiently small particles ($< T_e^* >$ large) the curve $< V_z >$-$< T_e^* >$ is a straight line when the representation is in logarithmic scale. We can write the algebraic equation for the lines in the mentioned representation and, as a result, we can obtain the formulation of the Settling Velocity, $< V_z >= \frac{K_s}{<T_e^*>^{0.19}}$, or, in dimensional terms,

$$< V_z^* >= \frac{K_s \Gamma^*}{R_v^* < T_e^* >^{0.19}} = \frac{K_s \Gamma^*}{R_v^*} \left[K_e \frac{18^5 \mu^{*5} \Gamma^{*2}}{\rho_p^{*5} g^{*4} \sin^4 \theta} d_p^{*-10} \right]^{-0.19} \tag{1}$$

with $< R_v^* >$ the radius of the vortex. The values of K_s for $\Gamma^* = 10^{-5}$, $3 \cdot 10^{-5}$, 10^{-4} m^2/s are: 1.0003541, 0.5371102, 0.2585931 s$^{0.19}$ (Rankine vortex) and 0.7445312, 0.3723925, 0.1978895 s$^{0.19}$ (Kauffmann vortex).

Transition Turbulence in a laboratory model of the Left Ventricle

Fortini S.[1] Querzoli G.[2] Marchetti M.[1] and Cenedese A.[1]

[1] Department of Idraulica Trasporti e Strade - University of Rome "La Sapienza"
stefania.fortini@uniroma1.it
[2] Department of Ingegneria del Territorio - University of Cagliari

The complexity of the left ventricular flows in the human heart is further increased in the presence of a diseased condition, such as unhealthy or prosthetic heart valves. Accurate diagnosis is fundamental in clinical practice which is focussed on the ability of early recognition of diseases; to this end measurements made in vivo (e.g. by echo-cardiography) can point out anomalies even if they are still not enough detailed to explain all the involved phenomena. Laboratory investigation can help assessing the origin of the anomalies observed in vivo by addressing the flow dynamics through the artificial valve. The objective of this study is to describe the changes of hemodynamic characteristics in the Left Ventricle due to variations in the transmitral flow: we document the velocity fields in a flexible, transparent ventricle model using an image analysis technique, i.e. Feature Tracking. Different transmitral flows, ranging from uniform inflow conditions to tilting-disk and bi-leaflet valves, are analyzed. In this paper we document vortices formation and vortices interaction; they are two important physical phenomena that dominate the filling and emptying of the ventricle. Moreover the flow is characterized by a strong variability in time with high remarkable velocity during the first diastolic peak. So, it is important to determine if and when turbulent characteristics arise during the average motion. This study was performed by the velocity variance analysis.

References

1. H. Reul, N. Talukder, E. W. Müller: Fluid Mechanics of the Natural Mitral Valve. J. Biomechanics **14**, 5 (1981)
2. D. Bluestein, E. Rambod and M. Gharib: Vortex Shedding as a Mechanism for Free Emboli Formation in Mechanical Heart Valves. J. Biomech. Eng. **122**, (2000)
3. O. Pierrakos, P. P. Vlachos: The Effect of Vortex Formation on Left Ventricular Filling and Mitral Valve Efficiency J. Biomech. Eng. **128**, (2006)

Experimental Study of Wake Structure Associated with Reduced Base Drag Using POD and LSE

Vibhav Durgesh[1] and Jonathan W. Naughton[2]

[1] University of Wyoming,
    Laramie, WY, 82070, USA vibhav@uwyo.edu
[2] University of Wyoming,
    Laramie, WY, 82070, USA naughton@uwyo.edu

An experimental study of the near wake of both a truncated wedge and flat plate has been carried out. Time-resolved base pressures and the near-wake velocity field have been measured for different boundary layer thicknesses at separation. Both the Proper Orthogonal Decomposition (POD) and Linear Stochastic Estimation (LSE) have been used to analyze the velocity data to determine the wake's structural content and its relationship to the fluctuating base pressure. The results indicate that, under certain conditions, thickening the boundary layer reduces base drag, which is accompanied by a change in the structure of the base flow. Specifically, thickening the boundary layer results in structures with lower peak vorticity that convect faster downstream. Dynamic reconstructions created with either a deterministic approach or LSE using the dynamic pressure measurements provide a very low-order descriptions of the flow. Such descriptions of the base flow (e.g. figure 1) are important for both fundamental understanding of the flow as well controlling the flow.

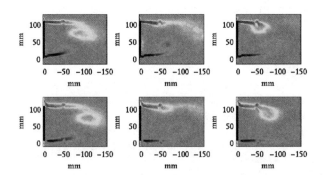

Fig. 1. Vorticity plot at three different times: top 3 figures are for the smooth fore-body surface, and the bottom 3 figures are for the rough fore-body surface.

On vorticity, vortices and material lines in turbulent channel flow

Anders Helgeland[1], B. Anders Pettersson Reif[2], Oyvind Andreassen[2] and Carl Erik Wasberg[2]

[1] University graduate Center, NO-2027 Kjeller, Norway andershe@ifi.uio.no
[2] Norwegian Defence Research Establishment (FFI), NO-2027 Kjeller, Norway.
   Bjorn.Reif@ffi.no Oyvind.Andreassen@ffi.no Carl-Erik.Wasberg@ffi.no

An advanced multifield visualization technique [Helgeland et al., *IEEE TVCG 2007 (to appear)*] is used to study the temporally evolving vorticity field in a fully turbulent developed channel flow. The objective is to gain further insight into various spatio-temporal characteristics of turbulent wall bounded flows. Techniques based on identifying vortex topology as compact tube-like objects are inherently unable to represent the kinematic behavior related to internal velocity shear. Velocity shear spawn regions of vorticity that are characterized by sheet-like structures, and these are dynamically very important. In this study vorticity field lines are employed to overcome this shortcoming. Multifield visualizations of vortices and vorticity sheets provide a complete instantaneous snapshot of the kinematic topology of the velocity field. The time evolution of these structures are considered to gain insight of the spatio-temporal characteristics. The difference between vorticity field lines and frozen-in material lines are quantified and it is for instance shown that the difference grows linearly in time (Fig 1).

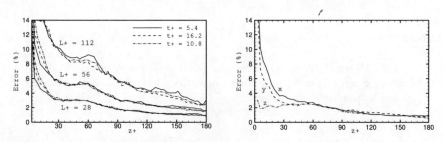

Fig. 1. Error measurements between vorticity field lines and material lines as a function of wall distance. Left: Error as a function of viscous length at the field line (L^+). Right: Error in the coordinate directions for $L^+ = 56$.

Orthonormal divergence-free wavelet analysis of cascading/backscattering process around coherent structures

Keisuke Araki[1] and Hideaki Miura[2]

[1] Department of Intelligent Mechanical Engineering, Faculty of Engineering,
   Okayama University of Science, Okayama, Okayama 700-0005 JAPAN
   araki@are.ous.ac.jp
[2] Theory and Computer Simulation Center, National Institute for Fusion Science,
   Toki, Gifu 509-5292 JAPAN

Nonlinear energy transfer of some isolated vortices which evolve from a thin shear layer is analyzed in terms of orthonormal divergence-free wavelet. We calculated numerically inter-scale, location-to-location wavelet nonlinear transfer defined by the integral

$$\langle j, \underline{l} | \mathbf{u} | k, \underline{m} \rangle := - \int \mathbf{u}_{j\underline{l}} \cdot (\mathbf{u} \cdot \nabla) \mathbf{u}_{k\underline{m}} \mathrm{d}^3 \underline{x}, \tag{1}$$

where j, k and $\underline{l}$, $\underline{m}$ are the scale and the location indices of wavelets, respectively. A graphical representation method for the wavelet nonlinear interactions developed in [1] reveals that the active nonlinear interactions are very closely distributed around the axis of rolling-up vortices irrespective of the forward or backward transfers.

References

1. Araki K, Miura H (2006) Orthonormal Divergence-free Wavelet Analysis of Nonlinear Energy Transfer in Rolling-Up Vortices. In IUTAM Symposium 2006 NAGOYA "Computational Physics and New Perspectives in Turbulence".

Experimental Investigation of Forced and Unforced Instabilities on the Cold Flow of a Swirl Burner

A. Lacarelle[1], C. O. Paschereit[1], D. Greenblatt[1] and E. J. Gutmark[2]

[1] Techniche Universität Berlin, Intitut für Strömungsmechanik und Technische Akustik, 10623 Berlin, Deutschland, arnaud.lacarelle@tu-berlin.de
[2] University of Cincinnati, Aerospace Engineering and Eng. Mech., 799 Rhodes Hall, PO Box 210070 Cincinnati, ephraim.gutmark@uc.edu

Swirling flows associated with vortex breakdown are commonly used to stabilize turbulent flames in modern combustion chambers. Among the advantages of swirl stabilized flows are compact flames and excellent mixing of the reactants. However, these flows are susceptible to so-called flow instabilities that are of a scale comparable to the burner dimensions (large-scale) and can take diverse forms, such as processing of the vortex core, helical instabilities, etc. Although such flows have been the subject of extensive experimental and computational investigations the instability mechanisms are still not well understood.

The objective of this investigation was to study the natural and forced instability on the cold flow of a swirl burner. It evidenced that the natural flow frequency propagates upstream of the burner, leading to velocity osillations at the fuel injection. Strong oscillation at the natural frequency ($St \sim 0.92$) was used to reproduce combustion instabilities induced by the vortex breakdown and which lead to strong velocity oscillations in the burner.

Experiments were conducted in a water test facility. An excitation mechanism could produce oscillations of the flow of up to 30 % of the mean burner velocity. A hydrophone located at the outlet of the burner provided a reference oscillating pressure signal that was used for phase-locked measurements. Phase averaged Particle Image Velocimetry (PIV) was used to capture the flow field at the burner outlet and showed a change in the structure between forced and unfoced case. Phase Averaged LDA measurement at the burner inlet were used to observe the impact of the natural instability and the excited flow on two velocity components at the fuel injection location.

Quantifying anisotropy in stratified and rotating turbulence using orthogonal wavelets

L. Liechtenstein, W.J.T. Bos and K. Schneider

MSNM-CNRS & CMI, Université de Provence, Marseille, France

Experiments as well as numerical simulations of anisotropic turbulence show different shapes of coherent structures, e.g. pancakes or cigares [1]. An understanding of these structures is of primordial importance for modelling rotating, stratified or magnetohydrodynamic turbulence. Even though we manage to vizualize, the quantification of anisotropy, related to these structures remains to date a challenge to fluid dynamicists.

In this work we discuss and investigate the possibility to characterize the anisotropy of a flow by using orthogonal wavelet decomposition of velocity. We will therefore use two properties of orthonormal wavelets [2]. First, they are directional, which allows to characterize the anisotropy of a flow (in figure 1 the directional energy is shown). Second, they allow to define scale dependent moments which can be used to quantify scale dependent higher order statistics in the different directions. In [3] we introduce the directional scale dependent flatness, which measures the spatial intermittency for different directions and we show that two-dimenionalization is not observed in rotating turbulence.

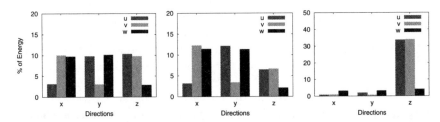

Fig. 1. Directional energy of the three velocity components u, v and w for isotropic (left), rotating (middle) and stratified turbulence (right).

References

1. L. Liechtenstein, F. Godeferd and C. Cambon. J. Turbulence, **6**, 24 (2005).
2. M. Farge, Annu. Rev. Fluid Mech., **24**, 395 (1992).
3. L. Liechtenstein, W.J.T. Bos, K. Schneider. Submitted (2007).

Variable Density Vortex Rings

Sonia Benteboula[1,2] and Ionut Danaila[1]

[1] Laboratoire Jacques-Louis Lions, Université Pierre et Marie Curie, Paris, France.
[2] LETEM, Université de Marne-la-Vallée, Champs-sur-Marne, France.

We numerically simulate the evolution of a variable density vortex ring resulting from the impulsive injection of a fluid into a quiescent surrounding of different temperature. We carefully investigate the influence of the jet-to-ambient temperature ratio $\alpha = T_j/T_a$ on the post-formation phase of the vortex ring. Two numerical codes using cylindrical [1] and, respectively, spherical coordinates [2] are considered to solve the Navier-Stokes equations for low-Mach flows. Figure 1 shows different vorticity evolutions for hot and cold jet injections. The baroclinic production term is responsible for the generation of a negative vorticity layer in the front of the vortex ring for the hot jet injection. The opposite effect is observed for cold jet injection. The ratio α is also found to considerably affect the integral characteristics of the vortex ring (circulation, impulse). We derive a simple model to describe these evolutions.

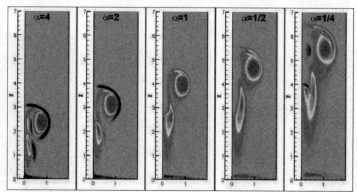

Fig. 1. Vorticity fields at $t = 10$ for different jet-to-ambient temperature ratios $\alpha = T_j/T_a$. The injection stops at $t = 6$.

References

1. S. Benteboula, PhD Thesis, Université Marne-la-Vallée (2006).
2. B. J. Boersma, In Ann. Research Briefs, Center for Turbulence Research (1999).

Behavior of Flow Structures inside a Round Buoyant Jet

Nobumasa Sekishita[1] and Hideharu Makita[2]

[1] Toyohashi University of Technology seki@mech.tut.ac.jp
[2] Toyohashi University of Technology makita@mech.tut.ac.jp

In a buoyant jet, coherent structures effected by buoyant force are observed not only in a plume region located far from the jet exit but also in a near field of the exit. Absolute instability of low-density jets was analyzed [1] and its result was verified experimentally [2]. The present paper aims to experimentally investigate flow structures in the near field of the exit of a heated jet.

A heated air jet was ejected vertically from a round exit of D =83mm in diameter. Velocity and temperature distributions were top-hat type with 2m/s at the jet exit. Temperature difference between the heated jet and ambient fluid was 0, 5, 30 K and its density ratio was 1.0, 0.98, 0.91, respectively. Flow patterns generated by smoke-wire method were taken by a high speed video camera. Velocity and temperature measurements were also conducted by using a thermal anemometer and an I-I type probe.

For unheated jets, a potential core is observed from the jet exit through about $4D$ as shown in Fig. 1. On the other hand, flow structure meanders inside of the heated jet. A turbulent region extends from both outside and inside area of the jet. As a result, the potential core is shorter in the buoyant jet than in the unheated jet. Moreover, its unstable flow structure was investigated by the wavelet analyses of velocity fluctuations.

This work was supported by the Japan Ministry of Education through grant-in-aid (No.18560166).

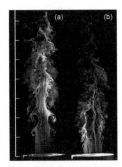

Fig. 1. Flow patterns of (a)unheated jet and (b)heated jet with 30K

References

1. P. Huerre, P.A. Monkewitz: Annu. Rev. Fluid Mech. 22, 473 (1990)
2. D.M. Kyle, K.R. Sreenivasan: J. Fluid Mech. 249, 619 (1993)

High Spatial Resolution MCCDPIV in a Zero Pressure Gradient Turbulent Boundary Layer

Chong Y. Wong[1] and Julio Soria[1]

Department of Mechanical Engineering, Monash University, Clayton Campus, Melbourne, Victoria, Australia 3800 chong.wong@eng.monash.edu.au

Multi-grid Cross-Correlation Digital Particle Image Velocimetry (PIV) in a nominally zero pressure gradient turbulent boundary layer was conducted in a 500mm-square water tunnel, with a PCO.4000 camera (4008*2672pixels), a Nikon 200mm f4 lens and a PK13 extension. 420 velocity fields/Re_θ were acquired at least 650 momentum thicknesses (θ) downstream of the tunnel exit plane (untripped). Reynolds number Re_θ (based on U_∞ and θ) is between 200 and 2000. Skin friction velocities ($u_\star$) were estimated by the Clauser method. Normalised mean velocity distributions are shown in Fig. 1. Except for the $Re_\theta = 209$ case, inadequate spatial resolution in the viscous sub-layer restricts the rest of the cases to model $U^+ = y^+$ (Eqn-1). $U^+ = \frac{1}{0.4}ln(y^+ - y_0^+ + \frac{1}{0.4}) + 5$ (Eqn-2) models the flow from $y^+ = 30$ to $y^+ = 200$. Here, $U^+ = \frac{U}{u_\star}$, $y^+ = \frac{yu_\star}{\nu}$ and $y_0^+ = 7.2$. In general, data in the overlap region agree well with $U^+ = \frac{1}{0.4}ln(y^+) + 5$ (Eqn-3). Results suggest that increased optical spatial resolution is needed to resolve velocities and structures below the logarithmic region.

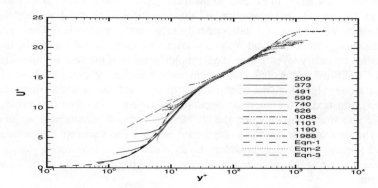

Fig. 1. Profiles of U^+ against y^+ for varying Re_θ.

Interaction of Quasi Two Dimensional Flow Field with Turbulent Boundary Layer as a Method of Investigating Drag Reduction of Polymer Additives

G.D.Roumbas, E.G. Kastrinakis, S.G. Nychas

Department of Chemical Engineering, Aristotle University of Thessaloniki, University Box 453, 54006 Thessaloniki, Greece
{ G.D.Roumbas, E.G. Kastrinakis, S.G. Nychas}@eng.auth.gr

Abstract

A combined experimental effort was undertaken to investigate the interactions, from the point of view of polymer drag reduction, between low dissipation vortical structures and polymers in a wall bounded turbulent flow. On the one hand, polymer induced drag reduction was investigated in a plain rotating disk and in a rotating disk with mounted Delta Wings (DWs) using two different molecular weights of poly(ethylene oxide) in distilled water. The required torque and the effected drag reduction vs the Reynolds number, shearing time and polymer concentration were measured. The DWs were mounted to introduce quasi two dimensional structures in the turbulent boundary layer of the rotating disk. The geometries used were a plain disk with a diameter of 290 mm and a disk of the same diameter with mounted DWs. The Reynolds numbers based to the diameter of the disks vary form 1.22×10^6 to 2.21×10^6. Poly(ethylene oxide) with molecular weights of 4×10^6 and 7×10^6 were used. On the other hand, to be able to measure velocity components and obtain an insight in the process of interaction of these vortices with the boundary layer, measurements were conducted independently in a wind tunnel, where DW vortices were interacting with a turbulent boundary layer on a flat plate. It was found that in the modified boundary layer region, the Reynolds stresses of the outward interactions, ejections, wallward interactions and sweeps were reduced compared to those of the flat plate alone. Moreover, the contributions of outward and wallward interactions increased compared to the ones for ejections and sweeps. The first two events are more likely to stretch the polymer; the interactions outward predominate in the wall region. The data suggest that the strengthened outward and wallward interactions by the DWs vortices in the disk boundary layer, could play a key role in the storing energy mechanism of the polymer and subsequently in the process of turbulent energy production.

Structure and Mean-Velocity Profile of Pipeflow

Matthias H. Buschmann[1] and Mohamed Gad-el-Hak[2]

[1] Institut für Luft- und Kältetechnik, Dresden, Germany
Matthias.Buschmann@ilkdresden.de
[2] Virginia Commonwealth University, Richmond, U.S.A. gadelhak@vcu.edu

One of the most intensively investigated turbulent flows is that through a straight axisymmetric pipe. We investigate some features of high-Reynolds-number pipeflows and connect them with the generalized log law.[1] First, we analyze the two regions—log and power—found in the mean-velocity profile by McKeon et al.[2] For this purpose, we examine the idea by Tennekes & Lumley[3] who suggested that in the wall-layer the mean flow could be split into first- and second-order inertial sublayers. We hypothesize that the power-law region coincides with the lower part of the first-order inertial sublayer and the log region with the second-order. To validate, we re-examine the borders of both regions employing the fractional difference (FD) calculated once for the log and once for the power law. Interestingly the lower border of the second-order inertial sublayer is in reasonably good agreement with the border between power and log regions. While the lower border of the first-inertial sublayer is found at $y^+ = 60$, which is slightly higher than originally proposed, the upper border of the log region ($\eta = 0.12$) lays much further out than the outer limit of the second-order layer originally assumed in [3]. The major implication from the existence of a second-order inertial sublayer in pipeflow is that the slope of the log law depends, however weakly, on the Kármán number. This in turn supports higher-order approaches such as the generalized log law. We therefore re-analyze McKeon's superpipe data with respect to this law. Employing the second-order solution resolves all profiles fairly well above $y^+ \approx 100\text{–}150$. Comparing the FD of the classical log law (with constant parameters) with the FD of the second-order solution clearly shows the superiority of the latter. To summarize, the higher-order approach proposed in [1] shows excellent agreement with pipeflow data. The power region proposed by McKeon et al.[2] does the same job as the offset parameter introduced by Wosnik et al.[4] into the log law to account for their mesolayer, or the additional $1/y^+$-terms of the generalized log law. A possible explanation for this behavior comes from the concept of first- and second-order inertial sublayer proposed in [3].

[1] *AIAA J.* **41**, pp. 40–48, 2003.
[2] *J. Fluid Mech.* **501**, pp. 135–147, 2004.
[3] *A first course in turbulence*, MIT Press, Cambridge, Massachusetts, 1972.
[4] *J. Fluid Mech.* **421**, pp. 115–145, 2000.

Detection of streamwise vortices in a turbulent boundary layer

J. Lin, J.-P. Laval, J.-M. Foucaut, and M. Stanislas

Laboratoire de Mécanique de Lille, CNRS UMR 8107, Blv Paul Langevin, 59655 Villeneuve d'Ascq, France. michel.stanislas@univ-lille1.fr

The wall region of a turbulent boundary layer contains different coherent structures which play an important role in the turbulence energy generation and transport. The objective of the present work was to study the characteristics of the turbulent flow in the near wall region, to observe and to quantify these coherent structures and to investigate their relations to provide a global organization model near the wall. A stereoscopic PIV experiment was carried out on a fully developed turbulent boundary layer flow in the wind tunnel of Laboratoire de Mécanique de Lille. [1]. This experiment was performed at large Reynolds number (Re_θ=7800) in 10 planes parallel to the wall at distances $14.5 < y^+ < 48$ with a spatial resolution of approximately 5 wall units in the plane. In order to quantify the statistics of the near wall structures a detection procedure was developed for each kind of structure. The procedure consists in choosing a suitable detection function and to apply a threshold to detect raw structures. Then, several morphological functions (dilatation, erosion) are used in order to clean the raw objects and reject degenerated patterns. This procedure removes the spurious structures and allows a detailed characterization of the remaining structures such as their size, occurrence or spatial organization. Evident links appear between the different coherent structures and streamwise vortices seems to play a significant role in the organization. They do not originate at the wall but in the buffer layer above the peak of turbulent kinetic energy. Looking at the spatial correlation between structures, a conceptual model of organization can be drawn. This model is in agreement with previous models proposed by several authors ([2],[3]) but is more quantitative as distances, sizes and intensities are provided.

References

1. L. Jie, Ph.D. thesis, Lille University Ph.D. thesis, 2006.
2. J. Carlier and M. Stanislas, J. Fluid Mech. **535**, 143 (2005).
3. R. Adrian, C. Meinhart, and C. Tomkins, J. Fluid Mech. **422**, 1 (2000).

Upstream Condition Effects on the Anisotropy of Rough Favorable Pressure Gradient Turbulent Boundary Layers

Raúl Bayoán Cal[1], Brian Brzek[2], Gunnar Johansson[3] and Luciano Castillo[2]

[1] The Johns Hopkins University, Baltimore, MD bayoan.cal@jhu.edu
[2] Rensselaer Polytechnic Institute, Troy, NY brzekb@rpi.edu/castil2@rpi.edu
[3] Chalmers University of Technology, Gothenburg, Sweden gujo@chalmers.se

Several experiments are carried out in the L2 wind tunnel facility in the department of Applied Mechanics at Chalmers University of Technology. A set of 5 different experiments is designed to isolate each external condition studied which are different magnitudes of roughness, 24 and a 60 grit sandpaper, upstream wind-tunnel speed, 5 and 10 m/s, and strength of favorable pressure gradient (FPG), 3.5 and 7 degrees.

The effects of the upstream and external conditions on the anisotropy tensor of a rough favorable pressure gradient turbulent boundary layer have yet to be reported. This parameter is given by $b_{ij} = \frac{\overline{u_i u_j}}{2k} - \frac{\delta_{ij}}{3}$. It is found that all the external and upstream conditions affect the anisotropy tensor in the outer region of the boundary layer. On the inner part of the boundary layer, the surface roughness is the only condition that promotes isotropy close to the wall even for the transitionally rough regime. This is observed in figure 1(a) where the b_{11} component of the anisotropy tensor is shown. In fig. 1(b), the production term, P_{12}, increases when the wind-tunnel speed decreases and conversely, when the favorable pressure gradient and surface roughness are increased.

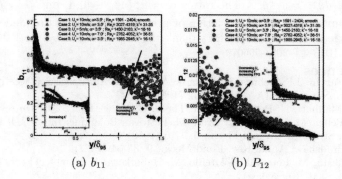

(a) b_{11} (b) P_{12}

Fig. 1. *The anisotropy tensor and production term for smooth and rough FPG data.*

Near-wall modeling of compressible turbulent boundary layers with separation

Margareta Petrovan Boiarciuc[1], Christophe Brun[1,2], and Michael Manhart[3]

[1] Laboratoire de Mécanique et d'Energétique/Polytech'Orléans, 8 rue Léonard de Vinci, 45072 Orléans Cedex 2, France margareta.petrovan@univ-orleans.fr
[2] Laboratoire d'Etudes Géophysiques et Industrielles/MoST, BP 53, 38041 Grenoble Cedex 9, France christophe.brun@hmg.inpg.fr
[3] Fachgebiet Hydromechanik, Technische Universität München, Arcisstr. 21, 80333 München, Germany m.manhart@bv.tum.de

Wall resolved Large Eddy Simulations of turbulent compressible attached and separating flow in a channel are considered for Reynolds and Mach number in the range $1200 - 4880$ and $0.4 - 1.5$,respectively. In addition to the standard friction quantities u_τ and T_τ [1], we determine pressure based quantities $u_p = \left|\frac{\mu}{\rho^2}\frac{\partial p}{\partial x}\right|^{1/3}$ [2] and $T_p = \frac{u_p^2}{2C_p}$, and define combined quantities $u_{\tau p} = \sqrt{u_p^2 + u_\tau^2}$ [3] and $T_{\tau p} = T_p + T_\tau$ [4]. A new near-wall scaling for velocity and total temperature is derived involving pressure gradient effects [5, 4, 7]:

$$U^*(y^*) = sign(\tau_w)\alpha y_1^{c*} + sign(\frac{\partial p}{\partial x})\frac{(1-\alpha)^{3/2}}{2}y_2^{c*2} \tag{1}$$

$$T_i^*(y^*) = T_w^* + sign(-q_w)\beta\sqrt{\alpha}Pr\,y_1^{c*} \tag{2}$$

$$+ sign(\frac{\partial p}{\partial x})\frac{\gamma}{\gamma-1}(1-\beta)Pr\left[\frac{\alpha\sqrt{1-\alpha}}{3}\frac{\mu_w}{\mu_1}y_3^{c*3} + \frac{(1-\alpha)^2}{12}\frac{\mu_w}{\mu_2}y_4^{c*4}\right]$$

with $U^* = U/u_{\tau p}$, $T_i = T + \frac{Pr}{2c_p}U^2$, $T_i^* = T_i/T_{\tau p}$, $y^* = \rho_w y u_{\tau p}/\mu_w$, and $y_n^{c*n} = \int_0^{y^{*n}}\frac{\mu_w}{\mu(T)}dy^{*n} = \frac{\mu_w}{\mu_n}y^{*n}$. The ratios $\alpha = u_\tau^2/u_{\tau p}^2 \in [0,1]$ and $\beta = T_\tau/T_{\tau p} \in [0,1]$ quantify which effect, friction or pressure gradient, is preponderant.

Acknowledgement: The authors thank the help provided by the HPC-Europa program RII3-CT-2003-506079.

References
1. C. Carvin, J.F. Debieve, A.J. Smits: AIAA 26th Aerospace Sci. Meeting (1988)
2. R.-L. Simpson: AIAA J., **21**, 142 (1983)
3. M. Manhart, N. Peller, C. Brun: Theor. Comp. F. Dyn., 2007 (in review)
4. N. Peller, M. Manhart, C. Brun, M. Petrovan Boiarciuc: In: *CEMRACS 2006*
5. B.S. Stratford: J.Fluid Mech., **5**, 1-16 (1959)
6. P. Bradshaw: Annu. Rev. Fluid Mech., **9**, 33-54 (1977)
7. C. Brun, M. Petrovan Boiarciuc, M. Haberkorn, P. Comte: Theor. Comp. F. Dyn., 2007 (in review)

Numerical Investigation of Turbulent Boundary Layer Relaminarisation

A.D.S. Borges[1,3], A. Silva Lopes[1,2], and J.M.L.M. Palma[1,2]

[1] CEsA — Centro de Estudos de Energia Eólica e Escoamentos Atmosféricos
[2] Faculdade de Engenharia da Universidade do Porto, Porto, Portugal
[3] Universidade de Trás-os-Montes e Alto Douro, Vila Real, Portugal
amadeub@utad.pt

Many boundary layer flows involve combinations and abrupt changes of surface curvature and streamwise pressure gradient, which can cause phenomena such as separation or relaminarisation. Relaminarisation (unlike transition to turbulence) is a gradual process, but is accompanied by drastic changes in the structure of the flow: the mean velocity profile departs from the logarithmic law, the boundary layer becomes thinner, the shape factor first decreases and then increases while the skin-friction increases first and then decreases.

The objective of this work was to study the turbulent boundary layer over a bump with $Re_\theta = 4030$, using the Large-Eddy Simulation (LES) and Detached-Eddy Simulation (DES) techniques, and verify if these methodologies were able to predict the friction on the top of the bump. However, the simulations predicted a friction 28% larger than measured in the bump summit and no evidence was found of the relaminarisation put forward in the experimental work [1]. Two parameters were used to check the possibility of relaminarisation: according to Sreenivasan [2], it can occur if $K = (\nu/u_\infty^2)(dU_\infty/dx) \geq 3 \times 10^{-6}$, while Patel [3] proposes that it is possible if $\Delta_p = -\nu(\partial p/\partial x)/\rho u_\tau^3 \geq 0.018$. In the flow over the bump, K never suggested that relaminarisation was possible, while Δ_p slightly surpassed the limit on the upstream side of the bump ($\Delta_p = 0.019$). However, this strong acceleration occurred before the hill summit, where the experimental work suggests the existence of relaminarisation, and did not remain for enough time (or length) for the flow to relaminarise [4].

References

1. D.R. Webster, D.B. DeGraaff and J.K. Eaton. *J. Fluid Mech.*, 320:53, 1996.
2. K.R. Sreenivasan. *Acta Mechanica*, 44:1, 1982.
3. V.C. Patel. *J. Fluid Mech.*, 23:185, 1965.
4. A.D.S. Borges. *Large-Eddy Simulation of the Turbulent Boundary Layer* (in Portuguese). PhD Thesis, 2007.

Turbulent flow over different groups of cubical obstacles

S. Leonardi[1], I.P.Castro[2] & P. Orlandi[3]

[1] Dep. Mech. Eng., University of Puerto Rico at Mayaguez, Puerto Rico
sleonardi@me.uprm.edu
[2] School of Engineering Sciences, University of Southampton, Southampton, UK
[3] Dip. Meccanica ed Aeronautica, University of Rome, "La Sapienza", Rome, Italy

Atmospheric boundary layers flow over a rough surface, composed of buildings, hills, valleys, vegetation. The present investigation extends previous studies (Coceal et al. 2006 and Orlandi & Leonardi 2006) by carrying out a set of DNS of flow over arrays of cubical obstacles with varying the plan densities (1:4 aligned and staggered, 4:25 and 1:9). The Reynolds number based on the bulk velocity and the obstacle height is $Re = 7000$, boundary conditions are periodic in the streamwise and spanwise directions, and free slip is applied on the upper boundary. The drag is almost entirely due to the form drag, the frictional contribution being negligible. The configuration $4 : 25$ is the one which maximizes the drag.

Fig.1 Roughness function (square) and roughness length (circle) as a function of the wall normal velocity rms.

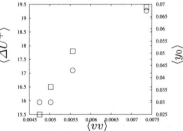

Scaled in wall units, the velocity profile can be expressed as $U^+ = 1/\kappa \log y^+ - \Delta U^+$ or $U^+ = 1/\kappa \log y/y_0$, where ΔU^+ and y_0 are the roughness function and roughness length respectively. The roughness function and the roughness length are plotted against the normal wall velocity rms at the crests plane in fig.1. Both quantities scale well with $\langle vv \rangle$ (v is the vertical velocity and angular brackets denote averaging on x,z and t). This confirms that the effect of roughness on the overlying flow is via the wall normal velocity fluctuations at the crests plane.

Orlandi P. & Leonardi S. (2006) *J. of Turb.* **7** (53), 1-22
Coceal, O., Thomas, T.G., Castro, I.P. and Belcher, S.E. (2006) *B. Layer Met.*, **121**, (3), 491-519

LES of transient turbulent flow in a pipe

Seo Yoon Jung[1] and Yongmann M. Chung[2]

[1] School of Engineering and Centre for Scientific Computing, University of
Warwick, Coventry, CV4 7AL, U.K. S.Y.Jung@warwick.ac.uk

[2] Y.M.Chung@warwick.ac.uk

Large eddy simulations are performed for a fully developed pipe flow that is
subjected to a linear acceleration in order to investigate the non-equilibrium
turbulence characteristics. The calculations are started from a fully-developed
turbulent pipe flow at $Re = 7000$. The flow rate increases linearly to a final
Reynolds number of $Re = 45200$ (see Fig. 1). The acceleration rate parameter
is $\gamma = \frac{D}{u_\tau} \left(\frac{1}{U_m} \frac{dU_m}{dt} \right) = 6.1$. These parameters are chosen to compare with the
experimental data of He & Jackson [2]. In the previous study [1], it was found
that, for a moderate increase in the flow rate, most popular turbulent models
are not capable of capturing this delay in the Reynolds stresses. The data
produced by the LES can be used to develop advanced turbulence models for
unsteady flow problems.

References

1. Y. M. Chung, M. M. Jafarian: *TSFP-4* pp. 319-324 (2005).
2. S. He, J. D. Jackson: *J. Fluid Mechanics* **408**, 1 (2000).

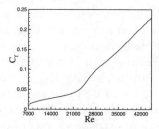
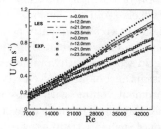

Fig. 1. Schematic of unsteady turbulent pipe flow, and time history of flow rate Q.

Investigation of a tripped
turbulent boundary layer flow
using time-resolved tomographic PIV

Andreas Schröder, Reinhard Geisler[1] and Dirk Michaelis[2]

[1] Deutsches Zentrum fr Luft- und Raumfahrt, Institut f. Aerodynamik und
   Strömungstechnik, Bunsenstrasse 10, 37073 Göttingen, Germany
   `andreas.schroeder@dlr.de, reinhard.geisler@dlr.de`
[2] La Vision GmbH, Anna-Vandenhoeck-Ring 19, 37081 Göttingen, Germany
   `dmichaelis@lavision.de`

1 Abstract

The spatial and temporal development of turbulent boundary layer flows is
governed by the self-organization of coherent structures like hairpin-like or
arch vortices and spanwise alternating wall bounded low- and high-speed
streaks. For a detailed analysis of the topologies of the wall normal fluid exchange namely of the four quadrants of the "instantaneous" Reynolds stresses
Q1 to Q4 a time-resolved 3D-3C measurement technique is desired. In this
feasibility study the tomographic PIV technique (Elsinga et al. 2006,Tomo-
graphic particle image velocimetry, Experiments in Fluids, Volume 41, Number 6, pp.933-947) has been applied to time resolved PIV recordings. Four
high speed CMOS cameras are imaging tracer particles which were illuminated in a volume inside a boundary layer flow at 4 kHz by using two high
repetitive Nd:YAG pulse lasers. The instantaneously acquired single particle
images of these cameras have been used for a three dimensional tomographic
reconstruction of the light intensity distribution of the particle images in a
volume of voxels (volume elements) virtually representing the measurement
volume. Each of two subsequently acquired and reconstructed particle image
distributions are cross-correlated in small interrogation volumes using iterative multi-grid schemes with volume-deformation in order to determine a time
series of instantaneous 3D-3C velocity vector fields with 250 sec time steps.
The measurement volume with a size of 34 x 19 x 35 mm was located near
the wall in a flat plate boundary layer flow with zero pressure gradients and
downstream of spanwise tripping wires. At a free stream velocity of U = 7 m/s
a turbulent boundary layer flow develops downstream of the tripping device.
For the first time spatio-temporal fluid exchange mechanisms can be studied
in 3D in this type of turbulent wall bounded flows by using experimental data.

Author Index

SPRINGER PROCEEDINGS IN PHYSICS